December 25, 1977

Chuck,

Friend & Colleague

Best to You & Jenn

As Ever

Jim Herberg

THE WORLD'S WORST WEEDS

Distribution and Biology

THE EAST-WEST CENTER—officially known as the Center for Cultural and Technical Interchange Between East and West—is a national educational institution established in Hawaii by the U.S. Congress in 1960 to promote better relations and understanding between the United States and the nations of Asia and the Pacific through cooperative study, training, and research. The Center is administered by a public, nonprofit corporation whose international Board of Governors consists of distinguished scholars, business leaders, and public servants.

Each year more than 1,500 men and women from many nations and cultures participate in Center programs that seek cooperative solutions to problems of mutual consequence to East and West. Working with the Center's multidisciplinary and multicultural staff, participants include visiting scholars and researchers; leaders and professionals from the academic, government, and business communities; and graduate degree students, most of whom are enrolled at the University of Hawaii. For each Center participant from the United States, two participants are sought from the Asian and Pacific area.

Center programs are conducted by institutes addressing problems of communication, culture learning, environment and policy, population, and resource systems. A limited number of "open" grants are available to degree scholars and research fellows whose academic interests are not encompassed by institute programs.

The U.S. Congress provides basic funding for Center programs and a variety of awards to participants. Because of the cooperative nature of Center programs, financial support and cost-sharing are also provided by Asian and Pacific governments, regional agencies, private enterprise and foundations. The Center is on land adjacent to and provided by the University of Hawaii.

East-West Center Books are published by The University Press of Hawaii to further the Center's aims and programs.

The
World's Worst Weeds

DISTRIBUTION AND BIOLOGY

LeRoy G. Holm

Donald L. Plucknett

Juan V. Pancho

James P. Herberger

AN EAST-WEST CENTER BOOK

from the East-West Food Institute

*PUBLISHED FOR THE EAST-WEST CENTER BY
THE UNIVERSITY PRESS OF HAWAII*

Honolulu

*To our mothers, wives, and children,
who must surely have wondered
how grown men could become so enraptured
with weeds!*

Contents

viii   CONTENTS

Preface

This is an inventory of the principal weeds of the world's major crops, with particular emphasis on their distribution, seriousness, and their known biology. From the time man first learned to plant seeds and harvest crops, weeds have been rather casually accepted as an inevitable nuisance. Only in recent years have we become aware of the staggering losses inflicted on a food-short world by destructive weeds. The origins of this book are deeply buried in this tradition of general acceptance of destruction by the world's worst weeds, and it is the authors' hope that this inventory may in some small way help the efforts of scientists and policymakers as they tackle a problem which can be solved. We have perfected the technology for building monstrous towers, not because they are fit places for human beings to live and work in, but because we know how to make elevators that will rise swiftly to the top. We have squandered our resources on the construction of supersonic jetliners, not because they fulfill a human need to be at a given place in the world in 2 to 4 hours, but simply because we know how to design and build them. Perhaps the crowning achievement of all science and technology to this time is that we have sent men to walk on the moon. But we cannot feed ourselves! When airborne, we have routine procedures which enable three or four young ladies to feed 100 people a complete meal and to clear away the dishes, all within an hour, while flying into or out of the richest or poorest countries in the world. But we have not mastered the means of providing the necessary meals for those who are earthbound! Have we got our priorities wrong?

Perhaps so, and as we seek to stay the terrible destructive forces of the world and search for ways to build a community of nations which can live together in love and trust, we ought to consider again some words written 2,000 years ago by Lucius Seneca, the Roman philosopher: "A hungry people listens not to reason, nor cares for justice, nor is bent by any prayers."

WEEDS! One of the most powerful forces to be reckoned with when planning for food production; and yet, strangely, a word almost without dimension in common usage. The same word may be used to describe the vegetation in a small rice paddy, or that in all of the rice fields of the world, or a mat of vegetation covering one of the world's largest rivers, or the plants which invade and capture vast expanses of grassland. We were unable to measure the word or give direction to our interests and concerns about the world weed problem until the need came into focus quite suddenly, for one of us, in the form of a question, during a night of mapping world weed vegetation in Rome almost a decade ago: "How many species of plants cause 90 percent of the food losses in agriculture and what are their names?"

As we began our work, the extent of these losses for the world and the steps needed to attend to them were clouded with uncertainty.

There were few among us who could name 10 books on weeds. In this volume we publish a list of about 300, and there are more. At that time we could ask "Are there 50, 500, or 5,000 species of plants which cause 90 percent of the weed loss in agriculture?" There were few who cared to guess. We recognized that we were dependent upon only 12 to 15 major crop species for a very large proportion of the world's food, but we had not asked whether, as we narrowed and refined the world's cultivated species, we had also singled out fewer weed species to match the very specialized cultural systems. We now can demonstrate that, of the 250,000 plant species, certainly fewer than 250 have become important weeds of the world. The world community has spent many millions of dollars on the biology and control of a few species of weeds which are of only secondary importance for world food production, yet several of the world's most destructive weeds still cannot be controlled in many of the crops where they are found. This fact engenders the more specific question: "Have weed scientists got their priorities right?"

Weeds have been man's constant companions; he is accustomed to seeing them everywhere. It is understandable, therefore, that hand-weeding, cultivation, and toleration were the systems employed in the fields from the beginning of agriculture until after World War II. But now we are certain of the waste of human energy, of power resources, and the loss of food quality and quantity resulting from weed competition on our farms and plantations. The costs are staggering. A look across the world may help us appreciate the magnitude of the problem. Rice supplies about 30 percent of all the food energy for mankind. Experiments at the International Rice Research Institute in the Philippines indicate that weeds may reduce rice yields on the average by 30 to 35 percent, and sometimes as much as 80 percent. In Australia, the annual loss caused by weeds to animal production alone has been estimated to exceed $A 100,000,000 (Meadly 1965). The crop losses due to weeds in India range from 10 to 80 percent and, even if the most conservative figure is used, the cost to agriculture is $U.S. 600,000,000 annually (Chakravarty 1963). One Latin American country has reported annual crop losses of $U.S. 450,000,000 because of weeds. A single species of poisonous weed was responsible for the death of 15,000 head of livestock (Furtick and Deutsch 1971). Annual crop losses of 15 to 20 percent have been reported for the United States and Canada. The estimated crop loss and the cost of controlling weeds in the United States in 1962 was 5 billion dollars (anon. 1965). It has been estimated that, if weeds were removed, the increased value of the wheat crop alone in the prairie provinces of Canada would amount to $U.S. 100,000,000 annually (Hay 1968).

From a simple beginning in which we made lists of all of the species we could find which were behaving as weeds, together with their locations, we soon realized that we must also attempt to summarize the known biology and the agricultural importance in order to bring an understanding of each species. Because a very large information file soon exceeded our ability to cope with all of the dimensions of the inventory, we turned at the appropriate time to the use of data-processing techniques. We believe that our entire system now holds 200,000 to 250,000 pieces of information about the weeds of the world.

The source of the information which came to be the backbone of the study was centered in the countries themselves, and we were wonderfully surprised at the large number of scientists across the world who truly understood their weed problems and wanted to communicate with others about them. We have traveled to most of the countries to visit these workers and to gather data. We have been in correspondence with more than 100 countries—some of them for over a period of almost a decade. We are genuinely grateful for the assistance of all of these men and women in the preparation of lists of weeds and ratings of their importance. Those portions of the major weed books and journals, abstracting journals, and the proceedings of many weed conferences which were appropriate to our studies have been entered into our data bank. Our procedures will allow us to return to the sources of almost all information in the text and on the maps. All of our data were finally sorted by species, crop, country, and importance for the several operations required in the preparation of these volumes.

What are the world's worst weeds? These data are presented that *you* may weigh the importance of the species of your interest and make comparisons between species. We believe that the first 18 species are arranged in the approximate order in which they are troublesome for the world. For the remaining species, we believe that the question of importance for the world has meaning only in the context of a specific region, climate, and crop. Perhaps it is enough to know that they rank among the world's worst weeds.

In spite of all our striving, either because we have not invested enough time or because we are not wise enough, we are unable to leave with you a formula or any suitable magic for arranging the world's worst weeds in an order of importance upon which everyone will be agreed. The species of weeds described in this volume have an important role in human affairs over much of the earth: they offer severe competition in cultivated fields and grasslands; are toxic to men and animals; may decrease the quality of wool, hides, meat products, and vegetable fibers; and may interfere with fishing, irrigation schemes, hydroelectric plants, and the movements of small and large vessels on navigable streams. The 18 species of weeds in group 1 are deemed to be the most troublesome for man in his agriculture and on waterways and they are presented in the approximate order of their importance for the world. The remaining species have been placed in group 2 and are presented in alphabetical order.

We believe that we have assembled and processed into meaningful form more data on the world weed problem than have been gathered up to this time, and we must emphasize that the importance that we have assigned to each species is a personal judgment made against this background of experience. Our judgments must remain as fragile as yours until weed scientists in a future time can gather similar information and make other, more perfect judgments.

This book, like most comprehensive works, could have used far more hands and many more years. There is so much more to do that the joy of accomplishment tempts one to go on and on. Another wonderful paragraph from the writings of Seneca describes our plight: "Nature does not reveal all her secrets at once! We *imagine* we are initiated in her mysteries, but we are, as yet, only hanging around her outer courts. Those secrets of hers are not opened to all indiscriminately—of one of them this age will catch a glimpse; of another, the age that will come after." Having received this glimpse of the order and arrangement of weed species across the climates and the crops of the world, we now must pause in our searching and record what we know and what we have seen so that weed science may be helped to recognize its priorities and to pursue the ideas and the needs which are most important to this hungry age.

With the omissions there will be errors as well, for our talents are limited and we confess to knowing, far better than anyone else, that our arms have been too short to reach over the whole world. We would count it a privilege if those who notice our mistakes will communicate with us about them, so that we may improve possible later editions. With your help, even our mistakes may be turned to use for we will know how to make corrections or add the information which is missing. Much of the writing for these volumes is being done in Hawaii, a land which ex-

isted in an uncharted sea and was unknown to the rest of the world until Captain James Cook and his crew came upon it less than 2 centuries ago. The maps he used to navigate to this part of the world were distorted and incomplete, those he left with us were full of inaccuracies; but, from the imperfect sketches of men such as this, the whole earth has now been charted in very great detail. We believe that the world will not soon have the resources to bring to absolute perfection such biological surveys as we have attempted here. The cost alone would be staggering. We hope our accuracy and completeness approaches 90 percent on an average across the world. Perhaps this is all we need to go forward with our work.

The tasks which we undertook were blessed at every step of the way with an openness and a willingness on the part of hundreds of people to help us. Our stations during this time were in the Food and Agriculture Organization of the United Nations (FAO), the University of Wisconsin, the University of Hawaii, and the University of the Philippines, but principal among them was the East-West Center in Hawaii. For many years we each worked alone at places which were scattered across the globe, with only an occasional meeting. At the time that we had completed our separate tasks, three of us were invited to come together for a year as Senior Fellows in affiliation with the Food Institute of the East-West Center to undertake the final preparation of this book. Through the Institute and the East-West Center units which support it, we were provided with a common room for the completion of the illustrations, maps, and text; with computer facilities for the updating and collation of our myriad data; with stenographic, clerical, and bibliographical assistance; with the added moral support of a genuine programmatic interest in integrated systems of crop pest management; and with an active linkage to our publisher, The University Press of Hawaii. Some of the most pleasant and helpful associations of this long journey were to be found in our work with The University Press of Hawaii.

These were the institutions, but it is to the individuals whose names appear below that we are indebted for so much of our good fortune. They were at times both architects and builders, for they were very able, and each became so genuinely interested that new insights were suggested and often the course of our work was changed. We wish to offer our warmest thanks:

To Barbara Kirchoff, Sylvia Lau, Margaret Sung,

and Mary Tugaoen who, in the critical stages of preparing this manuscript, performed a thousand tasks with a grace, ease, and wonderful good cheer that surely must be reflected in the pages of these volumes.

To Virginia Boehlke, Suzanne Heindl, Judy Herberger, Marian Holm, Fannie Lee Kai, Janet Lembcke, Sue Plucknett, Lynn Abe, and Arline Uyeunten for swiftly and carefully typing the thousands of lists and pages which were the background for this final copy.

To Sandra Klynstra of Wisconsin for her skill in providing the final design for all of our maps and for preparing so many of them. To Choy Ling Wang for continuing the work in Hawaii.

To the late Mr. Ernesto S. Calara, and to Avery Youn, Susan Nakagawa, and to Joanne Otsuka Ahana who prepared the illustrations of the plant species. To David Saiki for his help and leadership with this task.

To Larry Stevens, who gave more generously of his time than we had a right to expect, and whose genius at the computer weighs heavily in the successful completion of these volumes.

To Leo Rake, Rick Smith, and Dr. Sadiqul Bhuiyan for assistance at the outset with the preparation and photographing of distribution maps.

To Dr. Lee Ling, chief, Plant Protection Service, Food and Agriculture Organization of the United Nations, Rome; Professor Ova B. Combs, former head, Department of Horticulture, University of Wisconsin; Dr. Nicolaas Luykx, director, The Food Institute of The East-West Center, Hawaii; officials of the College of Agriculture, University of the Philippines, Los Baños; Dr. Wallace Sanford, chairman, Department of Agronomy and Soil Science and Dr. C. Peairs Wilson, dean of the College of Tropical Agriculture, University of Hawaii, who perceived that this work was worth doing, gave wonderful encouragement, and provided shelter, assistance, and peace of mind while we were about our tasks.

To Mr. John Fryer and his staff at the Weed Research Organization, Oxford, England, for assistance in the final preparation of the world list of books on weeds.

We wish to acknowledge a grant of funds from the United States Agency for International Development, through the International Plant Protection Institute at Oregon State University, for assistance during 1969 and 1970.

Finally, although our debt is far too heavy to express in the ordinary ways which are open to us, we must tell you that this first comprehensive summary of the world's weed vegetation rests heavily upon the kindness and the labor of hundreds of unnamed weed scientists and agriculturists with whom we worked as we made our headquarters at several institutions across the world, or whom we were privileged to visit as we traveled to gather the data. To these men and women we are truly grateful!

PART ONE
The Weeds

Introduction

For each of the important weed species we have attempted in these pages to present a summary of the world distribution, the known biology, and the agricultural importance. A line drawing and a description is included for each species, as well as a list of common names used in the countries where the weed is troublesome.

Those who are not acquainted with the distribution across the world of our worst weeds often seem confused that a plant which is an important weed in one area may be a valuable crop in another place. It frequently happens that a species introduced as a potential crop plant becomes a scourge of croplands if it escapes to behave as a weed. *Pennisetum polystachyon* was brought into Thailand as a possible pasture grass but escaped to become a terrible weed. Sugarcane growing was almost abandoned on one sugar plantation on the island of Kauai, in the Hawaiian chain, because of infestations of *Cynodon dactylon*, the world's number-two weed. Selected strains of this same species are, however, the most important pasture grasses in southeastern United States. It is true that the seeds or roots of many weeds may be used as food for humans or animals, or the stems and straw for thatching or papermaking, but this does not alter the fact that the same species may be very strong competitors for nutrients, light, and water when growing in croplands. Thus, a plant may sometimes be important for agriculture because we wish to grow it as a crop, while at other times and places the same plant may be important to agriculture because it behaves as a terrible weed which we are forced to bring under control *before* we can grow our crops.

In the early phases of data gathering we were puzzled by the words and phrases used across the world to express the importance of a weed, or to assign a relative rating among others in the same field. As our experience grew we began to perceive that we were being taught by those with whom we visited, worked, and corresponded that, although it goes unrecorded, there seems to be a common understanding about "serious" and "principal" weeds in a crop. When a worker has stated that a certain species ranks as one of the two or three most "serious" weeds in a crop, there is no question of his meaning. It is our experience that when he speaks of his "principal" weeds he is usually referring to about the five most troublesome species for his crop. Rarely does he name ten weeds, for example. A "common" weed is one that is very widespread in all the crops or regions of a country, requiring constant effort and expense to hold it at bay but never seriously threatening a crop. Every rating of weed severity recorded in this work was offered by a worker for his own country or by an author in a published document. We have not made estimates or provided ratings from the work or documents of other persons. It is with these terms, then, that we have tried to express our meaning, and it was in these ways and with these meanings that the work went forward. Others will surely find their own, perhaps more descriptive, methods. We have tried other systems, and we have often been advised that values for weed importance can be expressed only in numbers. After a very large experience in this work, however, we think it may be helpful to pose a question: "How would a rating of 6 differ from 4 or 8 in expressions from dozens of workers concerning hundreds of species in

50 to 100 countries across the world?" Would the point and purpose of all we seek to do here be made more clearly for *everyone* by the use of numbers, than with words such as we have chosen?

Our summary of the agricultural importance of a weed is often accompanied by a map depicting the major crops in which the species is a pest across the world. Where a species is shown to be a serious or a principal weed, we mean that a weed scientist of that country has given such a ranking for the species in that crop in some area of his country. This does not imply that a total evaluation of every weed species for every crop has been completed there on a national scale. There are perhaps only two or three countries in the world that may even make estimates at this high level.

There is one more matter which deserves comment as we seek to give expression to the relative significance of the weeds in our crops. This concerns the use of the terms "important weeds" and "problem weeds." *Imperata cylindrica* stands at the edge of every rubber and oil palm plantation in Southeast Asia, ready to invade when openings are created by removal of diseased trees, or whenever man relaxes his guard. *Agropyron repens*, another perennial grass, stands waiting at the fence line of every field in the corn growing belt of the United States. They must always be considered "important" weeds for these areas. During periods when the farms and plantations are under constant surveillance for new outbreaks, when there are annual expenditures for chemical and mechanical control measures, and when these measures are combined with the use of ground covers and rotations to suppress weeds, these two species *may not be problem weeds*. Should the areas be badly managed for only one or two growing seasons, however, these two "important" perennial weeds will quickly reappear and then become "problem" weeds. Such distinctions ought to be made in our writing and in our conversations so that we may understand one another clearly. Farmers and weed workers who may use several herbicides in weed control often assert that the worst weeds in the area are not "important." But this is to express only a kind of fleeting, ephemeral understanding of the ecology of our fields, for legislative restrictions or abrupt changes in farming practices which limit or even prohibit the use of herbicides will soon bring severe reinfestations. The weed species will again become a "problem" and this will serve to remind us that *they* are the "important" weeds of the area.

All of the drawings are original. They were made either at the University of the Philippines College of Agriculture by Mr. Ernesto S. Calara, or at the University of Hawaii College of Tropical Agriculture under the direction of D. L. Plucknett. Special efforts have been made to obtain drawings in which the habit and key features of each plant are emphasized. We have attempted to provide a description broad enough to assist the weed worker in recognizing the type that may be growing in his own land. The variations in morphology in the different climates and environments across the world make it impossible to depict an "average" plant of the species. For example, the maximum height of *Sorghum halepense* is said to be 50 cm in some parts of the world, and 3 m in others.

It was our task to try to comprehend the weed vegetation patterns in the fields of major crops across the whole earth, and the reader will soon observe that the particular directions in which we have probed represent the average of our collective biases about importance of species, principal crops, and the geography of food production. But the one unrelenting temptation from which we were never spared was the inclination to drift into emphasis on areas where data, points on maps, and biological information were easily available. It happens, however, that most of our people live in the warm areas of the world and by far the greater portion of our food is produced there. India, for example, is the leading producer of several major world crops. It is in these warm areas that more than 70 percent of the people, often entire families, must work in the fields in order to grow enough food simply for survival, and in many areas 60 to 90 percent of all income must go for food. It is for these reasons that we have tried, though not always successfully, to place greatest emphasis on the weeds of the warm regions, and to be most thorough in our search for data and in our analysis of those areas. It was here that information was most difficult to obtain, however.

We have tried to look across all of the continents in an attempt to show the *world distribution* of each species. At no time have we intended to provide detailed distribution maps for countries or even regions, for this must be done in an entirely different kind of program. We have missed some places and we earnestly seek your assistance in completing or correcting our maps. With close scrutiny of the several kinds of maps we have provided, you will soon share with us the fascinating discovery that the most serious weed species of a principal crop are very often found in every major production area for that crop, regardless of continent. A final word about distribution. As you find an area or two of the world where we have failed to mark the location of a particular species —and you surely will—we urge you to remember that the *pattern* of world distribution may still be valid as we have shown it. We would ask this question: "If the time and resources had been available to discover 10 to 20 percent more locations for the distribution maps of *Cyperus rotundus*, *Rottboellia exaltata*, *Imperata cylindrica*, *Eichhornia crassipes*, and *Convolvulus arvensis*, for example, would it significantly alter our impression of the world distribution of these species?" Perhaps the patterns shown here are sufficient for most species, and we may now spend our time in search of other kinds of knowledge we badly need if we are to bring these species under control.

In our summary of the biology of each weed we have tried to provide the information that will be most helpful in working with the species. We have not intended to review all of the world's literature for each species, but we believe that we have touched most of the major documents, and reference to these has been provided for those who wish to study in more detail. We have found it very difficult to summarize certain aspects of the biology of weeds on a world basis. For example, our present knowledge of seed dormancy and the role of phytochrome in seed germination demands that henceforth our research methods bring several factors of the environment under rigorous control, including some processes for which time must be measured in seconds. Our findings must then be interpreted within the context of these conditions, else the experiments may be meaningless. It is therefore likely, although we may wish it were not so, that many of the seed germination experiments reported in research journals cannot be understood because the methods of experimentation are no longer valid. For the seeds of most weed species the "new" experiments have not yet been done. In such matters we have tried to cite only the papers that might be meaningful, sometimes with tongue in cheek, and we hope the reader will appreciate the circumstances at the moment of writing. Many of these same arguments can be made concerning studies of photoperiod and flowering, as well as several other physiological and morphological events in the life of a species. The paucity of knowledge about most weed species remains one of the great challenges for weed scientists of the future. But something else about the status of our knowledge prompts us to issue a warning: In the interest of using our talents and our resources most efficiently, it seems a waste of time that often the same experiments are repeated over and over again across the world, by both students and senior researchers, because we have not tried or have not been able to communicate with one another. There are many things we *do already know* about our weed species and their control, and there are some experiments which *do not have to be performed again*.

We have frequent requests to provide literature citations and information about procedures and experiments for study of the life histories of weeds and for making weed surveys of countries or regions. The work of Penfound and Earle (1948) on *Eichhornia crassipes*, and the many life histories reported in the *Journal of Ecology* of Great Britain are excellent. For examples see Cavers and Harper (1964) and Sagar and Harper (1964). The weed survey of Canada's prairie provinces by Alex (1965), of Finland by Mukula (1969), of Sweden by Granstrom (1955), and of Taiwan by Lin et al. (1968) are all worthy of consideration.

GROUP 1

This section describes the 18 most serious weeds in the approximate order in which they are troublesome to the world's agriculturalists. Among them, *Cyperus rotundus*, *Cynodon dactylon*, *Echinochloa crusgalli*, *Echinochloa colonum*, *Eleusine indica*, *Sorghum halepense*, *Imperata cylindrica*, *Chenopodium album*, *Digitaria sanguinalis*, and *Convolvulus arvensis* are considered to stand apart from all other weed species for several reasons. Our data from personal visits and published reports reveal that workers have not only cited these species more often than other world weeds, but have also ranked them as the greatest troublemakers in the largest number of crops. For example, we have worked with more than 700 reports of *Cyperus rotundus*, with each report carrying up to 10 important pieces of information about the species, and with one-third of the reports ranking the weed as serious or principal. All of the above species carry from 200 to 500 citations of a similar nature. *Portulaca oleracea* is less often declared to be a serious weed but it should be emphasized that it is one of the three most frequently reported weeds across the world. The importance of *Eichhornia crassipes* is seldom ranked by workers and authors in a formal way for it seems to be understood that it is a threat wherever it is found in association with agricultural activity.

Cyperus rotundus is in 100 countries, one-half of the above species are in more than 60 countries, and all are in more than 50 countries. Several of the weed species are in more than 50 of the world's crops and all are in more than 30 crops. Exceptions are *Eichhornia crassipes*, which is a scourge of the world's major waterways; and *Avena fatua* (21 countries), which is mainly in cereals and vegetables so that its range of crops is not so great.

The remaining five weeds in the group, *Amaranthus spinosus*, *Amaranthus hybridus*, *Cyperus esculentus*, *Paspalum conjugatum*, and *Rottboellia exaltata* are about equal in importance. They are reported with great frequency, are widely distributed in the world, and have been ranked by workers as serious or principal weeds in more than one-third of the crops in which they are found. *R. exaltata* is worthy of special mention because it is said to be a serious or principal weed in one-half of the fields where it occurs.

❦ 1 ❧

Cyperus rotundus L.

CYPERACEAE, SEDGE FAMILY

Cyperus rotundus is the world's worst weed. It is a sedge, native to India, and is widely known in the world by the common names nutgrass, nutsedge, or purple nutsedge. It has very dark green leaves; a three-sided stem; grows to 100 cm on moist, fertile soils; and has an extensive subterranean system of rhizomes and tubers. The rhizomes can penetrate and pass completely through vegetable root crops. The outstanding characteristic of this plant is its prolific production of underground tubers that can remain dormant and carry the plant through the most extreme conditions of heat, drought, flooding, or lack of aeration. The inflorescence is reddish to purplish brown, hence the name purple nutsedge. The yellow or straw-colored inflorescence of a near relative, *C. esculentus*, has prompted the name yellow nutsedge for that species. The latter extends deeper into the cold regions of both hemispheres and is more tolerant of wet soils than is *C. rotundus*.

C. rotundus is reported to be a weed in 52 crops in 92 countries.

DESCRIPTION

C. rotundus is an erect, persistent, glabrous, *perennial* herb (Figure 1); *roots* fibrous, extensively branched, clothed with bent hairs; spreading by extensive, horizontal, slender *rhizomes*, which are white and fleshy and covered with scale leaves when young but which become brown, fibrous, or "wiry" when old; axillary buds do not appear on the rhizomes; rhizomes give rise at intervals of 5 to 25 cm to underground tubers which continue to proliferate, forming tuber chains that extend to a considerable depth in the soil; *tubers* irregularly shaped or nearly round, up to 2.5 cm in length, white and succulent when young, turning coarse fibrous brown or almost black with age, covered with papery scale leaves, which, when detached, result in leaf scars on the outer surface, buds germinating to form new plants; *aboveground stems* arise from a swelling on the rhizome that is referred to as a *basal bulb*, a *tuberous bulb*, or a *corm*, and this part forms a swollen or thickened plant base; *culms* erect, simple, smooth, triangular in cross section, 10 to 60 cm high, usually longer than the basal leaves; *leaves* grasslike, linear, acute, up to 50 cm long, 8 mm wide, smooth, shiny dark green, grooved on the upper surface, tubular and membranous where they grasp the three-sided stem, arising from very compact nodes in basal clusters in three rows, through the center of which the upright stem arises; *inflorescence* a loose *umbel*, terminal on the stem apex, simple or slightly compound, subtended by two to four leaflike bracts that are usually as long as or slightly longer than the flower-bearing rays, consisting of from three to nine slender, spreading three-sided peduncles of unequal length, near the ends of which are clustered the narrow spikelets; up to 30 cm; *spikelets* 0.8 to 2.5 cm long, 2 mm wide, 10- to 40-flowered, acute, compressed, red, reddish brown, or purple brown; *glumes* closely appressed, ovate, nearly blunt, reddish to deep brown or purplish brown, light green or yellow on membranous margin and the midvein, prominently three- to seven-nerved, 2 to 3.5 mm long; *calyx* and *corolla* absent; *stamens* three, an-

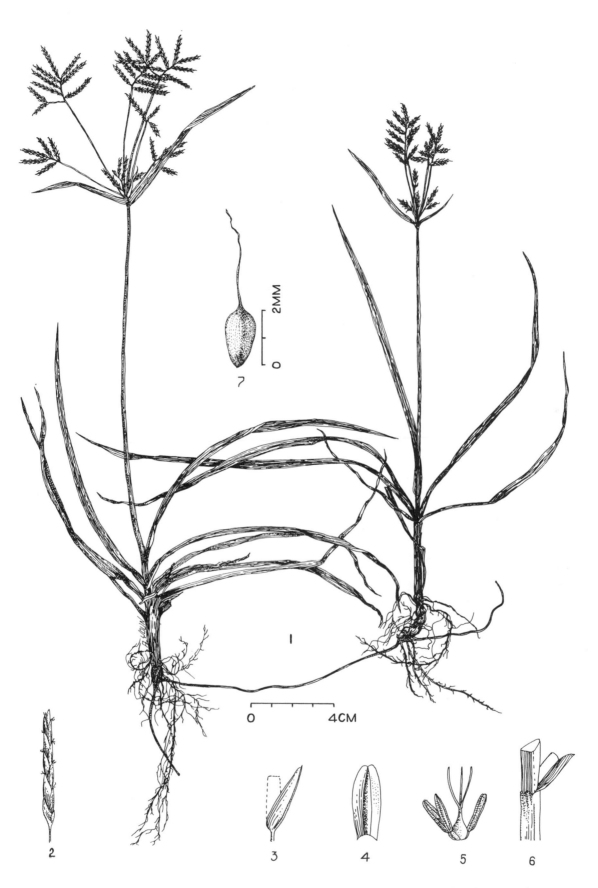

FIGURE 1. *Cyperus rotundus* L. *1*, habit; *2*, portion of inflorescence; *3*, flower with glume; *4*, glume; *5*, flower, glume removed; *6*, portion of leaf sheath and blade; *7*, achene.

thers 1 mm long; *stigmas* three, elongate; *style* three-branched, shorter than the nut; *nut (achene)* ovate or oblong-ovate, 1.5 mm long, about one-half the length of the glumes, three-angled, the base and apex obtuse, granular, dull, olive-gray to brown or black; covered with a network of gray lines.

This species is distinguished by its scaly or wiry tuber-bearing rhizomes; its red, reddish brown, or purplish brown inflorescence; and the mostly basal leaves, which are shorter than the inflorescence.

DISTRIBUTION

The distribution of *C. rotundus* is shown in Figure 2. It has been reported from more countries, regions, and localities than any other weed in the world. Its range at increasing latitudes in both hemispheres seems limited by cold temperatures; aside from this, however, it grows with reckless abandon in almost every soil type, elevation, humidity, soil moisture, and *p*H, and can survive the highest temperatures known in agriculture. Ranade and Burns (1925) reported that it cannot stand soils with high salt content. In spite of its vigor in most circumstances, the plant does not tolerate shade. When crops such as sugarcane and plantation trees close in the overstory and begin to shade the soil, the leaves of the weed yellow and die. The dormant tubers remain viable, however, and, as soon as an opening again appears, the sedge begins to infest the area.

Although there have been no studies on a regional or world scale to describe the ecotypes of *C. rotundus*, it is recognized that they do exist. Ranade and Burns (1925) have described types from India with the following variations in glume color: (1) yellowish white, (2) light red, (3) coppery red with metallic luster, and (4) dark red with a blackish tinge. Experiments were carried out with clonal material to show that the color variations are inherited and are not the result of different ages or environmental responses.

The species is found in cultivated fields, on roadsides, in neglected areas, and at the edges of woods, and it may cover the banks of irrigation canals and streams. As water becomes low, an entire stream or canal bed may support a population of nutsedge.

C. rotundus may be a serious problem in paddy rice in which the soil is puddled. In seasons of low water supply, when puddling cannot be done thoroughly or when the water supply fails after the rice has been transplanted, the nutsedge growing in the drying soil may choke the crop.

The tubers are moved in mud on the feet of men and animals and by being caught on machinery. They may be seen floating and drifting with the wind in flooded rice. They are also brought into new areas when streams flood after storms, and they are distributed in surface irrigation water.

ANATOMY AND MORPHOLOGY

C. rotundus is a perennial, grasslike herb with an unjointed, triangular, solid stem, and leaves in three ranks with closed sheaths and without ligules. An underground network of rhizomes produces tubers at intervals and those tubers near the surface give rise to aerial shoots. The tubers have an extensive root system that may reach deeply into the soil. Some principal studies of the life history of *C. rotundus* are those of Smith and Fick (1937), Honess (1960), Hauser (1962a), and Misra (1969). Contributions on the anatomy and embryology of the species have been made by Plowman (1906), Khanna (1965), and Wills and Briscoe (1970). The most comprehensive study of the species was made by Ranade and Burns (1925) in India a half-century ago.

Leaves

Leaves have a distinct midrib, are very dark green, and are 0.25 to 0.75 cm in width. Depending upon environmental conditions and the age of the plant, the leaves vary from 5 to 20 cm in length. The vessels of the leaf lead through the basal bulb into the underground rhizome system, and this large interconnecting aerial and subterranean vascular system probably remains intact for a growing cycle or season. The upper leaf surface has a waxy cutin without stomates. The lower surface is thinly cutinized and has many parallel rows of stomata. Wills and Briscoe (1970) have speculated that herbicides applied to the leaf surface must enter through the waxy upper surface, the stomates, or the thinly cutinized cells of the lower surface.

The leaves have parallel, collateral, vascular bundles with the xylem above the phloem. Each bundle is surrounded by a sheath having an inside layer of fibers and an outside layer with elongated, chlorophyllous cells. These are believed to make up the major portion of the photosynthetic tissue of the plant. Black, Chen, and Brown (1969), in a report on the biochemical basis for plant competition among weeds and crops, found this species to be among those plants with high photosynthetic efficiency by way of the C_4 dicarboxylic acid-carbon dioxide fixation pathway. It is the view of Black, Chen, and Brown that plants that are very competitive when

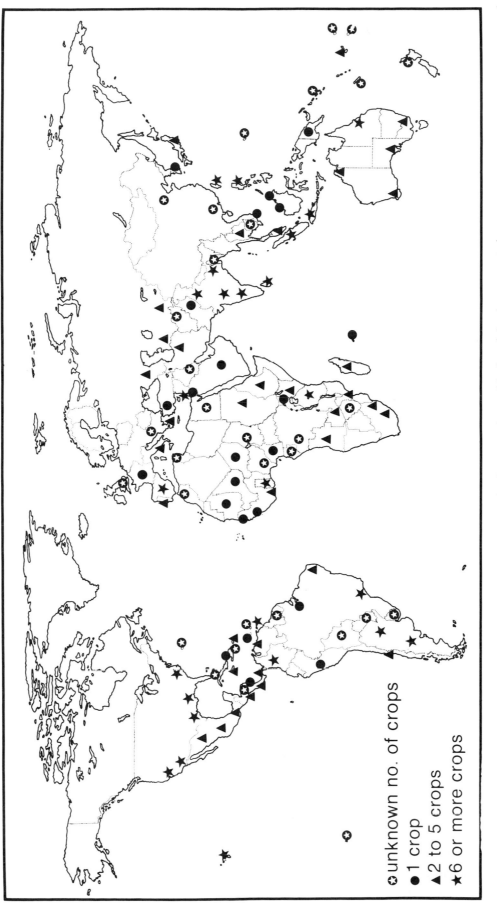

FIGURE 2. The distribution of *Cyperus rotundus* L. across the world where it has been reported as a weed.

unknown no. of crops
●1 crop
▲2 to 5 crops
★6 or more crops

temperatures rise and sunlight increases have the C_4 photosynthetic pathway in association with high concentrations of chlorophyllous tissue around the vascular bundles.

The leaves extend from the bulb in an infolded, triangular fascicle. Development begins at the outermost leaf and progresses inward to terminate with the seed-bearing rachis.

Inflorescence, Fruit, Seed, and Seedling

The flowering scape is a leafless stalk, solid and triangular, which pushes up from the center of the basal bulb. The reddish brown flower is a compound spike resembling an umbel and has been described in a previous section.

In the Cyperaceae the ovary is entire, erect, and sessile in the center of the flower, or it is sometimes placed on a short stalk which may be thickened or expanded to form a disk. It is one-celled and contains a solitary anatropous ovule, erect from the base. The dark brown trigonous achene is about one-half as long as the glume. The achene is really the fruit but, in the vast literature on the Cyperaceae, it is usually referred to as a "seed" and the term will be used in this discussion. The achenes are 1.5 x 0.8 mm, the seeds 1.2 x 0.5 mm, and the embryos vary from 100 to 300 microns. The seed is erect from the base of the fruit, free from the pericarp, and has a thin testa. The pericarp is thick and hard. The endosperm is made up of a peripheral oily layer with starchy tissue beneath. The small embryo is lenticular, globose, or ovoid, and it is situated within the base of this fleshy endosperm (Poiteau 1808).

Klebs (1881) investigated the process of germination and found that it was very uniform in several genera and species (*Scirpus lacustris*, *Cyperus iria*, *Cyperus papyrus*, *Isolepis Savii*, *Carex caucasica*, and *Kobresia caricina*). The embryo is somewhat different from that in grasses in that the lower end is always occupied by a radicle that is without any marked root-sheath development. The upper portion is filled by the cotyledon. As in the grasses, there is a distinct development of a cotyledonary sheath that encloses the clearly indicated first leaf.

In germination, only the cotyledon grows at first. The sheath elongates to break its coverings and to bend geotropically upward. The middle portion of the cotyledon then grows rapidly to pull the main root out of the seed. A circlet of long hairs develops at the base of the cotyledonary sheath before the root appears. These hairs take the place of the root in the early stages. The primary root gradually develops as the first leaf emerges from its sheath, and this root eventually shows considerable elongation. The first

adventitious root breaks from the base of the cotyledonary sheath. The endosperm is absorbed by a portion of the cotyledon, which then swells to fill the interior of the seed.

After the embryo has germinated, the apical portion elongates to form an absorbing organ through which food passes into the seedling. The basal end of the embryo passes through loosely arranged cells at the base of the seed; and just outside the coat a swelling appears and from this come the new root and shoot (Justice and Whitehead 1946).

Rhizomes

These organs are the means by which the plant spreads in all directions and it is through them that the food moves to the tubers. The young rhizomes are white, fleshy, and have an outer covering of scale leaves, whereas the old ones become brown and are often said to be "wiry." In cross section the young rhizome is seen to have an epidermis, a cortex, and a clearly defined endodermis surrounding the major part of the vascular system. The vessels inside the endodermis are unsheathed and loosely distributed around the outside edge of the central pith. In the cortex are scattered, small, immature, vascular elements surrounded by a sheath of small parenchymatous cells. These terminate in the scale leaves of the rhizome, whereas the central bundles are continuous throughout the bulb and tuber network. The change from a white, fleshy condition to a brown and wiry form comes as a result of the secondary, ligneous thickening of four to six layers of cortical cells exterior to the endodermis. The central vascular system thus becomes enclosed, and tissues outside this layer fall off to leave the highly resistant, wiry structure which persists in the soil.

The apical meristem of the rhizome is covered with several successive layers of scale leaves, each of which has grown through the apex of the previously matured leaf. The central anterior portion of the meristem is made up of inactive shoot and leaf primordia surrounded by actively dividing cells of new, growing, scale leaves. Behind this, unbranched provascular tissue develops into the rhizome vascular system. No evidence of axillary bud formation is found in actively growing rhizomes (Wills and Briscoe 1970). F. Andrews (1940) in Sudan could find no evidence that rhizomes connecting two tubers could give rise to new growth. He removed tubers from either one or both ends of the rhizomes and saw no germination from the nodes of any of the rhizomes.

Tuber

The new tubers are almost round and are white and succulent. They, too, darken, becoming almost black as they become very firm from the packed starch within. Mature tubers are 1 to 1.5 cm long and 0.5 to 1 cm in diameter. The papery scale leaves persist for a considerable time and leaf scars are evident when these leaves become detached. The tubers form at the end of the rhizome near the leaf primordia in the meristematic region. As development proceeds, the internodes cease to elongate and the parenchymatous cells inside and outside of the endodermis enlarge and accumulate much starch. The leaf primordia remain dormant. As the entire structure swells, the meristematic tissues branch into new areas and either become quiescent as dormant buds or immediately extend and grow off as new rhizomes (Wills and Briscoe 1970).

Basal Bulb

The swelling of the rhizome which bears the aerial shoot has been called a basal bulb, a tuberous bulb, and a corm. There seems to be no way to resolve this question at the moment, and the term "basal bulb" (which is by far the most frequently used term) will be used in this discussion. As a rhizome grows upward to the surface, it begins to swell as its tip approaches the soil surface. Formerly, basal bulbs were believed to develop only in an area 5 to 10 cm below the surface, but Hauser (1962*b*) has shown that they may develop at a depth of 20 cm. The basal bulb is formed from the meristematic region at the rhizome apex in much the same way as the tuber, except that now the leaf primordia give rise to shoots that emerge through the soil surface. A very short stem extends upward, with the leaves coming from the very compact nodes. The parenchymatous cells again enlarge (to give the bulbous form) and accumulate starch. The meristematic tissues branch off into new directions to send off rhizomes into the soil or more shoots to the surface (Wills and Briscoe 1970).

There are no data on the timing of the formation of the basal bulb with respect to the production of leaves, although some workers do feel that such formation may be triggered by leaf production because the enlargement quickly becomes a food storage center. The possibility exists that basal bulb formation may be initiated at the time the leaf primordia begin to grow and that the basal bulb fills with food assimilates from the photosynthetic process. Tubers placed on the soil surface either in light or dark formed basal bulbs after the development of a rhizome with two nodes. When placed in the light, however, tubers formed internodes so short that the basal bulb appeared almost as a shoot developing from the parent tuber (Standifer, Normand, and Rizk 1966; Standifer and Normand, unpublished*).

Root System

In Sudan Andrews (1940) dug pits more than 1.5 m deep to study the penetration of rhizomes and roots. He found most of the tubers growing within the top 15 cm of the soil, with a few penetrating as deep as 38 cm. The root system, however, extended to a depth of 135 cm in the heavy clay. There were many extremely fine, heavily branched roots at lower depths, and Andrews suggested that these may be very important in maintaining the water supply for tubers in the drier areas near the soil surface.

THE PHYSIOLOGY OF GROWTH AND DEVELOPMENT

Tuber Production

Valuable information on the ecology of *Cyperus rotundus* was revealed in the studies of Rochecouste (1956) on the role of this weed in sugarcane production in Mauritius. Cane is grown in three main zones which may be classified as subhumid (less than 1,250 mm of annual rainfall), humid (1,250 to 2,500 mm of rain), and superhumid (more than 2,500 mm), and the studies were made in canefields in each of the three zones. In the subhumid zone the soils were humic latosols of the Reduit bouldery clay type; those in the humid zone were humic latosols of the Richelieu bouldery clay type; and those in the superhumid zone were lithosols of the Rose-belle gravelly light clay type. *C. rotundus* was found to be much more serious in the humid zone than in the other two. The only perennial weeds which could suppress it were *Cynodon dactylon* and *Artemesia vulgaris*.

Plots 5,800 cm² were selected in areas with pure stands of nutgrass in the cane rows. The soil was removed at 7.5-, 15-, 22.5-, 30-, and 45-cm depths to determine the number of tubers in each layer. No tubers were found beyond the lowest level.

The total tuber production per pit in the humid zone exceeded that in the subhumid and superhumid zones by sixfold and fourteenfold, respectively. The total fresh weight of green tops and of tubers in the upper 45 cm of soil was almost 30,000 kilograms per hectare in the humid zone but only 12,300 and 7,400 in the subhumid and superhumid zones, respectively.

In the subhumid and the superhumid regions the weed tended to place a much larger proportion of its tubers in the upper levels of the soil. In the subhumid

* "Purple nutsedge: its characteristics and growth habits" (Department of Horticulture, Louisiana State University, Baton Rouge, United States, 1966).

zone 60 percent were in the top 7.5 cm and 90 percent were in the first 15 cm of the soil. In the superhumid area 80 percent were in the upper 7.5 cm and 95 percent were in the top 15 cm. Fifty percent were in the upper 7.5 cm and 75 percent in the top 15 cm of the soil in the humid zone and this, of course, means that 25 percent were at deeper levels where the tubers survive much longer and where they are very difficult to destroy with present control methods. Studies such as these help us to understand the mechanism by which this weed can offer such severe competition to our crops. The implications of these findings for water and nutrient reserves in the soil are discussed in the section on agricultural importance.

Andrews (1940) has contributed useful information about tuber production in the Gezira cotton area of Sudan. The soil of the Gezira is a heavy clay with low nitrogen and humus content, is strongly alkaline, and has low water permeability. Normally the bulk of the tubers are formed in the top 15 cm of soil with none found below 30 cm. If waterlogged soils and adjacent, cultivated, more permeable soils are compared, one finds slightly more tuber penetration in the drier soil.

In Trinidad, Honess (1960) found 50 to 85 percent of the tubers in the top 8 cm of a St. Augustine loam soil that was sampled at three different sites. The layer from 8 to 15 cm contained 10 to 35 percent, and it was believed that the higher figure resulted because plant refuse was burned regularly on one site, thus "burning out" the tubers in the soil near the surface. At two sites, 6 percent of the tubers were in the zone from 15 to 23 cm deep. At the third site (the only regularly cultivated area), 21 percent of the tubers were at this low level.

Rao (1968) in India found that one tuber could produce 99 tubers in 90 days and calculated that, on an area basis, this would mean 8 million per hectare in cultivated areas and 4.8 million per hectare in uncultivated areas. The maximum starch storage in uncultivated areas was found in tubers in the top 7.5 cm, whereas in cultivated areas it was found in tubers from the 25-cm level. In general, tubers in cultivated areas had comparatively lower levels of starch, regardless of soil depth.

Misra (1969), also in India, found most of the tubers growing in the top 10 cm of the soil with none below 30 cm, and he calculated an amount of 2,400 tubers per m² down to the 30-cm depth. This was seven times the number of shoots present on that area.

Misra (1969) reported that plants produced many more tubers at short photoperiods of 6 to 10 hours of daylight than at 12 hours or longer (see also Berger 1966). Ueki (1969) in Japan found 65 percent of the tubers in the first 3 cm of light soils. Smith and Fick (1937) in the United States found most of the tubers in the first 20 cm of the soil and found the tubers of greatest average weight at the 10-to-20-cm level.

In a 2-year experiment, Hauser (1962b) in the United States planted large plots with tubers which were equidistant at 0.9-m and 0.3-m intervals. About 75 percent of the tubers germinated in the clay loam soil. Emerged plants were counted at regular intervals and in October of each year soil samples were removed and sifted to a depth of 23 cm. In the second season, one-half of each plot was tilled and one-half of each tilled and untilled plot was covered with screen to give 72-percent shade.

Tubers sprouted in 7 to 10 days; many new basal bulbs and shoots were formed in 3 to 4 weeks; and the plant systems were overlapping at 5 weeks in the areas planted at 0.9-m intervals. Hauser pointed to the possibility that dormancy may be very low in the first few weeks of tuber growth. There was no flowering in the first 4 weeks, but flowering began at 6 and 8 weeks in plots planted at 0.9-m and 0.3-m intervals, respectively. At the end of the season, the number of aerial shoots in the plots with 0.9-m spacing was 5,800,000 per hectare and in the 0.3-m spacing it was 7,700,000; yet, the former wide spacing required only 11 percent of the amount of the original tubers required to plant the close 0.3-m spacing. In October of the first year, 11,000,000 and 7,000,000 bulbs and tubers per hectare had been produced in plots with 0.3- and 0.9-m spacings, respectively.

During the spring and early summer of the second season, the plants increased 66 percent and 84 percent in the 0.3-m spacing and 0.9-m spacing, respectively. In October of the second season, by comparison with the number of tubers formed in the soil in the year before, there was a threefold increase in the area of the 0.3-m spacing and a fourfold increase in the area planted with a 0.9-m spacing. One tillage at the beginning of the second season did not alter tuber and plant production significantly. Continuous shading gave a 10 to 57 percent reduction in the numbers of tubers and bulbs formed.

The nutsedge grew more vigorously in the first season. There also occurred wide differences between spacing variables, but Hauser felt that in the field there normally would not be great differences between plant and tuber densities at the end of the season, regardless of spacing in the original planting (see also Sierra 1973).

Tuber Dormancy

In the literature of *C. rotundus*, dormancy of the tubers and apical dominance in the tuber or the tuber chain frequently are discussed together. Apical dominance here is covered in the section on tuber germination.

Before discussing dormancy it is necessary to point out that many workers have found it difficult to study the formation and growth of tubers and basal bulbs because it is so hard to distinguish between them. The finding of Hauser (1962a) that basal bulbs may be produced as deep as 20 cm in the soil serves to emphasize that the sorting of these structures into two categories at any one moment after they have been excavated is a risky business at best. A basal bulb that is in a state of activation of leaf primordia that will reach toward the surface may not yet have produced visible evidence of what is about to happen. In this instance, the newly formed basal bulb located at the terminus of a rhizome may look very much like a newly formed tuber that is in a similar position. It is their activity that distinguishes them: the tuber remains temporarily dormant while the basal bulb differentiates into a new aerial shoot.

In one of the most detailed studies ever made on this species, Hauser (1962a) found no evidence of tuber formation in the first 4 weeks after the plant had emerged, but the number of basal bulbs had increased five times. In other words, all of the enlargements at the ends of the rhizomes were able to produce new plants and there seemed to be no dormancy. At 6 weeks there were still no tubers; at 7 weeks the first flowers appeared; at 8 weeks 94 percent of the plant units or systems had at least one flower, and now the first tubers appeared. The tubers formed on rhizomes at or near the youngest plant that was the first to grow from the original tuber planting. There were no tubers on plants that had already flowered and there were no chains, only single tubers, at this time.

At 10 weeks most tubers were still singles with an occasional pair connected. Some of the first-formed tubers now appeared to be mature and were brown to black in color. A four-tuber chain was found 3.5 months after the original tubers were planted, and at 4.5 months longer chains were seen.

Thus, in these experiments the first dormant tubers began to form sometime between 6 and 8 weeks, and this was also the period of early flower formation. Elsewhere, whether because of different climatic zones or different ecotypes of the species, flowering has been recorded in 3 weeks in Israel and

India and 4 weeks in Trinidad, with tuber formation occurring at 3 weeks in Hawaii, India, Puerto Rico, the southeastern United States, and Trinidad. The correlation between flowering and the first dormancy in the tubers needs to be explored elsewhere in other climates, for the phenology of the growing cycle will be quite different at increasing distances from the equator.

The mechanism of dormancy has been little explored in *C. rotundus*, although there are many experiments on breaking the rest period. Palmer and Porter (1959a,b,c) explored the possibility that dormancy of tubers in the soil was due to lack of aeration. They found that tuber-sprouting reached a very high level at high oxygen levels and low carbon dioxide levels. They also found that this tuber-sprouting suppressed apical dominance and that a greater percentage of tubers on a string or chain could produce shoots. Low oxygen and high carbon dioxide levels inhibited germination.

Berger (1966) searched for growth inhibitors that might be the cause of tuber dormancy. The most active compound was salicylic acid, which could be extracted from leaves and tubers that had sprouted. Although this chemical was not found in dormant tubers, Berger felt that it may be the major cause of seasonal dormancy. Short photoperiods induced flowering, tuber production, and the production of salicylic acid. Long photoperiods did the opposite (see also Jangaard, Sckerl, and Schieferstein 1971 and Agundis and Valtierra 1963).

Tuber Germination

Muzik and Cruzado (1953) described the apical dominance of *C. rotundus* and its tuber chains in the following way. First, the interconnected aerial and underground systems that arise during one growing season from one planted tuber must be considered as a unit. If we think of this unit as a long, slender stem that is itself without buds at maturity, that is interspersed at irregular intervals with swellings or enlargements which are crowded with buds (the tubers), and that terminates in a leafy shoot, then the behavior of this plant unit is comparable to the behavior of other higher plants in which apical dominance is so evident. The apical bud of a single tuber always sprouts first and this is still true for the top one-half of a tuber cut horizontally. On the bottom half, the bud nearest the cut surface germinates first for it was nearest the apical end in the intact tuber. The upper tuber exerts apical dominance over all other tubers in a long chain, but this control is not as strong as the dominance within one tuber. Tuber chains that are oriented in a horizontal position or that are turned upside down show sprouting of all tubers in the chains. The separation of a tuber from a chain removes it from apical dominance, and this fact has important implications for tillage operations that tear the plant units apart so that single tubers are distributed through the plow layer (see also Smith and Fick 1937).

Andrews (1940) in Sudan, where there are long periods in which soil moisture is very low, was interested in the minimum level at which sprouting could begin. He planted tubers in a heavy soil placed in containers which were almost sealed and in which soil moisture contents of about 10, 20, 30, 35, 40, 50, and 60 percent had been established. Eight days later there was no sprouting at 10-percent, very little at 20-percent, but much activity of buds and roots at 30-, 35-, and 40-percent soil moisture. From data obtained in other experiments he reported that dormancy was induced in tubers held for 2, 3, and 4 weeks in soils with moisture levels at 50 and 60 percent. The tubers were not killed and all germinated well when again placed in favorable circumstances. To study the point at which soil moisture becomes critical for plants with aerial shoots, he placed tubers in containers and watered them from below until the 10th day. As watering ceased, records of aerial growth were made daily; and as the plants ceased growing, the soil moisture content of some of the containers was measured. The moisture determinations were made at 3, 5, 7, 9, and 10 days in still other containers. The tubers were then resown to determine how many were still viable. From these trials he learned that, in the heavy soils of the Gezira area, the aerial growth of tubers with established root systems begins to wilt (and therefore ceases growing) when the soil moisture content falls below 20 percent. The maximum mortality of tubers comes when soil moisture is reduced to 8 percent.

Ueki (1969) found no germination in fields which were under water and suggested that, at the latitude of Japan, fields must be quite dry to obtain tuber sprouting. However, excess moisture does not necessarily destroy the tubers, for some that had been held in water for 200 days gave satisfactory germination when removed from the water and placed in suitable growing conditions.

Muzik and Cruzado (1953) planted tuber systems against a piece of glass that formed one side of a container filled with moist soil and planted others that they kept in complete darkness. They found extensive sprouting in the light but little in darkness. Ueki (1969) in Japan obtained similar results but he also showed that sprouting was about the same whether the light provided was red, green, blue, or white.

Shading of nutsedge foliage by taller crop plants was studied by Magalhaes (1967), who found that the jack bean, *Canavalia ensiformis*, greatly inhibits tuber formation. The beans quickly closed in the area to allow only 1-percent daylight to reach the weed. In tree orchards and plantations, top growth of *C. rotundus* begins to decline as heavy shading is provided by tree growth. This does not mean that the infestation has been destroyed, however, for, as soon as openings are made available by diseases or tree-cutting, the weed stand is quickly restored. Similarly, in sugarcane, the tall plants eventually shade out the top growth of *C. rotundus*, but the tubers only become quiescent and quickly reappear as later harvest and tillage operations move into the fields.

There appears to be no information on the effects of the photoperiod on tuber germination. The exposed buds of the tubers may prove to be sensitive as are the buds of trees.

Temperature is one of the important factors controlling germination of tubers. Japan is one of the coolest areas of nutsedge infestation, and Ueki (1969) has pointed out that the soil temperatures in upland fields in early May are about 15° C. The weed begins to appear about this time but comes on very slowly. The sprouting occurs over such a long period that he suggested that the fight against top growth is futile. In experiments he obtained 95-percent germination of tubers at 30° to 35° C, with no sprouting above 45° C or below 10° C. In incubation tests, tubers survived for 10 days, 12 hours, and 30 minutes at 45°, 50°, and 60° C, respectively. At –20° and –5° C they did not survive more than 2 hours. In Japan the northern limit of nutsedge is in a region where the average minimum atmospheric temperature is −5° C in winter. Confirming evidence has also been found in India that high temperatures favor tuber germination, and an advantage has been shown also for the use of alternating termperatures (Tripathi 1967). Sprouting of more than 50 percent of the tubers occurred between 13° and 40° C, but there was no germination at 50° C. At 40° C more than 80 percent of the tubers sprouted, but the germination was much prolonged as compared with other treatments. A remarkable stimulation of new growth was found at an alternation of 23° C (7 hours) and 31° C (17 hours). More than 80 percent of the tubers sprouted in 2 or 3 days.

In the United States Smith and Fick (1937) found

that tubers held at 50° C for more than 48 hours would no longer germinate, but exposure at −4° C for 8 hours did not impair viability.

The possibility that tillage operations can raise the dormant tubers to the surface where they are subject to extremes of temperature and drying has appealed to many workers. Some have also attempted simply to desicate the tubers by allowing fields to become as dry as possible.

Because *C. rotundus* is such a destructive weed, and because the tubers are generally regarded as the means by which this nuisance maintains itself and spreads, the weed is generally held to be well nigh indestructible. Smith and Fick (1937) were surprised to find that tubers left in the open in a laboratory very soon lost viability. There is general agreement that moisture content of the tubers is about 50 percent at harvest. Smith and Fick placed tubers in the following places during an experiment: (1) a laboratory desicator, (2) on the surface of dry soil in full sun, (3) in open air in a laboratory, and (4) in a storeroom which was more humid than the laboratory. The organs were desicated beyond recovery in 4 days in the sun, 16 days in the laboratory air or the desicator, and 32 days in the storeroom. In all cases the tuber moisture content was 15 percent at death, except that in sunlight it was 24 percent. In experiments approximating field conditions, tubers were planted at 5 and 10 cm in dry soil that was protected from rain but was exposed to direct sunlight. At 5 cm 80 percent of the tubers were killed in 8 days and all were dead at 12 days. All of the tubers at 10 cm were killed in 16 days. Andrews (1940) in Sudan, Ronoprawiro (1971) in Indonesia, and Ueki (1969) in Japan obtained similar results on tuber drying. Tripathi (1969*a*) in India placed tubers in open air at 22° to 26° C to study weight loss. Samples were taken each day to determine the remaining moisture content. The tubers contained 183 percent moisture (dry weight) at the outset and 65 percent at germination. After 2 days the moisture content had fallen by about one-third but there was no decrease in germination. After 4 days two-thirds of the moisture had been lost and germination was less than 40 percent. After 8 and 12 days moisture content fell to 30 and 15 percent, respectively, and germination was about 10 percent; beyond this time all tubers were killed.

The importance of an optimum level of oxygen and its significance in controlling morphological development are often implied in the literature, but, except for a general agreement on the need for some oxygen in the soil, these matters are far from clear (Palmer and Porter 1959*a*, Ueki 1969). Standifer,

Normand, and Rizk (1966) have speculated about the role of oxygen in tuber dormancy and basal bulb formation. They implied that sufficient oxygen may allow the terminal meristems of rhizomes to continue active growth, with the production of aerial shoots and the enlargement which we call a basal bulb. If oxygen is low, as may be the case in deeper layers of the soil, the terminal meristem of the rhizome may enlarge and provide a dormant tuber. This raises the question of the basal bulbs found rather deep in the soil by Hauser (1962*a*) and the need to know whether oxygen concentration has a significant role in the formation of such bulbs.

The hope for starvation of underground organs by constant removal of the tops has caused considerable experimentation. Misra (1969) prepared pot experiments in which the tops were clipped 0 to 16 times at intervals varying from 3 to 20 days. Tuber and basal bulb formation were severely inhibited by such treatments. The number of these organs was reduced by about one-third, one-half, and two-thirds when clipping was done at 20-, 10-, and 3-day intervals, respectively. In Israel, clipping every 2 weeks reduced tuber numbers by 60 percent and weight by 85 percent (Horowitz 1965).

A rhizome- and tuber-germinating technique reported by Thomas (1967*b*) may be useful for those engaged in studies on tuber germination. These vegetative propagules are carried on the farm by tillage implements and are moved into ditches by sheet and gully erosion to be distributed further by running water. Severe storms which inundate fields may bring in tubers which have been washed up elsewhere. *C. rotundus* comes to new localities with nursery stocks that often have been shipped long distances. Tubers are known to be caught up in the harvest and bagging of such crops as sweet and Irish potatoes.

Flowering and Seed Production

As the flowering stalk extends, the short, tightly clustered spikes of the inflorescence become longer and open up. The stigmas lengthen and protrude from the lower bracts of each spikelet. The umbel-like head is now above the leaves and continues to grow until anthesis. The early stigmas appear to curl and shrivel in 3 to 4 days and in another day or two anthesis begins. Flower-opening progresses upward from the bottom of the spikelet with two to four opening each day. On an average spikelet, therefore, some stigmas at the top of the spikelet will be receptive when anthers appear at the base. Much pollen is produced. The grains are smooth, light, and about

0.03 mm in diameter. The flowers are cross-pollinated mainly by wind. Honess (1960) in Trinidad reported that much of the pollen he examined was shrivelled and empty, and he believed much of it to be nonviable.

The most complete study of seeds of *C. rotundus* and their germination is that of Justice and Whitehead (1946). These two researchers initially had difficulty in locating stands of the weed that were producing seeds. Seeds were finally obtained from Alabama (United States) fields that had been cultivated in early spring. *C. rotundus* plants growing in perennial grasses, dry upland cornfields, or crops cultivated in late June or July had few seeds. Mature seeds were sometimes found in corn growing in moist river valleys.

Seeds tested in a greenhouse bench or in an incubator with alternating temperatures immediately after harvest did not germinate for 3 to 4 months. Germination continued to increase during the 13-month test. The highest germination was 15 percent but most lots averaged 3 to 5 percent.

An elaborate seed storage experiment was prepared and, at the outset, the seeds were separated into four categories according to their specific gravity. The two fractions of lowest weight contained seeds which were shrivelled and empty. In the third fraction some seeds had a small amount of endosperm but no embryo. Seeds of the heaviest fraction contained embryos and these were used in the storage experiment that was designed to hasten after-ripening by providing several different environments. The following storage conditions were tested over a 28-month period: (1) room temperature; (2) 2° C; (3) 10° C; (4) 2° C in darkness for 16 hours, then 20° C in light for 8 hours; (5) similar to (4) but with a dark temperature of 10° C.

Seeds were removed from the storage area and all were tested for germination at an alternating temperature of 20° C in darkness for 16 hours and 35° C in light for 8 hours on a substrate containing potassium nitrate. Seeds stored both at room temperature and at 10° C showed 1 percent germination at 4 months. Only seeds held at 2° C had not shown some germination by 7 months. All storage conditions brought some germination improvement with time, but alternating temperatures and dark and light gave best results. Some lots showed 10- to 18-percent germination but, of the 13,500 seeds used in all of the experiments, only an average of 4 percent germinated.

They confirmed a report from Ranade and Burns (1925) in India that heating the seed for 1 to 3 hours at about 60° C greatly increases germination. Only

about one seedling was established, on the average, out of all the seeds from each five inflorescences. No correlation was found between the presence of embryos and percent germination, for other factors obviously were involved in the stimulation or inhibition of the seeds. It was concluded that reproduction by seed is of little significance in the southern United States. It was also pointed out that, in the 5 years prior to 1946, not one seed of *C. rotundus* had been found as a contaminant in the samples of agricultural seeds tested in the Federal Seed Laboratory in Maryland or in the Alabama Seed Testing Laboratory.

The seed production of *C. rotundus* on the Gezira cotton scheme in Sudan is largely confined to the 3 months of the rainy season which extends from June into September (Andrews 1946). Tall flowering stems to 75 cm may be seen in August on low fields subject to flooding. The seeds are blown from the plant or soil surface and are distributed by small cyclonic winds. They infest the banks of irrigation canals, are thick at the water's edge, and if water is low may cover the bottom of the canal. Seeds produced on these plants obviously would be moved to the fields, for the seeds can float for a considerable time. Andrews strained 1,150 seeds from 3,500 cubic meters of irrigation water, and from this he calculated that each hectare could receive an annual supply of 5,000 seeds as the land was irrigated. He later sieved the tubers from the soil to study the *C. rotundus* plants that had emerged from the seeds. The total number of seeds in the field plots could not be counted, of course, because of their size, but it was shown that the plants which did emerge from seeds could resupply each square meter with 400, 800, 2,000, or 4,000 tubers if allowed to grow for 47, 78, 110, or 141 days, respectively. Most seedlings emerged from a depth of 1 to 1.5 cm and none came from the 2.5-cm level.

Seeds produced in Sudan would not germinate just after harvest. Germination increased with storage up to 7 years. Scarification of the seed with acid hastened germination. Andrews also showed that seeds planted in unsterilized soil gave much better results than those planted in sterile soil, and he concluded that microorganisms assist in the breakdown of the heavy fruit coat so that the embryo can be released.

A very low seed production in Trinidad and few viable seeds were indicated in Honess's (1960) studies. He found that no flowers were formed until seven leaves had been produced, with the peak of flowering coming at the nine-leaf stage. In the areas under study he showed that the seeds remained in the spike for a long time. Counts made when seeds were mature and then 2 months later showed that all seeds had remained in place.

Tripathi (1969*b,c*) found two flowering and fruiting periods in India: February to May and July to November. About 170 seeds were produced on each aerial shoot in May and 260 in November. Freshly harvested seeds gave 16-percent germination when incubated at 30° C and sprouting increased with age. Scarification in sulphuric acid for 15 minutes increased germination in some lots from 2 percent to 26 percent. This must certainly be one of the optimum areas of the world for seed production by *Cyperus rotundus*.

SUMMARY OF THE LIFE HISTORY

To undertake a life history of *Cyperus rotundus* at this moment seems a risky proposition because of the lack of understanding of many of the developmental processes, the contradictions in the literature, and the variations in the performance of the plant across the world. It is, however, the number-one adversary in the weed world, and the venture of mapping its growth cycle in light of our contemporary information may bring help to those who cannot see the major pieces of literature on the species but who must contest it in the fields. Parts of the growth cycle are well understood and some known facts of morphology and physiology always bring surprising agreement wherever and whenever they are studied. But there is much wasted effort, for many needless experiments have been carried out only to provide information on which substantial agreement has already existed for decades. There is an urgent need to coordinate efforts. There are new ideas at work in discovering the ways of *C. rotundus*, and there are many obvious voids and gaps in our understanding which will test the ingenuity of weed workers who are interested in the biology of the species. For all of these reasons it seems wise to try to draw together our present knowledge, that we may keep our direction as we seek to understand the form and function of this constant companion of man's crops.

Literature citations have been presented with the details of the research in the preceding sections and will not be repeated in the summary of the life history which follows.

Almost every new *Cyperus rotundus* plant originates from a tuber. Seed production, possible everywhere in the world, is unimportant. The sprouting tuber produces a rhizome which terminates as a green aerial shoot. In the process of emerging from the soil a swelling appears on the rhizome,

often near the surface but sometimes at a 20-cm depth. This enlargement, generally called a "basal bulb," is believed by some to be a corm and by others to be a specialized structure which may not be described with standard terminology. Removal of the basal bulb shortly after the plant has become established results in decreased growth of the aerial shoot. The ontogeny of the basal bulb and the factors in the environment that cause its initiation have been the subject of much speculation but little experimentation. Roots are formed on the basal bulb and rhizomes grow out from it for a distance of 1 to 30 cm horizontally before the tip turns up to produce a new aerial shoot with another basal bulb, or, alternatively, to form a subterranean tuber from which another rhizome appears at the apical end. Chains of tubers are thus formed, some deep in the soil. It is believed that there are no buds at the nodes of the rhizomes and that little or no propagation is possible from fragments of these organs. The apical bud of a tuber inhibits the buds below, and the tuber at the morphological apex of a rhizome tuber chain prevents sprouting in the chain. Rhizomes and tubers are white and fleshy when young and some become firmly packed with starch. On aging they darken, harden, and most of the tissue exterior to the endodermis of the rhizomes sloughs off to give a wiry structure which is resistant to desiccation and decay. For at least one season the vascular system is continuous from the aerial shoot throughout the entire plant unit.

The species is found on all soil types and from sea level to high mountains. It is sensitive to shade and grows in wet and dry soils and climates. In cool areas or in waterlogged soils it grows slowly, flowers little, and produces few tubers. It dies back, becomes quiescent, and is carried through extremely dry periods by the tuber system which draws its water from a root system that may penetrate deeply into the soil.

There is general agreement that the plant is stimulated to flower in short photoperiods of 6 to 8 hours, whereas the period from emergence to flowering varies from 3 to 8 weeks. There are some indications that short daylengths also stimulate tuber formation. Flowering is common everywhere, and many seeds are produced—although most of these are shrunken and have a shrivelled appearance. In some areas only one viable seed is formed for each four or five inflorescences, whereas elsewhere several good seeds may come from a single inflorescence. In general, however, the appearance of large viable seeds with plump embryos is rare. Seed germination seldom averages more than 1 to 5 percent.

The dormancy of the seed is difficult to describe because testing procedures are often incomplete and not uniform. Dormancy may be summarized by saying that few or no seeds will germinate at first maturity, but that storage—dry or moist—in soil or in laboratory containers allows for afterripening and will increase germination with time. Seeds that have been heated to 60° C for 1 to 3 hours or that have been stored for longer periods in conditions of alternating temperature and light and dark will provide the best germination for experimental purposes. The flowers are cross-pollinated, mainly by wind, and most pollen is shrunken and nonviable.

The seeds are distributed by wind or sheet erosion, transported in mud, or floated onto fields by flooding streams or with irrigation water. Most seedlings come from a 1- to 1.5-cm depth, and few can penetrate 2.5 cm of soil. Very few seeds of C. rotundus are found in commercial seedstocks.

Studies on tuber production and germination overshadow all else in the literature on C. rotundus. Most tubers are found in the top 15 cm of soil, rarely below 30 cm, and when planted at 90 cm are unable to grow to the surface. There are now indications that tubers do not form until flowering begins. The time from the emergence of a new plant until its flowering varies greatly across the world. Flowering is said to begin at the six- to eight-leaf stage in some regions. In other areas flowering has been reported 3 to 7 weeks after emergence, with first tuber formation at 3 to 8 weeks. The basal bulb and aerial shoot population may increase fivefold in the first 4 weeks after the first tuber is planted or begins growth. Starch reserves are highest in tubers near the surface in noncultivated land. In cultivated areas starch reserves are highest in tubers that are below the disturbed layer. There is speculation that, in the field, tubers deep in the soil remain dormant for lack of oxygen, but there have been no experiments to confirm this. A search is underway in several laboratories for growth regulators that may be involved in tuber dormancy.

Tubers dry quickly if they are detached from their rhizome and root system. The organs have about a 50-percent moisture content when they are dug, and they cannot survive when the level falls below 12 to 15 percent. Isolated tubers placed in air-dried soil in a field may not survive beyond 8 days. Experiments have shown that the tubers cannot survive more than 10 days at 45° C or 30 minutes at 60° C but nothing is known about the effects of high soil temperature for prolonged periods in the tropics. Experimental temperatures of -5° and -20° C killed tubers in 2 hours, but little is known about field survival in cold or temporarily frozen soils (Tripathi 1968b).

Tubers can survive long periods in very wet soils and at least 200 days under flood. They appear to become dormant but will germinate when again placed in favorable circumstances. Light promotes germination and a high temperature of 35° C is advantageous for experimental germination. In a heavy soil, very little sprouting can occur at a moisture content of 20 percent of the air-dried weight of the soil, but 30 and 40 percent permits good activity.

C. rotundus may produce up to 40,000 kilograms of subterranean plant material per hectare. There is recent evidence that the decaying organic matter from this and other perennial species may release toxic substances that can reduce the yield of crops. This chemical interaction between species may be an important factor in the ecological relationship of crop and weed in agricultural lands.

AGRICULTURAL IMPORTANCE

The information gathered on C. rotundus stands in such contrast to the data on all other species that a comparison seems in order. In the 7 years of searching, it was possible to select more than 700 entries to be added to the data from more than 90 countries, and many more reports were available. Each entry carries up to 10 details about the weed. About 200 of the entries concern such topics as taxonomy, biology, and distribution, and do not speak of a weed-crop relationship. The other 500 entries directly concern one or more crops and one-half of these were given a rank of importance by someone knowledgeable about weeds in his own country. Forty percent of all of these crop entries was given a ranking of principal, serious, or of being one of the three most serious weeds of a crop in a country. No other weed species approaches the seriousness recorded in these data by the individuals who are most qualified to judge the agriculture in their own country. No other weed has drawn the attention of man so frequently and prompted him to report on some aspect of its behavior.

Figure 3 portrays the importance of the weed in five of the world's principal crops. The picture is incomplete, for there was insufficient space to represent other major crops such as peanuts, sorghum, soybeans, and many plantation crops in which the weed is almost equally troublesome. It has been reported to be one of the three most serious weeds of corn in Ghana and the Philippines; of cotton in

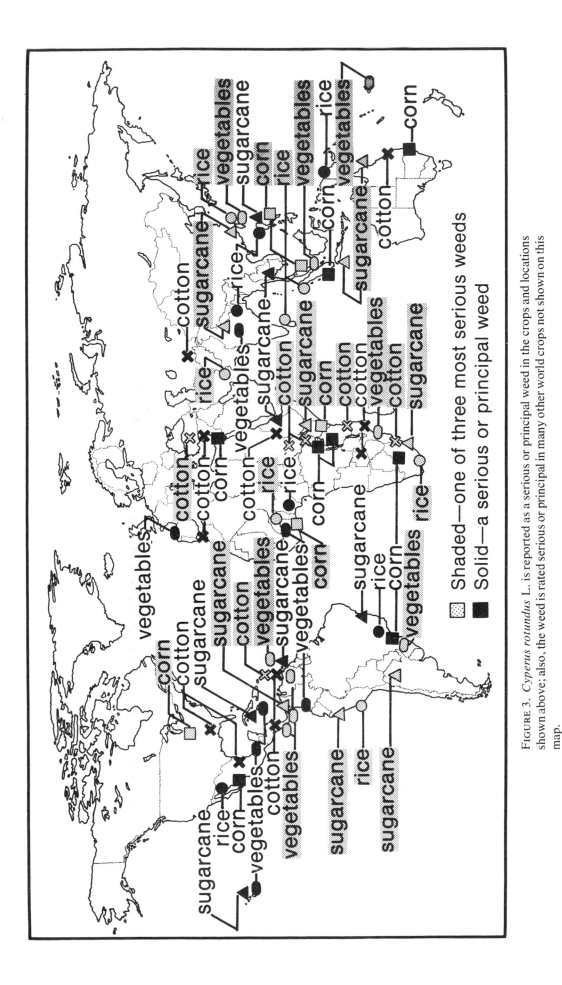

FIGURE 3. *Cyperus rotundus* L. is reported as a serious or principal weed in the crops and locations shown above; also, the weed is rated serious or principal in many other world crops not shown on this map.

Sudan, Swaziland, Turkey, and Uganda; of rice in Ghana, Indonesia, Iran, Peru, South Africa, and Taiwan; in sugarcane in Argentina, India, Indonesia, Natal, Peru, and Taiwan; and of vegetable crops in Brazil, Malaysia, Taiwan, and Venezuela. *C. rotundus* is a serious or a principal weed of corn in Australia, Brazil, Indonesia, Israel, Kenya, Malaysia, Mexico, South Africa, Tanzania, Uganda, and the United States; of cotton in Australia, Ethiopia, Israel, Mexico, Morocco, Mozambique, Nicaragua, the Soviet Union, Trinidad and the islands of the Lesser Antilles in the Caribbean, the United States, and Zambia; of rice in Brazil, Ceylon, India, Mexico, New Guinea, Nigeria, and the Philippines; of sugarcane in Australia, Brazil, Ethiopia, Hawaii, Jamaica, Kenya, Panama, the Philippines, South Africa, Thailand, and Trinidad; and of vegetable crops in Colombia, Costa Rica, Fiji, Ghana, Hawaii, India, Jamaica, Mexico, Mozambique, Panama, Spain, and Trinidad.

The above represent only about one-half of the entries on crops and countries in which the weed is of equal importance, but a recital of the others seems unnecessary. Some examples of the variety of problems and the nature of the difficulties caused by the weed may serve to underline the gravity of this weed problem in still another way.

Workers who have not had experience with *C. rotundus* in different crops and climates sometimes question the severity of this small weed on a world basis, but the relative size of biological organisms may tell us little about their ability to compete with one another. The array of mechanisms available to each organism in the mutual competition between them may be much more important than their size. One-celled microorganisms, for example, may destroy our largest plants and animals. The excellent experiments of Rochecouste (1956) on nutgrass in sugarcane in Mauritius have informed us about the manner in which ecological information can help us to understand the mechanisms of competition between crops and weeds. The details of his experiments and the data on production of green tops and of tubers under different climatic conditions were discussed in the section on the physiology of growth and development. Rochecouste has drawn our attention also to the way in which a plant of small stature, such as *C. rotundus,* can bring severe competition to a larger plant, such as sugarcane, a crop which may grow to 4 meters in height and which is harvested in greater bulk tonnage than any other world crop. He pointed out that, even in a humid region of more than 1,250 mm of annual rainfall, the production of 30,000 kilograms per hectare of green tops and of tubers of nutgrass can severely limit the availability of water in some seasons. His data show that in the humid zone the following quantities of fertilizer may be mobilized and stored in the weed: 815 kilograms per hectare of ammonium sulphate, 320 kilograms of muriate of potash, and 200 kilograms of superphosphate.

This competition in sugarcane has been confirmed in several places in the world. Experiments in Argentina, for example, have shown that in extreme cases the weed may reduce cane harvest by 75 percent and sugar yields by 65 percent (Cerrizuela 1965).

In experimental plots in Australia, sugarcane yields, even with cultivation, were reduced by 38 percent. It was suggested that the cause of the yield reduction is a competition for moisture at stooling time so that fewer canes are produced (Chapman 1966).

In Panama the propagules are distributed with each flooding of the river valleys, and some cultivated lands have been abandoned. Vegetable culture in the area is extremely difficult because the rhizomes pierce the roots and underground storage organs.

When irrigation was brought to the older parts of the Gezira cotton scheme in Sudan in 1925, *C. rotundus* increased very rapidly and the reduction of cotton yields brought considerable urgency to the task of finding control measures for the weed. If the weed was allowed to remain in corn fields in Colombia for the first 10 days of growth, the yield was reduced by 10 percent. If allowed to remain 30 days the yield dropped by 30 percent (Cruz et al. 1969).

In Australia and Italy tobacco-growing has been discontinued in some areas because the costs of hand-weeding and crop-reduction make it profitable no longer. In mulberries in Japan and lemons in Israel *C. rotundus* is an important competitor that can reduce tree growth. The weed is very competitive in coffee in Kenya, and some scientists believe that it may cause the death of coffee trees. In addition to competing for moisture, the weed withdraws and holds large quantities of nutrients in its massive underground system and these are not available to a current crop (Bhardwaj and Verma 1968).

Reports of control measures are legion in the long history of the struggle against *C. rotundus*. Until recently, of course, such measures were confined to tillage operations or rotations in which cultivated crops were included to provide disturbance of the underground system. The methods depended on a desiccation of the tubers at the surface or a starvation in the soil by isolation from their tops and root systems. All of these techniques are possible, of course, but most are not very practical because the land often cannot be used for cropping for two or three seasons. Smith and Mayton (1938, 1942) in the United States plowed or disked a sandy loam at intervals of 1, 2, 3, or 4 weeks for two entire growing seasons and found that a tillage operation each 3rd week reduced the tubers by 80 percent. The tubers increased when the interval was 4 weeks. These experiments were extended to several soil types from light to heavy with similar results being obtained. Tubers were most difficult to kill in heavy, wet clay. These methods, of course, would not allow the growth of a summer crop. Crops which can be planted in October and harvested in June, such as hay or winter grain, could be used in the scheme. Smith and Mayton also found that a clean-cultivated crop in the 3rd year helps to inhibit the weed population after the 2 fallow years. Sinha and Thakur (1967) repeated these experiments in India with similar results.

Recent research has turned away from cultural methods and cropping systems, the quest now being for translocated herbicides that will move to all parts of the plant unit. Parker, Holly, and Hocombe (1969) have reported recently on 10 years of testing for herbicides to control *C. rotundus*.

The knowledge that has been gained in the past half-century of the anatomy and physiology of the species must be taken into account for any method of control. To be effective, herbicides must be transported to the meristems of the most distant rhizomes and tubers. The fact that the vascular system of an entire plant unit remains intact for a growing season—sometimes even when all aerial structures have been destroyed—operates to the advantage of the use of translocated herbicides. The fact of apical bud dominance in both tubers and chains of tubers must be taken into account, however, when assessing the response to synthetic chemicals.

From the practical viewpoint, the knowledge that the tubers dry at a rapid rate is an important discovery. Machines which can turn up the tubers for exposure to the heat of the sun and for desiccation in the open atmosphere of the soil surface at regular intervals will surely reduce the tuber population very quickly. One should be forewarned, however, that a simple disturbance which allows the tuber to remain buried and still attached to roots and rhizomes will be of little value. Such a method may actually break some chains to give rise to new plant units, while also moving tubers to new locations in the field to grow again.

Recent evidence of a chemical interaction between *C. rotundus* and crop plants suggests a mechanism for yield reduction that has not yet been taken seriously in agriculture (Friedman and Horowitz 1970, 1971; Horowitz and Friedman 1971). In early experiments Friedman and Horowitz were able to show that an extract of soil incubated with pieces of tubers and rhizomes of *C. rotundus* would inhibit radicle growth of crop plants at germination. Later, rhizomes and tubers were allowed to decay in soil for periods of 1 to 3 months before barley was planted. Crop growth was inhibited 15 to 25 percent by the residues in the soil, and, in one instance, root growth in a light soil was reduced 40 percent. Horowitz pointed out that the root systems of crops grow into layers of soil that contain very large amounts of decaying residues from subterranean systems of perennial weeds. The release of inhibitory substances from the residues may have ecological significance if they do increase injury to crops in a manner which has not been considered previously.

The threat presented to all of agriculture makes domestic usefulness seem unimportant. In China the weed traditionally has been used for medicinal purposes, and a recent report from that country concerns its use in landscaping. It is sometimes used as a soil binder in India. It may be used as fodder where green matter is critically needed, although many animals prefer other vegetation. It quickly becomes fibrous as it ages. Pigs turn over the sod to get the starchy nuts.

C. rotundus is an alternate host of *Fusarium* sp. and *Puccinia canaliculata* (Schw.) Lagh. (Raabe, unpublished*); of abaca mosaic virus (Gavarra and Eloja 1970); and of the nematodes *Meloidogyne* sp. (Raabe, unpublished) and *Rotylenchus similis* (Cobb) Filip. (Filipjev 1936). *Tylenchus similis* Cobb has been reported by Muir and Henderson (1926) as

---

* "List of diseases found on the island of Hawaii in 1963" is a report compiled in 1963 by Dr. R. D. Raabe, formerly professor of plant pathology, College of Tropical Agriculture, University of Hawaii. This report was then enlarged by members of the Department of Plant Pathology to become a checklist of plant diseases in Hawaii. Professor A. Martinez graciously allowed us to use it while we were writing this book. His address is: Department of Plant Pathology, College of Tropical Agriculture, University of Hawaii, Honolulu, Hawaii 96822, U.S.A.

occurring in nutsedge tubers, but this nematode is not sufficiently destructive, however, to be materially helpful in eradicating the sedge.

The nutgrass moth, *Bactra truculenta*, which bores into the stems of *Cyperus rotundus*, showed promise for biological control in Hawaii in the early years after its introduction from the Philippines in 1925. As the populations of *Bactra* increased, so also did those of the insect *Trichogramma minutum*, which parasitizes the eggs of many moths and butterflies. So many of the eggs of *Bactra* were killed that biological control of nutgrass was never attained.

COMMON NAMES
OF CYPERUS ROTUNDUS

ARGENTINA
  castañuela
  cebollín
  cebollita
  chufa
  chufila
  cipero
  contra yerba
  juncea
  junquillo
  lengua de gallina
  negrillo
  paraquita
  pasto alemán
  pasto bolita
  pasto inglés
  tamascán
  totorilla

AUSTRALIA
  nutgrass

BANGLADESH
  dila
  motha

BERMUDA
  nutgrass

BURMA
  monhnyin-bin
  monhnyin-u

BRAZIL
  tiririca

CAMBODIA
  smao kravanh chrouk

CEYLON
  kalanthi

CHILE
  almendra de tierra
  chufa
  coquillo

COLOMBIA
  coquito
  cortadera

CUBA
  cebolleta

DOMINICAN REPUBLIC
  jonquillo

EASTERN AFRICA
  nut grass
  water grass

EGYPT
  se'd

FIJI
  nut grass
  soronakabani
  vucesa

FRANCE
  herbe-à-oignon
  souchet d'Asie
  souchet en forme
    d'olive

souchet rond

GERMANY
  Apotheker-Cypergras
  Asiatisches-Cypergras
  Runde-Cypergras
  Runde-Zyperwurzel

GREECE
  kupere

INDIA
  coco grass
  dila
  gantola
  korai
  mutha (Oriya)
  nagar motha

INDONESIA
  teki (Javanese)

IRAQ
  oyarslan

ITALY
  cipero

JAMAICA
  coco grass
  nut grass

JAPAN
  hamasuge

KENYA
  nut grass
  water grass

MALAYSIA
  rumput haliya hitan

MAURITIUS
  herbe-à-oignon

MEXICO
  cebollín

NATAL
  brown nutsedge

NEW ZEALAND
  nut grass

PAKISTAN
  deela
  dila
  notha

PERU
  coco
  coquito

PHILIPPINES
  balisanga (Ilokano)
  boto-botonis (Bikol)
  mala-apulid (Pampangan)
  mutha (Tagalog)
  sur-sur (Pampangan)

PORTUGAL
  junca de conta

PUERTO RICO
  coqui
  coquillo

SOUTH AFRICA
  red grass
  rooiuintjie

SPAIN
  castañuela
  chufa
  juncia

SUDAN
  seid

SURINAM
  adroe

TAIWAN
  hsiang-fu-tzu

THAILAND
  haew moo
  ya-haeo-mu

TRINIDAD
  nut grass

TUNISIA
  souchet

TURKEY
  topalak

UNITED STATES
  purple nutsedge

URUGUAY
  pasto bolita

VENEZUELA
  coquillo
  coquito
  corocillo

ZAMBIA
  nut grass
  nut sedge
  water grass

❧ 2 ❧

Cynodon dactylon (L.) Pers.

POACEAE (also GRAMINEAE), GRASS FAMILY

Cynodon dactylon, a perennial grass, may be the most serious weed of the grass family. It is a native of tropical Africa or the Indo-Malaysian area, but its range extends from lat 45° N to 45° S. It is one of the principal weeds of corn, cotton, sugarcane, vineyards, and plantation crops. More than 80 countries have reported it as a weed problem in 40 crops. Some strains are very useful pasture grasses, other strains are used to prevent soil erosion, and some make excellent lawns and playing fields.

DESCRIPTION

C. dactylon is a long-lived, prostrate, fine-leaved *perennial* grass (Figure 4) that spreads by strong, flat stolons and scaly rhizomes to form a dense turf; *stolons* root readily at the nodes; *culms* erect or ascending, 5 to 45 cm (rarely to 90 or even 130 cm) tall, wiry, smooth, sometimes reddish; *leaf sheaths* up to 15 mm long, shorter than internodes, smooth; *ligule* a conspicuous ring of white hairs; *blades* 2 to 16 cm long, 3 to 5 mm wide, smooth or hairy on upper surface; *inflorescence* of three to seven sometimes purplish spikes in one whorl, in a fingerlike arrangement (digitately), 3 to 10 cm long, or in robust forms spikes up to 10, sometimes in two whorls; *spikelets* 2 to 3 mm long, in two rows tightly appressed to one side of the rachis; *lemma* boat-shaped, acute with fringe of hairs on the keel, longer than the glume; *seed* (grain) very small, 1.5 mm long, oval, straw-colored to orange-red, free within the lemma and palea.

The distinguishing characteristics of this species are the ligule, which is a conspicuous ring of white hairs; the lemma, which has a fringe of hairs on its keel; and the often gray-green color of the foliage.

DISTRIBUTION AND HABITAT

This species probably originated in tropical Africa; however, Eurasia, the Indo-Malaysian area, and India have also been suggested as its home. It now grows throughout the tropical and subtropical areas of the world (Figure 5) and it extends into temperate zones along the coasts. In eastern Africa it is distributed from sea level to 2,200 m and in Hawaii, from the beaches to 1,250 m. It prospers in the sun and dies out as shade increases. It is a warm-season grass, makes little growth in cold weather, and is quickly damaged by frost. It especially prospers where temperatures approach 38° C.

The species is adapted to a wide range of soils from sand to heavy clay, but it thrives best on a medium to heavy soil which is moist and well drained. It will grow on either acid or alkaline soils and can survive under flood conditions or drought. In South Africa it often takes over wastelands or overgrazed veld.

In the tropics, it is found in areas of 600 to 1,800 mm of annual rainfall, but it is also a weed of arid lands where it thrives along rivers and in irrigated areas. It can withstand long dry periods but is unproductive on dry soils. In Ceylon it is planted on tank bunds to hold the soil. Near the Roosevelt Dam in Arizona (United States) it survives flooding at high water and provides grazing for cattle at low water.

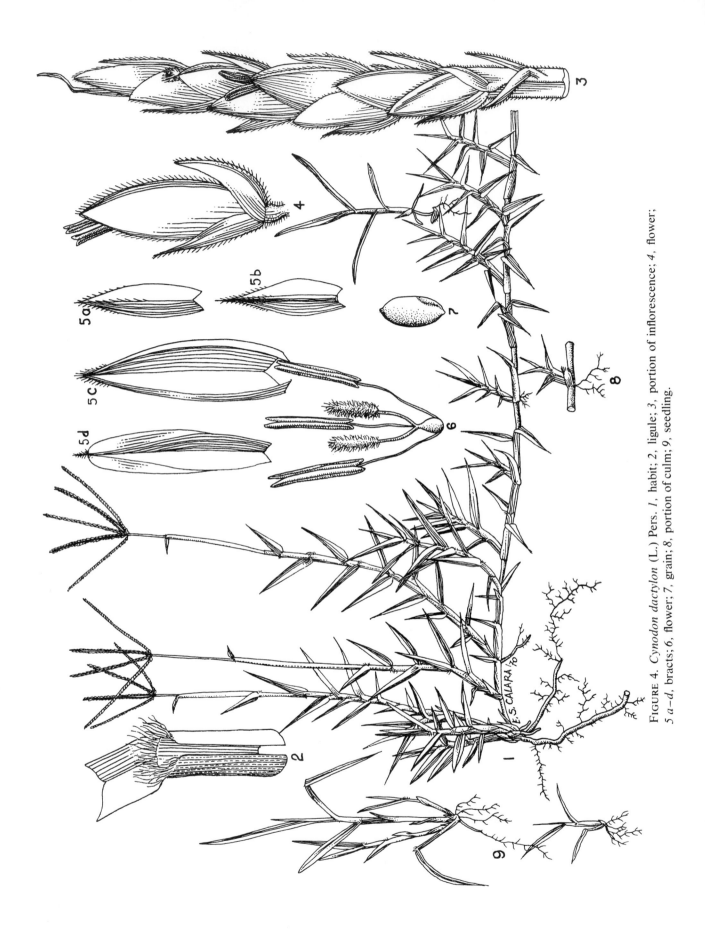

FIGURE 4. *Cynodon dactylon* (L.) Pers. *1*, habit; *2*, ligule; *3*, portion of inflorescence; *4*, flower; *5 a–d*, bracts; *6*, flower; *7*, grain; *8*, portion of culm; *9*, seedling.

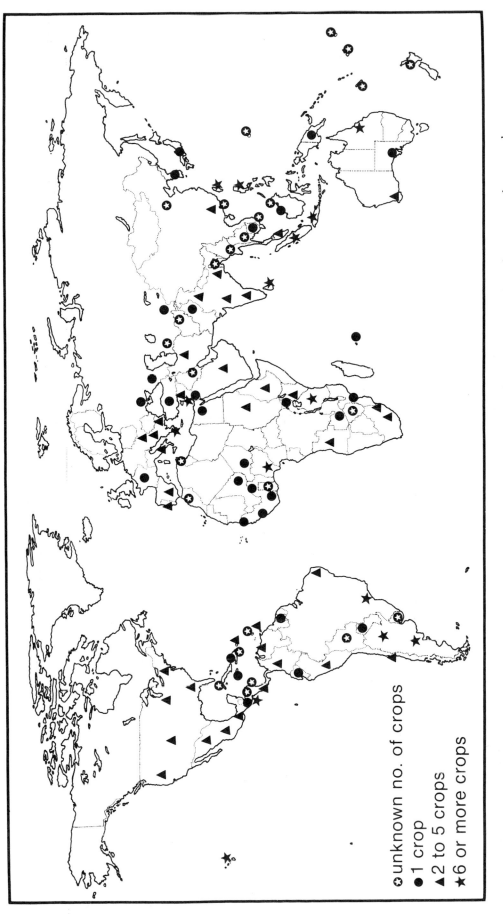

FIGURE 5. The distribution of *Cynodon dactylon* (L.) Pers. across the world in countries where it has been reported as a weed.

unknown no. of crops
1 crop
2 to 5 crops
6 or more crops

PROPAGATION

C. dactylon is a very sparse seed producer in most parts of the world; but good seed set may be obtained with some natural biotypes and improved varieties in Australia, India, and the southwestern United States. The seeds are very small, numbering about 4.4 million per kilogram. When ingested by cattle, the seeds remain viable and may show improved germination. Tests have established that seeds can survive at least 50 days of submergence in water.

The rhizomes and stolons provide the principal means of spreading. The vegetative plant parts catch in the mud on the feet of sheep and cattle at water holes. Blocks of floating sod on rivers and canals and rhizomes entangled in farm machinery provide means of distribution on land and water. Vegetative parts and seeds are known to move from port to port in the ballast of ships and in packing material. The seeds do not shatter easily when moved about in hay.

BIOLOGY

C. dactylon may have rhizomes which are quite superficial or very deep, ranging in soil depth from a few centimeters to a meter or more. This adaptation may be the principal factor in its ability to be an excellent pioneer weed, a weed of both waste places and arable lands, an inhabitant of several soil types, and a plant which can survive in extremes of climate.

The single bud of a rhizome or rhizome piece develops into a shoot, and the basal node of this shoot has lateral buds which give rise to tillers or rhizomes according to a well-defined pattern. Many have deep buds which can sprout from below the plow layer. As with some other perennial grasses, *C. dactylon* exhibits a seasonal pattern in the storage and depletion of its carbohydrate reserves. Where there is a winter season the reserve carbohydrates build up through autumn until midwinter. These reserves are stored in roots and rhizomes and are used in spring to support the growth of new shoots. The carbohydrate reserves then decrease until midsummer.

HABIT

C. dactylon is an example of a potential perennial which can behave as an annual and may fail to become perennial under the stress of long dry periods, overgrazing, or intensive weeding in the cropped phases of a rotation. In Sudan, where it normally has the habit of a stoloniferous perennial, it rarely sur-vives fallow periods in that form and may reappear during the rains as isolated tufts of two or three nodes only. In Sudan it is known as an all-season weed because it seems indifferent to season as long as it has water.

BIOTYPES

There are numerous natural strains or biotypes of *C. dactylon*. For example, 20 strains have been recognized in Hawaii and four in Mauritius. Rochecouste (1962*a*, 1962*b*), in an intensive study of the biotypes of Mauritius, described differences in the tendency to be upright or prostrate, the color of the plant and plant parts, and the length of the spikes. Hair group characteristics on vegetative organs were the best diagnostic characters for distinguishing biotypes. All had a very similar growth behavior. The tetraploid strains were more vigorous and one which had especially deep root penetration could withstand longer drought periods.

Through selection and hybridization, improved strains have become important pasture grasses (Harlan 1970). Dr. Glenn W. Burton of the United States Department of Agriculture and the University of Georgia led the way in breeding new types of *C. dactylon* for pasture, turf, and other uses (Burton 1966). *C. dactylon* is the principal pasture grass of India and of the southeastern United States. These strains are more resistant to grazing and trampling, and some are taller, larger leaved, more palatable, and more adapted to lighter soils and cooler temperatures. The strains are not suitable for pastures in rotation with cultivated land because they are too difficult to eradicate. Efforts are now being made to develop nonrhizomatous strains.

The plant parts develop hydrocyanic acid when allowed to wilt under some conditions. The percentage of toxic principle is high following a pronounced drought with high temperatures and following frosts. The toxicity of this species to cattle and horses has been demonstrated.

AGRICULTURAL IMPORTANCE

From Figure 6 it may be seen that *C. dactylon* is ranked among the three most serious weeds in sugarcane in Argentina, Colombia, India, Indonesia, Pakistan, and Taiwan; in cotton in Greece and Uganda; in corn in Angola, Ceylon, and Greece; in plantation crops in Kenya, Indonesia, and the Philippines; and in vineyards in Australia, Greece, and Spain.

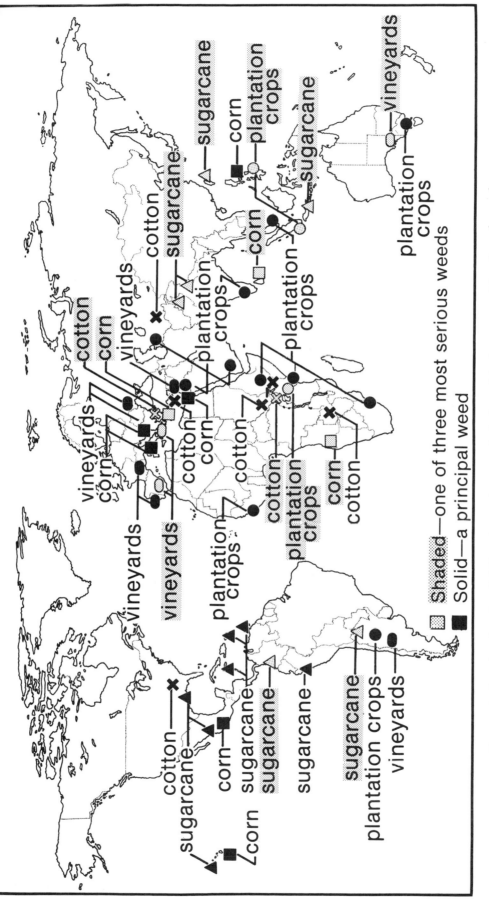

FIGURE 6. *Cynodon dactylon* (L.) Pers. is a serious or principal weed in the crops and locations shown above.

In addition it is a principal weed in sugarcane in Hawaii, Jamaica, Mexico, Peru, Puerto Rico, Trinidad, and the United States; in cotton in Israel, Kenya, the Soviet Union, Sudan, the United States, and Zambia; in corn in Hawaii, Israel, Italy, Mexico, the Philippines, and Yugoslavia; in plantation crops in Argentina, Australia, Ethiopia, Guinea, Kenya, Lebanon, Saudi Arabia, the Soviet Union, and Swaziland; and in vineyards in Argentina, France, Lebanon, Portugal, the Soviet Union, and Yugoslavia.

C. dactylon is also a troublesome weed in major world crops not shown in Figure 6. It is ranked among the three most serious weeds in rice, vegetables, and peanuts in Ceylon; in tobacco and sugar beets in Greece; in rice, rubber, and tea in Indonesia; and in coffee and rubber in Kenya and the Philippines. It is a principal weed in coffee, sisal, and tea in Tanzania; bananas, papayas, and pineapples in the Philippines; peanuts in Indonesia and Israel; rice in Brazil and India; sorghum in Israel and Italy; tea and tobacco in India; vegetables in Brazil, Colombia, Hawaii, and India; wheat in India and Jordan; bananas and orchards in Lebanon; coffee and rubber in Ethiopia; tea and orchards in the Soviet Union; and pineapples in Guinea. In addition, it is a common weed of taro, millet, flax, and pyrethrum in several places in the world.

The sugarcane industry in some parts of Hawaii was fighting for survival against this weed at one time. It is especially difficult to eradicate from ratoon cane because some rhizomes lie dormant under the cane stool and resume activity after harvest when environmental conditions become favorable.

C. dactylon is a species which should be carefully watched when preemergence herbicide programs for the control of annual weeds are initiated in vineyards and in similar cultures. When such competition is taken away from this perennial it may become dominant and provide a much more difficult weed problem.

With perennials there is always the question of using dry conditions to control a weed. W. Thomas (1970), working with material collected in Sudan, found that no buds survived on rhizomes stored 7 days in a light sandy soil which was air dry. He was able to show that, in general, the rhizomes are able to remain viable after suffering considerable water loss. On rhizomes which had a dry matter content of 42 percent, there were many buds which survived until the rhizome weight reached 50 percent of the original.

C. dactylon is an alternate host of *Puccinia cynodontis* Lacroix (Stevens 1925), *Ustilago* sp. (Raabe, unpublished; see footnote, *Cyperus rotundus*, "agricultural importance"), *Rhizoctonia solani* Kuehn (anon. 1960a), *Phyllachora* sp. (Chandrasrikul 1962); of the nematodes *Pratylenchus pratensis* (de Man) Filip. and *Meloidogyne* sp. (Raabe, unpublished; see footnote, *Cyperus rotundus*, "agricultural importance"); and the viruses which produce stripe disease of rice (Yamada, Shiomi, and Yamamoto 1956), barley yellow dwarf, and lucerne dwarf (Namba and Mitchell, unpublished*).

COMMON NAMES
OF CYNODON DACTYLON

ANGOLA
  usila
ARGENTINA
  chepica
  chepica brava
  grama
  gramilla
  gramilla blanca
  gramilla brava
  gramilla colorada
  gramilla forestal
  gramilla gruesa
  gramilla italiana
  gramilla perrera
  gramilla rastrera
  gramilla del tiempo
  gramón
  pasto bermuda
  pasto forestal
  pasto de perro
  pata de perdiz
  pie de gallina
  tejedora
  uña de gato
AUSTRALIA
  couch grass
BARBADOS
  Bahama grass

  Bermuda grass
  devil's grass

BRAZIL
  capim de burro
  grama rasteira
  grama seda

BURMA
  mye-sa-myet
  mye-sa
  mye-za-gyi
  myin-sa-myet

CAMBODIA
  smao anchien

CEYLON
  aruham-pul
  buha

CHILE
  pasto Bermuda
  pasto de gallina

COLOMBIA
Bermuda
  pasto Argentina
  pasto inglés

CUBA
  Bermuda
  hierba fina

* Dr. R. Namba and Dr. W. C. Mitchell are professors of entomology at the College of Tropical Agriculture, University of Hawaii, Honolulu, Hawaii 96822, U.S.A. They have kindly prepared a list of weeds that serve as alternate hosts for many important virus diseases of crops. We will refer frequently to this list in this work.

DOMINICAN REPUBLIC
grama fina de Bermudas

EASTERN AFRICA
Bermuda grass
couch grass
star grass

EGYPT
Negil

EL SALVADOR
barenillo
zacate de aguijilla
zacate de conejo

FIJI
balama grass
Bermuda grass
couch grass
kabuta

FRANCE
chiendent
chiendent dactyle
gros chiendent
herbes-des-bermudes
pied de poule

GERMANY
Echte-Hundszahn
Finger-Hundszahn

GREECE
agriada

GUATEMALA
grama Bermuda
gramilla
zacate de gallina

HAWAII
Bermuda grass

mahiki
manienie

INDIA
arugampul
devil's grass
doob
dub
duba
hariali

INDONESIA
djoekoet kakawatan
gigirinting (Sundanese)
grintingan
hoe maneek
sukit grinting (Javanese)

IRAN
cháir

IRAQ
thayyel

ISRAEL
yableet matsuia

ITALY
gramigna

JAMAICA
Bahama grass
dog's tooth grass

JAPAN
gyogishiba

LEBANON
shirch-un unjil
irk-en-najil

MALAYSIA
Bermuda grass

MAURITIUS
chiendent

MEXICO
agraisia
quick grass

MOROCCO
mor-chiendent

NETHERLANDS
hondsgras

NEW ZEALAND
Indian doab

NICARAGUA
zacate gallina

PAKISTAN
dub
khabbal
talla

PERU
grama bermuda
grama dulce

PHILIPPINES
babbalut (Bontoc)
galud-galud (Ilokano)
kawad-kawaran
kulatai (Tagalog)

PORTUGAL
grama

PUERTO RICO
ala quete queda
pelo de brujas
pepe ortiz
yerba bermuda

RHODESIA
couch grass

SAUDI ARABIA
nageel

SOUTH AFRICA
common quick grass

gewone kweekgras

SPAIN
grama
grama común
grama de España

SUDAN
nagila

SURINAM
tigriston

TAIWAN
gou-ya-gen

TANZANIA
couch grass
star grass

THAILAND
yah-phraek

TRINIDAD
Bahama grass
devil's grass

TUNISIA
chiendent

TURKEY
kopek disi ayrigi

UNITED STATES
Bermuda grass

VENEZUELA
pasto pata de gallina

VIETNAM
cò chỉ

YUGOSLAVIA
zubača

ZAMBIA
couch grass
kapinga

❧ 3 ☙

Echinochloa crusgalli (L.) Beauv.

POACEAE (also GRAMINEAE), GRASS FAMILY

Echinochloa crusgalli, an annual grass, is the principal weed of rice. A native of Europe and India, it has a range extending from lat 50° N to 40° S. It is a cosmopolitan weed that is troublesome in both temperate and tropical crops. Sixty-one countries report that it is a weed in 36 different crops. A closely related species, *E. crus-pavonis*, is reported as a weed in 10 countries, chiefly in rice. Some varieties are cultivated as cereals in the tropics and subtropics, whereas others are grown for coarse fodder and hay.

DESCRIPTION

E. crusgalli is a robust, tufted, *annual* grass (Figure 7); *culms* stout, erect, or decumbent, often rooting and branching near the base, to 1.5 m tall; *sheaths* smooth or often hairy; *ligule* absent; *leaf blades* linear with broad base and acute tip, smooth or with few hairs at the base, somewhat rough (scabrid) or smooth above, green, 5 to 50 cm long, 5 to 20 mm wide; *panicles* erect or nodding, green or purple-tinged, 5 to 20 cm long; *rachis* stout, nodes sometimes bearded; *racemes* numerous, 2 to 4 cm long (sometimes to 10 cm), spreading, ascending, sometimes branched; *spikelets* crowded, about 3 to 4 mm long, excluding the awns; *first glume* half as long as the spikelet; *second glume* and *sterile lemma* with short bristly hairs on the nerves, typically awnless; *awns* variable, mostly 5 to 10 mm (occasionally to 3 cm long); *grain* ovate, obtuse, tan to brown, with longitudinal ridges on the convex surface, usually 2.5 to 3 mm long, dorsally flattened. This is an ex-

tremely variable species which frequently has been split into various varieties and forms.

The absence of a ligule and the numerous racemes that are either spreading, ascending, or branched are distinguishing characteristics of *E. crusgalli*.

A closely related species, *E. crus-pavonis* (Figure 8), is much like *E. crusgalli* and is often difficult to distinguish from that polymorphic species. The rather soft, pinkish, or pale purple panicle with crowded spikelets with very long awns is characteristic of *E. crus-pavonis*.

Michael (1973), in collaboration with Dr. J. Vickery of the National Herbarium of New South Wales, has recently completed a long study of the *Echinochloa* species in the Asian-Pacific region. They have verified that the taxonomy of this important weedy genus is confused and that this confusion has resulted in difficulty in the interpretation of results of weed research. The *Echinochloa* species reported from several places in the world are discussed in their study and a partial clarification has been outlined for some of the species in Australia and other areas of the Pacific region. Dr. Michael has informed us that he wishes to receive herbarium material and seeds for growing from all parts of the Pacific region so that the work on these weedy species may be expedited. Desiring further examples from physiological and ecological research from several places in the world, he has pointed out that only when the identity of different species or forms of different species is known precisely can we understand their behavior in different regions and cli-

FIGURE 7. *Echinochloa crusgalli* (L.) Beauv. *1*, habit; *2*, ligule; *3*, spikelet, back view; *4*, spikelet, side view; *5*, spikelet, back view; *6*, flower, glumes excised; *7*, flower; *8*, grain; *9*, grain, cross section; *10*, seedling.

FIGURE 8. A, *Echinochloa crus-pavonis* (H.B.K.) Schult. *1*, habit; *2*, view of inflorescence and two views of spikelet; *3*, ligule. B, *Echinochloa crusgalli* (L.) Beauv.: view of spikelet and two views of inflorescence.

mates. This information is important for the accurate interpretation of herbicide trials as well.

DISTRIBUTION AND HABITAT

E. crusgalli is a native of Europe and India. It is now a common weed of most of the agricultural areas of the world, with the strange exception of Africa where it does not seem to be a problem (Figure 9).

In discussing the distribution, growth characteristics, and habitat requirements of this important weed, one must keep in mind that it exists in a number of races or ecotypes scattered over the world, and some of these may be distinct species (Yabuno 1966). Four definite varieties have been described in Japan and five have been described in the United States. Surely others exist in various parts of the world but we do not know at present whether any of the types are similar between regions.

This weed prefers wet soils and can continue to grow when partially submerged. On occasion it has been described as a weed of swamps and aquatic places. In drier soils it is not as tall and the yield of seeds and the numbers of panicles and tillers are reduced. It is normally found only at low and medium altitudes. It grows best in rich, moist soils with a high nitrogen content, but it can also thrive on sands and loamy soils.

Photoperiod is surely one of the most important factors governing the distribution and competitive ability of E. crusgalli. Experiments carried out in northeastern United States have shown that the plant will flower over a wide range of photoperiodic conditions. In 8- to 13-hour daylengths the plants passed into the flowering stage quickly and remained small in stature because vegetative development had been drastically reduced or had ceased altogether. In a 16-hour day, the plants were two times as tall and six times heavier; however, the number of leaves was the same in both cases, with those produced in long days being much larger. The number of panicles and tillers was larger in short days but the parts themselves remained small, whereas long-day plants went on to produce much larger panicles with more seeds. The rate of vegetative growth, as measured by height, appeared to be directly related to temperature, with slow extension of shoots in spring and very rapid growth in the heat of summer.

We have then a plant which normally appears to respond to short days by quickly flowering. When given favorable growing conditions and an extended period of long days it will produce very large, competitive plants which eventually flower and produce many seeds. In the temperate zone, at least, the understanding of its response to photoperiod makes possible some practical measures for control. Weed seedlings which emerge in spring and early summer have a long time in which to develop large plants before their vegetative growth is inhibited by the onset of the reproductive stage. Because this is also likely to be a period of optimum fertility and moisture supply, the plants may become large enough to be very destructive. Those which emerge in mid- or late summer, however, make only one-tenth or one-twentieth of the growth of seedlings that emerge in early season. This, together with the high susceptibility of E. crusgalli to crop shading, means that these late arrivals are not a serious problem. Therefore, because these plants grow in latitudes of decreasing daylength, it may be expected that the extensive vegetative growth found in long photoperiods will not be possible, the plant will become increasingly smaller, and fewer seeds will be produced (Vengris, Kacperska-Palacz, and Livingston 1966; HeJny 1957; and Li Sun-Zen 1962).

It appears then that some of the ecotypes can adapt to any photoperiod and that this adaptation may be an important factor in the plant's wide distribution as a cosmopolitan weed.

PROPAGATION

E. crusgalli reproduces and spreads by seeds. Much is known about the seeds for they have been the subject of studies in several places around the world. Reports of seed production vary from 2,000 per plant in the Philippines to 40,000 in Lebanon. Single plants in the United States have produced 5,000 to 7,000 seeds, and such production, in a weedy field, could result in a yield of 1,100 kg of weed seeds per hectare.

Arai and Miyahara (1960; 1962a,b,c,d,e) have conducted extensive studies on dormancy and seed germination. Generalizations about dormancy and storage viability are meaningless for these characteristics are peculiar to the ecotypes found around the world. In the Philippines some of the seeds will germinate immediately on harvest and within 3 months all seeds will respond. In Japan the dormancy period is 4 to 8 months and in the United States it is 4 to 48 months. Photoperiod and other environmental factors are known to influence the number of dormant seeds produced and the intensity of dormancy, and we may thus expect considerable

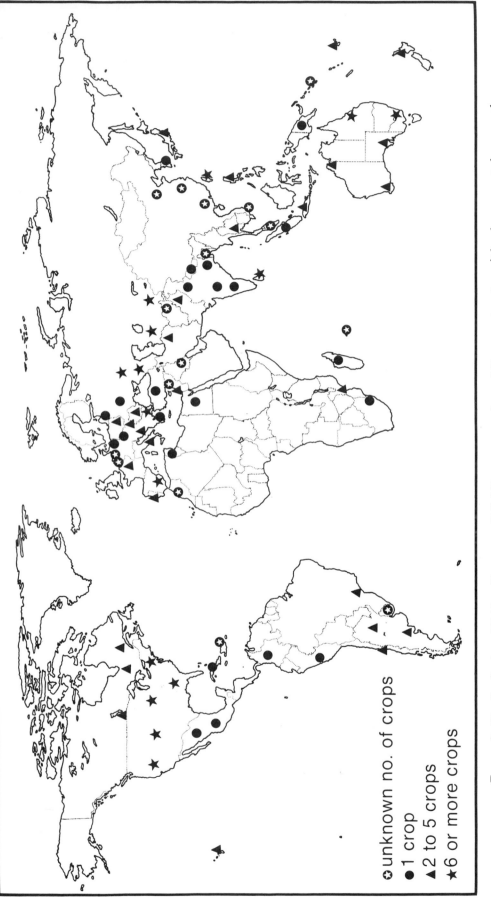

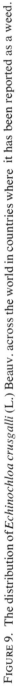

⊙ unknown no. of crops
● 1 crop
▲ 2 to 5 crops
★ 6 or more crops

FIGURE 9. The distribution of *Echinochloa crusgalli* (L.) Beauv. across the world in countries where it has been reported as a weed.

variation also with changing seasons and growing conditions in plants which thrive from the tropics to the temperate zone. Seeds were 100 percent viable after 7 years in dry storage and 33 percent viable after 1 year in soil storage in Germany. In the United States, 90-percent viability was found after 3 years' storage in either field or laboratory.

There is general agreement from several regions that the optimum temperature for germination is in the range from 32° to 37° C, and that germination falls off sharply below 10° and above 40° C. It is to be expected, therefore, that *E. crusgalli* seeds near the soil surface will germinate well on hot days.

In Germany, the optimum temperature for germination has been reported to be 20° C, and little response was found to changing soil moisture, light, *p*H, or treatment with salts. In the United States seeds in the light showed a tenfold increase in germination over those kept in the dark.

GROWTH CHARACTERISTICS

E. crusgalli is adapted to grow in wet soils. This is reflected in the influence of soil moisture content on seed germination and seedling emergence. Optimum moisture for germination in Japan varies with soil characteristics but is usually at 70 to 90 percent of maximum water-holding capacity. At this moisture content, maximum depth of seedling emergence was 10 cm, whereas in a water-saturated soil it was 3 cm, and when seeds were deeply submerged the distance was 2 cm. From elsewhere have come reports of emergence from 15 cm under favorable conditions. Elongation of seedlings was best at 30° C, was slow at 10° C, and ceased at 5° C. Elongation is more critically dependent on oxygen supply than is germination. At very low oxygen levels the seedling will elongate to about 1 to 2 cm and die. These are some of the requirements for early growth, but the plant's distribution indicates that it is quite able to adapt to varying environmental conditions (Noda and Eguchi 1965, Brod 1968). Anatomical and life history studies have been carried out by Dickerson (1964); Kacperska-Palacz, Putala, and Vengris (1963); Rahn et al. (1968); and Dunn, Gruendling, and Thomas (1968).

Compact soil favors germination and emergence. In experiments in the United States, germination was found to be 30 percent better in a compact soil than in a loose one. Maximum emergence was reached after the seeds were disturbed by a cultivation in mid-May and there was some emergence following each of the cultivations which were carried out at 3-week intervals. Emergence ceased in August. With continued emergence through the growing season more and more plants were present in the fields, and by August and September in temperate zones the weed problem was very noticeable. Jones (1933) studied the effect of water depth on growth of this weed and Dawson and Bruns (1962) studied emergence from several soil depths.

The first tillers in the field are formed 10 days after emergence. Normal plants produce about 15 tillers of various orders. Flowering in many varieties is photoperiodically controlled. Flowering takes place year-round in the Philippines, for example, and from June to November in Lebanon. For a discussion of the role of photoperiod in the distribution and competitive ability of *E. crusgalli*, see the section on distribution and habitat.

Workers in Germany found that a change from a close spacing of the plants from 7.5 cm to a generous distance of 60 cm resulted in a fivefold increase in dry weight and number of panicles and tillers. Plants grown in full sun had almost four times the dry weight and twice as many tillers and panicles as those grown in 50-percent shade. Plants responded to nitrogen, phosphorus, and potassium in that order.

WEED COMPETITION

The ecological requirements of *E. crusgalli* and rice are similar. The weed looks like rice in early stages and it is not uncommon that 10 percent of the weed plants in a rice field are introduced into the field while the rice plants are being transplanted. In direct-seeded rice, the weed germinates in about the same period of time (5 to 6 days). Relative growth rates depend upon the ecotype of the weed, the rice variety, and the growing conditions. In some areas such as the Soviet Union, *E. crusgalli* grows faster than rice from the outset, whereas in other areas (United States) the weed and crop grow at the same rate for the first 2 or 3 weeks, the weed becoming taller after that.

Because the weed germinates all through the season and because plants can produce new branches and tillers all through the season, *E. crusgalli* can emerge and develop rapidly in row crops wherever there are skips or misses in the field.

Experiments have shown that heavy stands of *E. crusgalli* may remove 60 to 80 percent of the nitrogen from the soil in a crop area. It has been shown in Japan that maximum competition for this element in rice fields occurs during the first half of the growing season. As is frequently the case, fertilizer applications favor the growth of *E. crusgalli* more than they do the rice crop. The fibrous root system of the weed overlays the rice roots, and the competition for nutrients is inevitable. A great deal is known about the effects of *E. crusgalli* on yield of rice and on the morphology of individual plants. In heavy competition, tillering in rice is reduced by 50 percent. Also, the number of panicles, the height of the rice plants, the weight of grains, and the number of grains per panicle are reduced. In Australia, infestations of this weed have caused yield reductions of 2 to 4 tons per hectare. Experiments in the United States have shown that one to five plants of *E. crusgalli* per square foot may reduce rice yields by 18 to 35 percent (Noda, Ozawa, and Ibaraki 1968; Swain 1967).

From the research of Lubigan and Vega (1971) in the Philippines we also know something of the competition offered by *E. crusgalli* at different periods during the growing cycle of lowland rice. At the two- or three-leaf stage, weed seedlings were transplanted into rice and allowed to remain for 7 to 40 days after the rice had been transplanted. At a weed density of 20 plants per sq m yields of rice were reduced 18 percent; 40 weeds caused a 30-percent loss; and, beyond that, an increased weed density resulted in no further crop losses. When weeds were placed in the field 7 days after the rice had been transplanted and allowed to grow to maturity, a density of 20 weeds per sq m reduced yields by one-third. Densities of 40, 60, 80, and 100 weeds gave further reductions. When weeds were introduced after 20 days and left until mature, the rice yield was reduced by 10 percent; increasing the weed densities as cited above caused about a 25-percent yield reduction. Weeds transplanted after 40 days and allowed to remain did not affect the yield of rice. From these results we may understand that under Philippine conditions the period from 7 to 40 days after rice has been transplanted is the time when barnyard grass offers the most severe competition. They believe that 15 weed plants per sq m will not result in reduced rice yields but that 20 plants per sq m represents the threshold where serious competition begins.

The yield of potatoes may be severely reduced by this weed, the degree of reduction depending on the density of weed plants and the time of their emergence. In experiments in sugar beets in Russia, yields were reduced by 85 percent in fields where *E. crusgalli* was the major species in a weed population of 500 plants per sq m. In sugarcane in Australia, *E. crusgalli* is one of three major grass weeds which

cause a severe problem before the cane closes in. After closure (summer and late autumn), the broad-leaved weeds are the major problem.

There is no known way by which *E. crusgalli* can be controlled with insect or disease organisms. Flooding remains an important control method. Fall plowing and shallow tillage encourage germination so that weeds can be killed before sowing. Clipping is ineffective, for the plants quickly regenerate. In the United States, rotations with rice, soybeans, and/or oats have been effective in reducing the levels of infestation. Alternating rice with pasture is used in Brazil and Australia (O. Williams 1957) to reduce *E. crusgalli* populations in succeeding rice crops.

AGRICULTURAL IMPORTANCE

Figure 10 indicates that *E. crusgalli* is ranked among the three most serious weeds in rice in Australia, Brazil, Ceylon, Chile, Greece, Indonesia, Iran, Italy, Japan, Korea, the Philippines, Portugal, Spain, and Taiwan; in cotton in Australia, the Soviet Union, and Spain; in corn in Australia and Yugoslavia; and in sugar beets in the United States. It is a principal weed of rice in Argentina, Colombia, Egypt, Fiji, Hungary, India, Nepal, Romania, the Soviet Union, and the United States; cotton in Iran, Mexico, Turkey, and the United States; corn in Italy, New Zealand, Romania, the Soviet Union, Spain, and the United States; sugar beets in Canada, Germany, Iran, Israel, and the Soviet Union; and potatoes in Bulgaria, Canada, Poland, and the United States.

In addition, *E. crusgalli* is a troublesome weed in major world crops not shown on Figure 10. It is ranked among the three most serious weeds in sorghum in Australia; peanuts and jute in Taiwan; sugarcane in Indonesia; and vegetables in Australia, New Zealand, Portugal, and the Soviet Union. It is a principal weed in citrus, orchards, soybeans, tea, tobacco, and vegetable crops in the Soviet Union; pasture, soybeans, and tobacco in the United States; sugarcane in Australia; vineyards in France; sunflowers in Argentina and Romania; vegetables in Bulgaria, Canada, and the United States; sorghum in Italy, the Soviet Union, and the United States; and in soybeans, sugarcane, and sweet potatoes in Taiwan. Finally, in several places in the world, it is also a common weed of cassava, taro, vegetable crops, bananas, coffee, tea, citrus, and millet.

E. crus-pavonis has been reported as the number-one weed of rice in Peru and number-two in rice in Brazil. It is a common weed of rice in Surinam

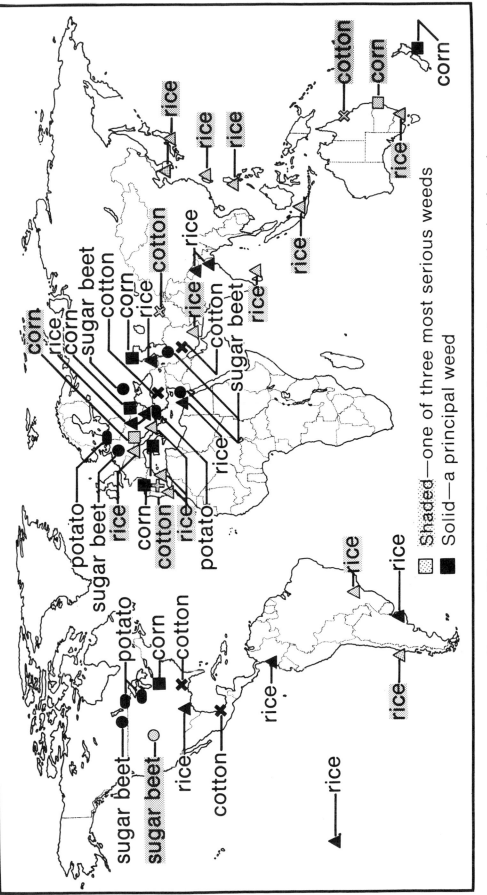

FIGURE 10. *Echinochloa crusgalli* (L.) Beauv. is a serious or principal weed in the crops and locations shown above.

and the United States. It has also been reported as a weed in potatoes, irrigated crops, wheat, and sugarcane.

E. crusgalli has been reported to accumulate levels of nitrate in its tissues high enough to be toxic to farm animals (Schmutz, Freeman, and Reed 1968).

E. crusgalli is an alternate host of *Piricularia oryzae* (Togashi 1942); and of the viruses producing tungro disease of rice (Watanakul 1964), dwarf disease of rice (Shinkai 1956), maize dwarf mosaic (Lee 1964), barley stripe mosaic, lucerne dwarf, oat pseudo-rosette, panicum mosaic, sugarcane mosaic, and wheat streak mosaic (Namba and Mitchell, unpublished; see footnote, *Cynodon dactylon*, "agricultural importance").

INFLUENCE OF CONTROL MEASURES

It has been shown in Poland that intensive hand-weeding on fields for 6 years caused a decrease in *E. crusgalli* populations, and that this species was replaced by *Galinsoga parviflora*. In New Zealand, continued use of phenoxy herbicides has resulted in the dominance of *E. crusgalli* as well as of species of *Paspalum* and *Panicum*. The extensive use of herbicides in rice and the resulting shift in weed types is best documented in Japan. About 1950 phenoxy herbicides came into use, being employed 30 to 40 days after the rice had been transplanted. Although broad-leaved weeds were greatly reduced, *E. crusgalli* increased. About 1957, pentachlorophenol came into use for control of this grass weed. More recently several other chemicals which are effective have come into use. In Japan and elsewhere the *Cyperus* species are now on the increase. In corn in Hungary, the continued use of herbicides for broad-leaved weeds over a period of 10 years has resulted in a significant increase in several weeds, including *E. crusgalli*.

COMMON NAMES
OF ECHINOCHLOA CRUSGALLI

ARGENTINA
arroz silvestre
capím
cresta gallo
grama de agua
jungle rice

AUSTRALIA
barnyard grass
BANGLADESH
shama
BRAZIL
barbudinho
capím arroz
BURMA
myet-ihi
CAMBODIA
smao bek kbol
CEYLON
kutirai-val-pul
martu
CHILE
hualcacho
COLOMBIA
barbarroja
paja de gallo
DENMARK
hønse-hirse
ENGLAND
chicken-panic grass
cock's foot
cockspur grass
FIJI
barnyard grass
FRANCE
crête-de-coq
ergot de coq
pattes de poule
pied-de-coq
GERMANY
Hahnenfuß-Fennich
Hühner-Hirse
INDIA
kayada
sawank
INDONESIA
djadjagoan
(Sundanese)
djawan (Javanese)
IRAN
seroof
ISRAEL
dochaneet
hatarnegoleem

ITALY
giavone
JAPAN
ta-in-ubie
KOREA
pi
LEBANON
cockspur
dhunayb
dinayb
water grass
MEXICO
pasto rayado
NETHERLANDS
hanepoot
NEW ZEALAND
barnyard grass
PERU
mijo japonés
PHILIPPINES
baobao (Ifugao)
bayokibok, daua-daua
(Tagalog)
dauana (Manobo)
lagtom (Bikol)
sasablog (Bontok)
PORTUGAL
milha maior
SPAIN
cerreig
pata de gallina
SWEDEN
hönshirs
TAIWAN
hsi-ye-ye-bai
THAILAND
hay kai mangda
ya-plong
UNITED STATES
barnyard grass
VIETNAM
cò lông vüt
YUGOSLAVIA
veliki muhar

❦ 4 ❧

Echinochloa colonum (L.) Link

POACEAE (also GRAMINEAE), GRASS FAMILY

Echinochloa colonum, an annual grass, is a native of India and is one of the serious weedy grasses of the world. Its range extends from lat 45° N to 40° S. It is a principal weed of rice culture. More than 60 countries have reported it as a weed problem in 35 crops. The grass is grazed by cattle and is sometimes cultivated in tropical Asia and Africa for the seeds which are made into flour. Young plants resemble rice. It is sometimes confused with *E. crusgalli* but differs from that species in that *E. colonum* is usually awnless.

DESCRIPTION

An erect or prostrate *annual* grass (Figure 11); *culms* to 70 cm in length, often in large tufts, rooting at the lower nodes; *leaf sheaths* smooth or hairy at the nodes, often tinged with red; *ligule* absent; *blades* rather lax, 3 to 25 cm long, 3 to 13 mm wide, occasionally transversely zoned with purple, smooth; *inflorescence* a panicle 5 to 15 cm long, appressed or ascending; *spikelets* 2 to 3.5 mm long, crowded, nearly sessile, awnless or shortly awned, with green or purple glumes; *lower lemma* minutely hairy on the surface, often with rigid hairs on the nerves; *grains* free, elliptic, wholly enclosed by the lemma and palea.

The absence of a ligule, the often purple-tinged leaf sheaths and blades, and the usually awnless spikelets are distinguishing characteristics of this species. *E. colonum* can be distinguished from *E. crusgalli* by the absence of awns on the spikelets.

DISTRIBUTION AND HABITAT

E. colonum is an important weed in five of the world's major crops which grow between lat 23° N and 23° S. From Figures 12 and 13 it may be seen that it is particularly troublesome in two areas. In the area below lat 30° N in Asia, in the warm parts of Australia, and in the Pacific Islands *E. colonum* is a serious weed in rice, sugarcane, corn, and sorghum. The second area is in the northern part of South America and the Caribbean where it prospers mainly in rice fields. Available reports indicate that the weed is seldom a problem in the Mediterranean areas of North Africa and Europe. It does not have the temperate zone range of *E. crusgalli* and is never reported in temperate cereals, fruits, or vegetables.

Because it is an annual, it grows rapidly during the rainy season or when water levels are on the rise and then dies out during the dry season. It is an all-season weed in the large irrigated Gezira cotton scheme of Sudan. In most crops it has the ability to germinate anytime during the growing season, and, for this reason, the first flush of weed seeds is often allowed to germinate before the crop is planted and these seedlings are destroyed by cultivation. Because it resembles rice in the seedling stage it is sometimes transplanted into the fields with the crop. Manual control is thus difficult in the early stages, and, by the time the weed is recognized and removed, the crop may be irreversibly damaged.

The weed is an excellent competitor and if rice culture is badly managed the crop may be forced out

41

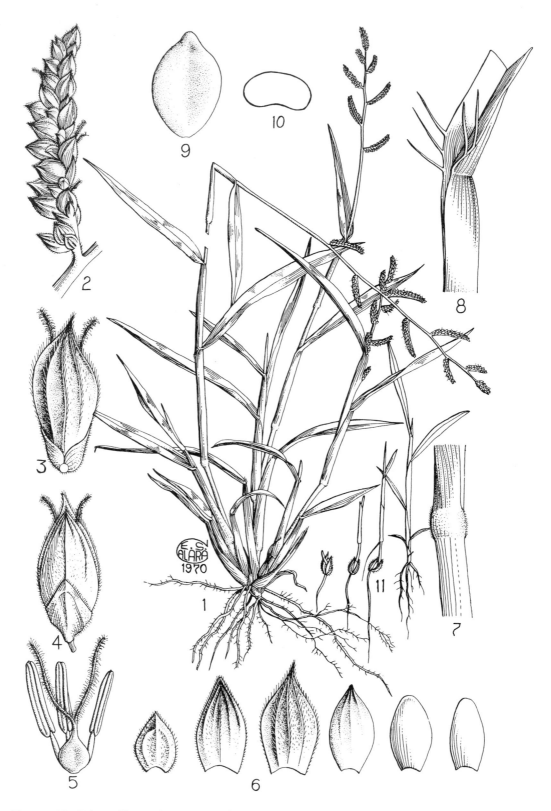

FIGURE 11. *Echinochloa colonum* (L.) Link. *1*, habit; *2*, inflorescence; *3*, spikelet, ventral view; *4*, spikelet, dorsal view; *5*, floret; *6*, bracts; *7*, portion of culm; *8*, portion of leaf base and blade; *9*, grain; *10*, grain, transverse section; *11*, seedlings.

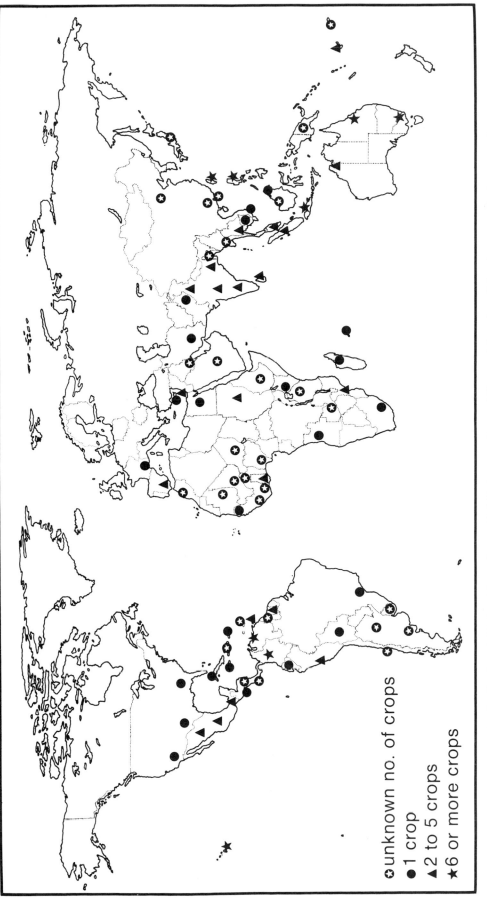

FIGURE 12. The distribution of *Echinochloa colonum* (L.) Link across the world in countries where it has been reported as a weed.

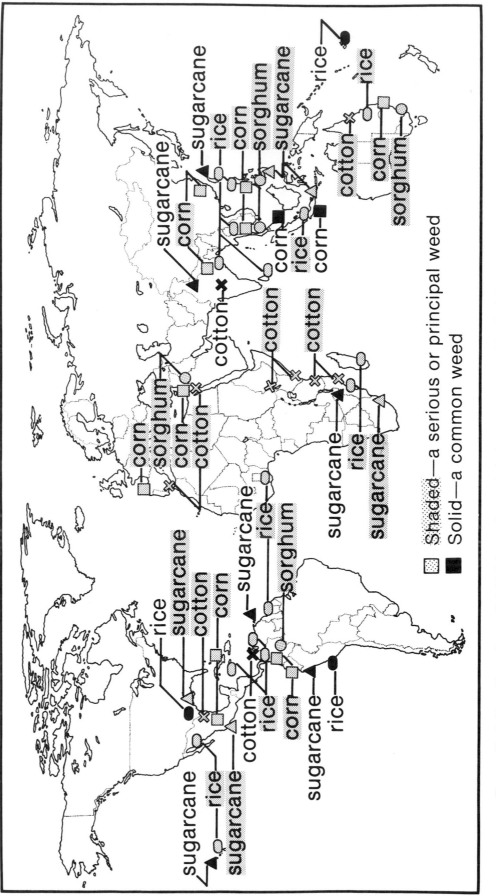

FIGURE 13. *Echinochloa colonum* (L.) Link is a serious, principal, or common weed in all of the crops and locations shown above.

by increasing numbers of this weedy plant. Its prostrate growth habit in early seedling stages—that of rooting at the nodes to gain space and assuming an erect posture when light is limiting—makes it a very competitive weed for most crops. For the effects of *Echinochloa* spp. on yield see the section on *E. crusgalli*.

PROPAGATION

One plant may produce thousands of seeds. Although it is an annual, it may be vegetatively propagated by production of new roots and shoots at the nodes when it is in a stage of prostrate growth. Some strains are reported to have a short period of dormancy following harvest but this dormancy disappears in less than 8 weeks of dry storage. Light is required for best germination. Flooding with 10 to 15 mm of water prevented germination in rice fields in Australia.

The seeds of *E. colonum* often enter rice fields with crop seeds or transplants. They may be transferred between fields on farm machinery and on mud on the feet, fur, feathers, and skin of rodents, birds, and larger animals including humans. In Australia, it is suspected that wild ducks may have been important in the initial distribution of the weed. Irrigated fields and rice paddies are often interconnected through a system of canals which is common to all and may, therefore, serve as a means of spreading weed seeds.

AGRICULTURAL IMPORTANCE

From Figure 13 it may be seen that *E. colonum* is ranked among the three most serious weeds in rice in Australia, Ceylon, Colombia, India, Mozambique, the Philippines, Surinam, Taiwan, and Venezuela; cotton in Australia and Spain; corn in Australia and Taiwan; sorghum in Australia; and sugarcane in Indonesia and the Philippines.

In addition, it is a principal weed in rice in Ghana, Hawaii, Indonesia, Jamaica, Madagascar, Mexico, and Thailand; cotton in Israel, Kenya, Mexico, Mozambique, Sudan, and Tanzania; corn in Colombia, Cuba, Ecuador, India, Israel, Mexico, the Philippines, Spain, and Thailand; sugarcane in Mexico, South Africa, and the United States; and sorghum in Colombia, Israel, the Philippines, and Thailand.

E. colonum is also a serious weed in major world crops not shown on the map. It ranked among the three most serious weeds in jute, peanuts, and rape in Taiwan and in vegetables in Australia. It is a principal weed in cassava in Thailand; bananas in Hawaii; beans in Mexico and the United States; peanuts in Colombia and Israel; irrigated crops, linseed, and safflower in Australia; sugar beets in Israel; bananas, cowpeas, mung beans, millet, papayas, peanuts, and soybeans in the Philippines; and in soybeans in Mexico and Taiwan. Finally, in several places in the world it is a common weed of abaca, coconuts, pineapples, tea, taro, sweet potatoes, and other vegetables.

E. colonum is an alternate host of *Piricularia* sp. (Chandrasrikul 1962) and *Meloidogyne incognita* (Kofoid & White) Chitwood (Valdez 1968); and of the viruses which produce abaca mosaic (Gavarra and Eloja 1964), hoja blanca disease of rice (Granados and Ortega 1966; Galvez, Thurston, and Jennings 1960), tungro disease of rice (Watanakul 1964), and sugarcane mosaic (Namba and Mitchell, unpublished; see footnote, *Cynodon dactylon*, "agricultural importance").

COMMON NAMES OF ECHINOCHLOA COLONUM

ARGENTINA
  arroz silvestre
  capím
  grama pintado
  pasto colorado
AUSTRALIA
  awnless barnyard
    grass
BARBADOS
  jungle rice
BRAZIL
  capituva
BURMA
  myet-thi
  pazun-sa-myet
  wumbe-sa-myet
CEYLON
  adipul
  gira-tana
CHILE
  hualcacho
COLOMBIA
  liendre de puerco
  paja de apto
EGYPT
  abu rokba

FIJI
  jungle rice
INDIA
  janguli
  kavada
  sawak
  sawank
  sharma
  water grass
INDONESIA
  djadjagoan leutik
    (Sundanese)
  roempoet bebek
    (Indonesian)
  toeton
  watoeton (Javanese)
IRAQ
  dahnan
ISRAEL
  dochaneet hashaleen
JAMAICA
  jungle rice
MALAYSIA
  jungle rice
  padi burong
  rumput kusa-kusa

MAURITIUS
  herbe de riz
  herbe sifflette

MEXICO
  arrocillo
  zacate de agua
  zacate pinto

NICARAGUA
  pata de conejo

PERU
  champa

PHILIPPINES
  dukayang (Ilokano)
  pulang puit (Tagalog)
  tumi (Bontoc)

PUERTO RICO
  arrocillo

SPAIN
  cerreig
  pata de gallina

SUDAN
  difera

TAIWAN
  máng-jì

THAILAND
  yah nok-see champo
  ya-plong

TRINIDAD
  jungle rice

UNITED STATES
  jungle rice

URUGUAY
  capím
  gramilla de rastrojo

VENEZUELA
  paja americana

✺ 5 ✺

Eleusine indica (L.) Gaertn.

POACEAE (also GRAMINEAE), GRASS FAMILY

E. indica, a tufted annual grass, is one of the serious weedy grasses of the world. There is no agreement on its place of origin but early records have come from China, India, Japan, Malaysia, and Tahiti. Its range extends from Natal in South Africa to Japan and the northern border of the United States. More than 60 countries report it is a weed problem in 46 crops. It is used for hay and silage in some areas of the world and is grown for seeds in Africa and Asia.

DESCRIPTION

This plant is an *annual* tufted grass (Figure 14); *culms* ascending, compressed, 5 to 60 cm high; *sheaths* keeled, smooth except for a few short hairs on the margins; *ligule* membranous with jagged edge; *blades* 5 to 15 cm long, 3 to 4 mm wide; *spikes* mostly two to seven, rarely one, 5 to 15 cm long, flat, straight or slightly incurved; *rachis* slender, margins slightly rough; *rachilla* separating at the joint above the glumes and between the florets; *spikelets* 2 to 4 mm long, three- to five-flowered; *glumes* rather membranous, the lower 1 to 1.5 mm long, pointed, rough on the keel, one-nerved, the upper 3 mm long, sharply pointed or tapering gradually to a point, with smooth keel, one- to five-nerved; *lemmas* similar in texture and shape to the glumes, ovate, sharply pointed or tapering gradually to a point, slightly rough on the keel toward the tip, the lateral nerves very slender, lemmas falling with the grain, *palea* somewhat rough on the keels; *grain* reddish brown to black, oblong-ovate, with conspicuous ridges.

The most important distinguishing characteristic of this plant is the windmill-like appearance of the inflorescence.

DISTRIBUTION AND HABITAT

E. indica is a weed problem mainly in crops grown in the warm areas of the world. It is seldom a serious weed outside the tropics of Cancer and Capricorn. From Figure 15 it may be seen that it is generally present in crops in southern Asia and the Pacific Islands, in eastern and southern Africa, and in northern Latin America.

It is one of six species of the genus that are widespread in the tropics, but, because of its broad tolerance to various factors in the environment, it can be found in diverse types of areas in the subtropics and the temperate zone. It grows well in open ground and so is found in lawns, pastures, and footpaths. It can stand much trampling. It is found in waste places and roadsides but prospers on arable land. It is present also in damp marshlands and is often most vigorous along irrigation field borders and canals; its vegetative growth is seriously reduced during dry seasons or when soil moisture is lacking.

PROPAGATION

E. indica reproduces by seeds. In the Philippines single plants may produce more than 50,000 seeds. In Rhodesia a maximum of 135,000 seeds per plant has been recorded, although the mean production of a population was 40,000 seeds per plant. In practical terms this means a possible production of 4,250

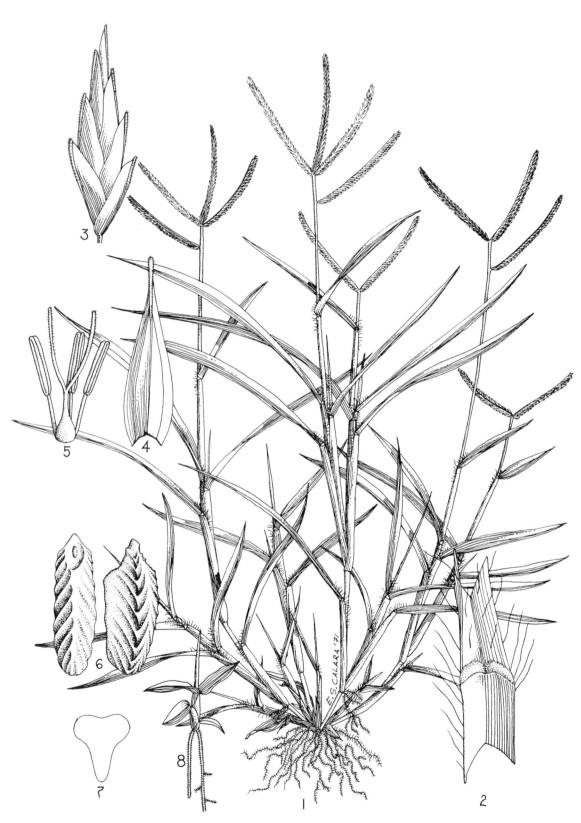

FIGURE 14. *Eleusine indica* (L.) Gaertn. *1*, habit; *2*, ligule; *3*, spikelet; *4*, bract; *5*, flower; *6*, seed, two views; *7*, seed, cross section; *8*, seedling.

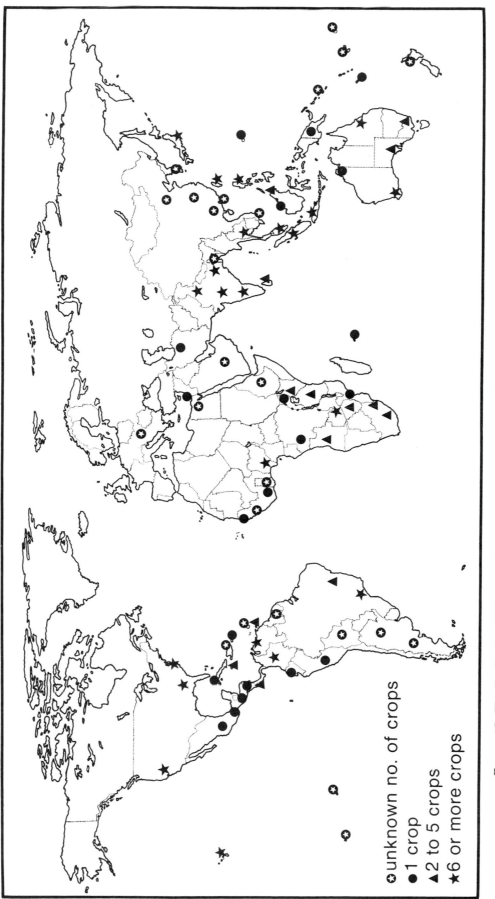

FIGURE 15. The distribution of *Eleusine indica* (L.) Gaertn. across the world in countries where it has been reported as a weed.

⊕ unknown no. of crops
● 1 crop
◀ 2 to 5 crops
★ 6 or more crops

kilograms per hectare or 5,000 million seeds, in a pure stand of this species. Given this weed's awesome capability for reproduction and its prevalence in both waysides and fields, one is not surprised to find that it is so widespread (Schwerzel 1970*a*). The seeds move by wind and in the mud on the feet of animals; they are ingested by wild and domesticated animals; and they move with man's commerce. The plant is a prominent adventive on rubbish heaps of ports and docks in both the tropics and the temperate zone.

Both fresh and 60-day-old seeds responded favorably in the United States to diurnal fluctuations in temperatures of 20° to 35° C and 25° to 40° C. Light was beneficial and the completeness of germination varied from 30 to 90 days, depending on the degree of maturity and the age of the seeds. Scarification of the seeds hastened germination. In other experiments, 6-day old seeds showed 90 percent germination with an alternating temperature of 20° to 35° C in both light and dark. Similar results were obtained in India, where it was also shown that prechilling wet seeds at 0° C for 40 days hastened subsequent germination carried out at 30° C (Andersen 1968).

BIOLOGY

In warm regions, *E. indica* grows and flowers at all seasons when moisture is sufficient. We have little knowledge of the variation in ecotypes across the world. In India, preliminary experiments with seeds from many places within the country have shown that the plant possesses a remarkable similarity of growth and morphology when cultured in a controlled environment. It has been suggested that the wide distribution of *E. indica* in India (approximately 3,300 kilometers from east to west, as well as north to south) is due to the species' wide ecological amplitude and not because it has formed distinct local populations.

We owe much of our knowledge of the biology of *E. indica* to studies done in India and Rhodesia. The work in India has shown that a variation in the length of the photoperiod has no effect on the time of flowering. The plant flowers at all daylengths between 6 and 16 hours but will not flower outside this range. Variation in photoperiod has a marked effect on vegetative characters. At the optimum photoperiod of 14 hours, the following were favorably influenced: number of tillers, number of nodes per tiller, length of the longest tiller, and length of the longest leaf. Longer photoperiods depressed growth of these plant parts rather sharply.

Drought and low temperature delay flowering and the growth of vegetative parts of the plant. Thus in India the reduced photoperiod of winter and conditions of low moisture and temperature all combine to cause smaller, more compact, plants whose flowering is much delayed.

A high light intensity produces an increase in number and lateral spread of tillers, number of inflorescences, and dry matter yield. As opposed to its prostrate habit in full sun, the weed tends to produce taller plants in the shade, but such shade also has a general suppressive effect on the growth of all plant parts. All of these characteristics together seem to indicate a tropical origin for this plant (Misra 1969, Singh and Misra 1969, J. Singh 1968).

The phenology of *E. indica* was studied in Rhodesia where plots were repeatedly dug or cultivated at 2-week intervals throughout the major part of the growing season. The seeds in the soil became more viable as the season progressed; and eventually more than 90 percent of the seeds in the plots completed their life cycle and the resulting plants produced large amounts of seeds. Of the six main Rhodesian weeds studied, *E. indica* was the second most prolific seed producer. When these experiments were repeated in a year of abnormally low rainfall, the seeds germinated more slowly, the percentage of seeds germinating in the plot declined more rapidly as the season progressed, and the time required for a complete cycle of growth was longer. Plants which germinated in December, the most favorable period for beginning growth in Rhodesia, flowered in 30 days and shed seeds in 70 days (Schwerzel 1970*b*). Elsewhere in the world a full cycle of growth may be as long as 120 to 180 days.

AGRICULTURAL IMPORTANCE

From Figure 16 it may be seen that *E. indica* is one of the three most serious weeds in corn in Angola, Malaysia, the Philippines, Taiwan, Venezuela, and Zambia; upland rice in Japan, the Philippines, Taiwan, and Venezuela; sweet potatoes in Hawaii, Japan, Malaysia, and Taiwan; and sugarcane in Indonesia, Taiwan, and Tanzania. It is a principal weed of corn in Colombia, Mexico, South Africa, and Thailand; of cotton in India, Kenya, Mozambique, Nicaragua, Nigeria, Rhodesia, Tanzania, Thailand, Uganda, the United States, and Zambia; of sugarcane in Peru, Puerto Rico, South Africa, and the United States; and of upland rice in Brazil, India, Indonesia, and Thailand.

E. indica is considered to be one of the three most

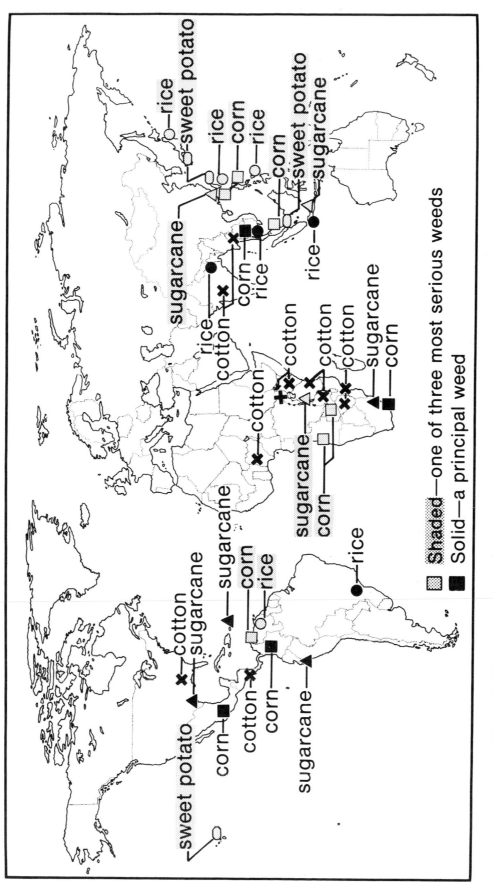

FIGURE 16. *Eleusine indica* (L.) Gaertn. is a serious or principal weed in all of the crops and locations shown above.

serious weeds of other crops which are not shown on the map: bananas, cassava, pineapples, rape, jute, soybeans, and peanuts in Taiwan; vegetables and papayas in the Philippines; peanuts, soybeans, sorghum, and vegetables in Malaysia; soybeans in Japan; peanuts in Indonesia and Zambia; and of vegetables in Venezuela. This species is reported to be a principal weed of abaca in Sabah; bananas in Hawaii; cassava in Thailand; cowpeas, millet, and mangoes in the Philippines; cacao in Brazil; coffee in Brazil and the Philippines; pineapples in Australia, Hawaii, the Ivory Coast, and the Philippines; sorghum in Colombia, the Philippines, Thailand, and Zambia; soybeans in the Philippines and the United States; peanuts in Gambia and the Philippines; rubber in Malaysia; vegetables in Brazil; tobacco in Trinidad; and wheat in South Africa.

For every crop listed above, one or more countries have also listed *E. indica* as a common, but not serious, weed. This means that it is widespread in crops and that some labor is required to control it. Elsewhere in the world the species is reported to be a common weed in many vegetable crops and in plantation crops such as oil palms, coconuts, rubber, tea, and coffee.

E. indica is believed by some to be palatable to grazing animals, although it may be a more satisfactory forage when it has been seeded with an annual legume. In other areas such as Ceylon, the crop becomes fibrous too early to be a satisfactory pasture grass. The species occasionally is grown for grain in Africa, India, and in the Orient. There are other members of the genus which are more satisfactory for this purpose, however.

In Australia and elsewhere it is known that young plants sometimes contain enough hydrogen cyanide to be responsible for the deaths of calves and sheep.

E. indica is an alternate host of *Helminthosporium* sp. (Chandrasrikul 1962) and *Piricularia oryzae* (Teng 1932); of the nematodes *Rotylenchus reniformis* Linford & Oliveira (Linford and Yap 1940), *Meloidogyne* sp., *Pratylenchus pratensis* (de Man) Filip. (Raabe, unpublished; see footnote, *Cyperus rotundus,* "agricultural importance"), and *Meloidogyne incognita* (Kofoid & White) Chitwood (Valdez 1968); and of the virus diseases which produce rice leaf gall, corn leaf gall (Agati and Calica 1950), tungro disease of rice (Watanakul 1964), sugarcane mosaic (Chona and Rafay 1950), maize dwarf mosaic (Lee 1964), groundnut rosette (Adams 1967), maize streak, and sugarcane streak (Namba and Mitchell, unpublished; see footnote, *Cynodon dactylon,* "agricultural importance").

COMMON NAMES OF ELEUSINE INDICA

ARGENTINA
gramilla
pie de gallina
AUSTRALIA
crowsfoot grass
BRAZIL
capím pé de gallina
BURMA
sin-ngo-let-kya
sin-ngo-myet
CAMBODIA
smao choeung tukke
CEYLON
bela-tana
tippa-ragi
COLOMBIA
grama dulce
grama de orqueta
pata de gallina
CUBA
pata de gallina
DOMINICAN REPUBLIC
pata de gallina
EASTERN AFRICA
crowsfoot grass
wild finger millet
EGYPT
negil
FIJI
crowsfoot grass
goosegrass
kavoronaisivi
wiregrass
GERMANY
Hartgras
Kreutzgras
GUATEMALA
pata de gallina
pata de gallo
HAWAII
wiregrass
INDIA
jangali maru
kodai
mandla
INDONESIA
djampang munding (Sundanese)
djoekoet djampang (Indonesian)
godong oela
lulangan (Javanese)
roempoet beloelang (Indonesian)
JAMAICA
fowlfoot grass
goosegrass
JAPAN
ohishiba
KENYA
wild finger millet
MALAYSIA
goosegrass
rumput sambou
MAURITIUS
chiendent
pattes de poule
MEXICO
pata de gallina
pata de gallo
yerba dulce
NEW ZEALAND
crowsfoot grass
NICARAGUA
yerba de camino
NIGERIA
gbegi
PERU
pata de gallina
PHILIPPINES
bag-angan (Bikol)
bakis-bakisan
bikad-bikad
bila-bila
palagtiki (Bisayan)
paragis
parangis (Ilokano)
sabung-sabungan (Tagalog)
sambali
PUERTO RICO
pata de gallina
RHODESIA
i gogadolo
mu kha
rapoko
SOUTH AFRICA
Indian goosegrass
Indiese osgras

SURINAM
mangrasie

TAIWAN
nyou-jin-tsau

TANZANIA
wild finger millet

THAILAND
yah teenka

TRINIDAD
fowlfoot grass
irongrass
yardgrass

UGANDA
kasibanti

UNITED STATES
goosegrass

VENEZUELA
guarataro

VIETNAM
cõ mân trâù

ZAMBIA
oxgrass
rapoko

❦ 6 ❧

Sorghum halepense (L.) Pers.

POACEAE (also GRAMINEAE), GRASS FAMILY

Sorghum halepense is a stout, erect, perennial grass that spreads by seeds and by long creeping rhizomes. It is a native of the Mediterranean area and its range as a weed extends from lat 55° N to lat 45° S. It is a principal weed of corn, cotton, sugarcane, and other crops from tropical to temperate climates. Fifty-three countries report that it is a weed in 30 different crops. On fertile soils the weed will spread to other agricultural crops and be very difficult to eradicate. Some strains make good hay. McWhorter (1971*a*) has provided an interesting historical account of this plant. He describes the origin of its most widely used common name, johnsongrass, and relates how it was introduced and spread throughout the United States until, by 1900, its destructiveness had become so alarming that the federal government took action to bring it under control.

DESCRIPTION

S. halepense is an aggressive *perennial* grass (Figure 17); *culms* erect, stout, 0.5 to 3 m tall, arising from extensively creeping scaly *rhizomes*; *leaf sheath* ribbed, smooth, or often hairy within at the junction with the blade, often with a waxy secretion at the base; *ligule* short, papery; *blades* alternate, smooth or rough on the edges, many-nerved, with conspicuous midribs, 20 to 60 cm long, 0.5 to 5 cm wide; *panicle* large, pyramidal, purplish, hairy, 15 to 50 cm long, often somewhat contracted after flowering, primary branches up to 25 cm long and then branched again; *racemes* 1 to 2.5 cm; *spikelets* usually in pairs although toward the tip of the

inflorescence they may occur in threes; *when spikelets in pairs*, the lower is usually sessile and the upper pedicelled, narrow, long, and stamen-bearing; *when spikelets in threes*, one (usually the middle) is sessile and perfect, the other two are pedicelled and staminate; *sessile spikelet* 4 to 5 mm long, green; *pedicelled spikelets* 5.5 mm long; *callus* sparsely bearded; *lower glume* silkily hairy; *upper lemma* unawned, sometimes awned in cultivated forms; *grain* nearly 3 mm long, oval, reddish brown, glossy, marked with fine lines on the surface.

The ribbed leaf sheath, conspicuous midrib, the large, purplish panicle, and the extensively creeping rhizomes are distinguishing characteristics of this species.

DISTRIBUTION AND HABITAT

S. halepense, a native of the Mediterranean region, has now become established as a formidable weed in most of the agricultural areas of the world. It seems best adapted to the warm, humid, summer-rainfall areas in the subtropics and not so well adapted to areas which are strictly tropical.

As a weed it is most serious from the Mediterranean through the Middle East to India, Australia and nearby islands, central South America, and the Gulf coast of the United States (Figure 18).

The species can grow on a variety of sites: on arable land, waste places, roadsides, and field borders. It occurs extensively along irrigated canals and at the edges of irrigated fields. Its general distribution in these areas is the result of water movement of

54

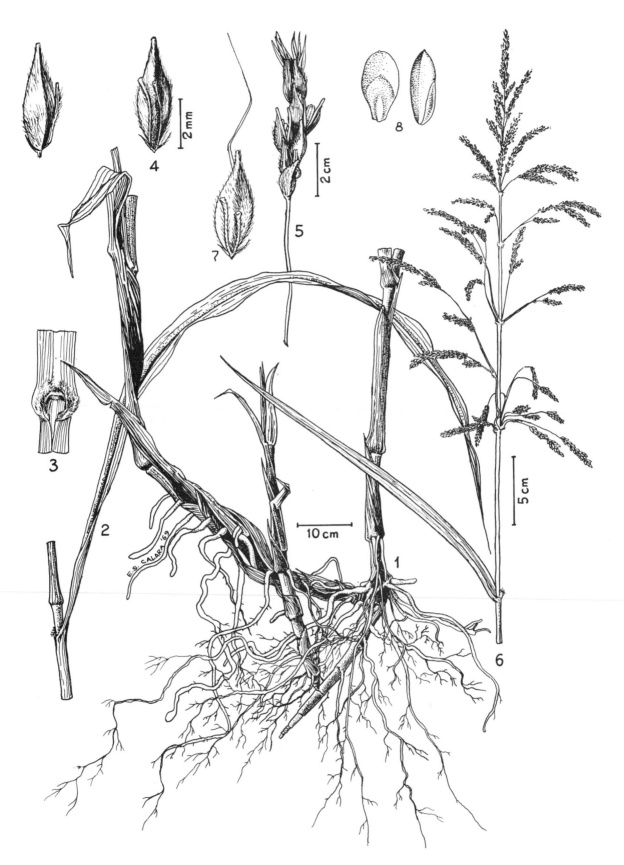

FIGURE 17. *Sorghum halepense* (L.) Pers. *1*, habit; *2*, blade; *3*, ligule; *4*, floret; *5*, spikelet; *6*, inflorescence; *7*, spikelet; *8*, grain, two views.

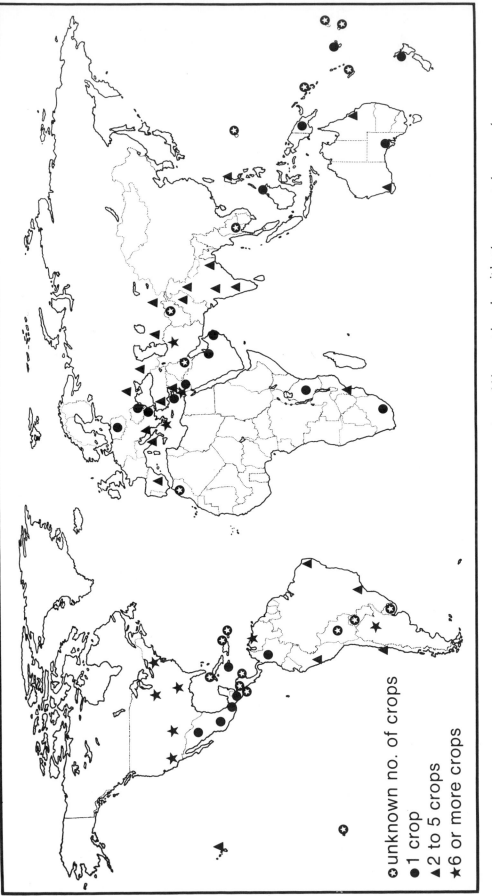

Figure 18. Distribution of *Sorghum halepense* (L.) Pers. across the world in countries where it has been reported as a weed.

the seeds, which readily fall from the head when mature.

Generalizations about habitat and growth physiology must first be qualified by noting that there are many ecotypes of this species. Studies have been made in the United States on 55 morphologically distinct vegetative types collected from many areas of that country and several other countries (McWhorter 1971c). The variability of ecotype response to herbicides has been studied by McWhorter (1971d).

The species has been classified as a short-day plant; if this is true of most ecotypes, its being a short-day plant may be one of the factors in its luxuriant vegetative growth during the long days in areas which are some distance north or south of the equator. In shaded conditions the seeds do not germinate and seedlings do not grow well. If one wishes to rid a field of seeds of this species, one should not use smother crops, for seed survival will only be prolonged. In temperate areas the tops are readily killed by frost. It has been reported from experiments that temperatures below 13° C tend to inhibit flowering.

PROPAGATION

The seeds of *S. halepense* provide the principal means of dissemination of this species. They travel with the wind, on water, are trapped on the coats of animals, are ingested by birds and cattle, and move in trade with dirty seed stocks and feed. After maturity, the seeds shatter readily from the spikelets. When plants are near irrigation systems, the seeds fall or are blown into the canals or furrows and float to new sites with the movement of the water. Seeds have remained viable after 30 months in the soil, after 7 years in dry storage, and after they have passed through the guts of animals.

Most ecotypes have a dormant period following maturity, but they afterripen in soil or storage so that it is not difficult to obtain seedling plants for experimentation. Taylorson and McWhorter (1969) studied the seed germination of 44 ecotypes and found that most were highly dormant. Pretreatment of seeds at 10° C for 2 weeks followed by a temperature shift to 40° C for 2 hours broke the dormancy. Without the cold treatment the seeds germinated best under conditions of continuous fluorescent light, a potassium-nitrate substratum, and alternating temperatures of 20° C to 35° C. Each of 12 ecotypes harvested in 2 different years germinated in a similar way, and the evidence suggests that seed

dormancy is largely imposed by the mechanical restriction of the seed coat.

Removal of the glume has given a 95-percent germination in some experiments with dormant seeds (Harrington 1916, 1917). It is believed that, in fields, the seeds from the current season have low germination but seeds which have lain in the soil for 1 year will sprout very readily. Seeds are known to germinate from the 15-cm depth, but most seedlings arise from the top 7 cm.

BIOLOGY

S. halepense is a very heavy seed producer, this being its principal means of distribution; however, its superior ability to compete with other plants and its persistence in the face of the most intensive control measures surely result from the long, very vigorous, and highly adaptable rhizome-root system which develops below the soil surface. Stamper (1957) has estimated that this species can produce a length of 600 km of rhizomes per hectare weighing 33 metric tons. Individual plants may produce 5,000 nodes in one growing season (Anderson, Appleby, and Weseloh 1960; McWhorter 1961a). In Yugoslavia a block of field soil measuring 1 sq m by 30 cm in thickness contained 1.2 kg of rhizomes which were 28 m in length and contained 2,000 buds. Much of our knowledge about the seasonal growth and morphological development of this plant concerns these underground organs (McWhorter, 1960, 1961a,b; Oyer, Gries, and Rogers 1959). In general, the system is made up of primary, secondary, and tertiary rhizomes. The primary structures are alive at the beginning of the growing season, providing buds for renewed growth. Extensions from the main rhizomes become the secondary structures which surface and give rise to new plants. Tertiary rhizomes, which grow out from the base of the plant at flowering time, are large, usually go deep into the soil, and usually continue to grow until the advent of cold or dry weather. These tertiary rhizomes produce new plants in the following season.

At the beginning of the season, either axillary or terminal buds may develop into new plants with aerial stems which subsequently develop crowns and tillers. It is known that crown formation comes more quickly on those plants which develop from vegetative parts than on those which come from seedlings.

A high temperature is necessary for renewed activity of rhizome buds after a dormant period. The plant grows rapidly and, within 60 days in the temperate zone, may develop sufficient rhizomes to give it a perennial character. Many types have an apical dominance of axillary buds by the terminal buds, so that early season cultivation may, in fact, result in a breaking up of the rhizome system to provide many small pieces from which several buds may sprout (McWhorter 1960).

The growth and development of 55 ecotypes from several countries have been described by McWhorter (1971c).

The primary rhizomes decay at the end of the season. In cool climates only a small percentage of the remaining roots and rhizomes are lost in winter. Rhizomes may be dried to 40 percent of their original weight without loss of viability, but the tissues will die if moisture content is reduced to 20 percent. Rhizomes may be found about 1 meter deep in the soil, but the greatest number are found in the top 20 cm.

The most detailed anatomical study of *S. halepense* is that by McWhorter (1971b). Over a period of 4 years hundreds of plants from greenhouse and field were examined at several different morphological stages. Because all of the material came from a nursery of ecotypes of this species, some very interesting comparisons were made of different strains. Their diameters varied from 60 to 150 microns in one ecotype to 100 and 230 in another type. The vascular bundles in culms were three times more plentiful in some types than in others. The number of stomata varied from 65 to 150 per sq mm in 10 different ecotypes. The average number of stomata on the upper leaf surfaces was 95 per sq mm, and the lower surfaces had about 20 percent more than that. These data contribute significantly to our understanding of the response of the species to its environment and indicate control measures of various kinds. They will be helpful as we attempt to explain the behavior of different ecotypes, particularly in their variable responses to herbicides.

Burt and Wedderspoon (1971) conducted laboratory experiments on temperature response and photoperiod with three strains of *S. halepense* from cool, intermediate, and warm areas of the United States. All strains grew well at 20° C but the type from the warm region grew much better at 35° C. A long photoperiod of 16 hours prevented flowering in all three strains. A 16-hour photoperiod, as compared to a 12-hour light period, reduced rhizome production in those strains from warm and intermediate regions. In a 12- and 8-hour photoperiod the types from cool and intermediate climates flowered, the former responding most rapidly. The strain from

a warm region did not flower under any photoperiod during the 11 weeks of the experiment.

In Mississippi (United States), Knight and Bennett (1953) studied the effects of 8-, 10.5-, 12-, 14-, and 16-hour photoperiods in experiments which continued for more than 12 weeks. Flowering occurred during all photoperiods, but seed-head formation was inhibited at 16 hours; and there was a tendency to produce poor seeds at the 14-hour daylength. The highest seed yields were obtained at 10.5- and 12-hour photoperiods.

The chromosome number of *S. halepense* is (2n=40); that of cultivated sorghums is (2n=20). Interspecific hybrids can result from crosses between *S. halepense*, *S. sudanense*, and cultivated sorghums; some of the resulting crosses may possess vigorous rhizomes (Tarr 1962).

TOXICITY

There are several reports that root exudates or extracts of fresh or decaying leaves, rhizomes, and roots can inhibit germination and seedling growth in several species, including clover, crown vetch, and in *S. halepense* itself (Abdul Wahab and Rice 1967; Friedman and Horowitz 1970). Although the ecological significance of this is not known for field conditions, it is known that *S. halepense* has a peculiar tenacity for holding an area once it has become established. In the United States the weed has often forced the abandonment of row crop culture on good soils because it became dominant. Studies on old field successions revealed that *S. halepense* plants which are present at the time a field is abandoned become important in the early stages of succession. They expand and persist for a long time in almost pure stands.

INFLUENCE OF CONTROL MEASURES

Some of the first and best evidence on the ecological shifts of weed species in field communities came from observing plantation crops after they had undergone continued treatment with herbicides—often with the same herbicide. *S. halepense* is a terrible weed of vineyards, and several countries have warned about the dangers of using herbicides for annual weed control without first taking precautions against the invasion of vineyards by perennial weeds after the competition of annuals has been removed. Greece, Australia, and the United States have reported serious infestations of one or more of the following: *S. halepense*, *Cynodon dactylon*,

Pennisetum clandestinum, and *Convolvulus arvensis*. In some areas, growers have had to return to hand tools and mechanical means to cope with these deep-rooted perennials. McWhorter (1971*d*) reported on the variable responses of different ecotypes to the same herbicide. Timmons and Bruns (1951) showed that 2 weeks was the longest effective interval between shallow cultivations for the eradication of the species in one season. Increasing the depth of shoot-cutting by duckfoot cultivator, plow, or shovel lengthened the effective interval between cultivations and, in most cases, reduced the number of operations necessary to eradicate the weed. No practical advantage was found in cultivating deeper than 7 to 10 cm.

AGRICULTURAL IMPORTANCE

From Figure 19 it may be seen that *Sorghum halepense* is ranked among the three most serious weeds in cotton in Greece, Mexico, and Venezuela; sugarcane in Argentina, Australia, Fiji, Pakistan, the United States and Venezuela; corn in Chile, Greece, the United States, and Yugoslavia; citrus in Mexico and Venezuela; and vineyards in Australia. It is a principal weed of cotton in Israel, Pakistan, Peru, the Soviet Union, Turkey, and the United States; sugarcane in Hawaii, India, South Africa, and the United States; corn in Israel, Italy, Mexico, Poland and Romania; citrus in Peru; and vineyards in Argentina, Greece, Lebanon, Spain, and Yugoslavia. In addition, *S. halepense* is a troublesome weed in major world crops not shown on Figure 19. It is a serious weed of alfalfa in Chile; rice in Venezuela; sugar beets in Greece; and wheat in Yugoslavia. It also has been reported as a principal weed of peanuts and sorghum in Israel; sugar beets and sorghum in Italy; orchards, peanuts, soybeans, and sorghum in the United States; vegetables in Argentina, Hawaii, Mexico, and the Soviet Union; peanuts in Pakistan; tea in the Soviet Union; abaca in Sabah; beans and rice in Mexico; upland rice in the Philippines; sorghum in Colombia; bananas and orchards in Lebanon; and orchards in Argentina and Turkey.

Finally, it is also a common weed in several places in the world in coffee, pineapples, barley, millet, pasture lands, potatoes, and sisal.

In contrast, there is no question of its usefulness for cattle feed in some areas. In Pakistan, for example, it is regarded as a palatable forage when properly managed and controlled for pasture or hay. Although the weed produces excellent hay in the southeastern United States, it may, under certain seasonal condi-

tions, accumulate prussic acid (hydrocyanic acid) in its leaves and stems. It may then be lethal to cattle grazing in pastures where it is growing. Periods of very dry weather and those following a first frost are regarded as especially dangerous in several parts of the world. In the United States it is believed to be one of the causes of hay fever. The species is a very strong competitor. Reports from several places in the world give these examples of crop reduction: 25- to 50-percent reductions in yield of ratoon crops of sugarcane; losses of 12 to 33 percent in corn, and losses of 300 to 600 kg per hectare of soybeans.

The weediness of *S. halepense* is due in great part to its adaptations for vigorous growth and for longevity. Most strains, however, quickly become sod-bound, often within 3 years, and the plants must be broken up to reestablish the stand. This disturbance and tearing of rhizomes into small fragments, when practiced in infested fields of row crops, may very well result in the establishment of greater populations of the weed.

S. halepense is an alternate host of *Botryosphaeria* sp. and *Puccinia purpurea* Cke. (Puckdeedindan 1966); and of the viruses which cause rice leaf gall, corn leaf gall (Agati and Calica 1950), stripe disease of rice (Yamada, Shiomi, and Yamamoto 1956), sugarcane mosaic (Chona and Rafay 1950), maize dwarf mosaic (Lee 1964), beet yellows, and wheat streak mosaic (Namba and Mitchell, unpublished; see footnote, *Cynodon dactylon*, "agricultural importance").

COMMON NAMES OF SORGHUM HALEPENSE

ARGENTINA
 canatillo
 canota
 canucho
 maicillo
 pasto ruso
 sorgo de Aleppo
AUSTRALIA
 johnson grass
BRAZIL
 capim massabara
CHILE
 sorgo de Aleppo
COLOMBIA
 pasto johnson
 sorgo maleza

CUBA
 Don Carlos
HAWAII
 johnson grass
INDIA
 barool
INDONESIA
 glagah rajoeng
  (Javanese)
 pangan (Indonesian)
IRAN
 ghiagh
ITALY
 cannarecchia
LEBANON
 hashishat-ul-faras

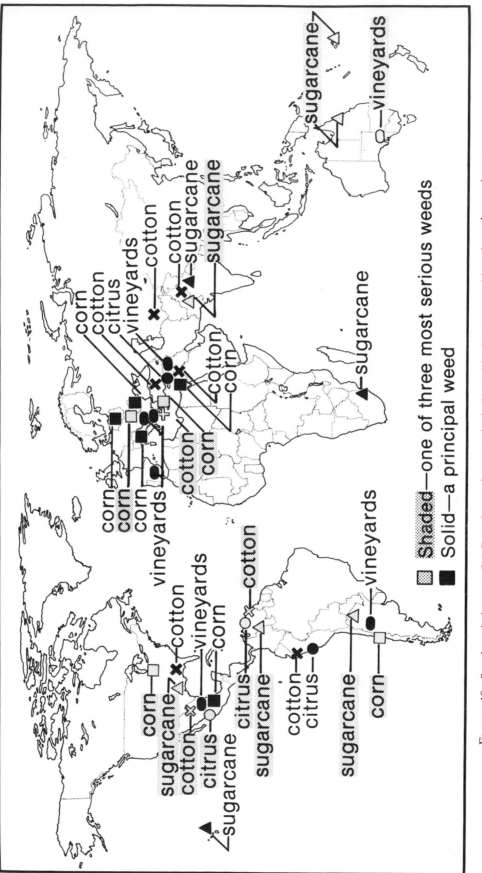

FIGURE 19. *Sorghum halepense* (L.) Pers. is a serious or principal weed in the crops and locations shown above.

MEXICO
  zacate johnson
NETHERLANDS
  Aleppobartgras
NEW ZEALAND
  johnson grass
NICARAGUA
  zacate johnson
PAKISTAN
  baru grass

PERU
  grama china
PHILIPPINES
  batad-bataran
  ngigai (Tagalog)
PUERTO RICO
  yerba johnson
SOUTH AFRICA
  johnson grass

SPAIN
  canota
SURINAM
  curaçaosche
THAILAND
  yah poeng
TURKEY
  gelis
UNITED STATES
  Aleppo grass

Arabian millet
Egyptian millet
evergreen millet
false guinea
johnson grass
Morocco millet
Syrian grass

YUGOSLAVIA
  kostan

❧ 7 ❧

Imperata cylindrica (L.) Beauv.

POACEAE (also GRAMINEAE), GRASS FAMILY

Imperata cylindrica, a native of the Old World, is a perennial grass forming long, hard, creeping, scaly rhizomes. The very attractive inflorescence is a dense, silvery white, fluffy, cylindrical spikelike panicle. The plant must be regarded as a major menace in high rainfall areas of the tropics although it is found as well in the warm temperate zone. It is on all of the continents and is the worst perennial grass weed of southern and eastern Asia. Seventy-three countries report that it is a weed in 35 crops which are as different in their cultural systems as tomatoes and coconuts. In many areas the grass enters the fields where shifting cultivation is practiced and may cause such fields to be abandoned early. This practice has caused the formation of great expanses of *I. cylindrica* in Asia and Africa. Two million hectares of rubber recently have been estimated to be seriously infested with this species in Malaysia. Fifteen to 30 million hectares are estimated to be covered with this species in Indonesia and 150,000 more hectares are invaded annually. In its very early growth stages the plant is palatable to livestock, and so the grass is often fired to stimulate growth of new sprouts. The plants make excellent thatching material and major efforts have been made to process them for papermaking in Africa, Asia, and southern Europe.

DESCRIPTION

I. cylindrica is a *perennial* grass that varies greatly in form, growing in loose to compact tufts with slender, erect culms from long, tough, extensively creeping scaly *rhizomes*. The inflorescence is branched but compacted into a dense, white, fluffy, cylindrical, spikelike head up to 20 or more cm long and 2 cm broad (Figure 20).

Culms erect or nearly so, 15 to 120 cm tall, rarely to 3 m, slender to moderately stout, sometimes very robust, one- to four-noded (rarely eight), unbranched, solid, smooth, glabrous; *sheaths* smooth or with margins ciliate (fringed with hairs), bearded at the mouth, the lower broad, loose to tight, overlapping, upper finally splitting into fibers to form a dense covering at bases of culms, the upper shorter than internodes, more often long-bearded at the nodes with silky hairs; *ligule* short, obtuse, or truncate, 0.5 to 1 mm long; *blades*, lower erect, or spreading and drooping, narrow, linear, gradually narrowing downward, sometimes to a stout, whitish midrib, blade tapering from above middle to a sharp tip, length varying greatly with habitat, from short to 150 cm or more in length, 4 to 18 mm wide, flat, involute or convolute, glabrous or hairy at the base, usually smooth except for scabrid (rough) margins, upper blades much shorter than lower; *panicles* spikelike, cylindrical, dense, 3 to 20 cm long (sometimes to 60 cm), 0.5 to 2.5 cm wide, silky, with silvery white or creamy hairs, speckled with purplish or brownish stigmas and orange or brown anther; *pedicels* unequal, 0.5 to 4 cm long; *spikelets* lanceolate to oblong, 3 to 6 cm long surrounded by silky hairs 10 mm long; *glumes* equal or the upper slightly longer than the lower, membranous, three- to nine-nerved, long hairs from basal callus and on lower part of back, lower lanceolate to oblong, upper

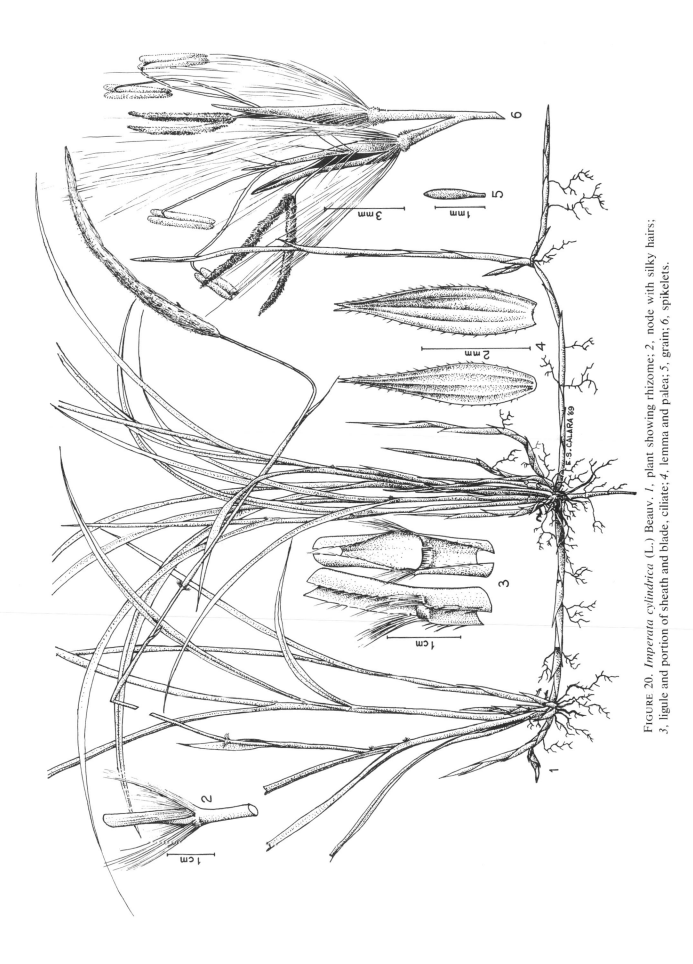

FIGURE 20. *Imperata cylindrica* (L.) Beauv. *1*, plant showing rhizome; *2*, node with silky hairs; *3*, ligule and portion of sheath and blade, ciliate; *4*, lemma and palea; *5*, grain; *6*, spikelets.

broader than lower; *lower floret* usually barren, *lemma* ovate-oblong to broadly oblong, acute to truncate, transparent, nerveless, ciliate, often toothed, 1.5 to 4 mm long, lemma of the upper floret lanceolate to oblong, acute, obtuse or toothed, ciliate, 1 to 3 mm long; *palea* usually very broad, toothed, ciliate, 0.8 to 2 mm long; *stamens* two; *anthers* 2 to 4 mm long on slender filaments; *grain* a caryopsis, single in each spikelet, oblong, brown, and 1 to 1.3 mm long.

Following the examination of a large number of specimens, Hubbard et al. (1944) placed them in five major groups. These varieties, for which they presented detailed descriptions, generally occupy distinct geographical regions that have been shown on a map prepared by those authors. The variety *major* is most widely distributed, extending from Japan and southern China through the Pacific Islands and Australia to India and eastern Africa. The variety *africana* is the next most widely distributed and is found from Senegal and Sudan southward through Africa. The variety *europa* extends from Portugal through southern Europe to the arid regions of Central Asia in the Soviet Union and Afganistan. It is generally distributed in the Middle East and North Africa. The variety *latifolia* is found only in northern India, and the variety *condensata* is found in Chile on the coastal region between lat 30° and 40° N.

Variety *major* has smaller spikelets than the other varieties. In Australia a few plants were found bearing two-flowered spikelets. This variety is distinguished from variety *africana* by having usually hairy nodes and slightly smaller anthers and spikelets. Variety *major* differs from variety *latifolia* by having generally smaller culms, leaves, and inflorescences. Variety *major* differs from variety *europa* in possessing smaller spikelets and anthers, hairy nodes, thinner and wider flat leaf blades, and less dense, much softer, more easily compressed inflorescences. Variety *condensata* most closely resembles variety *europa* but differs in having larger ligules, more finely pointed, flat, leaf blades; and shorter hairs on the spikelets.

DISTRIBUTION AND HABITAT

From Figure 21 it may be seen that *I. cylindrica* is very widely distributed in Australia, Africa, the southern half of Asia, and the Pacific Islands. In the New World it is found in Argentina, Chile, Colombia, Florida (United States) and the West Indies. It is present but not a serious weed in southern Europe and around the Mediterranean. Although normally confined to areas that are quite warm, it also is found in Japan and New Zealand at latitudes of 45° in both the northern and southern hemispheres.

Its habitat includes the dry sand dunes of shores and deserts as well as swamps and river margins. It grows in grasslands, in cultivated annual crops, and on plantations. It quickly enters abandoned farmlands, and it may be seen on railroad and highway embankments and on both deforested and reforested areas. It can withstand long dry spells on light soils and will tolerate waterlogging on heavy soils. It becomes established most quickly on medium to good soils and is less frequently a pest on poor soils. Although sometimes reported to be a weed of poor soils, it probably inhabits these areas because of lack of competition from better grasses which cannot be supported there. The species makes its maximum development in wet areas of good soils. If all other factors of the environment are suitable, *I. cylindrica* can occupy any soil type where sufficient moisture exists to support growth. The plant grows at altitudes to 2,000 m in several parts of the world and to 2,700 m in Indonesia. In eastern Africa it is most frequently present in areas with more than 1,000 mm of annual rainfall, whereas in Indonesia it does well in areas receiving 500 to 5,000 mm.

BIOLOGY

Coster (1932), Santiago (1965), Soerjani and Soemarwoto (1969), and Soerjani (1970) have studied the biology of *I. cylindrica*, and Hubbard et al. (1944) have provided some information in a general review of the species.

The species reproduces by seeds and by the extension of a very vigorous rhizome system. Aerial flowering culms arise from terminal or axillary buds of the rhizomes or from the basal portion of another aerial culm. Some individual plants produce flowers frequently, some never produce them, and others are intermediate between these conditions (Santiago 1965). Pollination occurs as the inflorescence expands; following pollination the callus hairs spread out to give the silvery white bushy appearance to the head. The varieties are widely distributed from north to south so that flowering times must be peculiar to a locality. Variety *europa* flowers from March to August near the Mediterranean. Variety *major* flowers all year in the Philippines and has been known to flower abundantly after a frost in the United States. Slashing, burning, defoliation, grazing, and the addition of nitrogen stimulate flowering.

I. cylindrica is a prolific seed producer and may

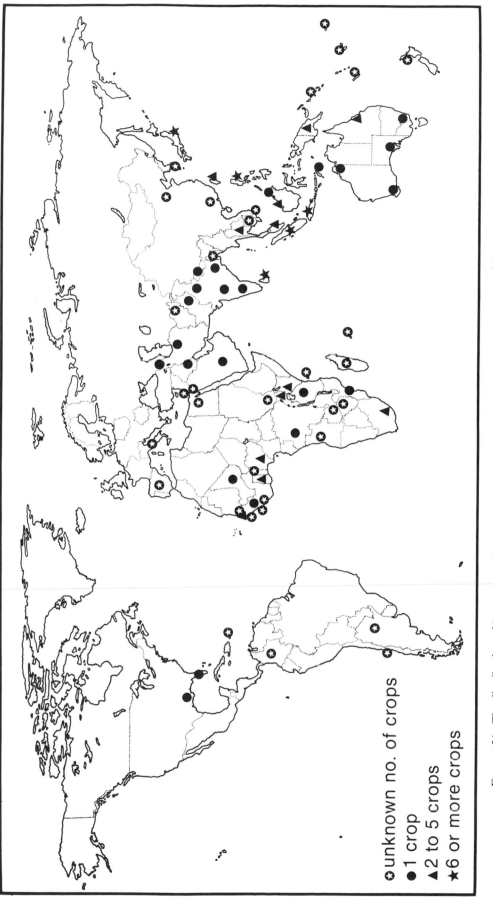

FIGURE 21. The distribution of *Imperata cylindrica* (L.) Beauv. across the world in countries where it has been reported as a weed.

give as many as 3,000 per plant. The plumed seeds may travel long distances over land and sea, but the average flight at inflorescence level is 15 m. The seeds have difficulty penetrating even a small forest barrier and seldom find their way to isolated clearings. They do, however, follow roads and railroads to interior areas. Santiago (1965) reported 95-percent germination within 1 week of harvest and the seeds were viable for at least 1 year.

The anatomy of the leaves, stems, roots, and rhizomes of the members of the genus *Imperata* and, in particular, of the varieties of *I. cylindrica* has been studied for almost a century. Information for *I. cylindrica* var. *europa* is given in the works of Kirchner, Loew, and Schröter (1904) and Duval-Jouve (1875); for variety *major* in the works of Hole (1911) and Vickery (1935); and for variety *latifolia* in the contributions of Hole (1911). Leaf size and shape vary greatly with the habitat so that a general account of anatomy and morphology is of little value. In areas of heavy cutting or grazing, for example, the leaf blades are small and the culms are short and slender, whereas in open grasslands the leaves are longer and wider with robust culms reaching to 120 cm. The roots and rhizomes possess modifications which help to conserve water within the central cylinder and to resist breakage and disruption when they are trampled or disturbed. The underground portions also have a remarkable resistance to the heat of fire. In variety *europa* the root has an outer cortex consisting of a band of sclerenchyma in which the large outer cells have greatly thickened outer walls. The cells of the endodermis show a marked thickening of the radial and inner walls. The rhizome has a band of sclerenchymatous fibers just below the epidermis, and each individual bundle in the pith is surrounded by sclerotic tissue.

New seedlings establish most quickly on the open ground of farms, plantations, and roadsides. The seedlings have no rhizomes for 4 weeks. The rhizomes are firmly rooted, white, and succulent, with scales that become papery on drying and then flake off to expose the internodes. Soerjani (1970) estimated that on 1 hectare there may be 4.5 million shoots, more than 10 metric tons of leaf material, and more than 6 metric tons of rhizomes. Roots may penetrate to 150 cm. Rhizomes may penetrate as deeply as 120 cm but most are in the top 15 cm of heavy soils and 40 cm of light soils. Soerjani and Soemarwoto (1969) tested short rhizome pieces for sprouting ability and found that 1- to 5-cm segments were the same. Rhizome pieces from 2 to 5 mm in diameter had about the same sprouting ability. Little

difference was found in the sprouting ability of 2-, 5-, and 10-cm cuttings buried at 0, 5, 10, and 15 cm in the soil. To study desiccation, Soerjani maintained rhizome segments at 100-, 80-, 70-, 47-, 19-, and 3-percent relative humidity for 1, 2, and 3 days. All rhizomes could sprout after the 1st day but none of those held at a humidity below 47 percent for 2 days or 70 percent for 3 days could sprout. Fresh rhizomes normally have a moisture content of about 80 percent.

Rhizome buds measuring 1 mm to more than 3 mm are about equal in sprouting ability and those which are smaller than 1 mm may not be viable. The orientation of the buds on the rhizome pieces influences the sprouting. Vertical rhizomes with the apical end up and horizontal pieces with the bud on top give best results. Sprouting is two to three times greater in light than in darkness. Buds of the rhizomes are white, brown, or dark brown, the latter having the lowest germinative capacity.

Little is known about the specific effects of light duration and intensity on the morphology and physiology or the cyclical growth of *Imperata*. It is generally recognized to be a light-loving plant which can be "shaded out" under a heavy canopy. Though the plant is generally weakened it does remain ready to invade areas which open up as a result of disease, storms, or man's activities. Soerjani (1970) found that even 50-percent shade did not eradicate the weed. From Nigeria have come reports of ecotypes which have a higher shade tolerance.

Soerjani (1970) reported that the optimum temperature for rhizome bud development in Indonesia is 30° C with no growth occurring at 20° or 40° C. In the United States, plantings made in Texas and Mississippi were killed out by low temperatures but those in Florida survived.

AGRICULTURAL IMPORTANCE

From Figure 22 it may be seen that *I. cylindrica* is ranked among the three most serious weeds in coconuts in Ceylon and Malaysia; oil palms in Malaysia; rubber in Indonesia and Thailand; and tea in Ceylon, India, and Indonesia. It is a serious weed in citrus in Malaysia and Thailand; oil palms in Colombia, Indonesia, and Nigeria; and rubber in Malaysia. It is a principal weed of citrus in Saudi Arabia; coconuts in Mozambique, New Guinea, and Zanzibar; oil palms in Dahomey; pineapples in Guinea; rubber in Ceylon and West Africa; and tea in Japan, Malaysia, Mozambique, and Uganda. Although these data are not shown on the map, *I. cylindrica* is also a serious weed in cas-

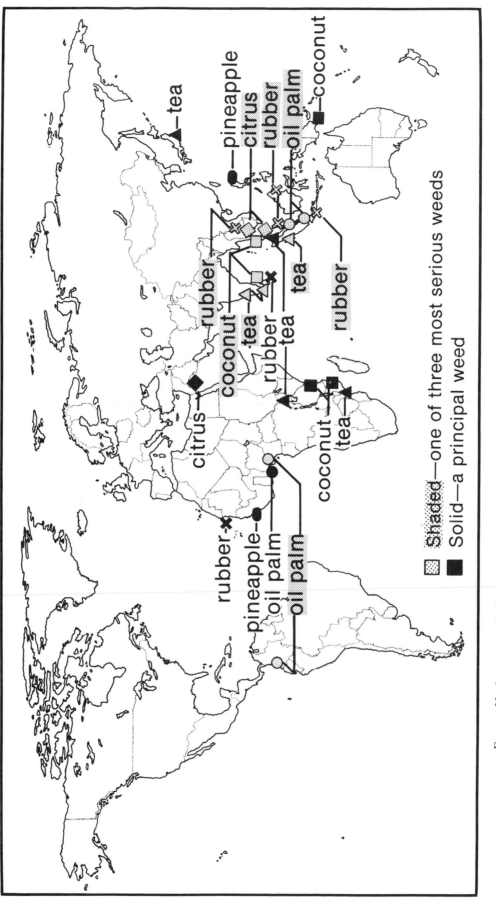

FIGURE 22. *Imperata cylindrica* (L.) Beauv. is a serious or principal weed in the crops and locations shown above.

sava, cinchona, corn, peanuts, upland rice, and sweet potatoes in Indonesia; of the teak and sal forest trees of India and Malaysia; of abaca in the Philippines; and of upland rice in Timor (Indonesia). It is also a principal weed of cotton and teak in Nigeria and of cotton in Ghana; of pastures in Australia and India; and of pineapples, sugarcane, and coffee in the Philippines, Taiwan, and Uganda, respectively.

In addition, *I. cylindrica* is found in rubber, tea, and pineapples in several other countries of Asia and Africa, and in abaca, cassava, coffee, corn, peanuts, upland rice, sugarcane, and sweet potatoes in many countries of Asia. Finally, it is in bananas in the Philippines and Thailand; barley in Iran; potatoes, soybeans, and vegetables in Japan; figs in Malaysia; roselles in Indonesia; and sisal in Kenya and Tanzania.

The weed has been so difficult to cope with in Asia that many studies have been made to examine the magnitude of losses after the crops have been in competition with *I. cylindrica*. Experiments in rubber plantations in Malaya show that, for the first 5 years after being planted, trees surrounded by the grass made about 50 percent of the annual growth of trees which were free from weeds or which had a legume ground cover of *Centrosema pubescens* (anon. 1937-1938). Most of the world's rubber supply comes from southern Asia and the weed is most serious there. On major plantations where the weed has been managed for many years through a combination of control measures, it is not regarded as a serious problem any longer. An annual budget allotment for *Imperata* control on many large plantations, however, is a recognition that this weed stands as a constant threat at the edge of every holding. Basnayake (1966) recently estimated that 2 million hectares of rubber are seriously infested with *I. cylindrica* in Malaysia. When rubber plantations were neglected in Sumatra in the early 1960s because of civil strife, *Imperata* quickly invaded; some managers later estimated that it would require 5 to 10 years to bring the grass again under control. For small holders the weed poses a very difficult problem.

Experiments similar to those described above also were conducted on coconut palms in Malaysia in an area of very heavy clay. Perhaps because of the heavy soil, the plots covered with *Imperata* made the best growth in early stages. It was suggested also that the grass provided shade for the newly planted coconuts in dry, hot weather and through transpiration helped to remove excess water from the heavy soil in the rainy season. In the plots which were free from weeds, the sun had baked the soil. In plots covered with *Centrosema pubescens*, trees were able to make good growth. At the end of several seasons, however, the trees in the weedy plots were underdeveloped, leaves were yellow, and growth was poor. The production of surface-feeding roots by the palms was inhibited in weedy plots and this tended to limit the uptake of nitrogen (Jagoe 1938). *Imperata* is reported to be harmful to coconut palms in Mozambique as well.

Coster (1932, 1939) reported that the growth of teak trees, *Tectona grandis*, in the 1st year was 13 cm in plots of *I. cylindrica* and 100 cm in plots where the weed had been removed. Often seedlings ceased growth, turned yellow, and the tops of some died in weedy plots. Similar harmful effects are experienced in Nigeria. The sal, *Shorea robusta*, which is second only to teak for lumber production in India, also suffers severe competition from *I. cylindrica* (Rowntree 1940).

In addition to providing tree crops with the best possible growing conditions so that they can develop a canopy quickly and thus shade out *Imperata* and striving to eliminate direct competition for nutrients, water, light, and space by weed plants, agriculturalists must also be aware of other factors if they are to produce an optimum crop. Where shifting cultivation is practiced, removal of natural shade creates conditions favorable to the growth of the *I. cylindrica*. Under natural conditions bush and tree vegetation would become reestablished and would shade and suppress the grass. But with the advent of dry weather the grass becomes dry and highly flammable and very often is ignited to give hot fires which destroy most other vegetation. The matted rhizomes of *I. cylindrica*, protected underground, regenerate quickly as the rains come and give rise to fresh growth which again takes over the area. If shifting cultivation is continued, the same thing reoccurs and soon very large areas are taken over by *I. cylindrica*. This practice causes the loss of agricultural lands and potential forest lands. The destruction of the forest organic layer affects the water-holding capacity of the area and, after burn-off, soil erosion may be severe until grass growth covers the area again. Dry *Imperata* produces a major fire hazard in cultivated plantations as well.

Locusts are serious pests in several parts of the world where extensive areas are covered with *I. cylindrica*; and there is speculation that these grasslands provide good breeding grounds. If this is true, serious economic losses may be attributed indirectly to the presence of this weed (Hubbard et al. 1944).

The underground rhizomes of the grass have sharp, hard points which actually enter the roots of rubber, coconuts, and pineapples and may grow within them for distances up to 60 cm. When these crop roots are destroyed, microorganisms may enter to affect other parts of the trees (Tempany 1951, MacLagan Gorrie 1950). The circumstances which surround the stunting of plantation trees and the symptoms these trees exhibit indicate that there may be a chemical interaction between the crop and weed species produced by compounds released from *I. cylindrica,* either from living tissue or from decaying residues.

This grass species is so widespread in India, Malaysia, the Philippines, and in many other areas that great efforts have been made to manage the grass in cultivated or natural stands so that it will remain in a palatable growth form. In stands which are undisturbed the grass becomes very coarse, the old leaves and culms hide the tender shoots from livestock, and light cannot penetrate to the new growth. Management techniques are generally directed toward maintaining the plant in a depauperate state by burning it about once a year and then grazing or cutting it sufficiently close to keep it short and fine-leaved. The growing point remains close to the soil surface and the outer leaves may be grazed without injury to this tip. The young shoots are badly nourished as a consequence of the loss of the outer leaves and a crop of fine grass about 5 to 10 cm tall results. If the grass is repeatedly cut or grazed the area will provide good pasture for many months. In some areas of Asia and Africa the crop is regarded as a valuable standby for stock during prolonged droughts.

In North Africa and the Middle East the grass is consumed by camels, goats, and sheep, but here again it is usually only the new growth that is eaten. Here also the Bedouins routinely fire the old coarse growth to encourage new, tender shoots. The grass is also used for grazing and is cut for hay in the arid and semiarid parts of the Soviet Union.

Studies of the nutritive value of *I. cylindrica* and of its usefulness as a tropical fodder grass date back almost a century. As with most grasses, the chemical composition varies greatly with the stage of growth, variety, soil type, and climatic conditions. In Indonesia the nutritive value of the variety *major* was found to be below the average of other grasses studied. This was too low for horses; and cattle developed scours when, for lack of other pasture, they were forced to feed on this grass. The animals also developed sore mouths, a problem which seems widespread whenever the grass is used beyond the very young stages. Eichinger (1911) collected data on the chemical analyses of 15 grass species from many places in the world. Corn, millet, rice, and sugarcane were included in the study. *I. cylindrica* was found to be very low in protein and ash, of average fat content, above average in carbohydrate content, and very high in crude fiber. Very similar results were reported by Pepa (1927) in analyses of *I. cylindrica* and four other forage grasses of the Philippines. The species in Malaysia was given a more favorable rating in a detailed study of tropical fodder grasses, pasture grasses, and other fodders. *Panicum*, *Pennisetum*, *Axonopus*, and *Paspalum* species, cowpeas, sweet potatoes, lucerne hay, and the cereals corn and rice were also included in the study (Georgi 1934).

In summary, many of the tropical pasture grasses are coarse, tall, and can be used for forage only in early growth stages. Many, including *Imperata cylindrica*, have several varietal forms and grow in a number of different habitats. The value of these grasses for forage may be dependent upon their stage of growth at cutting or grazing and upon a system of management which keeps the crop in a palatable condition. The judgment of worth in poor soils and in dry areas may hinge upon the presence or absence of a better fodder. It is not surprising, therefore, that there are varied reports on its usefulness across the world. There does seem to be general agreement that, in a depauperate state, *I. cylindrica* can provide worthwhile pasture. Pasture improvement research for some areas in recent years has provided grasses which are far superior, and in regions where these are utilized *I. cylindrica* usually is found only on poor pastures where the best grasses have been eaten out by overgrazing. For this species the injury to stock animals remains a problem everywhere. Some workers believe that even the young grass can cut the muzzles of grazing animals while the sharp tips of the older blades can injure their hoofs.

COMMERCIAL AND OTHER USES

In addition to being used occasionally as livestock feed, *I. cylindrica* serves man directly in several important ways. Hubbard et al. (1944) have reviewed its use as thatching for homes, public buildings, and temples, and for portable and temporary farm sheds in Asia and Africa. The authors summarized the costs of thatching and methods of construction and compared *I. cylindrica* with the widely used nipa palm *Nipa fruticans*. *Imperata cylindrica* is so highly regarded for thatching that it is planted and tended as a crop in many areas of Asia where cultivated and natural stands may be planted and hoed, and trees and brush removed to prevent shading. A normal management practice is to burn the first crop, allow cattle to graze the second crop while it is young and tender, and use the third and successive crops for thatch as the leaves become more supple.

For many decades serious attempts have been made to use *I. cylindrica* as raw material for papermaking. Hubbard et al. (1944) have reviewed the studies made in Africa, Asia, and southern Europe since 1900. There is at present no widespread commercial use of the grass for this purpose because there are other rough grasses that are more economical to process, and the problem of supplying paper mills with sufficient *I. cylindrica* of uniform quality from cultivated or natural stands is difficult. In the Philippines, for example, continued cropping brought reduced yields and invasion by grasses and woody plants that were not suitable for papermaking. All grasses are bulky and, therefore, expensive to transport. As is so often the case with both terrestrial weeds and waterweeds and vegetation, the construction of manufacturing plants for processing the raw material may be impractical at the site of greatest supply. Some workers believe that many of the areas of *I. cylindrica* production would yield far greater tonnage of raw material for paper if softwoods were grown on these areas.

I. cylindrica grows vigorously and, because it can thrive on many soil types and in a variety of habitats, it is frequently used as a soil binder. It quickly forms a sod and helps to prevent serious erosion where forests are clear-cut in high rainfall areas. It is used to stabilize canal and railroad embankments and may be a factor in flood control when used to hold riverbanks and earthen dams. It also serves as a sand binder for coastal sand dunes and moving sand hills in desert areas.

I. cylindrica is an alternate host of the rust *Puccinia refipes* Diet in Thailand (Chandrasrikul 1962). Vayssiere (1957) has reported on several polyphagous insects which attack both cultivated cereals and *I. cylindrica*.

COMMON NAMES
OF IMPERATA CYLINDRICA

AUSTRALIA
blady grass

BURMA
kyet-mei
thetke

CAMBODIA
sbauv

CAMEROONS
baya
ndongo limba
sosongo

CEYLON
darbai-pul
iluk
inanka-pilu

CHINA
mao-tsao
mau kan tso

CONGO
binkba
moto-moto
nianga

CYPRUS
xiphara

EASTERN AFRICA
swordgrass

EGYPT
beni el sham
deil el qott
halfa
hishka
sill

FIJI
gi

INDIA
chero
dabh (Hindi)
dharba (Telugu)
dhub
modewa gaddi
   (Telugu)
ooloo (Bengali)
siru
tharpai pullu (Tamil)
thatch grass

INDONESIA
alang-alang
eurih
lalang

IRAN
santintail

IRAQ
blady grass
cylindrical ha
halfa

IVORY COAST
nse

JAPAN
chi
chigaya
tsubana

KENYA
nyeki

MADAGASCAR
manevika
tena

MALAYSIA
alang-alang
lalang

MAURITIUS
lalang

NEW GUINEA
auturra
kawva
kunaigrass
kuru-kuru

NEW ZEALAND
imperata

NIGERIA
ata
ekan
isa
gasa kigere
soyo
speargrass
tibin
tofa
zarenshi

PHILIPPINES
buchid (Ivatan)
bulum (Ifugao)
gaon (Igorot)
gogon (Bikol)
goon (Bontoc)
ilib (Pampangan)
kogon (Tagalog,
   Ilokano)
parang

RHODESIA
ibamba
luwamba
silenge
silverspike

SOUTH AFRICA
bedding grass
mohlorumo
silverspike
um tente

SOVIET UNION
gyrjúk

kazarí
tiskan-čub

SPAIN
carrizo
cisca
marciega

SUDAN
doiya
mayani

SYRIA
halfa
meshian
sill
sulel

TAIWAN
bái-máu

TANZANIA
chiambi
motomoto

sanu

THAILAND
yah-ka

TUNISIA
dis

UNITED STATES
cogon grass

WESTERN AFRICA
dole
gombi

hada
nounour
soyo

VIETNAM
cò tranh

ZAÏRE
binkba
moto-moto
nianga

❧ 8 ❧

Eichhornia crassipes (Mart.) Solms

PONTEDERIACEAE, PICKEREL-WEED FAMILY

Eichhornia crassipes, a perennial, herbaceous, floating, aquatic weed is the scourge of the world's major rivers. It is a native of the Amazon basin. It is mainly a weed of the tropics and subtropics but extends to lat 40° N and lat 45° S in lakes and coastal areas where it can escape severe cold. In agriculture it is found only in paddy fields, but it stops up irrigation canals and impedes the flow of great rivers. It is a menace to human health, power supplies, and navigation.

DESCRIPTION

E. crassipes is a *perennial* aquatic herb (Figure 23); *stems* short, floating or rooting in mud, rhizomatous or stoloniferous, rooting from the nodes; *roots* long, sometimes dark because of their purple anthocyanin, pendant; *leaves* in a rosette; *petioles* spongy, in young specimens short and with a one-sided swelling or inflation but up to 30 cm long when older, tapering and narrowing from the bulbous base to the point of attachment with the lamina; *lamina* circular to kidney-shaped, glossy smooth, 4 to 15 cm long and wide, acting as a sail in the wind; *inflorescence* in *spikes* with about eight flowers, long peduncled, bibracteate, the lower bract with long sheath and small lamina, the upper almost entirely included within the sheath of the lower one, tubular with a small pointed tip (apiculate); *flower-bearing part of rachis* up to 15 cm or less long; entire *scape* may be 30 cm; *perianth* six-lobed, united below into a narrow tube, lilac, bluish purple or white, the upper lobe bearing a violet blotch with yellow center;

stamens six, three long, three shorter, attached to the tube; *capsule* membranous, three-locular, dehiscent, many-seeded, as many as 50 or so per capsule; *seed* ovoid, ribbed, 0.5 to 1 mm.

This species is distinguished by the almost one-sided swelling or inflation of the petiole, its long peduncled bibracteate spike, and its upper perianth blotched with yellow at the center.

BIOLOGY

The most comprehensive study of the biology of the water hyacinth was done by Penfound and Earle (1948). The mature plant consists of roots, rhizomes, stolons, leaves, inflorescences, and fruit clusters. The rhizomes produce all the roots, leaves, and reproductive structures. The reproductive tip of the rhizome, about 1.2 cm long, rests about 3.7 cm below the water surface. The stolons, which may be 5 to 45 cm long, are upright in dense stands and horizontal in open colonies. In open, sparse stands the leaf consists of a membranous ligule, a subfloat, a float (swollen midportion of petiole), an isthmus, and the blade which is merely a flattened portion of the petiole (Sculthorpe 1967).

In the succession of vegetation in freshwater systems, *E. crassipes* plays a dominant role by providing a floating platform for colonizing species of terrestrial, wetland, and aquatic plants (Wild 1961). These species form a floating marsh which increases in thickness until its base comes to rest on the bottom. A mat of medium-sized plants may contain 2,000,000 plants per hectare and the total wet weight

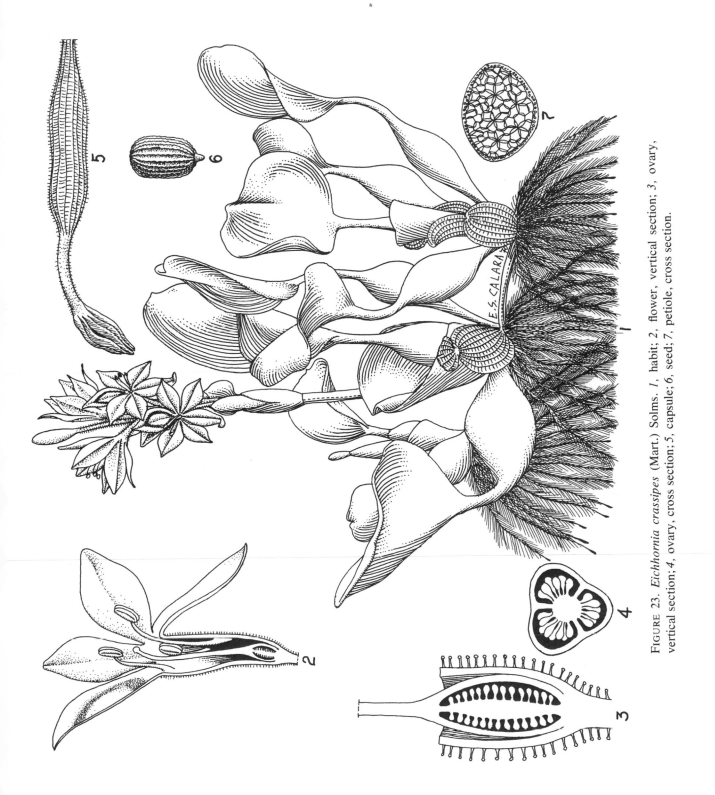

FIGURE 23. *Eichhornia crassipes* (Mart.) Solms. *1*, habit; *2*, flower, vertical section; *3*, ovary, vertical section; *4*, ovary, cross section; *5*, capsule; *6*, seed; *7*, petiole, cross section.

may be 270 to 400 metric tons per hectare.

The factors in the environment which influence the formation of floats on the petioles are not well understood, for the observations reported in the literature vary greatly. There are many reports that floats are not found on petioles of leaves when the plants are crowded as in a normal mat. Floats are formed on leaves of plants at the fringes of openings or borders of such mats. There are reports that floats do not form on plants rooted on land, but Mitchell (1972) has found and photographed petioles with floats on plants stranded for some months, as well as on plants arising from seedlings on mud banks. Shading inhibits float formation and Misra (1969) reported that floatless leaves are formed when the light falls below 500 foot candles, and death of the plant follows when light remains below 130 foot candles for very long. Thus we may have the situation that a small proportion of leaves in nature possess floats.

The plant cannot tolerate water temperatures above 34° C. Leaves are killed by frost but entire plants succumb only when the rhizome tip (just below the surface) is frozen.

The evapotranspiration rate is reported to be two to eight times that of evaporation from a free surface. The plant is 95-percent water. The period of survival for plants stranded or thrown on open ground may be several days and depends to some degree on the temperature and the amount of sun present. Mitchell (personal communication)* feels that humidity also has an important role, for he has observed that plants at the bottom of a heap, or those covered by soil, can survive at least 3 weeks. Plants rooted in the soil, if reflooded during active growth, will abscise just beneath the lowest living leaves and float to the surface; but if this occurs toward the season's end they remain at the bottom and perish.

DISTRIBUTION AND HABITAT

From Figure 24 it may be seen that *E. crassipes* grows almost everywhere in the agricultural world except in the northern reaches of the temperate zone. Because of his admiration for the flowers, man has assisted the spread of this plant by cultivating it in his pools and gardens. It is still for sale in several places in the world. His carelessness toward the

* Information obtained from Dr. D. S. Mitchell, Salisbury, Rhodesia, in 1972, concerning publications in preparation on ecological surveys of *Eichhornia crassipes* and *Salvinia* species, together with data on the taxonomic position of the latter genus.

cleanliness of his commercial craft on land and sea has also contributed to the movement of the plant. In Africa, fresh *E. crassipes* plants are used as cushions in canoes and as a means of plugging holes in charcoal sacks as they are transported from the bush. The plants catch on the sides and bottoms of river craft and thereby move with the commerce of the region.

Natural forces and events have also been important in the spread of this plant. The leaves are upright and serve as sails before the wind. From the nurseries in the swamps and backwaters, great islands of the weed may be flushed into a mainstream at flood time. Water hyacinths are found in rivers, lakes, ponds, reservoirs, sloughs, canals, and drainage ditches. In rapidly flowing streams they are commonly flushed out during high waters. They do not occur in water which has an average salinity greater than 15 percent that of seawater. When the species does get into brackish water, its leaves show epinasty and chlorosis, and finally death ensues.

Seeds can germinate on land, but plants can also strike root and survive. Size varies greatly in nature. On gravelly beaches the leaves may be only 1.2 cm long and plants may have only two flowers per inflorescence, whereas in open, aerated water leaves may be 120 cm long and there may be 40 flowers per inflorescence. It has been reported that giant plants produce few or no flowers.

Gay and Berry (1959), Parija (1934), Robertson and Thein (1932), Buyckx and Tas (1958), Heinen and Ahmed (1964), Sculthorpe (1967), Wunderlich (1964), Tabita and Woods (1962), Hearne (1966), and Parsons (1963) have reviewed the distribution and control of water hyacinth in several regions of the world.

PROPAGATION

The plant spreads by producing vegetative offshoots and seeds (Penfound and Earle 1948, Das 1969). The new offshoots are bound to the parent plant by strong stolons. The plants are separated by the action of wind, waves, and currents, as well as by deterioration of older, connecting stolons. In one experiment which started with two parent plants, 30 new offspring were vegetatively produced in 23 days and 1,200 at the end of 4 months. It has been estimated that 25 plants, under good conditions, can produce enough offspring to cover a hectare in one growing season in the temperate zone. Chadwick and Obeid (1966) completed an excellent study of the growth requirements of this species in Sudan.

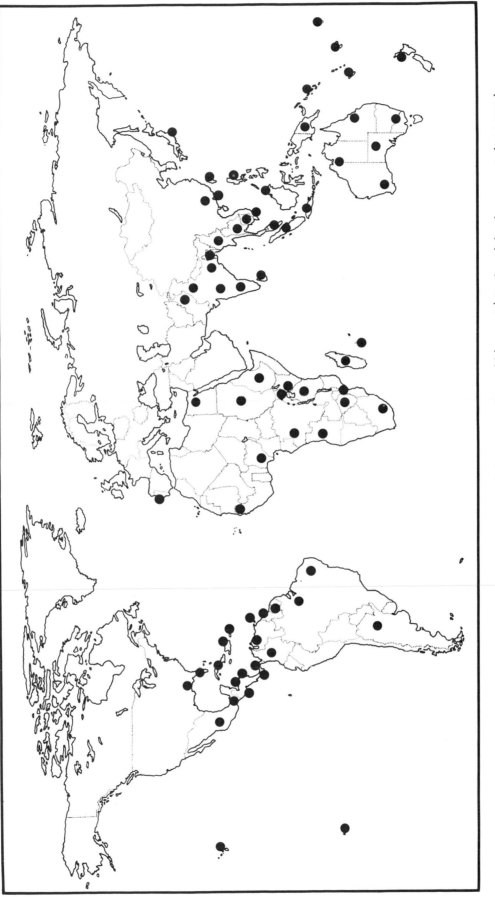

FIGURE 24. The distribution of *Eichhornia crassipes* (Mart.) Solms across the world in countries where it has been reported as a weed.

Seedlings may be seen on exposed shorelines, rotting logs, or on the debris of the floating mats. Seedlings may emerge 3 days after planting, will have two or three ligulate leaves in 10 days, and in 30 days will have seven or eight ligulate leaves plus one to three spatulate leaves in which floats are beginning to form. Within 60 days, most leaves have floats and small, new offshoots will be produced. The plants can produce a new leaf about every 3 days, and the number of leaves on mature plants tends to be constant because the old leaves at the bottom decay.

There is no agreement on the necessity for drying or exposure to air before seed germination takes place. It does appear that seeds must be scarified by physical, chemical, or biotic factors as a prerequisite for germination. Light is not necessary for germination.

Although the flowers seem to be very well adapted for insect pollination, such pollination occurred only rarely in plants studied in the United States. Soon after the plant flowers, the stems curve over into the water, or "wilt"; and it is during this period that self-pollination takes place. About 18 days after pollination, ripe seeds are shed. They may fall to the mat or bottom mud, and some of them have been known to survive for at least 15 years.

IMPORTANCE IN HUMAN AFFAIRS

E. crassipes is one of the world's worst weeds but it does not often bring its destructive force to bear directly on cultivated crops. It is a problem in some places in the world in paddy crops such as rice and taro. In Bangladesh some of the farming areas are covered with massive deposits of water hyacinths when the floods come down from the hills in the rainy season. Canals, streams, reservoirs, and lakes become unmanageable. Navigation becomes difficult or impossible because of the sheer mass of solid material on the water surface. Buoys are submerged and navigation channels are hidden. In times of flood, bridges may be pushed over. Fishing grounds are destroyed by shading and lack of oxygen as the weed cover becomes more dense; spawning areas are blocked; and the people of riverine communities cannot reach the fishing areas. For many communities this means the loss of the principal source of protein. Insect vectors of human and animal diseases seek harbor in the mats of this weed. Snakes and crocodiles take refuge in the weed cover and bring fear and injury to the people using the river.

For agriculture, the flow of water is impeded in irrigation canals, drainage ditches are closed, and great water loss occurs from the reservoirs and distribution systems. The weed clogs irrigation pumps and interferes with hydroelectric schemes.

There has been much outcry in the world about programs oriented toward the destruction rather than the utilization of water hyacinths. But these objections come from persons who do not have to face the challenge of drying great masses of the weed and who have little concern for the cost of transport, the means of finding a suitable product, or the economics of the entire process. Many efforts have been made in experiments and in pilot projects in several places in the world to find a use for the weed. A discussion of the constituents of the water hyacinth and of the plant's possible utilization, together with that of other aquatic weeds, has been made by Little (1968). In Asia the weed is used on a small scale for animal food, principally for pigs. It has been used also for cigar wrappers and as a growing medium for mushrooms (Alicbusan, personal communication*).

COMMON NAMES
OF EICHHORNIA CRASSIPES

ARGENTINA
  aquapey
  camalote
AUSTRALIA
  water hyacinth
BANGLADESH
  kachuripana
BRAZIL
  aquape
BURMA
  beda-bin
  ye-padauk
CAMBODIA
  kamplauk
COLOMBIA
  buchon
  lirio de agua
  tarulla
EASTERN AFRICA
  water hyacinth
FIJI
  bekabekairaga

  dabedabeniga
  jalkhumbe
FRANCE
  jacinthe d'eau
GERMANY
  Eichpilz
  Wasserhyazinthe
INDIA
  falkumbhi
  jalkumbhi
  kulavali
  neithamarai
  shoksamundar
INDONESIA
  bengkok
  etjeng gondok (Sundanese)
  etjeng padi (Indonesian)
  gendet
  wewehan (Javanese)
JAPAN
  hoteiaoi

* Information obtained from R. Alicbusan, Department of Plant Pathology, College of Agriculture, University of the Philippines, Los Baños, Philippine Islands, 1971.

MAURITIUS
  jacinthe d'eau
NETHERLANDS
  waterhyacint
NEW ZEALAND
  water hyacinth
NICARAGUA
  lirio de agua

PHILIPPINES
  water hyacinth
  water lily

PUERTO RICO
  flor de agua
  jacinto de agua
  pontederia

RHODESIA
  water hyacinth
SOUTH AFRICA
  water hiasint
THAILAND
  pak tob java
  phak top-cha-wa
  sawah

UNITED STATES
  water hyacinth

VENEZUELA
  laguner
  lirio de agua

VIETNAM
  lục-bình

❦ 9 ❧

Portulaca oleracea L.

PORTULACACEAE, PURSLANE FAMILY

Portulaca oleracea is an annual herb with succulent, fleshy stems that may grow erect or prostrate, depending on light conditions. It is frequently said to be a native of Europe but its succulent habit suggests that it is a desert or desert border plant and may have originated in North Africa. It is a weed of 45 crops in 81 countries. It was one of the early vegetables, and man carried its seed from place to place. It now is used widely as a food for pigs.

DESCRIPTION

P. oleracea is an *annual* herb (Figure 25), reproducing by seed and stem fragments on moist soil; *stems* succulent, smooth, fleshy, commonly prostrate, freely branched, commonly forming mats, arising from taproot, 10 to 60 cm long, smooth; *leaves* alternate or nearly opposite, often in clusters at the ends of branches, thickened, sessile, margins smooth with broad-rounded tips, 0.4 to 2.8 cm long, 0.6 to 2 cm wide; *flowers* yellow, sessile, solitary in the leaf axils or several together in the leaf clusters at the ends of the branches, 3 to 10 mm broad, including the five pale yellow petals (which open only on sunny mornings); *styles* four to six; *calyx* with lower portion fused with ovary, the upper part with two free sepals which are pointed at the tip and 3 to 4 mm long; *petals* and the six to 12 *stamens* appearing to be inserted on the calyx; *fruit* a globular, many-seeded *capsule*, 4 to 8 mm long, splitting open around the middle, the upper half (with two sepals on top) falling away like a lid; *seed* nearly oval, tiny, about 0.5 mm in diameter and length, the surface covered with

curved rows of minute wrinkles, black with a whitish scar at one end.

The fleshy, almost prostrate, reddish stem with leaves which are broad-rounded at the tips and the capsule which splits open around the middle (circumscisses) are distinguishing characteristics of this species.

DISTRIBUTION AND BIOLOGY

P. oleracea is one of the 12 noncultivated species which have been most successful in colonizing new areas (Allard 1965). It thrives in cultivated fields and gardens, barren driveways, waste places, and eroded slopes and bluffs up to 2,700 m elevation. The

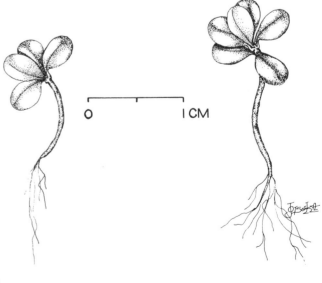

78

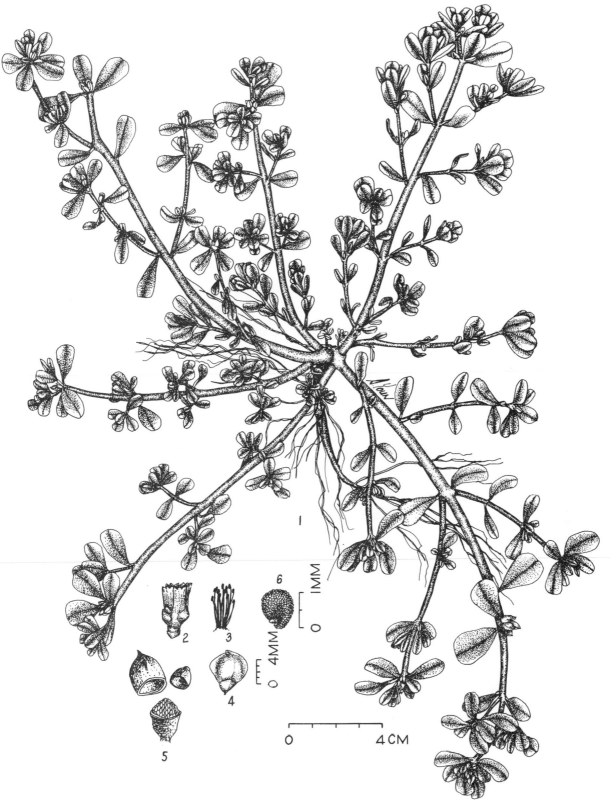

FIGURE 25. *Portulaca oleracea* L.
Above: *1*, habit; *2*, flower; *3*, stamens; *4*, fruit, undehisced; *5*, fruit, dehisced; *6*, seed.
Opposite: Detailed illustrations of the seedlings; *left*, two-leaf stage; *right*, four-leaf stage.

plant prefers an open habitat and, although it thrives in a rich moist soil, it does grow well on many soil types. It does not emerge until the soil is very warm.

The species has many ecological types and, as may be seen in Figure 26, it has become adapted to most of the agricultural areas of the world. A complete life cycle may be 2 to 3.5 months in the tropics or during the warmest seasons in the temperate zone. In cooler weather or in the rainy seasons the cycle may take 4 months. After being soaked, the seeds begin to germinate in 12 hours. At 30° to 40° C emergence is complete in 24 hours. At 10° to 20° C emergence may require 2 days. Early growth is slow but a rapid rate of vegetative growth begins at 15 days. At 1 month, when the plant has 10 or 12 leaves, flowering begins (Misra 1969). Flowering and fruiting begin in June or July in the northern part of the United States but may begin in April in the extreme south. In the tropics the plants are present and may flower all year round but in the temperate zone they are injured by the first frost.

Flowers open from 8 A.M. to 12 noon. Vegetative and reproductive growth continue together after flowering has started. Capsule maturation takes place 7 to 12 days after the first flowers open. The capsule turns pale yellow, the lid opens, and the seeds are dispersed. Seeds often germinate immediately, producing many new plants around the old ones. New seeds will not germinate in the dark but germinate well in continuous light. One-year-old seeds will germinate in the dark (Misra 1969, Dunn 1970).

The plants are day neutral and will flower well in a range of 4 to 24 hours of light; however, they do show a quantitative response to particular daylengths. There is an increase in flowering up to 12 hours of light, and a decrease beyond this point.

In full sun the plant is prostrate. As shade increases, the plant tends to grow more upright and to produce fewer leaves, flowers, and capsules.

The plant is extremely successful because it has many types and germinates well from 10° to 40° C (K. Singh 1968a). It exhibits good vegetative growth and flower production, as well as yielding viable seed, at 15° to 35° C, and flowers in photoperiods of 4 to 24 hours.

PROPAGATION

The seeds of P. oleracea are spread by wind, water, and with the seeds of crops; it is known also that some birds feed on them. The fleshy stems will root on contact with the soil, so that fragmentation with tools or machines may lead to movement and increase in plant numbers. Entire plants which have been lifted from the soil will survive for long periods and may reroot and become established in the same or a new area. Such plants can continue to ripen their seeds even though they have no root system.

The seeds of P. oleracea are very small. In the Philippines almost 10,000 seeds were counted on one plant. Under adverse conditions the plant may be cleistogamous (pollination and fertilization occurs within unopened flowers).

The seeds are reported to germinate both in light and dark. Although there is no agreement on the nature and extent of dormancy, we may be sure that some seeds will germinate promptly after maturation on the plant. Many experiments have shown that scarification, cold treatment, or a period of storage will increase the percentage of seeds that can germinate. A small portion of some seeds that had been in dry storage for 19 years germinated and produced normal plants. The seeds are favored by a high temperature for laboratory germination, this being consistent with experience everywhere in the field.

AGRICULTURAL IMPORTANCE

From Figure 27 it may be seen that P. oleracea is a serious weed of vegetables in Spain and Zambia; of corn in the Philippines and Spain; of cotton in Thailand and Turkey; and of rice in Taiwan. It is a principal weed of corn in Japan, Lebanon, Mexico, Taiwan, and Tanzania; of cotton in Australia, Greece, Israel, Kenya, Mexico, Nicaragua, the Soviet Union, Trinidad, Uganda, and the United States; of rice in Brazil, Colombia, Jamaica, Japan, and the Philippines; of sugarcane in Brazil, Indonesia, Peru, the Philippines, South Africa, Taiwan, and Tanzania; of potatoes in Brazil, Bulgaria, India, Japan and Mozambique; and of vegetables in Brazil, Canada, Japan, Mexico and the Philippines. It is a common weed of corn in India, Indonesia, and Israel; of cotton in Mozambique and Sudan; of rice in Indonesia; of sugarcane in Hawaii, India, Mozambique, and Zambia; and of vegetables in Venezuela.

Although this fact is not shown on the map, P. oleracea is a serious weed of coffee in Tanzania and of orchards in Spain. It is also a principal weed of peanuts in Australia, Indonesia, and Israel; of soybeans, sweet potatoes, and tea in Japan; of coffee in Brazil; of sorghum in Colombia; of wheat in India; of flax, sorghum, and wheat in Mexico; of rape in

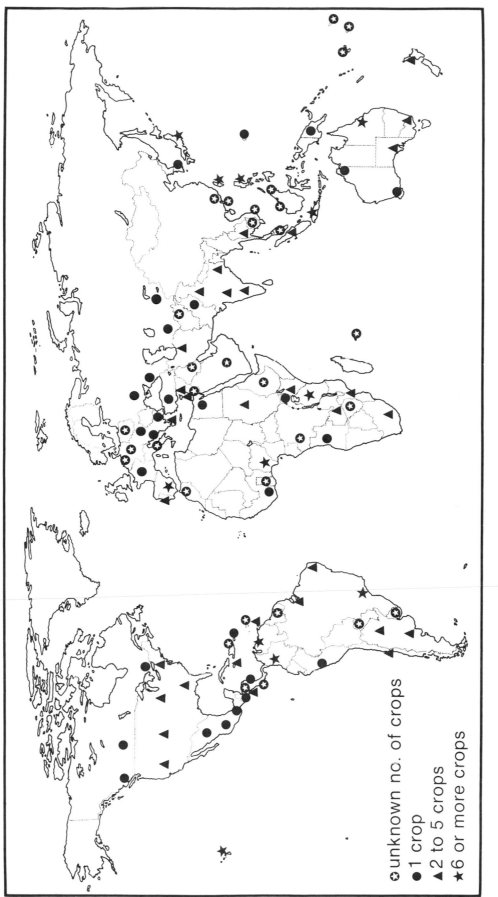

○ unknown no. of crops
● 1 crop
▲ 2 to 5 crops
★ 6 or more crops

FIGURE 26.   The distribution of *Portulaca oleracea* L. across the world in countries where it has been reported as a weed.

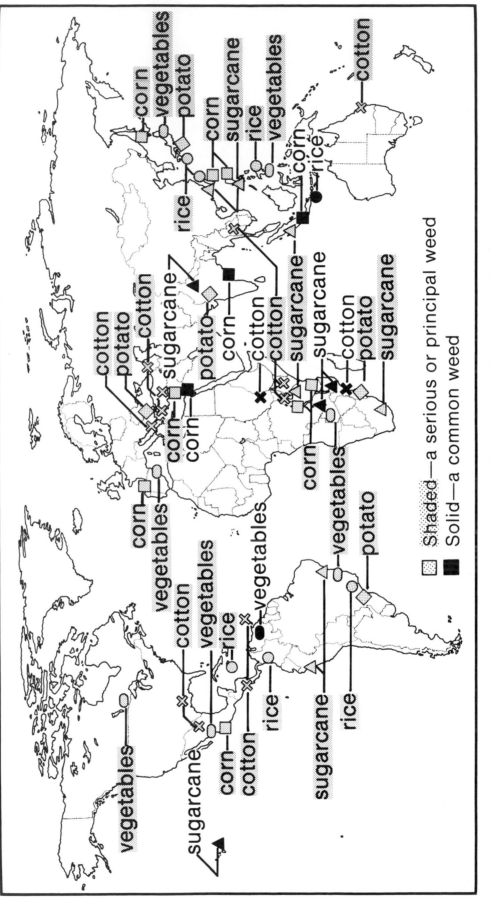

FIGURE 27. *Portulaca oleracea* L. is a serious, principal, or common weed in the crops and locations shown above.

Taiwan; of linseed and safflower in Australia; of sugar beets in Israel; of sorghum in Thailand; of tobacco in Trinidad; and of vineyards in the Soviet Union. It is also a common weed of bananas, citrus, coffee, fodder beets, millet, peanuts, roselle, sorghum, sugar beets, tea, vegetables, vineyards, and wheat in several other countries.

The variations in response to a single herbicide for a weed species growing in different geographical locations are not well understood. Kasasian (1967) made detailed observations on the responses of *P. oleracea* to the phenoxy herbicides in the Caribbean area. The species is susceptible to 2,4-dichlorophenoxyacetic acid in Barbados but not in Trinidad or Jamaica. It is more susceptible to 2,4,5-trichlorophenoxypropionic acid than to 2,4,5-trichlorophenoxyacetic acid in Trinidad, but the reverse is true for Jamaica.

The succulent leaves and stems accumulate toxic levels of oxalates and may cause sickness and death in livestock (Schmutz, Freeman, and Reed 1968).

P. oleracea is an alternate host of *Dichotomophthora portulacae* Mehr. & Fitz. (Mehrlick and Fitzpatrick 1935); of the nematodes *Meloidogyne* sp. (Raabe, unpublished; see footnote, *Cyperus rotundus*, "agricultural importance"), *Paratylenchus minutus* Linford (Linford, Oliveira, and Ishii 1949), *Rotylenchus reniformis* Linford & Oliveira (Linford and Yap 1940), *Heterodera marioni* (Cornu) Goody (Linford and Yap 1940), *Meloidogyne incognita* (Kofoid & White) Chitwood (Valdez 1968); and of the viruses which cause tobacco mosaic (Eugenio and del Rosario 1962), groundnut rosette (Adams 1967), anemone brown ring, aster yellows, beet curly top, chili veinbanding, clover big vein, tobacco broad ring spot, tobacco etch, and tobacco streak (Namba and Mitchell, unpublished; see footnote, *Cynodon dactylon*, "agricultural importance").

COMMON NAMES
OF PORTULACA OLERACEA

ARGENTINA
verdolaga

AUSTRALIA
pigweed
purslane

BRAZIL
beldroega
berdoega
salada de negro

BURMA
mya-byit
mye-byet

CAMBODIA
kbet choun

CANADA
pursley

CEYLON
genda-kola
(Sinhalese)

nilap-pachali
paruppuk-kirai
pulichchankirai
pulik-kirai
sikapu-pasali-pillu
(Tamil)

CHILE
verdolaga

COLOMBIA
verdolaga

EL SALVADOR
verdolaga

FRANCE
pourpier potager

GERMANY
Bergel
Gelber-Portulak
Portulak

GUATEMALA
verdolaga

HAWAII
pigweed
purslane

INDIA
baralunia
barinooni
jowar
khursa
kufa
kulfa
sanjatnatia

INDONESIA
gelang (Indonesian, Sundanese)
krokot (Javanese)

IRAN
khorfe

ITALY
erba porcellana

JAPAN
suberihiyu

KENYA
purslane

LEBANON
baklah
jinlah

MALAYSIA
gelang pasir
segan

MAURITIUS
pourpier

MEXICO
verdolaga

NICARAGUA
verdolaga

NIGERIA
esan omode (Yoruba)

PAKISTAN
kulfa
lunak

PERU
verdolaga

PHILIPPINES
alusiman (Bikol)
bakbakad (Ifugao)
dupdupil (Bontoc)
golasiman (Tagalog)
kantataba (Pangasinan)
ngalug (Ilokano)
ulasiman (Tagalog)

PORTUGAL
beldroega

PUERTO RICO
verdolaga

SOUTH AFRICA
porslein (Afrikaans)

SPAIN
verdolaga

SUDAN
rigla

SURINAM
gron posrin

TAIWAN
ma-chr-hsien

THAILAND
pak bia
phak-bai-yai

TRINIDAD
purslane
pussley

TURKEY
semiz otu

UNITED STATES
common purslane

URUGUAY
verdolaga

VENEZUELA
verdolaga

VIETNAM
rau sam

YUGOSLAVIA
mlada biljka

ZAMBIA
portulaca

❧ 10 ❧

Chenopodium album L.

CHENOPODIACEAE, GOOSEFOOT FAMILY

Chenopodium album, an annual, erect, rigid, pale green herb, is one of the most widely distributed species of weeds in the world and is one of the most successful colonizers as it moves into new areas. The plant is usually light green because it may have a waxy bloom or a white mealy pubescence. When growing in the open it may be tinged with red or purple. From both the Old World and the New World has come considerable evidence that in prehistoric times the seeds were harvested and stored for human consumption (Helbaek 1960). The weed has followed man into all of his settlements and agricultural areas, and it presently is found in 40 crops in 47 countries.

DESCRIPTION

C. album is an erect, rigid, pale-green *annual* herb (Figure 28) growing to 2 m in rich, moist soil, strongly tap-rooted; *stems* with ascending branches above, angular or ridged, smooth, often tinged with red or ribbed or striped with pink, purple, or yellow; *leaves* simple, alternate, ovate to lanceolate, petioled, without stipules, much longer than broad, 1.5 to 8 cm in length, 3 cm broad, no distinct lobes but with up to 10 shallow lobes, grayish green and mealy below, lower leaves may have goosefoot shape, upper may be linear and sessile, all leaves variable in shape; *inflorescence* a spiked panicle in leaf axils or at terminus of stems and branches, with small dense flower clusters crowded on branches; *flowers* perfect, green, small, sessile, in irregular spikes, without petals; *sepals* five, somewhat keeled and nearly covering the mature fruit, green with whitish membranous margin; *stamens* five; *pistil* one, with two or three styles; *ovary* one-celled; *fruit* an utricle (a seed covered by the thin papery pericarp which often persists); *seed* lens-shaped, with marginal notch, black, glossy, about 2 mm in diameter.

It is a light green plant with waxy bloom and can be distinguished by its inflorescence consisting of small inconspicuous flowers that are aggregated into dense clusters (glomerules) and by its fruit that is entirely enclosed in the perianth.

DISTRIBUTION AND HABITAT

C. album is found from sea level to 3,600 meters and from lat 70° N to more than lat 50° S. It grows in all inhabited areas of the world except in extreme desert climates. It is reported to be one of the 12 most successful colonizing species (Allard 1965). Coquillat (1951) suggested that it is one of the five most widely distributed plants in the world. It grows to large size in the long days of the temperate zone and it is there that it offers the most serious competition to crops.

It thrives on all soil types and over a wide range of pH values. It attains its greatest size on fertile, heavy soils but it can survive on coalpit heaps in England. Being a colonizing species it occurs on habitats that have been opened up by disturbances. Because the species has no special seed dispersal system, most of its seeds are deposited near the mother plant; such deposition causes it to grow in patches in crops. On waste ground and at margins one may find

FIGURE 28. *Chenopodium album* L. *1*, habit; *2*, flower, top view; *3*, flower, side view; *4*, flower, sepals removed; *5*, fruit; *6*, seed; *7*, seedling.

100-percent stands. The species is seldom seen as a single plant.

The seeds are commonly distributed as impurities in crop seeds. In one study in England *C. album* was found to be a contaminant in one-third of all the carrot seed samples examined at an official seed testing station (J. Williams 1963). The seeds were found in many cereal and pasture grass seeds but were most frequent in clovers, timothy, and Italian rye grass. Elsewhere in the world the seeds of this species often contaminate wheat seeds.

The world distribution of *C. album* is shown in Figure 29. In spite of the weed's extensive distribution in the cool regions, it also is sparsely distributed just along the equator. One is tempted to suggest that this limitation is due in part to the sensitivity of the species to photoperiod. During the long photoperiods of 16 to 18 hours in the temperate zone, the plant grows vigorously for a long time and attains great size before it is induced to flower by oncoming short days. Near the equator where the photoperiod is shorter, the species tends to flower earlier, remain somewhat smaller, and thus be less competitive in crops. Other annual weeds are the dominant species in such areas.

PROPAGATION

Propagation of *C. album* is always from seeds (Conley 1939). Large plants have produced 500,000 seeds. In a 5-year-old pasture in New Zealand 50 million seeds per hectare were found. When growing in crops such as potatoes and sugar beets, *C. album* may produce 13,000 seeds per plant.

There is no special system of seed dispersal so that most of the seeds drop to the ground. They are not buoyant, but surface water may wash them into ditches where they can be moved long distances. They also travel in dung and droppings of animals and birds. Fifty percent will pass unharmed through a pig, but the loss is greater in cows and bullocks. Twenty percent of the seeds in sparrow droppings are viable. The seed survives in ensilage.

Although seed longevity varies greatly with different conditions of storage in laboratory and soil, seeds have been known to survive from 30 to 40 years in soil.

There is considerable heteromorphy in the seeds so that some seeds are smooth, some are faintly striate, and others have a raised reticulum. The color of seeds on one plant may be black and shiny, brown, or brownish green. All of these characteristics of form and color may be found in the seeds of one plant

while, on the other hand, single individuals or whole populations may have only smooth seeds (J. Williams 1963). Many workers believe that there is a correlation between the amount of dormancy in a seed and its color or shape. It seems, for example, that brown seeds, which are usually few in number, have no dormancy.

Some of the seeds which fall from the plant are dormant, but all of the seeds from a given place or from different places in the world do not require the same conditions for breaking dormancy. The photoperiod at the time of seed formation influences the amount of dormancy in the seeds. Long photoperiods provide greater quantities of seeds, and a higher percentage of them are dormant. During one study in India, seeds were collected in January when 35 percent of them were still green. At this time 45 percent of all seeds gathered germinated readily, suggesting that dormancy is not necessarily present in the most newly formed seeds (Misra 1969).

In many places in the world one may expect to find about 35-percent dormancy in the seed at harvest. Low temperature treatments of 0° to 5° C will improve germination, as will alternating high and low temperatures. In general, maximum germination of *C. album* in India is found at 10° C, whereas in Canada 25° C is optimum.

Nitrates increase germination, and light often has been reported to be necessary. Some experiments in the temperate zone, however, suggest that there are types which germinate best in the dark.

Segments of the perianth frequently adhere to the seeds and, when dry, provide a thin filmlike covering which is difficult to see. A potent chemical inhibitor is contained in the perianth, and germination is greatly increased if the film is removed. In its floral parts only, the species accumulates high levels of a triterpene, pentacyclic oleanolic acid. The plant as a whole accumulates large amounts of ascorbic acid.

BIOLOGY

C. album exhibits great plasticity in its response to the environment, an environment which includes the proximity of neighboring plants. With good nutrition, good soil, and plenty of water, the weed may become more than 3 meters tall while growing in crops such as corn and sorghum. In waste places it is often small and depauperate, and may produce only 10 to 20 seeds per plant. When growing in very open places in various parts of the world, plants may be upright, or decumbent, or prostrate, or ascending.

In the tropics germination may be continuous;

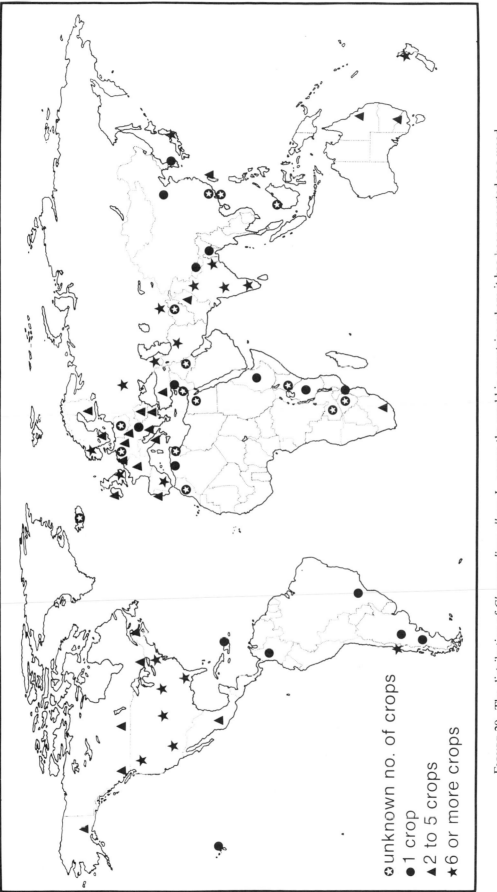

FIGURE 29. The distribution of *Chenopodium album* L. across the world in countries where it has been reported as a weed.

whereas in the temperate areas there is often an early- and a late-season germination, with the greatest flush at the beginning of the growing season. On the average the life cycle is completed in about 4 months, but the cycle often varies with the photoperiod of the season or location.

On spoil banks and waste places where drying is rapid the species survives well. Its leaves may show structural changes toward xeromorphy, to include a decrease in the number of stomata. It is killed early by frost in the temperate zone.

The species will flower at any daylength but an 8-hour photoperiod hastens flowering and maturity. In long days of 16 to 18 hours the plants are larger and produce more seeds, those seeds having a higher percentage of dormancy. It can flower as a very tiny plant in a very short photoperiod.

A slight, odorless, nectar secretion occurs at the base of the flowers. The time of anthesis varies from a few days to several weeks. It is believed that the flowers are mainly wind-pollinated, but the plant is often visited simultaneously by various insects —small bugs, aphids, and flies—that may assist in pollination (J. Williams 1963).

The plant has such a variable morphology in different environments that many hybrid forms have been suggested. Cole (1957, 1962), however, feels these must be viewed with caution. An interspecific isolating mechanism operates at or soon after fertilization. At one time the distinction between reticulate and smooth seeds was regarded as a basis for a specific distinction in the genus, but it is now known that extreme forms interbreed regularly and the full range of intermediates exists naturally. This has been emphasized by elaborate experiments in India in which various forms of the species were studied from north to south. It could be shown clearly that several physiologically distinct populations exist at various latitudes (Misra 1969).

The same studies also revealed that an increased density of plants resulted in shorter plants, fewer inflorescences, less seed production, and reduced dry weights. Seventy percent of full sunlight provided maximum vegetative and reproductive growth as expressed in dry weight of the plant and total weight of seeds produced. Low temperatures caused prolonged vegetative growth and gave greater seed production. Higher temperatures reduced the time from germination to flowering, decreased the length of inflorescences but increased the number of flowers per plant, and gave a higher dry weight.

Anatomical and life history studies have been reported by Artschwager (1920), Bharghava (1937), Gifford and Tepper (1961), and Gifford and Stewart (1965).

AGRICULTURAL IMPORTANCE

The origins of *C. album* are uncertain. The plant is frequently reported to be a native of Europe, but recent evidence from archeological diggings in Canada suggests that its seeds were stored and used by the Blackfoot Indians between 1500 and 1600 A.D., before European trade and goods had come to the area (Johnston 1962). This report of prehistoric occurrence and use is similar to those from the Soviet Union and elsewhere in eastern Europe.

In the centuries which followed, the species followed man into all of his settlements and agricultural areas. There is no special seed dispersal mechanism, and so the seeds must have been carried with man's goods. The plant seems to grow most vigorously in temperate and subtemperate areas. It is most frequently reported to be troublesome in sugar beets, potatoes, corn, and cereals wherever they are grown in the world. For example, it is one of the main weeds in wheat in the Great Plains of Canada and in Finland, whereas in India, Mexico, New Zealand, Norway, Pakistan, and South Africa it ranks among the top six weeds in importance for that crop.

From Figure 30 it may be seen that *C. album* is one of the three most important weeds in sugar beets in Iran, Italy, and Spain; vegetables in the Soviet Union; and corn in New Zealand and the Soviet Union. It is a principal weed of corn in Italy, Portugal, Romania, the United States, and Yugoslavia; of potatoes in Belgium, Bulgaria, Canada, Chile, Finland, India, New Zealand, Norway, the Soviet Union, Sweden, and the United States; of sugar beets in Algeria, Belgium, Canada, Czechoslovakia, England, France, the Soviet Union, Sweden, the United States, and Yugoslavia; and of vegetables in Alaska, Bulgaria, Canada, Finland, India, Ireland, Japan, New Zealand, Norway, Portugal, Spain, and the United States.

Although this fact is not shown on the map, *C. album* is reported to be the number-one weed in sunflowers in Iran, the number-two weed in barley in Finland, and the number-three weed in cereals in Alaska. It is also a principal weed in barley in Canada, India, and Norway; soybeans, tea, and upland rice in Japan; citrus, orchards, and vineyards in Spain; vineyards in France; cotton, soybeans, and strawberries in the Soviet Union; linseed in New Zealand; cotton, pastures, peanuts, and soybeans in the United States; rice in Mexico; tobacco in

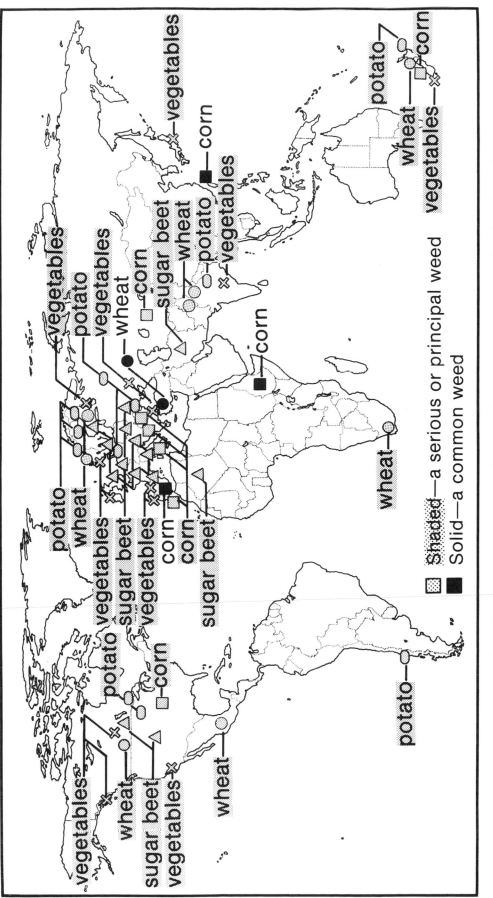

FIGURE 30. *Chenopodium album* L. is a common, principal, or serious weed in the crops and locations shown above.

Canada; cotton in Mexico and Turkey; flax in Germany and the United States; and sorghum in Italy, the Soviet Union, and the United States.

In many countries across the world it is reported to be a common weed of bananas, beans, cowpeas, hemp, peppers, pineapples, rape, sugarcane, and sweet potatoes.

The plant is used on most continents as a green vegetable. The seeds are dried and ground to make flour. In South Africa leaves are braised in oil and served with bread. Shanley and Lewis (1969) studied the nutritional value of 11 weeds of South Africa, including *C. album*. They determined protein and amino acid content of the leaves to estimate the importance of the leaves as a supplement to the cornmeal protein used in staple diets of Zulu and Indian communities. The levels of tryptophane and cystine in *C. album* leaves were found to be among the highest of all of the species tested. Lysine was at median level whereas methionine was low. The protein diet provided by mixing leaves of this species with cornmeal ranked sixth among the 11 weeds tested. Because of the general lack of appreciation of the importance of weed plants in the diets of many peoples of the world, we should point out here that 11 of the cornmeal-leaf mixtures in this study received markedly higher protein scores (Food and Agricultural Organization 1957) than did cornmeal alone.

When the plant is small it is eaten readily by livestock, but as it ages it becomes woody and unpalatable. If cut or trampled early the plants are unable to survive.

C. album does not offer the aggressive, highly destructive sort of competition that is characteristic of vigorous, perennial, tropical grasses. It produces so many seeds that it frequently emerges as a carpet in fields long cultivated. Upon entering new fields or new crops it first becomes patchy, but the stand densities are so heavy that crops may be smothered (Welbank 1959). It is so widespread, therefore, and stands of it can become so troublesome, that it must be reckoned with in world agriculture. Vengris (1955) showed that it competes strongly with corn for nitrogen, potassium, calcium, and magnesium. In competition studies with wheat and kale, it was recorded that the leaf area and dry weight of the crops were reduced by 40 percent in plots which received no nitrogen. When nitrogen was added, the competitive advantage was even more favorable to *C. album*, and these crop growth factors were reduced by 60 percent in wheat plots. Weed populations of

170 *C. album* plants per m² reduced the yield of sugar beets by 86 percent.

C. album is an alternate host of the viruses causing anemone mosaic, barley stripe mosaic, beet curly top, beet mosaic, beet yellows, brome mosaic, cucumber mosaic, dodder latent mosaic, hydrangea ring spot, potato mosaic, potato paracrinkle, potato X, radish mosaic, tobacco broad ringspot, tobacco etch, tobacco mosaic, tobacco ringspot, tobacco streak, turnip crinkle, and turnip mosaic (Namba and Mitchell, unpublished; see footnote, *Cynodon dactylon*, "agricultural importance").

COMMON NAMES
OF CHENOPODIUM ALBUM

ARGENTINA
  quinoa
  yuyo blanco
AUSTRALIA
  fat hen
BRAZIL
  ancarinha branca
CANADA
  fat hen
  goosefoot
  pigweed
CHILE
  campo
  quinqua del
  quinquilla
COLOMBIA
  cenizo
DENMARK
  hvidmelet gaasefod
EASTERN AFRICA
  fat hen
  goose foot
ENGLAND
  fat hen
  goosefoot
ETHIOPIA
  amadamddo
FINLAND
  jauhosavikka
FRANCE
  ansérine blanche
  chénopode blanc
  farineuse
GERMANY
  gemeiner Gänsefuss

  weißer Gänsefuss
INDIA
  bathua
  childhan (Marathi)
  dogstooth grass
  goosefoot
INDONESIA
  dieng putih (Javanese)
IRAN
  salmak
ITALY
  farinaccio
  selvatico
JAPAN
  akaza
  shiroza
MEXICO
  chual
NATAL
  fat hen
  goosefoot
NETHERLANDS
  luismelde
NEW ZEALAND
  fat hen
NORWAY
  meldestokk
PAKISTAN
  bathwra
  jhill
PORTUGAL
  catassol
RHODESIA
  fat hen

SOUTH AFRICA
  withondebossie
SPAIN
  cenizo

salado
SWEDEN
  svinmålla
  vitmålla

TAIWAN
  li
TUNISIA
  chénopode blanc

UNITED STATES
  lambs quarter
YUGOSLAVIA
  pepeljuga

❧ 11 ❧

Digitaria sanguinalis (L.) Scop.

POACEAE (also GRAMINEAE) , GRASS FAMILY

Digitaria sanguinalis, an annual grass, is a cosmopolitan weed that is troublesome in both temperate and tropical crops. It is a native of Europe and has a wide range extending from lat 50° N to 40° S. Fifty-six countries report that it is a weed in 33 crops. It is sometimes used for grazing and for hay.

DESCRIPTION

This plant is an *annual grass* (Figure 31), reproducing by seed, branching and spreading, often purplish; *culms* stout, usually decumbent at base, smooth, up to 120 cm long when prostrate, rooting at the nodes, flowering shoots ascending; *sheaths* densely long-hairy, especially the lower ones, rough to the touch, often more or less pilose; *ligule* membranous, flat at the top (truncate), glabrous, up to 2 mm high; *leaf blades* lax, 5 to 15 cm long, 5 to 10 mm wide, somewhat hairy; *inflorescence* in *spikes* 5 to 15 cm long, with three to 13 fingerlike segments, in whorls or whorllike at top of culm; *spikelets* 3 mm long, paired along one side of rachis, lance-shaped, pointed, one on a stalk 1.5 mm long, the other on a three-angled stalk 3 mm long, the outer glume 1.5 mm long, lance-shaped, the inner glume two-thirds the length of the spikelet, hairy, lance-shaped, pointed, three-nerved; *sterile lemma* strongly nerved, the lateral internerves appressed-pubescent, pale or grayish; *grain* about 2 to 3.5 mm long, 1 mm wide, elliptical, alternate on the branches of the inflorescence. The species is exceedingly variable and is easily confused with *D. adscendens*. The only constant character of diagnostic value to distinguish between the two species is the absence of minute spines on the nerves of the lower lemma in *D. adscendens*. The leaf blades are nearly always hairy in *D. sanguinalis*, glabrous but scabrid in *D. adscendens*. The ligule is usually more conspicuous in *D. adscendens*. The true *D. sanguinalis* is easily recognized by the presence of minute particles of silica (seen as minute triangular spines) on the lateral nerves of the lower lemma (examined under a high-powered hand lens).

PROPAGATION

D. sanguinalis is normally an annual but, because it can root at the nodes and form mats in moist soils, the plant exhibits perennial growth in some areas. In the United States, it has been shown that a single plant can produce 700 tillers and 150,000 seeds; whereas in the Philippines almost 2,000 seeds were counted on a single plant. The plant flowers all year in warm areas when it has sufficient moisture and, therefore, it produces an enormous number of seeds annually. In temperate areas, it produces seeds from early summer until the first frost. When the top growth is continually cut or grazed the plant can produce two to three large seed crops in 6 months in the temperate zone.

The plant is variable in different parts of the world, but available information indicates that there is only a short period of dormancy after the seeds are shed. This dormancy can be overcome by opening the seed coat or simply by allowing seeds to afterripen in dry storage. In other areas, however, those seeds pro-

FIGURE 31. *Digitaria sanguinalis* (L.) Scop. *1*, habit; *2*, ligules; *3*, portion of spike; *4*, spikelet, front view; *5*, spikelet, back view; *6*, spikelet, bracts removed; *7*, grain; *8*, grain, cross section; *9*, seedlings.

duced in 1 year will not germinate until the following spring. Seeds of the temperate zone will germinate satisfactorily at an alternating temperature of 20° to 30° C in the light, or at a constant temperature of about 35° C. Therefore, we find new seedlings in the field only where the soil can become very warm. Germination is often prevented in heavy sod or thatch because the dense shade holds the soil temperature down.

DISTRIBUTION AND BIOLOGY

Because of its ability to adjust to both tropical and temperate conditions, *D. sanguinalis* is found in an exceptionally large number of countries and crops (Figures 32 and 33). Some forms prefer moist regions at lower elevation, as found in Hawaii, whereas other types prosper in hot, dry fields. It is at home always in cultivated fields and gardens, in lawns, in trampled areas, and in waste places.

The plant prospers in high temperatures and often makes maximum growth when other plants may come under stress from heat and dry weather. The bunches or tufts increase in size by rooting at the nodes when they touch the soil. The branches frequently turn upward after rooting has occurred. If they germinate early enough so that a long growing season follows, single plants, if uncrowded and undisturbed, may easily cover areas 2 to 3 m in diameter. The very finely divided root system of such a plant may be 4 m in diameter at 10 weeks of age and 4.5 m at maturity. The roots can grow to a depth of 2 m. We need to know much more about the flowering habits of this weed in its various types. In the Philippines at lat 15° N the plant flowers all year round. In laboratory experiments in the United States the plant remained vegetative in a 14-hour photoperiod, but in a short day of 10 hours it began to flower. Once flowering was initiated, however, the plants continued to flower and set seed until frost. Flowering has a marked inhibiting effect on growth, and those plants which emerge early and have a long period of vegetative growth before beginning to reproduce will be much larger and more competitive than late-germinating plants. Plants turn purple quickly after a light frost (Peters and Dunn 1971).

AGRICULTURAL IMPORTANCE

From Figure 33 it may be seen that *D. sanguinalis* is one of the three most serious weeds in sugarcane in Argentina, Brazil, the Philippines, and Taiwan; in peanuts in Indonesia, Taiwan, and the United

States; in cotton in Spain, Swaziland, and Turkey; in corn in Portugal and Taiwan; and in sorghum in Taiwan. It is a principal weed of sugarcane in Australia, Cuba, India, and the United States; in peanuts in Brazil, Colombia, India, and Israel; in cotton in Israel and the United States; in corn in Brazil, Canada, Colombia, Israel, Italy, Mexico, New Zealand, the Philippines, Spain, and the Soviet Union; and in sorghum in Colombia, Israel, Italy, and the United States.

It is also a serious weed in the following crops not shown on the map: coffee and vegetables in Brazil; pineapples, rice, and vegetables in the Philippines; soybeans in Taiwan; and vegetables in Portugal and New Zealand. It has been reported as a principal weed in bananas in Hawaii; citrus in Australia; coffee in the Philippines; rice in Brazil, Colombia, India, Indonesia, and Thailand; soybeans in the United States; sunflowers in Romania; and potatoes in Brazil and the United States. It is also a weed of abaca, orchards, papayas, rubber, vineyards, flax, sisal, and sugar beets in several places in the world.

Because of this weed's prolific branching habit, the number of plants per unit area bears little relationship to the the total ground covered by mature plants. If only a few plants are present, they will spread rapidly until gaps in the vegetation are filled. Because *D. sanguinalis* is frequently short and not very obvious, it may be deceptively competitive. In experimental studies with several weed species on competition in corn, soybeans, peanuts, and sorghum in Taiwan, *D. sanguinalis* was found to be a very severe competitor. Regrowth of new plants after removal of the old ones was very rapid, and *D. sanguinalis* eventually became the predominant weed (Wang 1969). In an 8-year study in the United States of the competition by natural stands of weeds in peanuts, *D. sanguinalis* and *Richardia scabra* (a much-branched, spreading, broad-leaved plant) were found to be dominant among annual weeds, causing an average yield reduction of 20 percent. When once established in peanuts the weed is very difficult to control with mechanical cultivation alone. The United States has seen a marked increase in *D. sanguinalis* in corn fields in the past decade following the continued use of triazine herbicides. The weed is tolerant of these chemicals and, because it has been released from competition with other weeds, it spreads at an alarming rate.

Because the young plants are palatable to livestock, they are sometimes used for grazing and cut for hay.

D. sanguinalis is an alternate host of *Piricularia*

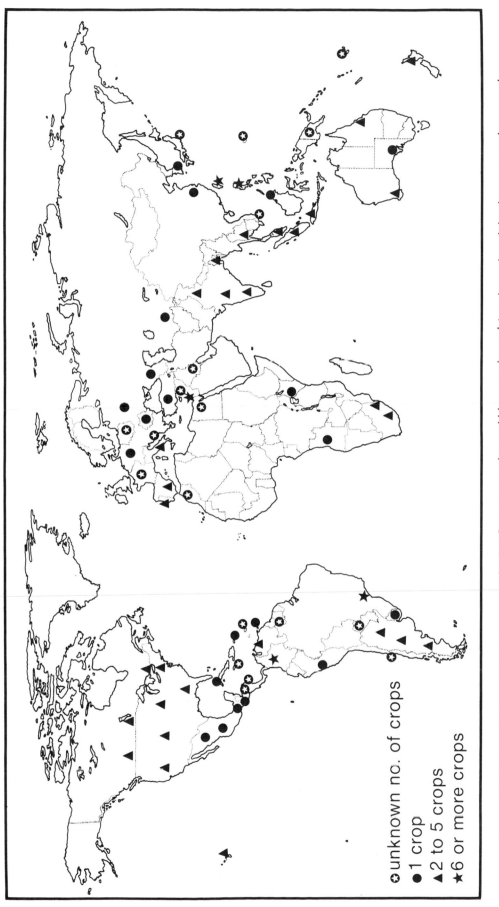

FIGURE 32. The distribution of *Digitaria sanguinalis* (L.) Scop. across the world in countries and locations where it has been reported as a weed.

○ unknown no. of crops
● 1 crop
▲ 2 to 5 crops
★ 6 or more crops

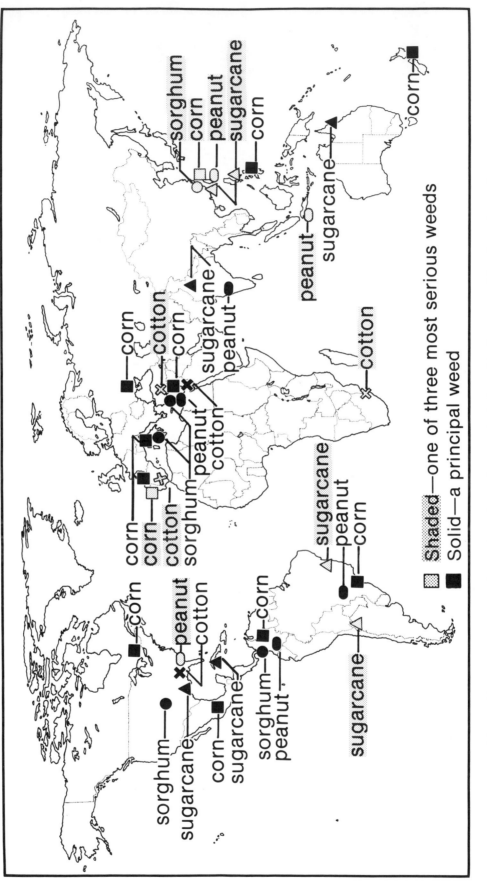

Figure 33. *Digitaria sanguinalis* (L.) Scop. is a serious or principal weed in the crops and locations shown above.

grisea (Cke.) Sacc. (Parris 1941), *P. oryzae* (Tullis 1941), and *Haplothrips melaleuca* (Bagnall) (Sakimura 1937); of the nematodes *Meloidogyne* sp., *Pratylenchus pratensis* (de Man) Filip. (Raabe, unpublished; see footnote, *Cyperus rotundus,* "agricultural importance" and *Meloidogyne incognita* (Kofoid & White) Chitwood (Valdez 1968); and of the viruses which cause abaca mosaic (Gavarra and Eloja 1970), barley stripe, lucerne dwarf, panicum mosaic, sugarcane mosaic, and wheat streak mosaic (Namba and Mitchell, unpublished; see footnote *Cynodon dactylon,* "agricultural importance").

COMMON NAMES
OF DIGITARIA SANGUINALIS

ARGENTINA
  gramillón
  pasto chato
  pasto cuaresma

AUSTRALIA
  summer grass

BRAZIL
  milha

BURMA
  byaing-chi-dauk
  myet-naya
  myet-sot

CHILE
  pata de gallina

COLOMBIA
  conejo
  gaudín
  guardarrocio

CUBA
  Don Juan de Castilla

FIJI
  large crabgrass

FRANCE
  manne-terrestre
  panic sanguin
  sanguinette

GERMANY
  Blut-Hirse
  Finger-Gras
  Mannahirse

GUATEMALA
  garrachuelo
  pasto cangrijo
  pata de paloma

HAWAII
  hairy crabgrass
  large crabgrass

INDONESIA
  djampang put
  djelamparan
  koenoe keo (Malay)
  suket djrempak
    (Sundanese)
  sunduk gangsir
    (Javanese)

ITALY
  sanguinella

JAPAN
  mehishiba

LEBANON
  ink-un-naqil
  tayyin

MEXICO
  frente de toro

  fresadilla
  pata de gallina

NEW ZEALAND
  summer grass

NICARAGUA
  manga larga

PERU
  digitaria

PHILIPPINES
  crabgrass
  pagpagai (Bontoc)
  saka-saka (Ilokano)

PORTUGAL
  milha digitada

PUERTO RICO
  pendejuelo

SPAIN
  pata de gallina

SOUTH AFRICA
  crab finger grass
  kruisgras
  kruisvingergras

UNITED STATES
  large crabgrass

VENEZUELA
  pendejuelo

ZAIRE
  hairy crabgrass
  large crabgrass

❧ 12 ❧

Convolvulus arvensis L.

CONVOLVULACEAE, MORNING GLORY FAMILY

Convolvulus arvensis, a prostrate or climbing perennial herb, may produce a root system that covers an area 6 m in diameter and extends to a depth of 9 m. It is a native of Europe and, although it is widespread in the world, it is most troublesome in cereals in the temperate zone. It extends from lat 60° N to 45° S. Forty-four countries report that it is a weed in 32 different world crops.

DESCRIPTION

A creeping or twining *perennial* herb (Figure 34); *stems* slender, smooth to hairy, 1 to 3 m long, twining or spreading over the soil surface; with a very deep *taproot* 0.5 to 3 m or more long, and cordlike and fleshy *rhizomes* which permeate the soil in all directions; *shoot buds* arise on these rhizomes which upon reaching the surface establish new crowns; *leaves* alternate, simple, long-petioled, margins entire, ovate-oblong, narrowing gradually upward to the rounded or blunt tip, the base squarish or the lobes pointing downward or concavely toward the stalk (sagittate), smooth to slightly hairy, up to 6 cm long and 3 cm wide; *flowers* perfect, regular, usually borne singly in the axils of the leaves; *flower stalk* one- to four-flowered, slender, up to 6 cm or more long, with two bracts 1 to 2.5 cm below the flower; *sepals* bell-shaped, 3 mm long, oblong, blunt; *corolla* funnel-shaped, white or pinkish, sometimes purplish or reddish stripes running from base to margin on the outside, 1.5 to 3 cm wide and long; *stamens* five, attached to the corolla; *pistil* compound with two threadlike stigmas, the ovary two-celled; *fruit* an ovate capsule, two- to four-seeded; *seeds* three-angled, ovoid, dull, dark brownish gray to black, coarsely roughened, 3 to 5 mm long, flat on one or two sides with the other side rounded, the basal scar rough with reddish depression at lower pointed end.

The arrow-shaped leaves, the flower stalk with two bracts below the flower, the deep taproot, the cordlike and fleshy rhizomes which penetrate the soil in all directions, and the white or pinkish funnel-shaped flowers are distinguishing characteristics of this species.

DISTRIBUTION

From Figures 35 and 36 it may be seen that *C. arvensis* is most troublesome as a weed in Europe, western Asia, Canada, and the United States, and it is a special problem in several crops grown widely in the temperate region. It is generally distributed over Africa and South America, Southeast Asia, and the Pacific Islands, but it is not a major weed in most of these areas. Although the plant can be found in waste areas, it can also grow in all kinds of cultivated lands. It prospers in dry or moderately moist soils and because of its deep root system can survive long periods of stress. It is not normally a weed of wetlands. It grows best on rich, fertile soils but persists on poor and gravelly soils as well.

BIOLOGY

The plant reproduces by seeds and by sending up new shoots from a deep and extensive underground

98

FIGURE 34. *Convolvulus arvensis* L. *1*, root system; *2*, habit with flowers; *3*, leaf variation; *4*, flower, showing five stamens of unequal length; *5*, capsule; *6*, seed, two views; *7*, seedlings.

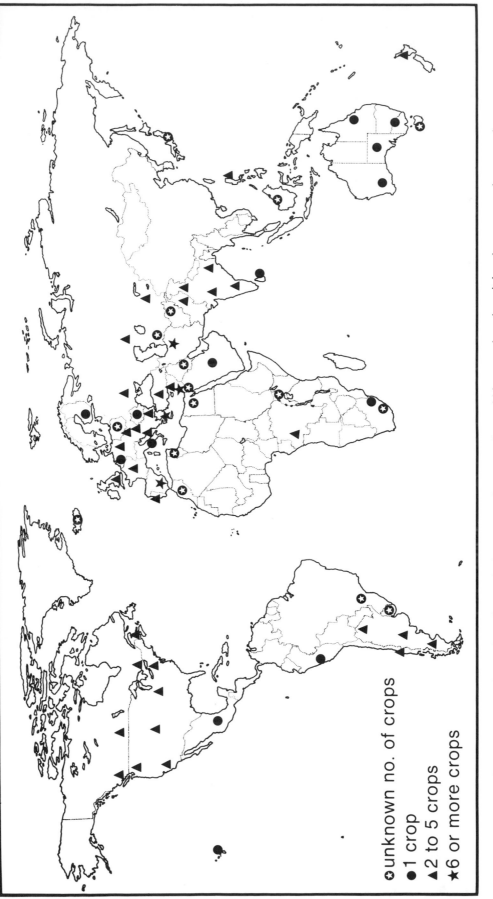

FIGURE 35. The distribution of *Convolvulus arvensis* L. across the world in countries where it has been reported as a weed.

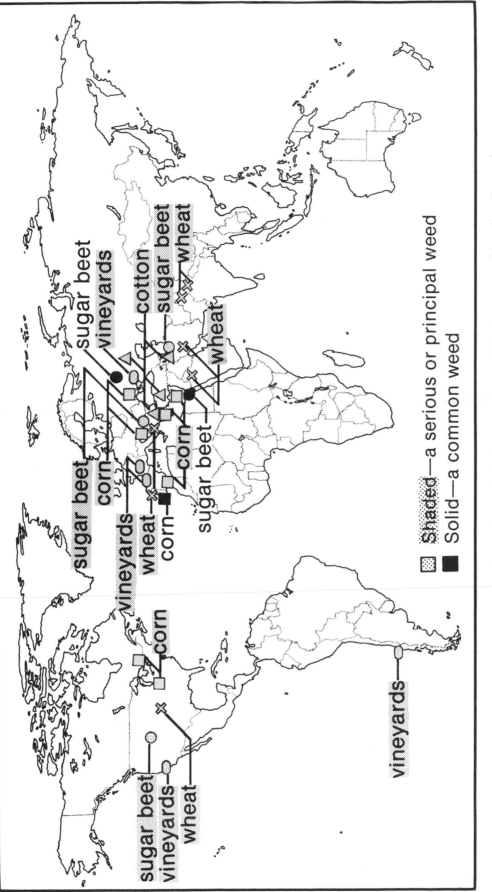

FIGURE 36. *Convolvulus arvensis* L. is a serious, principal, or common weed in the crops and locations shown above.

root system (Brown and Porter 1942). An excellent study on the nature and rate of development of the root system has been made by Frazier (1943a). Seeds are dispersed mainly by water, by movement with seedstocks in commerce, and by clinging to mud on farm vehicles. The seeds of *C. arvensis* will remain viable in the stomachs of some migrating birds for periods up to 144 hours, a span of time which would enable their distribution over thousands of kilometers (Proctor 1968). The seed has a hard coat and, when newly harvested, requires acid treatment for laboratory germination. For older seed, alternating temperatures of 20° to 30° C are required for germination. Timmons (1949) has shown in field studies that seeds of *C. arvensis* will remain viable in the soil for 20 years or more.

In a study of the role of shade and moisture in the competition of crops with this species, Stahler (1948) showed that, in the presence of ample moisture on fertile soil, both *C. arvensis* and spring-sown oats can develop normally as companion plants with neither apparently reducing the supply of essential growth elements to the detriment of the other. This is sometimes the case with corn also. Stahler believes, however, that when moisture stress is a factor, *C. arvensis* can compete successfully with practically all crop plants.

Competition for sunlight places the weed at a considerable disadvantage, and experiments have shown that, if adequate soil moisture is present, rye, soybeans, and sorghum will force it into abnormal growth and dormancy by shading (Bakke and Gaessler 1945, Stahler 1948). When the crops are harvested the bindweed resumes active growth without additional soil moisture. If adequate moisture is present, fall-sown rye and wheat intercept so much of the sunlight that *C. arvensis* develops poorly. Alfalfa has the ability to compete more successfully with this weed than does any other crop, if closed stands can be maintained. Summer-planted crops which will grow vigorously and provide early shade offer severe competition to the weed at a time when it is normally making its best growth. In summary, Stahler's data suggest that selected crops can be manipulated by proper agricultural practices in such a way that competition for light becomes the limiting factor in the development of *C. arvensis*.

Bakke (1939), Bakke, Gaessler, and Loomis (1939), and Bakke et al. (1944) studied the effects of cultivation on the root quantity and root reserves of *C. arvensis*. The fields used in the study had been infested for 20 years and the level of cultivation was comparable to that of a well-managed farm. Tillage resulted in a gradual and continuous reduction in total available carbohydrates, in a decrease in root quantity, and eventually in the death of the weed plants. The carbohydrates, when expressed as a percentage of dry matter, were two to four times more plentiful at lower than at upper levels of the soil. Total carbohydrates were greater at upper levels, however, because of the much greater quantity of roots near the top. The roots died first at the upper levels, which suggests that the conversion of starch to sugar and the translocation of the sugar from lower levels was not sufficiently rapid to sustain the roots in the upper soil horizons.

It is known that several ecotypes of this perennial weed exist, but variations over the world have not been identified or described. When grown under similar conditions the types respond differently to the same herbicide.

The most detailed anatomical study of *C. arvensis* ever made is that of Kennedy and Crafts (1931).

AGRICULTURAL IMPORTANCE

From Figure 36 it may be seen that *C. arvensis* is a serious weed of corn in Greece and Yugoslavia; of cotton in Greece; of sugar beets in Iran; and of wheat in Iran, Pakistan, and Yugoslavia. It is a principal weed of corn in Canada, Lebanon, the Soviet Union, Spain, and the United States; cotton in Iran, the Soviet Union, and Turkey; sugar beets in Greece and the United States; vineyards in Chile, France, Spain, the Soviet Union, and the United States; and wheat in India, Lebanon, Portugal, and the United States.

Although these data are not shown on the map, *C. arvensis* also has been reported as a serious weed of tea in Ceylon and of tobacco in Greece; and as a principal weed of cereals in Australia; citrus in the Arabian peninsula; flax in Germany; orchards in Belgium, the Soviet Union, and Spain; potatoes in India and the Soviet Union; pineapples in Hawaii; soybeans in the United States; vegetables in India, Lebanon, and New Zealand; and tea in the Soviet Union.

In addition, it is a common weed of corn in Portugal; sugar beets in Israel and the Soviet Union; cereals in Finland; citrus and vegetables in Spain; orchards in Yugoslavia; rice and tea in India; sorghum in Bulgaria; and pineapples in Swaziland. It is a weed of alfalfa, barley, beans, carrots, peas, peanuts, peppers, sugarcane, sunflowers, and strawberries in several other countries.

The root is used as a purgative in southern India.

C. arvensis is an alternate host of the viruses which produce potato X (Naperkovskaya 1968), tobacco streak, tomato spotted wilt, and vaccinium false bottom (Namba and Mitchell, unpublished; see footnote, *Cynodon dactylon,* "agricultural importance").

CONTROL MEASURES

C. arvensis is extremely difficult to eradicate once it has colonized an area. Its seeds remain in the soil for long periods, germinating sporadically. Experiments in the United States in which root samples were taken to a depth of 2 meters have shown that the weed may produce 1,300 kg of roots and rhizomes per hectare. This large reserve of root material, some of which survives at great depth, is the second reason for its tenacious hold on an area. (Frazier 1943*b*).

The most extensive study on the control of *C. arvensis* and its effect on the yield of crops was carried out by Phillips and Timmons (1954) at the Fort Hays Experimental Station in Kansas (United States) from 1935 to 1952. These 17 years included extremely wet and extremely dry seasons. Grain and forage yields of nine different crops grown under average farming conditions were reduced 20 to 80 percent because of the competition offered by this weed. Wheat and other small grains produced more nearly normal yields on infested lands than did sorghums and other summer-growing crops. With proper timing and implements the researchers were able to eliminate this weed in two seasons with careful fallowing practices. It was necessary to perform the operations in such a way that all aerial shoots were cut completely to prevent storage in the root system. When the tillage operations were carried out 12 days after each set of shoots emerged, the weed could be eliminated with an average of 16 cultivations.

Timmons and Bruns (1951) showed in another study that the longest effective interval between shallow hoeings was 2 weeks—and for duckfoot cultivation, 3 weeks—in order to eradicate this species. Increasing depths of shoot-cutting by duckfoot cultivator, plow, or shovel lengthened the effective interval between cultivations and in most cases reduced the number of operations necessary to eradicate the weed. No practical advantage was found in cultivating deeper than 7.5 to 10 cm. These methods gave control of the weed in two seasons.

An excellent study of the control of *C. arvensis* by a combination of cultivation, crop rotation, and her-

bicides was reported by Derscheid, Stritzke, and Wright (1970). In grain and forage crops they found that a weed stand of 25 shoots per sq m could be reduced by 90 percent with 2,4-dichlorophenoxyacetic acid alone or in combination with cultivation in 3 years in all rotations. These methods, of course, allow full use of the land during the period.

In Hungary, United States, and elsewhere it has been shown that continued use of the same or similar types of herbicides may remove several species of weeds, whereupon *C. arvensis* can become very serious for lack of competition.

COMMON NAMES
OF CONVOLVULUS ARVENSIS

ARGENTINA
  campanilla blanca
  campanilla europea
  campanilla perenne
  cien nudos
  correguela
  corrihuela
  enredadera europea
  enredadera flor blanca
  enredadera de la India
AUSTRALIA
  bindweed
BURMA
  kauk-yo-nive
CANADA
  European bindweed
  small-flowered morning glory
CHILE
  bocina
  correhuela enredadera
DENMARK
  ager-snerle
  ager-winde
EGYPT
  olleig
ENGLAND
  cornbine
  field bindweed
FRANCE
  liseron des champs
  petite vrillée
  petit liseron
GERMANY
  Ackerwinde

INDIA
  bhoomi chakra
    poondu (Tamil)
  chandvel (Marathi)
  hirankhuri (Bihari)
  pohi (Punjabi)
  pohli (Rajasthani)
IRAN
  pichak
IRAQ
  illake
  middade
ITALY
  erba leprina
LEBANON
  muddayd
MEXICO
  correhuela
  tripa de pollo
MOROCCO
  liseron des champs
NETHERLANDS
  akkerwinde
NEW ZEALAND
  field bindweed
NORWAY
  akervindel
PAKISTAN
  baker bel
  lehli
  naro
  verhi
PERU
  enredadera
PORTUGAL
  corriola

verdeselha

SAUDI ARABIA
madadi
oleik

SOUTH AFRICA
akkerwinde
SWEDEN
akervinda
THAILAND
phak-bung-ruam

TUNISIA
liseron des champs
TURKEY
roca ruskut
UNITED STATES
field bindweed

URUGUAY
campanilla blanca
corrihuela
YUGOSLAVIA
poponac

❧ 13 ❧

Avena fatua L. and other members of the "wild oats" group

POACEAE (also GRAMINEAE), GRASS FAMILY

It would be difficult to find a more serious group of weeds in cereals across the world than the collection of plants universally referred to as "wild oats." Few weeds are more difficult to eradicate from cereals. In any given locality the wild oats are those species or varieties which men do not seek to cultivate in the area today. In Limburg, Belgium, for example, *Avena strigosa* was cultivated before World War I, but it now is considered to be a weed of *A. sativa* and other cereals. In the upland districts of Wales and Scotland where the climate is not suited to *A. sativa* (the cultivated oats), *A. strigosa* is used as a crop; yet *A. strigosa* is regarded as a weed in the south of Great Britain. *A. barbata* is an important component of the range flora in the foothills of California in the United States. The most troublesome of all the wild oats, however, and by far the most widely distributed in the world is *Avena fatua*. The discussion which follows is concerned principally with this species, but we will refer also to the morphology, behavior, and distribution of other wild oat species when it is appropriate. *A. fatua* is an herbaceous annual, native to Eurasia, grows to 0.6 to 1.3 m, and looks very much like cultivated oats. Allard (1965) suggested that on a world basis it is one of the 12 most successful colonizers among the noncultivated plants. Hjelmquist (1955) believes that *A. fatua* and *A. strigosa* were cultivated in northern Europe before *A sativa*. The seeds shatter and fall to the ground before the cereal crop is harvested, and they persist in the soil for many years. Wherever plant husbandry requires the use of cleaned or certified seeds there is constant concern about contamination

with the seeds of *A. fatua*. This species is a weed of more than 20 crops in 55 countries.

DESCRIPTION

A. fatua is an erect *annual* grass (Figure 37); *root system* fibrous and extensive; *culms* stout, smooth or the lower ones soft-hairy, 30 to 120 cm tall, in small tufts; *leaf blades* flat, with broad base and acute apex, 7 to 40 cm long, 4 to 18 mm wide, rough; *sheaths* smooth or slightly hairy on the margins, especially in younger plants; *ligule* 1 to 4 mm, often irregularly toothed (dentate); *panicles* terminal, 15 to 40 cm long, loose, open, the axes slender, ascending, rough; *spikelets* two- to three-flowered, pendulous or drooping, about 2.5 cm long excluding the awn, each spikelet with two empty glumes within which are two or three florets, empty glumes longer than the florets; *glumes* smooth, finely lined (striate), acuminate; *lemma* two-toothed at the apex, glabrous or the lower part clothed with stiff, long, dark brown hairs; *awn* twisted in lower parts, about 3 to 4 cm long, upper parts bent sharply at right angles to the twisted parts, arising from the back of lemma, with a ring of hairs at the base and more or less appressed-pubescent with long stiff brownish hairs or glabrous; *rachilla* hairy, very obliquely jointed; *grain* 6 to 8 mm long, silky-hairy, especially near the base, the hairs longer above, white, pale yellow, brown, gray or black, enclosed by the flowering lemma and palca, all grains with a slanting, circular, depressed scar (also called a suckermouth) at the base, scar always with a circle of hairs.

FIGURE 37. *Avena fatua* L. *1*, habit; *2*, ligule; *3*, spikelet attached to rachis, note long awns; *4*, detached spikelet showing "suckermouth"; *5*, glume; *6*, glume; *7*, grain; *8*, grain, cross section.

The very obliquely jointed rachilla that leaves a visible scar at the base of the grain, the two-toothed hairy lemma apex, and the long, twisted and bent awns are distinguishing characteristics of this species. Wild oats shed seeds readily whereas cultivated oats do not.

DISTRIBUTION AND HABITAT

From Figure 38 it may be seen that *A. fatua* is present almost everywhere in the world where cereals are grown. The species is found as a weed in Iceland and Alaska; it is found at higher elevations at the equator; and it reaches into the southernmost agriculture of the southern hemisphere. The species is believed to have originated in Central Asia, a region which is much more arid than the cool, temperate regions of Europe and North America where it has achieved such notoriety in recent years. The species seems to be troublesome wherever cereals are grown at 375 to 750 mm of annual rainfall. Cold temperatures do not hinder the plant's growth and spread for it overwinters 1 to several years with a very persistent seed.

A. fatua is most frequently seen in grain fields, but it is in 20 other crops which are as variable as pastures, cotton, tea, peas, and vineyards, and it is found as well on roadsides, in fence rows, and on neglected lands. The fallen seeds are aided by an awn which is hygroscopic and twists and turns with changes of moisture to bury the seeds in cracks or under soil clods. They may then be held in place by the long hairs on the lemmas. The plant grows on a wide range of light to heavy soil types and succeeds on both acid and alkaline soils. It will tolerate soil pH down to 4.5.

The weed has been associated with the cultivation of oats and other cereals from the Early Iron Age dating back to 700 to 500 B.C. It sometimes formed a high proportion of the grain stored during that time (Jessen and Helbaek 1945).

There is a total of about 50 oats species, most of them having originated in the Old World. About 1,000 varieties of oats have been named for the cultivated crop. Only the varieties from two species, *A. sativa* L. and *A. byzantina* C. Koch, are of much economic importance (Coffman 1961).

A. sterilis, sometimes called "wild red oat," is second in distribution in the world. It has reddish culms and large lemmas that are covered with a dense growth of hairs. Both florets of the spikelet bear long, strong, twisted, and geniculate awns that adhere tightly to the kernel. The apex of the lemma is divided into four awn points (teeth) or bristles. A partial distribution is as follows: Afghanistan, Algeria, Arabia, Argentina, Australia, Crete, Egypt, England, France, Greece, Israel, Lebanon, Morocco, Peru, Portugal, the Soviet Union, and Tunisia. *A. sterilis* var. *ludoviciana* (Durien) Husnot. is a variable variety of the wild red oat which is regarded as a species, *A. ludoviciana*, in several places including England where much work has been done on it. It has brown, very hairy lemmas; long callus hairs; and long, strongly twisted, geniculate hairs. Its distribution parallels that of *A. sterilis*, but is much less extensive. *A. barbata* Brot., sometimes called "slender oat," is distinguished by its small, weak culms and its decumbent habit which it maintains for most of its life. It has narrower florets than *A. fatua*, each with an abscission scar at the base and two long bristle points at the tip of a very hairy lemma. When panicles begin to emerge, they are lifted up away from the decumbent plant by growth at the lower nodes. *A. barbata*, the third most widely distributed in the world, is never grown as a grain crop. A partial distribution is as follows: Argentina, Australia, Chile, France, Greece, Hawaii, India, Israel, Lebanon, Malta, Portugal, South Africa, the United States, and Uruguay. *A. strigosa*, called "sand oats," is rarely grown as a grain crop and is less widely distributed in the world; its distinguishing mark is the lancelike structure of the lemma which extends into two distinct points. It is generally distributed through Europe.

Barralis (1961), Lindsay (1956), Sexsmith (1967), and Stryckers and Pattou (1963) have described the various species and varieties of wild oats and their distribution in Belgium, Canada, and France. Stanton (1955) has published a lengthy bulletin on the description, history, and distribution of oats, including wild oats, in the United States.

BIOLOGY

Most countries which have a problem with *A. fatua* also have many variable types which seem to be neither wild nor cultivated forms, types which have arisen as crosses between *Avena* species. There are many Russian reports of the "reversion" of their cultivated varieties to wild forms and of the conditions which are conducive to this change. One of the important characteristics of *A. fatua* is the attachment of the seed to its minute stem by a joint. It is abscissed easily at this place and retains a circular scar at the base of the seed (often called a suckermouth). Cultivated oats lack this scar because they

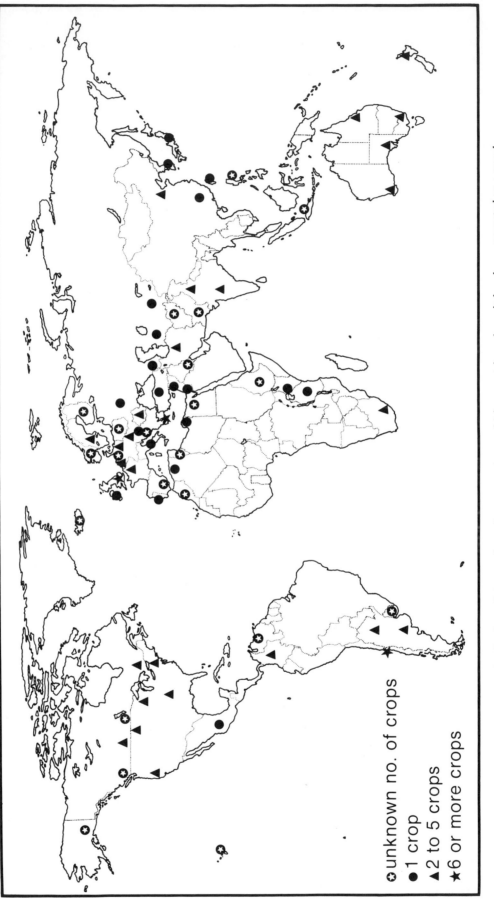

FIGURE 38. The distribution of *Avena fatua* L. across the world in countries where it has been reported as a weed.

depend on a fracture of the small stem for release. Each floret of the *A. fatua* spikelet has a twisted bristle. In the cultivated varieties of *A. sativa* there are no bristles or they are confined to only the lower floret and are usually straight. The panicles of wild oats are more loose and widely spread than are those in cultivated oats. Many cultivated oats can germinate immediately at harvest, whereas most wild forms have a dormant period.

Thurston (1959) made a very detailed study of the developmental morphology of *A. fatua* for its entire growing cycle. A great part of the literature on the biology of the species, however, deals only with the reproductive processes. This is understandable for seed dormancy is perhaps the prime reason for the difficulties encountered in the cereal fields of the world.

In the north temperate zone the plant begins to flower about early July. In Java, nearer the equator, it flowers all year. In another work in which she studied germination and growth of several wild oat species and types from different countries, Thurston (1963) in Great Britain found that the time of flowering differed more between countries of origin than between species. From the time it was sowed to 50-percent panicle emergence, *A. fatua* required 124 days to flower in plants from Russia but 97 days from Iraq. The time required for this process in *A. sterilis* was 93 days from Israel, 119 days from Crete, and 134 to 140 days from France. The effects of day-length were revealed in the same experiments when seeds of *A. fatua* from Australia were sown each 2 months from December to April. The first sowing began to flower in early May when plants were 60 to 70 cm tall and had four shoots each. The third sowing (4 months later) began to flower in late May when the plants were 30 cm tall and had two shoots each.

No generalizations can be made about the dormancy of the species of wild oats. There are surely several factors at work to make the seeds dormant, although attempts to explain these are often based on experiments to break the dormancy by working on only one or two of the factors; thus, the literature is filled with contradictions. If we add to this the knowledge that the seed dormancy may vary with the seed's position in the panicle and its age at the time of harvest, we may understand our failure to find ways to bring all seeds in the soil to simultaneous germination so that we may destroy them.

A. fatua and *A. sterilis* var. *ludoviciana* are two of the most serious weeds in cereals in England. Thurston (1960, 1963), working with these two species,

made extensive studies of dormancy and of the conditions required for germination in field and laboratory. The unripe seeds are viable and have no dormancy. The top seeds of the panicle may be germinable before the lowest seeds are free of the enveloping leaf sheath. About 95 percent of the seeds of *A. fatua* and only 50 percent of the variety *ludoviciana* are dormant at harvest. The former germinates mainly in spring and the latter in autumn and winter. In variety *ludoviciana* the seeds can make full development and establish their dormancy in 15 days. The first seed on the spikelet quickly loses whatever dormancy it had when it reaches the soil (and in dry storage as well) and is ready to grow in fall or winter. This resembles the postharvest dormancy of most cultivated oats. Thurston found that removal of the first seed of the panicle tended to reduce the dormancy of the second seed, and so forth. Larger panicles tended to have a larger proportion of viable seed and a lower level of dormancy than seeds of small panicles. In general, restricting water, food, or competition tended to increase seed dormancy in this variety. The differences in dormancy among seeds in different positions on the panicle were not as marked for *A. fatua* as they were for var. *ludoviciana*. In the seeds gathered from several countries peculiar patterns of dormancy appeared. In *A. sterilis* the first seed on the panicle had almost no dormancy and, thereafter, the seeds of the second, third, and fourth positions germinated in turn. Seeds from the third and fourth positions did not germinate until the 2nd and 3rd years after sowing.

Seeds of *A. barbata* from Malta gave 50-percent germination in the 1st year and fewer seeds germinated in 2nd and 3rd years. Only a few viable but dormant seeds remained at the end of the 3rd year. Seeds from Crete gave 50-percent germination in the 1st year, but no dormancy was present in seeds of plants from France or Australia.

A detailed study by C. Andrews (1967) describes the onset of embryo dormancy in *A. fatua*, the development of the caryopsis from fertilization to maturity, and the role of native growth substances during this period.

During the field trials in England, very little germination of *A. fatua* in summer was observed, so a fallow period during this season would not be profitable for wild oat control. Seeds of this species and *A. sterilis* var. *ludoviciana* can emerge from 23 cm. Each 2.5 cm of depth added 1 day of time from sowing to emergence. A higher percentage of the variety *ludoviciana* could emerge from the deep plantings, and fresh seeds of both types could penetrate more soil than seeds which had been dormant for 2 or 3 years (Thurston 1963).

In an effort to collect information about the effects of weather and climate on seeds of *A. fatua* after they have shattered and fallen to the earth, Thurston (1962*a*) coordinated cooperative uniform experiments in six countries of Europe. The seeds were allowed to remain on the soil surface from 0 to 3 months after harvest and then were stored dry in the laboratory from 0 to 6 months before testing. In general good agreement appeared among the countries that about 10 percent of the seeds germinated immediately after collection, with this figure rising to 50 percent by December or January. Viability was not affected by age or storage conditions. Many seeds were killed by winter weather when left outside in Denmark and Germany.

Kiewnick (1963, 1964) conducted some interesting research on the role of soil microflora in the life of the seeds of *A. fatua*. At 20° C and 100-percent relative humidity the presence of soil microorganisms reduced the viability by 23 and 35 percent in 3 and 6 months, respectively, as compared with seeds kept in a laboratory dish at 60 to 80 percent relative humidity. He felt that the seeds serve as a reservoir of saprophytic and parasitic soil fungi and play an active role in the life cycle of these fungi. The optimum soil moisture content for seed germination was 15 percent and the growth of fungi was most active at 50 percent. At a moisture content of more than 50 percent a secondary dormancy was induced in the seeds, and they became more susceptible to attack by soil microorganisms.

Seeds of *A. fatua* are more resistant to heat than most weed seeds and have survived in laboratory tests at 115° C for 15 minutes. A high temperature for storage purposes was unfavorable, however, for 50 percent died in 2 months when held at 27° C and 90 percent of those which survived became dormant. The seeds retained their viability and lost most of their dormancy at 10° C.

The role of the husk or the hull (consisting of lemma and palea) in the dormancy of the seeds is not understood. Atwood (1914), Ivanovskaya (1943), Hay (1962*b*) and others have had success in breaking dormancy by cutting or scorching the seeds, removing a piece of the seed coat above the embryo, or pricking the caryopsis through the hull. Johnson (1935) and Black (1959) could find no advantage in these treatments.

Atwood (1914) found dormancy was broken when more oxygen could be made to enter the seeds by puncturing the coats, scorching them, soaking them in potassium nitrate, or increasing the oxygen in the surrounding atmosphere. Mullverstedt (1963*a*) in Germany found *A. fatua* to be one of a group of species that require 6- to 8-percent oxygen in the atmosphere to begin germination, and 12 to 16 percent to reach 75-percent germination. A low level induced a long secondary dormancy, and a complete absence of oxygen killed the seeds. After 42 days, seeds held at 2-percent oxygen and 10° to 24° C showed 65-percent viability, whereas those held at 6° to 15° C were all viable. He concluded from his experiments that germination is directly related to the partial pressure of oxygen, and that the amount of oxygen required in soil for germination increases with temperature.

Cumming and Hay (1958) showed in the laboratory that white, blue, and infrared lights inhibited germination of partially dormant seeds (a secondary dormancy was induced by soaking the seeds in water). They point out that under natural conditions daylight may inhibit germination of partially dormant seeds on the soil surface, and burial of the seeds by tillage may bring about germination. Various explanations have been offered for increased germination in a spring following fall cultivation, but reports from several countries indicate that burial of the seeds by tillage may indeed be a way of getting an early crop of weed seedlings which then can be destroyed before the crop is planted.

Farmers often report that *A. fatua* seeds can survive from 50 to 75 years in the field. Experimental evidence does not confirm this, however, and most tests indicate that seeds in farm soils do not survive beyond 4 to 7 years. The exaggerated reports of longer life may have arisen because the fields were resupplied each year with contaminated crop seed stocks, thereby ensuring that a supply of *A. fatua* seeds be ever present (anon. 1961, Thurston 1961).

AGRICULTURAL IMPORTANCE

Figure 39 indicates that *A. fatua* is one of the three most important weeds in cereals in Argentina, Canada, England, and the United States, and it is the number-one weed in peas and beans in Greece. It is a serious weed of wheat in Australia and England; of wheat, barley, and oats in Greece; and of barley in England. It is a principal weed of cereals in Australia, Belgium, Ireland, and New Zealand; of wheat in Argentina, Canada, Chile, Iran, Mexico, South Africa, Turkey, and Sweden; and of barley and oats in Canada, South Africa, and Sweden. It is a principal weed of sugar beets in

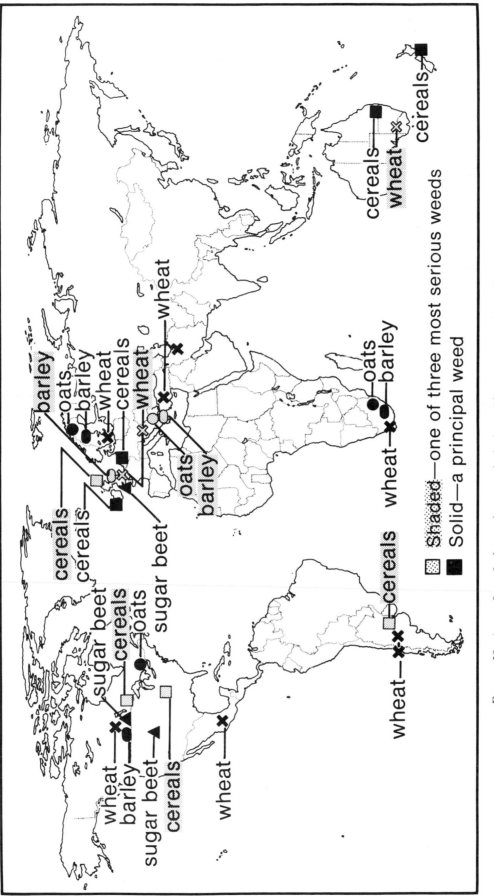

FIGURE 39. *Avena fatua* L. is a serious or principal weed in the crops and locations shown above.

Canada, England, and the United States; of peas and beans in Canada, England, and New Zealand; and of flax in Canada. It is also a weed of corn, cotton, orchards, pastures, potatoes, rape, sunflowers, tea, vegetables, and vineyards elsewhere in the world.

Summarizing world weed problems in wheat and other cereals, Hay (1970) estimated that *A. fatua* is probably the number-one weed. In the highlands of Colombia it is one of the four most important annual, grassy weeds in areas having 750 mm annual rainfall, acid soils, and fair to poor drainage. Recent surveys in Germany have indicated that modern crop husbandry methods have resulted in an increase of this weed. Ubrizsy (1968) in Hungary has reported on some excellent long-term experiments on the flora-changing effects of herbicides used on arable lands, pastures, and wastelands. When herbicides were used for 10 years without any annual rotation or alternation of chemicals, 30 to 40 species of weeds increased in the fields and *A. fatua* was among them.

A survey in Bavaria in Germany in the decade following 1949 revealed that more than 25 percent of the fields which had been planted to produce certified cereal seed were rejected because too many wild oats plants were present in the fields. Surveys in Argentina in the same period showed that *A. fatua* was one of the most common contaminants in cereal seed stocks. A cereal seed survey in the wheat belt of the United States and Canada showed that 54 percent of the seed stocks contained seeds of wild oats. A grain drillbox survey in North Dakota in the United States found an average of 15 seeds in each kilo of grain seeds. This would be about 1,500 seeds per hectare, and, because each plant can produce up to 250 seeds in a season, a heavy infestation could result in only 1 year. Wood (1954) made a survey of wild oat infestations in farmlands in the prairie provinces of Canada and the northern half of the wheat-growing area of the United States. He found that, out of 45 million hectares, 23 million had the weed present and 12 million were seriously infested.

One of the largest investigations ever undertaken to study economic losses due to weeds is that of Friesen and Shebeski (1960) in Canada. They obtained data in Manitoba on 28 species by using 150 fields of 8 hectares each. Wheat, oats, barley, and flax were studied. *A. fatua* was one of the four most important weeds in the study. They estimated the economic loss in cereals for Manitoba in that year to be 29 million bushels of grain valued at 32 million dollars.

Wherever agricultural practices are conducive to the growth of this weed it will enter fields with scarcely any regard for the type of crop. It is particularly damaging to grain because it can germinate later than the crop, will shed its seeds before harvest of the crop, thrives in the particular cropping system and sequence, infests the soil with seeds which have a very long life, and travels from field to field in commercial seed stocks. In addition to all these factors it is very competitive in the field. Many good competition studies have been made to measure the extent of loss. One may find losses ranging from 15 percent to 85 percent, depending on the crop, the area, and the level of infestation. Bowden and Friesen (1967), Allyn and Nalewaja (1968a,b,c,), and Thurston (1962b) carried out several types of competition experiments on different crops. Nalewaja and Arnold (1970) reviewed the work on weed competition by *Avena fatua*.

In an effort to study the behavior of weed seeds in the microsites in which they must germinate, Mullverstedt (1963b) in Germany made observations on the movement of seeds of *A. fatua* and other species during cultivation. Seeds were placed 5 cm in depth in a loam soil with particles which would pass through a screen with a mesh of 2.5 mm. When harrowed to a depth of 10 cm with tines set 3 cm apart, those seeds larger in size than the soil particles tended to move upward while smaller seeds moved downward. Twenty percent of the *A. fatua* seeds moved upward to within 1 cm of the soil surface. A second harrowing in the same direction moved the seeds upward still farther. A cultivation in the opposite direction tended to nullify the effects of the first operation. Light had no effect on germination so the presence of light cannot explain the improved activity of seeds near the surface. Experiments also showed that increased aeration of the loam soil stimulated germination in *A. fatua*, and Mullverstedt believes that this may be a major factor in the initiation of growth in seeds brought to the surface by cultivation.

Soil compaction is said to induce dormancy of wild oat seeds buried in the soil. Research in Canada indicates that nitrates in fertilizers applied early in the spring can break dormancy and cause increased wild oat infestation. Canadian farmers are interested in the implications of this for it would allow them to induce early germination of wild oat seedlings so that they can destroy the seedlings as they come up. Burning of stubble in the field tends to give higher germination of surface seeds in the next season. Long-term experiments indicate, however, that the large reserves of seeds buried in the soil are not affected by burning, so that no absolute decrease in weed seedlings is obtained (Viel 1963).

In Denmark, storage of *A. fatua* seeds for 2 months at a depth of 1 meter in sugar beet silage or waste resulted in the death of all seeds. In England, 2,000 seeds were fed to a calf and only 10 were viable on recovery (Thurston 1963).

Finally, it is estimated that 16,000 hectares of wild oats are used for hay in the interior and coastal valleys of California in the United States. Wild oats are screened from spring wheat and sold in carload lots as "feed oats." They are said to have 90 percent of the food value of cultivated oats.

COMMON NAMES
OF AVENA FATUA

ARGENTINA
  avena guacha
  avena negra

AUSTRALIA
  wild oat

CANADA
  wild oat

CHILE
  avenilla

COLOMBIA
  avena loca
  avena silvestre
  avenilla

DENMARK
  flyve-havre

EASTERN AFRICA
  wild oat

EGYPT
  zommeir

FINLAND
  wukkakaura

FRANCE
  avoine follette
  avron
  folle avoine

GERMANY
  Flughafer
  Windhafer

GREECE
  wild oat

INDIA
  jungoli jai

IRAN
  wild oat

IRAQ
  shoofan barri

ITALY
  avena
  avena selvatica

JAPAN
  karasumugi

JORDAN
  wild oat

MEXICO
  avena loca

NORWAY
  floghavre

PORTUGAL
  balanco

SOUTH AFRICA
  common wild oat
  gewone
  wildehawer

SPAIN
  avena loca

SWEDEN
  flyghavre

TURKEY
  yabani yulaf

URUGUAY
  avena mora
  balango
  cebadilla

VENEZUELA
  avena cimarrona

YUGOSLAVIA
  divlji ovas

❧ 14 ❧

Amaranthus hybridus L.

AMARANTHACEAE, AMARANTH FAMILY

Amaranthus hybridus is an erect, annual herb and is a native of North America. Commonly found in the Americas, it has also spread to Africa and south-central Asia. It has been reported as a weed in 27 crops in 27 countries. It is often used as a green vegetable in southern Africa and India.

DESCRIPTION

This plant is an erect, smooth, *annual* herb (Figure 40), often much-branched on the upper part, up to 150 cm tall; *stems* strongly ribbed, green to brownish or tinged red; *leaves* alternate, broadly lanceolate to ovate, up to 6 cm long, 3 cm wide, obtuse and shortly terminated by a short and sharp tip (mucro), tapering at the base into a slender petiole up to 5 cm long; *flowers* in dense clusters in terminal or axillary spikes, greenish, leafless at the tip, terminal spikes longer than the axillary, the staminate flower at the apex of the spike, the pistillate below, more numerous; *calyx segments* five, with !ong pointed tip; *fruits* split around the middle; *seeds* lens-shaped, 1 mm long, reddish brown to black, shiny. It can form a deep taproot. Its main distinguishing characteristics are its clusters of flowers in terminal and axillary spikes, the often reddish or brownish tinges of stems and leaves, and the ribbed stems.

BIOLOGY

A. hybridus grows widely in cultivated lands and waste places. Reproducing from its many seeds, hundreds of which may be produced by each flowering branch, the plant makes rapid early growth, competing quickly with crops. Seed germination appears to be stimulated by light, especially following a water-imbibition period of 72 hours in darkness. Germination of 1-year-old seeds was highest at alternating temperatures of 20° to 30° C, with or without light. Evidence for seed dormancy is conflicting; in some cases freshly harvested seeds did not germinate readily. Stands have been obtained in the field by sowing 0.6 to 1.3 cm deep after planting corn or soybeans (Andersen 1968).

DISTRIBUTION AND AGRICULTURAL IMPORTANCE

A. hybridus has been reported from 27 countries in four continents (Figure 41). It is reported most frequently from Latin America and North America, mostly as a weed of field or vegetable crops (Figure 42). It is reported as a principal weed in the following crops: sugarcane and sugar beets in Argentina; beans, corn, flax, potatoes, safflower, and wheat in Mexico; corn and vegetables in New Zealand; peanuts in the United States; peas in Brazil and New Zealand; cotton in Mexico and Rhodesia; and corn and soybeans in the United States. It is also reported as a weed of papayas, bananas, peanuts, garlic, onions, and barley.

A. hybridus has been reported to constitute as much as 40 percent of the weed population in peas in New Zealand. In Rhodesia it was found that, if *A. hybridus* plants were allowed to grow more than 2 weeks in cotton, yields were severely affected

114

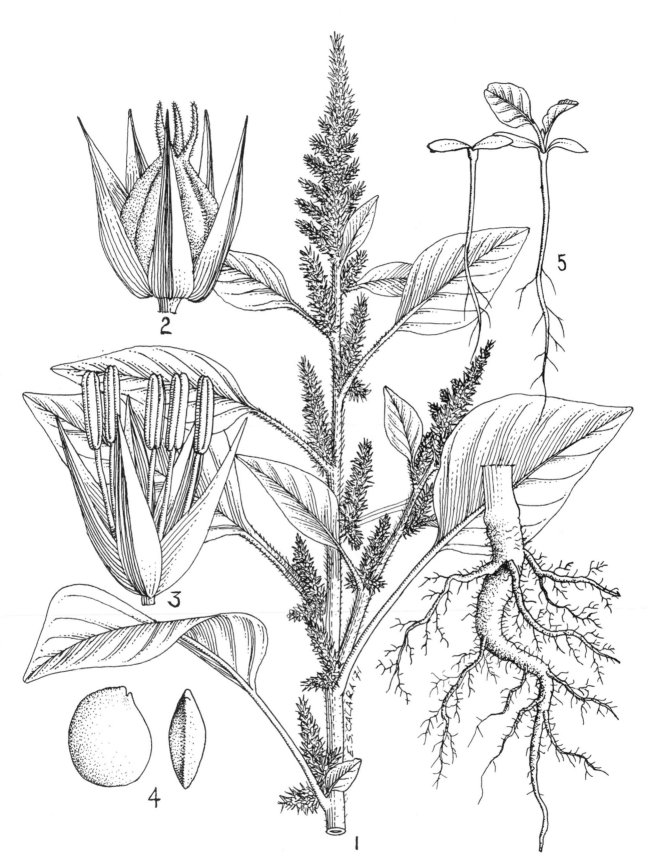

FIGURE 40. *Amaranthus hybridus* L. *1*, habit; *2*, female (pistillate) flower; *3*, male (staminate) flower; *4*, seed, two views; *5*, seedlings.

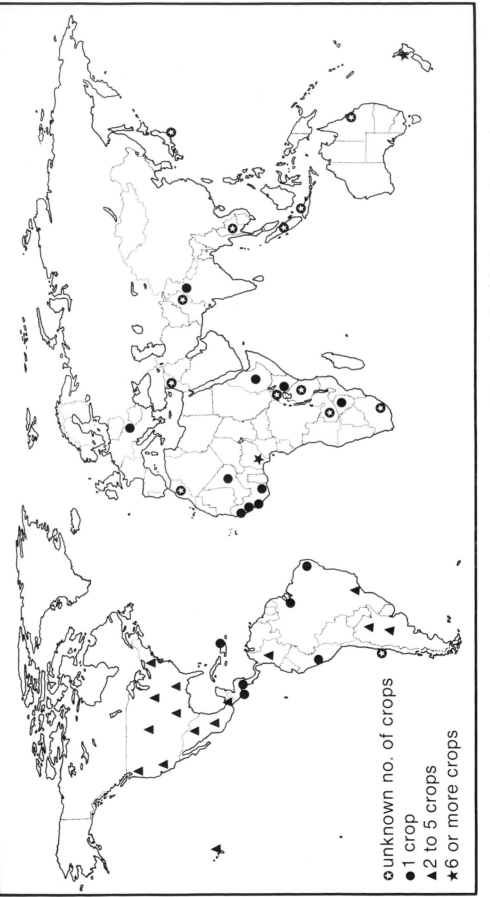

FIGURE 41. The distribution of *Amaranthus hybridus* L. across the world in countries where it has been reported as a weed.

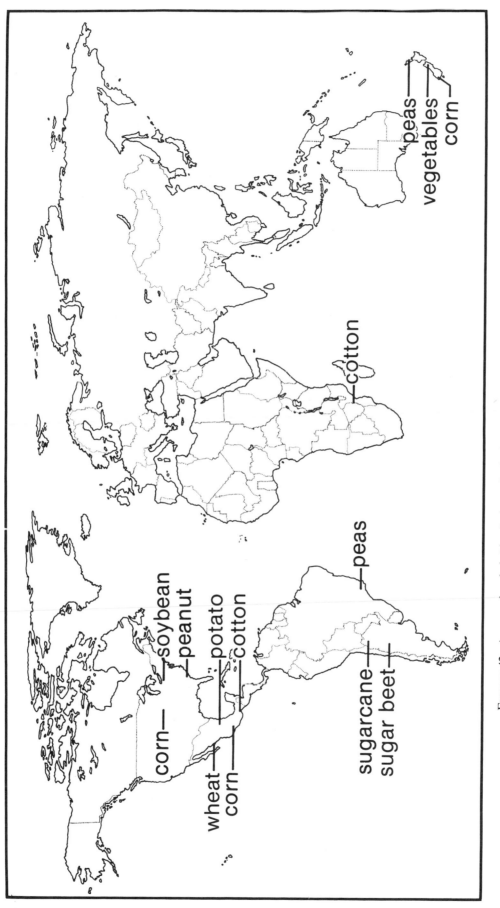

FIGURE 42. *Amaranthus hybridus* L. is a principal weed in the crops and locations shown above.

(Thomas and Schwerzel 1968). Very heavy stands of the weed can reduce corn yields by as much as 39 percent and soybean yields by 55 percent (Moolani, Knake, and Slife 1964). It grows taller than short-statured crops such as soybeans, causing losses in yields through shading. In taller crops it is able to compete because of its prolific seed production and rapid, early, seedling growth.

It may be an alternate host for the nematode *Meloidogyne* sp. (Raabe, unpublished; see footnote, *Cyperus rotundus,* "agricultural importance"); and for the tobacco mosaic virus (Namba and Mitchell, unpublished, see footnote, *Cynodon dactylon,* "agricultural importance").

Plants are used as green vegetables in South Africa (Shanley and Lewis 1969), India (Kanodia and Gupta 1968), and Rhodesia (Wild 1958).

COMMON NAMES
OF AMARANTHUS HYBRIDUS

ARGENTINA

ataco
bledo
caa-ruru

smooth pigweed
yuyu colorado

AUSTRALIA

slim amaranth

BRAZIL

caruru de espinho
caruru de folha larga

CHILE

moco de pavo
penacho
visnaga

COLOMBIA

bledo

EASTERN AFRICA

pigweed

EL SALVADOR

bledo
blero
lero

ETHIOPIA

alma

INDONESIA

bajem kejong (Javanese)
bajem merah (Indonesian)
singgang bener (Sundanese)

JAPAN

hosoaogeito

MEXICO

bledo
quelite

MOROCCO

queue de renard

NEW ZEALAND

red root

PERU

bledo
yuyo
yuyo hembra

RHODESIA

bwamanga (Ndao)
imbuya
imbuya jamabize (Ndebele)

SOUTH AFRICA

cape pigweed
kaapse misbredie

THAILAND

phak khom

UNITED STATES

smooth amaranthus

ZAMBIA

pigweed

❧ 15 ❧

Amaranthus spinosus L.

AMARANTHACEAE, AMARANTH FAMILY

Amaranthus spinosus, an erect annual, bears a pair of diverging, stipular spines almost as sharp and rigid as needles at the base of each leaf. They tear the flesh, their points breaking off in the hands of workers in sugarcane, cotton, and other crops. With minor exceptions its range as a weed extends from lat 30° N to 30° S. It is a native of tropical America. Forty-four countries report it as a weed in 28 world crops. The plant is avoided by most stock animals because of the spines. Leaves are sometimes used as vegetable greens.

DESCRIPTION

A. *spinosus* is an erect, much-branched, smooth, herbaceous *annual* growing to 120 cm (Figure 43); *stems* angled or with longitudinal lines or ridges, green or brown; *leaves* alternate, broadly lanceolate to ovate, discolorous, conspicuously veined beneath, up to 7 cm long, 4 cm wide, margins entire, the base tapering to the slender petiole up to 7 cm long, with a pair of straight spines up to 1 cm long at the base; *inflorescences* long, slender, terminal, with axillary spikes in clusters, greenish; *flowers* unisexual, straw-colored; *perianth segments* five, acuminate, upper third of spike being male with five stamens each, lower two-thirds female with three, rarely two, styles to the ovary; *fruit* one-seeded, opening by a line around the center; *seeds* reddish brown, lens-shaped, shiny.

The striated often reddish stem with two sharp, long spines at the base of the petioles, and the fruit which opens by a line around the center are distinguishing characteristics of this species.

PROPAGATION

A. *spinosus* is propagated by seeds which have a long viability and which are dispersed principally by wind and water. Seeds stored in glass containers for 19 years still gave 4-percent germination. Because the plant is quite variable, the seeds of some types will germinate within a few days after harvest if held at a high temperature, whereas others require 4 to 5 months of storage before they will germinate. Germination is satisfactory in both light and dark. Some types are known to produce 235,000 seeds per plant.

DISTRIBUTION AND BIOLOGY

Figure 44 indicates that A. *spinosus* behaves as a weed mainly in warm areas. It extends into the temperate zone in Japan and the United States. It is not reported to be a problem around the Mediterranean or in the Middle East. It is a problem weed principally around the Caribbean Sea, the west and south of Africa, around the Bay of Bengal, and in East and Southeast Asia from Japan to Indonesia.

The plant is found in cultivated fields, waste places, roadsides, garbage heaps, and abandoned fields. The plant will grow both in dry and wet sites but prospers when soil moisture levels are below field capacity. Waterlogging retards its growth. Maximum growth is obtained on soils which are high in organic matter, loamy in texture, and which have sufficient nitrogen.

The photoperiodic response is day-neutral and the plant may flower at daylengths between 8 and 16 hours. Overall optimum growth, however, is ob-

119

O
2MM O
IMM

4

3

2

O   4CM

FIGURE 43. *Amaranthus spinosus* L.
  Above: *1*, habit; *2*, seed; *3*, pistillate flower; *4*, utricle.
  Opposite: Seedlings.

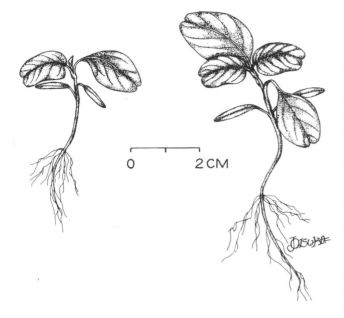

taincd at a daylength of 11 to 12 hours, and it is in areas with such daylengths that the earliest and most abundant flowering takes place. The plants do not grow well in shade or cool temperatures, because spine development and flowering are suppressed under these conditions. Conclusive evidence is lacking but there are indications that several different ecological populations exist in several regions of the world (Misra 1969).

AGRICULTURAL IMPORTANCE

A. spinosus is not often ranked the number-one weed or even as a principal weed in major crops grown on large acreages, but it is a common weed in many major crops around the world. There are frequent reports that it is troublesome, and sometimes serious, in specialty crops and those grown on smaller acreages. Figure 45 shows that it is reported to be the number-three weed in corn in the Philippines, a principal weed in that crop in Ghana, Hawaii, Mexico, and Thailand, and a common weed in Malaysia and Taiwan; in cotton it is ranked number-one in Thailand, principal in Nicaragua and the United States, and as a common weed in Mozambique; in peanuts it is a principal weed in Ghana, Hawaii, the Philippines, and the United States; and in sugarcane it is a principal weed in Brazil, South Africa, and Taiwan and a common weed in Hawaii, India, Indonesia, and Peru. In upland rice there are enough reports of competition on a world scale that we must consider it to be a common weed of this crop. Specifically, it is a principal weed of upland rice in

Mexico and the Philippines, and a common weed of upland rice in Ceylon, India, and Indonesia.

A. spinosus is a principal weed of the following crops (not shown in the figure): mangoes, sorghum, soybeans, and cowpeas in the Philippines; tobacco in Taiwan; beans in Mexico; vegetables in Malaysia; oil palms in Indonesia; papayas and sweet potatoes in Hawaii; mulberries in Japan; and cassava in Ghana. It is a common weed of bananas in Taiwan; oil palms in Nicaragua; pineapples in Hawaii; vegetables in Brazil, Ghana, Hawaii, India, the Philippines, and the United States; and of tea and jute in Taiwan. It is found in millet in the Philippines; coffee in Angola and El Salvador; and pineapples in the Philippines.

In 1973 in Hawaii, *A. spinosus* was implicated in a case of livestock poisoning when 39 dairy cows died after being fed green-chopped forage sorghum containing as much as one-fourth to one-third (by weight) of the weed. The poisoning was diagnosed as being caused by nitrate, and *A. spinosus* showed high nitrate levels. Hurst (1942) and Kingsbury (1964) both mention *A. spinosus* as a suspected poisonous plant.

A. spinosus is an alternate host of *Meloidogyne incognita* (Kofoid & White) Chitwood (Valdez 1968); and of the viruses which produce tobacco mosaic (Eugenio and del Rosario 1962) and groundnut rosette (Adams 1967).

COMMON NAMES
OF AMARANTHUS SPINOSUS

ARGENTINA	EASTERN AFRICA
ataco espinudo	spiny pigweed
AUSTRALIA	EL SALVADOR
needle burr	bledo
prickly amaranth	blero
BANGLADESH	huisquilite
katanata	FIJI
BRAZIL	spiny amaranth
bredo de espino	INDIA
caruru de espino	bajra
BURMA	chauli
hin-nu-nive-tsu-bauk	kantaneatia (Oriya)
tsu-gyi	kataili
CAMBODIA	spiny amaranth
phti banla	INDONESIA
COLOMBIA	baja badoeri
bledo espinoso	bajem duri
CUBA	(Indonesian)
bledo de espina	bajem eri (Javanese)

senggang tjutjuk
(Sundanese)

JAMAICA
prickly calau
wild calalu

JAPAN
haribiyu

MALAYSIA
bayam duri

MAURITIUS
brède malabar à pi-
quants
oseille
petit trèfle

trèfle

MEXICO
quelite espinoso

NATAL
thorny pigweed

NICARAGUA
bledo

NIGERIA
tete elegun
(Yoruba)

PERU
yuyo macho

PHILIPPINES
akum (Magindanao)

alayon (Ifugao)
ayantoto (Pampan-
gan)
gitin-giting
kalitis (Bisayan)
kalunai (Ilokano)
kilitis (Bikol)
orai (Tagalog)
tadtad (Bontoc)

PUERTO RICO
blero espinoso

RHODESIA
imbowa
mohwa-gura

thorny pigweed

SOUTH AFRICA
doring misbredie
thorny pigweed

TAIWAN
tsz-hsien

THAILAND
pak-khom-nam

UNITED STATES
spiny amaranthus

VENEZUELA
pira brave

VIETNAM
dên gai

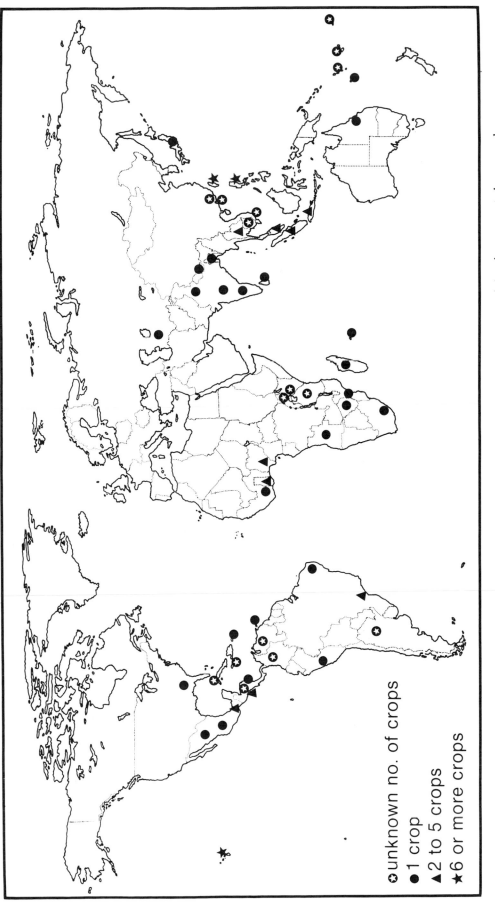

FIGURE 44. The distribution of *Amaranthus spinosus* L. across the world in countries where it has been reported as a weed.

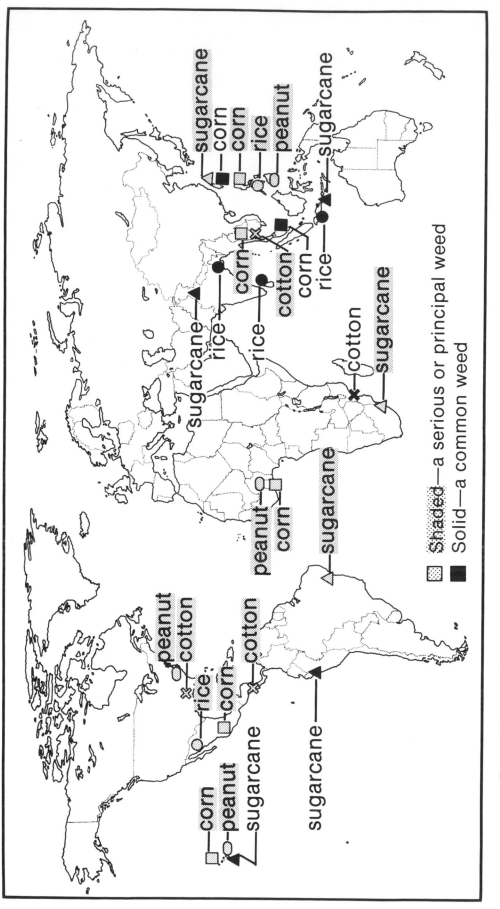

FIGURE 45. *Amaranthus spinosus* L. is a serious, principal, or common weed in the crops and locations shown above.

❧ 16 ❧

Cyperus esculentus L.

CYPERACEAE, SEDGE FAMILY

Cyperus esculentus is a light green perennial sedge having three-sided culms and growing to 1 meter in height. A basal bulb is formed by a swelling of the culm below the soil surface and rhizomes grow out from this bulb to terminate in single underground tubers. The species is often called yellow nutgrass or nutsedge in contrast to the purple nutgrass or nutsedge, *Cyperus rotundus*. The two species may grow in mixed stands and are difficult to distinguish before they have flowered. *C. rotundus*, however, produces rhizomes with several tubers in a chain as contrasted to the single tuber referred to above, and its subterranean parts are much darker. The plant propagates at a very fast rate. In one study it was shown that one tuber placed in a field produced 1,900 plants, almost 7,000 tubers, and covered an area about 2 meters in diameter within 1 year. The tubers are sweet, oily, and fleshy, and are used for human food. In some places the weed is cultivated to obtain the tubers for pig feed. *C. esculentus* is a weed in 21 crops in more than 30 countries.

DESCRIPTION

C. esculentus is an erect, *perennial* herb (Figure 46); *culms* simple, triangular, 30 to 80 cm tall, from perennial, tuber-bearing rhizomes; *leaves* in three ranks, mostly basal, about as long or longer than the culm, about 5 to 6 mm wide, with prominent midvein; *inflorescence* in more or less terminal umbels; *umbel* (an often flat-topped inflorescence whose pedicels and peduncles arise from a common point) subtended by unequal leaflike bracts varying from 5 to 25 cm long; *spikelets* yellowish brown or straw-colored, 1 to 3 cm long, of several flowers, flattened, two-ranked; *stamens* three; *style* three-cleft; *achenes* (fruit) three-angled, narrowing gradually from a square-shouldered apex toward the base, about 1.5 mm long, brownish gray to brown, covered with very small grains.

It is distinguished by its flattened, yellowish brown or straw-colored spikelets and its two-ranked, more or less terminal umbels.

DISTRIBUTION AND HABITAT

C. esculentus is a weed on all continents (Figure 47). It has the reputation of being a weed of warm areas, but in recent times the species seems to be advancing steadily into the cooler climates of the temperate zone. It grows at the equator but is now found in Alaska as well. At the present time it is most troublesome in eastern and southern Africa and in North and Central America. Frequent references may be found to ecotypes of the species, but no systematic classification of them anywhere in the world has been made.

The weed is found on low ground, moist fields, in heavily irrigated crops, along river banks and roadsides, and in ditches. It tolerates high soil moisture much better than does *C. rotundus*. There are reports that it can replace *Cynodon dactylon* and *Cyperus rotundus* in very moist places.

The species grows very well on all soil types,

FIGURE 46. *Cyperus esculentus* L. *1*, habit; *2*, spikelet; *3*, bract; *4*, seed, with closeup detail of surface reticulations; *5*, seed, cross section.

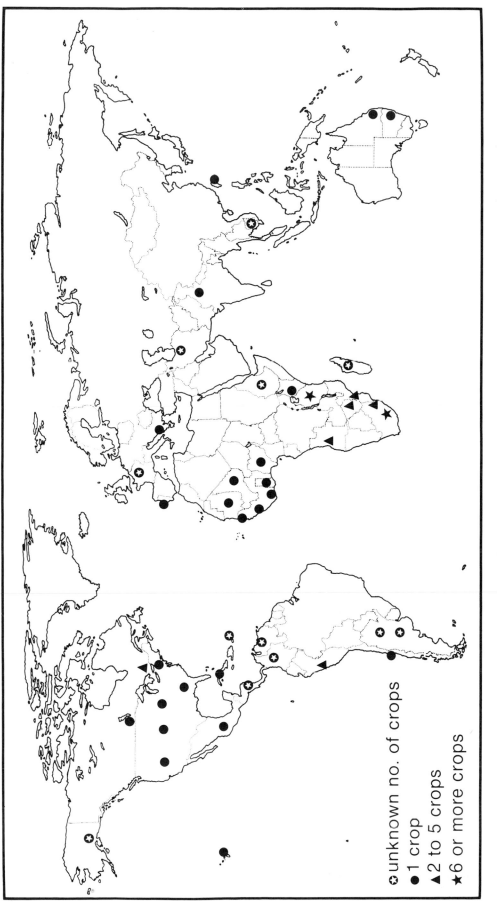

FIGURE 47. The distribution of *Cyperus esculentus* L. across the world in countries where it has been reported as a weed.

including black peat soils, and performs equally well at *p*H ranges from 5 to 7. It is quite intolerant of shade.

C. esculentus has shown a dramatic increase in the United States in the past decade, and it now infests more than 1 million hectares in the eastern half of the country alone. It grows in each of the states. Some changes in agriculture, such as herbicide programs to control predominantly annual weeds, less hand-hoeing, and a general tendency toward reduced tillage, may have contributed to its increase.

BIOLOGY

The morphology of *C. esculentus* during a full season of growth and development is most clearly understood from the experiments of Jansen (1971) in the United States. The life history presented in his paper is a synthesis of his own research and his observations with those of other workers.

In both *C. esculentus* and *C. rotundus* the basal bulb is the focus of the leafy shoot and the beginning of subterranean growth. It is the unit on which vegetative propagation ultimately depends. In plants grown from seed the basal bulb is first seen as a swelling at the junction of the mesocotyl and coleoptile. Apical growth from the basal bulb provides the leafy shoot and, under favorable circumstances, the scape and inflorescence. Rhizomes also originate from the basal bulb and, at a very short or a very long distance from the parent plant, the tip turns up to produce a secondary basal bulb. Still other rhizomes grow out to produce underground tubers which, upon sprouting, provide the first shoots in the following season. A tuber first produces one or more rhizomes from the apex and these grow upward to produce new basal bulbs. A repetition of these events makes it possible for the primary basal bulb to give rise to a series of secondary, tertiary, and higher order vegetative shoots, each in its turn producing more rhizomes and tubers.

We will use the terminology proposed by Jansen for certain of the vegetative structures to describe in greater detail the major events of this plant's life cycle. The rhizomes coming from the seedlings of *C. esculentus* originate as axillary buds at the first leafy node of the newly formed basal bud. In secondary bulbs they arise from the node of the prophyll (the first leafy structure of the new aerial shoot). The rhizomes are indeterminate, some growing to 60 cm in length and containing more than 30 internodes before differentiating at the apex to form either a tuber or a basal bulb. On the other hand, some

growth may be so short as to make it appear that the basal bulb has arisen directly from the tuber. Rhizomes have scale leaves (cladophylls) at the nodes and two or three cladophyll primordia enclosed in a nearly mature scale leaf at the terminal meristem. The branches of the rhizome originate in the buds of the axils of the older cladophylls.

When the tubers are formed elongation ceases at the rhizome tip, many small scale leaves are formed, and radial swelling takes place. At maturity the irregularly rounded tuber is covered with a single imbricated layer of scarified, dark-veined cladophylls and at the apex is a conical-shaped terminal bud surrounded by scale leaves. Short fibrous roots arise at the base of the scale leaves. The newly formed tuber is white and turns brown as it matures. The covering scale leaves darken as they become senescent.

C. esculentus (as does *C. rotundus*) develops basal bulbs very close to the soil surface but may form some at greater depths. The prophyll differentiates from the meristem at the rhizome tip in the same manner as a cladophyll but the prophyll grows to be two or three times longer. The leaf blades expand upon emergence and the sheath of the first typical leaf thickens at the base as the blade enlarges above. Each leaf sheath expands in turn. The leaves are inserted on a broad, conical stem which forms as the basal bulb continues to swell. Long fleshy roots are initiated at the base of the leaves.

From experiments in controlled-environment rooms and extending over a full growing cycle in a greenhouse, it was learned that a new leaf originates each 4.5 to 5 days and that each leaf has a sigmoid pattern of growth. At first, leaf development proceeds at a rapid rate but, shortly thereafter, the rate begins to decrease until within a week there is no growth at all. The total period of growth for a leaf varies from 24 to 40 days. Early leaves show a progressive decrease in length until the fourth or fifth leaf, subsequent leaves gradually becoming longer.

One of the earliest visible evidences of flowering is the appearance of a triangular, hollow, foliar tube through which the scape must elongate. As the scape begins to extend, floral structures begin to differentiate from meristems in axils of elongating involucral leaves. When the scape has elongated to one-third the length of the foliar tube, the inflorescences in the axils of the lower involucral leaves are macroscopic.

It is of interest that rhizomes may be differentiated from buds in axils of adaxial glumes. Though not often seen in nature, aerial tubers may form at the tips of such rhizomes. Other studies concerned with the biology and ecology of this species are those of Bell et al. (1962); Bendixen (1970); Garg, Bendixen, and Anderson (1967); Ito, Inouye, and Furuya (1968); and Taylorson (1967).

FACTORS INFLUENCING TUBERIZATION AND SHOOT PRODUCTION

It is to be expected that an adequate light intensity, good nutrition and soil moisture, and a suitable warm temperature will promote the general health and vigor of this species as it does with most plants. Certain levels of these, however, may be particularly influential in promoting the development of different organs of the plant. For example, Garg, Bendixen, and Anderson (1967) found that high levels of nitrogen, a long photoperiod (15 hours), and high levels of gibberellic acid (GA) inhibit tuberization. At high temperatures of 27° C and 33° C increases in tuberization occurred at very low nitrogen levels (the photoperiod was not specified for these experiments). At high levels of nitrogen and long photoperiods (14 and 15.5 hours) shoot formation was promoted. High temperatures favored shoot formation at 12.5- and 14-hour photoperiods, whereas GA had an inhibitory effect. As in many tuber-bearing plants, carbohydrate accumulation favors tuberization.

C. esculentus is particularly responsive to the length of the photoperiod. In the experiments of Jansen (1971) photoperiods began at 8 hours of daylight and extended by 2-hour increments to 24 hours. During a period of 3 months the number of rhizomes that developed from basal bulbs seemed to have no relation to photoperiod, but their age or order in the plant system was important. The original basal bulbs, for example, produced 15.4 rhizomes, whereas the basal bulbs and shoots of first, second, and third order shoots (may be called "peripheral" shoots) produced only 6.6, 3.7, and 4.0 rhizomes, respectively.

The number of rhizomes that differentiated into new shoots was markedly increased by the four longest photoperiods. At photoperiods of 14 hours or less no new shoots developed except those which had already been initiated. Although shoot development from parent rhizomes was favored by long photoperiods, there was a linear decline from 7 at 16 hours to 4.5 at 24 hours. The situation was quite different in rhizomes from peripheral shoots. None developed into a shoot at 16 hours but four were formed at 24 hours.

All of the indeterminate rhizomes present at the beginning of the experiment had formed tubers at the tips by the end of the experiment (3 months) if they had been exposed to short photoperiods of 8, 10, and 12 hours. Some were actually visible at 2 weeks and almost fully grown at 3 weeks. This was quite in contrast to the initiation of shoots in the long photoperiods. Tuber-formation diminished as photoperiod length increased, although some tubers were formed during long photoperiods following long delays. The number of new tubers produced, however, was unrelated to photoperiod at the end of the 3-month period. The reason is that, although the number of rhizomes was unaffected by photoperiod and although tuber formation was promoted in short photoperiods, the production in long photoperiods of many new shoots with their attendant new rhizomes which could produce tubers (though slowly, at long photoperiods) made the tuber number about the same in either short or long photoperiods after a long growth cycle (see also Bell et al. 1962, and Bendixen 1970).

TUBER DORMANCY

There is a tuber dormancy in *C. esculentus* which must be taken into account when one is attempting to cause tuber sprouting in order to destroy the new plants with herbicides or mechanical tillage. Reports of the level of dormancy vary a great deal because of differences in the history of the plant materials, times of harvest and testing, and methods of testing. Bell et al. (1962) found tubers harvested during the growing season in the eastern United States to be completely dormant in different types of germination tests at different temperatures. When held at 10° C for 1 month and then held at alternating temperatures of 20° to 25° C for 48 days, a germination of 42 percent was obtained. If held for longer periods the percentage of germination was even higher. Tubers which had overwintered in the field, however, showed a high percentage of sprouting in 7 days at the same alternating temperatures. In the north-central United States, Tumbleson and Kommedahl (1962) found 12-percent germination in fall-harvested tubers and 95 percent in spring-harvested tubers. They reported that storage at 3° C promoted sprouting and also increased the numbers of shoots per tuber. Washing freshly harvested tubers in cold water improved germination. Taylorson (1967) obtained better germination with unwashed tubers harvested from fields in the southeastern United States. He collected tubers monthly for 2 years and found them to be most dormant in late summer and fall with best sprouting in winter and spring. Mowing and disk-harrow cultivation sharply increased the sprouting of tubers from a stand which had previously been undisturbed, this occurring during a period when sprouting was normally low. Thomas (1967a) and Thomas and Henson (1968) working in England with plant material which originated in South Africa confirmed the results obtained by Tumbleson and Kommedahl (1962) that tubers held at 4° C increased markedly in germination. Washing in running water for 48 hours gave no advantage; scraping black tubers with sandpaper increased germination from 40 to 85 percent; and treating with hydrogen peroxide raised sprouting from 30 to 60 percent. Stoller, Nema, and Bhan (1972) have shown that tubers which differ fivefold in weight may germinate equally well but larger, more vigorous shoots are produced by large tubers. Sprouts were removed from germinated tubers and the tubers then replanted for a total of 3 germinations to simulate a condition which may obtain in fields which are tilled repeatedly. The proportion of tubers that regenerated new growth decreased after each germination. Multiple shoots were produced only in the first germination. The length and weight of shoots and weight loss by tubers decreased with each germination. More than 60 percent of the dry weight of carbohydrates, proteins, and oils in the original tuber was used up in the first germination. About 10 percent of the dry weight was consumed in the second and third germinations.

REPRODUCTION POTENTIAL FROM TUBERS

Tumbleson and Kommedahl (1961) have provided some interesting statistics on tuber production in *C. esculentus*. They worked with tubers which averaged 7 mm in diameter and 209 mg in weight. Moisture content was 40 to 60 percent. One tuber produced 36 plants and 332 tubers in the field in 16 weeks. In 1 full year this tuber had spread to a patch 2 m in diameter containing 1,900 plants and almost 7,000 tubers. A badly infested peat soil was sampled to 45 cm and calculations were made which indicated that there were 18 metric tons of tubers per hectare. In one area it was estimated that there could be more than 30 million tubers per hectare.

TUBER DISTRIBUTION AND BURIAL

In the eastern United States most tubers were found in the top 15 cm of the soil and firm tubers were rarely found below 23 cm. October and April samplings from fields showed 15 and 75 percent respectively, of tubers which were soft and presumed to be dead. There is obviously a high winter mortality in some areas (Bell et al. 1962). In the north-central United States 99 percent of the tubers were in the top 25 cm in a peat soil. An occasional tuber was found at 40 to 45 cm but none were found below that level. We assume, from a survey of many reports, that in most soils the greatest portion of tubers will be in the top 15 cm.

It is to be expected that soil type will greatly influence the depth from which tubers can emerge. Tumbleson and Kommedahl (1961) measured tuber emergence from several depths in sand, a sandy silt loam, and in a peat soil. Emergence was satisfactory from 8 and 15 cm in all soils. There was considerable emergence, though delayed, from 30 cm in peat soil. In greenhouse experiments 12 out of 100 tubers produced shoots after 4 weeks from a depth of 50 cm in peat soil. In the sandy silt loam soil one shoot emerged from a depth of 80 cm.

SEED PRODUCTION AND GERMINATION

When a photoperiod of 12 to 14 hours is reached in spring or late summer, inflorescences appear on all shoots which are disposed to flower in their particular growing conditions. Jansen (1971) found no flowers outside this range. Two or 3 weeks after plants were exposed to the above photoperiods, the foliar tubes were dissected and floral units were detected therein. In 3 to 4 weeks the inflorescences were elongating from the tube. Jansen suggested that the cool temperatures of spring may explain the general lack of flowering at that period.

The reports of seed production and viability are mixed across the world, but there seems little doubt that large quantities of good seeds can be produced under favorable growing conditions and that this is an important means of distribution of this species. Hill, Lachman, and Maynard (1963), in the eastern United States, demonstrated in experimental plots that one seedling could develop a plant system in one season capable of producing 90,000 seeds with better than 50-percent viability. Justice and Whitehead (1946) found one area of Maine (northeastern United States) where plants produced 1,500 viable seeds per inflorescence. While gathering seed lots from the north to the south of the United States, they found that viability varied from 50 to 95 percent.

There is little question but that seeds are dormant in most areas at maturity, but that the dormancy can be broken soon thereafter with appropriate methods.

Justice and Whitehead (1946) have conducted one of the most careful studies made on the seeds of *C. esculentus*. Dormancy was dissipated by storage in dry conditions at room temperature or in moist conditions at 10° C. After 4 months of storage, however, at alternating temperatures of 20° to 30° C, 80 percent of the seeds germinated.

Light has little effect on germination; 2-percent potassium nitrate is stimulatory; and scarification with sulfuric acid for 5 to 10 minutes improves sprouting slightly.

Bell et al. (1962) planted seeds at 0.6, 1.3, 2.5, 3.8, 5, and 6.3 cm in a sandy loam soil. Most seedlings came from 0.6 and 1.3 cm depths in 2 weeks. The only other seedlings to emerge came from 2.5 cm after some delay.

AGRICULTURAL IMPORTANCE

From Figure 48 it may be seen that *C. esculentus* is a serious or principal weed of sugarcane in Hawaii, Peru, South Africa, and Swaziland; of corn in Angola, South Africa, Tanzania, and the United States; of cotton in Mozambique, Rhodesia, and the United States; of soybeans in Canada and the United States; and of potatoes in Canada, South Africa, and the United States. Although these data are not shown on the map, *C. esculentus* is reported to be a serious weed of coffee in Kenya; cereals in Angola and Tanzania; and vegetables in Mozambique and the United States; and a principal weed of peanuts and sugar beets in the United States; pineapples in Swaziland; and sisal in Tanzania.

The species is also in sugarcane in places as widely distributed as Angola, Australia, Costa Rica, and Taiwan. It is in coffee, cotton, peanuts, rice, and tea in Tanzania and in corn in Rhodesia and citrus in Mozambique. It is also in tobacco in Australia; citrus in Peru; rice in Chile, India, Portugal, and the United States; and vegetables in South Africa and Tanzania.

The weed is troublesome not only because of reduced yields of crops but because of lowered quality as well. In some badly infested potato fields in the United States every potato tuber was found to have a rhizome running through it or into it. The nutgrass tubers become mixed with shelled beans. Costs are increased in several crops because of added cultivation, handweeding, herbicides, and wear on harvesting machinery as a result of the clumps of nutgrass which are pulled in during harvesting (Bell et al. 1962). It is one of the most important perennial weeds in the United States. A recent survey suggests that *C. esculentus* and *C. rotundus* together rank fifth in seriousness among all weed species and second only to *Agropyron repens* as perennial weed problems. It estimated that one million hectares are infested in the north-central and northeastern United States and the weed is still spreading. In these areas it is most troublesome in corn; in the south it is most troublesome in cotton and peanuts.

Tumbleson and Kommedahl (1961) found that a 4-year fallow period on peat soil in Minnesota (United States) reduced tuber numbers from 912 to 7 per 30 cm² or the equivalent of 21 to 1.6 metric tons per hectare. Tuber viability was reduced from 72 to 28 percent. Tubers rolled to the top in tillage operations dried quickly and, in 2 days, their germination was reduced by 80 percent. Bell et al. (1962) reported that viable tubers were reduced by 90 percent after a 2-year fallow period in an upland soil.

C. esculentus is frequently troublesome in irrigated areas because it can tolerate higher soil moisture levels than *C. rotundus*.

Wax et al. (1972) have recently reported a very effective system for controlling *C. esculentus* in soybeans. The procedure uses preplanting tillage, late planting of the crop, herbicides, and cultivation.

The tubers of *C. esculentus* contain 12 to 30 percent sucrose, 25 to 30 percent starch, some fatty acids, and almost 30 percent oil. They contain no alkaloids, caffeine, or asparagine. The plant is used both as a stimulant and as a sedative in Asia and Africa, but its value for human use is really dietetic rather than medicinal. Because of its sweet and nutty flavor it is cultivated in southern Europe, western Asia, and over much of Africa. It is sometimes roasted and used as a coffee substitute. Its nondrying oil resembles olive oil in quality.

The insect *Bactra verutana* Zeller is of interest for the biological control of *C. esculentus*. Keeley, Thullen, and Miller (1970) studied the insect in California (United States) and found it was able to suppress the growth of the weed, providing its numbers were large enough, the infestation occurred early enough, and that the attack was made on all newly emerging plants. In this particular area the weed began to grow in late February and early March, but the insect did not become active until early July. The long period of weed growth before the insects became plentiful may have limited their usefulness for biological control. Some injured plants survived to produce new vegetative plants and tubers.

It is an alternate host of the virus which produces lucerne dwarf (Namba and Mitchell, unpublished; see footnote, *Cynodon dactylon*, "agricultural importance").

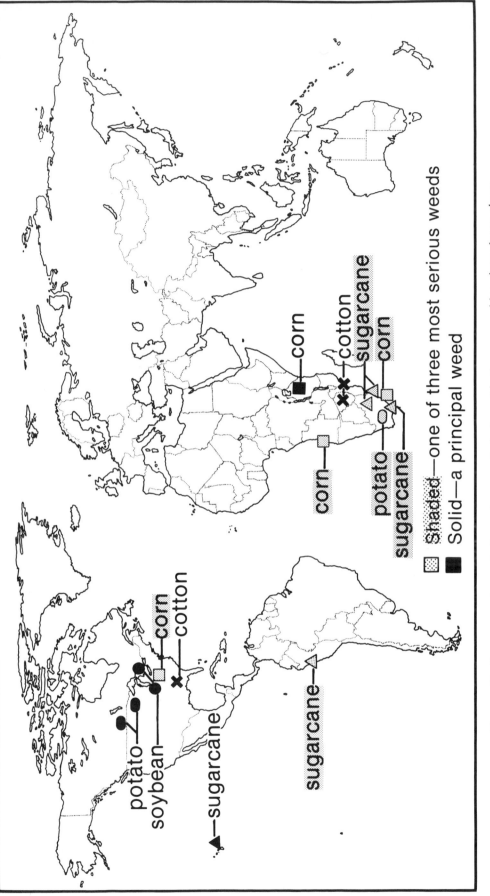

FIGURE 48. *Cyperus esculentus* L. is a serious or principal weed in the crops and locations shown above.

COMMON NAMES
OF CYPERUS ESCULENTUS

ANGOLA
  olonguesso

ARGENTINA
  chufa
  cotula
  junquillo
  tamascal

AUSTRALIA
  nutgrass

CAMBODIA
  phlong si moeum

CANADA
  yellow nut grass

CHILE
  chufa
  coquillo

COLOMBIA
  coquito

IRAN
  galee

KENYA
  water grass

  nut grass

MEXICO
  cebollín
  coquillo

NICARAGUA
  coyolillo

PERU
  coco
  coquito

PORTUGAL
  junquinha mansa

PUERTO RICO
  chapas

RHODESIA
  chupa (Shona)

  water grass

SOUTH AFRICA
  geeliuintjie
  uintjie
  yellow nutsedge

TANZANIA
  nutgrass

THAILAND
  haeo-thai

UNITED STATES
  yellow nutsedge

VENEZUELA
  chufa

✦ 17 ✦

Paspalum conjugatum Berg.

POACEAE (also GRAMINEAE), GRASS FAMILY

Paspalum conjugatum, a native grass of tropical America, may grow as an annual or a perennial. It has an extensive system of long, creeping, leafy stolons. It is principally a weed of the humid tropics. More than 30 countries report that it is a weed in 25 crops. It is suitable for grazing only when young. The seeds have been reported to stick in the throats of livestock and to choke the animals.

DESCRIPTION

This weed is an extensively creeping, stoloniferous, *perennial* grass (Figure 49); *stolons* leafy, to 2 m in length, rooting at the nodes, often reddish purple; *culms* reaching to 60 cm long, erect or ascending, the nodes smooth; *sheaths* compressed, ridged like the bottom of a boat (keeled), hairy on the margins and at the junction of the blade and stem; *ligule* short, appearing as if cut off at the end (truncate); *blades* lanceolate, acute, 5 to 25 cm long, 5 to 15 mm wide, flat, smooth or sparingly hairy, with rough or stiffly hairy margins; *racemes* two at the apex of culm, paired or nearly so, with rarely a third below, widely spreading, straight or somewhat arched, 4 to 15 cm long; *spikelets* ovate, acute, 1.5 to 2 mm, flattened, pale green, the margin fringed with long white silky hairs; *stigmas* white; *anthers* yellow. The plant is easy to recognize when in flower by the typical T-shaped inflorescence.

DISTRIBUTION AND AGRICULTURAL IMPORTANCE

P. conjugatum is a weed of the warm, wet areas of the world. Figure 50 shows it as a troublesome weed mainly in Central America, West Africa, and the islands and peninsulas of the Pacific and Southeast Asia. These are, for the most part, the humid tropics. It is a weed of wet lowland areas but in Hawaii and Ceylon it grows up to 1,875 m as well. It is found in waste areas and along trails and streams. Some members of this genus actually prefer an aquatic habitat. It is found in settled areas and cultivated fields, but more often it is common in natural and poorly managed pastures and in perennial or plantation crops where the soil is less frequently plowed or

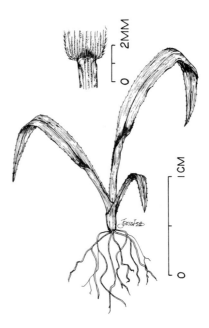

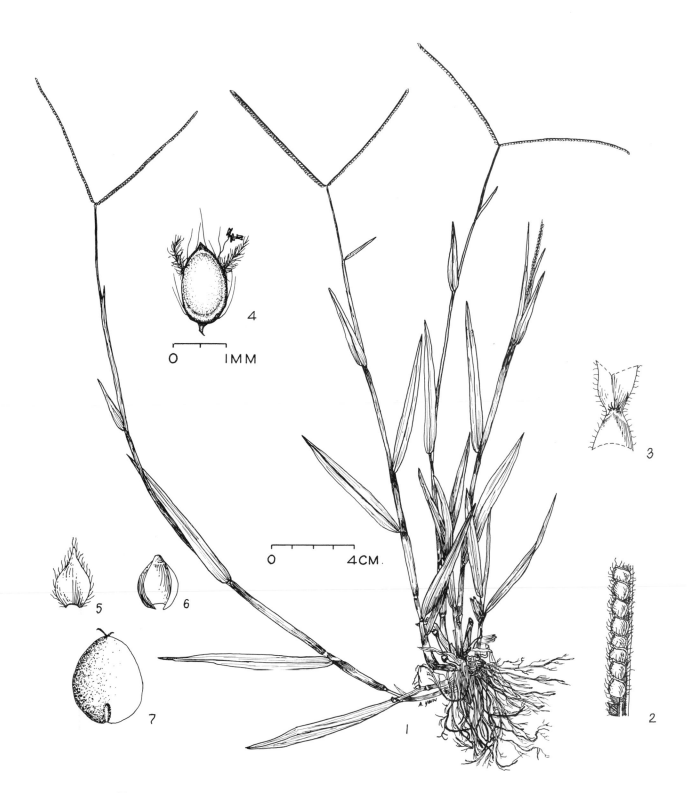

FIGURE 49. *Paspalum conjugatum* Berg.
Above: *1*, habit; *2*, portion of spike; *3*, ligule; *4*, spikelet; *5*, bract; *6*, bract; *7*, grain.
Opposite: Seedlings.

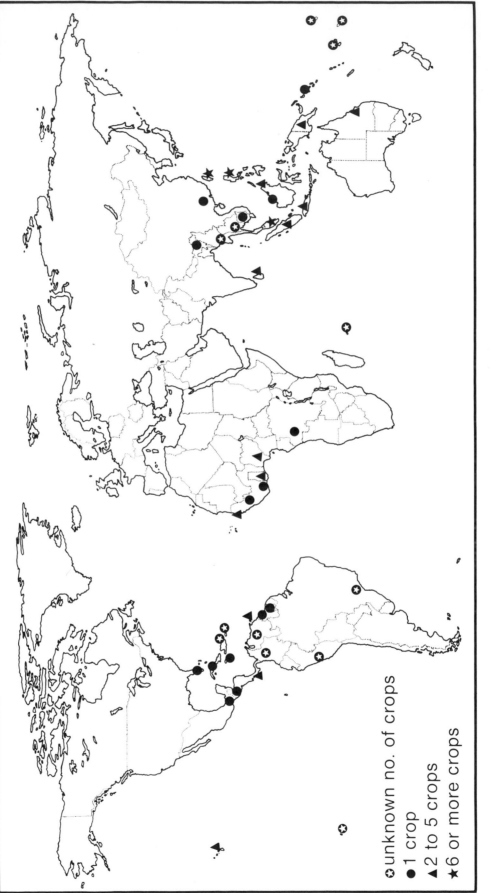

FIGURE 50. The distribution of *Paspalum conjugatum* Berg. across the world in countries where it has been reported as a weed.

tilled. The plant spreads quickly by means of its stolons and may soon form dense masses which can suppress or eliminate tree seedlings and herbaceous plants of all kinds. The plant will tolerate some shade and can make some growth on poor and acid soils.

Little is known of the biology of this weed. It can flower all year in the Philippines, and one plant can produce 1,500 seeds. It is also propagated and spread by vegetative pieces of stolons when machines are used for tillage.

From Figure 51 it may be seen that *P. conjugatum* is one of the three most important weeds in pineapples and coffee in the Philippines; coffee in Zaïre; rubber in Ceylon and Malaysia; oil palms in Malaysia; and tea in Ceylon, Indonesia, and Taiwan. It is a principal weed of pineapples in Hawaii and the Ivory Coast; coffee in Costa Rica and New Guinea; rubber in Indonesia and western Africa; tea in India; cacao in Ghana, New Guinea, and Trinidad; and oil palms in Costa Rica, Indonesia, and West Africa. It is not shown on the map, but *P. conjugatum* is also reported to be one of the three most serious weeds of bananas in the Philippines and coconuts in Malaysia. It is a principal weed of citrus and coconuts in Trinidad; bananas in Costa Rica; citrus in Malaysia; peanuts in Taiwan; papayas in the Philippines; abaca in Borneo; pastures in Australia and Hawaii; and of lawns and turf in Hawaii.

It is a weed of corn, tobacco, bananas, pineapples, sugarcane, cassava, and pastures in Taiwan; bananas, pastures, and sugarcane in Trinidad; cacao and vegetables in Malaysia; sugarcane, mangoes, bananas, and coffee in Mexico; corn, sorghum, macadamia nuts, sugarcane, and taro in Hawaii; peanuts, cassava, and rice in Indonesia; cowpeas, oil palms, and rice in Nigeria; taro, bananas, coconuts, and cacao in Samoa; abaca, rice, and sugarcane in the Philippines; rice in Cambodia and Surinam; and sugarcane in Australia.

P. conjugatum is an alternate host of *Cassytha filiformis* L. (Raabe 1965), *Leptosphaeria proteispora* Speg., *Sorosporium paspali* McAlp.

(Stevens 1925), *Halothrips gowdeyi* (Frank), and *Hoplothrips paumalui* Moulton (Sakimura 1937); and of the virus which produces sugarcane mosaic (Namba and Mitchell, unpublished; see footnote, *Cynodon dactylon*, "agricultural importance").

COMMON NAMES
OF PASPALUM CONJUGATUM

AUSTRALIA
  sour grass
BRAZIL
  grama comum
CEYLON
  sour grass
COLOMBIA
  horquetilla
CONGO
  sour grass
FIJI
  Thurston grass
  ti grass
  yellow grass
HAWAII
  Hilo grass
INDIA
  banhaptia (Assamese)
INDONESIA
  djampang pahit (Sundanese)
  djampang tjanggah
  paitan (Javanese)
  roempoet pait (Indonesian)
JAMAICA
  sour grass
MALAYSIA
  buffalo grass
MAURITIUS
  herbe créole
MEXICO
  grama

  pata de conejo
  trensilla
NIGERIA
  sour grass
PERU
  sour paspalum
  torourco
PHILIPPINES
  bantotan (Manobo)
  carabao grass
  kauad-kauaran
  kulape (Tagalog)
  sacate (Bisayan)
PUERTO RICO
  sour grass
  sour paspalum
SURINAM
  buta-grasse
TAIWAN
  buffalo grass
  Hilo grass
  mau-yin-chywe-bai
THAILAND
  ya-hep
TRINIDAD
  sour grass
UNITED STATES
  sour paspalum
VENEZUELA
  paja mala
ZAÏRE
  sour grass

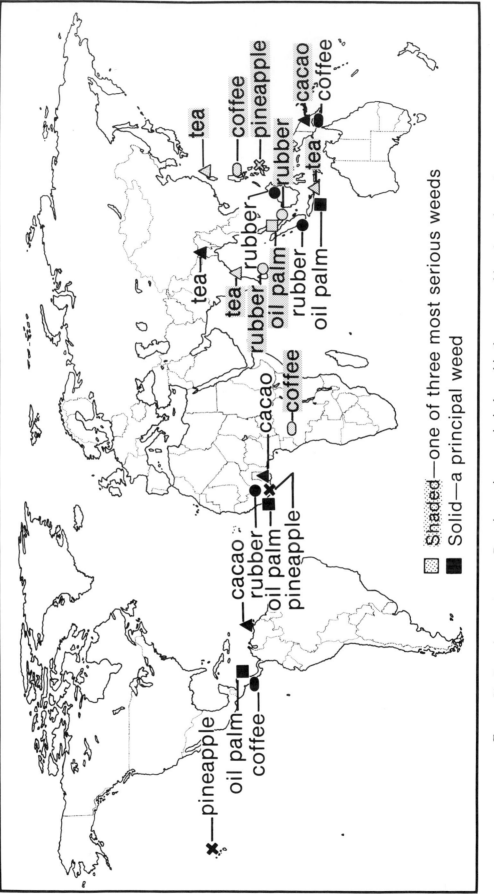

FIGURE 51. *Paspalum conjugatum* Berg. is a serious or principal weed in the crops and locations shown above.

❧ 18 ❧

Rottboellia exaltata L.f.

POACEAE (also GRAMINEAE), GRASS FAMILY

Rottboellia exaltata, a native of India, is an erect, strongly tufted, annual grass that may reach a height of 4 meters. It is an extremely vigorous competitor in upland crops and it is dreaded by hand-laborers because it has fiberglass-like needles on the sheath which penetrate the flesh, break off, and result in painful infections. It is a weed of 18 crops in 28 countries. Its rapid spread and destructiveness in recent years is cause for alarm.

DESCRIPTION

R. exaltata is a tall, erect, strongly tufted, *annual* grass (Figure 52) with stilt roots; *culms* obtusely angular, smooth, 1 to 3 m high; *leaf sheaths* with spreading, long, sharp, tubercle-based, siliceous, fragile, irritating hairs that break off on contact; *ligule* short, fringed with hairs (ciliate); *blades* linear-lanceolate, flat, with wide, white midnerve, 20 to 60 cm long, 1 to 2.5 cm wide, rough on both surfaces but sometimes smooth below, very rough along the margins; *inflorescence* a solitary raceme at the terminus of the culm and each branch of the culm; *spikes* cylindrical, 8 to 15 cm long, about 3 mm in diameter, mostly solitary, narrowed upward, glabrous, sheathed at the base, readily breaking into hard cylindric joints 6 to 7 mm long; *spikelets* sometimes sterile; *sessile spikelet* 5 to 7 mm, as long as the joint or distinctly shorter; *pedicelled spikelets* variable, 3 to 6 mm; *grain* obliquely ovate, strongly narrowed at apex, with flat back side and convex ventral side, 4 mm long, 2 mm wide.

The cylindric spikes and the sharp irritating sili-
ceous hairs on the leaf sheaths are distinguishing characters of this aggressive annual grass.

DISTRIBUTION AND HABITAT

R. exaltata is a weed of warm season crops but its habitat varies widely across the world. In many areas it is prominent in open, well-drained places and is one of the important species in old field successions. In Rhodesia it is present at altitudes up to 1,600 m; in Kenya and Tanzania to 1,900 m; and in Madras, India, it is reported above 2,300 m. In South Africa, however, it frequents wet places and, in Madras, may even grow in shallow water. In some regions it requires sunny or moderately shaded places, whereas in others it is found in thickets or teak forests. Finally, it is common on contour banks and roadsides and its importance as a weed of several cultivated world crops is increasing. In a detailed survey of the weeds of arable lands in Rhodesia (P. Thomas 1970), the weed was recorded only on heavier soils—clay to sandy clay loams. In these main heavy-soil cropping areas it is one of the most serious weeds in corn, cotton, peanuts, and soybeans. It is not a serious problem on lighter soils in Rhodesia. The weed is most troublesome between 800 and 1,300 m elevation. Thomas believes that the main limiting factor below 1,300 m is rainfall; above this elevation it is temperature.

Except for those weed problems with *R. exaltata* reported from the United States and Australia, all others have been reported from lat 23° N to 23° S (see Figure 53). Actually, the locations are all within the

FIGURE 52. *Rottboellia exaltata* L. f. *1*, habit; *2*, seedling; *3*, ligule; *4–5*, joints with pedicelled and sessile spikelets; *6–7*, lower and upper glumes, both of sessile spikelet; *8–9*, lower and upper glumes, both of pedicelled spikelet; *10*, grain.

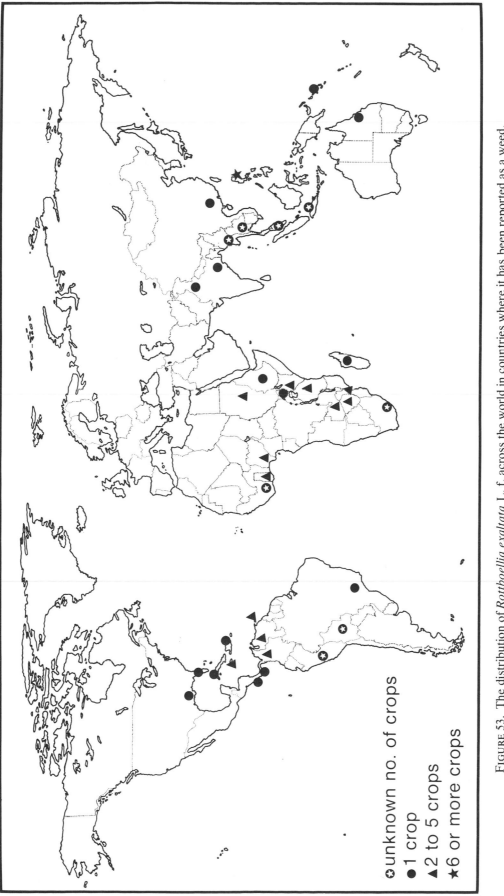

FIGURE 53. The distribution of *Rottboellia exaltata* L. f. across the world in countries where it has been reported as a weed.

20° C isotherms in the Northern and Southern hemispheres.

BIOLOGY

In eastern Africa the species is one of the primary colonizers of disturbed land. In Trinidad it is one of the important fast-growing annuals which finally become dominant and take over from early perennial colonizers such as *Cynodon dactylon* and *Cyperus rotundus*; it tends to develop well even in the driest parts of the island. It is a very vigorous plant and often exists in pure stands in old, cultivated fields and near habitations (Richardson 1963).

Reproduction is by seeds. In the Philippines the plant flowers all year round, and a single plant may bear 2,200 seeds. In Rhodesia dense stands have produced 665 and 590 kg per hectare of seeds in consecutive seasons (Schwerzel 1970*a*). In some regions the individual plant produces seeds continuously once flowering has set in. Seeds are spread by water, in poorly cleaned crop seeds, and by harvesting machines.

The dormancy of the seeds and the germination habit vary a great deal across the world. In some areas the seeds germinate together at the beginning of the season, so that cultivation and chemical controls at this time may be more efficient. In other areas the seeds are said to have a deep dormancy, sometimes lasting for years; in old fields this gives a staggered germination throughout the season or year. Deep-plowing to bury the seeds may extend their life span. There are reports that nondormant seeds can emerge from 15-cm depths. At other times it seems necessary to expose the seeds to the weathering action of sun and air to prepare them for germination when moisture conditions become suitable.

In Rhodesia the germination has been studied in seasons of normal and dry rainfall (Schwerzel 1970*b*). In a normal year 75 percent of the plants were found to produce seeds; the greatest seed production came from plants which emerged 2 weeks after the start of the growing season, the seeds produced becoming less viable as the season progressed. In seasons of low rainfall, each plant had a longer life span; seeds in the soil germinated more slowly; and the number of seedlings which emerged declined more rapidly than they would have in the normal season.

Perhaps there is a unifying explanation that can bring order out of these reports. If the plant does indeed produce seeds which are nondormant as well as seeds with varying degrees of dormancy (as is the case with most weed species) and if the plants produce viable seeds continuously through the season, then both immediate germination and staggered germination must be dealt with if this weed is to be controlled.

The weed is a very serious competitor in sugarcane in the Philippines and studies there have provided some information on the biology of the species. Fernandez (unpublished)* studied growth and development in pot experiments in upland soil. When seeds were planted at a 2-cm depth the coleoptiles were visible in 4 to 5 days. The additional time required for the appearance of successive leaves after emergence of the coleoptiles was: first leaf, 1 day; second leaf, 3 to 4 days; third leaf, 6 days; fourth leaf, 9 days; and fifth leaf at 14 days. First tillers appeared at about 3 weeks after planting (four- to five-leaf stage), one to five tillers per day were produced in the early life of the plant, tillering continued for 44 days, and the average plant finally had about 100 tillers per plant.

The first external sign of flowering was an enlargement of upper internodes and the formation of a wide angle between the mother tiller (which usually grew upright) and all other tillers. This was followed by flag leaf formation and eventually by the emergence of the tip of the inflorescence. Spikelets emerged over a period of 15 days. Pollen was shed 4 to 9 days after the emergence of spikelets.

Seed maturation was first indicated by a change from green to brown on the portion of the spikelet where "snapping-off" would take place. The first 8 to 12 spikelets "snapped off" 2 to 4 days after the last spikelet of the inflorescence had emerged. Others fell within 2 weeks. The maturation period for all spikelets covered about 1 month. It was found in these experiments that the spikes had nine to 34 spikelets, the plants produced 270 to 950 spikes each, and 10,500 to 16,500 seeds were harvested from each plant. About one-half of the seeds did not fill under these conditions.

In still other experiments (Fernandez, unpublished) seeds were sown monthly for an entire year and it was demonstrated that sowings early in the year (February, March, and April) provided the longest period between planting and flowering and the plants produced were almost twice as tall as those from later sown seeds. Tillering occurred at 13

---

* "Studies on *Rottboellia exaltata*. 1. Biology. 2. Control" is the title of a paper presented by Dr. D. Fernandez at the Philippine Sugar Technologists Convention that was held in Manila, 14 August 1963. Dr. Fernandez's address is: Department of Agricultural Botany, College of Agriculture, University of the Philippines, Los Baños, Philippine Islands.

days from seeds sown in April and at 29 days from those sown in September. September plantings produced plants with 18 tillers, whereas those planted in June gave 100. Plants emerging in March required 154 days to flower as compared with 47 for plants started in October. The greatest quantities of seed were harvested from plants with the longest periods of vegetative growth.

AGRICULTURAL IMPORTANCE

This species is presently most destructive in eastern Africa, the Philippines, and along the shores and islands of the Gulf of Mexico and the Caribbean area. Many research reports express alarm over this weed. It seems to be a recent adventive in many crops and areas and is causing concern because of its destructiveness and rapid spread. In some areas in northern South America it is responsbible for the abandonment of cropland. Its rapid growth and spreading habit make it very competitive in the early season. If the weed growth is advanced, the workers in some areas refuse to go into the fields for weeding, because the fiber-glass-like needles in the area of the leaf sheath can penetrate hands and clothing and result in painful infections.

Rottboellia exaltata is one of three most important weeds found in sugarcane, corn, and upland rice in the Philippines (see Figure 54). It is a serious weed in sugarcane in Jamaica, Mozambique, and Tanzania; corn in Ghana, Rhodesia, and Zambia; cotton in Rhodesia and Zambia; and peanuts and soybeans in Rhodesia and Zambia. It is a principal weed of sugarcane in Trinidad and the United States; soybeans in the Philippines; and peanuts in Sudan. The number-one weed of sugarcane in Zambia, it is also a weed of sugarcane in Kenya, Madagascar, and Venezuela; corn in Colombia, Nigeria, Tanzania and other countries in eastern Africa, and Venezuela; peanuts in Trinidad; and cotton in Ethiopia, Mozambique, Sudan, Uganda, and Venezuela. It is a weed in one or more of the following crops in Cuba, Ghana, Jamaica, the Philippines, Trinidad, and Venezuela: bananas, cassava, citrus, cowpeas, papayas, peanuts, pineapples, rice, and sorghum.

Although attempts have been made to use it for pasture in many areas, there have been conflicting reports of its desirability. It was introduced into the Caribbean for grazing. Although it is used for hay in Africa, it is also suitable for green fodder and silage. Cattle and horses appear to relish it at all times, even when it is in seed. In Ceylon it is considered to be dangerous for stock because the stiff hairs may lacerate the animals' mouths and intestines. It is not liked by stock animals in Angola, but here the stems are very coarse and fibrous as compared with the plants of East Africa. In India most grazing animals avoid it because of the sharp hairs. The water buffalo will eat it anytime and other stock will use it if there is little else available.

In East Africa it is said that a 2-year fallow period will bring the plant under control. The area is first burned to destroy the seeds on the surface. A shallow plowing is then done to stimulate germination of seeds. Finally deep plowing buries the seedlings and many more of the seeds in the soil. Fallowing is continued until the land is clean.

Some believe that it cannot be controlled by cultivation alone. During experiments in Rhodesia the grass made such remarkable growth on arable land early in the season that the resulting dense mass of vegetation suppressed other weeds that had started more slowly. In this case, however, only two cultivations were required for effective control of *R. exaltata* (Smartt 1961).

In March 1972 a Ph.D. thesis entitled "Studies of the biology of *Rottboellia exaltata* Linn f. and of its competition with maize" was completed by P. E. L. Thomas (P.B. 222A, Salisbury, Rhodesia). Copies of this thesis are lodged at the University of London, the University of Rhodesia, and at the Weed Research Organization, Oxford. Microfiche is available on loan or for 50 Rhodesian cents from Dr. Thomas.

R. exaltata is an alternate host of rice leaf gall virus and corn leaf gall virus (Agati and Calica 1950).

COMMON NAMES OF ROTTBOELLIA EXALTATA

AUSTRALIA
kokoma grass

CUBA
sancarana

EASTERN AFRICA
guinea-fowl grass

INDONESIA
bandjangan (Javanese)
djoekoet kikisian (Sundanese)

PHILIPPINES
aguiñgay (Tagalog)
annarai (Ivatan)
bukal
gaho (Bikol)
girum

nagei (Bontoc)
sagisi (Ilokano)

RHODESIA
guineafowl grass
kokomo grass
shamva grass

TRINIDAD
corn grass

UNITED STATES
Raoul grass

VENEZUELA
paja peluda

ZAMBIA
jointed grass
mulungwe
shamva grass

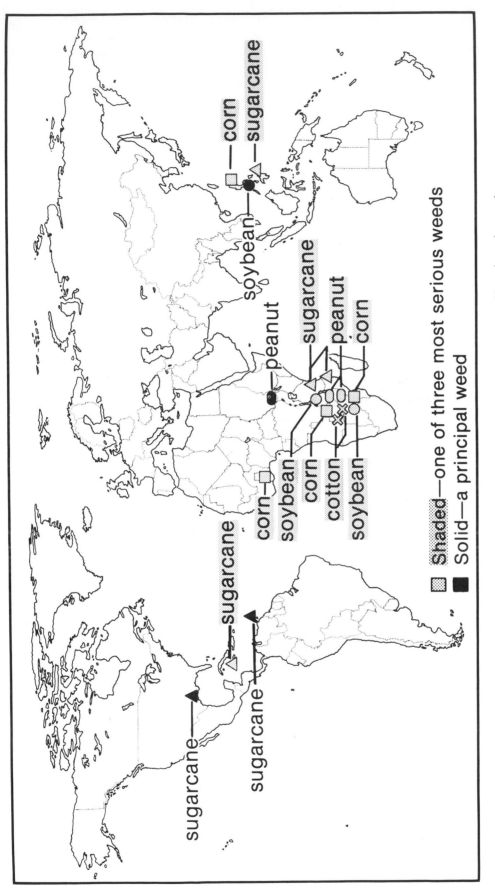

FIGURE 54. *Rottboellia exaltata* L. f. is a serious or principal weed in the crops and locations shown above.

GROUP 2

The species of weeds in this group are next in importance and are troublesome for man in cultivated crops, pastures, and waterways. Some are confined to one or two regions of the earth, others are limited to specific groups of crops. Those which are widely distributed seldom compete with crops as severely as do those of the weed species in group 1. The weeds are arranged in alphabetical order for there is at present no meaningful way to rank them in order of importance. Perhaps it is enough to know that they are among the world's worst weeds.

☙ 19 ❧

Ageratum conyzoides L. and *Ageratum houstonianum* Mill. (= *A. mexicanum* Sims)

ASTERACEAE (formerly COMPOSITAE), ASTER FAMILY

Ageratum conyzoides, an annual, is a native of tropical America. Its range as a weed extends from lat 30° N to 30° S. Forty-six countries report that it is a weed in 36 different crops. It has escaped from the New World, where it is usually an ornamental, to become one of the most common weeds in the warm regions of the world. It is one of two really important weeds in a genus of about 300 species. A closely related species, *A. houstonianum*, can easily be mistaken for *A. conyzoides*. Some of the reports of the weedy nature of *A. conyzoides* may actually be confusing this plant with *A. houstonianum*.

DESCRIPTIONS

Ageratum conyzoides

A. conyzoides is an erect, branching, *annual* herb (Figure 55) that may be 5 to 7.5 cm tall or 60 to 120 cm tall at flowering, depending on environmental conditions; sometimes decumbent at the base with branches forming roots; *roots* shallow, fibrous; *stems* erect, cylindric, the nodes enlarged, nodes and young parts clothed with rather long, partly crispy hairs; *leaves* opposite, soft, 2 to 10 cm long, 0.5 to 5 cm wide, on hairy petioles 0.5 to 5 cm long, sometimes broadly ovate, triangular-ovate, or rhomboid-ovate with a subcordate, rounded, or narrowed acute base and an acute or obtuse, sometimes acuminate, tip, sparsely long-hairy on both surfaces, glandular on the lower surface, leaf margins toothed (serrate), the surfaces rough, veins prominent; *crushed leaves* have a characteristic odor; *inflorescence* terminal or often axillary, made up of

several branches, each bearing a number of flower heads which are arranged in showy, more or less flat-topped clusters; *flower heads* light blue, white, or violet, 5 mm across, 4 to 6 mm long, with 60 to 75 flowers; *bracts* surrounding the flower heads linear, in two or three rows, sparsely hairy or smooth-toothed (serrate) in the upper part, acuminate, acute at the tip, 3 mm long, green with pale or reddish violet top; *head clusters* (corymbs) small, on peduncles 5 to 17 mm long; *heads* made up of 60 to 75 *tubular flowers* which are not (or are scarcely) exserted above the rows of bracts forming the involucre; fruit an *achene*, black, ribbed or angled, 1.5 to 2 mm long, with a small white cap at lower end,

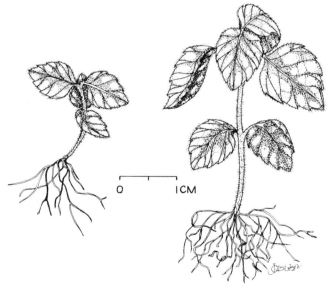

146

FIGURE 55. *Ageratum conyzoides* L.
Above: *1*, habit; *2*, head; *3*, bract; *4*, flower; *5*, achene.
Opposite: Seedlings.

glabrous or sparsely hairy, pappus of five chaffy bristles, white to cream-colored, lance-shaped, with upward-turning spines, slightly longer than the achene.

Ageratum houstonianum

A. houstonianum is an erect or creeping-at-base and rooting *annual* or short-lived *perennial* herb (Figure 56), from 15 to 90 cm tall; *stems* cylindric, sometimes branched, covered with white hairs which spread or open at an angle from the stem, slightly purple-tinged; *leaves* opposite, on 0.5 to 6.5 cm long petioles, subtriangular or deltoid, top leaves ovate, with a notched (subcordate), flattened (truncate), or rounded base passing acutely into the petiole, obtuse or almost acute at the apex, both surfaces with hairs which spread or open at an angle from the surface, 2 to 10 cm long, 1.5 to 5.5 cm wide; *crushed leaves* have a characteristic odor; *inflorescence* terminal or often axillary, made up of several branches each bearing a number of flower heads which are arranged in showy, more or less flat-topped clusters; *head clusters* (corymbs) small, dense to very dense, branches with white hairs spreading or open at an angle from the surface; *peduncles* 2 to 10 mm long, with spreading or open white hairs; *flower heads* blue or violet, 5.5 to 7 mm long, with 75 to 100 flowers; *bracts* surrounding the flower head very hairy, some hairs long, some hairs short and glandular, smooth on the margins, tapering to the needlelike (acicular) top; *flower heads* made up of 75 to 100 flowers, the styles of which are exserted above the rows of bracts making up the involucre; *fruit* an achene, angled, covered with hairs, but not densely hairy, 1.5 to 2 mm long, pappus shorter than the corolla.

A. conyzoides and *A. houstonianum* can easily be confused. However, distinguishing characteristics of the species are: *A. conyzoides* has pale blue, white, or violet flowers, smaller flower heads with fewer flowers per head, and nearly glabrous and toothed bracts of the involucre; *A. houstonianum* has blue flowers, larger flower heads with 75 to 100 flowers per head, very hairy bracts of the involucre with smooth margins and needlelike tips, deltoid leaves, and spreading or open hairs on the flower head branches and peduncles, both surfaces of the leaves, and on the stems.

DISTRIBUTION AND BIOLOGY

A. conyzoides grows in a wide range of arable crops including, to some extent, grasslands in many tropical and subtropical countries. It sometimes extends into the temperate zone. It is also found in wastelands and on roadsides, in both light and heavy soils, and in dry areas—though it prefers a moist habitat (Figure 57).

All of the nonweedy species of the genus *Ageratum* are confined to the New World. The extreme adaptability of *A. conyzoides* has made it a very successful colonizer in many regions of the world, and it colonizes rapidly in disturbed or cultivated areas. Its life cycle may be completed in less than 2 months; it can flower when as few as two pairs of leaves have been formed; and it may be seen as a plant up to 90 cm tall with hundreds of flower heads in favorable conditions, although it can also exist as a tiny plant with a single flower when it is crowded or when it is growing under extremely wet or dry conditions (Baker 1965).

A. conyzoides appears to have no photoperiodic requirement. It will flower at low or high night temperatures, produces an economical amount of pollen in its small seedheads, and is self-pollinated. It may produce 40,000 seeds per plant and in some areas one-half of the seeds will germinate shortly after they are shed. Seeds are mainly spread by wind and water and will germinate under a wide range of conditions (Baker 1965).

Although we know little of the biology of *A. houstonianum*, we feel that it probably is similar to that of *A. conyzoides*, for the two species are often found in the same habitat. *A. houstonianum* probably produces more seeds than *A. conyzoides* because it has more florets per head.

Because *A. conyzoides* and *A. houstonianum* are easily mistaken for each other, some reports of the weedy nature of one or the other may have confused them. Therefore, careful study should be made to determine which species is present in a country and is acting as a weed.

AGRICULTURAL IMPORTANCE

A. conyzoides and *A. houstonianum* are among the weeds most commonly seen in the warm regions, but they do not have the destructive capacity of large, vigorous weeds such as *Sorghum halepense* or *Panicum maximum*. Because they can flower all year in many areas and because they produce such large quantities of seeds which germinate readily, either of the two species can form thick carpets of plants which compete with crops for nutrients and moisture. When a stand of these weeds is destroyed, another quickly takes its place.

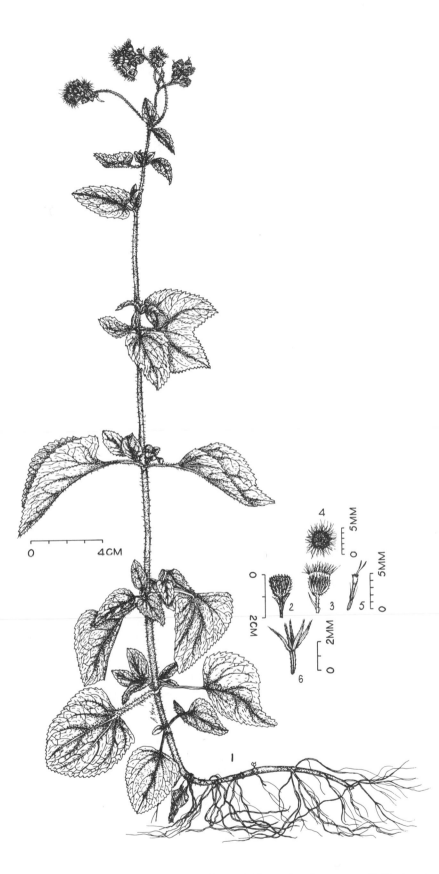

FIGURE 56. *Ageratum houstonianum* Mill. *1*, habit; *2–4*, flower head, note needlelike bracts of the involucre; *5*, ray flower; *6*, achene with chaffy pappus of five bristles.

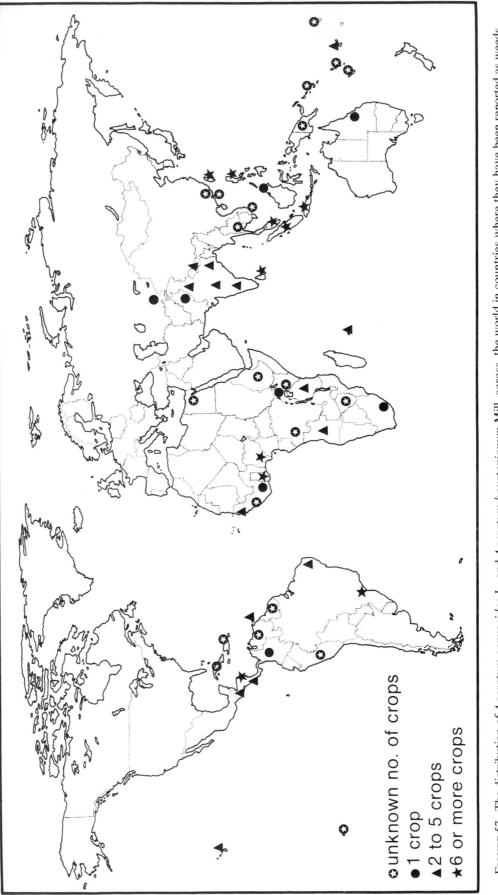

FIGURE 57. The distribution of *Ageratum conyzoides* L. and *Ageratum houstonianum* Mill. across the world in countries where they have been reported as weeds.

A. conyzoides is often a weed of plantation crops and pastures, but it can occur in short-term crops as well (Figure 58). It is a serious weed of corn in Ceylon and Ghana; peanuts, upland rice, and chilies in Ceylon; sugarcane in Taiwan and the Philippines; bananas, cassava, and tea in Taiwan; and cassava in Ghana. It is a principal weed of tea in Ceylon, India, Indonesia, and Mauritius; peanuts in Ghana and Indonesia; sugarcane in Australia; corn in Nigeria and the Philippines; upland rice in Indonesia and the Philippines; pineapples in Taiwan; millet, soybeans, and cowpeas in the Philippines; cotton in Uganda; overgrazed wetland pastures in Hawaii and of pastures in India and Australia; cacao in Brazil; potatoes in Colombia; sweet potatoes in Malaysia; wheat in Angola; and oil palms in Nigeria. It also has been reported a weed of abaca, taro, cinchona, coconuts, rubber, coffee, citrus, tobacco, sisal, wheat, sorghum, and several vegetable crops in several countries.

A. conyzoides is widespread in arable crops in eastern Africa and is found there from sea level to 3,125 m. In Australia it germinates late in the wet season after the sugarcane can no longer be cultivated and flourishes even after the cane provides a heavy cover. The plant is sometimes cultured as an ornamental, both in gardens and for use of the cut flowers in homes (Pope 1968). A herb medicine is prepared from the plant in Nigeria.

A. houstonianum is reported as a weed of tea, cinchona, and coffee plantations in Indonesia. It is one of the most serious weeds of sugarcane in Australia because in dense stands it prevents a good cane fire, causes difficulty and increased costs in both manual and mechanical harvest, and hinders surface drainage in the wet season (Young 1962).

A. houstonianum grows from 200 to 1,650 m in fields and along water courses and roadsides in Indonesia. It is often found in open degraded pastures and grasslands, especially in the wet tropics.

A. conyzoides is an alternate host of *Cassytha filiformis* L. (Raabe 1965), *Cercospora agerati* Stevens, and *Puccinia conoclinii* Seym. (Stevens 1925); of the nematodes *Meloidogyne* sp., *Pratylenchus pratensis* (de Man) Filip. (Raabe, unpublished, see footnote, *Cyperus rotundus,* "agricultural importance"), *Rotylenchulus reniformis* Linford & Oliveira (Linford and Yap 1940), *Aphelenchoides fragariae* (Ritz.-Bos) Christie (Sher 1954), *Meloidogyne incognita* (Kofoid & White) Chitwood, *M. javanica* Treub, *M. arenaria* Chitwood, and *M. arenaria thamesis* Chitwood (Valdez 1968);

and of the viruses which produce spotted wilt (Sakimura 1937), anemone mosaic, and tobacco leaf curl (Namba and Mitchell, unpublished, see footnote, *Cynodon dactylon,* "agricultural importance").

A. houstonianum can be an alternate host of *Pseudomonas solanacearum* (Pegg and Moffett 1971).

COMMON NAMES OF AGERATUM CONYZOIDES

AUSTRALIA
billy goat weed
blue top
Mother Brinkly
winter weed
BRAZIL
erva São João
mentrasto
CEYLON
goat weed
hulantala
mulanchala (Sinhalese)
poon-pillu (Tamil)
EASTERN AFRICA
goat weed
FIJI
goat weed
botebotekoro
sogovanua
HAWAII
ageratum
INDIA
billy goat weed
gundhuabon
mahakaua
INDONESIA
bandotan (Indonesian)
bandotan leutik (Sundanese)
berokan (Javanese)
KENYA
goat weed
MALAYSIA
ruput tahi-ayam
white weed
MAURITIUS
herbe de bouc

NATAL
billy goat weed
NIGERIA
goat weed
imiesu (Yoruba)
tamasondji bata
PHILIPPINES
asipukpuk (Pangasinan)
bahu-bahu
bahug-bahug
budbuda (Igorot)
bulak-manok (Tagalog)
kakalding (Bontoc)
kamabuag (Ivatan)
kolokong-kabayo (Tagalog)
kulong-kogong-babae (Bikol)
RHODESIA
billy goat weed
SURINAM
boko-boko-wiwiri
TAIWAN
hwo-hsiang-ji
TANZANIA
white weed
THAILAND
ya-tabsua
ya-sap-raeng
TRINIDAD
herbe à femme
tropic ageratum
UNITED STATES
tropic ageratum
VIETNAM
cò cút-heo (bò xít)

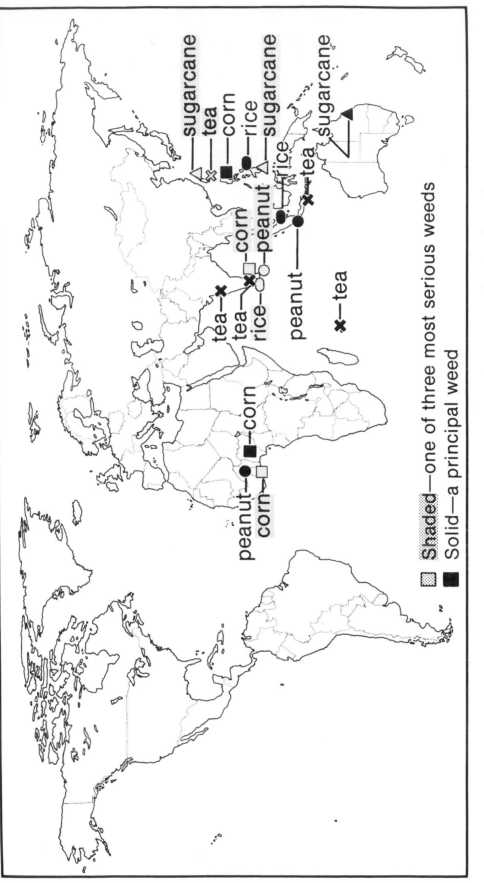

FIGURE 58. *Ageratum conyzoides* L. is a serious or principal weed in the crops and locations shown above.

❧ 20 ❧

Agropyron repens (L.) Beauv.

POACEAE (also GRAMINEAE), GRASS FAMILY

Agropyron repens is an aggressive perennial grass, native to Europe, that spreads by seeds and rhizomes. It is believed by some to be the most serious perennial weed of the cooler regions of the north temperate zone. It is particularly troublesome in the northern United States, southern Canada, and over most of Europe. The inflorescence arrangement is like that of wheat and the species was originally called *Triticum repens* by Linnaeus. Because the species is cross-pollinated, there is great variability among plants grown from seed; however, a persistent, vigorous rhizome system is the principal source of man's difficulty with the species. The plant is closely associated with human activity and is reported to be a weed in 32 crops in more than 40 countries. Two names are most commonly used: couchgrass and quackgrass. The former term is used for many grasses and so the latter will be used as a common name in this text.

DESCRIPTION

A. repens is an erect, *perennial* grass (Figure 59); *culms* erect or curved at the base, 30 to 100 cm tall or more; yellowish or straw colored, cordlike, creeping, extending laterally to 150 cm, 5 to 15 or 20 cm in depth, about 3 mm in diameter, often numerous enough to form a tough, tangled mass, the tips being often very sharp; at each *node* there is a tough, brownish sheath, giving a scaly appearance, branches arising at nearly every node, fibrous roots arise from the rhizome nodes; *leaf sheaths* hairy on the lower leaves, smooth or slightly soft-hairy above; *ligule* membranous, ragged, 0.5 mm long; *auricles* in pairs at the bases of the leaves, small; *blade* soft, relatively flat, often with whitish bloom when growing in drier conditions, dark green in humid conditions, rough or sparsely hairy above; *inflorescence* in dense or lax *spikes*, somewhat like the spikes of wheat, 5 to 25 cm long; *spikelets* compressed, 0.5 to 2.2 cm, usually four- to six-flowered; *glumes* three- to seven-nerved, usually awn-pointed; *lemma* mostly 8 to 10 mm long, the *awn* bristlelike, from less than 1 mm in length to as long as the lemma; *palea* obtuse, as long as the lemma, rough on the keels; *seeds* enclosed in the flowering glumes, elongated toward the slender, short-awned tip, broadest below the middle and tapered to the blunt base; *grain* (caryopsis) 4 to 5 mm long, slender, brownish, pointed at the base, topped by a ring of hairs.

A. repens may be distinguished by its prominent pale yellow or straw-colored rhizomes which have a tough brownish sheath at each joint, giving the rhizomes a scaly appearance.

DISTRIBUTION AND HABITAT

Figure 60 indicates that *A. repens* is present in all of the major agricultural areas of the north temperate zone. It is a weed of Alaska and extends above the Arctic Circle in Norway. Almost every country of Europe has quackgrass in its croplands. It grows in many of the north temperate countries of Asia, and is a principal weed of coffee in the higher areas of New Guinea. The scattered reports from Central and

FIGURE 59. *Agropyron repens* (L.) Beauv. *1,* habit; *2,* ligule; *3,* spike; *4,* spikelet, front view; *5,* spikelet, back view; *6,* spikelet, bracts removed; *7,* flower.

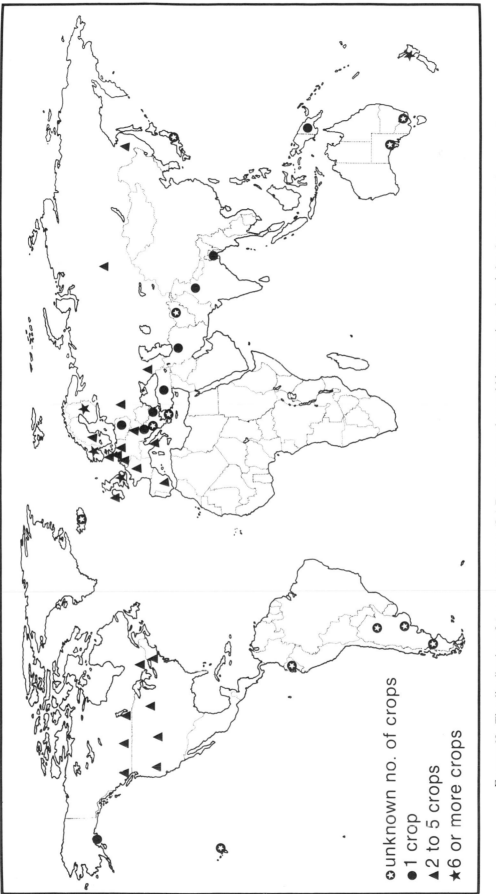

FIGURE 60. The distribution of *Agropyron repens* (L.) Beauv. across the world in countries where it has been reported as a weed.

South America suggest that, although unreported as a troublesome weed, *A. repens* may grow in the cooler mountain valleys of several countries there. The general absence of this perennial from the tropics and from most warm regions, however, invites a physiological explanation. Some workers have speculated that it cannot succeed without a cold dormant period. Those who have worked with the species are impressed with the vigorous vegetative growth which seems to follow the cold winter period. Hakansson (1969d) found that growth was generally suppressed by high temperatures in a short photoperiod and speculated that perennation of the weed may be difficult, or even impossible, in the tropics. In any case, *A. repens* is unable to compete with weeds better adapted to the tropics.

The species is found in arable land, in lightly grazed pastures, and in waste places. It will inhabit sites where trees and shrubs are not continuous, the edges of fields and roads, and the margins of rivers. It is a serious problem of many orchards of the North.

The weed can grow on many soil types from light to medium to heavy but prefers the latter. It is found on organic soils and chalk, but cannot tolerate a low *p*H or exposed rock faces. When it colonizes a new area it first appears in clumps; in the next season it may spread to form larger patches; and, if undisturbed, goes on to form pure stands.

Buried rhizomes and roots are seldom harmed by freezing, but, if they are raised to exposure on the surface of the soil by tillage, they are easily killed by freezing and thawing. If undisturbed the plants can stand long periods of drought.

BIOLOGY

The studies of the morphology and physiology of this perennial weed have followed the traditional pattern whereby researchers plant rhizomes of varying lengths at different depths, compare the plant units derived from seeds or vegetative parts, and make observations on the effects of light intensity, photoperiod, temperature, soil texture, and soil moisture content. Even though such systematic exploration of the stages of growth has rewarded us with knowledge which has been useful in planning control measures, it still is generally recognized that large areas remain to be explored. Some workers believe it to be the most troublesome perennial weed in the cool regions of the world's temperate zones. We are becoming increasingly aware that the species is extremely variable across the world and that most of the plant types are very sensitive to changes in

climate and to major factors in the immediate environment. This means, of course, that no "standard" quackgrass plant or habit of growth can be described.

Clones in the United States and Canada may be either light or dark green in color and hairy or smooth. There is large variation in plant height and length of the spikelet; outer glumes and lemmas may be blunt or may have awns of varying length; and in some clones the spikelets remain together at harvest whereas others they break up into individual florets (Raleigh, Flanagan, and Veatch 1962). Several of these variations have been found in plants in Great Britain and Norway, where, in addition, some forms are erect whereas others are prostrate, leaf colors are blue-green to yellow-green, and the dates for beginning growth and flowering are quite different between clones. Plants from northern regions tend to be prostrate in habit and to flower late. Varietal names have been given to several of the forms (Palmer and Sagar 1963).

DEVELOPMENT OF PLANT UNITS FROM RHIZOMES AND SEEDS

Hay (1962a) in Canada, Palmer and Sagar (1963) in England, and Raleigh, Flanagan, and Veatch (1962) have reviewed portions of the life history of *A. repens* and some of the literature for their regions. Sawhney (1965) has reported on the anatomy of the rhizomes of this species. A very important contribution to our knowledge of the shoot apex of *A. repens* was made by Sharman (1947). Although the internal structure of the shoot apices had been much studied, he attempted to describe the morphological changes during the ontogeny of the apex or, in his term, "the biology of the region." The most comprehensive investigation of the growth and development of *A. repens* was reported in a series of nine papers from Sweden (Hakansson 1967; 1968a,b; 1969a,b,c,d; 1970; and Hakansson and Jonsson 1970). The planting materials which they used are typical of those found in the central and southern parts of their country. Laying a foundation for their many experiments, they first described the developmental morphology with time during the plant's first two seasons of growth. With concomitant chemical analyses they were able to determine the stage at which the net loss of dry matter was most severe; for it is at this point that the plant units should suffer the greatest setback if disturbed in any way.

When rhizome pieces were planted in early autumn, some buds produced new shoots which reached the soil surface before the winter freeze. If planted later the shoots surfaced the following spring. After the first aerial shoots (primary shoots) had developed three or four foliage leaves in spring, lateral shoots began to develop from nodes below and close to the soil surface. Some of these, usually those closest to the soil surface, developed into tillers (becoming aerial branches of the primary shoot), whereas others grew as new rhizomes. The young rhizomes were more plentiful and appeared when there were three leaves on the aerial shoot. Tillers, on the other hand, were not often seen until the primary shoot had four leaves. Lateral buds on these new rhizomes, together with those on the piece originally planted, produced branches late in the summer and thereafter branched repeatedly in a dichotomous habit.

In late May and early June aerial shoots elongated into culms with only a few producing flowering spikes. In late June and early July the apical buds of new, young rhizomes began to turn up to the surface to form new aerial shoots. The process continued until October, and some of the new shoots survived the subsequent winter. The oldest aerial shoots perished in winter. Very few aerial shoots developed from lateral buds of the rhizomes.

Initiation of aerial shoots, whether during the natural procession from one season to the next or following tillage and burial of whole plant units, was always accompanied by a decrease in dry matter in the plant as a whole. In a normal cycle of events, the minimum dry weight was reached in spring when new aerial shoots had about two developed leaves. Hakansson believes that up to this point a net import of food reserves into the shoot was necessary for the establishment of a new plant unit. Shortly thereafter, when three or four leaves were visible, a net export of food assimilates left the shoot. This was the time at which further development of the plant unit began. It was during the brief period of minimum dry weight, when the plants were showing about two developed leaves, that the plant seemed least able to withstand extremes in the environment or disturbance by man. It is at this moment that tearing and burial are used most effectively for control purposes.

Summarizing the many types of investigations with these clones of *A. repens* from Sweden, Hakansson wrote that the general morphology and physiology of the relationship between reserve food and growth renewal is quite similar whenever old aerial shoots have been destroyed or removed; for, when that occurs, the plant must reorient its growth in the direction of new aerial shoots as soon as possi-

ble, and it must produce tillers and new rhizomes as soon as food reserves can be made available. Whether the new aerial shoots come from undisturbed quackgrass stands, from rhizomes of various lengths planted at increasing depths, from burial operations during tillage, or from new plants originating from seed, the minimum dry weight of the plant falls to its lowest point at about the time two leaves have developed. With further leaf production and food manufacture the plant goes on to produce branches on both rhizomes and aerial shoots.

After the experimental results were obtained, Hakansson made extensive field observations to determine how many of the new rhizome branches formed during a season eventually would grow into new aerial shoots. A survey in late September in 15 untilled stubble fields in south and central Sweden revealed that from 20 to 100 percent of the tips of rhizomes had already appeared above the surface. Very few aerial shoots came from lateral rhizome buds. In another year a survey was made in mid-May on 18 fields, most of which had been fall-plowed but had received no spring tillage. The tips of all rhizomes examined had by this time produced aerial shoots with three or four leaves. Again, no lateral shoots (either rhizomes or tillers) were observed on any aerial shoot possessing two leaves or less. In these fields, a relatively large number of lateral rhizome buds had developed vertical shoots growing directly toward the surface or had already emerged and formed aerial shoots. This tendency to produce aerial shoots from lateral rhizome buds was most pronounced in those fields which had been plowed in autumn, but there was considerable variation from place to place in any given field as well as between fields.

In the United States, aerial shoots arising from rhizome pieces planted in the previous fall produced new rhizomes and tillers by mid-May. The rhizomes averaged 25 cm in length by the end of May. During the first 2 weeks in June the rhizomes grew 20 to 25 cm weekly. During the 3rd week average growth was 13 cm and a few of the rhizome apices had turned up to produce new aerial shoots. During the last week in June the average growth was 8 cm and most of the apices had turned up to produce aerial shoots; immediately behind each new shoot a new rhizome was extending horizontally.

The excavation in September of one entire plant from these experiments provided some interesting statistics about *A. repens*. The diameter of the rhizome spread was 3.3 m, 14 rhizomes from the original plant had grown to a total length of 135 m, 206 aerial shoots had been produced by the system, and 232 additional growing points were found on rhizomes in the soil. Thus, a total of 440 branches had arisen from the parent plant during one growing season (Raleigh, Flanagan, and Veatch 1962).

In the milder climate of Great Britain, a different pattern of growth was found for the clones growing there (Palmer and Sagar 1963). The aerial shoots were formed in autumn and progressed to the two- or three-leaf stage. Active growth of the shoots began in March and, by late March and early April, new rhizomes and tillers had been formed at the base of the shoots. Lateral buds on rhizomes produced in the previous season formed new horizontal branches in July and thereafter branched repeatedly. These lateral buds may have been activated earlier, however, if the terminal bud of the rhizome had been detached prematurely or if the apex had come up against a stone or other hard object. If the rhizome is severed from the parent plant, the terminal bud and several lateral buds respond by turning up to produce new aerial shoots. The terminal and lateral buds of the rhizomes were never seen to develop into aerial shoots while still attached to the parent plant, this lack of development being quite different from that seen in the clones studied in Sweden. Horizontal rhizome growth continued without influence from daylength or flowering until extension was terminated sometime between mid-August and mid-October by senescence of the shoots or unfavorable growing conditions. In Great Britain, rhizomes in pure stands of A. repens may live for several years. The aerial shoots form three primary tillers but normally only the two basal ones grow vigorously. After forming five aerial shoot leaves, a large number of the shoots produce inflorescences which emerge in June or July. In a mature stand of quackgrass only the original primary shoots flower. When there is more room and "clumping" is possible, flowering takes place on both primary and secondary shoots. In earlier experiments Palmer (1958), working in England, reported that all rhizomes grew horizontally beneath the soil in spring and summer, and the tips turned up in autumn to produce a few transitional leaves and generally one small aerial shoot leaf. The latter pierced the surface and growth of the shoot proceeded slowly in winter with two or three mature aerial leaves being present by spring. He found that seven lateral buds normally grew out from the base of the primary shoot in the spring. Three of these became tillers and the remainder became rhizomes. Plants growing in open communities (in contrast to mature stands of quackgrass) produced the secondary rhizomes and tillers in the first season. The rhizome growth seems to be independent of the flowering process.

This then summarizes the general growth and development of the whole plant, and a more detailed description of behavior and response of the individual structures is in order.

Growth of Rhizomes

The rhizomes of A. repens play important roles in the spread of the species and in its persistence in agricultural lands in the face of intensive cropping. As a result there have been more investigations on the rhizome and its behavior than on any other organ or any other portion of the plant's life history. The depth of the rhizomes has been studied in almost every area where the weed grows; and everywhere and on all soil types it has been shown that most rhizomes occur in the upper 10 to 15 cm. Hakansson (1967; 1968a,b) placed rhizomes of different lengths on the surface and at 10 different depths down to 30 cm in a sandy soil. He found that the greatest rate of emergence of shoots and the highest overall production came from plantings at the 2.5- to 7.5-cm level. The range of this optimum depth for regrowth of rhizomes became greater, however, as the longer rhizomes were planted. Only a few shoots emerged from 4- to 8-cm rhizome pieces buried below 15 cm, the rate of emergence for all sizes of rhizome pieces decreasing rapidly below this level. A few shoots emerged from 32-cm rhizome pieces buried at a 30-cm depth. Aerial shoots arising from rhizomes at great depth were much reduced in weight, and their subsequent rhizome production was decreased. This may be explained in part by the great delay in emergence, a delay which greatly reduced the total photosynthetic time available to the plant unit for the season. More aerial shoots came from deeper levels in heavy clay soils than in sandy soils. Kephart (1923), Raleigh, Flanagan, and Veatch (1962), and Vengris (1962) in the United States and Kraus (1912) in Germany obtained similar results, although the latter had some plants emerge from a 40-cm depth.

Rhizomes planted at the surface showed poor shoot production, with the original piece being broken down by the end of the season. The planted rhizome pieces showed good survival when buried anywhere in the top 10 cm of soil. In general, survival of rhizome material decreased below this depth. Large, plump rhizomes with much stored food are most likely to succeed in placing shoots above the soil so that they may eventually photosynthesize and return food assimilates to the underground organs for storage. Short rhizome pieces may exhaust their food supplies before new shoots can reach the surface but, also, a gradual deterioration process moves into the rhizomes from both cut ends and this also contributes to the more rapid demise of small pieces. Grummer (1963) had an interesting report of 30-percent production of aerial shoots from buds on rhizome pieces only 3 mm long.

The soil level at which the greatest concentration of rhizomes is located is an important consideration when the depth of plowing or tillage is to be selected for control purposes. Hakansson pointed out that this may vary with different clones, soil types, and the season. He planted rhizomes in a sandy soil at 5, 10, 15, and 20 cm in autumn and left them without disturbance for a year. The growth for all of the pieces placed at 5 and 10 cm occurred entirely in the upper 10 cm of the soil. Eighty-seven percent of the growth from pieces planted more deeply was also in the upper 10 cm of the soil. Thus, as the aerial shoot produced from a rhizome at any depth begins the process of secondary rhizome production from basal buds (on the shoot), it tends to concentrate these secondary rhizomes in the top 5 to 10 cm of soil. This new production, coupled with the decreased survival of rhizomes deeply placed (as discussed previously), helps to explain the concentration of rhizome material at the surface even by the end of 1 year.

Most workers are aware that tips and lateral buds can be sources of new growth following disturbances as well as during the beginning of new growth cycles. It may not be so well known, however, that research in different areas of the world has shown that the subterranean parts of the aerial shoots have a remarkable ability to survive in times of environmental stress, these parts perhaps being as important as the new rhizomes in the regeneration of new plant units.

The common practices of mowing or otherwise removing the aerial portions of perennial weeds or of disturbing the entire plant are aimed at the reduction of carbohydrates and the weakening of the whole plant unit. When such treatments are properly timed, there may be a withdrawal of food reserves from the subterranean parts to provide energy and building blocks for the replacement of the aerial structures.

There are reports from many places in the world recommending that the rhizomes be torn or that the tops be mowed when leaves have reached a certain length. Other recommendations base control of this weed on intervals of elapsed time between disturbances to give the most efficient reduction in weed infestation with the minimum expenditure of farm labor and power. All techniques must be adapted to the climate, the soil type, the intensity of the infestation and other factors. Two examples from the work of Hakansson (1969a) and Turner (1969) will suffice to illustrate the respective responses of the weed in Sweden and England.

Hakansson buried (turned under) plant units to a depth of 7.5 cm at 1-, 2-, and 4-week intervals for 8 weeks. Burials once each week produced almost no shoots of any size above the surface between reburials. With burial at 2-week intervals, shoots varying from 3 to 9 cm were visible before the three first reburials, but none were seen for the fourth or fifth. At the 4-week interval, shoots 3 to 21 cm long were present before reburials. Many of the buried shoots died, but buds sprouted at their bases to give new shoots and rhizomes. A count of shoot production in the spring following these experiments showed that even the least effective burials had reduced the shoot production by 80 percent. If the quackgrass was broken into small pieces before burial the survival rate was much reduced. Turner (1969) studied the effect of clipping on carbohydrate loss in the rhizomes. Removing the foliage to the ground level each 10 to 14 days caused the maximum loss of carbohydrates, and an application of nitrate, an ammonium salt or urea sometimes hastened the loss. Clipping tops at 14- or 21-day intervals permitted little production of new rhizomes. These results are in agreement with those data obtained from investigations by Dexter (1936), Johnson and Dexter (1939), and Hakansson (1969a). The latter pointed out that clipping at 2 cm above the soil surface was much less effective in stunting the quackgrass than cutting at the soil surface. There was generally produced an increase in tillering which varied with the type of cutting regime.

The conditions which influence the direction of growth of the rhizomes were studied by Palmer (1962) in England. Rhizomes grown experimentally in a horticultural loam mix, vermiculite, or river sand tended to angle downward at about 5° to 10° from the horizontal. Opaque screens placed over the top did not influence growth. The orientation of the rhizomes seemed to be geotropically determined, and there was no evidence for a physiological depth-sensing mechanism. Observations in natural sites showed a similar pattern but with the plants inclining slightly less downward. Grummer (1963) in Germany confirmed a slight downward inclination of rhizome growth if the soil was loose, but found a tendency for the rhizomes to grow closer to the horizontal plane in more compact soils. In loose alluvial soils some rhizomes descended to 40 cm. Where litter was heavy on the surface rhizomes traveled freely on the soil surface. The behavior of clones in other areas of the world and on other soil types has not been investigated.

Plants Grown from Seeds

The period of vulnerability for plant units derived from seeds already has been briefly described, but there are some other characteristics of the developing seedlings which are worthy of special mention. In early stages, plants originating from seeds are really no different than are those originating from rhizomes. The new plants of *A. repens* from seeds are weaker, the leaves are more slender, and a longer time is required to produce the first tillers and rhizomes than when the plant units have been derived from a rhizome piece (Hakansson 1969a, 1970). When newly formed seeds were planted, only a few germinated in autumn and the largest number emerged in early May. The appearance of new tillers and rhizomes on these plants was 3 to 5 weeks later than it was for plants derived from rhizome pieces. Although the time between developmental stages was increased, the general pattern of development was again similar to that described above. Plants from seeds tended to have a lower number of rhizome branches and, therefore, had a lower number of new aerial shoots produced by rhizome branches by the time the growing season was over in autumn.

Seedling plants were buried (turned under) at 14, 20, 28, 32, 35, and 41 days after seed-sowing. They were placed at 1.5, 2.5 and 4 cm in the soil. The plants could not regenerate and produce new shoots until the 41-day period, when the primary plants had reached a stage at which rhizomes several centimeters long had developed.

From experiments which were carried into a second season Hakansson observed that the first aerial shoots to be seen in spring came from shoots which had overwintered without significant injury, from buds either underground or very near the surface, or from the remaining vertical parts of old shoots which had suffered some deterioration the season before or during winter. The new shoots also originated, of course, from the rhizome apices which had turned upward and from lateral buds of horizontal rhizomes. All of these shoot types produced new tillers and rhizomes at their basal buds in May, and subsequent development paralleled that of the first season (see also E. Williams 1970a,b).

Effects of Light and Temperature

Hakansson (1969d) grew plants from rhizomes at temperatures of 2°, 5°, 10°, 15°, 20°, 25°, 30°, 35°, and 40° C. The photoperiod was 20 hours and the light intensity was 6,000 lux. At 2° C 50 percent of the rhizomes had emerged from a 2-cm depth in 125 days. The most rapid emergence and increase in dry weight of shoots and total plants was at 25° to 30° C. The rate of leaf development was greatest at 20° to 25° C, and this was also the optimum temperature for the development of subterranean parts of the plant. In general, higher temperatures favored the tops and lower temperatures favored underground parts. A temperature of 35° C was harmful to the plants and there was no development at 40° C.

Plants were tested for the effects of shade by being grown in the greenhouse at 13° to 21° C (night and day averages) and 11° to 17° C in an 18-hour photoperiod. Light intensities were 6,000 to 15,000 lux at noon, and this was reduced to 2,500 to 6,500 lux in shaded areas. Shading caused reduced dry matter production at both temperatures. Underground parts were retarded more severely than aboveground parts by the shading.

From the results of a 2-year experiment in outdoor plots, E. Williams (1970a) in England compared growth of *A. repens* plants under shade (46 percent of daylight) with those grown in the open for an entire season. In the 2nd year some plots were shaded for 2 months in the early part of the season (mid-May to mid-July); others were shaded for a similar period in late season. Continuous shading in both years cut the total dry weight of rhizomes produced by 50 percent. The rhizomes which were produced in shade had a 25-percent dry matter content as compared with 30 percent for rhizomes grown in full light. Shading early in the season reduced shoot numbers at midsummer, but when shade was removed the shoot numbers were comparable to unshaded plants by fall. By mid-July, early shading had significantly decreased shoot weights and reduced the dry weight of rhizomes by 50 percent. By the end of the season, both early and late shading had produced a small reduction in both shoot and rhizome dry weight, but plants kept in shade for the entire first half of the season (early shading) had made a remarkable recovery.

In short photoperiods *A. repens* is decumbent, producing more shoots than in long photoperiods (E. Williams, 1971a). Long photoperiods favor rhizome formation more than do short photoperiods (see also McIntyre 1967 and Hakansson 1969d). Optimum conditions for rhizome growth include bright light, long photoperiods, and high nitrogen levels. The effects of nitrogen and photoperiod are additive.

Effects of Soil Moisture

Hakansson and Jonsson (1970) studied the responses of rhizomes planted in a sandy loam (with about 10-percent humus content) at different soil moisture levels. Rhizomes normally have a moisture content of 60 to 80 percent, and in these experiments they lost viability at about 16 percent. Rhizomes survived for 4 weeks with little injury when the soil moisture content was held at 4 percent of the dry weight of the soil. This was found to be near the wilting point of this soil in a test in which sunflower plants were grown. At 2-percent soil moisture there was heavy damage to rhizomes in 1 week and all were killed at 4 weeks. No visible development of buds took place except for very slight extension of buds at 4-percent soil moisture. At 6 percent there was shoot emergence which was very much delayed, and at 8 percent shoot development was normal.

It was learned during these experiments that old rhizomes which were taken in November and which were dark colored and somewhat decayed were much more susceptible to drying than were the new, vigorous rhizomes of the season. When the two types of rhizomes were taken in midsummer, however, the older ones were in good condition and resisted drying, whereas the young, immature ones dried out easily.

Dexter (1937 and 1942) found large seasonal variations in the susceptibility of the rhizomes to drying. When the moisture content was reduced to 40 percent of total fresh weight in July the rhizomes were killed; when reduced to 30 percent in very early spring or late fall, a large number survived; and when moisture content was reduced to 20 percent in November and December, some rhizomes survived. He also found that rhizomes from fertilized sod sprouted more readily and were more susceptible to drying and to attack by microorganisms. We may conclude, therefore, that tillage operations performed in very early spring may find rhizomes most susceptible to desiccation.

Bud Dormancy

The presence of a "late spring" or "summer" bud dormancy as reported by Johnson and Buchholtz (1961, 1962) from the northern United States is not in agreement with the behavior of buds in the researches of Hakansson (1967) in Sweden and Turner

(1966) in England. This question will be difficult to resolve until similar clonal plant materials are used in experiments carried out at the same or a series of locations with careful control of fertility and moisture levels as well as other factors in the environment. Johnson and Buchholtz planted rhizome samples from an established sod on agar in flasks each week for a period of 2 years to study bud activity. They were unable to surface-sterilize the rhizomes without injury, and as a result used only agar in the medium because contamination was serious when nutrients were added. Nevertheless they reported a decreased activity from rhizome buds from mid-April to June, a "dormant" condition in June, and then increased activity for the rest of the summer. Hakansson, on the other hand, transplanted rhizomes into soil at several times during the season and found no bud dormancy to hinder the early emergence of new shoots. In all cases at least 60 percent of the buds developed shoots at least 0.5 cm in length. McIntyre (1965, 1967) in Canada has shown that high levels of nitrogen can prevent the onset of dormancy in rhizome buds, and Palmer (1958) has shown that buds with an apparent winter dormancy were easily stimulated to growth by higher fertility levels and increased light intensity. Meyer and Buchholtz (1963) reported optimum temperatures of 20° to 25° C for rhizome bud growth but also noted activity at a high temperature of 35° C. These and observations in Sweden make it appear that summer dormancy caused by high temperatures is not a control problem. Hakansson suggested that the "late-spring" dormancy referred to above occurs at the period when many carbohydrate and nitrogen reserves have been used for spring growth; buds placed on a medium without nutrients at such a time may be inactive because they have exhausted their supplies of organic and inorganic materials needed for growth. A similar situation may arise after persistent tillage to exhaust the rhizomes. Buds which are inactive for these reasons are probably not dormant.

There is a dormancy, or an inhibition, in quackgrass buds which does govern growth. This is the apical dominance of the bud at the rhizome apex. Such dormancy is readily broken by cutting the rhizome but it is quickly reestablished by the first bud to develop on the new rhizome piece.

Seed Production

Flowering in *A. repens* is very responsive to soil fertility, soil moisture, light intensity, crowding by other plants, and so forth. The first flower initials may be seen on meristems from mid-April to early May in the northern hemisphere, with flowers appearing about the second half of June. Once flowering has begun the process may be continuous through September. Certain primary shoots in a stand remain vegetative for an entire season. In England, only primary shoots flower in mature stands of quackgrass, but when the weed grows in clumps in more open areas both primary and secondary shoots produce inflorescences.

The species is often reported to be self-sterile, but careful tests with enclosed spikes, both in greenhouse and field, have shown that there is some self-fertilization and a small amount of seed can be produced without cross-pollination. Palmer and Sagar (1963) have suggested a possible explanation for the general view that the plant is self-sterile. Over a period of several years, very large areas may be covered with plants having their origin in the same clone as vegetative multiplication takes place. This is especially true in cultivated areas where broken rhizomes are moved some distance each year. If the plant tends to be self-sterile, the only seed which could form would be found near the margins of the clone where pollen is received from other nearby clones. Wind is the principal vector for cross-pollination.

Because of clonal isolation, poor growing conditions, and a lack of understanding of proper conditions for germination, the reports of seed production, dormancy, and viability are very mixed in the literature. Recent studies have resolved some of the conflicts, however, and it is now quite generally accepted that seed production can be high in one place and poor in an area which is not far distant yet in the same region. Viability may be high or low depending on growing conditions, and the seeds of the species have little dormancy at harvest. As with the small seeds of many species, the seeds of *A. repens* have a lower viability than do large, plump seeds (Kolk 1962). There is often uncertainty about the effects of removing the seedhulls before germination tests. In tests that gave poor germination of hulled seed, a closer examination showed that many seeds had been injured during hull removal. It may be of some advantage to remove the hulls if conditions are not advantageous for germination (E. Williams 1968).

The caryopses mature in late summer and early autumn. Most seeds will germinate as soon as they are mature if properly treated. Korsmo (1925, 1954) found immediate germination of 90 percent of mature seeds in Norway, found 30° C to be better than 20° C, and discovered that light is advantageous. It is now widely accepted that alternating temperatures of 30° C and 20° C (16 hours in darkness) give very good germination for most seeds. E. Williams (1971*b*) found no advantage of light in germination. When constant temperatures were used, the older seeds germinated best. In summary, it seems that the seeds of *A. repens* can germinate to some degree whenever circumstances are favorable. In parts of the United States and Great Britain some seeds germinate in autumn and may produce small aerial shoots. In Canada, Sweden, Russia, and other areas of lower temperature, most of the seeds germinate in spring. There is germination delay whenever seeds are buried too deeply, when moisture is lacking, when soils are waterlogged, or when temperatures are too low. Under these unfavorable circumstances, it is not correct to assume that the seeds are in a period of dormancy; for, in most cases, they will sprout readily when placed in favorable circumstances.

A careful study of seed production by *A. repens* in fields in England and Wales was made by Williams and Attwood (1971). More than 250 samples were taken, most of them being from winter and spring cereals but some also from 11 other crops and from a fallow area. Ninety-five percent of the samples had some viable seeds, one-third had over 15, one-third had six to 15, and one-third had less than five viable seeds per spike. The average of all lots was 13 viable seeds per spike.

Most seeds emerge from shallow soil depths. Korsmo found that 50 to 80 percent came from the top 1 cm of soil in Norway. None grew from 7 cm. In the United States, a few seeds planted in a sandy loam emerged from 10 cm, but most of them came from the top 5 cm.

Seeds of *A. repens* stored inside have been reported to survive 1 to 6 years. Viability gradually decreases with age. The strong sensitivity to (or the need for) alternating temperatures in germination disappears as the seeds become older. It now seems likely that most stored seeds will germinate at anytime if given the proper conditions.

In a final report on the long-term Duvel buried-seed experiment (Toole and Brown 1946), the survival of *A. repens* seeds was given as follows: 20 cm burial—21-percent germination at 1 year with no further germination; 50 cm burial—73-percent germination in 1st year and 1 percent at 2 years; 100 cm burial—67-percent germination in 1st year, 19 percent at 3rd year, and 2 percent at 10 years.

Bruns and Rasmussen (1958) carried out long-term

survival studies on seeds of weed species placed in flowing freshwater. About 3 percent of the *A. repens* seeds survived for 3 months and none survived for 27 months.

In a survey of seed-stock contamination with *A. repens* in United States, samples of oats were collected from the drill boxes of planting machines. Sixty of 268 samples contained quackgrass but only 22 of the lots contained viable seeds. This points up a problem in the interpretation of data on quackgrass seed, for although many unfilled caryopses look normal they will not germinate. Sterile florets terminate the end of each spikelet and, in some clones, all the spikelets break off as units. In others the individual florets break off and are distributed separately. Many of the florets found in the above survey contained ergot bodies. The weed seeds were also found in spring wheat seed stocks, and a germination test of 6-percent viability was recorded for all of the material found in the planter boxes. However, after the sterile florets had been blown out of the samples, germination was 49 percent (Raleigh, Flanagan, and Veatch 1962).

Nutrition

The knowledge which is most valuable to us if we are to control perennial weeds of all kinds is concerned with factors of the environment and the role of these factors in morphogenesis. For example, perennial grasses which spread by rhizomes and stolons often respond to changes in the environment by changing the ratio of aerial to underground structures produced, by flowering more or less readily, or by shifting bud dormancy as light intensity, temperature, moisture, photoperiod, or other conditions are altered. Quackgrass is no exception, and research on stimulus and response by buds on rhizomes and by the basal portions of aerial shoots has provided information which can be very helpful in controlling the species. There is general agreement across the world that the weight of new aerial shoots and underground parts produced in one season is increased as rhizomes of greater thickness or length (thus greater weight) are planted. When growing in pure stands, *A. repens* displays a tendency toward more vigorous growth if soil fertility is increased. Good nutrition is often more favorable to the production of dry matter by the aboveground plant parts than by the underground organs. The number of flowering culms is also greater on fertile sites and in areas where larger or longer rhizomes have been planted.

The most detailed studies on the morphogenetic effects of nutrition on the direction of growth in *A. repens* were reported by McIntyre (1965, 1967) in Canada. Plants were started from pieces of 1-year-old rhizomes and were held at 210, 10.5, 5.25, and 2.6 ppm of nitrogen in sand nutrient cultures. One of McIntyre's principal findings was that the nitrogen supply greatly influenced tillering and rhizome bud activity. At the lowest level of nitrogen, tiller emergence was almost completely suppressed. From this it follows that the production of secondary rhizomes, i.e., those arising from tillers, would also be reduced. At 2.6 and 10.5 ppm of nitrogen the rhizome buds ceased to grow after reaching a length of 3 to 6 mm, which is close to the size normally found on plants in the field. At a level of 210 ppm, however, the rhizome buds reaching back to the fifth and sixth nodes from the apex increased in size and, finally, all produced lateral branches several cm long. The high level of nitrogen in the medium had thus completely eliminated the onset of dormancy. After a refinement of technique it was shown that the bud in the axil of leaf one on the basal part of the shoot was very responsive. At a high nitrogen level of 315 ppm, most buds at this location produced tillers; whereas at a low level (2.1 ppm) the buds were transformed into rhizomes. Buds which had already contained rhizome initials could be induced to form tillers by increasing the nitrogen supply. High daytime temperatures of 27° C favored tiller production; low temperatures of 10° C favored rhizome production. Reducing the daylength from 18 to 9 hours promoted tillering and prevented rhizome production. Reduction of the light intensity from 4,000 to 2,000 ft-c seemed to promote the turning up of rhizomes to form new shoots, but did not otherwise alter the pattern of bud development. McIntyre pointed out that, in these experiments as well as in those of other workers, the conditions which increase or may be expected to increase the accumulation of carbohydrates will tend to promote rhizome development. Long days, low temperatures, and low nitrogen are among these conditions. Other aspects of nutrition are discussed in connection with bud dormancy and crop yield reductions (see also Dexter 1937, 1942; Hakansson 1967; Johnson and Buchholtz 1958; and E. Williams 1971a).

AGRICULTURAL IMPORTANCE

Figure 61 indicates that *A. repens* is a serious weed of cereals in England, the Soviet Union, and the United States, and of corn in the United States. It is a principal weed of cereals in Alaska, Canada, Finland, Norway, and Spain; sugar beets and fodder

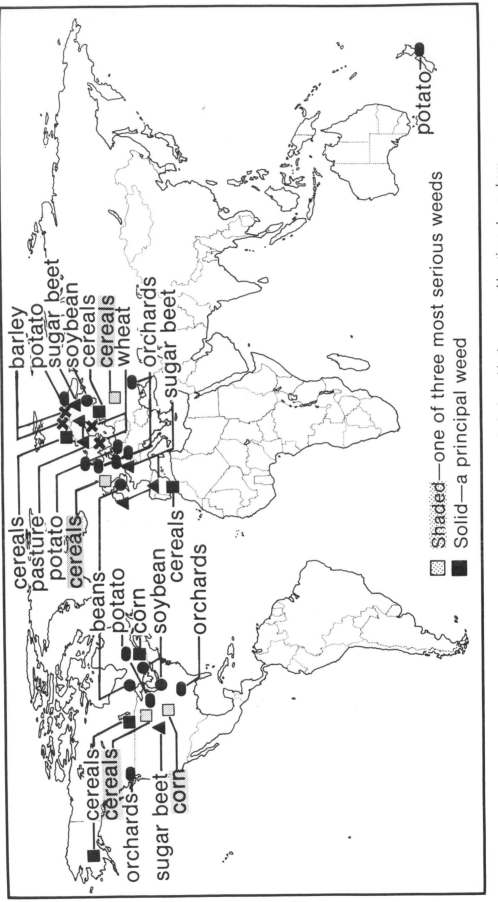

FIGURE 61. *Agropyron repens* (L.) Beauv. is a serious or principal weed in the crops and locations shown above.

beets in Belgium, Finland, Ireland, Spain, and the United States; potatoes in Belgium, Canada, Finland, New Zealand, Norway, and the United States; orchards in Belgium, Canada, Germany, Norway, the Soviet Union, and the United States; wheat in Finland and Norway; barley in Finland and Norway; vegetables in Canada, England, Ireland, Norway, and the United States; corn in Canada; beans in Canada and England; pastures in Finland and Sweden; and soybeans in Canada, Finland, and the United States. Although these facts are not shown on the map, *A. repens* is also a principal weed of alfalfa in Spain; coffee in New Guinea; tobacco in Canada; vineyards in the Soviet Union; and fruit in Canada and the Soviet Union. It is a weed of many other crops and countries in the temperate regions of the world.

There are persistent reports from farmers and research workers alike that "quackgrass land" may have strange effects on crops. It is claimed that poor stands of crop seedlings, symptoms of nutrient deficiency, and wilting are seen on soils with heavy infestations of quackgrass. Kommedahl, Kotheimer, and Bernardini (1959) showed that stands and dry weights of alfalfa, flax, wheat, oats, and barley were reduced on fields which had been previously infested with quackgrass. They also found that ground rhizomes added to soil or water extracts of these organs added to soil in the greenhouse inhibited crop growth. In later experiments Ohman and Kommedahl (1960) reported that the toxic substance was present in roots, rhizomes, stems, and leaves of both old and young plants, and that there was evidence of different levels of toxicity between plants belonging to different clones.

Welbank (1960) studied the effects of leachates of living quackgrass plants on the growth of wheat, barley, kale, and tomatoes, and also on the growth of seedlings of rape and turnips when they were placed on moist filter paper resting on the roots of a living plant of *A. repens*. No evidence was found of a secretion of toxic substances from the roots of the intact plant and, in some instances, a slightly better growth of the crop actually resulted. Soils in which rhizomes and roots of the weed had been incubated were sometimes strongly toxic, inhibiting or preventing the germination of seeds and early growth of rape seedlings. In later experiments (Welbank, 1963*a*) the production of toxins from decaying plant parts was studied in different soil types. Production of inhibitors by *A. repens* was no greater than that by *Agrostis tenuis*, another rhizomatous grass, or by *Lolium multiflorum* or *Medicago lupulina* which are frequently used as green manure crops. Previous history of the plant material influenced the amount of toxicity which later developed. For example, rhizomes which had decayed in waterlogged soils produced more toxins than those which had been killed by freezing or drying. Soil type had little influence on degree of toxicity. Aqueous extracts of quackgrass rhizomes inhibited root growth in seedlings of peas and wheat in the experiments of LeTourneau and Heggeness (1957). As the concentration was increased, wheat coleoptile growth was retarded and finally seed germination was inhibited altogether.

Obtaining evidence for a chemical interaction of living or dead quackgrass (or for any other plant) with the crop in the field always has been difficult because the reaction must finally be explored in a soil with the microflora present. Chemical compounds released from living or dead material may react with the soil solution and be altered or adsorbed or may enter a chain of reactions mediated by one or more soil organisms, finally to emerge with different chemical structures. Beclouding all of these issues is the possibility that large quantities of organic material—as from vigorous, perennial weeds—may initiate a burst of activity from soil microorganisms which themselves may release inhibitory or stimulatory compounds and change the ecology of the microsite for both crop and weed. Ohman and Kommedahl (1964) have estimated that 42 metric tons of fresh material and 8 metric tons of dry material may be present in and on each hectare of some fields. The extensive need for nitrogen and other nutrients by microorganisms during great periods of decay of organic matter may evoke temporary symptoms of nutrient deficiency in crops, and these periods of deficiency may very well be critical in the life of the developing plant. The variables and the complexities in such investigations are thus quite difficult to overcome. These problems of method, however, do not negate the mounting evidence across the world that certain species cannot tolerate one another and in many cases cannot follow one another in the same field. The fact of chemical interaction in the ecology of natural or cultivated sites must now be taken into consideration whether we understand the mechanism or not (Holm 1971).

With these considerations in the background, we may investigate other studies which shed further light on the relationship of *A. repens* to crop plants. Ohman and Kommedahl (1964) set out to explore more fully the role of nutrition and the differences between living and dead organs in these interactions. Alfalfa and oats, grown in a nutrient solution in which quackgrass rhizomes had been planted at the same time, showed no harmful effects from the weed. When the crops were planted in an established culture of *A. repens* they became severely stunted; yet bioassays showed no evidence of toxic substances being given off by the living, intact plants. The competition for nitrogen and other nutrients by older, but not younger, quackgrass seems to explain the harmful effect of the weed in this instance. During experiments in which alfalfa and oats were planted in pots and ground rhizomes of *A. repens* were added to the soil, the crop plants again displayed chlorosis and, when dried and weighed, a reduction in weight. There was no reduction in stand. The authors point out that field tillage may cause clumping rather than uniform distribution of rhizomes in the soil, and these foci of intensive decay activity may be part of the explanation for uneven stands of some crops. To explore the possibility that the reduced growth may have been due to a deficiency of nitrogen because of a change in the carbon:nitrogen ratio during decay of the rhizome material, Ohman and Kommedahl added several types of amendments to the soil before planting the crop. Cellulose, ground rhizomes, rhizome extracts, and the rhizome material after extraction all caused severe chlorosis and reduced dry weight in alfalfa. Adding ammonium nitrate to the pots completely overcame the effects of the soil amendments. It was further shown that when large amounts of quackgrass or crop plant residues were incorporated into the soil, phytotoxic extracts were obtained only when the soils were saturated and decomposition was anaerobic.

Bandeen and Buchholtz (1967) examined the competitive effects of *A. repens* on corn in the northern United States. The easy assumptions found in the literature that all plant competition in nature is for nutrients, light, and water do not hold very well for this crop-weed relationship. Quackgrass is widely recognized for its deleterious effects on corn, yet the morphology of the crop plant, the fact that it is planted in some areas about the time that the rhizome buds of *A. repens* are said to become dormant, and the usually favorable moisture supply in spring in the northern United States all make it improbable that competition for water and light are of much consequence. It was found in a 2-year experiment on a clay loam soil that a drastic competition occurred between the weed and crop, so that the height and yield of the corn were reduced and the time required for the crop to reach maturity was delayed. The weed (shoots and rhizomes) took up

110, 18, and 68 kilograms per hectare of nitrogen, phosphorus, and potassium annually, respectively, during the two seasons. Of this total, 55, 45, and 68 percent, respectively, were removed by the weed before mid-July. It was prior to this period that the corn made rapid growth and development and required a high level of nutrition. High levels of fertilization, however, did not overcome the effects of quackgrass competition. In later experiments Buchholtz (1968) used a split-root technique for the study of competition between corn and quackgrass. The experiments were made in a quackgrass sod area where corn plants made little progress even with the addition of a plentiful supply of water and nutrients. He found that the crop plants made much more satisfactory growth if nutrients were made avilable to a few of the attached corn roots by placing the roots in a separate plastic container of liquid nutrients located just beside each plant. Is it possible that metabolites from the quackgrass rhizomes may interfere with the process of nutrient uptake by corn roots? Is there enough rhizome material in the soil to be influential? Buchholtz estimated from field soil samples that some areas may have 4.5 to 6.8 metric tons dry weight of rhizomes in each hectare.

Kommedahl et al. (1970) also found that plant heights and seed yields of oats, corn, and soybeans were lower when grown on field plots which had been in quackgrass sod for 2 years, as compared with areas which had been in oats or in fallow for 2 years. The areas received ammonium nitrate at 280 kilograms per hectare and the crops responded differentially, depending on the previous use of the field. In general it may be said that the addition of nitrogen only partially corrected the reduced yields caused by previous quackgrass infestations. Wheat was grown on the quackgrass plots in a 2nd year of cropping, and the crop was reduced in height and yield as compared with wheat grown on areas previously planted in oats, corn, soybeans, or left fallow.

Efforts to control this perennial weed by mechanical means must necessarily vary with the area (Bylterud 1965, Lowe and Buchholtz 1951). Bylterud pointed out that in Scandinavia the crop harvest is often so late that little cultivation is possible in the autumn. Spring comes so quickly that planting must be done immediately for most of the crops which are grown in such a short season. Because the weather is cool and the soils are moist, there may not be much weakening of quackgrass tilled after harvest in the fall or before planting in spring. As labor becomes scarce and more expensive farmers tend to add more tractors and to raise only a monoculture of cereals.

Thus, *A. repens* tends to increase because it has good growing conditions year after year during the yellow-ripening stage of the grain and onward through harvest until the ground freezes. It is understandable that efforts have turned toward a combination of mechanical methods and herbicide treatments in recent years.

In drier areas and in regions where the cold is severe, such as in some areas of the United States and Canada, an alternative method is sometimes used when there is no crop in the field and when the weather of spring and fall allows time for tillage and drying of the rhizomes. In such places significant reductions in quackgrass infestations may also be realized through freezing injury. The quackgrass is torn up repeatedly to bring large numbers of rhizomes to the top where they dry out. Just before the ground freezes a final portion of the rhizome material is brought to the surface where it is killed by freezing and thawing during the winter.

Other workers have emphasized the value of weakening the quackgrass by shading (Palmer 1958, Cussans 1968). The experiment of E. Williams (1970*a*, 1971*a*) reminds us, however, that, although shading in the early season cut the dry weight of rhizomes by one-half, removal of the shade allowed an almost complete recovery of seasonal rhizome growth during the last half of the season. This may approximate the situation for quackgrass in cereals which open up late in the season, thereby allowing the quackgrass to resume vigorous growth.

In summary it should be pointed out that *A. repens* is still considered by some to be the worst perennial weed of the cooler parts of the northern temperate zone, and it is obvious that much work remains to be done on it. Herbicides have given successful control in a few crops infested with quackgrass but in some cases the perennial weed has been replaced by serious annual weeds. Historically, the intent of a fallow period to clear "quackgrass land" was based on the premise that rhizomes could be desiccated or the weed would become exhausted as a result of using up its reserves. Drying requires a few hours or a few days, and the susceptibility of the rhizomes to drying varies a great deal with the time of year (Dexter 1937, 1942). In many areas in the spring and fall when crops are absent from the field the soils and weather are moist and there is no drying.

Hakansson (1967; 1968*a,b;* 1969*a*) has shown that in Sweden there is a period in late spring, and also a period during regeneration from tillage operations, when the food reserves of the weed fall to a minimum level. It is at this moment that the next tillage operation is most effective for survival after tillage is low. There is no certainty, however, that the reserves of quackgrass change greatly during the year in other areas of the world. The investigations of Arny (1928) and Tildesley (1931) are of interest here.

We have discussed previously the fact that the small rhizome pieces are short-lived because both cut ends deteriorate. With this in mind, Fail (1954, 1959) explored a third alternative for control. He recognized as well the fact of apical dominance and the theoretical need to reduce all rhizome pieces to one bud segments so that they might escape the inhibiting influence of the terminal bud. If each bud then grows, it may be destroyed at the next cultivation. In experimental trials he was able to show that three to six rotary cultivations at 21-day intervals give complete eradication. In one experiment the combined lengths of all rhizome pieces in an area 30 square centimeters were 300, 95, 40, and 13 cm before each of four succeeding rotovations. After the fourth operation no rhizomes were alive. Hay (1962*a*) has confirmed these results for Canada. Rotary cultivation cannot be used on a large scale, but the idea is a good one and awaits the development of a practical means for chopping rhizomes into ever smaller segments in all of the kinds of crops and fields infested with *A. repens*.

In Tanzania, *A. repens* was chewed to nullify the effects of arrow poison, and the macerated material was applied to the wound. A liquid preparation from roots and rhizomes is used in some countries because of its diuretic properties. The diuresis is probably due to the glycolic acid content. The plant has many medicinal uses in Guatemala. Some rhizomes contain 11 percent mucilage.

Damage to barley by wheat bulb fly larvae has been reported in areas where the original infestation in the area was on *A. repens* in the fields (Shaw and White 1965).

It is an alternate host of the bromegrass mosaic virus (BMV) (Slykhuis 1967).

COMMON NAMES OF AGROPYRON REPENS

ARGENTINA
  grama menor
  gramón
CANADA
  couchgrass
  quack grass
  quickgrass
  quitchgrass

  scutchgrass
  twitchgrass
DENMARK
  almindelig kvik
ECUADOR
  grama
ENGLAND
  couch

quack grass
twitch-grass
FINLAND
  juolavhena
FRANCE
  chiendent rampant
  froment chiendent
GERMANY
gemeine Quecke
kriechende Quecke
Päde
Quecke
GREECE
  creeping couch-grass
ITALY
  caprinella
dente canino
NETHERLANDS
  kweek
  kweekgras
NEW ZEALAND
  couch
NORWAY
  kveke
SPAIN
  grama de Europa
SWEDEN
  kvickrot
  vitrot
UNITED STATES
  couch grass
  quackgrass

❧ 21 ❧

Anagallis arvensis L.

PRIMULACEAE, PRIMROSE FAMILY

Anagallis arvensis, a small but highly variable and adaptable winter or summer annual herb, is a native of Europe or the Mediterranean region. It has spread throughout the world as an unnoticed companion of European emigrations, probably in dirty crop seeds. It is primarily a weed of pulses, cereals, vegetables, and small oil seed crops. It is a weed of 22 crops in 39 countries. There are at least two subspecies of *A. arvensis*, subspecies *arvensis* (also known as subspecies *phoenicia*) and subspecies *foemina*, and many botanical varieties. Most are differentiated on the basis of flower color which can range from bright red through blue or white. The plant is very interesting physiologically because it has specific and rather complex flowering and germination requirements. Although small, the plant can germinate during cool weather and make early growth in reduced sunlight, quickly occupying open sites in cultivated fields. It has long been known as a poisonous plant if taken internally, especially for dogs, horses, and man, and has been reported to cause dermatitis in some individuals if the leaves or stems are handled. The poisonous principle is thought to be a saponin.

DESCRIPTION

This weed is a procumbent, ascending or sometimes erect, glabrous, *annual* herb (Figure 62); fibrous *roots*; *stems* quadrangular, weak, gland-dotted, diffusely branched from the base, 10 to 40 cm long; *leaves* opposite or in whorls of three, ovate to oval, 5 to 25 mm long, margins entire, obtuse to somewhat acute at tip, sessile to clasping at base,

both leaf surfaces glabrous, bottom dotted with black glands; *flowers* solitary, axillary on thin peduncles 1 to 5 cm long, erect in flower, curved down in fruit; *calyx* persistent, with five lanceolate, acuminate lobes 3 to 5 mm long; *corolla* deeply five-parted, 8 to 14 mm across, a little shorter than the calyx, glandular ciliate on the margins, scarlet or salmon, red, purple, pink, rarely blue or white in color, depending upon subspecies or varieties; *stamens* five, attached to base of corolla, clothed with numerous purple-fringed (villous) filaments; *fruit* a membranous globose capsule 3 to 5 mm across, the top falling off as a lid; *seed* about 1 mm long, three-angled, brown, finely pitted.

Other *Anagallis* species are also weeds. *A. arvensis* L. can be differentiated from *A. pumila* Sw. by the following characters: *A. arvensis* has opposite leaves and usually salmon, pink, red or blue flowers (rarely white), borne singly, whereas *A. pumila* has alternate leaves and white flowers, which often form in leafy racemes.

DISTRIBUTION

A. arvensis is a native of Europe and has spread throughout the world, mostly as a result of the movements of European emigrants. It was probably carried in dirty crop seed stocks and is now found as a weed in 39 countries, from Argentina in the southern hemisphere to Germany in the northern hemisphere (Figure 63). Although the plant occurs in the tropics, mostly at higher elevations, it grows best in temperate regions at low elevations. It is most

169

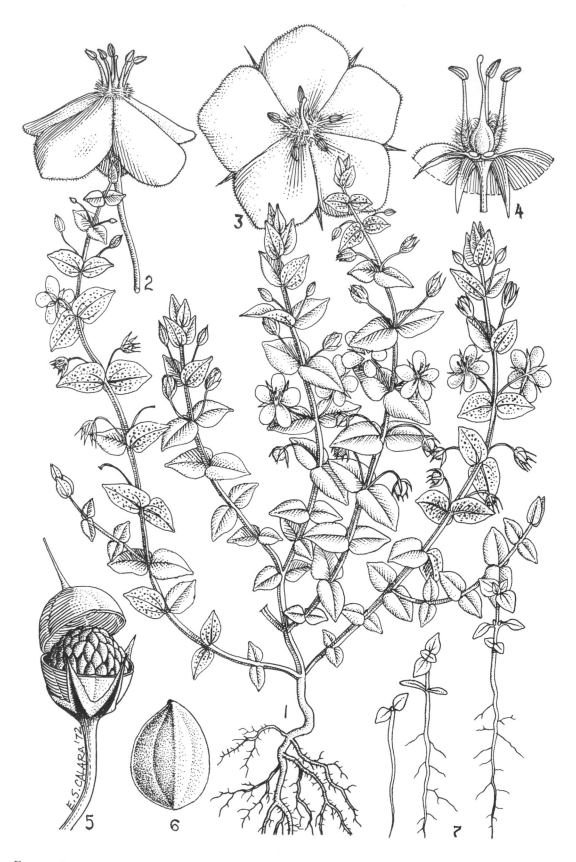

FIGURE 62. *Anagallis arvensis* L. *1*, habit; *2*, flower, side view; *3*, flower, top view; *4*, flower, petals removed to show hairy stamens below; *5*, capsule; *6*, seed; *7*, seedlings.

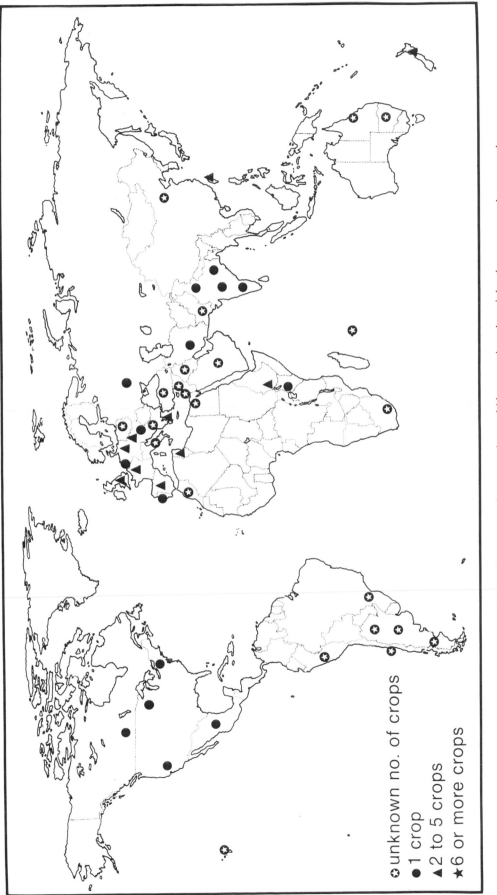

FIGURE 63. The distribution of *Anagallis arvensis* L. across the world in countries where it has been reported as a weed.

widely distributed in Europe, the Mediterranean, and Central Asia.

BIOLOGY

There are three major factors to consider in the biology of *A. arvensis*, namely: (1) the numerous forms and varieties which exist, depending upon growing conditions and genetic types, (2) its specific germination requirements, and (3) its specific flowering requirements and behavior.

Comprehensive reviews of *A. arvensis* relating to flower induction and behavior (Ballard 1969) and general biology (Misra 1969) have been completed.

A. arvensis is a branched, low-growing or nearly prostrate, annual herb which is extremely variable, both in regard to taxonomic varieties as well as in response to environmental conditions. For many years several subspecies or varieties have been used in physiological and genetic experiments. For this reason and because so many forms of the species exist, we feel that a list of some of the variant subgroups might be useful.

The species has been divided into at least two subspecies: subspecies *phoenicia* (Scop.) Schinz & Keller, which usually has a scarlet corolla; and subspecies *foemina* (Mill.) Schinz & Thell. with a blue corolla. Taylor (1955) proposed that subspecies *phoenicia* should be classified as subspecies *arvensis*. There are also several varieties and forms of subspecies *phoenicia* (*arvensis*?): variety *phoenicia* Scop., scarlet corolla; variety *carnea* Schrank, salmon or flesh-colored corolla; variety *pallida* Hook f., pale or white corolla; variety *lilacina* Alefeld, lilac corolla; and variety *coerulea* Gren. & Godr., blue corolla. Other varieties reported include *arvensis*, *latifolia*, *azurea*, *decipiens*, and *vinacea*. Chromosome numbers for some of the varieties in Israel are as follows: *arvensis*, n=20; *coerulea*, n=20; *foemina*, n=20, and *latifolia*, n=40. It is believed that *arvensis*, *coerulea*, and *foemina* are tetraploids and that *latifolia* is an octoploid (Kollmann and Feinbrun 1966). In Poland chromosome numbers were reported as 2n=40 for the following seven varieties: *arvensis*, *azurea*, *carnea*, *lilacina*, *pallida*, *decipiens*, and *vinacea* (Svĕřepová 1968). Apparently polyploidy has played a role in giving rise to the multiplicity of forms in this variable species.

A. arvensis grows in cultivated fields, gardens, lawns, meadows, pastures, waste places, and along roadsides. The specific name, *arvensis*, means "growing in cultivated land." It is found on a wide range of soils, mostly under disturbed conditions. It has been observed that *A. arvensis* grows frequently near coastal areas. The reasons for this have not been fully explored; however, Wacquant and Passama (1971) found that roots of the weed absorb more potassium and sodium and less calcium than other species grown under the same conditions.

The plant grows well inside crop fields as well as in the open along field borders and lightly shaded crop areas but is seldom found in dense shade. Light intensity has been considered a factor in its habitat preference. At 50 percent of full sunlight after 10 days of age, shoot length, leaf number, leaf area, and flower number were greater than with 100 percent or with 20 percent of full sunlight (Misra 1969). Greatest leaf area was obtained after 22 days at 100-percent sunlight. These data indicate that less light is required for maximum growth at early growth stages of the plant than at later stages. Only 50 to 68 percent of full sunlight is required to obtain maximum leaf production during the first 10 days. At this point little mutual shading occurs but, later, as the plants develop more leaves, mutual shading results, so that the plants now require more sunlight for maximum growth.

Misra (1969) studied shoot length, root length, number of nodes, number of leaves, and shoot dry weights of plants grown from seeds collected from lat 24° to 27° N in India. All of these factors differed significantly with site, highest values generally being recorded for plants grown from seeds collected at lat 25° to 26° N.

A summer annual in the colder climates, *A. anagallis* is usually a winter annual in warm regions. It propagates by seeds which are produced in large quantities. Yields as high as 900 seeds per plant have been observed in the field in England, whereas in the greenhouse as many as 250,000 seeds per plant have been reported (Ballard 1969). Seeds may remain viable in the soil for as long as 10 years (Champness, 1949). Harper (1960) reported 2,480 *A. arvensis* seeds per square meter in a soil with a history of 8 years of cultivated cropping and 1 year of ley before sampling.

Germination requirements of *A. arvensis* are so intricate that they guarantee the continued propagation of the species. All factors involved in breaking dormancy of the seeds are those which ensure that the plant will have best chance for survival and the best possible growing conditions. Also, the requirements for breaking dormancy are interrelated, so that no one factor should trigger a large germination response during unfavorable growing conditions.

The seeds have some dormancy. There is evidence that seeds may contain a water-soluble germination inhibitor, and different seed sources may vary greatly in degree of dormancy (K. Singh 1969). Germination increased when seeds were washed for 24 to 48 hours in running water, and a water extract of the seeds inhibited seed germination of test species, including *A. arvensis* (Misra 1969). Dry seeds contain very little inhibitor. Freshly collected seeds are often partially dormant, for periods as much as 1 year or more; for example, 1- and 13-month-old seeds gave 17 and 85 percent germination, respectively. Longer periods of dormancy result when seed formation and maturation occur during periods of low temperature. After maturation, factors which help to overcome dormancy include: increased temperatures; dry storage, at either higher or lower temperatures; increasing age of seed, washing of seeds, rainfall, light, and soaking in gibberellic acid. Once dormancy is broken, germination occurs in 6 or 7 days with seeds which have imbibed for 3 days. Misra (1969) proposed that *A. arvensis* seeds may contain two types of inhibitors, one which is water-soluble and another which is responsive to light. One year after harvest 100-percent germination can be obtained with light. Soaking seeds in gibberellic acid substituted for light requirement for germination.

Early work in Australia showed that germination in the blue-flowered subspecies *foemina* (subspecies *arvensis*?) was increased by red light but was inhibited by far-red light (Grant Lips and Ballard 1963). Misra (1969) obtained 11- to 15-percent germination in the blue-flowered variety *coerulea* with a single exposure to white, red, or far-red light 24 hours after the seeds were moistened. Seeds will not germinate in the dark. Red light penetrates soil to greater depth than does blue light, and far-red penetrates more than red. Possibly a sensitivity to red light might ensure better growing conditions below the soil surface for the young seedling, but seeds buried too deeply will remain dormant because their light requirements are not being met.

A. arvensis seeds germinate best at moderately low temperatures, with the optimum being about 7° to 20° C (Andersen 1968). Pandey (1968) reported an optimum at 15° C. Germination can occur below 2° to 5° C (Singh 1968*b*). This capacity to germinate at low temperatures is not unexpected for, as a winter annual or high elevation plant in subtropical or tropical areas or as a summer annual in the temperate region, the plant germinates and grows during low-temperature periods. In India the seeds germinate in October or November at the beginning of the cold

season (November to February) when air temperatures are 10° to 25° C. These temperatures also occur in early spring in most colder climates, and for this reason *A. arvensis* makes an early start in spring. Although it is a long-day plant, it will flower at 10- to 11-hour daylengths during low-temperature periods such as those which prevail during the cool seasons in the subtropics. Plants growing during these cool periods are small; and in India may flower with only four or five pairs of leaves as early as December. Vegetative growth and flowering continue until nearly April. Late seedlings, though stunted, produce flowers very early with the onset of longer days and warmer temperatures. Seeds are shed in April and May, remaining in the soil during the wet, warm summer season. As was pointed out before, higher temperatures and the leaching action of rainfall will reduce dormancy in the freshly matured seeds. Seeds mature in 6 to 8 weeks, but can germinate 3 weeks after they are formed in the capsule (Ballard 1969).

Ballard (1969) has pointed out that *A. arvensis* is a long-day plant. For flowering to occur, the plant can be sensitive to daylength at the cotyledonary leaf stage, but becomes less sensitive with age. Flower initiation depends upon two processes, one in leaves and the other in meristems. Exposure to 1 long day can cause flowering and, when conditions are suitable, anthesis occurs about 5 to 6 weeks after the first long day. Some seeds may be germinable 3 weeks later, though they usually mature in 6 to 8 weeks. If proper conditions for flower initiation do not continue, the plant ceases flowering and reverts to vegetative growth. If suitable conditions for induction of flowering do not take place, the plant can grow very large. For this reason some lower nodes may produce flowers, whereas later nodes may not. When seeds are vernalized, flowering may take place somewhat earlier. Critical daylength is generally about 12 hours at temperatures above 10° C, but at very low temperatures the plant will flower at very short daylengths. This explains its ability to flower and produce some seeds during winter in the tropics and subtropics.

Flowering dates for the plant have been reported as follows: Lebanon, February to September; United States, May to September (seed, June to October); Germany, mid-June to early August; England, April to August; and India, November to February.

As one might expect in such a variable plant, growing conditions can affect the plant in a number of ways. Ballard (1969) pointed out that leaf and stem size and other plant characters are greatly altered by environmental conditions. In India increasing nitrogen fertilization from 0 through 44, 65, 112, and 132 kilograms per hectare increased the number of capsules per plant, seeds per capsule, seed weights, and germination of freshly harvested seed. The rate of 132 kilograms per hectare of nitrogen seemed to bring about increased dormancy (Misra 1969).

The plant is killed by prolonged freezing temperatures but it can withstand nine daily half-hour exposures to -10° C (K. Singh 1968b).

The plant has sometimes been called the "poor man's weatherglass" because the flowers begin to close as bad weather approaches.

AGRICULTURAL IMPORTANCE

A. arvensis is a weed in 22 world crops, mostly of cereals, pulses, oil seeds, and vegetables. It is a principal weed of sugar beets in France, wheat in India and Iran, flax in the United States, and potatoes in India. It is a common weed of tobacco in India, of wheat in Portugal, of peanuts, rice, and sweet potatoes in Taiwan, and of wheat, oats, and barley in Greece. It is reported also as a weed in beans, corn, vegetables, vineyards, carrots, linseed flax, sugarcane, peas, onions, and upland rice.

The ability of A. arvensis to compete with crop plants has been questioned, for it is a small weed and appears not to be able to compete seriously with larger plants. Tripathi (1968a) compared the competition of A. arvensis, Asphodelus tenuifolius, Euphorbia dracunuloides, and Trichodesma indicum in wheat and gram (Phaseolus aureus). Of the four species Anagallis arvensis was least competitive in the two crops. Welbank (1963b) found that, of 16 weed species studied, A. arvensis was least competitive in kale and wheat. Sinapis arvensis was most competitive, followed by Polygonum convolvulus and Alopecurus myosuroides. These two studies indicate that Anagallis arvensis is not very competitive; however, experiments by Bornkamm (1963) may point out the nature of competition of this weed. Sinapis alba, Bromus secalinus, Agrostemma githago, and Anagallis arvensis were grown alone and in all combinations, and growth rates of all weeds were recorded. Using the values for each weed grown alone as a "growth optimum" for that weed, he found that one species could reduce the growth optimum of another if the two were grown together. He proposed that speed of invasion is a very important factor in determining dominance among species. In earlier experiments Bornkamm (1961) grew A. arvensis var. azurea in pots at 15, 30, 60, or 100 percent of full sunlight. Yield decreased as sunlight intensity decreased, but the relative growth of A. arvensis increased with decreasing light when grown with Sinapis alba or Agrostemma githago.

The ability of Anagallis arvensis to compete at lower light intensities, to make early growth in spring, and to develop local populations may be important in its behavior as a weed. A. arvensis germinates during cool weather and makes early growth before many other plants begin to grow. It can, therefore, invade open ground early in the season. As discussed previously, it can make good seedling growth with only 50 to 68 percent of full sunlight. Therefore, it may be most competitive during the early growth stages of crops, a period during which crop losses can occur if weed competition is serious enough. Varieties or subspecies may vary in their ability to compete with crops. Out of 27 principal weeds of the Catania plain in Spain, Viola (1947) ranked variety coerulea third and variety phoenicia fourth in importance.

A. arvensis has long been blamed for cases of poisoning in man and dogs, and especially in horses if taken internally. However, Kingsbury (1964) reported that poisoning cases are difficult to authenticate. He tells of work in Australia where, under certain conditions, sheep fed with the weed at about 2 percent of their body weight were killed in 2 days but that, at other times, no toxicity was demonstrated. Symptoms included depression, anorexia, and diarrhea. Internal hemorrhaging of hearts, rumens, and kidneys resulted. It has been blamed for the deaths of calves in Pennsylvania (United States).

The plants are known to have a sharp, bitter taste. Georgia (1938) reported that cattle reject the plant in pastures, thereby allowing it to increase because of selective grazing. According to Arnold (1944) the plant may not be of great importance as a poisonous plant for man because of its bitter taste and limited acceptability. He did report that some individuals may develop dermatitis from handling the leaves or stems. Watt and Breyer-Brandwijk (1932) reported that the plant roots contain a saponin, cyclamin, which is known to be toxic in other plants and causes gastroenteritis in dogs and horses.

A. arvensis is sometimes used in herbal medicine (Watt and Breyer-Brandwijk 1932).

COMMON NAMES
OF ANAGALLIS ARVENSIS

AUSTRALIA
scarlet pimpernel

CHILE
pimpinela azul
pimpinela escarlata

DENMARK
rød arve

EASTERN AFRICA
pimpernel

EGYPT
'ain el-gamal

ENGLAND
scarlet pimpernel

FRANCE
morgeline
morgeline d'été
mouron des champs
mouron rouge

GERMANY
Acker-Gauchheil
Feld-Gauchheil

GREECE
pimpernel

HAWAII
poisonous pimpernel

INDIA
biliputi (Punjabi)
krishnaneel
pimpernel

IRAN
pimpernel

IRAQ
poor man's weather-
glass
rmaimeeneh
scarlet pimpernel

ITALY
bellichina
mordi-gallina

LEBANON
adhan el far el nabti
lubbayn
scarlet pimpernel
zaghilah

MAURITIUS
mouron
pimpernel

MOROCCO
mouron des champs

NORWAY
rødarve

PAKISTAN
bili booti

PERU
pilpis

PORTUGAL
murrião
murrião vermelho

SOUTH AFRICA
pimpernel
scarlet pimpernel

SPAIN
murajes

SWEDEN
rödarv

TAIWAN
hwo-jin-gu

TUNISIA
mouron des champs

TURKEY
tarla farekulagi

UNITED STATES
common pimpernel
poison chickweed
poisonweed
poor man's weather-
glass
red chickweed
scarlet pimpernel
shepherd's clock
wink-a-peep

YUGOSLAVIA
vidovcia

❧ 22 ❧

Argemone mexicana L.

PAPAVERACEAE, POPPY FAMILY

Argemone mexicana, a herbaceous annual plant, is easily recognized by its yellow flowers, bluish green color, and the waxy marbling of the leaves. Before flowering, the plant could easily be mistaken for a thistle. It belongs to a family of about 450 species which are found mostly in the subtropics and the temperate regions of the northern hemisphere. *A. mexicana*, however, is native to tropical America and is now pantropical. It is not an aggressive competitor of cultivated crops but has long been recognized as a plant which is highly toxic to poultry, sheep, cattle, horses, and to human beings as well. It is a weed of 15 different crops in 30 countries.

DESCRIPTION

A. mexicana is a smooth, thistlelike, prickly, *annual* herb (Figure 64), 30 to 100 cm tall, with yellowish sap, on a firm taproot; *stem* pithy with scattered prickles, smooth to slightly pubescent; *leaves* alternate, *lower leaves* crowded in a rosette, petioled, higher ones sessile, somewhat clasping the stem, variable in shape, usually unevenly margined, covered with powdery bloom, producing a bluish green color, 6 to 20 cm long, 3 to 8 cm wide, the prickles scattered along the margins and on the undersurface; *flowers* sessile or on a short pedicel; *calyx lobes* with acute horn below the apex, few prickled; *petals* obovate, bright yellow, pale yellow, or creamy white, 1.5 to 5 cm wide; *stamens* numerous; *stigma* dark red, three- to six-lobed; *capsule* 2.5 to 5 cm long, 2 cm in diameter, crowned with the persistent style, with rounded ribs, separated by grooves, prickly, dehiscing from the apex to one-third of its length; *seeds* globular, 1.5 to 2 mm in diameter, with a fine network of veins (reticulations), black-brown with a prominent hilum.

Its yellow sap, its prickly lobed leaves, the dark red and three- to six-lobed stigma, and its reticulated seeds are the distinguishing characteristics of this species.

DISTRIBUTION AND BIOLOGY

A. mexicana is adapted to a very wide range of habitats. In Mauritius it is most plentiful in the subhumid regions, whereas in Puerto Rico it is most prevalent in the semiarid northern regions. Though often seen at sea level, it is a weed at an elevation of 2,900 m in Tanzania. It is found on many soil types, on cultivated ground, in pastures, in fence rows, on stony ridges, along roads, in waste places, and on bare soil. It seems to prefer moist sites, growing freely along rivers in Australia and following sandy riverbeds in southern Africa.

Much of the seed crop falls near the parent plant, sometimes resulting in almost pure stands along roadsides and in waste places. In crops such as sugarcane, however, the weed is well adapted and grows in association with other annuals and perennials.

The world distribution of *Argemone mexicana* is shown in Figure 65.

The species is propagated by seeds, and the physiology of seed production and germination in the field varies throughout the world. Mauritius re-

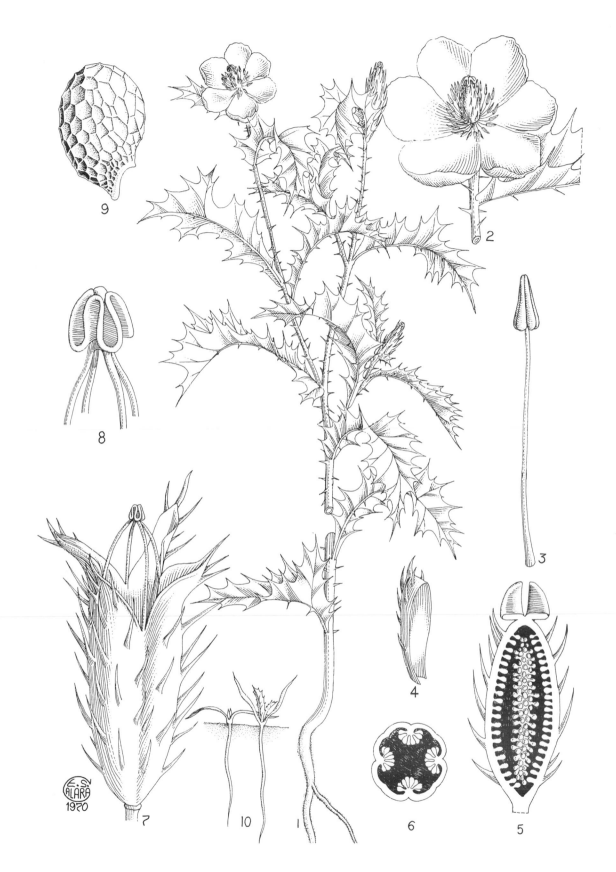

FIGURE 64. *Argemone mexicana* L. *1*, habit; *2*, flower; *3*, stamen; *4*, sepal; *5*, ovary, vertical section; *6*, ovary, cross section; *7*, fruit, dehisced; *8*, stigma, attached to style; *9*, seed; *10*, seedlings.

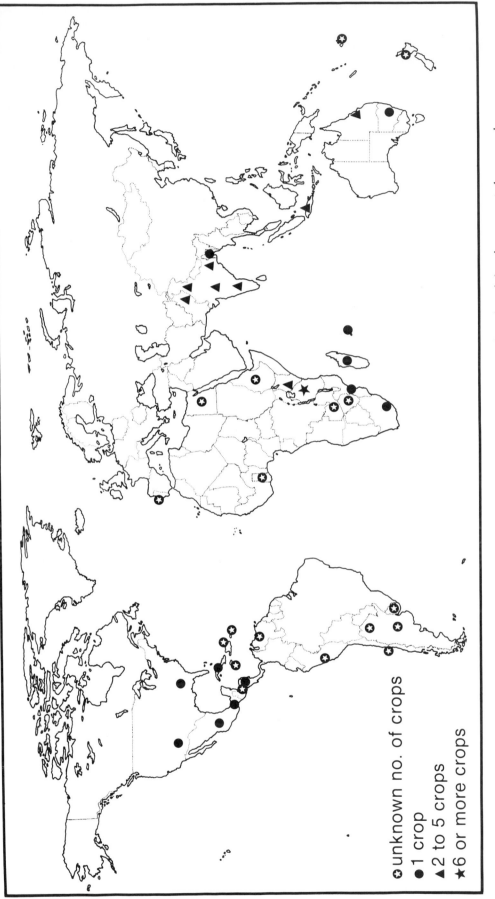

○ unknown no. of crops
● 1 crop
▲ 2 to 5 crops
★ 6 or more crops

FIGURE 65. The distribution of *Argemone mexicana* L. across the world in countries where it has been reported as a weed.

ports the greatest seed production with an average of 60 to 90 capsules per plant with 300 to 400 of the black, oily seeds in each capsule. In this region there is one generation per year. The seeds in the soil remain dormant in the hottest summer weather and begin to germinate in the late summer with the advent of cooler weather. The full development of the plant comes in winter and seeds are shed continuosly during this season (Rochecouste and Vaughan 1960). In eastern Africa, on the other hand, germination is continuous, even in the driest season. In Australia (New South Wales) the weed is present and green all the year round.

The seeds are dormant at harvest. Laboratory tests in several areas have shown that germination will proceed if the proper treatment is applied a few weeks or months after harvest. Seed treated with dilute sulfuric acid for 30 minutes showed 10-percent germination. A storage temperature of 4° to 32° C for 24 days gave 80-percent germination as did a high temperature treatment, under moist conditions, for 24 hours.

Although the plant is able to produce 6 to 9 metric tons of fresh weight per hectare when growing undisturbed, it is not one of the most aggressive weed species when growing on cultivated land. It is perhaps more widely known for its toxicity to all animals including human beings. The plant is a near relative of the opium poppy and many medicinal properties have been attributed to the yellow sap and the oil of the seeds. Two alkaloids, berberine and protopine, have been isolated from the sap.

In some grazing areas the plants are selectively avoided, so that not many deaths occur among large animals. It is reported that sheep will die of starvation rather than eat green plants of this species. However, chafed hay containing the weed causes a violent cholic in horses, and seeds mixed with feed grains often poison poultry.

In some areas the plant is attacked by a bacterial wilt, *Xanthomonas papavericola*, which periodically checks the spread of the plant.

AGRICULTURAL IMPORTANCE

A. mexicana is a principal weed of beans and corn in Tanzania, cereals in Australia and India, cotton in Nicaragua, potatoes in India, tobacco in Argentina and Puerto Rico, and wheat in Pakistan. It is also a weed of beans, coffee, peanuts, and sorghum in Tan-

zania; sugarcane in Australia, India, Mauritius, and South Africa; corn, beans, and cereals in eastern Africa; sisal in Madagascar; tobacco in the Philippines; vegetables in India; cotton in Morocco; and wheat in India.

It is a noxious weed in all parts of Queensland, Australia. The seed is a contaminant of alfalfa seed in Argentina.

The herbicides now available can give fairly good control of seedlings if applied early in the season, but the seeds continue to germinate and those which emerge later in the season become fully developed and a problem at harvest.

COMMON NAMES
OF ARGEMONE MEXICANA

ARGENTINA
 cardo amarillo
 cardo blanco
 cardo santo
 Mexican prickly
  poppy
AUSTRALIA
 Mexican poppy
BURMA
 khy-ya
CHILE
 cardo blanco
 cardo santo
HONDURAS
 cardo santo
INDIA
 agara (Oriya)
 bharband
 brahamadandi
 bramandandu
 kantakusama (Oriya)
 katelisatyanasi (Punjabi)
 Mexican poppy
 prickly poppy
 satyonasi
INDONESIA
 droedjoe
 tjelangkringan
  (Javanese)
JAMAICA
 Mexican poppy
 Mexican thistle

prickly poppy
yellow poppy
MAURITIUS
 chardon du pays
 Mexican poppy
 prickly poppy
MEXICO
 chicalote
NETHERLANDS
 stekelpapaver
NICARAGUA
 cardo santo
PAKISTAN
 kanderi
 kundiari
 sialkanta
PERU
 cardo santo
PUERTO RICO
 cardo santo
RHODESIA
 umjelemani (Ndebele)
SOUTH AFRICA
 Mexican poppy
TANZANIA
 Mexican poppy
UNITED STATES
 Mexican prickly
  poppy
URUGUAY
 cardo santo
VENEZUELA
 cardo santo

❦ 23 ❧

Axonopus compressus (Sw.) Beauv.

POACEAE (also GRAMINEAE), GRASS FAMILY

Axonopus compressus, a creeping, perennial grass, is native to tropical America. It may be recognized by the blunt, rounded tips of its leaves and a flower which has two branches at the apex. It is sometimes confused with *Paspalum conjugatum*. It is mat-forming and thus can squeeze out almost all other species. It is most serious in perennial plantation crops. In appropriate areas it is a valuable forage or lawn grass. It is reported to be a weed of 13 crops in 27 countries.

DESCRIPTION

A. compressus is a creeping, stoloniferous, *perennial* grass (Figure 66); *flowering culms* erect, 15 to 16 cm high, solid, laterally compressed, the stolons strongly branched, rooting at each node; *leaf sheaths* strongly compressed, finely hairy along the outer margin, otherwise smooth, the nodes densely pubescent, *ligule* very short, fringed with short hairs; *blade* lanceolate, flat, 4 to 15 cm long, 2.5 to 15 mm wide, with broadly rounded base and blunt apex, often fringed with hairs; *inflorescence* with slender peduncles, two to four, seldom eight, developing successively, the secondary and succeeding inflorescences remaining hidden inside the sheath but ultimately projecting beyond the sheath (long-exserted); *peduncle* smooth, bearing at its apex two slender, one-sided spikes, usually 5 to 8 cm long, often with a third below them, rarely a fourth; *spikelets* oblong, rather acute, 2 to 2.5 mm long, 1 to 1.25 mm wide, pale green or tinged with purple, solitary on alternate sides of rachis and forming two

rows, ciliate on the margins; lower glume absent, the upper as long as the spikelet with two nerves near each margin; *anthers* yellowish white or slightly tinged with purple; grain (caryopsis) yellowish brown, about 1.25 mm long.

It is commonly distinguished by the short ligule which is fringed with short hairs and by the inflorescences which develop successively. The secondary and succeeding inflorescences remain hidden inside the sheath but ultimately become long-exserted.

BIOLOGY AND DISTRIBUTION

Figure 67 indicates that *A. compressus* is troublesome as a weed mainly along the equator. It is seldom reported to be a weed problem in the subtropics or the warm temperate regions, and is even considered to be a valuable pasture grass for some soils in those areas. It needs sun but will tolerate shade. It is found in lawns and, indeed, is a valuable lawn cover for some areas. It is found in gardens, waysides, and waste places.

The species grows best if the soil is rich, the ground is moist, and the climate is slightly humid so that the plant is not subjected to extreme periods of low humidity. It is found in the tropical rain forest areas of Fiji, at all elevations in Ceylon, and under the shade of mature rubber trees in Southeast Asia. Like most plants, it performs best in fertile areas but also grows remarkably well on poor sandy soils if moisture is present. In such places it is more prosperous than *Cynodon dactylon*, although the latter is

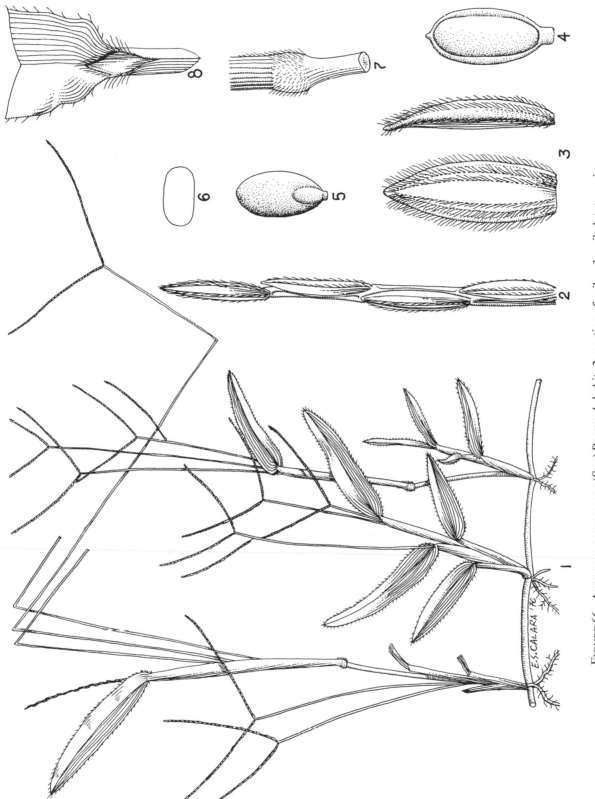

Figure 66. *Axonopus compressus* (Sw.) Beauv. *1*, habit; *2*, portion of spike; *3*, spikelet, two views; *4*, floret; *5*, grain; *6*, grain, cross section; *7*, portion of culm with node; *8*, ligule.

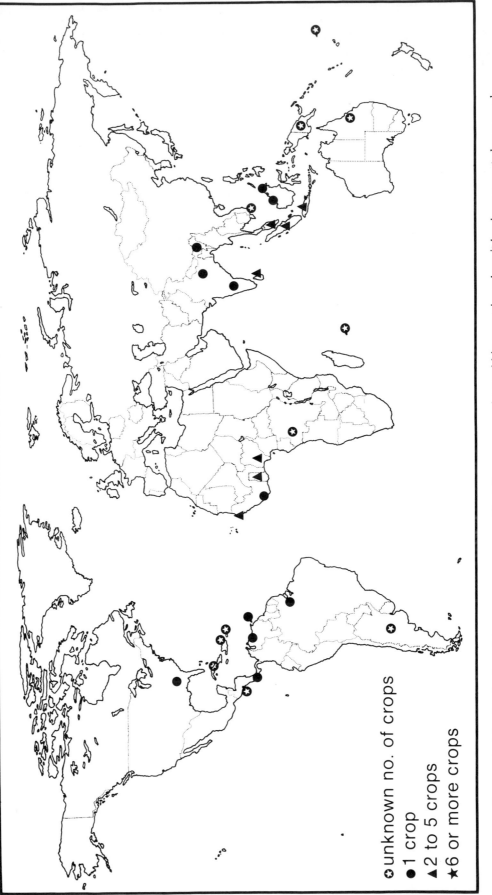

FIGURE 67. The distribution of *Axonopus compressus* (Sw.) Beauv. across the world in countries where it has been reported as a weed.

more tolerant of drier sites and will survive longer under drought conditions. As compared with most other tropical and subtropical grasses, however, it is considered very useful for it can tolerate poor drainage and even seasonal flooding and can make a valuable contribution to fodder resources in dry seasons when pasture may be scarce. It grows over a wide range of soil pH and likes a compact soil. In loose soils stock trampling tends to improve the growth of *Axonopus compressus*.

The plant can reproduce by seeds or by vegetative parts. To plant an area such as a lawn or pasture, one may spread chopped stolon segments on the surface and work them in. Under favorable conditions the plant may flower continuously, although in the northern extremes of its range flowering begins in early spring and continues until first frost. The majority of the stem material is above the ground and probably cannot survive temperatures below -12° C. *Cynodon dactylon*, however, can survive such a temperature because it has a much more extensive underground system which is protected from the cold. *Axonopus compressus* greens up better than *Cynodon dactylon* in a mild winter and thus can be grazed earlier in the spring and later in the fall.

A young plant of this perennial grass begins growth in a circular patch. If it has no serious competition the patch may reach a size of 0.6 to 1 m in diameter in one season. It often crowds out all other weeds and grasses. The weed is quite easily destroyed by plowing and cultivation.

AGRICULTURAL IMPORTANCE

This species is most troublesome in plantation crops which are grown very near the equator in humid regions (see Figure 68). It is one of the three most serious weeds of citrus, coconut, and oil palm in Malaysia and of rubber in Brazil, Ceylon, and Malaysia. It is a principal weed of cacao, citrus, and oil palm in Ghana; of coffee in Costa Rica; of rubber in Ghana and Sabah; of oil palms and rubber in western Africa; of sugarcane in India; of oil palms and tea in Indonesia; and of dry-land crops in Sarawak. It is also a weed of cacao in Malaysia and Venezuela, corn in Ghana and Indonesia, oil palms

and rice in Nigeria, peanuts in India and Indonesia, rice and rubber in Indonesia, and sugarcane in Trinidad.

The species is excellent for pastures and for lawn turf in appropriate areas. It withstands close grazing, has good food value, and its adaptability to extremes of soil type and moisture conditions makes it valuable forage. The most important pasturage for stock in Ceylon is under coconut trees, and here *A. compressus* becomes the dominant, mat-forming grass. It is a major pasture grass under coconut trees across the world (Plucknett 1973). The palatability varies with the ecotype or variety and the conditions under which it is grown. In most areas it is readily eaten by cattle, but in some places the animals will not touch it until they are hungry and then they thrive on it. It is used in lawns in hot areas because it makes its best growth in the warm season and tends to be dormant during cool periods.

A. compressus is an alternate host of *Rhizoctonia solani* Kuhn (anon. 1960*a*).

COMMON NAMES
OF AXONOPUS COMPRESSUS

ARGENTINA
  pasto chato
  pasto jesuita
  zacate amargo
AUSTRALIA
  carpet grass
BANGLADESH
  shial kata
CEYLON
  potu-tana (Sinhalese)
  sappu pul (Tamil)
DOMINICAN REPUBLIC
  cardo santo
EL SALVADOR
  grama
FIJI
  American carpet grass
INDONESIA
  djoekoet pahit

  djukut pait (Sundanese)
  roempoet pait (Indonesian)
NIGERIA
  carpet grass
PHILIPPINES
  carpet grass
PUERTO RICO
  tropical carpet grass
SOUTH AFRICA
  carpet grass
THAILAND
  ya-hep
TRINIDAD
  carpet grass
  flat grass
  savanna grass
VENEZUELA
  baracoa

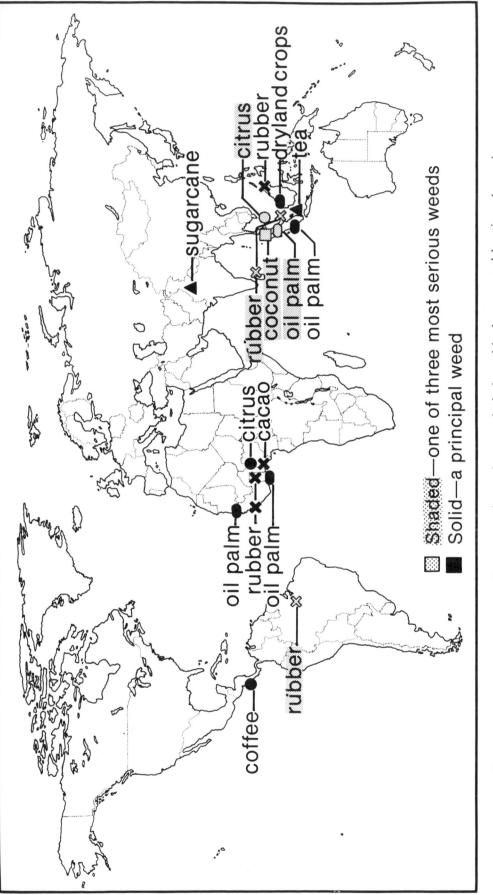

FIGURE 68. *Axonopus compressus* (Sw.) Beauv. is a serious or principal weed in the crops and locations shown above.

❧ 24 ❧

Bidens pilosa L.

ASTERACEAE (also COMPOSITAE), ASTER FAMILY

Bidens pilosa is an annual which originated in tropical America but is now spread throughout the warm regions of the world. The weed is easily recognized by the elongated burlike fruits that bear recurved or hooked bristles and that have played an important part in its spread. It is troublesome in both field and plantation crops and is reported to be a weed of 31 crops in more than 40 countries. The leaves are sometimes used for human food in either fresh or dry form.

DESCRIPTION

B. pilosa is an *annual*, erect herb (Figure 69), growing to 150 cm; *stems* and branches marked with parallel lines or ridges, smooth, green or with brown stripes, types found in some regions may have small, inconspicuous, white hairs on the stem; *leaves* opposite, petioled, pinnate, (arising on opposite sides of the midrib of the leaf) usually with three (sometimes five) ovate, acute leaflets, the upper leaflet usually largest, up to 9 cm long, 3 cm broad, margins sharply serrate, sparsely hairy to smooth on both sides, the upper and lower surfaces may have different green colors; *inflorescence* a capitulum (congested head of flowers), yellow, terminal, 7 mm in diameter, on peduncles 5 cm long; outer *involucral bracts* (ring of bracts at the base of the flower head) oblong or more or less spoon-shaped, ciliate, shorter than the florets; *fruit* (achene) blackish, about 11 mm long, narrow, ribbed, sparsely bristled to smooth; *pappus* a ring of awns (two to four) with recurved barbs, 3 mm long.

DISTRIBUTION AND BIOLOGY

The generic name *Bidens* refers to the barbs of the fruit, suggesting "two-toothed." This weed belongs to a group which is often spoken of as the beggar ticks, stick tights, or Spanish needles. The plant is in the Asteraceae (Compositae), the largest family of flowering plants, and one which contains many of the weed species which are widely distributed in the world. The spread and colonization of areas by these species can be attributed in part to very effective pollination arrangements and to special adaptations which allow the distribution of its fruits by workers, animals, wind, and water. Several *Bidens* species are found rather far north in the temperate zone but *B. pilosa* prefers the warm regions. It can usually be seen at all seasons in the tropics but it grows most actively in the warmer and wetter parts of the seasons (Figure 70).

The weed is found in gardens, on cultivated land, in open waste places, and along roadsides. It mixes easily with annuals and perennials in different types of plant communities. It is reported in crops at increasing elevations in several countries but the final altitude at which it can grow depends, of course, on the climate of the region.

The plant is easily recognized in the out-of-doors by its collection of black, barbed fruits radiating in all directions from a common receptacle. Very young plants have strap-shaped cotyledons and purple-tinged hypocotyls. Single plants have yielded 3,000 to 6,000 seeds. Many of the seeds germinate readily at maturity. This makes it possible to have three to

185

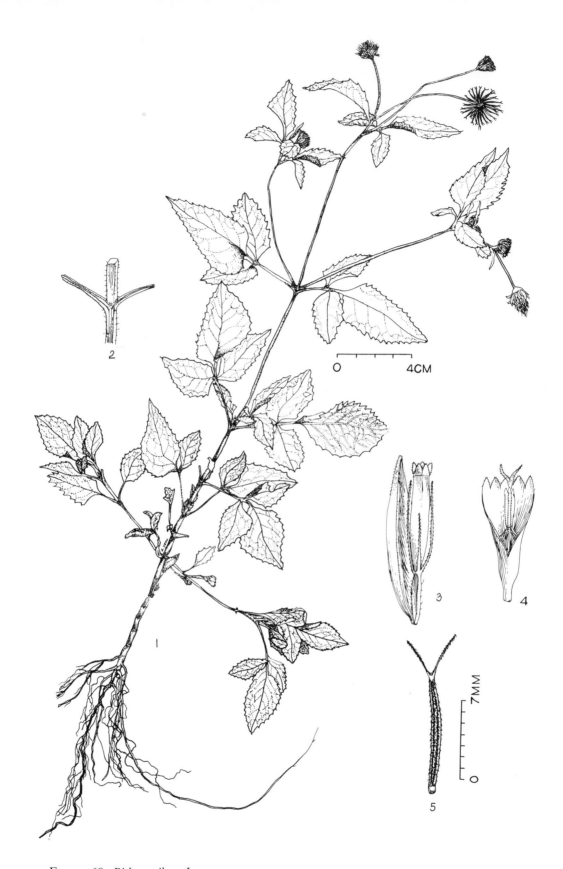

FIGURE 69. *Bidens pilosa* L.
  Above: *1*, habit; *2*, portion of stem; *3*, ray flower; *4*, disk flower, partly opened; *5*, achene.
  Opposite: Seedlings.

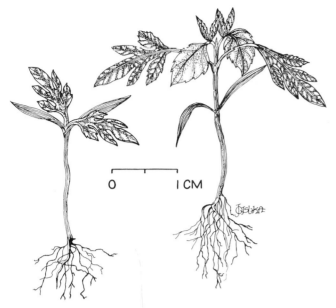

four generations per year in some areas. Light and good aeration are required for germination. In one experimental test 35 percent of the seeds germinated 1 week after harvest, whereas in another there was 60-percent germination. Seeds which are 3 to 5 years old may give 80-percent germination (Rochecouste and Vaughan 1959).

AGRICULTURAL IMPORTANCE

Figure 71 shows *B. pilosa* as a principal weed of sugarcane in Brazil and Mexico; corn in Mexico and Mozambique; coffee in Brazil and Mexico; tea in India; cotton in Peru and Swaziland; potatoes in Colombia, Mexico, and Mozambique; and citrus in Mexico and Venezuela. It is a serious weed of sorghum in Hawaii, and a principal weed of vegetables in Brazil and Venezuela and of bananas and beans in Mexico. In addition, it is a common weed of coffee and pyrethrum in Tanzania; corn in Ghana and Hawaii; cotton in Mozambique; tea in Mozambique and Taiwan; sugarcane in Argentina, Mauritius, Taiwan, and Venezuela; beans in Brazil; cassava in Ghana and Taiwan; peanuts and vegetables in Ghana; pineapples in the Philippines and Swaziland; pastures in Australia; and coconuts in Trinidad. It is also found in oil palms, papayas, rice, rubber, alfalfa, and tobacco.

From Figures 70 and 71 it may be seen that Latin America and eastern Africa have reported the most serious infestations of *B. pilosa*. It is believed by some to be one of the most important annual weeds of eastern Africa. In plantation crops in which herbicides have been used to remove perennial grasses, this weed often returns to become dominant. In tea in southern India, the advent of the southwest monsoon signals a heavy growth of annual weeds from July to September with the dominant weed in many places being *B. pilosa*. Following *B. pilosa*, another annual, *Ageratum conyzoides*, begins to take over.

The species is an introduced weed of South Africa and is one of the first herbs to come up after the spring rains in Natal. The Zulus and the Indians eat it when no more palatable herbs are available. The leaves are boiled with a small amount of water and eaten alone or with cornmeal (Shanley and Lewis 1969). An analysis of nitrogen, protein, and amino acid content of 11 crops and herbs (among them were three common vegetable crops—*Colocasia antiquorum*, *Curcurbita pepo*, and *Ipomoea batatas*; a common weed—*Chenopodium album*; three members of the genus *Amaranthus*; and *Bidens pilosa*) showed *Bidens pilosa* to be one of the least nutritious.

Bidens pilosa is an alternate host of the fungi *Cassytha filliformis* L. (Raabe 1965), *Cercospora megalopotamica* Speg. (Stevens 1925), and *Uromyces bidenticola* (P. Henn.) Arth. (anon. 1960a); of the nematodes *Meloidogyne* sp. (Raabe, unpublished; see footnote, *Cyperus rotundus*, "agricultural importance") and *Paratylenchus minuta* Linford (Linford, Oliveira, and Ishii 1949); and of the viruses which cause spotted wilt (Sakimura 1937), ground rosette (Adams 1967), and aster yellows (Namba and Mitchell, unpublished; see footnote, *Cynodon dactylon*, "agricultural importance").

Other fungi found on *Bidens pilosa* in Mauritius were *Cercospora bidentis* Tharp., *Entyloma bidentis* P. Henn., and *Uromyces bidenticola* (P. Henn.) Arth. (Rochecouste and Vaughan 1959).

COMMON NAMES
OF BIDENS PILOSA

ANGOLA
  olokosso
ARGENTINA
  amor seco
  espina de erizo
  picón
AUSTRALIA
  cobbler's pegs
BARBADOS
  Spanish needle
BRAZIL
  picão
  picão preto

CHILE
  amor seco
  asta de cabra
  cacho de cabra
COLOMBIA
  cadillo
  masquia
  papunga chipaca
DOMINICAN REPUBLIC
  margarita silvestre
  romerillo
EASTERN AFRICA
  black jack

HAWAII
  beggar-ticks
  Spanish needle
INDIA
  cobbler's peg
  dipmal (Marathi)
  phutium (Gujarati)
INDONESIA
  adjeran harenga (Sun-
    danese)
  djaringan ketul
    (Javanese)
JAMAICA
  Spanish needle
JAPAN
  ko-sendangusa
KENYA
  black jack

MAURITIUS
  herbe villebague
MEXICO
  acetillo
  amor seco
  hierba amarilla
  rosilla
NATAL
  black jack
NEW ZEALAND
  cobbler's pegs
PANAMA
  arponcito
  cadillo
  sirvulaca
PERU
  amor seco

  cadillo
  pega-pega
  perca
PHILIPPINES
  dadayem (Ivatan)
  nguad (Igorot)
  pisau-pisau
  puriket (Bontoc)
PUERTO RICO
  margarita
  margarita silvestre
  romerillo
RHODESIA
  nyamaradza (Shona)
SOUTH AFRICA
  black jack
  gewone knapsekerel

TAIWAN
  hsien-feng-tsau
THAILAND
  yah koen-jam khao
TRINIDAD
  Spanish needle
  railway daisy
UNITED STATES
  hairy beggar-ticks
URUGUAY
  amor seco
VENEZUELA
  cadillo rocero
WEST INDIES
  railway daisy
ZAMBIA
  black jack

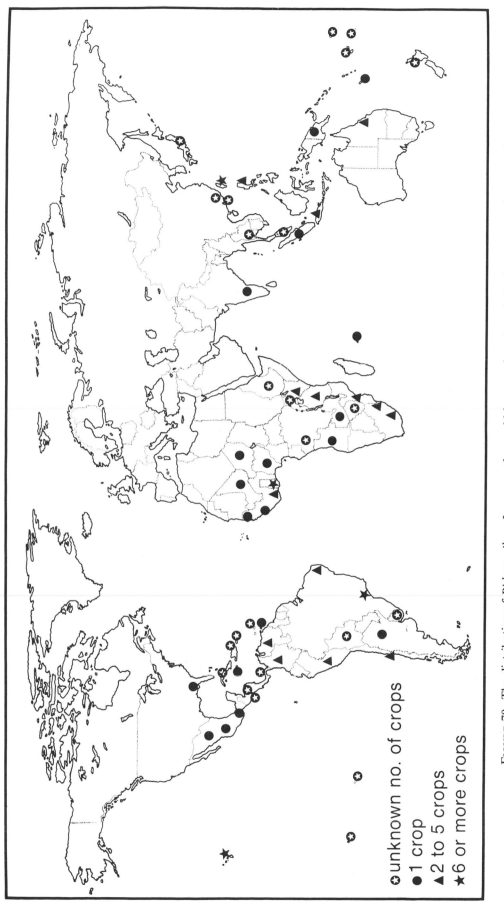

FIGURE 70. The distribution of *Bidens pilosa* L. across the world in countries where it has been reported as a weed.

unknown no. of crops
● 1 crop
▲ 2 to 5 crops
★ 6 or more crops

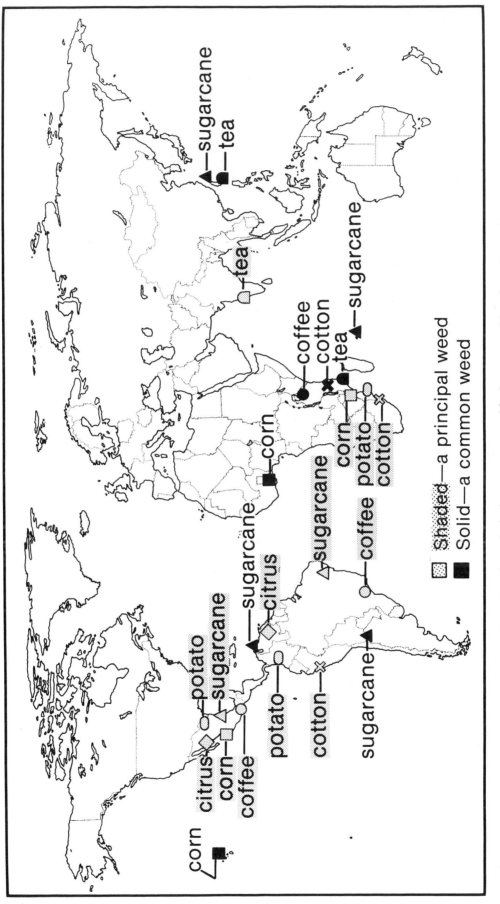

FIGURE 71. *Bidens pilosa* is a principal or common weed in the crops and locations shown above.

❧ 25 ❧

Brachiaria mutica (Forsk.) Stapf (= *Panicum purpurascens* Raddi)

POACEAE (also GRAMINEAE), GRASS FAMILY

Brachiaria mutica is a coarse, trailing, perennial grass that roots at the nodes. It is native to tropical Africa and is now widely distributed both as a weed and as a fodder grass in most humid tropical and subtropical areas. Thirty-four countries report that it is a weed in 23 crops. It is a valuable fodder grass both for soilage and for pasture in many parts of the world because it is able to withstand prolonged droughts, as well as being able to grow luxuriantly in swampy, nearly waterlogged lowlands.

DESCRIPTION

B. mutica is a stout, spreading, *perennial* grass (Figure 72) with culms growing up to 2.5 m and stolons reaching out to 5 m; *roots* fibrous, sometimes spreading from short rhizomelike stolons; *culms* decumbent to ascending, rooting at the base, stoloniferous, nodes densely covered with long soft hairs (villous); *sheaths* villous or smooth on upper portion with densely hairy (pubescent) collar; *ligule* short, long-ciliate; *blades* 10 to 30 cm long, 5 to 15 mm wide, flat, smooth; *panicle* 15 to 30 cm long, densely flowered branches somewhat separated, subracemose, ascending to spreading, 2.5 to 10 cm long; *spikelets* subsessile, 3 to 5 mm long, elliptic, five-nerved, glabrous, occasionally with light purple tinge; *grain* minutely transversely wrinkled. The most important distinguishing characteristics of this plant are the long vigorous stolons and the hairy nodes.

DISTRIBUTION AND AGRICULTURAL IMPORTANCE

B. mutica is mainly a weed of wet areas although it is adaptable to a very wide range of moisture conditions. It will flourish in wet valley bottoms, swamps, poorly drained and seasonally flooded land; yet it is drought-resistant. It responds readily to both irrigation and high soil fertility conditions. It is somewhat shade-tolerant and for that reason can become very serious in both tree and field plantation crops. It can be especially valuable as a fodder resource in dry seasons when it is needed most. It is very susceptible to frost and thus cannot move into the cool regions of the temperate zone (Figure 73). It grows in wet areas at all elevations in Ceylon (Senaratna 1956).

It frequents the edges of watercourses and often enters sugarcane and other crops by this pathway. It can take root on banks, encroach into small streams and canals, and end by clogging and impeding water flow and traffic. Under high moisture conditions a dense, pure stand may develop into a stolon mat 1 meter in depth.

Of all the cultivated fodders it is perhaps the only one that can grow well in brackish water. Like many tropical grasses with a stool-forming or trailing habit, *B. mutica* will not persist under continual close grazing and trampling. It is best when grazed in rotation and may need cutting and mowing at intervals to remove the coarse growth of canes and stolons left by stock. The protein content decreases and dry

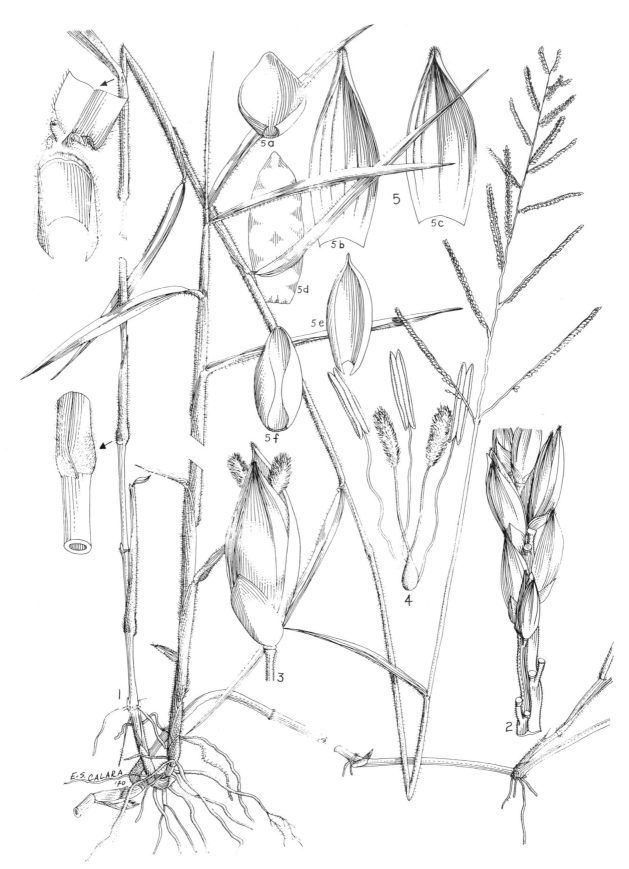

FIGURE 72. *Brachiaria mutica* (Forsk.) Stapf. *1*, habit, showing detail of sheath and node (center) and ligule (top); *2*, portion of inflorescence; *3*, spikelet; *4*, flower; *5 a–f*, bracts.

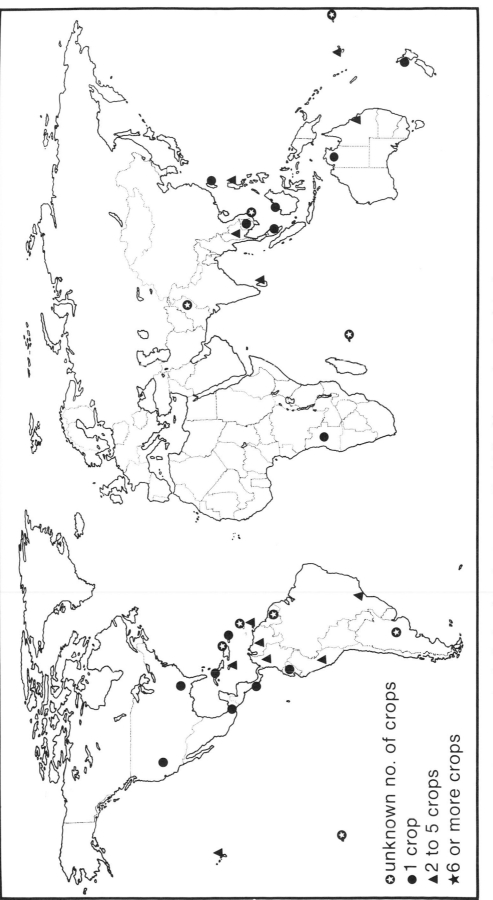

○ unknown no. of crops
● 1 crop
▲ 2 to 5 crops
★ 6 or more crops

FIGURE 73. The distribution of *Brachiaria mutica* (Forsk.) Stapf in countries where it has been reported as a weed.

weight goes up as the interval between cuttings increases.

A major disadvantage of the species is that it may invade cultivated land to become a serious pest. Figure 74 depicts it as a serious weed in citrus in Thailand and a principal weed in sugarcane in Jamaica, Peru and Puerto Rico; citrus and tobacco in Trinidad; rice in Australia and Fiji; pineapples and sugarcane in the Philippines; and oil palms in Colombia. It is a common weed of sugarcane in Hawaii and rice in Ceylon, Peru, and the Philippines.

Although not shown on the map, it is a serious weed in bananas in Fiji and the West Indies; in orchards in Thailand; and a principal weed in coconuts in Trinidad. It is also present in sugarcane in Angola, Australia, Brazil, Colombia, and Trinidad; rice in Cambodia, Colombia, New Zealand and Thailand; bananas in Brazil and Jamaica; rubber in Ceylon and Malaysia; citrus in Jamaica and the United States; taro and papayas in Hawaii; corn in the Philippines; coffee in Fiji; beans in Brazil; cacao in Venezuela; cotton, linseed, and safflower in Australia; and pastures in Ecuador and Thailand.

B. mutica rarely sets seeds. It is usually propagated by stolon cuttings which sprout and take root rapidly, sending out long creeping stolons which readily root at the nodes. Because the spread or persistence of the weed is dependent upon the presence of some vegetative material, good land preparation to kill existing vegetation coupled with use of herbicides to prevent spread from field borders can help to control it in cultivated lands.

B. mutica is an alternate host of *Cassytha filliformis* L. (Raabe 1965), *Helminthosporium* sp. (Parris 1936), *H. sacchari* (B. de Haan) Butl. (Raabe, unpublished; see footnote, *Cyperus rotundus,* "agricultural importance"), *Pythium arrhenomanes* Drechs., *P. rostratum* Butl. (Sideris 1931a), *P. artotrogus* (Mont.) By. (Sideris 1932), *Sclerospora graminicola* Schroet. (Raabe, unpublished; see footnote, *Cyperus rotundus,* "agricultural importance"), *Uromyces leptodermus* Syd. (Stevens 1925), *Piricularia oryzae* (McIntosh 1951; Paje, Exconde, and Raymundo 1964), *Chirothrips mexicanus* Cwfd. (Sakimura 1937), and of the rice leafhopper *Thaia oryzivora* (Leeuwangh and Leuamsang 1967).

COMMON NAMES
OF BRACHIARIA MUTICA

ARGENTINA
  caprim
AUSTRALIA
  panicum grass
  Para grass
BRAZIL
  capim de Angola
CAMBODIA
  smao barang
CEYLON
  diya-tana (Sinhalese)
  tanni-pul (Tamil)
COLOMBIA
  pasto Para
CUBA
  hierba bruja
DOMINICAN REPUBLIC
  yerba páez
FIJI
  Mauritius grass
  Para grass
HAWAII
  California grass
  Para grass
  tall panicum
INDONESIA
  djukut malela (Sundanese)
  kalandjana (Javanese)
  roempoet inggris
  rumput malela (Indonesian)
  soeket babanggalaan
JAMAICA
  Para grass
MAURITIUS
  Para grass
MEXICO
  zacuti Para
PERU
  gramalote
PHILIPPINES
  Para grass
PUERTO RICO
  malojillo
SURINAM
  Para grass
THAILAND
  yah koen
TRINIDAD
  Para grass
UNITED STATES
  Para grass
VENEZUELA
  hierba del Para

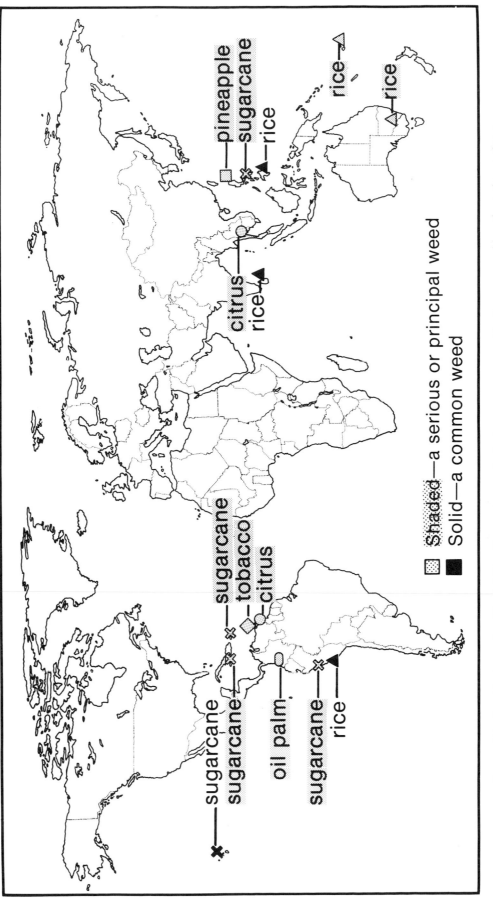

FIGURE 74. *Brachiaria mutica* (Forsk.) Stapf is a serious, principal, or common weed in the crops and locations shown above.

❧ 26 ❧

Capsella bursa-pastoris (L.) Medic.

BRASSICACEAE (also CRUCIFERAE), MUSTARD FAMILY

Capsella bursa-pastoris is a small herbaceous annual, winter annual, or biennial with rosette leaves, and is a native of Europe. It is mainly a weed of temperate areas, although it is found at higher elevations in the tropics. It has been reported to be a weed of 32 major crops in 50 countries. It is a cosmopolitan weed and is commonly found in cereal crops, pastures, cultivated lands, roadsides, gardens, and lawns. The specific name, *bursa-pastoris*, refers to the shape of the fruit which is triangular or shaped like a purse; hence, the weed's most common name is shepherd's purse.

DESCRIPTION

C. bursa-pastoris is a small, erect, *annual* or *biennial* herb (Figure 75); *stems* erect, simple or branched from the base; 5 to 80 cm tall, angled and striate, glabrous or sparsely gray hairy, hairs simple or branched, pale green to straw-colored, from a thin, sometimes branched *taproot*; *leaves* alternate, the *basal leaves* petioled, commonly coarsely deeply lobed, rarely entire, 15 cm long, 4 cm wide, with a large terminal lobe, spread in a rosette form, the *upper stem leaves* alternate, sessile, toothed and clasping the stem, much smaller, rarely to 8 cm long, 1.5 cm wide, lanceolate to oblong, entire to dentate or with earlike lobes; *flowers* white in long terminal racemes, the rachis and pedicel elongating as the seeds ripen; *calyx* pinkish or green; *corolla* white; *petals* four, obovate, 2 mm long; *seedpod* (silique) triangular, notched at the apex, of two valves, flattened at right angles to the partition, valves fall

leaving many small seeds attached to thin, membranous, elliptical septa; *seeds* about 1 mm long, oblong and flattened, dark reddish yellowish brown or sometimes yellowish brown, with two longitudinal grooves separating the seed face into three almost equal parts, surface minutely roughened.

The species is extremely variable in size, fruit, and leaf form, but it can be distinguished by its long terminal racemes, its triangular seedpods which are flattened at right angles to the partition, its small white flowers, and by the toothed leaves of the rosette.

BIOLOGY

C. bursa-pastoris reproduces by seeds. Although an annual (or usually biennial) in temperate regions, the plant may also act as a biennial in warmer countries or in the subtropics. It is not commonly found in tropical areas except at higher elevation. In Kenya it is found from 1,600- to 2,300-m elevation, and in Colombia it is reported as a weed of the cooler highlands.

C. bursa-pastoris flowers during April and May in Lebanon and in August to December in South Africa. The seeds may be very long-lived in the soil, and this, coupled with its high seed production, may explain its ability to survive and persist as a weed.

Much work on the seed biology of this plant has been conducted in England by Popay and Roberts (1970*a,b*). In their laboratory studies they found a low temperature treatment, especially below 10° C, for imbibed seeds to be necessary. When this condi-

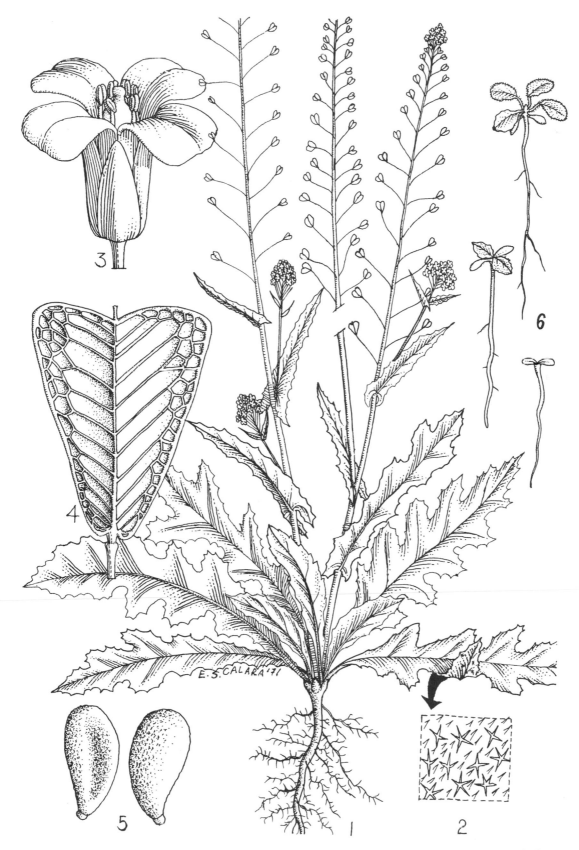

FIGURE 75. *Capsella bursa-pastoris* (L.) Medic. *1*, habit; *2*, undersurface of leaf; *3*, flower; *4*, fruit or seedpod (silique); *5*, seed, two views; *6*, seedlings.

tion was met, germination occurred when the seeds were exposed to light. Low temperatures (below 15° C) aided in breaking dormancy; higher temperatures (especially 25° to 30° C) and alternating temperatures, speeded germination of nondormant seeds. Rate of germination increased with fluctuating temperatures. Storage of seeds for as long as 2 years did not reduce dormancy. Darkness, as well as high carbon dioxide and low oxygen levels, suppressed seed germination at temperatures of 4° to 25° C; however, upon being returned to light and normal gas concentrations, the seeds again began to germinate. Sowing of nondormant seeds below 2 mm of moist soil inhibited germination. It appeared that nitrates and nitrites could substitute for the low temperature required to break dormancy; in addition, nitrates and nitrites and fluctuating temperatures exercised a synergistic effect.

Popay and Roberts (1970a) suggested that the combination of low temperatures to break dormancy followed by the required light may account for early spring growth of this weed. Also, the combination of factors necessary for both degree and rate of germination probably accounts for the ability of the plant to germinate throughout the growing season. Later germination is probably influenced by increasing temperatures, drying or wetting of soil, and by the amount of nitrate present in the soil. The varying conditions necessary for germination probably ensure a constant reservoir of seed in the soil. As one might expect, stirring or mixing of the soil by tillage exposes more nondormant seeds to light at the surface, thereby completing the germination requirement under moist soil conditions.

DISTRIBUTION AND AGRICULTURAL IMPORTANCE

C. bursa-pastoris is more a weed of the temperate zone than of the tropics (Figure 76). If present in the tropics or subtropics, it is usually more important at higher elevations. It is commonly found in waste places, cultivated lands, pastures, lawns, and gardens. It is reported to be a weed of light, cultivated soils in Australia (Maiden 1920).

It is a principal weed of flax in Germany; onions in Ireland and the Soviet Union; peas in Bulgaria, Greece, and Ireland; pastures in Alaska; barley, oats, wheat, potatoes, and vegetables in Norway; and vegetables in Belgium and Canada. It is reported to be a common weed of barley and wheat in Finland and Nepal; orchards in Yugoslavia; peas in New Zealand; sugar beets in the Soviet Union; sugarcane

in Taiwan; and wheat in Jordan. In addition, it is a weed of sugarcane and onions in Australia; vineyards in Australia, Chile and Greece; rice in Korea; hemp in the Soviet Union; cereals in England, Greece, and New Zealand; wheat in Colombia, France, Germany, Greece, Italy, Portugal, and Spain; barley in Colombia, England, and Greece; rye in Greece; sugar beets in England, Italy, New Zealand, and Spain; cowpeas and beans in Iran; vegetables in England, Iran, Japan, and New Zealand; strawberries and tobacco in New Zealand; pastures in Colombia, England, New Zealand, and Spain; potatoes in Australia, Colombia, England, and New Zealand; citrus in the United States; and alfalfa in Colombia.

C. bursa-pastoris has been blamed for the deaths of horses and cattle in Australia, deaths which occurred when the plant's fibrous stems and fruits formed ball-like obstructions in the bowels of these animals (Whittet 1968). The plant is sometimes used medicinally in Argentina (Marzocca 1957).

It is an alternate host of diseases of other Cruciferae such as *Cystopus candidus* (Maiden 1920) as well as of the following viruses: anemone mosaic, aster yellows, beet curly top, beet mosaic, beet ring spot, beet yellows, cabbage black ringspot, cabbage ring necrosis, cauliflower mosaic, tobacco broad ringspot, tobacco mosaic, tobacco ringspot, turnip crinkle, turnip mosaic, clover big vein, cucumber mosaic, potato yellow dwarf, radish mosaic, and turnip yellow mosaic (Namba and Mitchell, unpublished; see footnote, *Cynodon dactylon*, "agricultural importance").

COMMON NAMES OF CAPSELLA BURSA-PASTORIS

ARGENTINA
  bolsa de pastor
AUSTRALIA
  shepherd's purse
BRAZIL
  bolsa de pastor
CANADA
  shepherd's purse
CHILE
  bolsita de pastor
  mastueros
COLOMBIA
  bolsa de pastor
DENMARK
  almindelig hyrdetaske

EASTERN AFRICA
  shepherd's purse
EGYPT
  kees el-raat
ENGLAND
  shepherd's purse
FINLAND
  lutukka
FRANCE
  bourse-à-pasteur
  bourse-de-capucin
  boursette
GERMANY
  gemeines
    Hirtentäschel

Hirtentäschel
Hirtentäschelkraut
Säckel
Täschelkraut

IRAN
khorjinak

ITALY
borsa pastore

JAPAN
nazuna

MOROCCO
bourse-à-pasteur

NETHERLANDS
herderstaschje

NORWAY
hyrdetaske

PORTUGAL
bolsa de pastor

SOUTH AFRICA
herderstassie

SPAIN
bolsa de pastor

SWEDEN
lomma

TAIWAN
jì-tsài

TUNISIA
bourse-à-pasteur

TURKEY
coban cantasi

UNITED STATES
shepherd's purse

URUGUAY
bolsa de la pastora

VENEZUELA
mostaza silvestre

WESTERN AFRICA
chamberaka

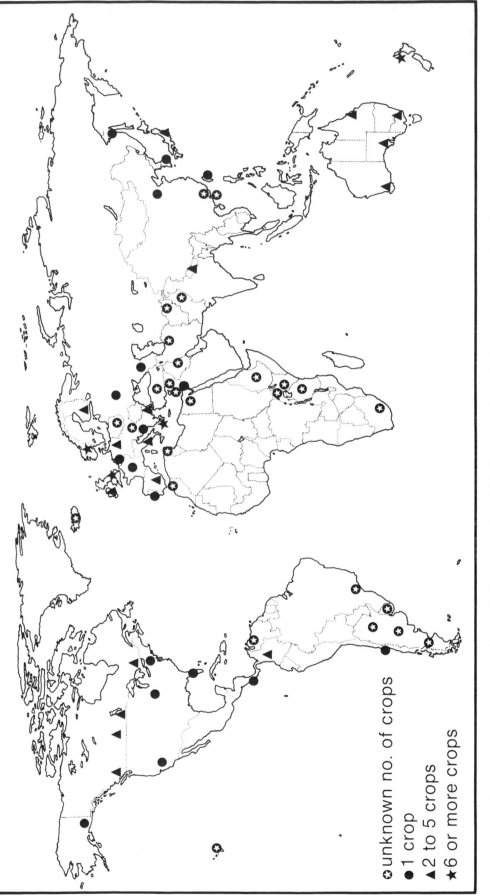

○ unknown no. of crops
● 1 crop
▲ 2 to 5 crops
★ 6 or more crops

FIGURE 76. The distribution of *Capsella bursa-pastoris* (L.) Medic. across the world in countries where it has been reported as a weed.

❧ 27 ❧

Cenchrus echinatus L. (= *C. pungens* HBK, *C. brevisetus* Fourn., *C. viridis* Spreng., and *C. quinquevalvis* Ham. ex Wall.)

POACEAE (also GRAMINEAE), GRASS FAMILY

Cenchrus echinatus, a native of tropical America and a weed of the tropics and subtropics, is one of a dozen or more species in the genus which are undesirable. These grasses are commonly named sandbur, burgrass or sandspur. The common name refers to the fact that these grasses are adapted to porous sandy soils and to the spiny burs of the seed heads which easily detach from the racemes and become attached to clothing or animal hair. The spiny burs can cause painful or annoying injuries to human skin and can contaminate feeds and hay, thus reducing its palatability and acceptability for animals. Seeds are easily disseminated because they adhere to clothing, fur, machinery, tires, or other surfaces. *C. echinatus* has been reported as a weed of 18 crops in 35 countries, mostly in cereals, pulses, vineyards, plantation crops, and pastures. When young it can provide good forage for cattle.

DESCRIPTION

C. echinatus is an *annual* grass that is erect or decumbent at the base, somewhat tufted, branched, (Figure 77); *roots* fibrous; *culms* 25 to 90 cm tall, the lower part often prostrate, compressed, rooting at the nodes, often reddish at the nodes; *leaf sheaths* smooth or with few stiff hairs on the margins on the upper portion; *ligule* 0.5 to 1.7 mm long, with marginal hairs; *blades* smooth to hairy, 5 to 30 cm long, 0.5 to 1 cm wide, flat, lower surface smooth, upper side rough, slightly hairy at the base; *inflorescence* a *spike*, dense, cylindric, 2.5 to 10 cm long, about 1 cm in diameter; the *burs* five to 50 or

more, not crowded, almost sessile, globular, densely arranged, 3 to 6 mm in diameter, 5 to 10 mm long; the *spines* or *bristles* 2 to 3.5 mm long, usually turning purple with age, sometimes straw-colored, the basal bristles numerous, usually turned downward, inner bristles attached below the middle of the bur, bristles united for about one-half their length to form a deep cup, hairy, bristles irregular in size and thickness; *spikelets* two to four (usually three) in each bur, about 5 to 7 mm long; *stamens* three; *grains* ovoid, 1.6 to 3.2 mm long, 1.3 to 2.2 mm wide.

This species is distinguished by the large, spiny burs which are easily detached from the flowering spike. The burs are covered with numerous, sharp bristles which are usually turned downward. The spiny bristles usually turn purple with age and may be strong enough to penetrate shoe leather. There is a variety in Hawaii, variety *hillebrandianus*, in which the plants are softly hairy throughout rather than just at the base of the leaves.

DISTRIBUTION AND BIOLOGY

C. echinatus grows from about lat 33° S to 33° N, mostly in the tropics or subtropics. It is found most frequently in tropical America where it is native, but it has also spread to Australia, Nigeria, and Southeast Asia (Figure 78). It has been reported as a weed in 35 countries.

An annual reproducing by seeds, *C. echinatus* produces from five to 50 or more spiny burs on each raceme. When mature, these fall to the ground and produce new plants. If brushed or touched the burs

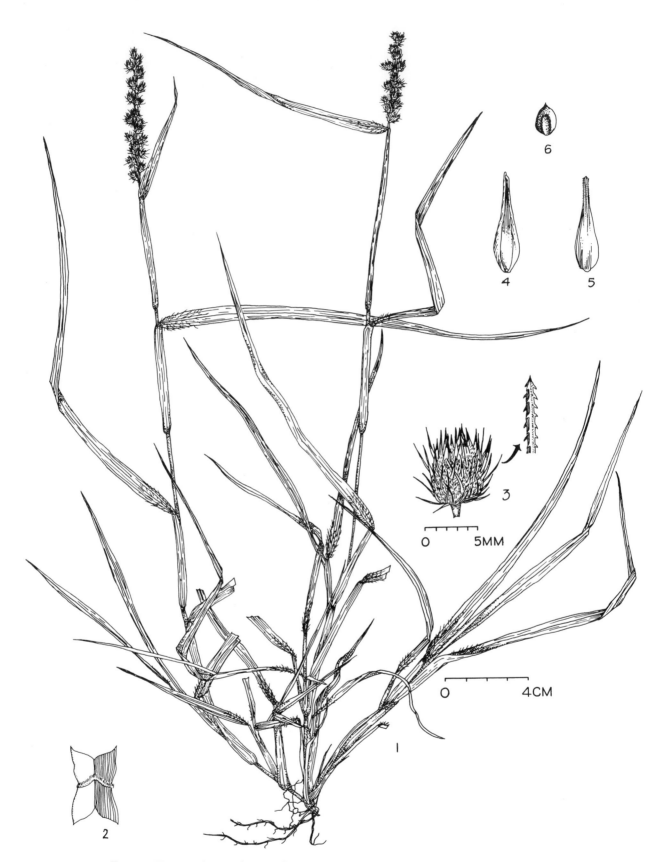

FIGURE 77. *Cenchrus echinatus* L.
Above: *1*, habit; *2*, ligule; *3*, bur, with detail of ligule; *4*, spikelet, dorsal view; *6*, seed.
Opposite: Seedling.

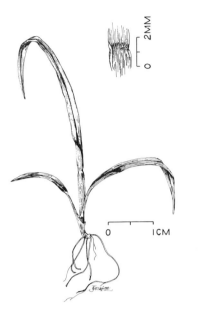

cling to clothing, fur, feathers, or other surfaces, often traveling long distances as unwanted wayfarers. Undoubtedly the primary method of dispersal for this weed is its spiny bur. The plants flower throughout most of the year in the moist tropics.

C. echinatus can grow in many habitats. Often associated with low moisture and with sandy or light, well-drained soils of the lowland tropics, the plant can spread rapidly in moderately moist regions. Under dry conditions the plant is short-lived, stays very small, and produces few burs; under moist conditions it may be long-lived, can grow very large, and produces many burs. It colonizes bare or open ground, quickly filling gaps in crop stands. It is found in cultivated fields, abandoned fields, pastures, field margins, along roadsides, in lawns, on sand dunes, along river sands, and in sandy soils along beaches. The plants can withstand repeated defoliation in lawns, in frequently harvested forage crops, or in pastures.

AGRICULTURAL IMPORTANCE

C. echinatus is a weed in many crops in many countries (Figure 79). It is one of the most serious weeds of corn and upland rice in the Philippines; corn in Venezuela; and pastures in Fiji. It is a principal weed of corn in Brazil and Hawaii; sugarcane in Australia and Peru; peanuts and beans in Brazil; cotton in Peru; alfalfa, papayas, peanuts, and sweet potatoes in Hawaii; pastures in Australia and Hawaii; and pineapples in the Philippines. It is also a weed in the following areas and crops: orchards,

vineyards, fruits, cereals, coffee, onions, vegetables, bananas, coconuts, and soybeans.

C. echinatus is important not only because of its spiny burs but also because it competes with upland crops. It can grow rapidly under moist conditions to produce thick strands which compete readily with field crops for both light and moisture, often smothering other vegetation. The ready dispersal of the spiny burs can spread the plant widely and quickly.

C. echinatus is also a problem because the spiny burs can penetrate the skin of animals or the feet and hands of humans. The plant becomes a nuisance, therefore, in recreational areas along beaches and pathways, on playing fields, or in other places where persons might walk barefooted. In hand-harvested sugarcane the burs can cause painful punctures in the hands and feet of field workers, thereby increasing the cost and decreasing the efficiency of harvest. In forage crops or pastures the burs cause injury to the eyes, mouths, and tongues of animals. The quality, acceptability, and value of hay, chopped fresh forage, or pasture which is heavily infested with the burs may be sharply reduced. Burs in wool will reduce the quality and price of the fleece.

If grazed when young, *C. echinatus* can provide valuable forage, especially in dry, sandy areas. Close grazing is recommended to prevent flowering and production of burs.

COMMON NAMES
OF CENCHRUS ECHINATUS

ARGENTINA
  cadillo
  cadillo correntino
  roseta
AUSTRALIA
  mossman river grass
BRAZIL
  arroz bravo
  capím carrapicho
CEYLON
  bur grass
  kuvenitana
COLOMBIA
  cadillo carretón
    morado
CHILE
  cadillo
CUBA
  guizazo

EL SALVADOR
  mozote
FIJI
  bur grass
  se bulabula
GUATEMALA
  cadillo
  mozote de caballo
  rosetilla
HAWAII
  konpeito-gusa
  sandbur
  'ume'alu
HONDURAS
  cabeza de negro
JAMAICA
  bur grass
MAURITIUS
  herbe à cateaux

MEXICO
   abrojo

PANAMA
   pega-pega

PERU
   cadillo

PHILIPPINES
   cauit-cauitan

PUERTO RICO
   abrojo

TAHITI
   piri-piri

THAILAND
   sandbur
   ya-bung

UNITED STATES
   southern sandbur

URUGUAY
   cardillo
   pasto camelo
   pasto roseta

VENEZUELA
   cadillo

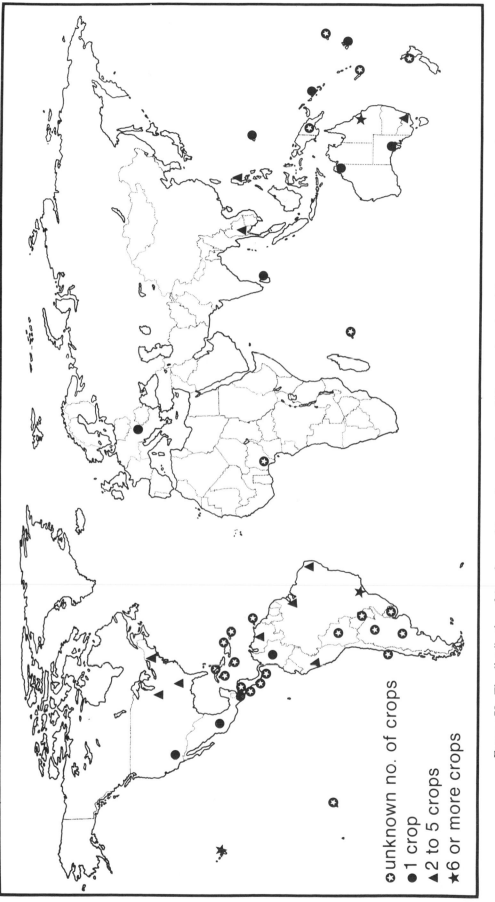

FIGURE 78. The distribution of *Cenchrus echinatus* L. across the world in countries where it has been reported as a weed.

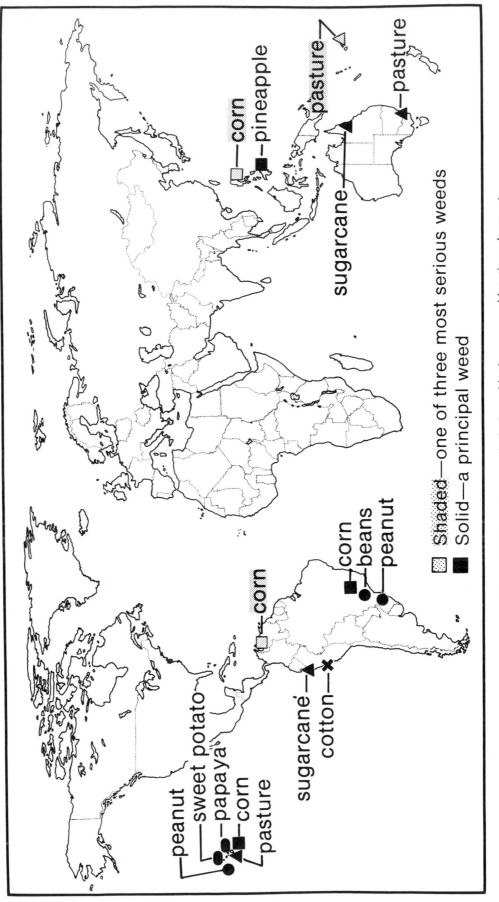

FIGURE 79. *Cenchrus echinatus* L. is a serious or principal weed in the crops and locations shown above.

❦ 28 ❧

Ceratophyllum demersum L.

CERATOPHYLLACEAE, HORNWORT FAMILY

Ceratophyllum demersum is a fragile, annual or perennial, algallike herb that spends its entire life beneath the surface of the water. Although its origins are uncertain, it has been identified in Pliocene deposits in northern Europe. It is one of the hydrophytes among the angiosperms that has become completely adapted to a life beneath the surface of the water. It grows submerged or afloat and has no roots but possesses modified leafy branches that are rhizomelike and may serve for anchorage and for the absorption of nutrients. It is one of the delicate floating or submerged freshwater hydrophytes with its extreme latitudinal range made possible by the uniformity of the environment to which it has become adapted. Its wide distribution over the world remains a mystery, for its seeds sink to the bottom and the plant is quickly killed by seawater.

DESCRIPTION

C. demersum is a *submerged*, rootless, densely leafy, *perennial*, freshwater plant (Figure 80), *stems* often much-branched, forming large masses, flexuous, the internodes 1 to 3 cm long; *leaves* in whorls of seven to 10, 1 to 4 cm long, the base gradually or abruptly thickened above the middle, the segments spreading when submerged, collapsing out of water, twice or thrice forked with linear segments; *flowers* very small, sessile in the leaf axils, each surrounded by minute bracts, a real perianth not present; *flower bracts* nine to 12, linear, transparent, with numerous short, brown lines; *female flowers* with a minute ovary and simple style; *male flower*, anthers white, stamens sessile; *fruit* black, 4 to 5 mm long, with three spines, the apical spine (style) 11 to 12 mm long, soft, the two basal ones straight to slightly recurved, 9 to 11 mm long, one-seeded.

This submerged plant is distinguished by the presence of one apical and two basal spines on the fruit and its rootless branched stems with forked leaf segments.

DISTRIBUTION AND HABITAT

The native origins of *C. demersum* are unknown, although it has been an inhabitant of the earth at least since the Pliocene period. Some believe it came from the temperate zone, but doubt has been cast on this view by observations that it requires almost tropical temperatures for the maturation of its fruits.

The plant is an inhabitant of shallow ponds, tanks, lakes, and stagnant water. It tends to seek sheltered sites in standing or slowly flowing water where bottoms are much silted. There are some records of growth in 10 m of water. It is easily killed by very high light intensities and may be shattered if the water is violently disturbed. Surveys in the southwestern United States found the weed to be present in lakes having a pH of 7.1 to 9.2, and being most plentiful where the pH was 7.6 to 8.8.

Although it is most widely distributed in the freshwaters of the temperate zones, and mainly in the northern hemisphere, it is truly a cosmopolitan species (Figure 81). It requires for its life cycle that remarkable uniformity of temperature, light, mois-

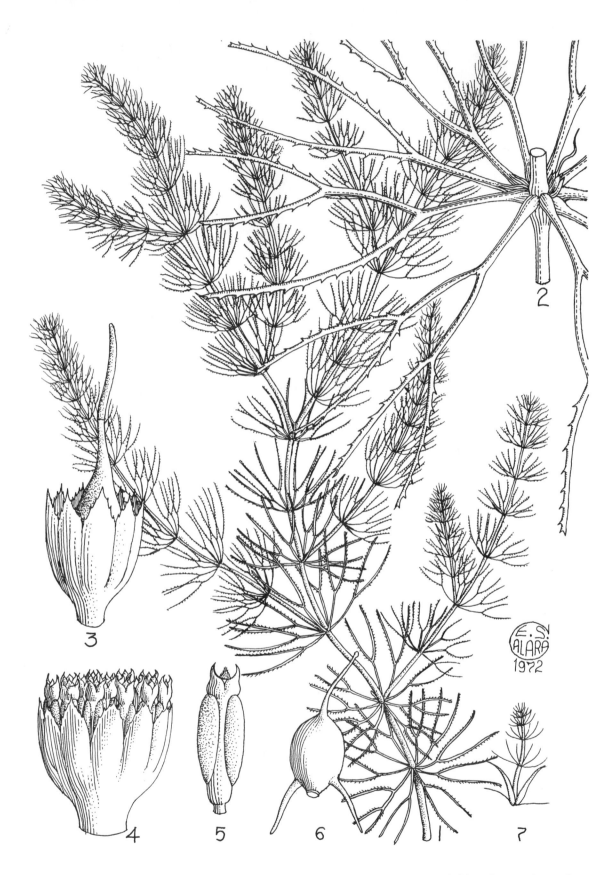

FIGURE 80. *Ceratophyllum demersum* L. *1*, habit; *2*, whorl of leaves; *3*, pistillate flower; *4*, staminate flower; *5*, stamen; *6*, fruit; *7*, seedling.

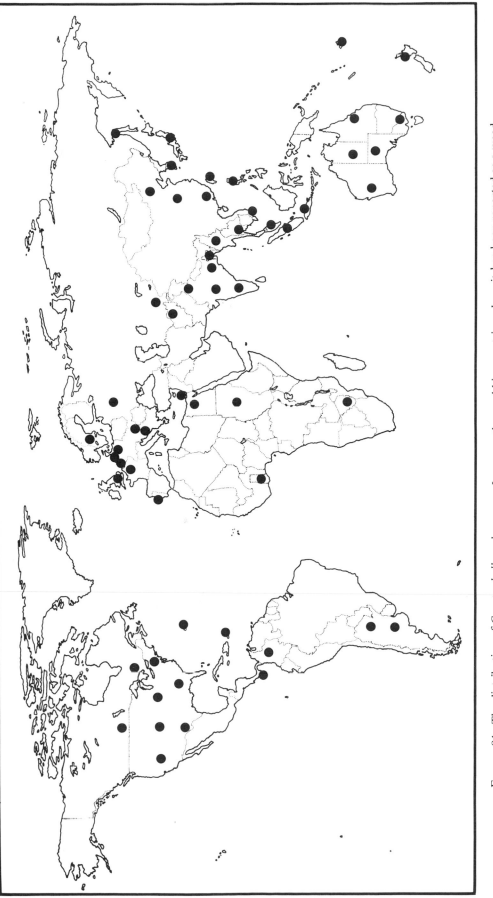

FIGURE 81. The distribution of *Ceratophyllum demersum* L. across the world in countries where it has been reported as a weed.

ture, and nutrients found at the shores of freshwater bodies which do not permit ice formation all the way to the bottom. There are several submerged or floating hydrophytes which exhibit remarkable latitudinal range because of the uniformity of such an environment.

C. demersum is very sensitive to freezing and is said to be unable to survive if encased for only a few days in ice. Suspended solids in the water, which may seriously reduce sunlight penetration, will drive the weed out but it returns when the water is again clear.

The means by which the species has spread so widely in the world remains a mystery. Its seeds quickly sink and become lodged in silty bottoms. Its vegetative fragments quickly will dry if taken out of the water. The pollination process can only be carried out in quiet waters. The plants are quickly destroyed by seawater, yet the species is present in Bermuda and Fiji.

Birds are known to fly with pieces of the plant for short distances. Strong winds, strong currents, foraging animals, or any violent disturbance of the water may tear the plants into many pieces, each of which is a viable source of a new plant.

As with most of the weeds, this species may serve both men and animals or may be a source of problems to them. The plants may support large populations of shore-living fishes but, in nutrient-rich waters, they may become so thick that light and oxygen become limiting and a fishery may disappear. The species is, nevertheless, an important part of the food chain for aquatic creatures. Its achenes are sought by diving ducks.

The world has seen a dramatic increase in the rampant growth of aquatic weeds in recent decades, weeds which are often detrimental to the inhabitants of streams and lakes. Such weeds also seriously interfere with many of the activities of man by which he benefits from these natural resources. This increase in weeds has come about because new species have been introduced in new areas; many artificial lakes have been created; thousands of kilometers of irrigation canals have been opened; recreational activities have tended to move the plants about; and the effluent from cities, factories, and fields has provided an abundance of nutrients which can promote the rapid growth of aquatic species. *C. demersum* has become more weedy for all of these reasons, and it now interferes with water flow, plugs the gates of irrigation systems, and may interfere with fishing and water transportation. A sudden explosion of the weed in New Zealand in 1965 caused the shutdown of some electric power plants.

BIOLOGY

C. demersum is a member of a small single genus in the family *Ceratophyllaceae*. The plant is without roots or stomata and seems perfectly adapted to spend its entire life beneath the surface of the water. The radicle does not enlarge or elongate during seed germination, and no adventitious roots are formed. Because the plant is rootless, it anchors itself with leafy branches which grow into the mud. These basal lateral branches develop very finely divided leaves which appear as white threadlike segments called "rhizoid shoots." These shoots also assist with the absorption of nutrients. The plants may break off and float to the surface or in shallow water may grow to the surface; in either case they often become entangled with other filamentous plants (Sculthorpe 1967).

Each plant is about 88 percent water; there is a complete absence of lignification; and one-third of the plant consists of air spaces. The species is less fragile than one might expect, however, for it is more cuticularized than most plants. It does not collapse and dry as rapidly as do many submerged species when removed from the water. The plant has mucilage-containing hairs on its leaves and stamens, but the function of these hairs is not known.

In the temperate zone, the stems which are produced in one season will decay in June or July of the next year.

This plant not only lives beneath the surface for its entire vegetative cycle, but even the stigmas of the flowers remain submerged and pollen is conveyed to to them by the water. The minute, monoecious flowers are born singly in the axil of one leaf of the whorl. There are about 12 stamens in each male flower. At maturity, the stamens become detached and form small terminal floats; these floats enable them to rise to the surface where they dehisce. The pollen has a specific gravity higher than water and gently sinks, whereupon it comes in contact with the stigmas of the flowers below. When mature, the fruits sink at once. The achenes have a hard, durable cover and they may be found in the mud at the bottoms of ponds and lakes or in marshes from which water has receded. The seedlings germinate in the mud and upon reaching 8 cm some may rise to the surface (Arber 1963).

Some of the submerged and floating hydrophytes

of the cool, temperate regions do not have persistent rootstocks, rhizomes, or tubers to carry them from one season to the next. Certain of these form very specialized types of "winter buds" which often seem to consist mainly of leafy material. In autumn, *C. demersum* becomes brittle and proceeds to form one of the simplest of these resting vegetative bodies. As days shorten and temperatures fall, the apices of lateral shoots cease to elongate and they bear tight clusters of dark green leaves which contain much starch and possess an increased amount of cuticle. Thus, a sort of dormant apex forms which will remain attached, or may fall, and will rest until temperatures rise and days begin to lengthen in spring. In winter, those on the bottom are often weighted down by annelids, aquatic molluscs, and insects.

Hill (1967) reported on the favorable influence of nitrogen and phosphorus on *C. demersum*. He concluded that these elements were extracted from the products resulting from the biologically active cycles of the lake itself as well as from the runoff from farmlands. Goulder and Boatman (1971) confirmed that this species is a nitrophilous plant which requires high inorganic nitrogen levels in the surrounding water at least part of the year. Wilkinson (1963, 1964) studied the effects of light intensity and quality, daylength, and temperature on *C. demersum* in controlled experiments. He found that the plants could grow slowly at 2 to 3 percent of full sun, could grow rapidly at 5 to 10 percent, and required less than 2 percent of the amount of red light normally present in sunlight. He concluded that control of the plant with shade is not feasible. In studies on photosynthesis in this species, Meyer (1939) and Meyer and Heritage (1941) have reported on a diurnal cycle for the process and on the effect of turbidity of the water.

BIOLOGICAL CONTROL

The preference of the grass carp (*Ctenopharyngdon idella*) for *Ceratophyllum demersum* as a source of food has been reported from four widely separated places in the world. This fact indicates that useful biological control systems may someday be worked out for many similar species of weeds now on the increase. *Ceratophyllum demersum* was much preferred in tests in England and the Ukraine area of the Soviet Union. In India, Bhatia (1970) arranged tests in which the fishes, allowed to choose between seven species of plants, consumed 100 to 150 percent of their body weight daily. *C. demersum* was the third most popular species behind *Najas minor* and *Hydrilla verticillata*. Penzes and Tolg (1966) in Hungary ran feeding tests on 22 species of weeds and reported that one of the best liked was *Ceratophyllum demersum*. There was no feeding below 14° C, only soft plants were taken at 16° C, and there was intense feeding at 20° C.

COMMON NAMES
OF CERATOPHYLLUM DEMERSUM

ARGENTINA	PHILIPPINES
cola de zorro	arigman
AUSTRALIA	inata (Tagalog)
hornwort	linamum (Magindanao)
INDIA	
sivara	
INDONESIA	TAIWAN
ganggang (Javanese, Sundanese)	common horn weed
	hornwort
JAPAN	jin-yu-tzao
matsumo	morass weed
MALAYSIA	
kantjil	THAILAND
NETHERLANDS	sa-rai-hangma
hornblad	
NEW ZEALAND	UNITED STATES
hornwort	common coontail

❧ 29 ☙

Chromolaena odorata (L.) R. M. King & H. Robinson (= *Eupatorium odoratum* L.)

ASTERACEAE (also COMPOSITAE), ASTER FAMILY

Chromolaena odorata, a perennial, is a diffuse, scrambling shrub that is mainly a weed of plantation crops and pastures of southern Asia and western Africa. It forms a tangle of bush from 3 to 7 meters in height when growing in the open. Native to Mexico, the West Indies, and tropical South America, it was spread widely by early navigators. It is a weed of 13 crops in 23 countries.

Chromolaena was previously included in the genus *Eupatorium* (King and Robinson 1970). It is distinguished by the rather consistent pattern of many rows of phyllaries (bracts) that are progressively longer and give a markedly cylindrical appearance to the head, by the three prominent veins on the leaves, and by the pungent smell emitted by the crushed leaves.

DESCRIPTION

This weed is a spreading, sometimes scrambling, tangled, thicket-forming, *perennial shrub* growing from 3 to 7 m high (Figure 82); *taproot* deep and massive; *stems* yellowish, shortly hairy or nearly smooth, round, with fine longitudinal lines, spreading, profusely branched, herbaceous when young, tough and semiwoody when older; *leaves* opposite, ovate-deltoid, acuminate, 6 to 12 cm long, 3 to 7 cm wide, margins toothed, the teeth pointing forward (serrate), otherwise entire, dark green, sparsely hairy to smooth, conspicuously three-veined; *petiole* 1 cm or more long; *flower clusters* in more or less terminal and axillary pedunculate flat-topped inflorescence, 10- to 35-flowered, the outer flowers

opening first (corymb); *involucre* cylindric, the bracts to 1 cm long, 3 mm in diameter, straw-colored to greenish in five to six rows closely overlapping; *corolla* five-lobed, with two-branched stigma projecting above; *florets* pale mauve, pale blue, or whitish, protruding from the involucre; *fruits* (achene) narrow, linear, angled, brown or black with short white stiff hairs on the angles, 5 mm long; *pappus* white, of rough bristles, 5 mm long. As the species name implies, the leaves have a pungent odor when crushed.

DISTRIBUTION AND HABITAT

C. odoráta can grow on many soil types but seems to prefer well-drained sites. It must have sunlight or partial shade, for it cannot survive in forests or plantations after the canopy has closed in. It is found on cultivated lands, abandoned or neglected fields, wastelands, and along forest trails, fence rows, roadsides and banks. Because it loves the sun it invades clearings rapidly, forming brush thickets.

Its agressiveness is due to its rapid growth rate, its profuse branching habit, and to its very prolific annual seed production; in addition, some workers feel that the weed may produce chemicals which can inhibit adjacent plants. It has been suggested that the nonsusceptibility of the plant to attack by insects may be due to oils which have insect-repellent properties.

During seasons when soil moisture is plentiful, slashing and burning may result in the weed's making rapid regeneration from the deep perennial tap-

212

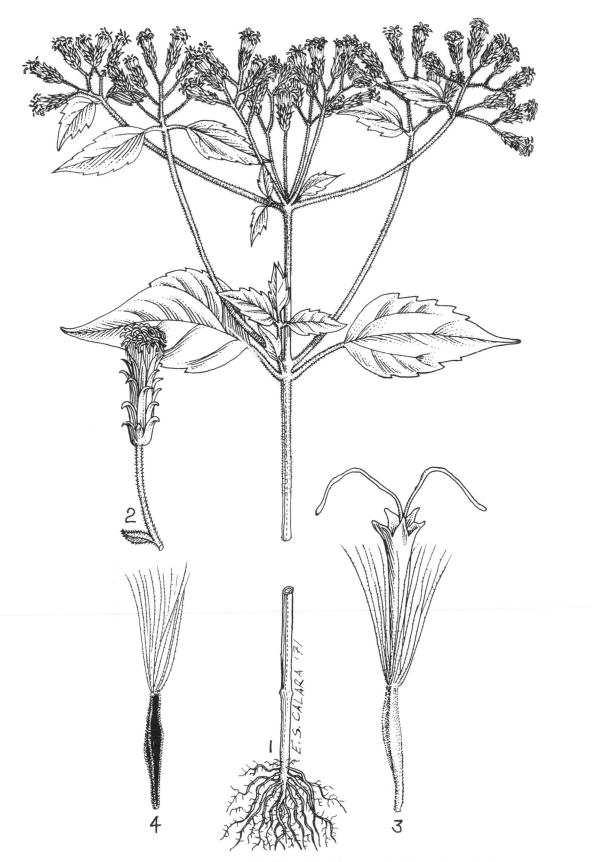

FIGURE 82. *Chromolaena odorata* (L.) R. M. King & Robinson. *1*, habit; *2*, flowering branch; *3*, flower; *4*, achene.

roots. In some areas large plants may represent a significant fire hazard when they are dry.

Flowering is controlled by photoperiod so that in a given region all plants tend to flower together. In Cambodia the plant does not flower until after the rainy season ends in November (Litzenberger and Ho Tong Lip 1961). The plant is almost entirely spread by seeds which are produced in great numbers and which are airborne. Seed germination occurs mainly after the rainy season. Little is known about the propagation or the reproductive physiology of this species; however, the development of the embryo and the female gametophyte has been described (Ghosh 1969).

The genus *Chromolaena* is essentially tropical but does extend into higher altitudes. The geographical distribution of *C. odorata* is shown in Figure 83.

AGRICULTURAL IMPORTANCE

C. odorata is ranked among the three most serious weeds of coconuts in Ceylon; rubber in India, Indonesia, and Malaysia; and oil palms in Nigeria and Sarawak. It is a principal weed of abaca in Sabah, of coconuts in Trinidad, and of rubber and pineapples in Ceylon. In addition, it is found in oil palms and tobacco in Malaysia; cotton, corn, and rubber in Thailand; rubber, tea, teak plantations, and vegetables in India; coconuts and pastures in the Philippines; sugarcane in Trinidad; pastures and rice in Nigeria; and upland rice in Indonesia.

In the Philippines, where it has been recently introduced, it spreads in pastures like wildfire. It is believed to be responsible for the poisoning of livestock at certain times of the year, but this observation has not been confirmed in the laboratory. Large tracts of pastures have been abandoned as a result of its invasion. On some occasions whole villages have been deserted because of the nonproductivity of the land infested with this weed (Pancho and Plucknett 1971).

C. odorata is one of the most faithful followers of human activity in Southeast Asia (anon. 1967c). It has spread at an alarming rate in Ceylon (Salgado 1963). It is very competitive and can replace grasses in open lands.

It is found by the sides of logging roads far into the forests and comes up in every clearing, waste place, and roadside. In experiments at the Rubber Research Institue of Malaya this weed, when compared with ground covers of small grasses or recommended legumes, has been found to depress the growth of rubber trees (anon. 1967c). It is regarded as a heavy feeder on nutrients which are then locked up in rather large quantities of slow-rotting litter.

Biological control measures are being studied for this weed in Trinidad (Cruttwell 1968). Biological control of other *Chromolaena* species is being used with some success in Hawaii (Nakao 1969) and Australia (Auld 1970).

It is reported in Thailand to be an alternate host of the leaf spot, *Cercospora* sp. (Pukdeedindan 1966).

In Nigeria thickets of *C. odorata* harbor wild pigs and rodents which cause damage to crops (anon. 1959).

Attempts to use it in India for manufacturing a good quality paper were not too successful (Bhat and Karnik 1954).

In Cambodia *C. odorata* is considered to be promising as a green manure crop and as a means of reducing nematode-fungus attacks in black pepper production (Garry 1963). Some very interesting experiments were conducted in this country to determine the plant's efficacy as a mulch and as a green manure crop for improving rice, cassava, and black pepper yields (Litzenberger and Ho Tong Lip 1968). Using a rate of 20 metric tons of fresh *C. odorata* per hectare, Litzenberger and Ho Tong Lip estimated that the nutrients added to the soil by the green manure were 112 kg nitrogen, 12 kg phosphorus, and 87 kg potassium per hectare. Higher rice yields were obtained with *C. odorata* than with farm manure and chemical fertilizers. Fish were killed by the *C. odorata* green manure treatment. When the green material was applied to a rice field—either on the soil surface after plowing, on the soil surface before plowing, or into the water of the flooded rice field after transplanting—no differences in yield were found. In cassava, *C. odorata* green manure yields were 12 tons per hectare higher than those obtained from the untreated plots. Black pepper vine growth was improved by mulching with *C. odorata*. This improvement was attributed to control of a nematode (*Heterodera marioni*) and, indirectly, as a result of control of the nematode, to prevention of infection by root-rot fungi, *Pythium complectans* or *P. splendens*. The successful and economical use of *Chromolaena odorata* as a green manure plant might keep it under some control in the absence of suitable biological or chemical control measures.

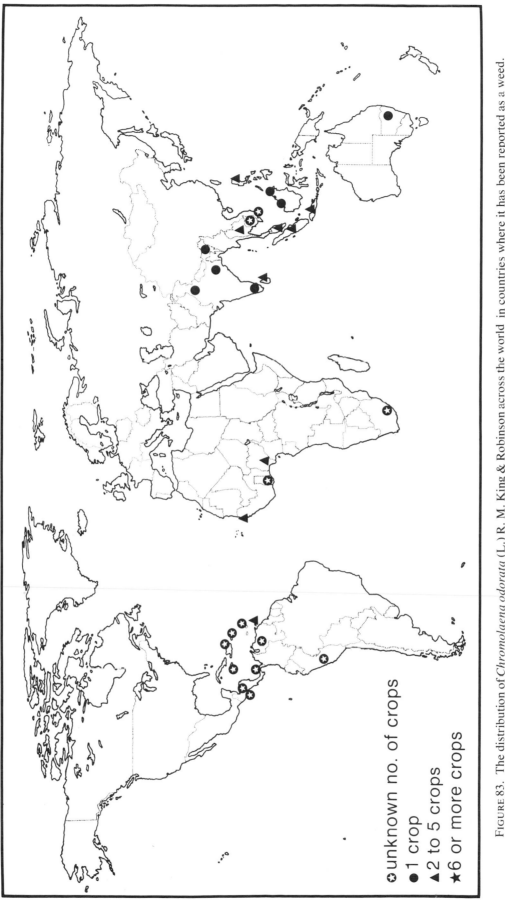

FIGURE 83. The distribution of *Chromolaena odorata* (L.) R. M. King & Robinson across the world in countries where it has been reported as a weed.

COMMON NAMES
OF CHROMOLAENA ODORATA

CAMBODIA
  tontrean khet
CEYLON
  Siam weed
DOMINICAN REPUBLIC
  rompesaraguey
EL SALVADOR
  chimuyo
HONDURAS
  crucito

  rey del todo
INDIA
  sam-solokh
INDONESIA
  Siam weed
JAMAICA
  archangel
  bitter bush
  Christmas rose
  hemp agrimony

  jack-in-the-bush
MALAYSIA
  Siam weed
  pokok Tjerman
NIGERIA
  Siam weed
PANAMA
  hierba de chiva
  paleca
PHILIPPINES
  agonoi
  huluhagonoi

PUERTO RICO
  santa maria
SIKKIM
  asloke lata
THAILAND
  sap sua
  ya-su'a-mop
TRINIDAD
  bitter bush
  Christmas bush
  jack-in-the-bush
VIETNAM
  co hoi

❧ 30 ❧

Cirsium arvense (L.) Scop.

ASTERACEAE (also COMPOSITAE), ASTER FAMILY

Cirsium arvense, a native of Europe or temperate Asia, is an herbaceous perennial growing to more than 1 meter in height and having spiny leaves and an elaborate subterranean system of roots and rhizomes from which new shoots are produced at irregular intervals. Small fragments of the underground material give rise to new plants. One plant can colonize an area several meters in diameter during the first one or two seasons of growth. The plumed achene is assumed by some to be the mechanism of long-distance dispersal, but other evidence indicates that many plumes are broken from achenes which remain in the head. It constitutes the worst thistle problem of many cool and warm temperate regions. In pastures it reduces forage consumption, for cattle will not graze near either tall or spreading plants because of the sharp spines on the leaves. It is a weed of 27 crops in 37 countries.

DESCRIPTION

C. arvense is an erect, *perennial* herb (Figure 84); with extensive creeping, white or yellowish, horizontal *rhizomes* up to 5 m or more long, the vertical roots 2 to 5 m in depth, with an extensive system of fibrous absorbing rootlets; *stems* erect, grooved, 40 to 120 cm tall, nearly glabrous or slightly hairy when young, increasingly hairy with age, arising from numerous buds on the horizontal rhizomes; *leaves* alternate, oblong, or lanceolate, usually with crinkled edges and spiny-toothed margins, very irregularly lobed, terminating in a spine, hairy beneath or smooth when mature, upper leaves sessile but only slightly decurrent, the narrowed base continuing down the stem beyond the point of leaf attachment, giving the impression of a spiny stem; *heads* dioecious (male and female flowers in separate heads and on different plants), in corymblike clusters, terminal and axillary, 2 to 2.5 cm in diameter; *involucre* 1 to 2 cm high, bracts numerous, overlapping, spineless; *staminate heads* oblong; *pistillate heads* ovoid or flask-shaped; *receptacle* bristly, chaffy; *achene* oblong, smooth, shiny, finely grooved lengthwise, curved or straight, more or less four-angled, flattened, apex with a characteristic conical point in the center, 2.5 to 4 mm long, light to dark brown, surmounted by a pappus; *pappus* plumose, of white or rather brownish, feathery hairs, 2 mm long, easily separating from the achene (deciduous), leaving a small projection at the apex of the achene.

This species is distinguished by its horizontal branching rhizomes; its dioecism; its small, almost spineless, heads; and by its characteristic of growing in circular patches with each patch usually consisting of only one clone and, in some cases, of only female or male plants.

DISTRIBUTION

This species is most frequently reported to be a weed problem in the temperate zone of the Northern Hemisphere where it is very widely distributed (see Figure 85). It is a weed of the temperate zone of the Southern Hemisphere in South America, Africa, Australia, and New Zealand. There is some indication that related species such as *Cirsium vulgare*

217

Figure 84. *Cirsium arvense* L. *1*, habit; *2*, head; *3*, flower; *4*, achene; *5*, seedlings.

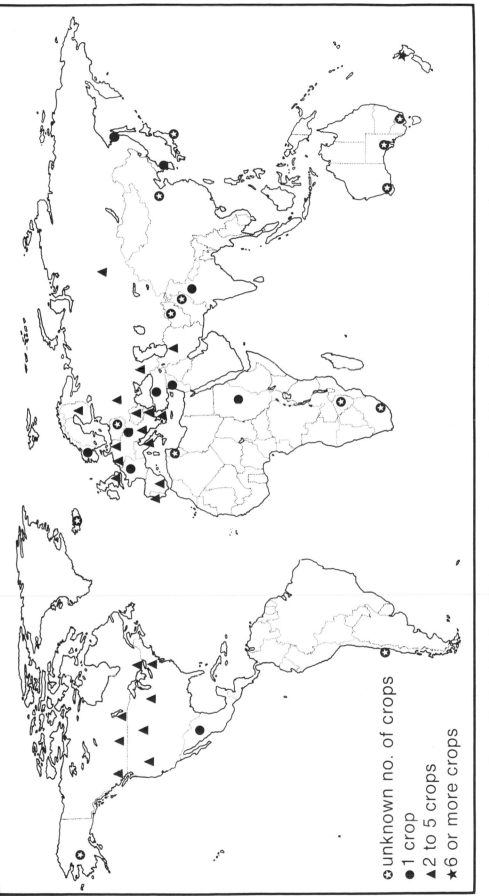

○ unknown no. of crops
● 1 crop
◀ 2 to 5 crops
★ 6 or more crops

FIGURE 85. The distribution of *Cirsium arvense* L. across the world in countries where it has been reported as a weed.

may be found more frequently as weedy species in the southern regions. *C. vulgare* is a biennial and does not reproduce vegetatively. *C. arvense* is rarely seen as a weed near the equator.

The species is adapted to areas where summer temperatures are moderate and where rainfall is not too heavy. Many of the serious weed problems are in areas receiving 450 to 900 mm of rainfall per year. A high water table severely limits root growth; the plant can survive on a wide range of soil types and moisture conditions and will tolerate a 2-percent salt content. It is less common or often absent from very light, dry soils. It is most competitive on deep, productive, well-aerated soils which do not become too warm.

The plant grows in many crops, is serious in pastures, is found on ditch banks, at the edge of woods, and on open, neglected land. A survey in mid-America reported it growing near sugar beet factories and yards, stockyards, grain elevators, railroad yards, greenhouses, and cement plants. The seeds move with packing straw.

BIOLOGY

Several excellent life histories of *Cirsium arvense* have been recorded by Bakker (1960), Detmers (1927), Hayden (1934), and Hamdoun (1967). Sagar and Rawson (1964) described the development of the plant from seeds and from fragments of roots and stems.

It is a late-emerging, small-seeded plant which establishes itself from seed only when environmental conditions are favorable within narrow limits and when competition is at a minimum. If one individual prospers, however, many seeds are produced and some may be carried long distances to colonize new areas. Fragmentation of roots and stems can produce dense stands which encroach upon new territory by vegetative propagation. These characteristics, coupled with the development of many ecotypes, make *C. arvense* one of the worst of the thistles and one which causes serious yield losses in many crops.

The species varies greatly in appearance and growth characteristics in Europe and the United States. The most extensive study on ecotypes was made by Hodgson (1964) who gathered 10 distinct ecotypes from the western United States and described and photographed them. Some leaves had smooth margins; others had deeply lobed and spiny margins. Some of the lobed leaves, with different patterns, varied from long to short in length. Some leaves were narrow, others were very broad. Flower color varied from pale blue to purple and one variety had white petals. Emergence in the spring extended over a period of 2 weeks. The heights of the plants and the shoot and flower development were also variable. Flowers were usually dioecious but mature seeds occasionally were found in staminate heads. J. Williams (1966) gathered seeds of several species of *Cirsium* from Europe and studied their response to several different dormancy-breaking treatments. These species were present: *Cirsium oleraceum, C. canum, C. eriophorum, C. vulgare*, and *C. arvense*. With the exception of *C. vulgare*, various biotypes were present in all species. Often the same biotypes could be found in widely separated areas. Palaveeva-Kovachevska (1966) has identified and described five varieties of *C. arvense* in Bulgaria.

Moore and Frankton (1969) have provided a key to 26 native and introduced species of *Cirsium* in Canada and the United States east of long 100° W. Systematic position, morphology, and chromosome numbers are presented.

The seedlings usually start slowly and are quite sensitive to competition from crops or other weeds. They grow poorly if they are shaded. Thus they often make a beginning on disturbed grazing areas, on open areas which are not cropped, or on neglected places such as ditch banks. As the seedling grows the hypocotyl elongates and breaks the surface. If the cotyledons stay together and are not caught in the soil and if they move straight up, the plant will be vertical. If the cotyledons are caught in sticky, wet soil or by a stone or other obstruction, the plant will be arched. When small, the seedlings resemble many other *Asteraceae*. The first foliage leaves are round to ovate with regularly spaced, coarse, marginal hairs which become spinose as the margins of the leaf become serrate and then lobed (Hayden 1934).

By the time the plant has developed two foliage leaves the branched root system is up to 15 cm long and the main root has begun to thicken. As food accumulates, lateral shoots from the main root grow obliquely upward or proceed horizontally for some distance before arching upward to emerge from the soil. The main vertical root or its branches produce two kinds of structures: the first consists of roots which generally grow horizontally but sometimes arch downward or upward. These branch again and again to provide a complex interlaced system which may run very deep but which initiates new roots only. The other structures are stems which come from the original vertical root system and grow upward, or come from buds on arched, ascending or descending root branches. Both types of shoots pierce the soil to form rosettes at the surface. These root-borne shoots must be regarded as rhizomes because histologically the structure is stem tissue.

The first aerial shoots do not flower (Bakker 1960). Other shoots will produce flowers and mature seeds in the year of the initial seed germination. The seedling plants can reproduce vegetatively when they are 7 to 9 weeks old. Shoots from root and stem fragments can penetrate the remains of a turf or a vigorous stand of rye grass. Subsequent growth may be retarded, however, if the grass provides very much competition. Early in the season the plants are not very responsive to applications of nitrogen; but later in the season they do show some response, with shoot growth being stimulated more than root growth (Hamdoun 1967). All tops die when winter comes. The primary root dies 3 to 4 months after germination and seedling growth and, at this time, adventitious roots form in the lower stem. Two to 4 months after germination rhizomes may be produced from the basal leaf axils of the stem.

Hayden (1934) has provided a remarkable reconstruction of the subterranean system of *C. arvense*. With her co-workers she excavated from a dense thistle patch all of the root system contained in a block of soil 30 cm thick, 3 ⅓ m long, and more than 2 meters deep. All sections were marked so that all parts could be replaced in order in the laboratory. The growth of the whole underground system terminated at the depth of her excavation because the water table was encountered at a clay layer. The upper 15 cm of the soil had been cultivated early in the season and it then became very hard because of dry weather. Some shoots found beneath the surface were curled and twisted from the effort to penetrate. Below this region the soil was friable and Hayden learned that the first 1.5 meters were interlaced with vertical and horizontal roots, whereas the first meter contained the horizontal rhizomes which eventually arched upward to provide new plants. At the ends of the roots and on the nodes of the shoots were many fibrous branches. The previous cultivation had stimulated the growth of many buds in this system. The depth of the subterranean system has been reported at 3 to 6 m in the United States, 2.25 m in Sweden, and 5.5 m in the deep, black soils where cotton is grown in the Soviet Union. While excavating root systems of 2-year-old plants in the western United States, Hodgson (1968) found 54 percent of

the roots in the 7- to 23-cm layer of soil, 30 percent at 23 to 38 cm, and 16 percent at 38 to 53 cm. Of the total mass of root material found in the upper 50 cm of the soil, 84 percent was in the top 38 cm.

Hamdoun (1967) reported that 1-year-old plants in England may expand and produce enough shoots to occupy an area almost 2 meters in diameter with an underground system extending to 70 cm. In the Netherlands plants may spread to an area 4 to 5 m in diameter in 2 years. A spread of 6 m in a single season has been recorded in the United States. Salisbury (1961) in Great Britain reported an instance of growth to 12 m in a like period. Old, exhausted, brittle, vertical roots in a colony become hard and brown, then soften and become black, and finally disintegrate. The running, horizontal roots die off a little on the end of origin each season while the other end is extended as new plants are formed; thus, it is very difficult to know the age of the underground parts.

During the period when the most recent polders were being formed from the former Zuider Zee in the Netherlands, *Cirsium arvense* and *Tussilago farfara* made their appearance early as the water subsided; in some areas they then became very troublesome. Bakker (1960) made extensive studies of the life histories of these weeds and the effects of the environment on their growth and development in the field. In this circumstance the initial infestation had to be from seeds which had been carried in either by wind or water. In experimental areas he found that the seedlings were extremely sensitive to density of plant stand, light, soil aeration, and soil moisture conditions. The greatest percentage survival of seedlings was in the plots of lowest density. On one occasion he recorded a very high mortality in plots of high density during a drought at the end of June. Watering them brought only partial success in carrying them through the drought. Some seedlings were washed out of the soil by rain and many died soon after emergence, presumably from a fungus infection. This is consistent with field observations that the highest density of seedlings on the polder seems to be in the bare soil.

PROPAGATION

Surely one of the causes of the spread of this species through a field or to adjacent fields is the distribution of fragments of stems and roots during tillage operations. In northern Germany an invasion of rats in an area of permanent pastures once resulted in the storage of vegetative portions of *C. arvense* in their underground tunnels. It was not unusual to see an otherwise "clean" pasture become thickly populated with thistles in 1 year.

Upright shoots or rhizomes are able to produce roots and buds at all nodes, and so when plants are broken one new shoot may come from each node. Horizontal and vertical roots, however, may produce buds and roots at any point, and so they have the potential to produce many more new plants per unit of mass than the stem fragments which have buds at intervals of 2.5 to 5 cm at the nodes. Root pieces which are in good condition can be cut into pieces 1.2 cm long and 0.3 to 0.6 cm in diameter to provide 95-percent production of new plants. Longer pieces containing more stored food give larger, more vigorous plants in a shorter time. When fragments are taken at the end of the flowering season, only 5 to 10 percent of the pieces produce new plants. Plants emerge from root and stem pieces in about 15 days (Hayden 1934, Prentiss 1889). Underground roots and stems which are fragmented during tillage possess sufficient food reserves to survive for more than 100 days and, without replenishment of food supplies, may then produce new shoots and roots. In Canada, a smooth-leaved variety, *integrifolium*, is able to propagate vegetatively at double the rate of a common spiny-leaved variety, *horridulum*. Hamdoun (1967) studied the regenerative capacity of the plant at four times during the year in England. Shoots were produced more vigorously in April than in July, October, or December. He found that roots were formed mostly at the apical ends of the segments and shoots were formed toward the basal ends. Drying by exposure to a relative humidity of 58 percent for 1 day killed all stem segments, as did an exposure for the same period to a temperature of -4° C. Root segments survived 5 weeks at 0° C. The optimum temperature for shoot production from segments was 15° C. Shoots were able to emerge from a root fragment of 2.5 cm when it was buried at 50 cm. To summarize, we may say that vegetative reproduction cannot proceed if there is competition for space, if plants are shaded, or if the soil is too wet or poorly aerated.

The reproduction by seeds and their subsequent dissemination to new areas is an interesting story. *C. arvense* is a long-day plant. Most ecotypes flower well at 18 hours of daylight but will not flower at 8 or 10 hours. Most plants are dioecious and are pollinated by insects; their seeds are disseminated by flight of the plumed achene, by water, by dirty crop seeds, by sticking to clothes and farm vehicles or adhering with mud to fur and feathers or to the feet of men and animals.

The flowers are on a common stem axis in clusters of 100 called a "head." All flowers are perfect, each bearing stamens and a pistil, but functionally they are either staminate (pollen-bearing) or pistillate (seed-bearing). The pistils of the staminate flowers are abortive and the stamens of the pistillate flowers are not functional. The normal stamens are often the color of the corolla whereas the aborted ones are purple. There are many disagreements in the literature about the possibility that some flowers may be perfect. Detmers (1927) grew plants from seed to seed production and found only dioecious plants. Now that we have available many more studies of the flower and have some knowledge of the variability among the ecotypes of the plant, we have become aware that in most places in the world the majority of the flowers are imperfect but that occasionally perfect flowers are produced. Hodgson (1968) found staminate plants bearing both pollen and seeds. Isolated colonies of pistillate plants thus bear no seeds and produce achenes which are thin and shriveled. Hayden (1934) suggested that seed production can take place when colonies of plants of the different sexes are within 60 to 90 m of one another—a distance that can be worked by insects. In the Netherlands, Bakker (1960) found the best seed production to be at a distance of 20 m but found that some seeds were produced when the distance was 50 m. In his populations male and female plants occurred in a ratio of 1 to 3. Perfect flowers are selfed by the opening and closing of flowers at dawn and dusk. The female florets have a strong scent resembling vanilla. Male florets are not strongly scented.

In a survey in northern Iowa (United States), Hayden (1934) found that about 50 percent of the heads on the plants bore seeds, each head had about 100 flowers, and there was an average of 46 seeds per head. The flowers at the fringes matured earliest. Mature seeds could be found about 10 days after the flowers opened (Derscheid and Wallace 1959, Derscheid and Shultz 1960). Many mature seeds are destroyed by the larvae of several insects. The number of seeds produced per flowering stalk varies but as many as 40,000 have been recorded. Hayden (1934) obtained 95-percent germination of new seeds. Kolk (1962) found that the age of the seeds influenced their response to light. Young seeds germinate well in bright daylight and old seeds respond best to weak light. High temperatures almost consistently retard the germination of weed seeds, although *C. arvense* germinated best at 30° C. Seeds stored in wet sand during the first winter germinate better than those kept in dry storage. Bakker (1960) also found good germination at the high temperature, but germination was also good at an alternating temperature of 10° to 28° C. This approximates conditions on the polder in spring. At temperatures of 15° to 20° C high light enhanced germination. The seeds germinated in a clay loam which was at 40 to 50 percent of its water-holding capacity and was well aerated. When the polders were first drained there was little aeration of the surface layers so *C. arvense* was not able to be established for a time.

Kollar (1968) studied the germination and viability of the seeds after they had matured in winter wheat at various depths in the soil. Some seeds matured and dropped before the wheat harvest. Fifteen percent of these germinated on the soil surface. When seeds were buried at depths of 0.25 to 1.5 cm, only 4 percent emerged from the most shallow depth. Bakker (1960) also found that the seedlings can penetrate only a very thin layer of soil. *C. arvense* seeds may be viable after 20 years of storage in the soil. Viability is greater when the seeds have been buried at 100 cm than at 55 or 20 cm (Toole and Brown 1946). Seeds will germinate after 4 years of storage in water.

Small (1918) elaborated on the flight of the plumed seeds of the *Asteraceae*. Many assume that the flight of the plumed achenes of *C. arvense* is the chief source of its long distance dispersal, but few studies have been made to determine whether these are only the plumes or whether the achenes are attached. After examining the plumed achenes carefully, Bakker (1960) found that many plumes broke off and drifted away while the achenes remained in the head. Until more information is available, we may assume that the seeds are distributed by all of the conventional means available to most weed seeds. Distribution of these seeds in crop seeds is a serious source of infestation on all continents. Feeding trials have been conducted with several species of birds to determine whether the seeds may move long distances in the guts of migrating birds. All results indicate that the seeds are destroyed after ingestion.

AGRICULTURAL IMPORTANCE

Combinations of tillage, herbicides, and rotations have been used recently in areas where wheat, temperate-zone grains, and pastures are infested with *C. arvense*. These methods have been so successful that the plant has not been reported as a serious or principal weed of late. It still ranks as a troublesome species and a constant battle is required to hold it at bay. It is a weed of barley, flax, millet, oats, rye, sorghum, wheat, and other cereals in the following countries: Bulgaria, England, Finland, Germany, India, Iran, New Zealand, the Soviet Union, the United States, and Yugoslavia. It is a problem weed in pastures and perennial forage crops in Canada, the United States, and much of northern Europe. It is a weed of beans, peas, and other vegetables in places as widely separated as Bulgaria, Iran, and New Zealand. It is a weed of vineyards in southern Europe; orchards in Canada, the Soviet Union, the United States, and several places in northern and central Europe; of rape in New Zealand, sunflowers in Romania and Sudan; and of corn in New Zealand, Romania, the Soviet Union, the United States, and Yugoslavia.

The weed is generally respected for the trouble it can cause in cereals and pastures; but the variety of crops it infests and the seriousness of the weed when uncontrolled, though growing in widely separated locations, cannot be overemphasized. It is believed to be the most common, troublesome thistle in Great Britain. It is the most common perennial weed in spring cereals in Finland. It is widely established in New Zealand and Australia. In 1952 it infested more farm acreage than any other weed in the four northwestern states of the United States. Its invasion of the newly drained polders of the Zuider Zee in the Netherlands has been discussed. The weed is increasing in importance as a weed in vineyards in southern France. The seed is reintroduced to fields as an impurity in seed stocks of small grains, alsike and white clovers, alfalfa, and several other forage crops.

Derscheid and Wallace (1959) reported wheat yield reductions of 15, 35, and 60 percent with stands of 2, 12, and 25 shoots per 0.8 sq m, respectively. Higgins (1967) reported that severe infestations of *C. arvense* reduced potato yields by 75 percent. The size and quality of the potatoes were affected, and the weed vegetation interfered with harvest operations. Schreiber (1967) studied losses and gains in alfalfa production under several densities of competition from *C. arvense*. In an elaborate experiment, 22,000 vegetative portions were placed in the field to provide plants in proper spacing for grazing trials, mowing, and herbicide treatments. Plots were grazed for 4-day periods two, three, or four times yearly. Some plots were mowed immediately after grazing. Records were kept of animal consumption and forage harvested by mowing. Total production loss was 16.5 metric tons per hectare of alfalfa during the 4 years where thistle stands had been planted two plants per 30 cm², and animal consumption had been reduced by 10.5 metric tons per hectare at this level of competition. Mowing the sward to 5 cm height after each grazing period practically eliminated the thistles in 4 years, and alfalfa production was increased by 13.9 metric tons per hectare for the period.

In many areas the proper timing of tillage operations has resulted in good control. The food reserves of the thistle are at their lowest point about 4 to 6 weeks after the first plants have emerged in spring. This is the very early bud stage (Hodgson 1971). The area should be plowed, disked, and then cultivated to a depth of 8 to 10 cm when the first shoots appear; repeated cultivations should be made at about 21-day intervals for the rest of the season. In many areas this program will eliminate almost all of the plants and the remainder can be destroyed by tillage during the following spring. Competing crops can inhibit thistle growth severely, and winter wheat and barley which germinate quickly in spring (*C. arvense* emerges late) serve this purpose well. Alfalfa recovers more quickly than does the thistle after being mowed and thus is an effective competitor. Phenoxy herbicides are used widely on this species, and, if applied at the early bud stage or when the plants are growing actively, will eliminate the weed in 2 to 3 years.

Experiments have been conducted in Canada on use of the insect *Altica carduorum* for biological control of *Cirsium arvense* in Canada, but the data collected so far are not encouraging.

COMMON NAMES
OF CIRSIUM ARVENSE

AUSTRALIA
  Canada thistle
  creeping thistle

CHILE
  cardo

DENMARK
  ager-tidsel
  mark-tidsel

ENGLAND
  corn

FINLAND
  pelto-ohdake

FRANCE
  chardon des champs
  cirse des champs
  sarrette des champs

GERMANY
  Ackerdistel
  Acker-Kratzdistel
  Feldkratzdistel

INDIA
  thistle

ITALY
  scardaccione
  stoppione

JAPAN
  ezonokitsuneazami

NETHERLANDS
  akkervederdistel

NORWAY
  akertistel

SOUTH AFRICA
  Kanadese dissel

SPAIN
  cardo

SWEDEN
  akertistel

TUNISIA
  cirse des champs

UNITED STATES
  Canada thistle

YUGOSLAVIA
  palamida

❧ 31 ❧

Commelina benghalensis L., *Commelina diffusa* Burm. f. (= *C. nudiflora sensu* Merr., *non* L.), and *Murdannia nudiflora* (L.) Brenan (= *Commelina nudiflora* L., *Aneilema nudiflorum* [L.] Wall., and *Aneilema malabaricum* [L.] Merr.)

COMMELINACEAE, SPIDERWORT FAMILY

Commelina benghalensis, *C. diffusa*, and *Murdannia nudiflora* are fleshy, herbaceous, creeping annuals or perennials, and are natives of the Old World tropics. Their sprawling, creeping habit, coupled with their ability to root readily at the nodes, causes them to be problem weeds in perennial as well as vegetable and cereal crops. *Commelina benghalensis* is reported as a weed in 25 crops in 28 countries, *C. diffusa* is reported in 17 crops in 26 countries, and *Murdannia nudiflora* is reported in 16 crops in 23 countries. Although livestock will eat them, these plants have too high a moisture content to be of much forage value.

DESCRIPTIONS
Commelina benghalensis

C. benghalensis is a somewhat fleshy or succulent creeping *annual* or *perennial* herb (Figure 86); *stems* creeping-ascending, 15 to 40 cm long, branched, rooting at the nodes; *leaves* ovate or elliptic, acuminate, 3 to 7 cm long, 1 to 2.5 cm wide, the base narrowed into a petiole; *bracts* subtending the flower (spathe) funnel-shaped, about 1.5 cm long and wide; *flowers* lilac or blue, the petals 3 to 4 mm long, anterior petal sometimes white, long-pedicelled at pollen-shedding; two anterior ovary cells two-ovuled, posterior ovary cell one-ovuled; two anterior cells of *capsule* open when mature (dehiscent), two-seeded or empty; *seeds* ribbed-rough (rugose), grayish brown, sometimes appearing sugarcoated, 2 mm long. The plant has white, burrowing rhizomes which can produce subterranean flowers and seeds. It may be distinguished from other blue-flowered species by the short flower stalk which does not extend above the spathe, by the partially joined spathe margins, and by the reddish brown hairs on the leaf sheath (Ivens 1967).

Commelina diffusa

C. diffusa is a smooth or sparsely hairy *annual* or

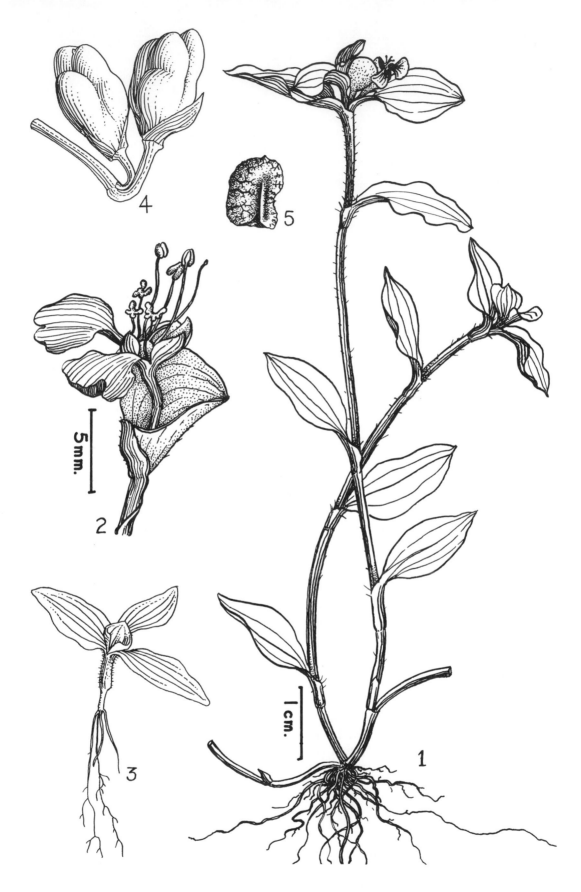

FIGURE 86. *Commelina benghalensis* L. *1*, habit; *2*, flower; *3*, seedling; *4*, fruits; *5*, seed.

perennial herb (Figure 87); stems creeping, ascending above, branching below; leaves lanceolate to broadly lanceolate, 4 to 6 cm long, 1 to 2 cm wide, gradually acute to acuminate; bract subtending the flower (spathe) broad, rounded or shallowly heart shaped at base, gradually tapering above to a rather acute apex, 2 to 3 cm long, 1.5 to 2 cm wide when unfolded; flowers blue, fertile stamens three, sterile stamens (staminodes) two, rarely three; capsule three-celled, five-seeded; seeds reticulate-ribbed, ridged on one side, and finely reticulated.

It is distinguished by its longer spathe with acute apex, its lanceolate to broadly lanceolate leaves, and by the seeds which are finely reticulated and ridged on one side.

Murdannia nudiflora

This species is a slender creeping herb (Figure 88), rooting at the nodes, annual or perennial, ultimate branches ascending, 15 to 30 cm high; leaf sheaths with long shaggy hairs sometimes smooth; leaves oblong-lanceolate, acute, 3 to 7 cm long, 1 to 2 cm wide; inflorescence axillary or terminal, peduncled, the short-lived flowers clustered near the ends of few branches, petals oval-obovate, rounded, pink-purplish or violet, about 4 to 6 mm in diameter; fertile stamens three; sterile stamens (staminodes) three, rarely two, one, or none, which are essentially trilobed at the apex; filaments conspicuously bearded; capsule subequal, dehiscing between the partitions, 4 to 6 mm; seeds smooth to coarsely reticulate.

It may be distinguished by the more or less regular flower and the conspicuously bearded filaments.

BIOLOGY AND DISTRIBUTION

Commelina benghalensis, C. diffusa, and Murdannia nudiflora, weeds of moist places, reproduce both by seeds and by stolons. They are weeds of the tropics and subtropics (Figures 89, 90, 91). One plant of Commelina benghalensis may produce as many as 1,600 seeds (Pancho 1964). The plants of all three root readily at the nodes of the creeping stems, and will do so especially when broken or cut. Cultivation may cut or injure these weeds, but stem cuttings may lie for several days to several weeks on the soil surface, finally to revive, take root, and grow again. The ability to grow rapidly from vegetative cuttings makes these weeds especially difficult to control in field areas. In Rhodesia C. benghalensis has been reported to produce underground stolons or branches with reduced leaves as well as closed modified flowers which set viable seed (Wild 1958). Such reproduction, of course, makes control more difficult.

Several species of Commelina are classified as wetland hydrophytes, plants which take root in water-saturated soils (King 1966). Although the classification holds true for the three species being considered, these plants, once well rooted, can live for extended periods after the soils have dried out. All three weeds grow best under conditions of high soil moisture and fertility. However, they can persist in sandy or rocky soils—even under fairly dry conditions—and will grow rapidly with the onset of rains.

C. benghalensis, C. diffusa and Murdannia nudiflora are found mainly as perennials in tropical and subtropical lowlands. They can also grow as annual plants in some temperate zone countries. In rice and other lowland crops they may be almost subaquatic, growing readily along dikes and on banks of irrigation ditches. They can withstand flooding and waterlogged conditions. They are present in cultivated lands, field borders, wet pasturelands, gardens, roadsides, and in waste places. They can become the dominant species in pastures.

AGRICULTURAL IMPORTANCE

The importance of Commelina benghalensis, C. diffusa, and Murdannia nudiflora relates to their great persistence in cultivated lands and to the difficulty with which they are controlled. Because they are well adapted to moist, perhaps swampy and even waterlogged, conditions, they compete easily with crop plants, making rapid vigorous growth. The plants form dense, pure stands, smothering out other plants, especially low-growing crops such as vegetables, pulses, or cereals. In pastures they grow rapidly over desirable grasses and legumes, competing with them for light and nutrients.

Figure 92 shows the ocurrence of Commelina benghalensis in important world crops. C. benghalensis is one of the three most important weeds in coffee seedling plantings in Tanzania and is one of the principal weeds of corn in Kenya. It is a principal weed of sugarcane in South Africa; coffee in Kenya and Tanzania; upland rice in India and the Philippines; cotton in Kenya, Mozambique, and Uganda; tea in India; wheat in Angola; and soybeans in the Philippines.

It is also reported as a common weed of rice in Ceylon; sugarcane in India, Mozambique, and the Philippines; cassava in Taiwan; corn in Angola,

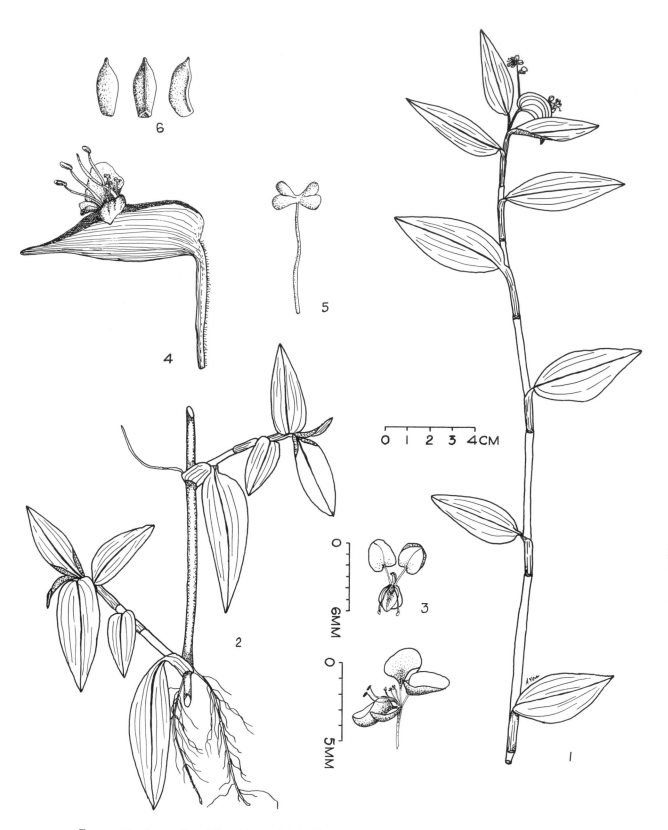

FIGURE 87. *Commelina diffusa* Burm. f. *1*, habit; *2*, portion of stem with leaf sheath; *3*, flowers, note petal shape; *4*, flower, with bract; *5*, sterile stamen; *6*, seed, three views.

FIGURE 88. *Murdannia nudiflora* (L.) Brenan. *1*, habit; *2*, node and leaf sheath; *3*, inflorescence; *4*, seed, two views; *5*, seedlings.

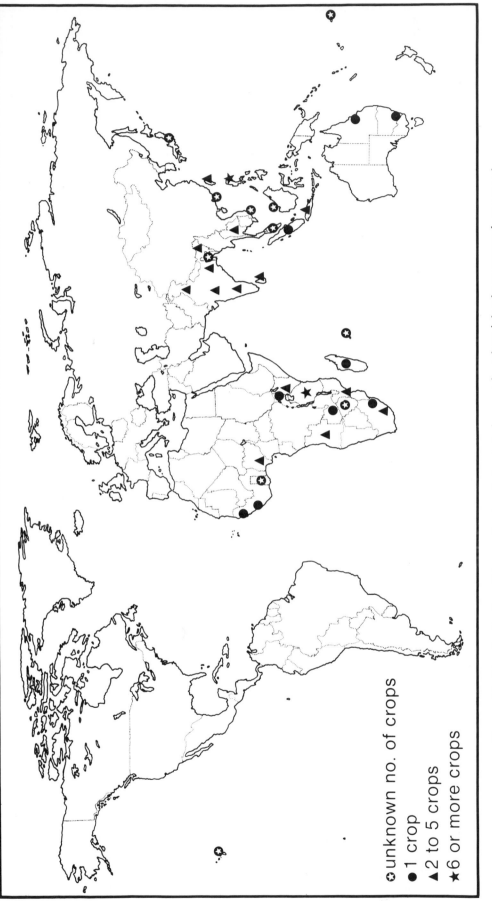

FIGURE 89. The distribution of *Commelina benghalensis* L. in countries where it has been reported as a weed.

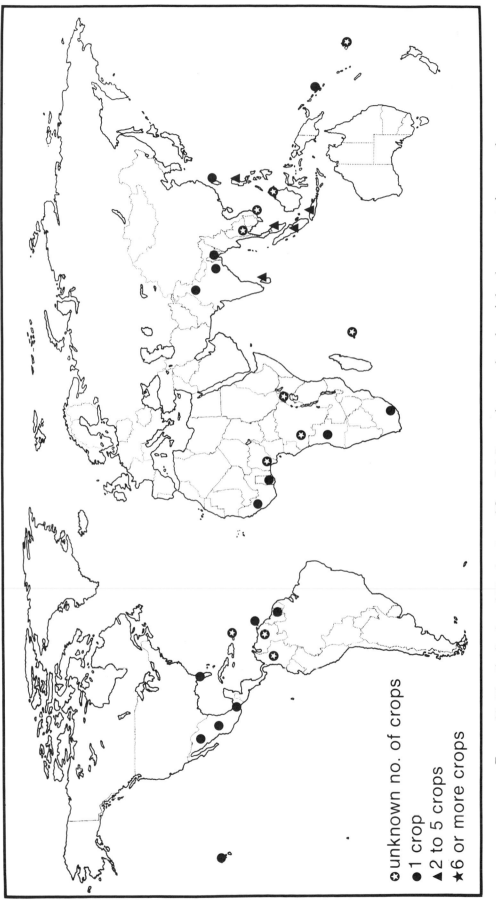

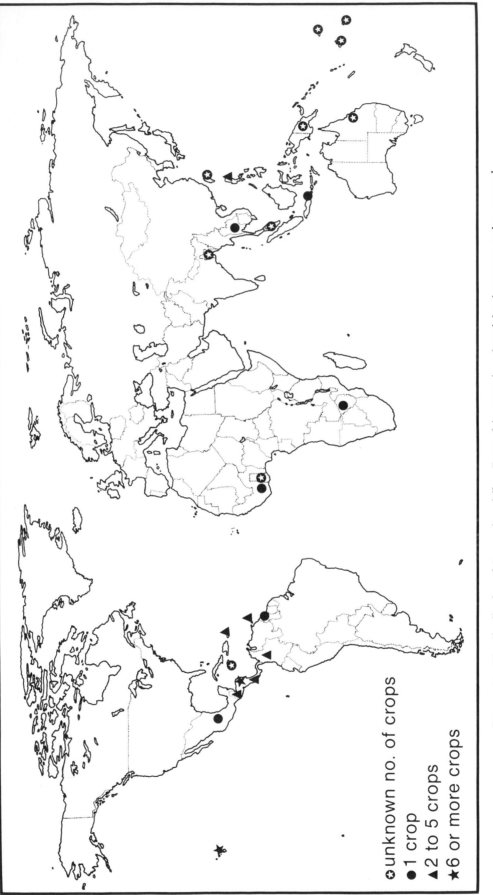

FIGURE 91. The distribution of *Commelina diffusa* Burm. f. in countries where it has been reported as a weed.

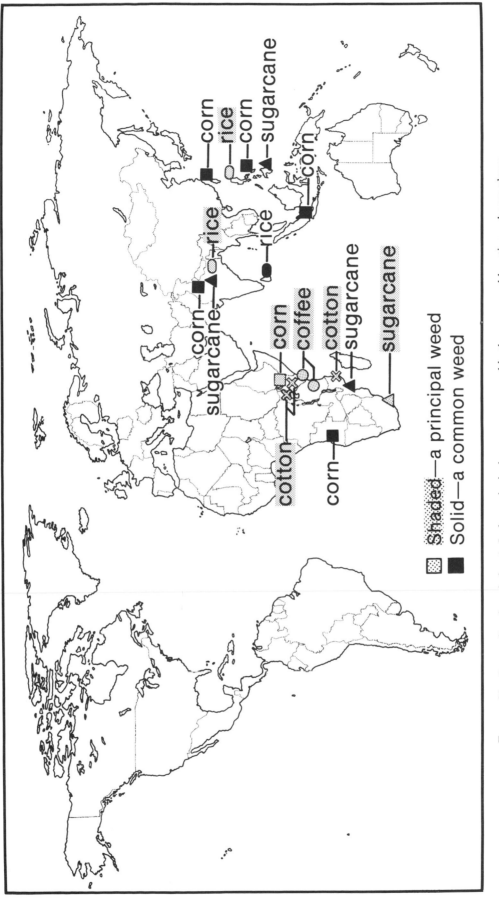

FIGURE 92. *Commelina benghalensis* L. is a principal or common weed in the crops and locations shown above.

India, Indonesia, the Philippines, and Taiwan; peanuts in India and the Philippines; pineapples in Taiwan and Swaziland; cowpeas and sorghum in the Philippines; tea and citrus in Mozambique; and roselles in Indonesia. It is a weed of barley, jute, sisal, beans, pastures, sweet potatoes, vineyards, and cereals in many countries of the world.

C. benghalensis can be an alternate host of the nematode *Meloidogyne incognita* (Valdez 1968) and of the groundnut rosette virus (Adams 1967).

Commelina diffusa is a principal weed of bananas in Mexico and Hawaii; of beans, citrus, coffee, and cotton in Mexico; of papayas in Hawaii; of sugarcane in Puerto Rico; and of sorghum in Thailand. It is a common weed in pineapples and sugarcane in Hawaii. It also has been reported as a weed in corn and vegetables in Mexico; bananas, papayas, and pineapples in the Philippines; rice in Bali, Colombia, and Mexico; sugarcane in Mexico and Trinidad; taro and pastures in Hawaii; and coffee in Costa Rica.

C. diffusa is an alternate host of *Cuscuta filiformis* L., *Cuscuta sandwichiana* Choisy (Raabe 1965), and *Meloidogyne arenaria* Chitwood (Valdez 1968); and of the viruses which produce cucumber mosaic and spotted wilt (Raabe, unpublished; see footnote, *Cyperus rotundus*, "agricultural importance").

Murdannia nudiflora is reported to be a principal weed of peanuts in Indonesia; tea in Ceylon; and bananas, citrus, and coffee in Mexico. It is a common weed of lowland and upland rice in Ceylon and Indonesia, of corn in Indonesia, and of oil palms and rice in Malaysia. It has also been reported as a weed in cocoa and rubber in Malaysia; tea in Indonesia; pineapples in Guinea, Hawaii, and the Philippines; sugarcane in Angola, Hawaii, Indonesia, Natal, the Philippines, and Taiwan; taro and pastures in Fiji; rice in India, Surinam, Malaysia, and the Philippines; taro in Hawaii; and corn, rice, sugarcane, and vegetables in Mexico.

M. nudiflora is an alternate host of *Pratylenchus pratensis* (de Man) Filip., *Meloidogyne* sp. (Raabe, unpublished; see footnote, *Cyperus rotundus*, "agricultural importance"), *M. arenaria* Chitwood (Valdez 1968), *Pythium arrhenomanes* Drechs. (Sideris 1931*b*); and of the viruses which cause cucumber mosaic (anon 1960*a*) and the southern celery mosaic (King 1966).

The plants are sometimes used as famine vegetables in India and for poultices and animal fodder in Indonesia, Malaysia, and Africa (Burkill 1935). Years ago in Hawaii *Commelina diffusa* plants were cut and fed as green forage to dairy cattle (Pope 1968).

COMMON NAMES

Commelina Benghalensis

ANGOLA
  ndakala
BANGLADESH
  kanaibashi
BURMA
  myet-cho
EASTERN AFRICA
  wandering Jew
INDIA
  kanasiri (Oriya)
  kanchara
  kankaua
  kena (Marathi)
  konasimalu
    (Assamese)
  krisnaghas
  mankawa
INDONESIA
  gewor (Indonesian)
JAPAN
  tsuyukusa
KENYA
  wandering Jew
MAURITIUS
  herbe aux cochons

NATAL
  wandering Jew
PHILIPPINES
  alikbangon (Tagalog)
  bias-bias (Pampangan)
  kuhasi (Ivatan)
  kulkulasi (Ilokano)
  sabilau (Bisayan)
RHODESIA
  wandering Jew
SOUTH AFRICA
  Bengaalse commelina
TAIWAN
  ju-ye-tsai
TANZANIA
  wandering Jew
THAILAND
  pak prarb
UGANDA
  Mickey Mouse
ZAMBIA
  wandering Jew

Commelina diffusa

BANGLADESH
  manaina
COLOMBIA
  siempreviva
  suelda
HAWAII
  dayflower
  honohono-kukui
  wandering Jew
INDONESIA
  brangbangan
    (Javanese)
  gewor lalakina (Sundanese)
JAMAICA
  water grass
JAPAN
  shima-tsuyu-kusa

MEXICO
  carutillo
  empanadilla
NICARAGUA
  cohitre
PHILIPPINES
  alikbangon (Tagalog)
  bangar-na-lalake
    (Ifugao)
  kolasi (Ilokano)
PUERTO RICO
  cohitre
THAILAND
  phak-prap
TRINIDAD
  water grass
UNITED STATES
  spreading dayflower

Murdannia nudiflora

MALAYSIA
  common spiderwort

rumput kupu-kupu
rumput sur

MAURITIUS
  herbe aux archons
MEXICO
  cohitre
  comelina
  maclalillo

PHILIPPINES
  alikbangon (Tagalog)
  bangar   na   lalake
    (Ifugao)
  katkatauang (Bontoc)
  kohasi (Ivatan)

  kolasi (Ilokano)
SURINAM
  gadodede
THAILAND
  phak-prap
UGANDA

  Mickey Mouse
  vanda
UNITED STATES
  spreading dayflower
VENEZUELA
  suelda con suelda

❧ 32 ☙

Cyperus difformis L.

CYPERACEAE, SEDGE FAMILY

Cyperus difformis is a small, tufted, annual sedge that is a native of the Old World tropics. It has been reported as a weed from lat 35° S to 45° N. A prolific seed producer which grows mostly in lowland or flooded areas, it is a weed of rice in 46 countries. However, it is sometimes reported to be a weed in upland rice and in other upland crops such as bananas, sugarcane, tea, and corn.

DESCRIPTION

C. difformis is a tufted, smooth, *annual* or infrequently *perennial* sedge (Figure 93), 10 to 75 cm high; *stems* rather weak or tender, triangular, slightly winged, smooth, 1 to 3 mm thick; *roots* numerous, fibrous, reddish; *leaves* linear, few, weak, rather abruptly acuminate, smooth or slightly rough at the top, usually shorter than the plant is high (often two-thirds of the height), 2 to 5 mm wide, sometimes reduced to sheaths; *sheaths* tubular united, the lower sheaths straw-colored to brown; *inflorescence* in dense, globose, umbellate *heads*, simple or compound, 8 to 15 mm across, with numerous (more than 40) stellately spreading spikelets, the inflorescence rather loose, simple or compound, subtended by one, two, three, or four, variable, leaflike *bracts* which are contracted or spreading, one may be almost erect, up to 25 cm long; the *umbel rays* 1 to 5 cm long, some sessile, some long peduncled; *spikelets* linear to oblong-linear, compressed but slightly swollen, obtuse, 2.5 to 8 mm long, 1 to 1.25 mm wide, 10 to 30 flowered, remarkable for the numerous, very small (0.6 to 0.8 mm), almost orbicular, concave, very obtuse *glumes* which occur in two opposite rows, closely packed, at first a rich brown, dirty green, or blackish, but becoming variegated by their pale yellow or almost white margins, with a green keel; *stamens* one or two; style branches three; *achenes* elliptic to slightly obovate, acutely subequally triangular, lightly pitted, shining, straw-colored, yellowish brown or pale brown, about 6 mm long.

This species may be distinguished by the dense, globose heads composed of many radiating spikelets and by the many small orbicular, obtuse glumes which are a rich brown at first but later become variegated by their pale yellow or almost white margins.

DISTRIBUTION AND BIOLOGY

C. difformis is widespread throughout southern Europe, Asia, Central America, North America, Africa, and the islands of the Indian and Pacific oceans (Figure 94). It is mainly a weed of the tropics and subtropics but can be found from lat 35° S to 45° N; it has been reported as a weed in 46 countries.

The plant normally grows in flooded or in very moist soils. It is primarily a weed of paddy or flooded rice. It is frequently found in small pools, along rivers, canals, and streams, in open wet places, and in grassy swamps. It grows best in rich, fertile soils,

FIGURE 93. *Cyperus difformis* L. *1*, habit; *2*, portion of culm; *3*, spikelet; *4*, bract; *5*, flower; *6*, achene; *7*, achene, cross section; *8*, seedling.

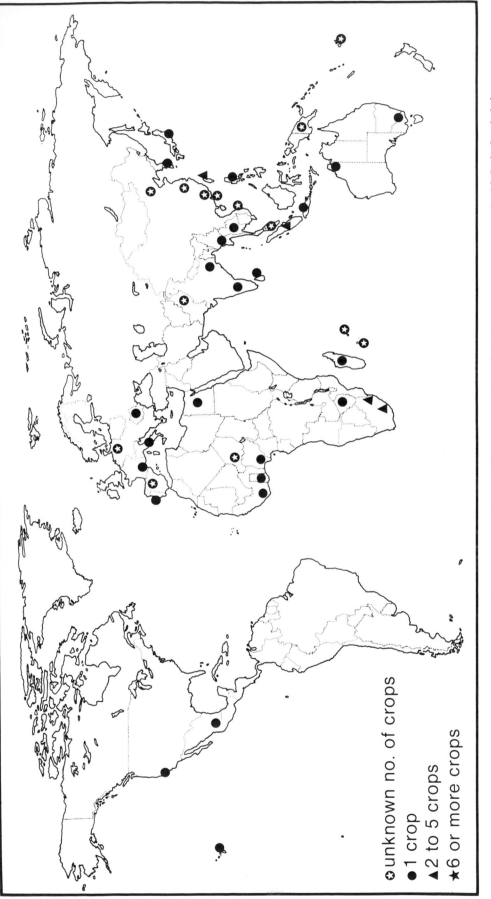

○ unknown no. of crops
● 1 crop
▲ 2 to 5 crops
★ 6 or more crops

FIGURE 94. The distribution of *Cyperus difformis* L. across the world in countries where it has been reported as a weed, chiefly in flooded rice.

but can grow in poorer sandy or clay soils of unused lands or in fallow rice fields.

C. difformis is not often found in upland soils but may be a problem weed on lowland soils used for upland cropping. It can grow from sea level to over 2,000 m in Madras, India, and has been reported from sea level to 1,400 m in Indonesia.

There is evidence that the plants cannot stand deep flooding; indeed, deep-water management has been used to control them in some cases.

Usually annual, *C. difformis* is propagated by seeds which are produced in large quantities. In an early study in Italy, Jacometti (1912) reported that one plant could produce 50,000 seeds with about 60-percent germination. Having heavy seed production and massive seedling densities, the plant quickly covers the ground and becomes dominant. The sedge may become a major weed in rice when herbicides are used which control weedy grasses but which do not kill sedges like *C. difformis*.

An indication of the potential rapidity of spread of *C. difformis* can be obtained from work in Nigeria (Vaillant 1967) where plants completed a vegetative cycle in 1 month in June, and produced a new generation of weeds in August. Growing in densities of 480 to 1,200 plantlets per square m, it was the most numerous sedge present.

During field surveys in Taiwan (Lin, 1968), *C. difformis* was found to be the most prevalent sedge present in paddy rice. It was fifth in fresh weight production behind *Monochoria vaginalis*, *Echinochloa crusgalli*, *Sagittaria trifolia*, and *Fimbristylis miliacea*.

In tropical areas the plant apparently can flower and produce seeds all year long provided sufficient soil moisture is present. The plant flowers and fruits all through the year in Hawaii and the Philippines. It has been reported to flower and fruit following rains in India, to flower during June to October in Burma, and to flower in June in Vietnam. All of these countries are subject to a monsoonal climate.

AGRICULTURAL IMPORTANCE

C. difformis is a widespread and serious weed of paddy rice. It is important because it produces abundant seed and rapidly becomes a dominant weed in the fields. Although it is not a tall shading weed like *Echinochloa crusgalli*, it does produce abundant plants per unit area, thereby forming dense, solid mats of vegetation in the young rice crop. The weed appears to be more of a competitor for water and nutrients than for light, and the plant can produce a very large fresh weight yield per hectare. Also, the ability of the plant to complete a vegetative and reproductive cycle within a month or so makes it especially competitive in a crop which requires at least 90 or more days to reach maturity.

C. difformis has been reported as a serious weed of rice in Australia, Italy, Japan, Madagascar, Taiwan and the Philippines; and as a principal weed of rice in Ghana, Mexico, Nigeria, Portugal, Romania, and Swaziland. It is ranked as a common weed in rice in Ceylon, India, Indonesia, Thailand, and the United States. It is a weed of upland rice in the Philippines; lowland rice in Korea; bananas, corn, sugarcane, tea, and pastures in Taiwan; lowland taro and pineapples in Hawaii; and rice in Egypt, France, and South Africa.

COMMON NAMES
OF CYPERUS DIFFORMIS

AUSTRALIA
  dirty Dora
JAPAN
  tamagayatsuri
  umbrella plant
KOREA
  albangdongsani
PHILIPPINES
  baki-baki
  ballayang (Ilokano)

  gilamhon (Bisayan)
PORTUGAL
  negrinha
TAIWAN
  chyou-hwa-hau-tsau
THAILAND
  kok ka-narg
UNITED STATES
  small-flowered umbrella plant

📚33📚

Cyperus iria L.

CYPERACEAE, SEDGE FAMILY

Cyperus iria, a native of the Old World tropics, is an annual, tufted, tall, herbaceous sedge that is mainly a problem in the rice fields of Asia. It is reported to be a weed in 17 crops in 22 countries.

DESCRIPTION

C. iria is an *annual*, herbaceous sedge (Figure 95); *roots* yellowish red, fibrous; *culms* tufted, 20 to 60 cm high; *leaves* linear-lanceolate, usually all shorter than the culm, 3 to 6 (occasionally 8) mm wide, the margins somewhat rough on the upper part; *sheath* enveloping the culm at base, membranous; *inflorescence* simple or compound, usually open, up to 20 cm long; *leaf bracts* (involucre) three to five (occasionally seven), the lower one longer than the inflorescence; *spikes* elongate, rather dense; *spikelets* erect-spreading, crowded, six- to 24-flowered, 5 to 13 mm long, 1.5 to 2 mm wide, yellow; *stigmas* three; *glume* broad-ovate, 1 to 1.5 mm long; *fruit* a small achene, slightly shorter than the glume, the terminal half broader than the base (obovate), brown, triangular in cross section, 1 to 1.5 mm long.

This species is distinguished by its yellowish red, fibrous roots; its yellowish, usually open, inflorescence; and by the lowest bract of the flower always being longer than the inflorescence.

DISTRIBUTION

There is no biological information available on *Cyperus iria*. From Figure 96 it may be seen that the weed is most often found in an area from Japan south through the Pacific Islands and Australia and west through India. Outside Asia it has only been reported in southern and western Africa and in the United States. It is mainly a weed of open, wet places. It spreads by seeds. In the Philippines, one large plant may produce about 5,000 seeds.

AGRICULTURAL IMPORTANCE

C. iria is principally a weed of rice in the world. It is one of the three most important weeds of rice in Ceylon, India, and the Philippines; it is a principal weed in Indonesia and Japan; a common weed in Fiji, Thailand, and the United States; and it is a weed of rice in Cambodia, Korea, Senegal, Swaziland, and Taiwan. It is a principal weed also of tea in Japan, vegetables in Malaysia, pastures in India, peanuts in Indonesia, and soybeans in the Philippines. It is a common weed of bananas in Taiwan and corn in Malaysia.

In Taiwan it is also a weed in pineapples, sugarcane, cassava, tea, and corn; in Japan, a weed of soybeans, vegetables, and sweet potatoes; in Indonesia, a weed of sugarcane, corn, and soybeans. It is a weed of sugarcane in Australia.

COMMON NAMES OF CYPERUS IRIA

BANGLADESH	yellow sedge
barachucha	INDONESIA
FIJI	djekeng (Javanese)
sedge	JAPAN
INDIA	kogomegawatsuri
morphula	umbrella sedge

240

KOREA
  chambangdonsani
MALAYSIA
  grasshopper's
  cyperus

PAKISTAN
  khana
PHILIPPINES
  alinang
  paiung-paiung

taga-taga (Bikol)
okoiang (Bontoc)

TAIWAN
  sha-tsau

THAILAND
  kok huadaeng
  ya-kok-sai
UNITED STATES
  umbrella sedge

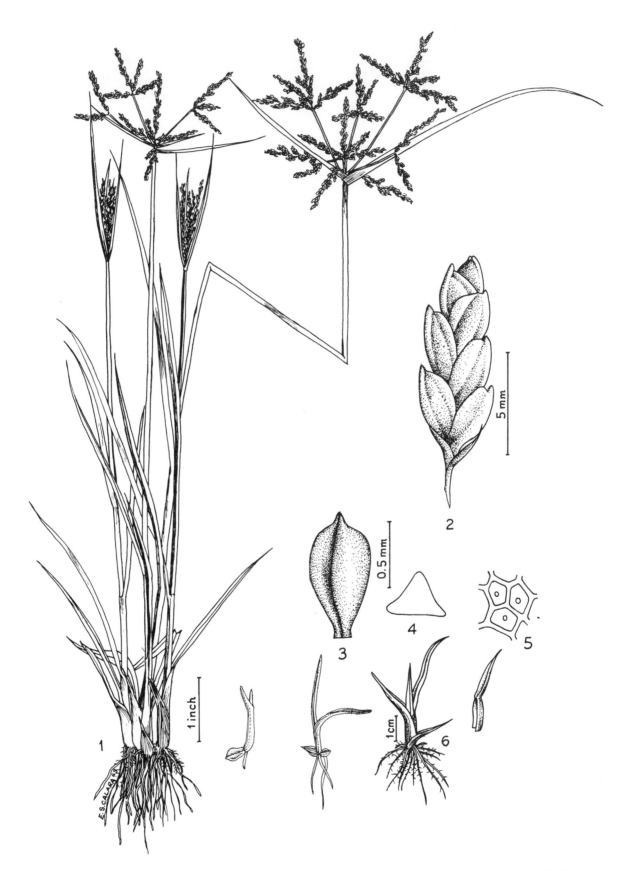

FIGURE 95. *Cyperus iria* L. *1*, habit; *2*, spikelet; *3*, achene; *4*, achene, cross section; *5*, detailed surface of reticulations; *6*, seedlings.

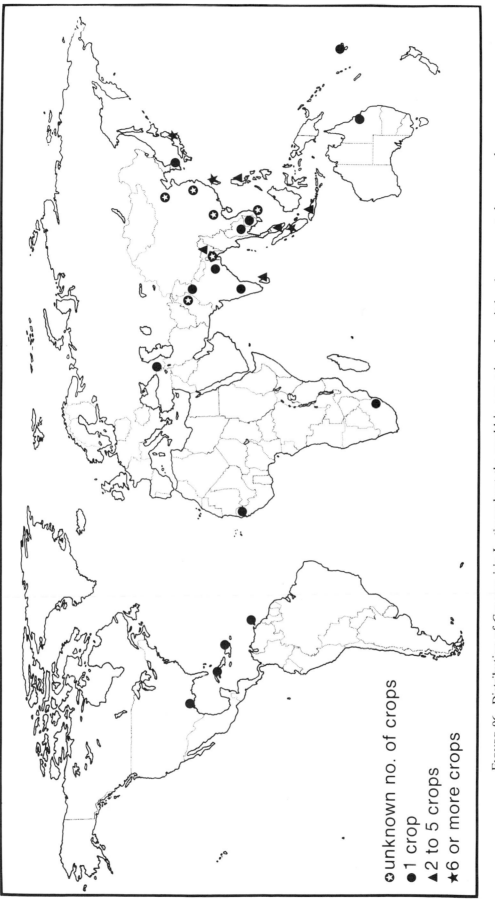

Figure 96. Distribution of *Cyperus iria* L. throughout the world in countries where it has been reported as a weed.

❧ 34 ❧

Dactyloctenium aegyptium (L.) Beauv.

POACEAE (also GRAMINEAE), GRASS FAMILY

Dactyloctenium aegyptium, an annual grass, is a native of the Old World tropics where it is a troublesome weed in corn, cotton, sugarcane, and peanuts. Its principal range as a weed is from lat 15° N to 15° S. Forty-five countries report that it is a common weed in 19 world crops. It is sometimes used for cattle forage. Its seeds have been used for human consumption on three continents in times of want. It is sometimes toxic to men and animals.

DESCRIPTION

D. aegyptium is a spreading to slightly ascending, *annual* or sometimes *perennial* grass (Figure 97), *culms* compressed, spreading with ascending ends, rooting at the nodes, branching, commonly forming radiate mats, often 20 to 40 cm high, sometimes up to 1 m; *sheaths* compressed, smooth; *ligule* a short membrane with a jagged edge ending in hairs; *blades* flat, smooth, or sparsely covered with soft hairs (pilose), with hairs swollen at the bases on both surfaces, particularly on the margins and nerves beneath; *inflorescence* in coarse spikes two to seven (rarely one), arranged in a fingerlike fashion (digitate), 1 to 3 cm long, 6 to 8 mm wide; *spikelets* 2.5 to 3 mm long; *glumes* one-nerved, somewhat rough on the keel, the lower 1.5 mm long, the upper 2.5 mm long with an equally long rigid awn; *lemma* 3 mm long with a stiff sharp point or tip (mucro), often recurved; *palea* shorter than the lemma; *grain* (caryopsis) about 1 mm long, obovate, with loose pericarp, irregular, transversely ridged, orange-brown.

The species can easily be distinguished by its two to seven coarse, digitate spikes; its lemma with stiff, sharp, often recurved tips; and the transversely ridged, orange-brown grains.

DISTRIBUTION AND HABITAT

D. aegyptium is one of nine species of this genus found mainly in the tropics but with some extensions into warmer parts of the temperate zone (Figure 98). It grows to lat 45° N in the United States but is not a problem weed there. It is a weed of cotton near the Tropic of Capricorn in Mozambique. So little is known about the developmental physiology and morphology of the species that it is difficult to understand the factors which govern its distribution. It is a weed of both arable land and waste places near the sea and, in Malaysia, is found in sandy soils of the lowlands and on sandy beaches at the sea. It prefers light, dry soils in Java and grows mainly on sandy areas in Sudan. Despite its preference for a habitat with light soils and low moisture, it has been reported to be an important weed in many countries in the humid tropics.

PROPAGATION

D. aegyptium is an annual grass which reproduces mainly by seeds but which has creeping or spreading stems which root at the lower nodes. Although the weed flowers all year round in warm regions, no research has been reported on its response to photoperiod. It can be a heavy seed producer for, in one

244

FIGURE 97. *Dactyloctenium aegyptium* (L.) Beauv. *1*, habit; *2*, ligule; *3*, spikelet; *4*, glumes, upper and lower; *5*, grain.

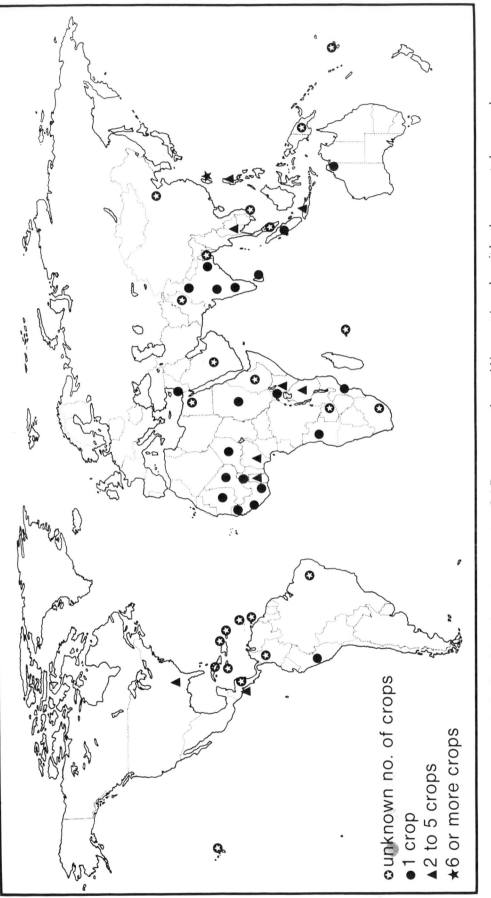

FIGURE 98. The distribution of *Dactyloctenium aegyptium* (L.) Beauv. across the world in countries where it has been reported as a weed.

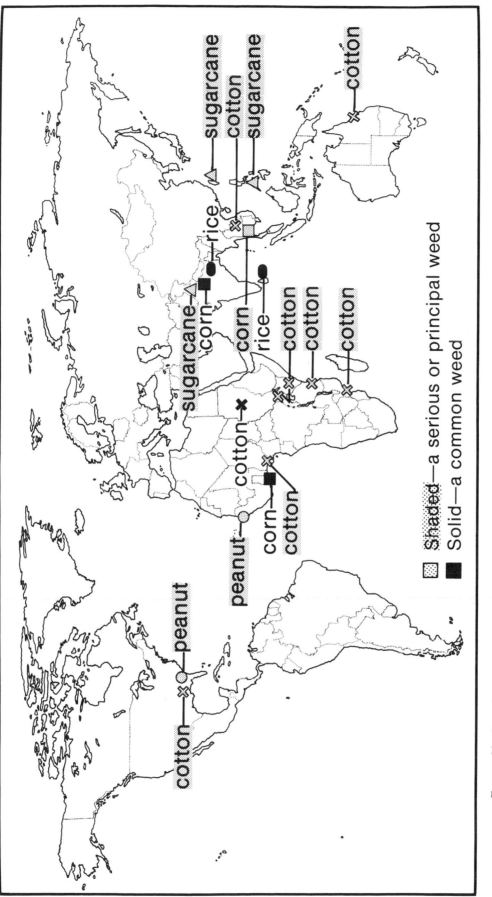

FIGURE 99. *Dactyloctenium aegyptium* (L.) Beauv. is a serious, principal, or common weed in the crops and locations shown above.

experiment, one plant produced 66,000 seeds. One sample of seeds which had been in dry storage in a glass container for 19 years still had 5-percent viable seeds. In the Ord River region, a tropical area in northern Australia, this species is a principal weed of cotton. Seed studies have shown that 50-percent germination can be obtained 5 months after harvest. At 10° C there is no germination; at 20° C germination is 80 percent; at 30° C, germination falls to less than 10 percent.

AGRICULTURAL IMPORTANCE

From Figure 99 it may be seen that *D. aegyptium* is a serious weed of cotton in Thailand and a principal weed in the following crops and locations: cotton in Australia, Kenya, Mozambique, Nigeria, Tanzania, Uganda, and the United States; sugarcane in India, the Philippines, and Taiwan; peanuts in Gambia and the United States; and corn in Thailand. It is a common weed of corn in Ghana and India; of rice in Ceylon and India; and of cotton in Sudan.

Although not shown on the map, it is a common weed of rice in Indonesia, and a weed in the following crops and locations; rice in Nigeria, the Philippines, and Thailand; pineapples in Taiwan and Thailand; coffee in Kenya and Tanzania; and tea in Taiwan. It is a weed of bananas, papayas, cassava, citrus, sweet potatoes, and millet in countries of Africa, Asia, and Central America.

Although the weed is readily eaten by cattle and poultry, it has been rated only fair to poor as a forage. It seems to be little used as a food grain by humans except in emergencies. In Africa the seeds are extracted to treat a kidney ailment. There are reports of toxicity to human beings and livestock following ingestion of seeds or leaves.

D. aegyptium is an alternate host of the viruses which cause rice leaf gall and corn leaf gall (Agati and Calica 1950) and sugarcane mosaic (Namba and Mitchell, unpublished; see footnote, *Cynodon dactylon,* "agricultural importance").

COMMON NAMES
OF DACTYLOCTENIUM AEGYPTIUM

AUSTRALIA
  coast button grass
BANGLADESH
  kakpaya ghash
BARBADOS
  crowfoot grass
BRAZIL
  mão de sapo
BURMA
  didok-chi
  myet-le-gra
CEYLON
  puta-tana
COLOMBIA
  estrella del mar
  paja de palma
  tres dedos
EASTERN AFRICA
  crowfoot grass
EGYPT
  naim el salib
  rigel el herbaya
HAWAII
  beach wiregrass
INDIA
  madana
  makra
INDONESIA
  sapadang babi
    (Indonesian)
  soeket dringoan
    (Javanese)
  tapak djalak (Sundanese)

JAMAICA
  crowfoot grass
KENYA
  crows foot
LEBANON
  naim el-salib
  rigal-ul-harbayah
MALAYSIA
  Egyptian finger grass
MAURITIUS
  chiendent
  pattes de poule
PERU
  pata de gallina falsa
PHILIPPINES
  damong balang
  krus-krusan
PUERTO RICO
  crowfoot grass
  yerba de egipto
SUDAN
  um assabia
TAIWAN
  ai-ji-jr-shu-tsau
THAILAND
  ya-pak-khwai
TRINIDAD
  crowfoot grass
UNITED STATES
  crowfoot grass
VIETNAM
  co chan ga

❧ 35 ❧

Digitaria adscendens (HBK) Henr.
(= *D. chinensis* Horn., *D. henryi* Rend.,
and *D. marginata* Link)

POACEAE (also GRAMINEAE), GRASS FAMILY

Digitaria adscendens is a nearly prostrate, mat-forming, annual grass that is said to be a native of Taiwan. It is a weed of 22 crops in 19 countries and is a problem in plantation crops, cereals, pulses, and vegetables. Although growing mostly in the tropics and subtropics, it is found as a weed from about lat 35° S to 40° N. It is a common wayside weed which is sometimes used for animal feed.

DESCRIPTION

D. adscendens is a tufted *annual*, sometimes *perennial*, grass to 60 cm high (Figure 100); *culms* 50 to 100 cm long, decumbent or geniculate, often rooting at lower nodes and copiously branched, forming dense mats, usually compressed, few- to many-noded; *leaf sheath* loose and thin, smooth or with tubercle-based hairs, often bearded at the base; *ligule* membranous, appearing as if cut off at the end (truncate), smooth, up to 1.5 mm; *blades* lanceolate to linear, somewhat rounded or gradually tapering at the base, flat, 4 to 20 cm long, 2.5 to 12 mm wide, weak, smooth or thinly hairy at mouth, with narrow thickened, often wavy, margins; *inflorescence* in *racemes*; *rachis* of racemes three-angled, distinctly green-winged; *racemes* two to nine, rarely more, subsessile, subdigitate, solitary or conjugate or sub-whorled along short main axis, ascending to spread-ing, green or somewhat purplish, usually forming compact delicate heads, 5 to 15 cm long, finely short hairy at base; *spikelets* appressed, 1.5 to 3 mm long, more or less acuminate; *first glume* a minute ovate scale; *second glume* ovate-lanceolate, acute, usually more than half the length of the spikelet, three-nerved, with rows of fine hairs between nerves and along margins; *sterile lemma* lanceolate, five- to seven-nerved; *fertile lemma* as long as the sterile lemma, one-nerved; *fruit* (grain) as long as the spikelet or slightly shorter, 1.5 to 2 mm long, oblong-lanceolate, acuminate, yellowish or purplish brown at maturity.

This species can be recognized by the absence of spreading hairs on the unevenly spaced nerves of the lower lemma and the three-angled and distinctly green-winged rachis of the racemes. *D. adscendens* may be confused with *D. sanguinalis* which differs in having rough nerves on the lower lemma, sparsely to densely hairy leaf blades with spreading tubercle-based hairs, evenly spaced nerves of the lower lemma (those of *D. adscendens* are usually unevenly spaced), and a less conspicuous ligule.

BIOLOGY AND DISTRIBUTION

D. adscendens is mainly a weed of the tropics and subtropics but does grow as a weed within the range

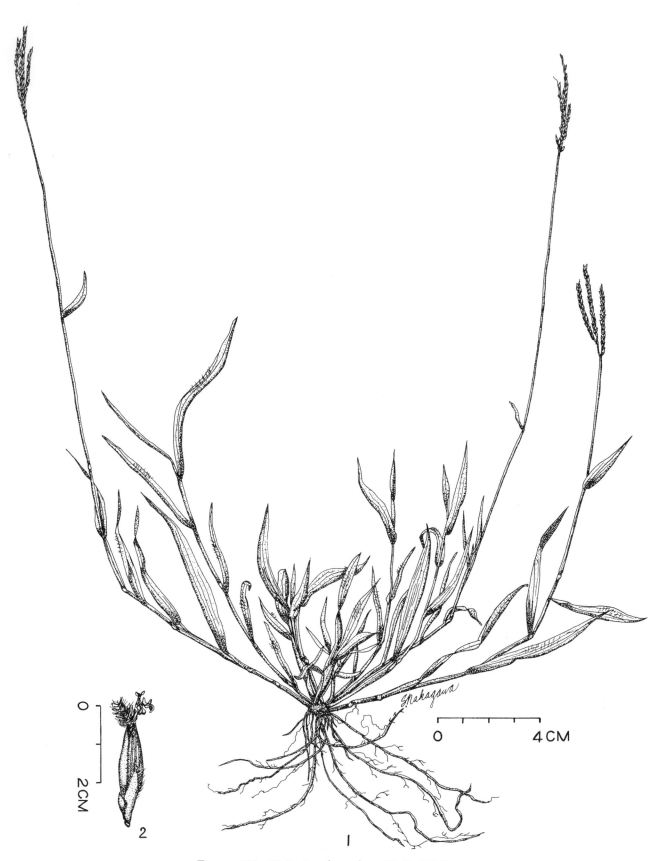

0

2CM

0   4CM

2

FIGURE 100. *Digitaria adscendens* (H. B. K.) Henr.
Above: *1*, habit; *2*, spikelet.
Opposite: Seedling, detail of the ligule.

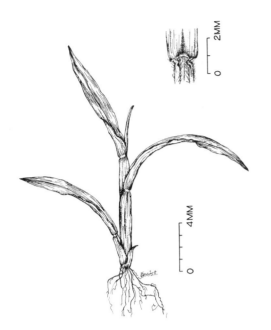

from lat 35° S to 40° N (Figure 101). Although found in Hawaii and Jamaica, it usually grows in countries of the Old World. It is usually found at low elevations and in warm regions. It is called summer grass in Australia where it is troublesome in field crops of the coastal areas. It frequently grows in lawns, lowland pastures, field crops, plantation crops, open ground, and disturbed areas. It grows from sea level to 1,600 m in Indonesia. Thriving in moist areas, it can become a major problem in irrigated lands.

An annual or rarely perennial under ideal year-round growing conditions, it reproduces by seeds. The plants may also spread and reproduce by rooting at the nodes and, in this manner, can form thick, nearly prostrate mats 0.6 m or more in diameter. The plants produce many seeds and several seed crops may be produced each year if grazing or mowing occurs.

Under field conditions, the seeds germinate in the spring or summer in more temperate climates or with the onset of rains in tropical areas.

During laboratory experiments in Australia (van Rijn 1968) germination was found to be 40 percent or more within a temperature range of 25° to 40° C, but was only about 10 percent at 20° C. No germination resulted at 10° and 50° C. Highest germination occurred at 27° C. Germination was 79 percent when fluctuating temperatures of 20° to 35° C were used. The seeds apparently do not germinate readily after maturity; for example, 7 months' storage was neces-

sary before 50-percent germination was obtained.

In Japan an incubation period of at least 45 days was required before the seeds began to germinate, and seeds collected from 100 plants in a random population showed wide variations in germination, dormancy, and in response to storage (Matumura, Takase, and Hirayoshi 1960).

AGRICULTURAL IMPORTANCE

D. adscendens is important because it competes aggressively with other plants by spreading and crowding. Its heavy seed production coupled with extensive rooting at the nodes makes it a threat in moist, warm conditions under heavy irrigation or incident rainfall. It can tolerate defoliation and may become dominant in lawns and playing fields.

It is a principal weed in plantation and field crops throughout the world (Figure 102). It has been ranked as the most important weed of tea, upland rice, corn, soybeans, vegetables, beans, potatoes, and sweet potatoes in certain areas in Japan. It is considered a serious weed of tea in Taiwan, vegetables in Malaysia, and cotton in Australia and Mozambique. It is a principal weed of pineapples in Australia; cassava, cotton, and sorghum in Thailand; corn in Ceylon, Malaysia, and Thailand; peanuts in Ceylon and Gambia; chilies, rice, and tea in Ceylon; sorghum and sweet potatoes in Malaysia; and cassava in Taiwan. It is also a weed in bananas, sugarcane, rubber, citrus, rice, and papayas.

The grass is sometimes grazed by cattle.

D. adscendens is an alternate host of the viruses which produce rice stripe disease (Shinkai 1955) and black-streaked dwarf disease of rice (Shinkai 1957).

COMMON NAMES
OF DIGITARIA ADSCENDENS

AUSTRALIA
    summer grass
CEYLON
    arisi pul (Tamil)
    guru tana (Sinhalese)
HAWAII
    Henry's crabgrass
INDIA
    suruwari
INDONESIA
    suket djrempak (Sun-

danese)
    sunbuk gangsir
        (Javanese)
JAPAN
    mehishiba
MALAYSIA
    bamboo grass
TAIWAN
    jya-ma-tang
THAILAND
    yah-tin-nok

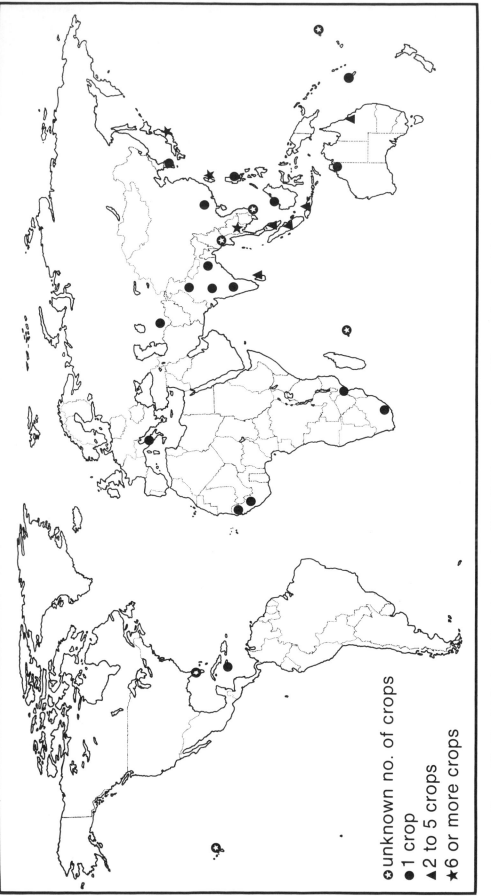

FIGURE 101. The distribution of *Digitaria adscendens* (H. B. K.) Henr. across the world in countries where it has been reported as a weed.

⊙ unknown no. of crops
● 1 crop
▲ 2 to 5 crops
★ 6 or more crops

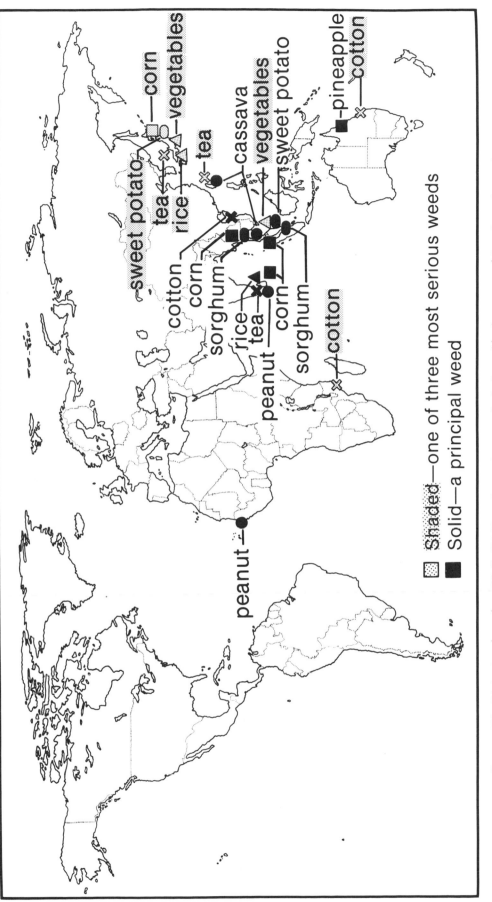

FIGURE 102. *Digitaria adscendens* (H. B. K.) Henr. is a serious or principal weed in the crops and locations shown above.

❧ 36 ❧

Digitaria scalarum (Schweinf.) Chiov.

POACEAE (also GRAMINEAE), GRASS FAMILY

Digitaria scalarum is a rhizomatous, creeping, perennial grass native to eastern Africa and may be the worst weed of the major crops of that area. It is not reported to be a troublesome weed of crops anywhere in the world except in this region. Even though it is palatable for cattle when young, it is not a productive enough grass to use for grazing.

DESCRIPTION

D. scalarum is a *perennial* grass (Figure 103) that creeps extensively by means of well-developed, usually branched *rhizomes*; *culms* are 20 to 50 cm high, simple, or branched near the base; *sheath* loosely to fairly densely hairy with minutely tubercle-based hairs; *ligule* conspicuous, membranous; *blades* 3 to 7 mm wide, expanded; *racemes* up to 9 cm long, two to 12, arranged subdigitately on a central axis, racemes rather loose, usually unequal in length, often divided and compound; *spikelets* usually in pairs on unequal pedicels, or some on the raceme, solitary, up to 2.5 mm long, acute, glabrous, spikelets toward the base may be reduced; *lower glume* a small scale, *upper glume* and *lower lemma* about equal in length and width, both longer than the upper floret and prominently nerved, the glume three- to five-nerved or frequently with an additional fainter nerve near each margin, the lemma seven-nerved. The branched, well-developed rhizome and the unequally pediceled spikelets are distinguishing characteristics of this species.

DISTRIBUTION AND BIOLOGY

From Figure 104 it may be seen that the distribution of the species as a weed is limited to eastern and southern Africa. It is distinctly set apart among the world's important weeds in having such a regional habitat.

In eastern Africa, *D. scalarum* is widely distributed in the moister regions from sea level to 3,125 m and is the most important of the rhizomatous grass weeds. Some feel it is the most troublesome of all weeds of eastern Africa. In South Africa it has been introduced into the Transvaal and Natal, and has also been planted in the Cape of Good Hope peninsula where it forms a thick turf on mountain slopes.

The leaves are softer than those of *Cynodon dactylon* and are a much darker green. The plant differs also from *Cynodon* in bearing a conspicuous membranous ligule at the junction of the leaf sheath and blade. The species is closely related to *Digitaria decumbens* (pangola grass) which is an unimportant grass in its home on the Pangola River in South Africa, but which is becoming one of the important pasture grasses of the tropics and subtropics, including Trinidad, Florida, Hawaii, Australia, Brazil, Jamaica, and Puerto Rico.

Little is known of the biology of *D. scalarum*. Harker (1957) and Huxley and Turk (1966) studied seed production and germination of the species in Uganda. For three seasons, beginning in 1953, collections were made from plants which were seeding

254

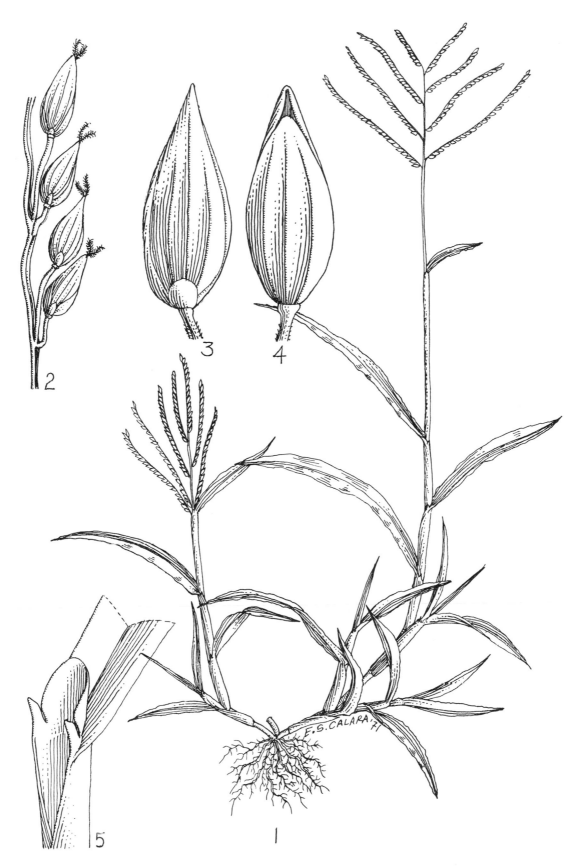

FIGURE 103. *Digitaria scalarum* (Schweinf.) Chiov. *1*, habit; *2*, spike; *3*, flower, back view; *4*, flower, front view; *5*, ligule.

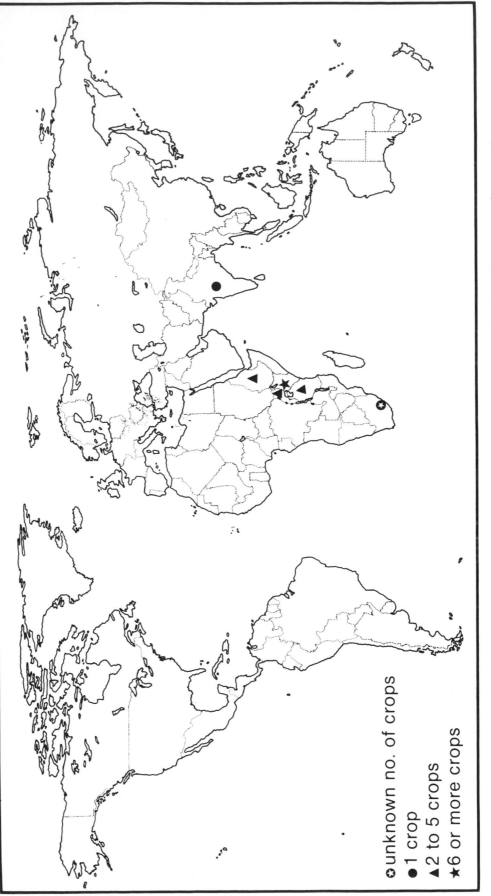

⊙ unknown no. of crops
● 1 crop
▲ 2 to 5 crops
★ 6 or more crops

FIGURE 104. The distribution of *Digitaria scalarum* (Schweinf.) Chiov in countries where it has been reported as a weed.

profusely. The seed lots gave quite variable germination but, in general, their germination rate was 1 to 7 percent 3 to 5 weeks after harvest. At 6 months some lots were giving 50-percent and at 18 months 80-percent germination. Seeding was so prolific that an average of 26,000 seeds per square m was produced. Even 1-percent germination means that a single crop may contribute three viable seeds to each square dm of a field. Some researchers have found that germination is retarded if the seed coat is intact. In still other tests scarification of the seed coat had no appreciable effect.

The plant appears to become dominant on lands after soil fertility has declined. From Kenya there are frequent reports of *D. scalarum* replacing *Pennisetum clandestinum* when soil nutrient levels become low.

AGRICULTURAL IMPORTANCE

It is most troublesome in the crops of Ethiopia, Kenya, Tanzania, and Uganda. *D. scalarum* is said to be the number-one weed in coffee in Kenya and cotton in Uganda. It is a serious weed of coffee in Ethiopia, cotton and sisal in Kenya, and sugarcane in Tanzania. It is also a problem weed in tea, pineapple, pyrethrum, forest nurseries, flax, and wheat in these areas.

The growth and yield of crop plants is greatly reduced where the weed occurs. Coffee plants may be completely killed by a severe infestation. In Kenya heavy infestations have caused losses of more than 2 metric tons per hectare in a sisal crop cycle (Richardson 1965). As with many rhizomatous grasses, this one is difficult to control with machinery. It can be severely reduced by sufficient cultivations but, once it has become established in a crop, it is impossible to destroy all of the rhizome pieces even by forking. In perennial crops the rhizomes mix with the crop roots and may severely damage them; and attempts to dig the weed can further disrupt the crop's root system.

COMMON NAMES
OF DIGITARIA SCALARUM

EASTERN AFRICA
  blue couch
  couch grass
KENYA
  couch grass
  sangare

TANZANIA
  couch
  thangari

UGANDA
  lumbugu

❧ 37 ❧

Eclipta prostrata (L.) L.
(= *E. alba* [L.] Hassk.)

ASTERACEAE (also COMPOSITAE), ASTER FAMILY

Eclipta prostrata is a herbaceous plant that is a native of Asia. It may grow either in prostrate or erect form, is most troublesome in low areas or regions of high rainfall, can endure saline conditions, and is confined to the warm areas of the world. It may be either annual or perennial. It is reported to be a weed of 17 crops in 35 countries across the world.

DESCRIPTION

E. prostrata is a subherbaceous *annual* or *perennial* plant (Figure 105); it may be prostrate or grow up to 90 cm in erect form; *stems* with white, appressed hairs from bulbous bases, occasionally purple, branched; *leaves* opposite, simple, sessile, up to 12 cm long, 2.5 cm wide, lanceolate to elliptical, margins entire or faintly toothed with sharp, appressed, straight, and basally swollen hairs on both surfaces, usually acute, nerves prominent; *inflorescence* a head, axillary or terminal, white, subglobose, 0.5 to 1 cm in diameter, heads often numerous; *peduncle* thickened at the top, appressed, hairy, 0.5 to 7 cm long; *involucral bracts* five to six, ovate-acuminate, appressed, hairy, in two rows; *fruits* (achene) brown or black, 3 mm long, compressed, with more or less thickened margins, densely covered with small, round, protruding bodies, crowned with small beak; *young fruits* bristly at the apex; *pappus* absent. It may be distinguished by the white flowers; the two-rowed, involucral bracts; and the absence of a pappus.

DISTRIBUTION AND AGRICULTURAL IMPORTANCE

Eclipta alba is the name most widely used in the world literature for the species, but the correct name is *Eclipta prostrata*. Although it is an important weed in world crops, little is known of its biology. Its success is surely due to its adaptability to changing environmental conditions. It is usually found on poorly drained, wet areas; along streams and ditches in marshes; and on the dikes of rice paddies. However, it is also common in lawns and in upland conditions where rainfall is about 1,200 mm or more.

It can grow under wet, saline conditions but is often a weed of drier sites in plantation crops. The name *prostrata* infers that it is a low, creeping plant, but it may also grow to almost a meter in erect form. In South Africa it grows as an annual and in India, as a perennial. In the Philippines it flowers all year round and has been known to produce more than 17,000 seeds per plant.

E. prostrata has general distribution over the world (Figure 106). It is on all continents and tends to remain in the tropics and subtropics. It has never been reported to be a serious weed, but it is troublesome in several crops. It is a principal weed of sugarcane and flax in Taiwan. It is a common weed of corn in Indonesia, Taiwan, and Thailand; lowland rice in India, Indonesia, the Philippines, Taiwan, and Thailand; upland rice and sugarcane in India; and roselles and sugarcane in Indonesia. It is a weed of taro,

FIGURE 105. *Eclipta prostrata* (L.) L. *1*, habit; *2*, achene; *3*, ray flower; *4*, disk flower; *5*, involucral bract; *6*, flower; *7*, undersurface of leaves; *8*, seedlings.

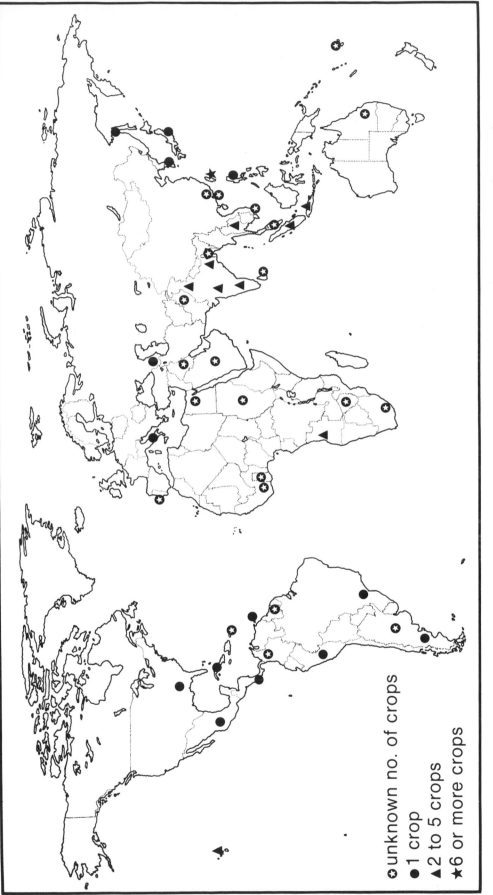

FIGURE 106. The distribution of *Eclipta prostrata* (L.) L. across the world in countries where it has been reported as a weed.

lawns, papayas, bananas, and sugarcane in Hawaii; rice in Japan, Portugal, and the United States; bananas and soybeans in Taiwan; sugarcane in Angola, Peru, and Trinidad; sisal in Angola; sorghum in Taiwan and Thailand; vegetables in the United States; soybeans in Indonesia; and cotton in India and Thailand.

E. prostrata is an alternate host of the nematode *Meloidogyne incognita* (Kofoid & White) Chitwood (Valdez 1968 as *E. alba*).

COMMON NAMES
OF ECLIPTA PROSTRATA

ARGENTINA
erva potão
yerba-de-tago

AUSTRALIA
white eclipta

BANGLADESH
keshuti

BURMA
kyeik-hman

COLOMBIA
eclipta
yerba-de-tago

EGYPT
sada
sowweid

FRANCE
éclipte droite

HAWAII
false daisy

INDIA
bhangra
ghuzi
gutakalagra
karisaranganni
mochkand

INDONESIA
daon sipat
goman (Javanese)
urang-aring
(Indonesian)

JAPAN
takasaburo

MALAYSIA
aring-aring
white heads

MEXICO
anisillo

PAKISTAN
daryai buti

PERU
florcita

PHILIPPINES
higis manok

PUERTO RICO
eclipta blanca

SOUTH AFRICA
eclipta

SUDAN
tamar el ghanam

SURINAM
loso-wiwiri

TAIWAN
jan-chang

THAILAND
ka-meng

TRINIDAD
congo lala

VIETNAM
cò múc

❦§ 38 ᴈ❦

Equisetum arvense L.
and *Equisetum palustre* L.

EQUISETACEAE, HORSETAIL FAMILY

Equisetum arvense and *E. palustre*, perennial fern allies, have conspicuous nodes (joints) which are easily separated, fluted, prominently ridged stems, and whorled branches which give the whole plant a horsetail appearance. They are dreaded everywhere because they are so poisonous to cattle, sheep, and horses. They do not often reproduce by spores but rather by very extensive rhizome systems; under some circumstances they can produce underground tubers which may serve as vegetative propagules. The plants are most often found in cereals and grasslands, but they are also reported in 25 other crops across the world. Because the stems may become encrusted with siliceous crystals, the plants have been named "scouring rushes."

DESCRIPTIONS

Equisetum arvense

E. arvense is a *perennial*, creeping, herbaceous fern ally (Figure 107); *rhizomes* attached to small tubers, deep-seated and long-running, sending out new shoots each year; *stems* annual, erect or decumbent, 15 to 60 cm tall, hollow, jointed, with sheaths at the joints; *fertile stems* unbranched, 10 to 25 cm high, terminating in a cone which may be 2.5 to 10 cm, the cone formed of shield-shaped, stalked scales from which *spores* are produced; stem with large, easily separated joints, up to 8 mm thick, sheaths 14 to 20 mm long, with large, partly united teeth, 5 to 9 mm long, flesh-colored, yellowish or

brownish, fertile stems appear in spring; *sterile* or vegetative stems tough or wiry, branched, 20 to 100 cm high, smaller joints, with whorls of numerous four-angled green branches ("leaves") arising at the joints, *sheaths* cup-shaped, 5 to 10 cm long, gradually widening upward, teeth dark brown to light tan, persistent, free or partly united, 1.5 to 7 mm long, thin, dry, membranaceous; *branches* 10 to 15 cm long, sterile stems appear in late spring, last until frost; "*leaves*" reduced to mere scales united by their margins and sheathing the stem joints; this plant does not flower but reproduces by *spores* which germinate to produce a minute, rarely seen sexual stage (gametophyte) from which the spore-bearing plants develop; *spores* globose, with a pair of elators for dissemination, pale green to yellow.

The hollow, jointed, spore-producing cone-bearing stems and the scalelike leaves sheathing the stem joints of the sterile stems are the distinguishing characteristics of this species.

Equisetum palustre

E. palustre is a *perennial* creeping, herbaceous fern ally; *rhizomes* deep (up to 150 cm) and long, growing both horizontally and vertically, circular in cross section when young, seven- to eight-angled when old, surfaces smooth, glossy, white near tip, tan farther back, other parts dark brown or dark purple or black, no tubers; *fertile* and *sterile stems* similar, emerge simultaneously, annual, erect or strongly ascending, sometimes prostrate, at most 20

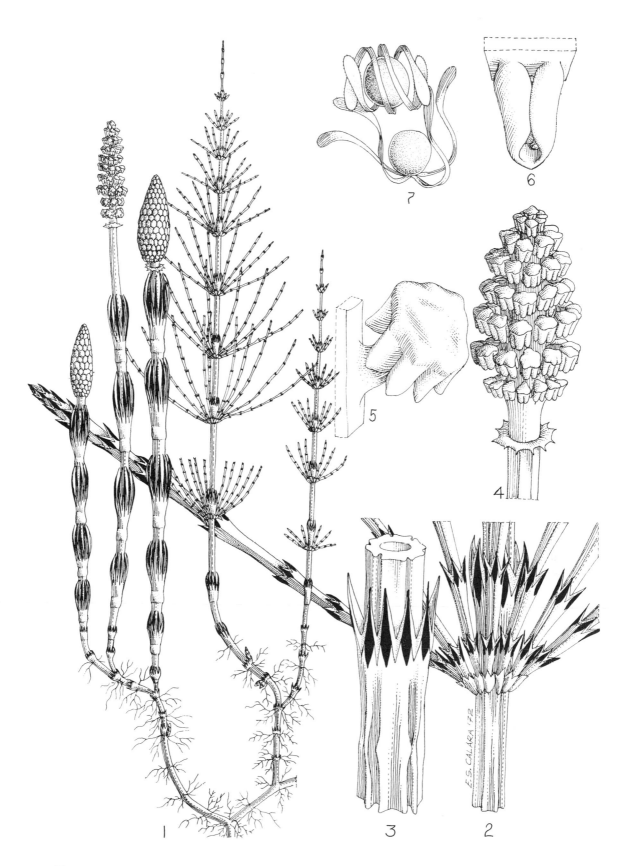

FIGURE 107. *Equisetum arvense* L. *1*, habit, fertile and sterile stems; *2*, node, with leaves and lateral branch; *3*, stem section at node; *4*, strobilus; *5*, sporangiophore; *6*, sporangium; *7*, spore with elators.

to 100 cm high, unbranched to highly branched, at most 2 to 5 mm thick with seven (rarely up to 12) grooves, stems hollow with small central cavity; *leaf sheaths* including teeth 5 to 18 mm long, green, close to stem, teeth 1.5 to 35 mm long, green with white margins, upper part black, strongly three-angled, seven (rarely eight) in number; *branches* simple, usually not more than 6 cm (occasionally 10 to 25 cm) in length, five- (sometimes four- to six-) angled, symmetrically attached to the stem nodes, three or more per node, ascending, straight or arched; *basal sheath* of branches brown or black; *spore-bearing cone* (strobilus) appears in midsummer to late summer, usually at tips of branches, 0.5 to 2.5 cm long; *spores* spherical, minute.

E. palustre may be distinguished from *E. arvense* by the absence in the former of tubers on the rhizomes, by sterile and fertile stems being similar in appearance and emerging at the same time, and by the fertile spore-bearing cones which usually appear at the tips of the stems but sometimes appear at the tips of branches.

DISTRIBUTION AND HABITAT

E. arvense is spread around the world (Figure 108), but it grows most readily in the temperate zones. It is principally a weed of North America, Europe, and the Soviet Union. We have long been aware that the genus *Equisetum* is a very old one. Fossil materials have been identified from the Eocene carbonaceous shales in the state of Wyoming (United States). One of the most northerly locations in which it is found is the area between Brooks Range and the Arctic coastline in Alaska. *E. palustre* is found there in the tundra and the boggy meadows near the coastal plain.

These species are especially adapted to open grasslands. They prefer sandy, gravelly, loose soils and are found on railroad and highway embankments, on streambanks, in open woods, and in ditches. In the United States they are found at altitudes up to 2,750 meters.

BIOLOGY

Equisetum arvense

Studies made during the 19th and 20th centuries dealing with the developmental anatomy of *E. arvense* have been cited and reviewed in part by Golub and Wetmore (1948a,b). There are 30 species of Equisetaceae and all are in the single genus *Equisetum*. They are mostly rank, silica-encrusted plants found in wet places. Several of them are poisonous if ingested by animals. They are characterized by articulated shoots with fluted internodes which separate very easily at the nodes. Several have whorled branches. Sizes range from *E. scirpoides* which is about 2 cm to *E. giganteum* which may be 10 m tall.

E. arvense is often seen as a plant about 30 cm tall. In the northeastern United States the primordia for the fertile shoots are laid down and the buds develop in summer but the plant does not emerge until the following May. Green vegetative shoots, which were also formed during the previous season, emerge about 1 week later from smaller buds on the rhizomes, either terminal or axillary. Terminal buds are larger than axillary buds and give rise to a more vigorous vegetative growth. As their distance increases from the apex, the lateral buds become less vigorous; those most removed may remain dormant.

The growth of sterile shoots from rhizome axillary buds may be completed by the 3rd or 4th week in May; when the sterile shoots are 8 to 12 cm in height, they may produce whorls of slender, lateral branches. It is this arrangement which resembles a horse's tail. The leaves are inconspicuous and consist of a small set of teeth closely appressed to the nodal sheath in a whorllike arrangement. In *E. arvense*, therefore, the shoots are without functional leaves. This explains in part why the plants cannot tolerate much shading.

In the Equisetaceae the young sporelings are erect but, once established, they soon become rhizomatous and quickly send lateral branches downward. They establish successive, layered, horizontal rhizome systems at about 30-cm intervals as growth continues downward. Golub and Wetmore (1948a,b) found five such layers by digging to a 2-m depth and noted that the system extended beyond that point. It is recommended that land in the Soviet Union which is endangered by this species be deep-plowed once each season, this action preventing rhizomes from reaching below the plow layer and thus making it possible eventually to eliminate the species.

All root development beyond the primary root takes place at the bases of lateral branch buds, both on rhizomes and erect shoots.

Kiselev and Sinyukov (1967) studied shoot development in the northern part of the Soviet Union in cultivated (winter wheat) and noncultivated heavy loam soil. The period of active consumption of food reserves for spore formation took place from 23 April to mid-May. From mid-May to mid-August the plant accumulated and stored carbohydrates. Life

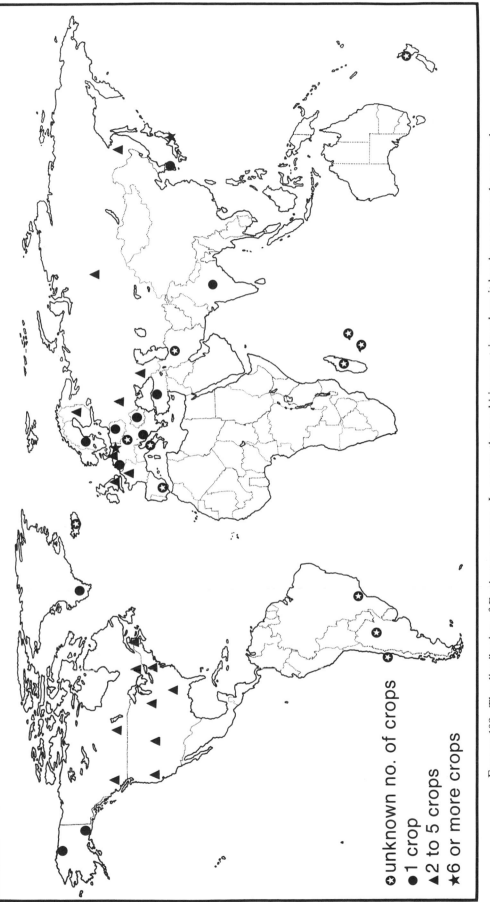

○ unknown no. of crops
● 1 crop
▲ 2 to 5 crops
★ 6 or more crops

FIGURE 108. The distribution of *Equisetum arvense* L. across the world in countries where it has been reported as a weed.

cycles were shorter on noncultivated land. Spore production normally took place in spring but, on noncultivated land, a secondary period of spore release was sometimes seen in August. The density of shoots on cultivated land was lower; shoots were taller; tuber formation was delayed 40 days; the dry weight of rhizomes showed a steady decrease; and rhizome penetration was deeper (2 m) than in noncultivated areas. In the latter areas the bulk of the rhizomes were found in the top 40 cm of soil.

Because of the depth and expansiveness of its rhizome system, with its possibility for storage of reserves, the plant has a surprising regenerative capacity. For example, in the state of Vermont in the United States during 1936 severe floods occurred which covered 1,600 hectares of agricultural land with silt to a depth of 3 cm to 2 m in some areas. Of all the species which were buried, *E. arvense* proved itself most capable of survival, sometimes emerging through 1 m of silt.

Much has been written about the plant's toxicity to animals but the possibility of toxicity to the surrounding vegetation is also suggested by the high level of alkaloids which it contains. Zelenchuk and Gelemei (1967) studied the effects of water extracts taken from this species on seed germination and early growth of some of the meadow grasses of the Soviet Union. When tested with 12 other species, *E. arvense* displayed the strongest inhibitory effect on seed germination and seedling vigor after it had been applied to 30 species of grasses.

Our comments on the biology of *E. arvense* may be closed with an interesting note on this species' exceptional ability to accumulate gold in its tissues. Plants in general have the capacity selectively to absorb certain minerals from the soil and to accumulate in their tissues a higher concentration of these minerals than that found in the ambient soil. The soil and ground water of a region contain recognizable amounts of metals which are characteristic for that area, and gold, though nearly insoluble, will dissolve sufficiently to be recorded on the highly sensitive instruments now available. From various studies we have indications that *E. arvense* may be able to accumulate greater amounts of gold than does any other plant, 4½ ounces of gold per ton of fresh plant material having been recorded. Mining engineers look upon it as an "indicator" plant rather than as a commercial source of gold (Benedict 1941). Nemec, Babicka, and Oborsky (1937) have confirmed these observations on gold accumulation in Czechoslovakia. They found 0.0063 percent gold in the ash

of *E. arvense* and just a slightly less percentage in *E. palustre*.

Equisetum palustre

Mukula (1963) and von Kries (1963) have recently reviewed and cited the principal studies on *E. palustre* from the 18th century on and have reported on their own studies of the biology of the species. In an extensive survey in Finland, Mukula found this species in 1 percent of the hay fields and in 8 percent of the cereal fields where its density averaged one to five plants per 10 sq m. But these data did not include the open ditch drains which are quite close together and which are necessary in 85 percent of the agricultural fields of the country. When animals are allowed to graze selectively they try to avoid this weed but in cut hay there is little way they can do so. In Central Europe most of the grasslands are used for grazing but in Finland 80 percent of the hay is cut and dried, thus making *E. palustre* a dangerous species. Tests conducted in Finland demonstrated that 2 grams of the dried weed in the daily fodder of cattle will cause milk production to fall. Larger amounts cause lack of appetite, diarrhea, and general indisposition of the animals. Horses were not as sensitive as cattle but over long periods they too could be seriously poisoned.

The plants of *E. palustre* are extremely variable and taxonomic treatments of the species tend to be confusing. Mukula believes that this is due in part to the species' extreme sensitivity to environmental changes. For example, damage to the growing point, shade, trampling, spring frosts, and air currents near roadways can cause the otherwise erect plant to become prostrate and bushy. Heavy shading by crop plants or very wet soils delay or almost completely inhibit branching.

An interesting example of the variable morphology of *E. palustre* is the case in which plants, connected in a common rhizome system and growing at various distances from a highway, closely resembled three different varieties of the species which had been described in the literature. Plants immediately at the road had one form; at some distance another; and at the bottom of the ditch along the road, near the level of the standing water, still another form. Upon being dug up, these varied forms were found to be connected by a common rhizome system!

The principal means of spread is by the growth of the vigorous rhizome system. Spores live only a few days or weeks after they have been released from the strobili. The germinability of the spores was demonstrated in the early 1800s, but the prothalli are so delicate that successful establishment of new individuals is believed to be uncommon in the outdoors. In artificial cultures antheridia and archegonia develop from the prothalli in this species and fertilization may take place. If fertilization does not occur, the female prothalli may later produce antheridia in addition to archegonia and self-fertilization may follow, with an embryo developing inside the archegonium. The embryo requires several months to develop to a point where it can survive in a difficult environment. During these months the prothalli may be killed by direct sunlight (although they do require light); they are very susceptible to diseases; and they cannot survive drying, flooding, or freezing. They have been reported to survive and grow into new plants on the shores of the Rhine and Rhone rivers in Central Europe. Reproduction by means of spores is believed to be impossible in the rigorous climate of Finland.

By contrast, the perennial rhizome system is extremely hardy. Rhizomes may be 30 m in length and descend 5 m into the soil in Finland. Over a period of a few years the growth from one plant may spread the weed a distance of 100 m. This is not to say that all plants in such a system will still be connected, for it is known that the individual rhizomes live only a few years. The buds formed at the nodes will become lateral branches or aerial shoots.

Rhizomes must be in contact, or nearly in contact, with the groundwater, and this requirement confines the species to stream and ditch banks and boggy places. The plant can survive in dry soil if the surface layer is loose and porous and if deeper portions of the rhizome system are in contact with sufficient moisture. The plant will not tolerate firm, packed soils and rhizomes may travel long distances in soft subsoil before surfacing in porous soils at the edges of the hard places. Vigorous growth of aerial shoots is promoted by removal of other competition and this, in turn, favors a more ambitious extension of the rhizome system. Fragments of shoots and rhizomes can survive for long periods. Although these fragments are moved about by water and machinery, they seldom produce roots and, therefore, such movement does not constitute an important mechanism of dissemination of the plants.

Many species of *Equisetum* have shortened, swollen, rhizome internodes, consisting of sclerotinized tissue. These internodes are called tubers and, being resistant to drying, serve for storage and reproduction as well as distribution. Tuber formation is re-

ported from Central Europe and Norway but has not yet been observed in Finland. Because all tubers are capable of flotation and rooting, they serve to spread the species when they are released into new, favorable areas.

Sonneveld (1953) in the Netherlands surveyed 1,000 fields in an effort to establish some relationships between the growth of *E. palustre* and several factors of the environment such as land use, soil type, and fertility. The weed was found more often on fields which had been mowed repeatedly (29 percent) than on those which had been under continuous direct grazing (8 percent). It was on 29 percent of the wet soils, 9 percent of "normal" soils, and 5 percent of dry soils. It grew on 31 percent of the fields poor in potassium but on only 3 percent where this element was abundant. It was on 22 percent of the fields low in phosphorus but on 9 percent where phosphorus was in good supply. In summary, the weed was rare on well-drained, grazed pastures which had been supplied with manure rich in phosphorus and potassium.

The weed's toxicity to animals is, of course, the principal concern of farmers. Holz (1957) reported that spraying the weed with phenoxy herbicides reduced the level of toxic compounds in the weed by 50 to 100 percent within 2 to 4 weeks. Many workers have since explored this idea and also the possibility of control of the species with the chemical. Such herbicides will seriously inhibit the plant and reduce shoot numbers but, depending on the location and cropping system, several years of continual treatment may be required for practical results.

In many agricultural systems, controlling the water table is not a method which can be used, for often the crops will suffer more than the weed. Tillage of the proper type and heavy shading by appropriate crops may be effective when used continually for several years. Complete removal of the tops for 3 or 4 years should wipe out the weed. Votila and Mukula (1961) have reported that spring plowing for 3 years in succession gave almost complete control of the weed.

AGRICULTURAL IMPORTANCE

E. arvense is reported to be a weed in more than 25 crops of the world but it is seldom ranked as a principal weed. This species, often called field horsetail, is reported in crops much more often than *E. palustre* which, in turn, is known as marsh horsetail. The greatest threat from both species is, of course, the harm that may be done to animals. It is interesting, nevertheless, to see the wide range of crops in which *E. arvense* thrives. The following are samples of the various crops which are disturbed in some of the countries most bothered by the weed. In Japan it is found in barley, corn, oats, rape, rice, rye, soybeans, sweet potatoes, tea, vegetables, and wheat; in the Soviet Union, in cereals, citrus, corn, millet, orchards, pastures, potatoes, and soybeans; in Germany, in barley, carrots, flax, forest nurseries, potatoes, and sugar beets; in Canada, in flax, pastures, peas, and tobacco; in Finland, in cereals and orchards; in England, in cereals, vegetables, and potatoes; in France, in orchards, cereals, and corn; and in the United States, in corn, pastures, potatoes, and vegetables.

Animals which have eaten the equisetums are said to have "equisetosis," and several different compounds are reported to be the cause of this illness. The toxicity is reported from Japan to India through the Soviet Union and Europe to North America. An interesting report from the Soviet Union (Cupahina 1963) suggests that some *Equisetum* species are rated as satisfactory forage in the eastern coastal Maritime Territory.

Thiaminase, an enzyme found in the species, is said to produce vitamin B_1 deficiency in monogastric animals through the destruction of thiamine. Sheep and cattle are more susceptible to poisoning by the green plant but horses suffer more from hay. Hay containing the weed causes illness in 4 or 5 weeks and the first symptom is a weakness, especially in the hindquarters. This is followed by a lack of coordination, excitability, rapid pulse, trembling, and difficulty in breathing; death finally ensues, preceded by convulsions and coma. Massive doses of thiamine have sometimes save the afflicted animal.

The toxic compound in *E. palustre* is an alkaloid called "palustrine" or "equisetine." As stated previously, 2 grams of air-dried *E. palustre* in the daily fodder will cause a decrease in milk production. A toxicity much like this is reported from Argentina.

A further note of interest about the horsetail species concerns the name "scouring rush" which is common in some regions. Because of the siliceous deposits on the stems, the plants are used to clean or "scour" pots and pans and silver articles.

COMMON NAMES
OF EQUISETUM ARVENSE

CANADA
  common horsetail
  field horsetail
  horse pipes
  mare's tail
  snake grass

CHILE
  helecho

FINLAND
  peltokorte

FRANCE
  prêle des champs

GERMANY
  Acker-Schactelhalm

INDONESIA
  bibitungan

  (Sundanese)
  rumput bitung
  (Indonesian)
  tropongan (Javanese)

IRAN
  horsetail

JAPAN
  sugina

NEW ZEALAND
  field horsetail

TURKEY
  at kuyrugu

UNITED STATES
  field horsetail

YUGOSLAVIA
  rastavič

❦§39§❧

Euphorbia hirta L. (= *E. pilulifera* L.)

EUPHORBIACEAE, SPURGE FAMILY

Euphorbia hirta is a very small, prostrate, reddish colored, annual weed with milky sap. A native of tropical America, it has been reported from 47 countries as a weed of arable lands, grasslands, gardens, and waste places. Plants are sometimes used in herbal medicine.

DESCRIPTION

A rough-hairy, *annual* herb (Figure 109), 15 to 30 cm tall; *stems* much-branched from the base, with ascending or spreading, often reddish branches, clothed with brownish crisp hairs, and having milky sap; *leaves* opposite, two-ranked, elliptic-oblong to oblong-lanceolate, 1 to 5 cm long, 0.5 to 2.5 cm wide, minutely toothed, the teeth pointing forward (serrulate), often blotched with purple in the middle, hairy on the veins, the base oblique with small *stipules* on the axil; *cymes* collected in dense axillary or terminal clusters, one to two on a 4- to 15-mm-long stalk, subglobose, greenish yellow; *flowers* unisexual, staminate and pistillate in the same involucre which is cup-shaped, four-toothed at the top, the teeth alternating with minute glands; *capsule* three-angled, hairy, about 1.5 mm long, which breaks into three segments when mature; *seeds* red-brown, 0.5 to 1 mm long, slightly transversely ribbed or wrinkled when dry.

The main distinguishing characteristics of this weed are the milky sap of the stem and the sometimes purple-blotched leaves with toothed margins. It has a taproot.

DISTRIBUTION AND BIOLOGY

E. hirta is a weed of the tropics and subtropics, especially of Central America, South America, tropical Africa, tropical Asia, and the Pacific Islands (Figure 110). It is found in both moist and dry environments in cultivated lands, in poor or degenerate grasslands and pastures, and in lawns, gardens, and waste areas. It is an early colonizer of bare ground in many situations, especially under damp or irrigated conditions.

The plant reproduces by seeds which are produced abundantly, as many as 2,990 per plant (Pancho 1964). In a month or less the plant can produce seeds which germinate readily and dense stands can develop in a short time. There are reports, however, that the seeds will not germinate unless they have been leached to remove an inhibitor. When the seed pods mature they explode, throwing the seeds some distance from the plant. In most tropical or subtropical areas the plant flowers year round. Because the prostrate plant can tolerate mowing, it can become important in lawns or turf.

AGRICULTURAL IMPORTANCE

E. hirta is reported to be the most important and numerous weed of pineapples in Taiwan. It is a principal weed of cacao in Ecuador, sugarcane in South Africa, rape in Taiwan, corn in Nigeria, and citrus in Trinidad. It was listed by Rochecouste (1967) as being a dominant weed in sugarcane fields before the

269

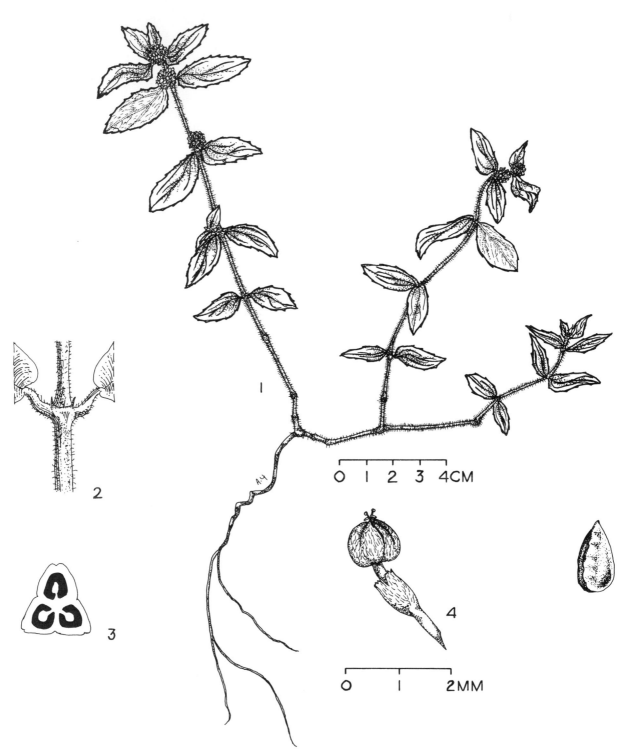

FIGURE 109. *Euphorbia hirta* L.
Above: *1*, habit; *2*, portion of stem to show stipule; *3*, ovary, cross section; *4*, flower; *5*, seed.
Opposite: Seedlings.

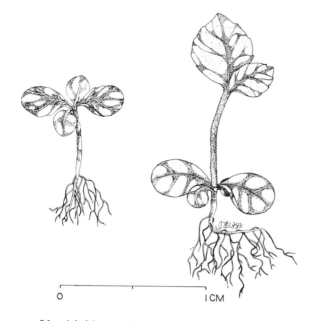

0   I CM

use of herbicides. It is a common weed of cassava and peanuts in Ghana; lowland rice in Indonesia; bananas in Taiwan; sugarcane in India, Mauritius, Taiwan, and Trinidad; upland rice in Ceylon and India; corn in Ghana, India, Indonesia, and Thailand; pineapples in Hawaii; cotton in Thailand; and peanuts and potatoes in India. In addition, it is a weed of sugarcane in Angola, Brazil, El Salvador, Indonesia, Mauritius, Mexico, Peru, and the Philippines; tea in Ceylon and Taiwan; rice in Surinam and Thailand; upland rice in Indonesia and the Philippines; sorghum in Indonesia and Thailand; corn in El Salvador, Hawaii, Mexico, and Taiwan; peanuts in Hawaii, Indonesia, and Trinidad; cotton in El Salvador, Kenya, Mozambique, and Sudan; and vegetables in Hawaii, Mexico, and the Philippines. It also has been reported to be a weed in abaca, papayas, millet, soybeans, sweet potatoes, sisal, and tea.

E. hirta is an alternate host for the nematodes *Rotylenchus reniformis* Linford & Oliveira (Linford and Yap 1940) and *Meloidogyne incognita* (Kufoid & White) Chitwood (Valdez 1968); as well as for the crop virus which causes tobacco leaf curl (Namba and Mitchell, unpublished; see footnote, *Cynodon dactylon,* "agricultural importance"). It also has been reported to be a host for two vectors of groundnut rosette virus (Adams 1967) and a host for the rust *Aecidium tithymali* Arth. (Puckdeedindan 1966).

The plant may have slightly poisonous properties and is useless for livestock feed. It sometimes is used in native medicines in Fiji (Parham 1958), Malaysia,

the Philippines (Burkill 1935, Quisumbing 1951), Brazil, and Indonesia (Burkill 1935).

COMMON NAMES
OF EUPHORBIA HIRTA

AUSTRALIA
asthma plant
BANGLADESH
bara dudhia
BURMA
mayo
CAMBODIA
tuk das khla thom
COLOMBIA
pimpinela
EASTERN AFRICA
asthma weed
ECUADOR
mal casada
EL SALVADOR
golondrina
GUATEMALA
sábana de la virgen
HAWAII
garden spurge
HONDURAS
golondrina
INDIA
baridhudi
chitakuti (Oriya)
dhuli
dudhani (Marathi)
INDONESIA
daoen bidji katjang
gendong anak (Indonesian)
koekon-koekon
nanangkaan
nangkaan (Sundanese)
patikan (Javanese)
JAMAICA
spurge
JAPAN
shima-nishiki-sō
MALAYSIA
ara tanah
hairy spurge
MAURITIUS
Jean Roberts

MEXICO
hierba de la golondrina
NATAL
red euphorbia
NIGERIA
buje
PERU
lechera
PHILIPPINES
bambalinag (Ifugao)
botobotonis
bugayau
gatas-gatas (Bisayan, Tagalog)
magatas (Pampangan)
maragatas (Ilokano)
paliak (Subanon)
patik-patik
soro-soro (Bikol)
tairas (Ivatan)
tawa-tawa
teta (Bontoc)
RHODESIA
asthma weed
pillpod spurge
SOUTH AFRICA
red euphorbia
rooi euphorbia
SURINAM
mirki-tite
TAIWAN
ru-tzu-tsau
THAILAND
namnom ratcha-si
TRINIDAD
milkweed
UGANDA
aksasandasanda
UNITED STATES
garden spurge
VIETNAM
cỏ sũa lông

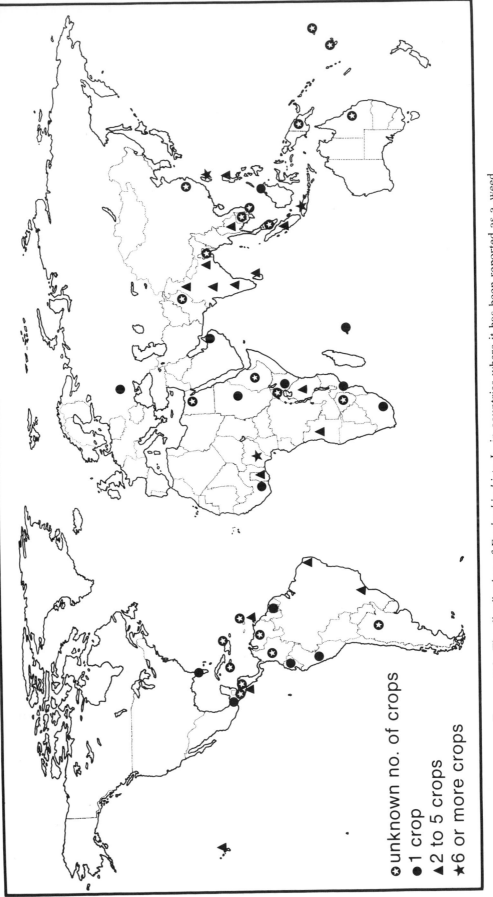

○ unknown no. of crops
● 1 crop
▲ 2 to 5 crops
★ 6 or more crops

FIGURE 110. The distribution of *Euphorbia hirta* L. in countries where it has been reported as a weed.

⋘§ 40 ৶⋙

Fimbristylis miliacea (L.) Vahl (= *F. littoralis* Gaudich.) and *Fimbristylis dichotoma* (L.) Vahl (= *F. annua* Roem. & Schult., *F. diphylla* [Retz.] Vahl, *F. longispica* Steud., *Scirpus annuus* All., and *S. dichotomous* L.)

CYPERACEAE, SEDGE FAMILY

Fimbristylis miliacea is a sedge that has become increasingly troublesome in paddy rice in recent years. It is a native of tropical America. The seeds in the soil germinate all through the growing season, so that plants may escape even the best weed control programs. The plant can grow as an annual or perennial. It has been reported to be a troublesome weed in 21 countries. Another sedge, *F. dichotoma*, is a weed in 21 countries in paddy crops, old rice fields, ditches, lawns, open wetland pastures and meadows, roadsides, cultivated lands, and along forest margins.

DESCRIPTIONS

F. miliacea is a tufted, erect sedge that may grow as an *annual* or *perennial* (Figure 111); it has a fibrous *root* system; *culms* slender, 40 to 60 cm tall, four-angled and somewhat flattened; *leaves* 1.5 to 2.5 mm wide, up to 40 cm long, basal leaves half as long as culm, linear, threadlike and stiff, two-ranked, with sheaths; *leaf bract* shorter than inflorescence; *inflorescence* a rather lax and diffuse compound *umbel*, 6 to 10 cm long, globose or sub-globose, 2.5 to 4 mm long, 1.5 to 2 mm wide, round or acute at apex, reddish brown, the lower scales fall early; *stigmas* three-branched, rarely in a few flowers two-branched; *anthers* yellow; *glumes* ovate, brown, about 1 mm long, spirally arranged, membranous, obtuse, the green midvein or keel broad; *seed* (achene) white, yellowish, less than half the length of the glume, three-angled, biconvex, broadest above the middle, very finely warty, somewhat sugarcoated.

F. dichotoma is a quick-growing, erect, tufted *annual* or *perennial* sedge (Figure 112), 10 to 75 cm

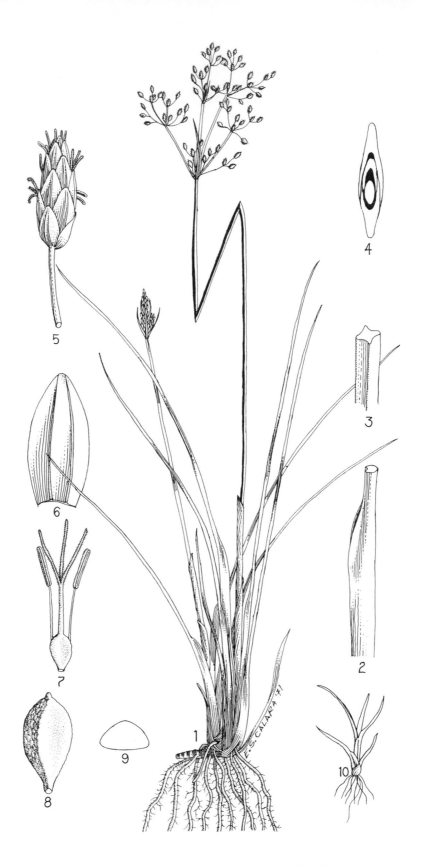

FIGURE 111. *Fimbristylis miliacea* (L.) Vahl. *1*, habit; *2*, sheath, basal portion; *3*, portion of culm; *4*, culm, basal cross section; *5*, spikelet; *6*, bract; *7*, flower; *8*, achene; *9*, achene, cross section; *10*, seedling.

FIGURE 112. *Fimbristylis dichotoma* (L.) Vahl. *1*, habit; *2*, spikelet; *3*, longitudinally grooved fruit (achene).

tall; *roots* fibrous, black, wiry; *stems* angular, smooth, compressed below the inflorescence; *sheath* margin membranous, with a small ligule; *leaves* numerous, basal, flat, abruptly acuminate, glabrous or more or less pubescent, rough to the touch, green or glaucous, 1.5 to 5 mm wide, about one-half the length of the plant; *inflorescence* loose or dense, with few to numerous spikelets up to 20 cm long, some spikelets borne on a short stalk (pedicel), others sessile; two to three of the *pedicels* which bear the clusters of spikelets 25 to 60 mm long, the shorter ones but a few mm long; lower *involucral bracts* shorter to somewhat longer than the inflorescence; *spikelets* ovoid to oblong-ovoid, pointed, brown, many, 3 to 10 mm long, 2.5 to 3 (rarely 5) mm wide; *glumes* brown, 1.5 mm long, spirally arranged, broadly egg-shaped to almost round, with a short point, with five to seven nerves near the keel, keel may be green; *stamens* one to three; *style* short, thick, hairy, two-branched at the tip; *fruit* (achene) obovate to broadly obovate, 0.8 to 1.1 mm long, 0.8 to 1 mm wide, dry, hard, with longitudinal grooves, white, pale, or rarely brownish.

F. miliacea can be distinguished by its stiff, threadlike, two-ranked leaves; the finely warty, somewhat sugarcoated three-angled seeds; the leaf bracts which are shorter than the inflorescence; and the three-branched style. *F. dichotoma* can be distinguished by its obovate two-sided or slightly compressed seeds with longitudinal grooves, its two-branched style and its three-ranked leaves.

BIOLOGY

Fimbristylis miliacea is very competitive in rice. Because it is a prolific seed producer, it soon becomes widespread when it enters a new area of rice production. In the Philippines it flowers all year round and in one experiment was shown to produce 10,000 seeds per plant. In many places the seeds seem to have no dormancy period and thus, when ripe, will germinate very quickly if moisture is present. In the Philippines the experimental germination of weed seeds over a 3-year period from a quantity of soil obtained from a rice field revealed that *Fimbristylis* emerged in all periods and constituted 70 percent of all the seedlings which appeared (Vega and Sierra 1970). In Trinidad the seeds have been found to have a dormant period which was broken only by desiccation in laboratory experiments.

In Malaysia *F. miliacea* is the first sedge to emerge after the rice has been transplanted and is the first

sedge to recover after plowing (Burkill 1935).

The most helpful work on the biology of this species was done in Japan. The emergence of seedlings was studied in both experimental and field plots in early, ordinary (planted at the usual time for rice in the area), and late plantings of rice in southwestern Japan (Noda and Eguchi 1965). The seasonal variations in germination and the effects of soil moisture were also studied.

The greatest numbers of seedlings of *F. miliacea* emerged in the early plantings of rice. In plantings made at the usual time, emergence was about 20 percent of that in early plantings. Late plantings produced still fewer seedlings. Moreover, in all plantings, the species had the unique ability to keep seedlings emerging in the field throughout the entire culture period. It had few peaks of emergence during a given culture period. This accounts in part for its increasing importance as the use of herbicides becomes more widespread. A single treatment will affect only the crop of seedlings present at the time, and the many plants which appear later can escape the herbicide. This permits both competition and seed production. As other species are removed with herbicides, *F. miliacea* can compete more effectively.

In the same experiments in Japan, the seedlings were observed as they emerged from: (1) fields where the water level was held 3 cm above the surface, or (2) at the soil surface, or (3) 20 cm below the soil surface. In the latter case a considerable fluctuation of soil moisture content was observed with soil moisture being, on the average, below field capacity. The result of these trials was that *F. miliacea* had many more seedlings than did any other weed emerging in the driest soil. There were twice as many seedlings coming up from the dry soil as from the submerged soil and this was true in both the early and usual planting seasons. The differences were not great in late plantings.

These results agree with data taken from rice fields in Surinam where it is recognized that a shortage of irrigation water may signal a heavy germination of the species, which may then become dominant and may seriously injure stands and yields of rice. Keeping a layer of water (15 cm) on rice in Surinam is said to suppress germination of *F. miliacea* completely. In Vietnam *F. miliacea* is commonly found in the acid soils of rice fields which are not used in the dry season (Ho-Minh-Si 1969).

We may expect, therefore, to find the weed in damp, open, waste places where it may not establish itself well in submerged conditions but may compete heavily following germination during dry periods or during low water conditions.

F. miliacea is eaten by cattle, the seeds passing through their digestive tracts mostly undigested and germinating near the droppings. For this reason it is known as *rumput tahi berbau* or buffalo-dung grass in Malaysia (Burkill 1935).

Studies of the growth of root systems in rice fields in India have shown that roots of *F. miliacea* spread much more rapidly than do the roots of rice. The roots of the weed extend vigorously in all directions, growing between the rice roots and eventually surrounding them and competing seriously with them for nutrients. In some cases the roots of the weed penetrate and enter the rice roots. *F. miliacea* is, therefore, one of the most harmful weeds in rice fields in India.

F. dichotoma appears to be more adapted to upland conditions than is *F. miliacea*, although *F. dichotoma* also grows well in poorly aerated soils with high moisture. The plant reproduces by seeds. It is an exceedingly variable species, producing many seeds which fall to the ground when mature, germinate quickly, and produce seedlings after rains. The plants flower all year round in the Philippines. They grow from sea level to 1,500 m in Indonesia. In Malaysia they may be eaten by cattle, and the undigested seeds may be moved thereby to new locations (Burkill 1935). Their height may vary widely depending upon whether they are cut or mowed.

DISTRIBUTION AND AGRICULTURAL IMPORTANCE

F. miliacea is becoming one of the serious weeds of rice in Asia. Its increase with the advent of herbicide use has been discussed in the biology section. From Figure 113 it may be seen that it appears all around Asia and the Pacific Islands—wherever rice is grown. The only other regions reporting it as a troublesome weed are located in the Carribean area.

F. miliacea is reported to be one of the three most serious weeds in rice in Ceylon, Guyana, India, Indonesia, Malaysia, Surinam, and Taiwan. It is a principal weed of rice in the Philippines, Thailand, and Trinidad. It is found in taro in Hawaii and in rice in Cambodia, China, Japan, Korea, and the United States. Although *F. miliacea* is mainly a weed of

paddy crops, it is also found in bananas in Taiwan, abaca in the Philippines, sugarcane and corn in Indonesia and Taiwan. It is reported to be one of the most prevalent weeds in sorghum in Malaysia.

The distribution of *F. dichotoma* is shown in Figure 114. It is a serious weed in Fiji, Hawaii, India, Indonesia, Malaysia, and Nigeria. In Indonesia it can be troublesome also in teak and arable lands. It is common and often troublesome in pastures. Other countries which report it as troublesome are Japan and Taiwan. Crops in which the plant may be found include roselles, pineapples, taro, rice, upland crops, pastures, and rangelands.

In Malaysia *F. dichotoma* grows in fallow rice fields and is considered to be a rather poor green manure crop. Cattle may also graze on it. It has been used to make inferior mats in the Philippines (Burkill 1935).

COMMON NAMES

Fimbristylis miliacea

BANGLADESH
 bara javani
 joina
BURMA
 mônhnyin
CAMBODIA
 kak phnèk kdam
 smao
INDONESIA
 adas-adasan
  (Javanese)
 bawagan
 bebawangan (Sundanese)
JAPAN
 hiderikō

KOREA
 barambaneulgiji
MALAYSIA
 lesser fimbristylis
 rumput bukit
 rumput keladi
 rumput kurau
 rumput ta hi berbau
PHILIPPINES
 ubod-ubod
TAIWAN
 mu-shih-tsau
THAILAND
 agor
 yah nuad maew
 yah nuad pladouk

Fimbristylis dichotoma

HAWAII
 fimbristylis
 futaba-tentsuki
 futabo-tentuki
 tall fringe rush
MALAYSIA
 rumput kepala lalat

 rumput para-para
 rumput purun batu

TAIWAN
 pyau-fo-tsau

UNITED STATES
 two-leaf fimbristylis

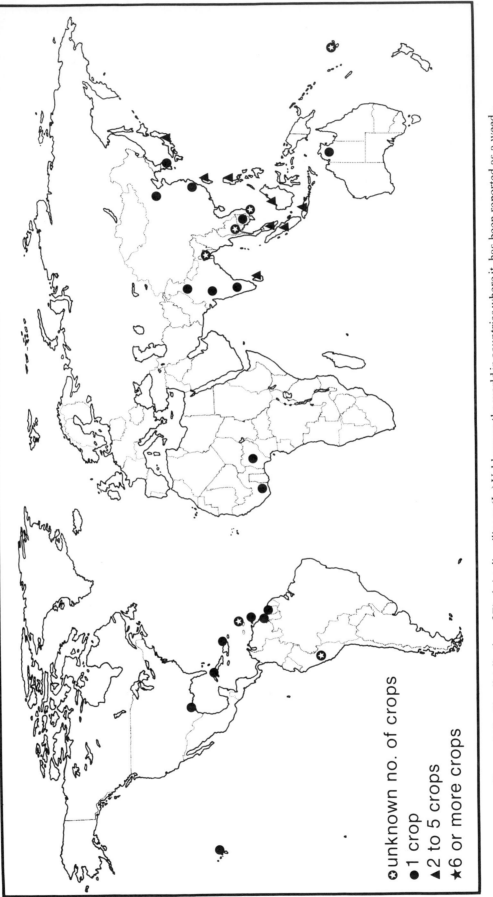

FIGURE 113. The distribution of *Fimbristylis miliacea* (L.) Vahl across the world in countries where it has been reported as a weed.

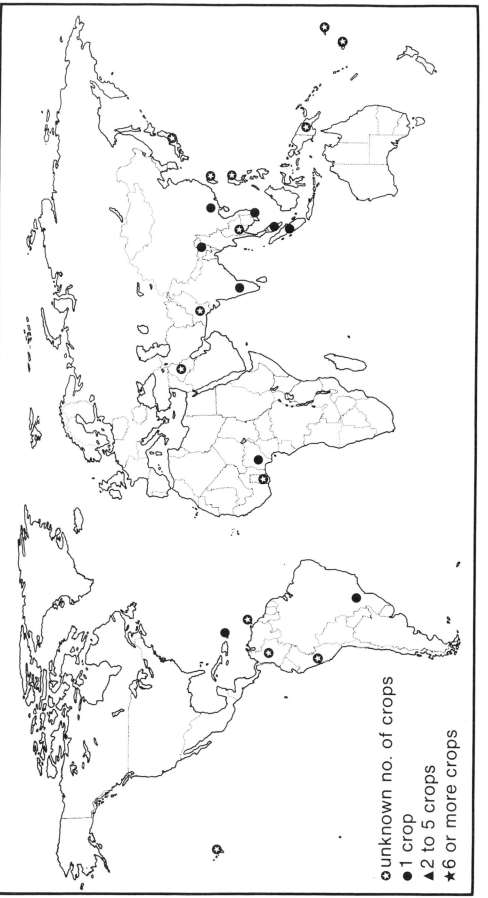

FIGURE 114. The distribution of *Fimbristylis dichotoma* (L.) Vahl across the world in countries where it has been reported as a weed.

41

Galinsoga parviflora Cav.

ASTERACEAE (also COMPOSITAE), ASTER FAMILY

Galinsoga parviflora is a low, slender, branched, annual herb that is a weed in 38 countries, both in temperate and tropical areas. A native of tropical America, it is a common weed under moist conditions in arable lands, gardens, and waste places.

DESCRIPTION

G. parviflora is an erect, slender, soft, often branched, *annual* herb, (Figure 115); 10 to 90 cm tall; *stem* with short hairs on the upper part, sometimes with gland hairs, otherwise smooth; *leaves* opposite, simple, three prominent nerves, sparsely rough-hairy, ovate or ovate-oblong, shallowly toothed, the teeth pointing forward (serrate), with wedge-shaped (cuneate) base and tapering blunt or pointed apex, 1 to 6 cm long, 6 to 12 mm wide; *petiole* 2 to 15 mm long; *head* small, 5 to 8 mm in diameter, 5 mm long, glandular-hairy, the *ray flowers* white, four or five, three-lobed, the *disk flowers* yellow, tubular, five-lobed; *bracts* of two series, membranous; *fruit* (achene) black, angled or the outer ones flat, widening upward, sparsely hairy, 1.5 mm long; *pappus* of numerous persistent scales, about 1.5 mm long. The root system is usually shallow and fibrous. The distinguishing characteristics of this weed are the usually five, small, white, outer (marginal) florets and the yellow, central disk florets, plus the absence of a pappus on the seeds of the ray flowers.

BIOLOGY

G. parviflora reproduces by seeds. Under favorable conditions it may be one of the first plants to germinate and make rapid growth in the early, warm, growing season, thus competing quickly with crop plants. A prolific seeder, often producing several thousand per plant, it can complete a life cycle (germination to shedding of seeds) in as little as 50 days. Viable seeds may be produced when the plant is only a few cm high. These seeds may germinate readily after falling to the ground; however, germination may be retarded if the seed coat remains intact. Light, as well as alternating temperatures, may favor germination. Temperatures of 10° to 35° C appear to be suitable for germination.

The plants grow best in moist conditions and can become very numerous under irrigation or in year-round rainfall in the tropics. Although it is a weed of the early growing season, the plant is very susceptible to frost injury and is often the first to succumb in the fall in Canada. The seeds can be dispersed by wind, or by clinging to clothing or the hair of men and animals.

DISTRIBUTION AND AGRICULTURAL IMPORTANCE

G. parviflora is a weed of over 32 crops in 38 countries. It grows from about lat 54° N to 40° S (Figure 116). It is reported as being the most important weed of cotton in Uganda and is a serious weed of wheat in Angola. In Figure 117 it can be seen that it is a principal weed of beans in Canada, Colombia, and Mexico; coffee in Brazil, Mexico, South Africa, and Tanzania; corn in Angola, Ethiopia, Mexico, and Tanzania; potatoes in Belgium, Mozambique, and Poland; and vegetables in Belgium, Brazil, Hawaii, the Philip-

280

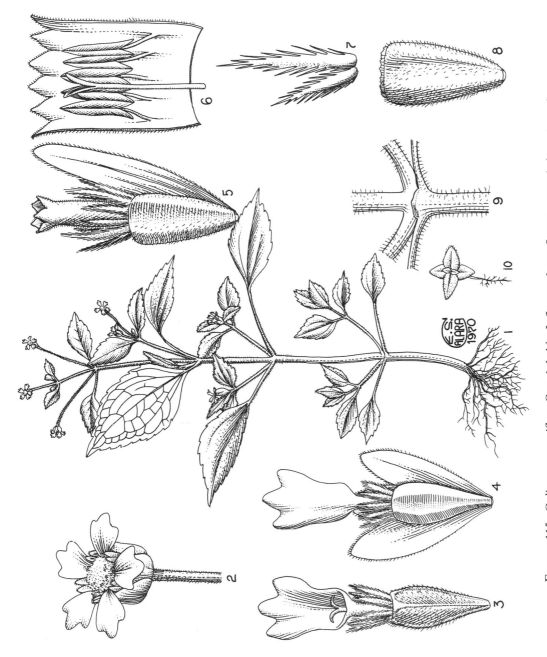

FIGURE 115. *Galinsoga parviflora* Cav. *1*, habit; *2*, flower; *3*, ray flower, dorsal view; *4*, ray flower, ventral view; *5*, disk flower; *6*, corolla opened to show stamens with anthers joined to form a cylinder around the style (syngenesious); *7*, pappus; *8*, capsule; *9*, portion of a twig; *10*, seedling.

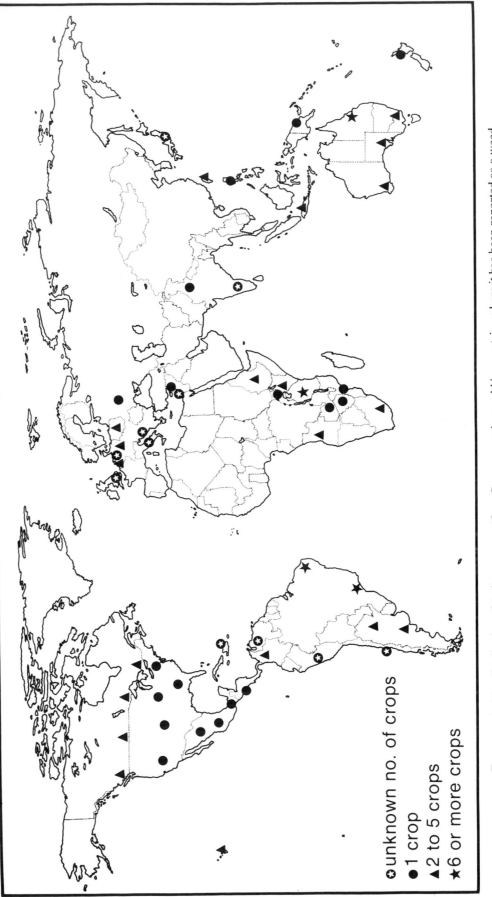

○ unknown no. of crops

● 1 crop

▲ 2 to 5 crops

★ 6 or more crops

FIGURE 116. The distribution of *Galinsoga parviflora* Cav. across the world in countries where it has been reported as a weed.

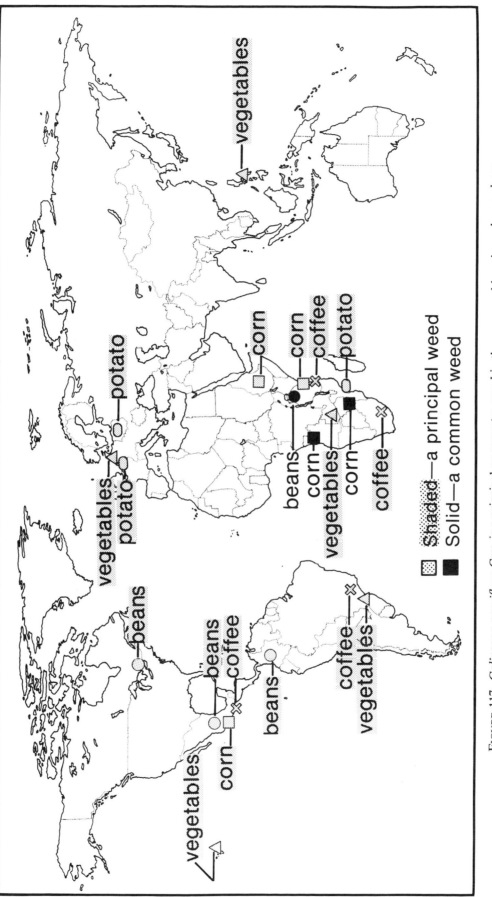

FIGURE 117. *Galinsoga parviflora* Cav. is a principal or common weed in the crops and locations shown above.

pines, and Zambia. It is a common weed of beans in Tanzania and corn in Rhodesia. Although not shown on the map, it is a principal weed in the following crops and locations: bananas, citrus, and sugarcane in Mexico; field beets in Belgium; onions, peas, and strawberries in Brazil; pyrethrum in Tanzania; sorghum in Taiwan; sugarbeets in Belgium and Germany; and wheat in Ethiopia and Tanzania. It is a common weed of strawberries in Australia; sorghum in Tanzania; garlic in Brazil; and roselle in Indonesia. In addition, it has been reported as a weed in the following crops: barley, cereals, peanuts, rice, and tea.

G. parviflora is mainly a weed of field crops where it is important because of its ability to develop quickly in the early growing season and to provide competition for small or emerging crops. It has been singled out in Taiwan as a weed found always in sorghum. Its ability to produce seeds and complete its life cycle quickly enables the weed to become especially abundant in seeded or small statured crops; for example, it has been reported to constitute one-third of the weeds in corn during the early season in Ethiopia. Also it is ranked third in frequency of distribution in wheat in Angola. It is one of three dominant weeds in coffee seedlings in Tanzania, one of two most important weeds in strawberry beds in Brazil, and one of the most important weeds in peas in Brazil.

G. parviflora can be an alternate host of the nematodes *Meloidogyne* sp. (Raabe, unpublished; see footnote, *Cyperus rotundus*, "agricultural importance") and *Heterodera schachtii* Schmidt (Vinduska 1967). Virus diseases associated with the weed are cucumber mosaic, tobacco mosaic, tomato aspermy (Namba and Mitchell, unpublished; see footnote, *Cynodon dactylon*, "agricultural importance"), and spotted wilt (Sakimura 1937). It is also a host of the fungus *Ascochyta phaseolorum* (Alcorn 1968).

COMMON NAMES
OF GALINSOGA PARVIFLORA

ANGOLA
okalume
onglo

ARGENTINA
albahaca silvestre
picao bravo
saetilla

AUSTRALIA
chick weed
potato weed
yellow weed

BRAZIL
botão de ouro
picão branco

CAMBODIA
ciliate galinsoga
quick weed

CHILE
pacuyuyo

COLOMBIA
guasco

DOMINICAN REPUBLIC
yerba boba

EASTERN AFRICA
gallant soldier
macdonaldi

ENGLAND
gallant soldier

ETHIOPIA
abadabbo

GERMANY
Gängelkraut
kleinblütiges Franzosenkraut
kleinblütiges Knopfkraut
Knopfkraut

HAWAII
galinsoga

INDONESIA
balakatjioet losih (Sundanese)
bribel (Javanese)

ITALY
galinsoga

JAPAN
hakidamegiku

KENYA
macdonaldi

MEXICO
rosilla chica

NATAL
quick weed

NETHERLANDS
knopkruid

NEW ZEALAND
galinsoga

PAKISTAN
khanna

PERU
chuminca
pacuyuyo

PHILIPPINES
galinsoga

RHODESIA
gallant weed
kew weed

SOUTH AFRICA
knopkruid
small-flowered quick weed

SWEDEN
tandgängel

TANZANIA
galinsoga

UGANDA
kofume

UNITED STATES
small-flowered galinsoga

VENEZUELA
canilla de blanca

ᵉᵍ42ᵍᵉ

Galium aparine L.

RUBIACEAE, MADDER FAMILY

Galium aparine is a decumbent, slender, annual herb native to North America. It is a widespread weed, found on all continents, and grows best in the temperate regions where it is important in cereal crops and rice. Recent studies indicate that one-third of the drained winter paddy fields of Japan are dominated by this weed. It is a weed in 19 crops in 31 countries.

DESCRIPTION

G. aparine is a decumbent, slender, *annual* herb (Figure 118); *roots* branching, short, shallow; *stems* wiry, weak, straggling, four-angled, jointed, smooth, but with short, downward-pointing, bristly hooks along the ridges, 60 to 150 cm long; *leaves* in whorls of six to eight at a node, simple, narrow, oblanceolate to linear, 2 to 7 cm long, bristle-pointed, rough, sparsely hairy, entire but with margins with backward-pointing, strong-hooked prickles, one-veined; *inflorescence* a cyme; *flowers* perfect, axillary, in clusters of one to three on a peduncle; *calyx* not present; *corolla* white, 1 to 1.5 mm in diameter with four blunt petals, borne on short peduncles attached to the nodes; *stamens* four on the corolla, alternate with the lobes; *fruit* burlike, prickly, covered with stiff bristles, composed of two spherical halves, separating at maturity into two seedlike, indehiscent, one-seeded carpels; *seed* warted with short sharp spines, with a deep pit on one side of the point of attachment with the other half of the fruit, 2 to 3 mm in diameter, gray-brown.

It may be distinguished by its square stems which have short downward-pointing bristly hooks on the angles, its whorled leaves, and the paired spherical fruits with stiff hooked bristles, borne on straight stalks.

DISTRIBUTION AND HABITAT

G. aparine is a member of a large genus whose members are essentially found in the temperate zone but yet may be found as important weeds at higher elevations in the tropics (Figure 119). *G. aparine* may be found with monotonous consistency in wheat, barley, oats, and rye throughout the world. It grows on moist lands in meadows, pastures, rich woodlands, thickets, seashores, waste grounds, and fence rows. In meadows it may be found in bare places near bordering woodlands. If drainage is satisfactory it grows well on both loam and sandy soils. In Europe it is found from Portugal to the Soviet Union and from Great Britain to Italy. In North America it grows in Alaska, British Columbia, across the wheat belt of Canada, in Newfoundland, and is generally distributed across the United States. In South America it is a problem mainly in Argentina, Chile, and Uruguay. In Asia it extends from Pakistan to mainland China and from Japan to New Zealand. It is one of the serious weeds of drained winter paddy fields in temperate Japan. The species has been reported as a weed in cereals in Tunisia and at higher elevations in Ethiopia.

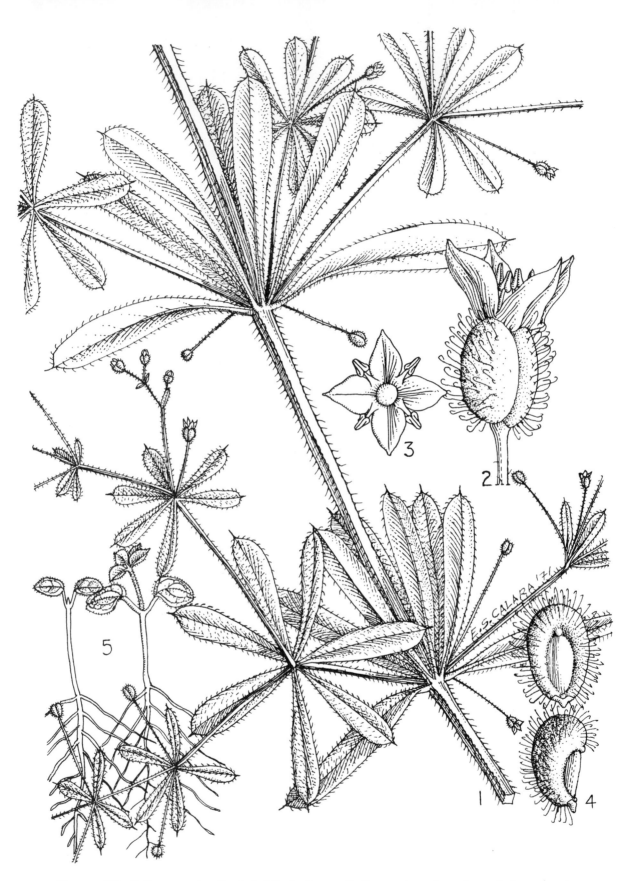

FIGURE 118. *Galium aparine* L. *1*, habit; *2*, flower; *3*, flower, top view; *4*, seed, two views; *5*, seedlings.

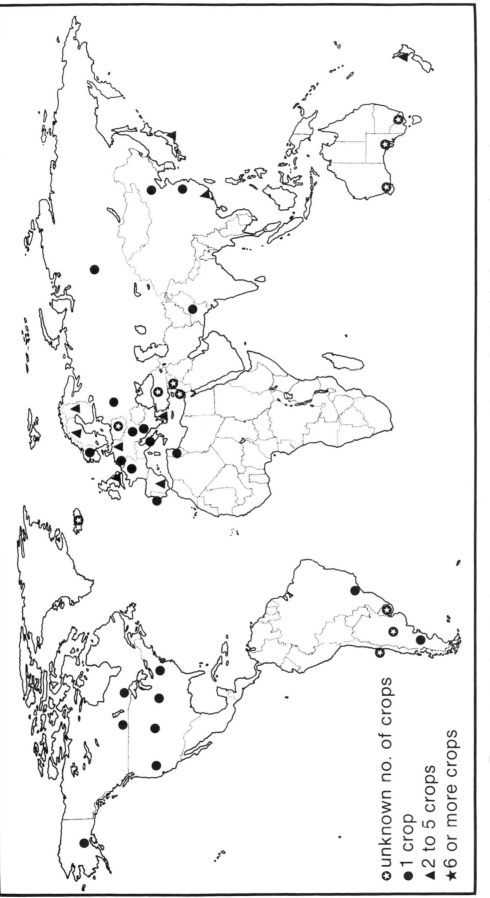

FIGURE 119. The distribution of *Galium aparine* L. across the world in countries where it has been reported as a weed.

PROPAGATION

G. aparine reproduces by seeds which are moved by wind, water, and soil tillage implements. Small hooks on the seed coat provide a special mechanism for attachment to fur, bags, and clothing. Viable seeds have been recovered from the droppings of cattle, horses, pigs, goats, and birds.

Harper (1965) has pointed out that the inadvertent importation of the seeds of weeds with the seeds of crops, through carelessness, is a critical matter in agriculture for it brings into the field species accustomed to growing with the crop. Under these conditions the weeds are provided with an environment which again favors their multiplication. If the weed seeds are small, crop seed stocks with a relatively high percentage of purity can still contaminate a field. For example, when cereals are sowed at a rate of 125 kilos per hectare, *G. aparine*, if present at 1 percent by weight, will be distributed at the rate of 50 per sq. m.

In England, Brenchley and Warington (1930) have shown with long-term experiments that natural dormancy in the soil is usually less than 2 years, and this affords an opportunity to clean the field rather quickly if seeding can be prevented. It is also believed in Germany that viability in the soil is limited to 2 or 3 years.

Seeds will not emerge from a depth of 4 cm if placed in a heavy, firm soil. If buried at 10 cm in light soils they will emerge 7 to 12 days later and will flower and fruit later than seeds placed in the upper few cm of the soil. In addition to the main root, plants which emerge from deep in the soil give rise to supplementary, adventitious roots formed on the hypocotyl (Hanf 1941).

In Sweden and Germany, freshly harvested seeds have been reported to germinate very well (von Hofsten 1947, Lauer 1953, Sjostedt 1959). Germination was found to be best in the dark during the 1st year; high levels of light at any time retarded germination; low levels permitted some seeds to begin growth. A higher percentage of seeds germinated if they had been planted in soil or if they had been in contact with soil extract. Seeds placed in dry storage developed a long-lasting dormancy.

Ueki and his colleagues in Japan have made important contributions to our knowledge of the germination and early growth of *G. aparine* (Ueki unpublished;* Ueki, 1965; Ueki and Shimizu 1967a,b; Arai 1961; Arai, Chisaka, and Ueki 1961; Noda et al 1965). Concerning dormancy, the most important findings under Japanese conditions were

that: (1) seeds which are harvested and dried for 1 month go into complete dormancy for 1 year or more, although dormancy gradually diminishes with long-term dry storage, and (2) a period of afterripening for about 1 month at 30° C while the seeds are moist, this to be followed by exposure of the seeds to a temperature of 10° C, permits a high rate of germination. Large seeds seem to have the deepest dormancy. In arriving at these conclusions, however, Ueki and his colleagues learned a great deal more about the seeds. Histological and physiological changes were recorded during the termination of dormancy (Ueki and Schimizu, 1967a). Rates of germination were higher when seeds were given 5 to 10 minutes of light as compared with unbroken darkness. Scarification, hulling, rubbing, puncturing, and soaking in enzymes, kinetin, thiourea, and nitrate solutions were ineffective for breaking dormancy. Soaking seeds in 1,000 ppm of gibberellic acid caused an increase in germination (Ueki and Shimizu 1970).

In the field, 20 percent of the seeds were able to emerge from a depth of 5 cm, with maximum emergence from 8 cm. A soil moisture content of 30 to 50 percent was best for germination. The optimum temperature for growth of buds and roots after germination was 20° C. Seeds stored in compost piles were killed when temperatures exceeded 50° C.

Useful information resulted from trials in which seeds were buried in flooded soils and in drier upland areas. In submerged soils, dormancy was terminated in 10 days but seeds were no longer viable if left for 20 days. In upland soil, termination of dormancy occurred in about 45 days but the speed of germination varied with the soil moisture content. The *p*H of the soil had little influence on germination.

BIOLOGY

The polyploid complex of the strictly annual and predominantly self-pollinating members of *G. aparine* are among the most successful colonizing plants in the angiosperms. Members of the group have reached a particularly extensive distribution in the floras around the Mediterranean Sea and in the Middle East. As weeds, however, they have followed man over most of the earth (Ehrendorfer

* "Breaking of weed seed dormancy by chemicals" is a paper presented at the United States-Japan Joint Seminar for New Biochemical Approaches to Pest Control: Biochemical Approaches to Weed Control held at North Carolina State University, Raleigh, North Carolina during March 1969.

1965). In the case of *G. aparine* we would like to point out that one must be cautious in using the word "native," or in assigning periods of introduction for a given place. This species is one of a considerable number of weeds recorded from Paleolithic or Mesolithic deposits in northern Europe (Salisbury 1961). This tells us only that they were present prior to Neolithic man's agricultural activities and that they were, therefore, not dependent upon artifically created habitats. Their proven antiquity does not necessarily mean that present-day populations were derived from those early representatives. Many present-day weed species probably have been repeatedly replenished and reintroduced into any given area. In some instances new and different morphological and physiological strains may have replaced those already present. Thus, the weeds of the present may be quite different from weeds of that same species growing in an earlier time. Therefore, to suggest that a plant is "native" to a given region is to engage in speculation unless all studies have been completed to trace the plant's continuous presence back to an earlier time. This helps to explain why so many of our weed species are said to be "native" to two and often three continents, even though there often is no scientific basis for the assignment of such a term.

For those who must contest with *G. aparine* in cereal fields, the studies of Noda et al. (1965), can be helpful in understanding the growth and development of the species in relation to control measures. Soil-treatment herbicides controlled some weeds of the paddy fields but allowed an increase of *G. aparine* in Japan. Emergence over a long period helps the plant to escape herbicides. A period of dry, warm weather conditions which lasted for several years in temperate Japan also gave optimum growing and fruiting conditions for the weed. In this area, *G. aparine* emerges from mid-November to February, with the peak about the middle or end of December. Peak germination usually follows the initial germination by about 40 days. In some years there is another period of germination in February and one in March. The plant matures from mid-May to the end of June. In wet soils some seeds germinate and die before they emerge. Some which are deep in the soil are delayed in germination and these may be the source of plants which emerge through the season. Noda found that the maximum number of seeds in the field emerged from a depth of 8 to 15 mm but the average for all was 18 mm. The greatest depth was 33 mm. By comparison, *G. aparine* seeds emerged from relatively greater depths than other weeds, such as

Stellaria uliginosa, *Polygonum aviculare*, and *Alopecurus aequalis*, which were also present from mid-November to February.

Because of low winter temperatures, the shoots elongate very slowly until growth becomes more vigorous in mid-March and reaches a maximum in April when flowering begins. Branching of the top begins in late February and reaches a maximum in mid-March.

When the seed germinates, the seminal root grows to a length of 3 cm before the aerial portion of the plant breaks the soil surface. When the leaves first appear, the roots have reached a length of 5 to 6 cm. The roots continue to grow more rapidly than the tops and this is believed to be one of the reasons why *G. aparine* plants which are small in size may prove to be surprisingly hard to kill with herbicides. Further, the root system of *G. aparine*, as compared with the root systems of several other weeds species present at this season, spreads extensively through the entire plow layer. The other weeds tend to have a large proportion of their root systems very near the surface so they are unable to use the water and nutrients of a large area as does *G. aparine*.

AGRICULTURAL IMPORTANCE

The use of phenoxy herbicides in several crops, but especially in cereals, has brought a shift in weed species in several parts of the world. This has been one of the contributing factors toward an increase in seriousness of *G. aparine* in barley, rape, rye, and wheat in Japan. The weed is also widely distributed in cereals, vegetables, small fruits, and sugar beets in England, France, Germany, Spain, and several other countries in northern Europe. It is a troublesome weed of wheat in Italy and the Soviet Union; soybeans in Yugoslavia; cereals, small fruits, vegetable crops, and pastures in New Zealand; cereals and vineyards in Greece; winter season crops in Pakistan; cotton and wheat in China; cereals in Tunisia; and coffee in Brazil. A close relative, *Galium spurium*, is a problem in northern and eastern Africa. Though found in several crops, it is most troublesome in cereals. It interferes with harvesting, encourages lodging, and sometimes smothers an entire crop. *G. spurium* is similarly resistant to phenoxy herbicides.

A study of weed surveys and herbicide field trials conducted in Germany in the 15 years following 1950 revealed an increase in the number of weeds in several crops and especially of *G. aparine* which was found in 25 to 50 percent of all sugar beet fields.

An examination of the purity of seed stocks in Europe for the period 1950 to 1957 as compared with 1960 and 1961 revealed that *G. aparine* seeds had increased in wheat and barley and decreased in oats and rye. Seeds of this weed are common in root and vegetable crop seed.

Koch (1964) and Mullverstedt (1963*b*) have made important contributions to our understanding of the effects of tillage on weed and crop seeds and seedlings both in the microsites in the soil and on the types and numbers of weeds in the fields over the long term. Weed seeds equal to the size of the soil particles were moved toward the surface by a harrowing operation, while those which were smaller were moved downward. A second harrowing carried out in the same direction intensified the movements described above. A second harrowing carried out in the opposite direction, however, tended to nullify the effects of the first one. In one experiment the seeds of *G. aparine* were planted at a depth of 5 cm in a loam soil whose particles had passed through a 2.5-mm sieve. When the area was harrowed to a depth of 25 cm with tines placed 3 cm apart, 3l percent of the weed seeds were found in the top 1 cm of soil. In further experiments it was shown that harrowing had enhanced germination because it had increased aeration. The increased probability of exposure to light was not a factor in improved germination.

As expected, seedlings were uprooted or buried when they were harrowed, hoed, or cultivated with a chisel-tined cultivator in the field. Experiments carried out over a period of several years in cereals demonstrated that harrowing the soil before the crop had emerged increased the numbers of *G. aparine* plants in the fields. Some weed species remained the same and others decreased in number as a result of this operation. In general, harrowing or cultivating before the three-leaf stage harmed the crop, as did hoeing to a depth greater than 4 cm. In most cases, cultivation gave slight increases in yield and shallow hoeing was most safely carried out after tillering.

Preventive measures for reducing populations of *G. aparine* in paddy and upland fields in Japan have been summarized by Ueki (1965) as follows:

1. If at all possible, flood both paddy and upland fields to kill the seeds. Where this is not possible use crop rotations or use deep plowing to bury the seeds and decrease the numbers of plants which can produce seed.
2. Use a straw mulch at germination time.
3. Apply a systemic herbicide to the soil before the weeds emerge.
4. Apply a contact herbicide or a systemic herbicide shortly after emergence of the weed.

It has been shown in the Soviet Union that a toxic chemical can be washed from the aerial parts of this weed or extracted from the plant. This chemical will inhibit the growth of the tops of 1-year-old oak seedlings (*Quercus robur*) but will stimulate the growth of their tap roots (Matveev and Timofeev 1965).

The oat race of the stem eelworm can infect *Galium aparine* and thus will survive on the plant from one crop to the next (anon. 1966).

COMMON NAMES OF GALIUM APARINE

AUSTRALIA
  cleavers
  goose grass
  robin-run-over-the-hedge
CANADA
  cleavers
CHILE
  lengua de gato
DENMARK
  burre-snerre
ENGLAND
  cleavers
  goose grass
  robin-run-over-the-hedge
FRANCE
  gaillet grateron
  galium grateron
GERMANY
  Klebkraut
  kletterndes Labkraut
  Klebern
GREECE
  bed straw

ITALY
  aparine
  attacca-mani
  attacca-veste
JAPAN
  yaemugura
NETHERLANDS
  kleefruid
NORWAY
  klengemaure
SPAIN
  lapa
SWEDEN
  snärjmåra
  snärj gräs
TUNISIA
  gaillet grat
UNITED STATES
  bedstraw
  catchweed
  cleavers
  goose-grass
  scratch-grass
YUGOSLAVIA
  divlji broc

❦ 43 ❧

Heliotropium indicum L.

BORAGINACEAE, BORAGE FAMILY

Heliotropium indicum, a hairy annual herb, is native to the Old World tropics and has become a weed in the Caribbean region, Africa, and southern Asia. It is readily distinguished by its curled, one-sided terminal or near-terminal spikes, the flowers of which open first at the base. Often associated with rich moist soils of the lowland tropics, it is a weed of crops such as rice, corn, soybeans, bananas, sugarcane, cassavas, and abaca. It is a weed of 15 crops in 28 countries. It is frequently found in fallow paddy fields, shallow pools, or ditches which dry periodically, and along roadsides, streams, and watercourses.

DESCRIPTION

H. indicum is an erect, robust, coarse, succulent, *annual* herb (Figure 120); strong primary *taproot*; *stem* branched, 10 to 75 (occasionally 100) cm high, sometimes deeply grooved, clothed with rather dense or sparse, large, coarse, white hairs (hirsute); *leaves* opposite or alternate, 3 to 15 cm long, 2 to 10 cm wide, ovate to oblong-ovate, with dense long white hairs on both surfaces, the lower surface pubescent, acute or acuminate, margin with shallow undulating teeth, the base narrowing and extending down along the *petiole* to form wings on both sides, the leaves prominently veined on the lower surface and with a coarsely rough and grooved upper surface; *inflorescence* in simple spikes, terminal or opposite a leaf, 3 to 10 (rarely 30) cm long, cattail shaped but curled at the tip, the flowers all on one side in two rows, one on each side of a flattened axis,

densely clustered, the lower ones opening first; *calyx* deeply five-lobed, the unequal segments narrowly lanceolate, finely hairy; *corolla* tubular, five-lobed, 3 to 3.5 mm wide, segments rounded, toothed, pale lavender or blue to nearly white, about 5 mm long, the tube light purple; *stamens* five; *ovary* four-celled; stigma ending in a conical disk; fruit 4 to 5 mm long, broadly ovate, dividing into two halves when mature, each half two-celled, beaked, outer cell one-seeded, the inner large, swollen, empty.

It may be distinguished by the curled terminal or leaf-opposed spikes with the flowers all on one side, the lower ones opening first.

DISTRIBUTION AND BIOLOGY

H. indicum, a native of the Old World tropics, has spread to become a weed in the Caribbean region of Central and South America, eastern and western Africa and southern Asia (Figure 121). It is usually associated with the moist rich soils of the lowland tropics near rivers and lakes, on dikes and roadsides, and in waste places, fallow rice fields, pools, ditches, or muddy soils which dry out periodically. The plant prospers in open sunny areas. It may also be found in moist sandy soils or in shallow swamps.

An annual, *H. indicum* reproduces by seeds which are produced in quantity on the long, curled, one-sided spikes. The flowering season of a plant is very long, and basal flowers of the spike mature and fall to the ground while new flower buds are still developing at the apex (Tadulingam and Venkatanarayana 1932). The plants flower all year round in the Philip-

291

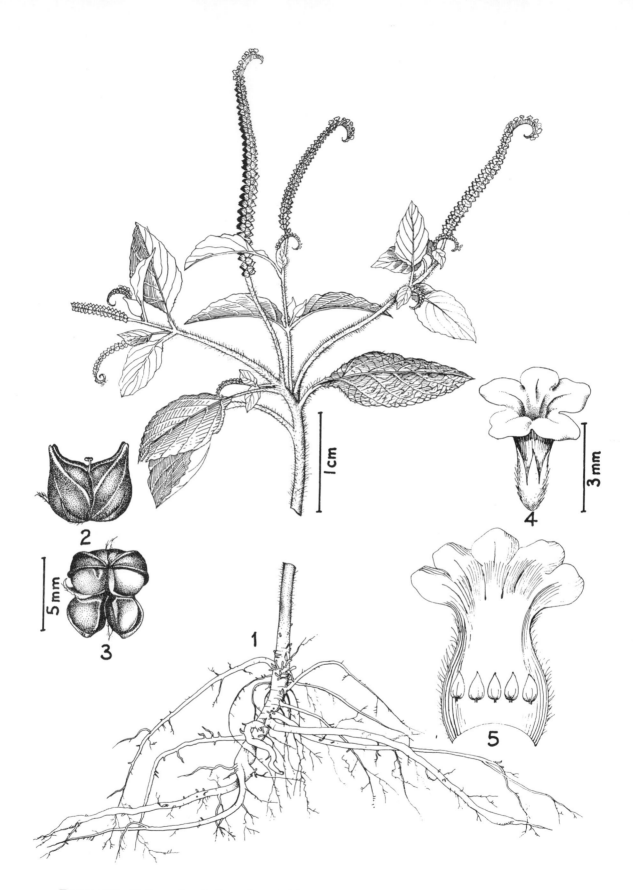

FIGURE 120. *Heliotropium indicum* L. *1*, habit; *2*, fruit; *3*, fruit, opened to show nutlets; *4*, flower; *5*, corolla, opened to show stamens.

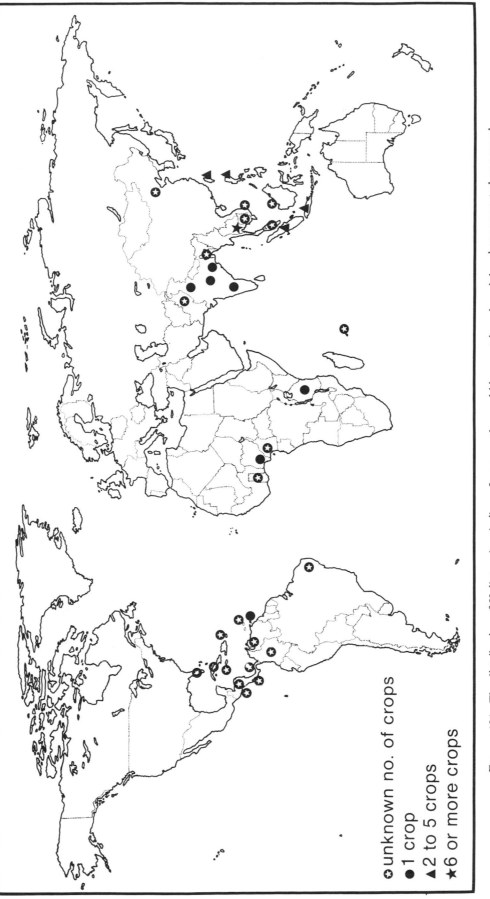

⊛ unknown no. of crops
● 1 crop
▲ 2 to 5 crops
★ 6 or more crops

pines. There is little information on the biology of the weed.

AGRICULTURAL IMPORTANCE

H. indicum has been ranked in some areas as the second most important weed in sugarcane in Indonesia and also as a weed in that crop in the Philippines, Taiwan, Tanzania, and Trinidad. It is a common weed of lowland rice in Indonesia and a weed of lowland rice in Thailand. It is a weed of peanuts in Indonesia; corn in Indonesia, the Philippines, Taiwan, and Thailand; vegetables in India; cassava in Nigeria and the Philippines; abaca, beans, cotton, papayas, pineapples, upland rice, soybeans, and tobacco in the Philippines; and bananas in Taiwan. The weed is present in many crops, probably competing mostly for moisture and nutrients and, to a lesser extent, for light. Its ability to grow in periodically wet or partially flooded situations enables it to grow and hold space in crops under conditions where weeds sensitive to excess water could not endure.

COMMON NAMES
OF HELIOTROPIUM INDICUM

BANGLADESH
hatisur

BARBADOS
white clary

BRAZIL
crista de galo

BURMA
sin-letmaung-gyi

CAMBODIA
pramoy damrey

COLOMBIA
heliotropo silvestre

COSTA RICA
largatillo

EL SALVADOR
borraja de la tierra
hierba de alacrán
pico de zope

HONDURAS
cola de alacrán

INDONESIA
bandotan lombok
djingir ajam (Sundanese)
gadjahan
tlale gadjah (Javanese)
tusok konde (Indonesian)

INDIA
hatisundha

JAMAICA
erisipelas plant
scorpion weed
turnsoles
wild clary

MAURITIUS
herbe papillon

NETHERLANDS
wilde hiliotroop

PAKISTAN
devil weed
ounth chara

PANAMA
cola de alacrán
flor de alacrán

PHILIPPINES
buntot-leon (Tagalog, Bikol)
pengnga-pengnga (Ilokano)
punta elepante

PUERTO RICO
cotorrera
heliotropo

TAIWAN
gou-wei-chung-tsau

THAILAND
yah nguang-chang

TRINIDAD
clary
erisipelas plant
scorpion weed
wild clary

VENEZUELA
rabo de alacrán de playa

VIETNAM
vòi voi

❧ 44 ❧

Ischaemum rugosum Salisb.

POACEAE(also GRAMINEAE), GRASS FAMILY

Ischaemum rugosum, a native of tropical Asia, is a vigorous, aggressive, annual grass of the warm, humid tropics. It is easily recognized by its spiral awns and by the prominent transverse ridges on the lower glume of the spikelet. It is reported to be a weed in 26 countries and, although found predominantly in rice, is also found sometimes in sugarcane. The grass is eaten by cattle.

DESCRIPTION

I. rugosum is an *annual* grass (Figure 122); *culms* 60 to 120 cm high, branched and purplish, erect or ascending; *sheaths* fairly loose, and pilose on margins near tips, densely soft-hairy on the nodes; *ligule* 1 to 7 mm, fused with the auricles; *blades* linear-lanceolate, sparingly pilose on both surfaces, 20 to 30 cm long and 9 to 12 mm wide; *inflorescence* in paired racemes 5 to 10 cm long, firmly adpressed together at the base but gradually separating, with one flat side each; *spikelets* 3 to 4.5 mm long, lower spikelet sessile, upper spikelet stalked, obtuse, awned; *lower glume* of sessile spikelet strongly transversely ribbed, hard, with a green, herbaceous, ovate tip, about 6 mm long, yellowish, keels unequally winged above, one keel of lower glume of stalked spikelet with a wide crescent shaped wing; *awn* 1.8 to 2.5 cm long, slender, spirally twisted in the lower half; *seed* 2.5 to 3 mm long, 1 mm wide, triangular in cross section, oblong, small tip (mucro), slightly concave near point of attachment.

The species is easily recognized when in flower by the prominent transverse ribs or ridges on the lower glume of the spikelet. These are distinctive and can be seen with the naked eye. Also, the prominent spinal awns and the tufted hairy nodes of the culm are excellent diagnostic characters for this weed.

DISTRIBUTION AND BIOLOGY

I. rugosum is a weed of low country, swamps, and paddy fields. In the Philippines it is found in wet grasslands at sea level but also grows up to 2,400 m. From Figure 123 it may be seen that it is mainly a problem in warm humid regions often near the equator. It is in northern South America, western Africa, southern Asia, and the Pacific Islands.

In the Philippines it has been shown that some plants may produce 4,000 seeds, although in Surinam it is believed that it produces few seeds. It flowers and fruits in February and October in Fiji. The seeds have a dormant period after they have matured.

AGRICULTURAL IMPORTANCE

I. rugosum is one of the most serious weeds in rice in Ceylon, India Madagascar, and Thailand, and it is one of the two most serious weeds in rice in Fiji and Surinam. It is also a principal weed in rice in Brazil, Ghana, Peru, and the Philippines. It is a weed of rice in Cambodia, Guinea, Liberia, Sarawak, Senegal, and Trinidad. It is a weed of sugarcane in Peru and Trinidad.

In India this vigorously growing plant is troublesome because its vegetative habit is similar to that of

295

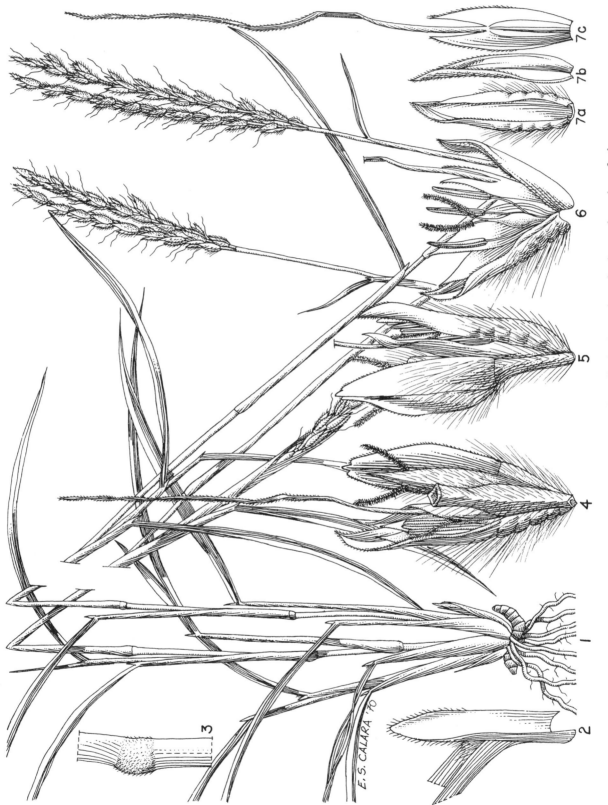

FIGURE 122. *Ischaemum rugosum* Salisb. *1*, habit; *2*, ligule; *3*, portion of culm to show the tufted hairy node; *4*, spikelet, front view; *5*, spikelet, back view; *6*, spikelet, partly opened; *7a–c*, bracts.

E. S. CALARA '70

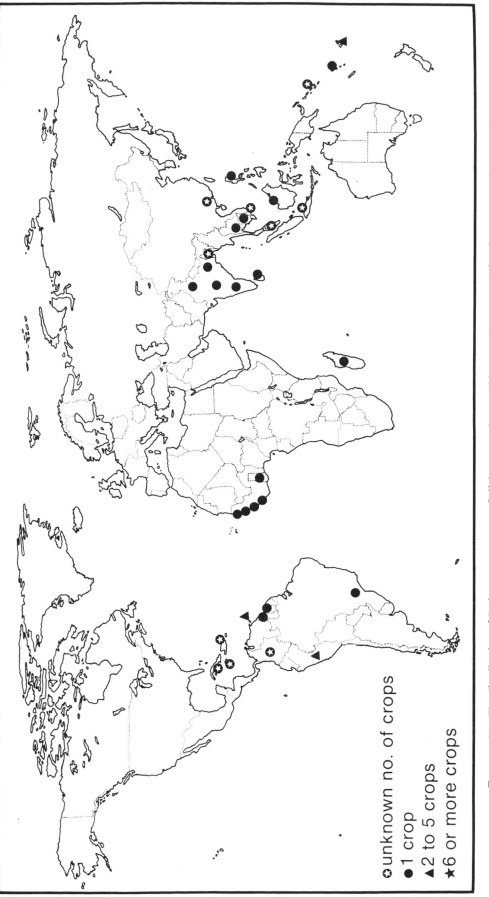

FIGURE 123. The distribution of *Ischaemum rugosum* Salisb. across the world in countries where it has been reported as a weed.

rice, so that it is not recognized as a weed and removed at the time of hand-weeding. The compact racemes with their flattened, awned spikelets are quite distinguishable from those of rice, but the panicles of *I. rugosum* cannot be seen before head formation in rice.

The difficulty in recognizing the weed during hand-weeding also has been reported from Surinam (Dirven and Poerink 1955). Here the weed is so vigorous that it severely limits rice growth. Much of the weed vegetation of the rice fields was native to the swamps before rice was finally planted and cultivated. *I. rugosum* was not native to the swamps, and it is believed that this weed came in and was spread with impure seedstocks of the rice. The spread of the weed and the severity of competition may depend on the differences in ecotypes of the weed which may exist in the several parts of the world. Also the variety of rice being sown or the type of rice being cultured may influence the spread of the weed. For example, some countries report that the weed matures with the rice and its seed is harvested and carried off with the crop. Elsewhere it has been reported that the seed ripens and shatters immediately so that a constant supply of seeds falls back into the soil to result in ever-increasing stands of *I. rugosum*. The grass is eaten by cattle. It is most palatable during its early growth.

I. rugosum is an alternate host of the viruses causing rice leaf gall and corn leaf gall (Agati and Calica 1950) and of *Piricularia* sp. (Chandrasrikul 1962).

COMMON NAMES
OF ISCHAEMUM RUGOSUM

BANGLADESH
moraro
BRAZIL
capim macho
BURMA
ka-gyi-the-myet
CAMBODIA
smao srauv
CEYLON
kudukedu (Sinhalese)
COLOMBIA
trigillo
DOMINICAN REPUBLIC
yerba de papo
FIJI
co muraina
muraina grass
INDIA
kaddukken pillu

(Tamil)
mararo (Bengali)
INDONESIA
blemben (Javanese)
djoekoet randan
(Sundanese)
PERU
mazorquilla
PHILIPPINES
daua (Subanun)
gulonglapas (Pangasi-
nan)
tinitrigo (Tagalog)
SURINAM
Saramacca grass
THAILAND
ka-du-ai-nu
yah daeng

❧ 45 ❧

Lantana camara L.

VERBENACEAE, VERBENA FAMILY

Lantana camara is a spreading, thicket-forming, woody shrub with multicolored flowers. It is a native of the Americas, grows in temperate areas, and has escaped to become a widely distributed and serious weed throughout the tropics. Forty-seven countries have reported that it is a weed in 14 crops. It has infested millions of hectares of natural grazing lands. In some areas of India its invasion of cultivated lands led to the shifting of several entire villages. Leaves and seeds are poisonous for stock.

DESCRIPTION

L. camara is a *perennial*, erect shrub (Figure 124) or, in shady places, is a straggling shrub, strongly odorous, 2 to 5 m tall, branched; *stems* four-angled, armed with recurved prickles; *leaves* opposite, ovate to ovate-lanceolate, 2.5 to 10 cm long, 1.75 to 7.5 cm wide, acuminate, the margins crenate to dentate, rough above, hairy below; *petiole* 2 cm long; *flowers* axillary and terminal in dense, almost flat-topped, peduncled or short spikes, 2.5 cm in diameter, generally yellow and pink on opening but changing to orange and red, sometimes blue or purple, individual flowers tubular, 9 mm long; *petals* four; *peduncle* 2.5 to 7.5 cm long; *fruit* a drupe, globular, dark purple to black, 6 mm across at maturity, borne in clusters; *seed* one, about 1.5 mm long.

This species is distinguished by the strong unpleasant odor of its leaves when crushed, its four-angled stems with recurved prickles, and its flat-topped multicolored flower clusters.

DISTRIBUTION AND BIOLOGY

From Figure 125 it may be seen that *L. camara* can grow from lat 45° N. to 45° S., although it is most serious in the Caribbean area, in eastern Africa, South Africa, southern Asia, Australia, and the Pacific Islands. It is a weed of cultivated land, fence lines, pastures, rangelands, and waste places. It thrives in dry and wet regions and often grows in valleys, mountain slopes, and coastal areas. It spreads by seeds which are readily eaten by birds and carried long distances. The plant flowers all the year in many warm countries. It is somewhat shade-tolerant and, therefore, can become the dominant understory in open forests or in tropical tree crops.

L. camara is known to exist in several varieties or forms throughout the world. Eighteen forms have been described from Australia alone. Attempts are often made to separate the varieties on the basis of flower color. The lantanas are widely cultivated as ornamentals and most are considered to be varieties of *L. camara*.

L. camara is one of the weeds which has been studied most widely in efforts to find biological controls and, indeed, was the first weed for which biological control was studied and reported. This work was conducted by Perkins and Koebele in Hawaii in 1902 (Nakao 1969). The work goes back to more than 70 years ago. Two insects—*Teleonemia scrupulosa*, a lace bug which eats leaves, and *Ophiomyia lantanae*, a seed-destroying fly—have received the most attention. Today perhaps the most

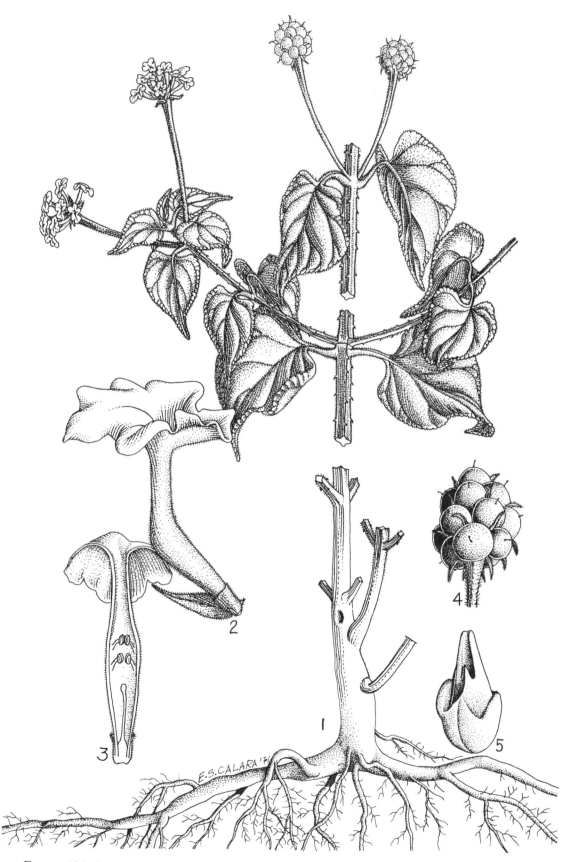

FIGURE 124. *Lantana camara* L. *1*, habit; *2*, flower; *3*, flower, longitudinal section; *4*, fruits; *5*, seed.

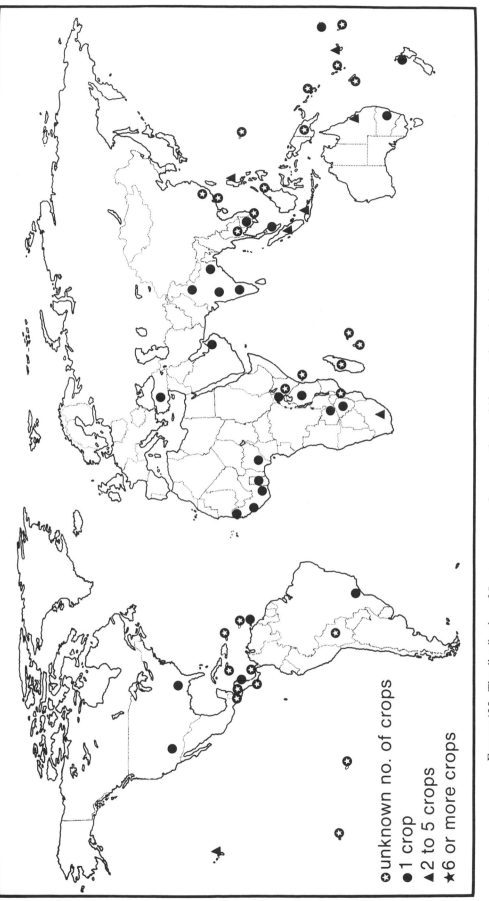

FIGURE 125. The distribution of *Lantana camara* L. across the world in countries where it has been reported as a weed.

important biological control is the lantana defoliator caterpillar, *Hypena strigata* (Nakao 1969). Scores of other insect species have been studied in Hawaii, Australia, and eastern Africa. It has been reported that biological controls are effective in the dryer parts of southern Queensland in Australia and in Hawaii. Some believe that *L. camara* may already be under stress in these areas, and the added pressure from the insects may keep the weed under control.

AGRICULTURAL IMPORTANCE

L. camara is mainly a weed of plantation crops and pastures. It is ranked among the three most serious weeds in coffee in Indonesia; and is a principal weed of oil palms in Nigeria, coconuts in Fiji and Trinidad, and cotton in Turkey. It is troublesome in pastures in Australia, East Africa, Fiji, Hawaii, India, the Philippines, Rhodesia, South Africa, and Zambia. It is a weed of bananas in Samoa; pineapples in the Philippines; sugarcane in Australia, India, and South Africa; tea in India and Indonesia; rubber in Malaysia; cotton in Nicaragua; and rice in Indonesia. In pastures it forms dense thickets which shade out and encroach upon desirable pasture plants. With time it can form pure stands over large areas, thereby rendering the land useless for pasture. It is especially pernicious in steep, mountainous, rocky rangelands.

It is estimated that nearly 4 million hectares in Australia and almost 160,000 hectares in Hawaii are infested with *L. camara*. It was first introduced in Madagascar in 1900 and now covers thousands of hectares. In India, this species was first recognized as a pest in forested areas near Madras in 1893. By 1917 it had invaded 2,000 hectares and by 1941 it was a serious problem on 40,000 hectares (Chakravarty 1963). For many countries there are no estimates of the sizeable areas which have been taken over by this weed. The seeds are carried by birds, and the plant, once it has become established, quickly closes over open areas where it forms dense, thorny thickets. In this way this species has taken over much natural grazing ground in Asia and Africa. Conventional control methods such as burning, slashing, and digging result in the regrowth of an even larger number of shoots. *L. camara* thickets are potential breeding places for rats, wild pigs, insects, and diseases. In eastern Africa the weed provides a favorable environment and breeding areas for the tsetse fly.

The leaves and seeds are toxic to many animals. Photosensitivity, gastrointestinal disturbances, and death have followed the ingestion of these plant parts by sheep and calves.

L. camara is an alternate host of *Diaporthe* sp., *Physalopora fusca* Stevens (Stevens and Shear 1929), *Ascochyta phaseolorum* (Alcorn 1968), *Pratylenchus pratensis* (de Man) Filip. (Raabe, unpublished; see footnote, *Cyperus rotundus*, "agricultural importance"), *Hoplothrips flaviceps* Jones, and *Thrips tabaci* Lind. (Carter 1939).

This species has been detrimental to sandalwood forests in India, both through the competition it offers and through its role in the spread of the sandal spike disease (Chakravarty 1963).

COMMON NAMES
OF LANTANA CAMARA

AUSTRALIA
lantana
BARBADOS
red-flowered sage
BRAZIL
cambara de espinho
CAMBODIA
ach mann
EASTERN AFRICA
lantana
tick berry
FIJI
kauboica
lantana
HAWAII
lantana
INDIA
bands
guphul (Assamese)
nagaairi (Oriya)
phullaki
putus
tantbi
INDONESIA
boenga pagar
kembang satik
kembang telek
saliara (Sundanese)
tahi agam
   (Indonesian)
telekan (Javanese)
JAMAICA
wild sage
MALAYSIA
bunga tahi ayam

prickly lantana
MAURITIUS
vieille fille
NEW ZEALAND
lantana
NICARAGUA
cuasquito
NIGERIA
lantana
PHILIPPINES
bahug-bahug
lantana (Tagalog)
sapinit
PUERTO RICO
cariaquillo
RHODESIA
chiponiwe (Shona)
lantana
tick berry
SOUTH AFRICA
lantana
TAHITI
tatara moa
THAILAND
pha-ka-krong
red-flowered sage
white sage
TRINIDAD
red-flowered sage
white sage
UNITED STATES
lantana
VIETNAM
thôm õi

❧ 46 ❧

Leersia hexandra Sw.

POACEAE (also GRAMINEAE), GRASS FAMILY

Leersia hexandra is a tall, perennial, tufted aquatic or swamp grass that reproduces by seeds or creeping rhizomes. It is a native of tropical America. It is widely distributed as a weed only in the thermal tropics but its natural range carries it into temperate climates. Because it has a high crude protein and a low fiber content, it is valued as an important forage grass in several regions of the world. It is a weed of six crops in 23 countries.

DESCRIPTION

L. hexandra is a *perennial* grass (Figure 126); *culms* hollow, usually creeping and rooting at the base, the upper part erect, slender, with fine longitudinal lines, 30 to 120 cm tall; *leaf sheath* with thickened, fleshy, cufflike base (sheath-node) which is densely covered with white-rough hairs bent or turned over backward; *ligule* 4 to 9 mm, smooth, sometimes thin and rather stiff and dry (scarious); *blades* tapering at base, acute, rough on both sides, rolled at night or when dry, 15 to 30 cm long, 4 to 6 mm wide; *panicles* erect or nodding, 5 to 12 cm, spreading or contracted, the branches slender, naked at base; *spikelets* in two series, overlapping, with a small knoblike projection (callus) at the base of the spikelet, 2.5 to 4.5 mm; *outer bracts* white or purple between green nerves, five-nerved, the midnerves comblike, the inner bract much narrower, acute, three-nerved; *fruit* (caryopsis) oblong, but seldom produced.

The hollow slender culms with fine longitudinal lines and the outer bracts of the spikelets with comb-like (pectinate) projections extending from the midnerves are distinguishing characteristics of this species.

DISTRIBUTION AND BIOLOGY

L. hexandra is a member of a small genus that is found in a variety of moist habitats across the world. It is a salt-shy grass found on permanent moist or marshy habitats, along irrigation ditches and other watersides, in humid thickets, in ponds, in rice fields, and on moist arable lands. Present reports of its behavior as a weed originate only in the tropics and are mainly concerned with rice production (Figure 127). The species can extend into temperate climates whether by reason of latitude or the presence of high valleys or tablelands in the tropics, and it has, therefore, a most interesting distribution and adaptation which will be discussed further in the section on agricultural importance.

Little is known of the biology of the species. The plant is ricelike in some habitats and in several areas is known as "rice grass." The leaf blades are narrowly linear which, when the plant is growing on upland soils, may roll at night or when dry. When this species is growing in water, it may have floating, flexuous branches several meters in length.

The plant can reproduce by seeds but is easily propagated by division of rhizomes. The culms are often creeping and, if they are cut into pieces and spread on wet soil, they root at every node.

In Malaysia the species does not set fruit but the ovary is sometimes attacked by the fungus

FIGURE 126. *Leersia hexandra* Sw. *1*, habit; *2*, spikelet; *3*, flower; *4*, ligule; *5*, portion of culm to show node; *6*, grain; *7*, grain, cross section.

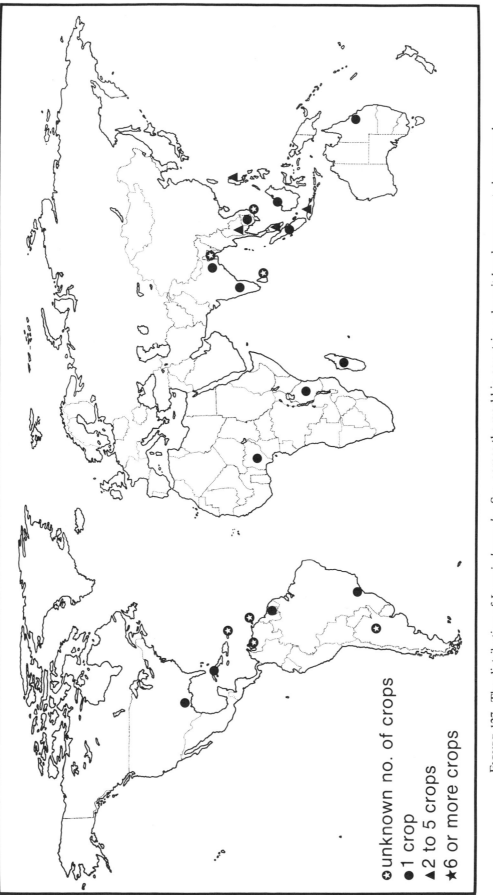

unknown no. of crops
1 crop
2 to 5 crops
6 or more crops

FIGURE 127. The distribution of *Leersia hexandra* Sw. across the world in countries where it has been reported as a weed.

Testicularia leersiae. When this occurs, the ovary expands and looks like a grain.

AGRICULTURAL IMPORTANCE

This perennial grass is most troublesome in rice. It has been reported to be a serious weed of rice in Guyana and Brazil and a principal weed in Madagascar, the Philippines, and Sarawak. It is also a weed of rice in Cambodia, India, Indonesia, Malaysia, Nigeria, the Philippines, Surinam, and Thailand. It is a serious weed of tea in Indonesia. It is a weed of corn in Indonesia; rubber in Malaysia; and sugarcane in Australia, the Philippines, and Tanzania. The plants can also grow into drains and impede the flow of water.

In many areas of the world it is highly regarded as a fodder crop and is palatable to most animals. In the Philippines it is grown on irrigated fields or near tidal rivers just about at mean tide level or slightly above. Here the fields are watered by open flooding as the tide comes in and are drained as the tide recedes. It is highly prized in this country as forage for horses. In Ceylon it is readily eaten by cattle and in India by water buffalo. It is used in pastures for cattle and for cavalry horses in Indonesia. The grass grows nearly all over India and Burma, extending from the lower Himalayas, Kashmir, Nepal, east to Burma and south to Madras. Deep in the Congo there are tablelands covered with grasses west of Lake Tanganyika. These tablelands are used only for grazing cattle, with *L. hexandra* being one of the most important grasses in the moist soils of the region. In Angola it is one of the most satisfactory forage grasses during the cool season. The species was found to have the highest crude protein (12.4 percent) and lowest crude fiber content (30.9 percent) of seven grasses studied from the fallow rice fields of Surinam. It has also produced one of the highest forage yields (Dirven 1962; Dirven, Dulder, and Hermelijn 1960). The savannas of Venezuela cover about 15 percent of the country and are used for cattle production. *L. hexandra* has the highest protein content of all the grass species in this grassland.

The grass was eaten by the manatee (*Trichechus manatus*) in trials on the clearing of irrigation canals and ditches in Guyana (Allsopp 1960).

L. hexandra was found to be susceptible to leaf yellowing, a virus disease of rice in India (Raychandhuri, Mishra, and Ghosh 1967), and to the fungus *Piricularia* sp. in Thailand (Chandrasrikul 1962).

COMMON NAMES
OF LEERSIA HEXANDRA

AUSTRALIA	NETHERLANDS
swamp rice grass	rijstgras
BANGLADESH	PHILIPPINES
araila	barit
BURMA	PUERTO RICO
thaman-myet	arrocillo rosada
INDONESIA	SURINAM
benta (Javanese, Indonesian)	alesi grasie
djoekoet lameta (Sundanese)	THAILAND
kalamenta (Indonesian)	yah-sai
	VENEZUELA
MALAYSIA	lamedora
tiger's tongue grass	VIETNAM
	cõ bắc

❦ 47 ❧

Leptochloa panicea (Retz.) Ohwi (= *L. filiformis* [Lam.] Beauv.) and *Leptochloa chinensis* (L.) Ness

POACEAE (also GRAMINEAE), GRASS FAMILY)

Leptochloa panicea and *L. chinensis* are annual grasses that are native to tropical Asia. *L. panicea* has spread to tropical America and Africa. Although both are weeds of rice, they also are important in upland crops. *L. chinensis* is a weed of 11 crops, including sugarcane, vegetables, sweet potatoes, soybeans, cotton, and corn in 15 countries. *L. panicea* is a weed of 10 crops, including sugarcane, cotton, corn, soybeans, peanuts, and pastures, in 19 countries. Both are found in wet or marshy areas; however, *L. chinensis* appears to be more tolerant of flooded situations. *L. chinensis* is grazed by livestock.

DESCRIPTIONS

Leptochloa panicea

L. panicea is a strongly tufted, *annual* grass (Figure 128); *culms* erect or branching, geniculate below, smooth, 40 to 100 cm tall or often shorter; *leaf sheaths* smooth to sparsely hairy, often with thin, long, tubercle-based hairs on the upper portions; *ligule* of lower leaves 1.5 to 3 mm, irregularly toothed (dentate); *blades* flat, linear-lanceolate, thin, acute, smooth or slightly rough above, otherwise smooth and sparsely long-hairy, from 3 to 10 mm wide, 5 to 45 cm long; *inflorescence* a panicle, with an axis 6 to 30 cm long, straight, with longitudi-

nal lines (striate), slightly rough in part; *panicle* often reddish or purplish, somewhat sticky, of numerous slender racemes; *racemes* solitary or in clusters of two to five, ultimately wide-spreading, 5 to 10 cm long, on an axis mostly about one-half the entire length of the culm; *spikelets* two- to three-, rarely four-flowered, 1.3 to 2.5 mm long, rather distant on the rachis; *pedicels* of spikelets 0.3 to 0.7 mm long; *glumes* acuminate, longer than the first floret, with scattered harsh hairs on the midnerve, upper glume slightly longer (1 to 1.8 mm); *lemmas* awnless, hairy on the nerves, 1.5 mm long; *paleas* hairy on the nerves; *grain* (caryopsis) broadly oblong, smoothly or finely reticulated, reddish brown or brown, 0.7 to 0.8 mm long.

This species may be distinguished by its irregularly toothed ligule, its two- to three-, rarely four-flowered spikelets which are rather distant on the rachis, and by the tubercle-based hairs which often occur on the leaf sheaths.

L. chinensis is similar to *L. panicea* but differs by having a longer inflorescence (10 to 40 cm), *spikelets* which are four- to six- (often five-) flowered, and a *ligule* which is deeply divided into hairlike segments.

Leptochloa chinensis

L. chinensis is a strongly tufted *annual* or *perennial* grass (Figure 129); *roots* fibrous; *culms*

307

FIGURE 128. *Leptochloa panicea* (Retz.) Ohwi. *1*, habit; *2*, ligule; *3*, spikelets and axis of raceme; *4*, close-up of florets.

FIGURE 129. *Leptochloa chinensis* (L.) Nees compared with *Leptochloa panicea* (Retz.) Ohwi.
A, *L. chinensis: 1*, plant; *2*, ligule; *3*, spikelet; *4*, grain. B, *L. panicea: 1*, spikelet; *2*, inflorescence.

erect or geniculately ascending, 12 to 120 cm high, smooth, leafy; *leaf sheath* smooth; *ligule* 1.2 to 2 mm, membranous, deeply divided into hairlike segments; *blade* linear, acute, rough on the upper surface, otherwise smooth, 6 to 32 cm long, 4 to 9 mm wide; *inflorescence* a panicle, the axis 10 to 40 cm long, straight, marked with longitudinal lines (striate), slightly rough; *racemes* solitary or two to four together, ultimately widely spreading, 1 to 10 cm long; *pedicels* of spikelets 0.5 to 0.75 mm long; *spikelets* four- to six- (often five-) flowered, 2.5 to 3.5 mm, often purplish; *glumes* unequal, with a small sharp point, the upper larger, with scattered harsh hairs on the midnerves; *lemma* with submarginal lateral nerves, higher lemmas successively smaller, all lemmas with appressed hairs along nerves; *palea* smooth or hairy along the nerves; *grain* (caryopsis) brown, smoothly or finely roughly reticulated (rugose), 0.7 to 0.8 mm long.

DISTRIBUTION AND BIOLOGY

L. panicea and *L. chinensis* are both natives of tropical Asia. Today *L. panicea* is the more widespread of the two, being found in tropical and subtropical America, Africa, and Asia (Figure 130). *L. chinensis* has a more limited distribution, being confined mostly to southern and southeastern Asia, and northwest as far as Japan.

Both *L. panicea* and *L. chinensis* are associated with wetlands, swamps, or streams in open lowland regions. *L. chinensis* appears to grow more frequently under marshy or somewhat wetter conditions than does *L. panicea*. *L. chinensis* can grow in heavy or light soils, along streams and watercourses, in marshy grounds and in lowland rice fields. *L. panicea* cannot withstand extremely dry or extremely wet soils, is frequently associated with heavy soils, and is found in waste places, swampy areas, gardens, roadsides, disturbed soils, rice fields, along streams, and in teak forests. In Java, *L. chinensis* occurs from sea level to 900 m, while *L. panicea* occurs from sea level to 200 m. *L. panicea* can grow in open sun or in light shade.

Both species reproduce by seeds. *L. panicea* is an annual, and *L. chinensis* may be an annual or sometimes a perennial when suitable growing conditions exist. Little is known of the germination or other biology of these weeds. Dissemination of both species has probably resulted from impure rice seedstocks; however, because the grains of both species are much smaller than rice, they should not be difficult to separate from rice during cleaning.

L. chinensis can reproduce by division of the culm clumps or rootstocks following tillage or incomplete weeding.

AGRICULTURAL IMPORTANCE

Although both *L. panicea* and *L. chinensis* are best known as weeds of rice, they are both important in other crops as well. They are important as weeds because after they have germinated they make abundant growth under nearly waterlogged or flooded conditions. In keeping with its ability to grow under somewhat wetter conditions than *L. panicea*, *L. chinensis* is listed more frequently as a weed of rice. It is a serious weed of sugarcane in Indonesia. In lowland rice it is a serious weed in Indonesia and Thailand; a principal weed in Swaziland; a common weed in Ceylon, Malaysia, the Philippines, and Taiwan; and a weed of that crop in India and Japan. In upland rice it is a principal weed in the Philippines; a common weed in Ceylon, Indonesia, and Taiwan. It is a common weed of vegetables, sweet potatoes, and soybeans in Japan, and of sweet potatoes, peanuts, bananas, tea, and sugarcane in Taiwan; as well as a weed of cotton and corn in Thailand.

L. panicea is a principal weed of rice in Colombia and Mexico; corn in Colombia, Ecuador, and Mexico; cotton and soybeans in Mexico; and of peas in Taiwan. It is a common weed of cotton in Colombia and is a weed of the following crops and countries: rice in Argentina, the Philippines, and Trinidad; corn in Argentina and Mexico; cotton in Argentina and the United States; sugarcane in Colombia, Jamaica, Peru, and Puerto Rico; peanuts in Colombia; and pastures in Taiwan.

Both weeds compete with crops for nutrients, space, and light in lowland rice; and, in addition, compete for moisture in upland crops. Their ability to grow in both flooded and upland conditions makes them problem weeds. In Mexico *L. panicea* competes with cotton and soybeans in spring and with rice and corn in summer.

Because their seeds are small, careful seed cleaning in rice can remove these weeds from crop seeds as well as from those grains sold for human food.

L. chinensis may be a fairly good fodder grass for livestock, and animals will eat it.

L. panicea is an alternate host of *Xanthomonas oryzae* Uyeda & Dowson (Dalmacio and Exconde 1967 as *L. filiformis*).

COMMON NAMES

Leptochloa panicea

COLOMBIA
paja colorado
paja de burro
paja mona
plumilla

EL SALVADOR
cola de buey

JAMAICA
sprangle top

MEXICO
zacate salado

UNITED STATES
red sprangletop

Leptochloa chinensis

INDONESIA
bebontengan
(Javanese)
timoenan (Javanese)

JAPAN
azegaya

PHILIPPINES
malay-palay

THAILAND
ya-yang-khon
ya-dock-kao

VIETNAM
cò duõi phung

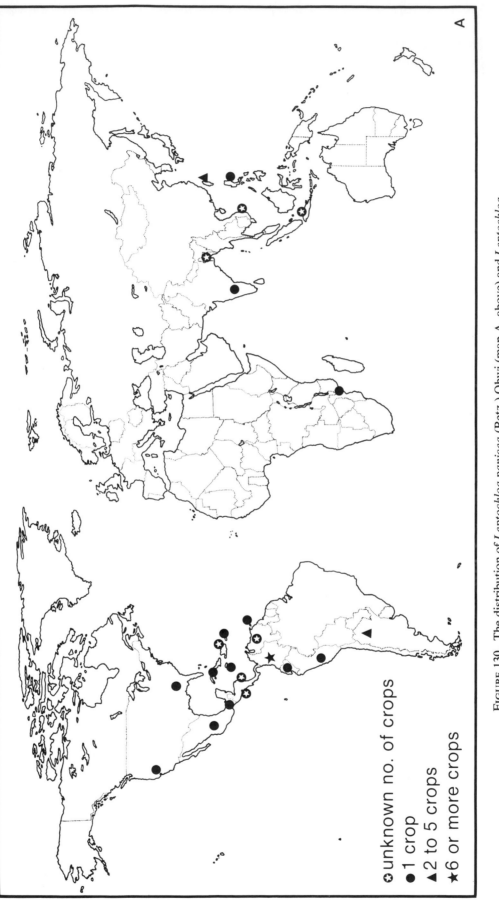

A

unknown no. of crops
● 1 crop
▲ 2 to 5 crops
★ 6 or more crops

FIGURE 130. The distribution of *Leptochloa panicea* (Retz.) Ohwi (map A, above) and *Leptochloa chinensis* (L.) Nees (map B, facing page) across the world in countries where they have been reported as weeds.

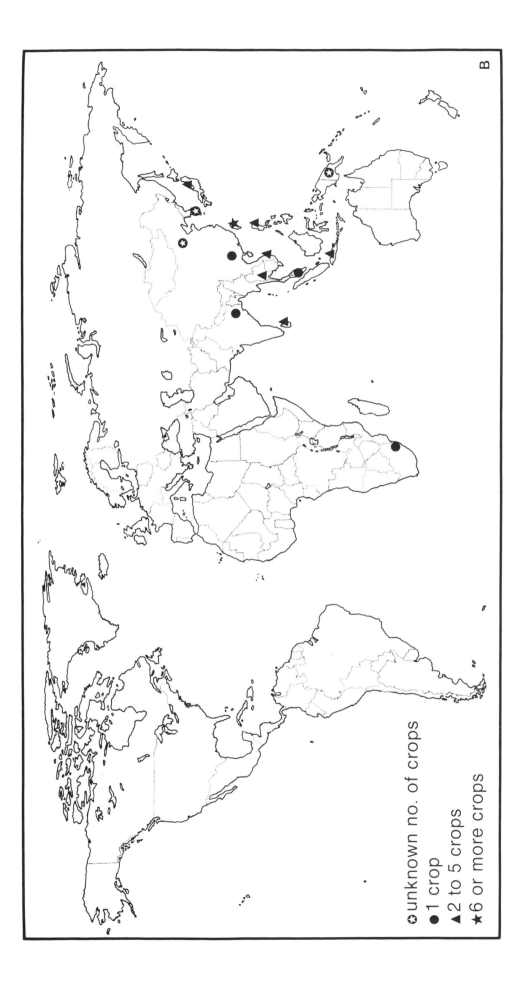

⊗ unknown no. of crops
● 1 crop
▲ 2 to 5 crops
★ 6 or more crops

B

❦48❧

Lolium temulentum L.

POACEAE (also GRAMINEAE), GRASS FAMILY

Lolium temulentum, known since ancient days as darnel, is probably also the tares of the Bible, of which it was written, "But while men slept, his enemy came and sowed tares among the wheat, and went his way" (Matt. 13:25). It is chiefly a weed of wheat and of other small grains. A native of the Mediterranean region, the weed has been disseminated throughout the world as an impurity in cereal seed stocks. It is a weed of 14 crops in 38 countries. The weed has long been notorious as a poisonous plant because its seeds, when milled with wheat, cause flour to become gray and bitter, sometimes poisoning those who eat the bread. Animals are also poisoned when they eat the seeds mixed with other grain. The poisoning appears to be caused by a fungus, *Endoconidium temulentum* Prill. & Del., which attacks the grain, within which mycelia develop. An alkaloid, temuline, and a glucoside, loline, that may be produced by the fungus have been found to cause the poisonous reactions. Symptoms of poisoning in both men and animals include blurred or impaired vision, giddiness, partial stupor, upset stomach, and diarrhea. The specific name *temulentum*, meaning drunken or intoxicated, refers to the common sympton of the poisoning.

DESCRIPTION

L. temulentum is a robust, densely or loosely tufted, *annual* grass (Figure 131); fibrous *roots; culms* 30 to 130 cm high, erect or only slightly spreading, often in clumps, two- to four-noded, smooth, usually rough below the inflorescence; *leaf sheaths* overlapping, smooth below or somewhat rough above; *ligule* 1 to 2 mm, membranous, obtuse; *blades* 5 to 7 in number, narrowly linear, 10 to 40 cm long, 3 to 15 mm wide, acute at the apex with smooth or rough margins; *inflorescence* a stiff, erect, flattened spike; *spikes* two-ranked, 10 to 25 cm long, with the sessile spikelets two-ranked in depressions on alternate sides of the rachis, rachis portions between spikelets as long or longer than the spikelets; *spikelets* four- to 10-flowered, much-flattened, oblong, up to 2.5 cm long; *florets* plump; *flowering glumes* rounded, bifid at the apex, with or without awns; *lower glume* absent or usually suppressed and very small except in the terminal spikelet; *upper glume* about 2.5 to 3 cm long, longer than the spikelet, usually as long as or longer than the uppermost lemma, rigid, with an acute tip; *lemmas* become swollen and hard, with *awns* from a notch between two short lobes; *awn* up to 2 cm long; *grain* 4 to 8 mm long, narrowly obovate, with a tuft of hairs at the apex, dull, yellow-brown to orange-brown.

The long bristle at the tip of the outer bract of each floret, the bifid apex of the flowering glume, and the involute margins of the young leaf are characteristics of this species.

DISTRIBUTION AND BIOLOGY

A native of the Mediterranean region, *L. temulentum* has spread widely across the temperate world wherever wheat or other temperate cereals, especially winter cereals, are grown (Figure 132). Although a weed of the temperate zone, it does grow

FIGURE 131. *Lolium temulentum* L. *1*, habit; *2*, portion of spike; *3*, glume; *4*, ligule.

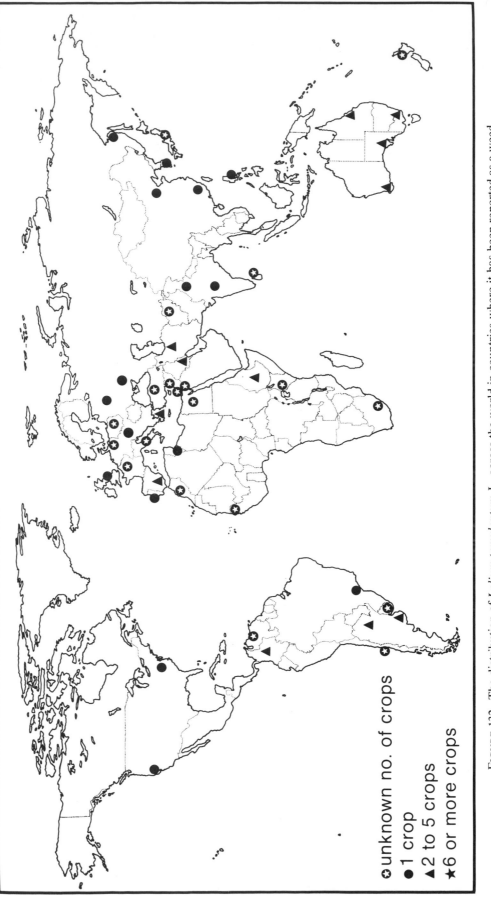

FIGURE 132. The distribution of *Lolium temulentum* L. across the world in countries where it has been reported as a weed.

at high elevations in the tropics; for example, it is sometimes found in crops and waste lands in Kenya between 2,000 to 3,000 m. In addition to being commonly found in grain fields, it can be found in cultivated fields, pastures, roadsides, and waste places. It may be found on a wide range of soil types.

Dissemination of *L. temulentum* has been mainly through dirty seed stocks, and this species may be one of the most widespread of poisonous plants because of its association with man's oldest cultivated crops, the cereal grains (King 1966).

L. temulentum, an annual, reproduces by seeds that are remarkably similar in size and weight to the grains of wheat and other small-seeded cereal crops. The seeds appear as hard, slim grains of wheat. These similarities in seed characteristics have made separation of the crop and weed seeds very difficult. However, this is not the whole story for, although wheat is easily differentiated from *L. temulentum* after emergence of the panicles, plants in the vegetative stage closely resemble one another. This makes hand-weeding difficult. The life cycle of the weed closely approximates that of several cereal crops, and both winter annual and summer annual strains occur.

The seeds are able to germinate within a very wide temperature range; for example, Andersen (1968) reported experiments in which nearly complete germination occurred within a range of 2° to 35° C. Little seed dormancy seems to occur, and vernalization may not be required for germination. However, flowering may be speeded by vernalization in winter annual strains (Evans 1969).

Evans (1969) reviewed the flowering behavior of *L. temulentum*, especially those factors which induce flowering in the species. The plants are self-fertile because of lack of exsertion of the stigmas, giving rise to uniform populations. Many strains of the grass are known, and some of these have been used in detailed physiological studies. It is a long-day plant, and some strains will flower after exposure to 1 long day. With some strains under long-day conditions, the inflorescence may develop when only a few leaves, as few as four, have formed. Evans points out that there are three leaf primordia in the embryo; thus, in plants which flower with four leaves, even the primordial leaves may respond to long days. Older plants require a shorter induction period than do younger plants. Spikes develop more rapidly with increasing temperature and daylength. Under short days (8 to 9 hours or so) more than 30 leaves may form before flowering results, and plants may remain vegetative for 120 days or more.

The plant flowers during April and May in Lebanon and in late June and July in the United States; and seed formation occurs in July and August in the United States.

POISONOUS NATURE OF THE WEEDS

Although *L. temulentum* has been notorious for its poisonous properties since ancient times, information concerning the poisonous nature of this plant is conflicting. It is agreed that poisonings are associated with ingestion of the seeds rather than of the leaves or stems. Indeed, the leaves and stems can be grazed and made into hay or other feed forms without injury to animals, provided that seed formation has not occurred.

Poisoning has been reported in man following the eating of bread made from wheat, barley, or rye flour in which *L. temulentum* seeds were present at milling. Such flour has been reported to be gray in color and bitter to the taste. When the seeds are malted with barley, the resulting beer or ale is reputed to cause intoxication very suddenly, indicating that the toxic effect is still present. Human poisoning has been reported from South Africa in cases where the seeds were ground up in household cereal meals.

Poisoning reports on animals often conflict; however, it is generally agreed that animals such as dogs, horses, swine, cattle, and sheep that have been fed grains in which the seeds are present may suffer mild to severe symptoms of poisoning, sometimes resulting in death. Georgia (1938) reported that poultry and cattle can eat the seeds without exhibiting bad effects, but Blohm (1962) recommended that poultry not be fed the seeds. Watt and Breyer-Brandwijk (1932) cited reports that dogs, sheep, and horses are affected but that swine, cattle and poultry are not.

Some symptoms of mild poisoning reported include: stupor and movements resembling drunkenness, stiffness, lassitude, and slow movement. Higher doses may result in ringing in the ears, nausea, and impaired vision; sometime later, violent abdominal pains and diarrhea may result. Paralysis may result in swine (Blohm 1962). If a lethal dose has been taken, all symptoms are accentuated, and convulsions and delirium may occur before death (Forsyth 1968). If death occurs, it is usually preceded by several days of suffering (Blohm 1962). In less severe cases, several days may be required for the animal to recover.

Seeds are thought to be poisonous only when they are infected with a fungus, notably *Endoconidium temulentum* Prill. & Del. However, another fungus,

Chaetonium kunzeanum Zopf., has also been blamed for toxicity of infected grains. Guerin (1899) stated that *E. temulentum* was not the fungus involved. Most reports, however, link *E. temulentum* with occurrences of toxicity. Mycelia of the fungus grow within the seeds of *Lolium temulentum*.

The compounds believed to be involved in the toxicity of the grains are temuline, an alkaloid, and loline or loliin, a glucoside. Watt and Breyer-Brandwijk (1932) listed some of the toxic properties of these two compounds. Both appear to cause stomach disorders, and pain and paralysis of the brain and nervous system.

In summary, there appears to be no doubt that sometimes *L. temulentum* seeds may cause toxic effects in man and certain animals if eaten in prepared human foods or in contaminated livestock feeds. Animals may also suffer poisoning if they graze on seeds in the spikes or heads. There seems to be no question that seeds free of the fungus are safe. Georgia (1938) reported that poisoning may be most dangerous when the plant grows on wet soils or during a wet season. Certainly factors favoring growth of the fungus would seem to make the toxicity more serious; however, there seem to be no critical studies on incidence and occurrence of the toxic conditions. For further reading on the subject, see Pammel (1911), Hurst (1942), Muenscher (1939), and Kingsbury (1964).

AGRICULTURAL IMPORTANCE

L. temulentum as a poisonous plant may not be as important today as it was in the past. This is due in part to improved techniques of harvesting and cleaning cereal grains (King 1966), as well as in improved techniques in weed control. Forsyth (1968) stated that the weed was formerly common in wheat fields in England but now has disappeared from that country as a result of careful efforts by farmers, millers, and others involved in agriculture or cereal grain processing. However, in countries where grain harvesting or cleaning techniques are poor, the plant still poses a threat.

In addition to its poisonous properties, *L. temulentum* is a problem weed in cereals, flax, and other small-seeded annual crops. Impure grain is always undesirable because the contaminants reduce crop quality and market price. Seeds of *L. temulentum* in linseed flax reduce the value of the seed flax and permit loss of flax grains during the extensive cleaning process. Wheat, barley, or rye containing seeds of this weed are not suitable for flour or other products.

The plant is ranked as a principal weed of barley, beans, peas, and wheat in Ethiopia; lowland rice in the Philippines; and vegetables in Spain. It is a common weed of cereals in Tunisia and of flax in Argentina. It is a weed of barley in Greece, Iran, and Iraq; cereals in Argentina, Australia, and India; oats in Greece; potatoes in Colombia and the Soviet Union; sugar beets in Spain; pastures in Australia; wheat in Argentina, China, Colombia, Greece, India, Iraq, Iran, and Portugal; and of winter season crops in India.

Control of *L. temulentum* should be related to preventive measures: only clean crop seeds must be sown and seed formation must be prevented by harvesting infested fields before the weed seeds have matured. Fallow cultivation or crop rotation of cereals or grain crops with cultivated crops will help to reduce the weed population by tillage during the seedling stage. This approach can be especially useful because this annual weed must rely for survival on seed formation and a constant seed supply. Diligence in seed cleaning can also assist in preventing traffic in commerce of weedy grain.

COMMON NAMES
OF LOLIUM TEMULENTUM

ARGENTINA
  cizaña
  joyo
  trigollo
AUSTRALIA
  darnel
  drake
BRAZIL
  alho bravo
CHILE
  ballico
COLOMBIA
  ballico
DENMARK
  giftig najgraes
EASTERN AFRICA
  darnel
ETHIOPIA
  inkerdad
EGYPT
  zawan
FRANCE
  ivraie enivrante

GERMANY
  Taumellolch
GREECE
  darlet
HAWAII
  darnel
INDIA
  darnel
IRAN
  gij dane
IRAQ
  annual darnel
  darnel
  rwaitah
  tares
ITALY
  gioglio
  liglio
JAPAN
  darnel
  poison darnel
LEBANON
  darnel

tares
zuwan

NEW ZEALAND
darnel

NORWAY
svimling

PORTUGAL
joio

SOUTH AFRICA
bearded cheat
cheat
darnel
drabok
dronkgras
SPAIN
vallico
SWEDEN
dårrepe

TUNISIA
ivraie
TURKEY
delice

UNITED STATES
annual ray-grass
bearded darnel
cheat
darnel

ivray
poison darnel
poison rye-grass
white darnel

URUGUAY
joyo

VENEZUELA
cizaña

❧ 49 ❧

Mikania cordata (Burm. f.) B. L. Robinson, *Mikania scandens* (L.) Willd., and *Mikania micrantha* HBK

ASTERACEAE (also COMPOSITAE), ASTER FAMILY

Mikania is a genus of about 250 species of herbaceous or slightly woody vines. Because the plants are quite variable they are often difficult for workers to identify; but the most commonly known species are *M. cordata*, which is native to the Old World and widely distributed there; *M. scandens*, which is confined to North America; and *M. micrantha*, which is native to tropical America but which some workers have reported in Asia and the Pacific Islands. By far the most important weed of the three is *M. cordata*, a species which is troublesome in plantation crops, especially in tree crops such as tea, rubber, coffee, coconuts, cacao, and oil palms. It is a weed of 10 crops in 23 countries. It is a rapid-growing, rampant vine which smothers young tree crops and other plants. It cannot tolerate heavy shade. It contains substances which inhibit the growth of other plants and which depress nitrification rates in the soil. Clearly more taxonomic work is needed on the *Mikania* species of the Old World and the Pacific Islands in order to verify reports of distribution and correctly to identify these weeds.

DESCRIPTIONS

Mikania cordata

M. cordata is a fast growing, creeping or twining, *perennial* vine (Figure 133); *stems* branched, pubes-

cent to glabrous, ribbed, from 3 to 6 m long; *leaves* opposite, cordate or triangular-ovate, blade 3 to 12 cm long, 2 to 6 cm wide, on a slender petiole 1 to 8 cm long, base broadly cordate, tip acuminate, margins crenate, dentate, or entire, surfaces nearly glabrous, three- to seven-veined from base; *flowers* in small heads in open, nearly flat-topped (corymbose) panicles; axillary and terminal *heads* 6 to 9 mm long, four-flowered; *involucral bracts* four, obtuse or acute, 5 to 6 mm long, glabrous or subglabrous with one additional smaller bract about 3 mm long; *corolla* white or yellowish white, about 5 mm long; *anthers* bluish gray or grayish black; *style* white; *fruit* an achene, linear-oblong, 2 to 3 mm long, five angled, blackish brown, glandular; *pappus* of 40 to 45 bristles, about 4 mm long, white at first, reddish afterwards. Native to tropical regions of Southeast Asia and East Africa.

Mikania scandens

This weed is an herbaceous, *perennial* vine; *stems* branched, slightly four-angled, sparsely to densely pubescent; *leaves* opposite at swollen nodes, mostly triangular, 4 to 12 cm long, 2 to 7 cm wide, on a petiole 1 to 10 cm long, base sagittate, cordate, or hastate, tip acuminate, margins entire or with few teeth, surfaces glabrous, three- to seven-veined from base; *flowers* in small heads 5 to 7 mm long, crowded

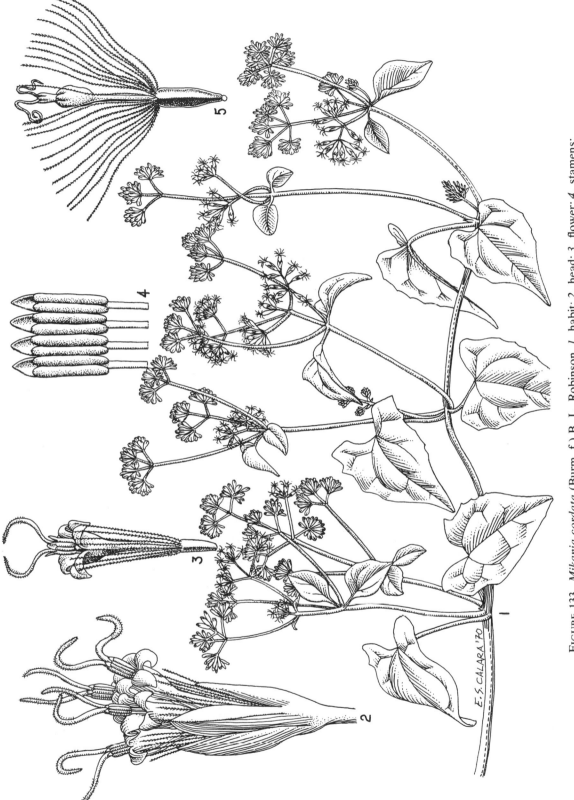

FIGURE 133. *Mikania cordata* (Burm. f.) B. L. Robinson. *1*, habit; *2*, head; *3*, flower; *4*, stamens; *5*, achene, with the attached flower.

E.S. CALARA '70

in roundtopped, lateral and terminal, modified pani-cles (corymbs); *involucral bracts* four, linear-lanceolate, 4 to 5 mm long, attenuate, often purplish tinged, with an additional smaller bract; *corolla* pink, pale purplish, or rarely white; *fruit* an achene, oblong, 1.5 to 2.5 mm long, brownish black, five-angled; *pappus* of 30 to 35 bristles 2 to 4 mm long, white or sometimes purple. Native of temper-ate to subtropical eastern North America, as far north as New York (Robinson 1922).

Mikania micrantha

M. micrantha is a fast growing, *perennial*, creep-ing or twining plant; *stems* branched, pubescent to glabrous, ribbed; *leaves* opposite, thin, cordate, triangular, or ovate, blade 4 to 13 cm long, 2 to 9 cm wide, on a petiole 2 to 8 cm long, base cordate or somewhat hastate, tip acuminate, margins coarsely dentate, crenate, or subentire, both surfaces glab-rous, three- to seven-nerved from base; *flowers* in heads 4.5 to 6 mm long, in terminal and lateral openly rounded, corymbous panicles; *involucral bracts* four, oblong to obovate, 2 to 4 mm long, acute, green, and with one additional smaller bract 1 to 2 long; four flowers per head; *corollas* white, 3 to 4 mm long; *fruit* an achene, linear-oblong, 1.5 to 2 mm long, black, five-angled, glabrous; *pappus* of 32 to 38 soft white bristles 2 to 4 mm long. Native of South America, Central America, and the Caribbean reg-ion, it is now reported to be in Asia and the South Pacific (Parker 1972).

Robinson (1934) briefly pointed out the differences in the *Mikania* complex as follows: the inflorescence is habitually looser and more paniculate in *M. micrantha* than in *M. scandens*; the inflorescence in *M. scandens* is mostly crowded with round-topped corymbs. The phyllaries (bracts) of the heads are acute in *M. micrantha* rather than attenuate as in *M. scandens*. *M. micrantha* seems never to show the purplish coloration which is nearly always present in *M. scandens*. Although the leaves vary greatly in contour both in *M. scandens* and in *M. micrantha*, they tend on the whole to be more sharply angled and triangular-sagittate or -hastate in *M. scandens* and more oval, cordate, and merely crenate in *M. micrantha*. *M. cordata* may be distinguished by the following characteristics: 40 to 45 reddish pappus bristles, corollas white, and heads 7 to 7.5 mm long (usually longer than those in *M. micrantha*, which are 4.5 to 6 mm long).

Another species which is sometimes reported as a weed is *M. cordifolia* (L.) Willd., an herbaceous or soft-woody twining plant with hexagonal stems; with densely gray-pubescent, round-ovate leaves with acute or acuminate tips; and with a cordate base and open sinus. It is native to South America.

DISTRIBUTION

Most species of *Mikania* are native to America. The most widely distributed species, however, is *M. cordata* which is probably native to Southeast Asia and East Africa and is widespread in those areas and in the Pacific Islands (Figure 134). The species is confined to the tropics where it is frequently found in young secondary forests, forest clearings, aban-doned or uncared-for clearings, forest margins, sec-ondary regrowth areas, thickets, ravines or moun-tain slopes, along roadsides and water courses, fal-low lands, and in plantation tree crops.

M. micrantha is widespread throughout South America and Central America, but is a weed in only a few places (Figure 134). It is usually found in damp lowland clearings or open areas but it may grow at an elevation of 2,000 m or more. In Bolivia it has been observed at 3,000 m. It grows along streams and roadsides as well as in or near forests. Parker (1972) reported that this species has been identified from several countries of Asia and he suggested that it may be the most aggressive species in some parts of that continent. He reported *M. micrantha* from Assam and Kerala in India, and from Bangladesh, Ceylon, Malaysia (both West Malaysia and Sabah), and Indonesia. Other countries or loctions of *M. micrantha* reported to the authors of this book dur-ing its preparation include the Cook Islands, Samoa, Fiji, India, Jamaica, and Thailand. Because of the questions regarding the presence of *M. micrantha* in Asia, we have not shown the weed in Asia in Figure 134.

M. scandens is found mostly along the Atlantic and Mexican gulf coasts in the United States. It extends as far north as New York and as far west as Texas. It sometimes grows in inland areas. The plant frequently is found in woods, thickets, and swamps. It is seldom reported as a weed. In the past it was thought that *M. scandens* was found in Asia, and reports of its presence have come from Ceylon, In-donesia, Malaysia, Mauritius, the Philippines, and Pakistan as well as from the Congo and Ghana. However, it is probable that these weeds were plants of *M. cordata*.

Another species, *M. cordifolia* (L.) Willd., be-haves sometimes as a weed. A densely pubescent plant, it is widespread in Central America, Mexico, South America, and the West Indies.

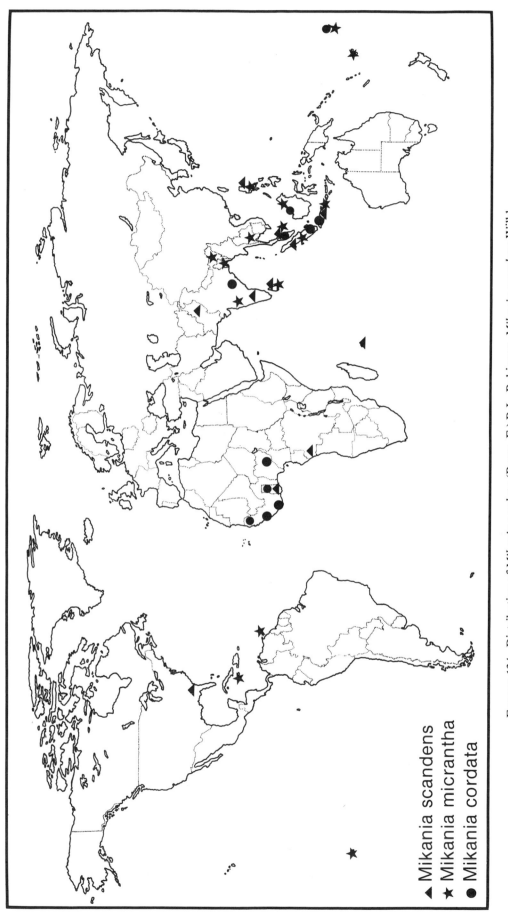

Mikania scandens
Mikania micrantha
Mikania cordata

FIGURE 134. Distribution of *Mikania cordata* (Burm. F.) B.L. Robinson, *Mikania scandens* Willd., and *Mikania micrantha* H.B.K. across the world in countries where they have been reported as weeds. The identity of *Mikania* species is much confused across the world. The locations above were provided by contemporary weed scientists for their countries.

PROPAGATION

M. cordata propagates by seeds and by rooting at stem nodes. It does not have rhizomes but is perennial. Seeds are produced in large numbers on the masses of rampant twining or creeping stems. The feathery parachutelike pappus of white or tawny hairs makes it possible for the wind to spread the seeds for long distances.

As important as seed propagation is for *M. cordata*, rooting from stem nodes may be of equal significance. A small stem fragment with at least one node can sprout and take root when it comes in contact with the soil. Fragments or sections pulled or cut from cropland during handweeding fall to the ground and take root; therefore, handweeding or cultivation without destruction or drying of stems may do little to control the weed. Craig and Evans (1946) reported an alarming rate of spread of the weed in Mauritius, a spread due mainly to the movement of stem materials by streams and floods; from this incident the authors concluded that vegetative propagation is the most important means of dispersal on that island.

Dispersal by humans is important. The luxuriant growth is frequently cut as green fodder for cattle and other livestock. Dropped fragments or uneaten and undecayed stem materials may take root in new locations. When the plants grow near a river or stream, they are sometimes cut and thrown into the water and thereby are transported elsewhere. Craig and Evans (1946) reported that *M. cordata* was cultivated for feeding deer on a hunting preserve in Mauritius.

A major factor in its dispersal in the Old World may have been its use by tropical plantation managers as a cover crop or smother crop to keep out other weeds (Bamber 1909, Burkill 1935, King 1966). This was a recommended practice in Ceylon, Indonesia, Malaysia, and probably elsewhere in countries dependent upon plantation tree crops.

BIOLOGY AND HABITAT

M. cordata grows most frequently in places receiving high rainfall, probably 1,500 mm or more; prefers rich, damp soil; rarely grows in dry areas; and thrives in open, disturbed places. For that reason it is common in young secondary forests, in forest clearings, in plantation tree crops, fallow or neglected lands, and along rivers and streams, waste areas, steep hillsides, and even mountainsides from whence winds probably spread the seeds to new

areas. The species will grow in partial shade but cannot tolerate dense shade (Macalpine 1959). Burkill (1935) stated that the plant takes up considerable quantities of potassium. Craig and Evans (1946) studied production of green material and dry matter and major nutrient uptake of a representative infestation of *M. cordata* in Mauritius. The fresh weight yield was 41.7 metric tons per hectare, the dry weight yield was 3.72 tons per hectare. Nutrient concentrations in the dry matter were nitrogen, 1.72 percent; phosphorus, 0.33 percent; and potassium, 3.86 percent. Converted to a field area basis, *M. cordata* removed 179 kg of nitrogen, 12 kg of phosphorus, and 139 kg of potassium per hectare.

Caum (1940) and Craig and Evans (1946) studied *M. cordata* growth and development. When the seedlings or small plants which start from stem sections begin to grow, they spread very quickly over the ground. If vertical supports such as crops, trees, rock walls, fences, and so forth are available, the plants quickly twine and climb on them. In the absence of a vertical support the plants creep over the ground and soon form a dense cover of entangled stems bearing many leaves. Once established the rate of spread of the weed can be alarming. Craig and Evans (1946) tell of a 10 hectare corn field which was covered with the weed to a depth of 0.6 to 1 m deep after a 6 months fallow period. In cultivated crops the plants twine about the stems of the crops and climb to the uppermost part of them, covering and smothering them; if unchecked the weed may do this as well to trees and shrubs. Caum (1940) described large breadfruit trees (*Artocarpus altilis* [Park.] Fosberg) which were being killed by *Mikania cordata* in Samoa, and he told of coconut plantations which had to be abandoned because of the incursions of this weed. Abandoned banana, coconut, or other tree crop lands are quickly taken over and the crop choked out. Perhaps the habit of growth is best expressed in the words often used to describe the plant (species?): "rampant," "dense tangle," great confused tangle," "smothering," "swamping." A common name for *M. cordata* in the Pacific Islands is "mile-a-minute." Because of its abundant, smothering growth habit *M. cordata*, if undisturbed, often spreads in great massive circular patterns. In Mauritius the plant became widespread over more than half of the Flacq district in 17 years, infesting 17,380 hectares of the total 29,440 hectares in that time.

M. cordata produces abundant axillary or terminal flowers in compound flat-topped panicles. Flowering occurs throughout the year in most areas, but may reach a seasonal peak during the summer. The flowers are fragrant, their scent somewhat resembling that of vanilla.

AGRICULTURAL IMPORTANCE

M. cordata is probably the most important weed in the genus *Mikania*. Figure 135 indicates that *M. cordata* is usually a weed of plantation crops and may be most serious in rubber and tea. Not shown in this figure are the data that *M. cordata* is a common weed of tea in Ceylon; a weed of coffee in the Congo; of coconuts in Ceylon, Fiji, and Malaysia; of cacao in Ghana and Malaysia; of oil palms in Malaysia; and of pastures, tree crops, and bananas in Fiji.

M. cordata is important because of its rapid growth and competitive habit and because of its ability to grow from even the tiniest stem fragments; these characteristics make it almost uncontrollable by handweeding or by mechanical means. For this reason the weed must be controlled by applying herbicides or by cutting and destroying all of the stems. The latter is accomplished by drying the stems, burning them, or feeding them to cattle. The weed is especially troublesome in young plantings of coconuts, oil palms, bananas, or cacao, for, if uncontrolled, it will cut out all sunlight by choking and smothering the crop plants. The weed competes with tea seedlings in nurseries and also with the mature crop (Dutta 1965), but it also makes plucking difficult when it grows over the tops of bushes (Kasasian 1971). In young rubber trees or oil palms it invades the cover crop areas, competing seriously with the leguminous cover plants (anon. 1967*a,b*, 1964; Watson, Wong, and Narayanan 1964*b;* Riepma 1962, 1965; Wong 1966). Rubber and oil palms eventually shade out the *M. cordata* and, where the weed is not controlled, the ground may then become bare and subject to erosion. This is especially true in hilly areas.

In Malaysian oil palm plantations which were unfertilized, a ground cover of *M. cordata* resulted in smaller fruit clusters than those which had been produced under a leguminous cover (Gray and Hew 1968).

Although considered more of a problem in perennial tree crops, *M. cordata* also can be a problem in cultivated crops such as sugarcane. The weed's active growth, its twining habit, and the tremendous quantity of green matter it produces make it a problem in this crop just at the time when weed control is most difficult.

There is evidence from Malaysia and Indonesia that the effects of *M. cordata* on a crop may extend beyond the normal competition for nutrients, light, and soil moisture. It was observed that the growth and girth of rubber trees were reduced when *M. cordata* was the cover plant or when it became dominant (Mainstone and Wong 1966, R. Wong 1964, Wycherley and Chandapillai 1969, Seth 1971). Rubber trees in areas where *M. cordata* was dominant showed lower nitrogen and phosphorus levels in their leaves, less rooting, and smaller canopies. This resulted in spite of the fact that *M. cordata* covers produced less dry matter and took up less nutrients than other covers used. In pot incubation experiments, Guha and Watson (1958) found lower rates of nitrification in soils mixed with leaves and stems of *M. cordata*. In later studies lower nitrate-nitrogen levels were found under the *M. cordata* cover than under legume or grass covers (Watson, Wong, and Narayanan 1964*a*). It was difficult to explain the effects of *M. cordata* covers as resulting from competition for nutrients or water alone.

R. Wong (1964) found that *M. cordata* contains substances which inhibit the growth of rubber trees, tomato plants, and tropical kudzu (*Pueraria phaseoloides*). Water extracts (1 and 2 percent) of oven-dried stem and leaf materials from *Mikania cordata* significantly depressed dry weight and nitrogen and phosphorus content of tomato seedlings as compared to water extracts of *Paspalum conjugatum* or of tropical kudzu. Petroleum ether extracts of *M. cordata* reduced the dry weight of tomato seedlings. Water extracts of both root and green materials caused a marked reduction in dry weight of tropical kudzu seedlings, an increase in nitrogen and a marked decrease in phosphorus and calcium in the legume seedlings. In studies of the effect of *M. cordata* on *Fomes lignosus*, a fungus disease which causes white root disease of rubber, it was found that water extracts inhibited the growth of the fungus. Root extracts were most active if they were followed by leaf and stem extracts. Increasing concentrations gave increasing inhibition. Petroleum ether extracts could be separated further into hot-water-soluble and -insoluble fractions, the former being the most active. The hot-water-soluble fraction was more active than fractions of water extracts of the same roots. Extracts of *M. cordata* significantly reduced the rate of nitrification in incubation pots. The inhibitory compounds were found to be phenolic and flavenoid substances. It is not known whether these compounds are released from the roots during growth.

M. cordata extracts did not affect ammonification

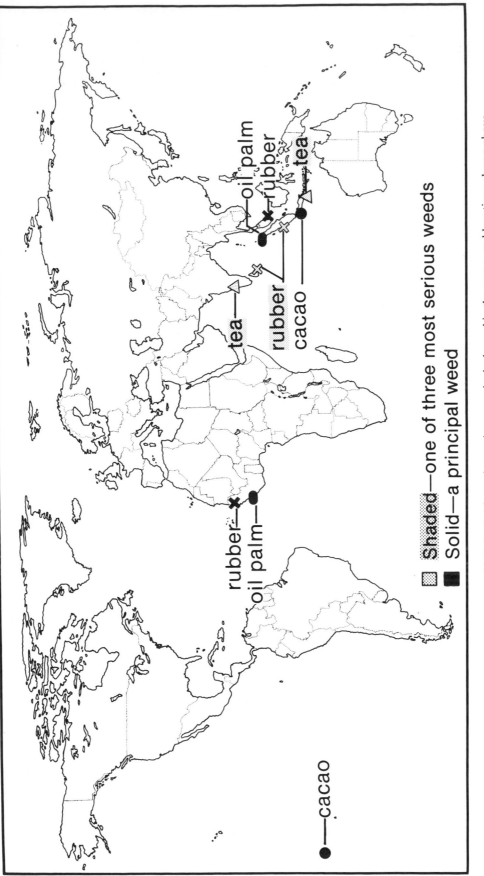

FIGURE 135. *Mikania cordata* (Burm. f.) B. L. Robinson is a serious or principal weed in the crops and locations shown above.

but did suppress nitrification. This finding seems to relate to data obtained from earlier studies in which lower nitrate-nitrogen levels were found under *M. cordata* than under other cover plants. It was observed that the leaves of young rubber plants growing in a ground cover of *M. cordata* showed low nitrogen levels; however, when these same trees were older, their leaves showed near-normal nitrogen levels. This suggests that, as increased shading occurs and *M. cordata* is shaded out, nitrification becomes less and less affected by the weed. Therefore, inhibitory effects of *M. cordata* may be most severe when it is in vigorous, active growth. It does appear that *M. cordata* has both direct and indirect effects on rubber and other crops.

M. cordata has been reported to be susceptible to parasitic attack by *Cuscuta chinensis* in Ceylon (King 1966) and to *C. australis* in Fiji (Parham 1958) and Malaysia (anon. 1960*c*).

M. cordata is used as a cure for snake and scorpion bites in South Africa (Watt and Breyer-Brandwijk 1962), as a remedy for body itch in Malaysia, and as a poultice for wounds in Java (Burkill 1935).

COMMON NAMES

Mikania cordata

CEYLON
  gam-palu (Sinhalese)
  kadugannawa creeper
  kehel-palu (Sinhalese)
  loka-palu (Sinhalese)
  pulun-taliya (Tamil)
  tuni-kodi (Tamil)
  vatu-palu (Sinhalese)

FIJI
  mile-a-minute
  wa bosucu
  wa butako

MALAYSIA
  akar lupang

  akar ulam tikus
  cheroma
  mile-a-minute
  seletpattungau

MAURITIUS
  liane marzoge
  liane Pauline
  liane raisin

SAMOA
  fue saina
  fue sega
  mile-a-minute

SOUTH AFRICA
  kamele

Mikania scandens

UNITED STATES
  climbing hempvine

Mikania micrantha

PERU
  camotille

Mikania cordifolia

DOMINICAN REPUBLIC
  mata finca
  verdolín

UNITED STATES
  climbing hempvine

❦§ 50 ᔈ❧

Mimosa invisa Mart.

MIMOSACEAE (also LEGUMINOSAE), MIMOSA FAMILY

Mimosa invisa is a shrubby, herbaceous, spreading plant that becomes woody with age. A native of Brazil, it is a legume and a member of a large genus with most of the species in the New World. Depending on the site, it may grow as a biennial or a perennial. The most outstanding characteristic of the species is the presence of recurved spines or thorns on its stems. It is found in waste places, but is becoming increasingly troublesome in plantation crops and pastures, thriving in arable crops as well. It is a weed of 13 crops in 18 countries.

DESCRIPTION

M. invisa is an erect, climbing, ascending or prostrate, *biennial* or *perennial* shrub (Figure 136) that often forms a dense thicket, the root system strong, often woody at the decumbent base; *stems* conspicuously angular throughout the length, up to 2 m tall with many randomly scattered recurved spines or thorns 3 to 6 mm long; *leaves* bipinnate, 10 to 20 cm long, moderately sensitive to the touch; pinnae four to nine pairs; *leaflets* 12 to 30 pairs, sessile, opposite, lanceolate, acute, 6 to 12 mm long, 1.5 mm wide; *inflorescence* a head, one to three in the axils of leaves, on stalks 1 cm long, hairy, about 12 mm in diameter; *corolla* united at least at the base (gamopetalous), pale pink; *stamens* twice as many as the petals; *fruit* a pod, spiny, three- to four-seeded, borne in clusters, linear, flat, 10 to 35 mm long, splitting transversely into one-seeded sections which separate at grooves or seams (sutures); *seeds* flat, ovate, 2 to 2.5 mm long, light brown.

It may be distinguished by its angular stems with recurved thorns or spines and by its stamens which are twice the number of the petals.

DISTRIBUTION AND BIOLOGY

M. invisa is sometimes called the giant sensitive plant to distinguish it from its near relative, *Mimosa pudica*, the sensitive plant. *M. invisa* is only moderately sensitive to touch. It has a very robust growth and scrambles over other herbage by means of spiny stems which form spreading, tangled masses or thickets of undergrowth up to 1.5 or 2 m in height. Animals may die if they become trapped in such thickets and men may be severely injured.

The plant flowers all year round in the Philippines. It reproduces by seeds enclosed in spiny pods well adapted to being carried by animals. Very young seedlings a few weeks old can produce viable seeds and some will germinate immediately. Some seeds, however, remain in the soil for years before germinating. The species is found in moist waste places, in plantations, pastures, and cultivated grounds. A thornless type (variety *inermis* Adelb.) has been described in Indonesia and the Philippines. *M. invisa* is reported to be troublesome as a weed mainly in Southeast Asia and the Pacific Islands (Figure 137). There is a single report of its occurrence from South America (Argentina) but it is not yet a problem in western Asia, Africa, Europe, North America, or Central America.

328

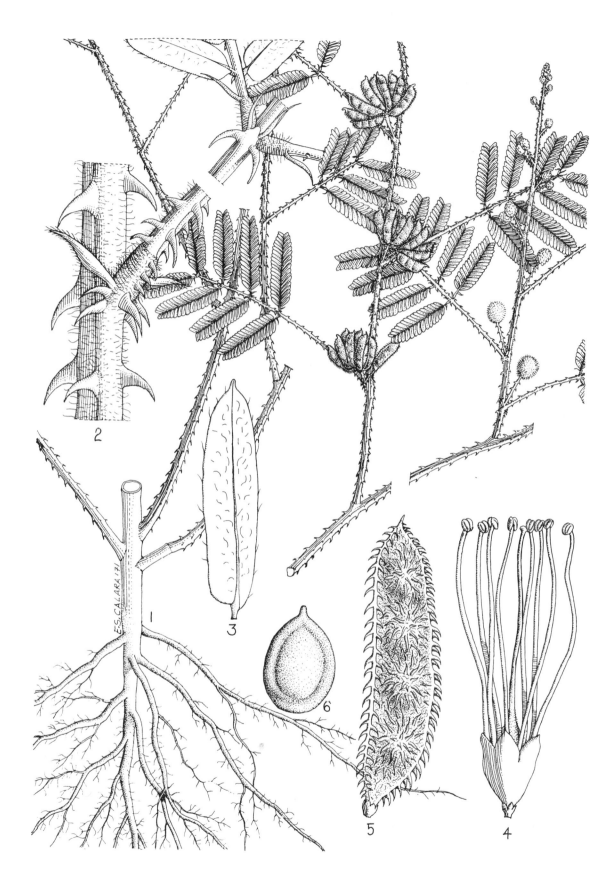

FIGURE 136. *Mimosa invisa* Mart. *1*, habit; *2*, stipules; *3*, leaf, enlarged; *4*, flower; *5*, pod; *6*, seed.

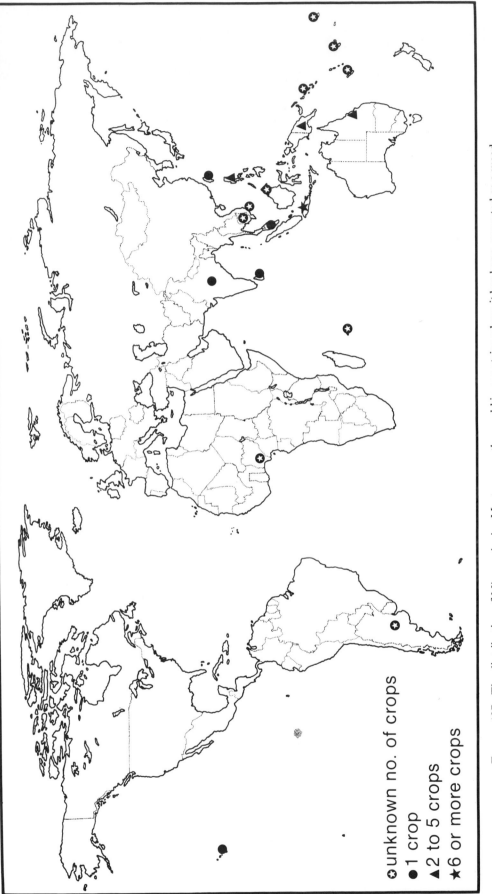

FIGURE 137. The distribution of *Mimosa invisa* Mart. across the world in countries where it has been reported as a weed.

unknown no. of crops
● 1 crop
▲ 2 to 5 crops
★ 6 or more crops

AGRICULTURAL IMPORTANCE

M. invisa is a principal weed of rubber and coconuts in New Guinea, of rubber in Indonesia, of sugarcane in Taiwan, and of sugarcane, pastures, and tomatoes in the Philippines. It is also a weed of sugarcane and pastures in Australia, of upland rice, cassava, soybeans, corn, and tea in Indonesia, of coconuts in Ceylon, of rubber in Malaysia, and abaca and pineapples in the Philippines. In plantations and arable lands in Fiji and the Philippines it is potentially one of the worst weeds ever introduced to the islands.

Vacant rice fields are used as pasture in many places in Southeast Asia; but, once the new rice has been planted, roadsides, coconut plantings, and unused areas must be used for grazing animals. If *M. invisa* invades such areas, the animals will not step on the spiny plants or eat them and, because the species now has no other plant competition, it becomes very difficult to eradicate. In cultivated fields it is rapidly distributed by machinery. The use of mechanical harvesters in sugarcane will make the problem even more severe in that crop. In Asia the plant is sometimes used as a cover crop. It grows about 1.5 to 2 years and is a good soil cover during this period because of its spreading habit and because it adds nitrogen to the soil through symbiotic nitrogen fixation. When the plant dies and dries up it is a serious fire hazard. Just before the rainy season it is sometimes burned to stimulate the germination of a good seedling crop. The plants are allowed to grow and fix nitrogen for a time and then are plowed under as they become spiny. In countries where sugarcane is hand-cut, fields infested with *M. invisa* are difficult to harvest because the thorns puncture and lacerate the hands of the workers. Owners may have to pay a premium to the workers to induce them to harvest infested fields, and the sugar mill may deduct a penalty for impure cane if weed fragments are not removed prior to milling.

When growing vigorously it can hold its own against *Imperata cylindrica*, the worst perennial grass of southern Asia. In many places, however, when the life cycle of *Mimosa invisa* is over, the perennial grass quickly reestablishes itself as light again becomes available.

COMMON NAMES
OF MIMOSA INVISA

AUSTRALIA
  giant sensitive plant

CAMBODIA
  banla saet

FIJI
  giant sensitive plant
  wagadrogadro levu

INDONESIA
  borang

  djoekoet borang
    (Sundanese)
  pis koetjing (Indonesian)

PHILIPPINES
  makahiang lalake

UNITED STATES
  giant sensitive plant

VIETNAM
  cõ trinh nũ móc

∽§51∂∾

Mimosa pudica L.

MIMOSACEAE (also LEGUMINOSAE), MIMOSA FAMILY

Mimosa pudica is a perennial in the warm regions of the world. It is a native of tropical America and, because it is often grown as an annual ornamental plant, it is now widespread in the world. It is easily recognized because the petiole drops and the leaflets close when the plant is touched or disturbed. It has been reported to be a weed in 22 crops in 38 countries.

DESCRIPTION

M. pudica is a much-branched *perennial* herb (Figure 138), slightly woody at the base; with either an upright or low trailing habit, from 20 to 100 cm in height; *stems* stiff with thorns and scattered prickles on internodes, reddish brown or purple; *leaves* normally with one or two compound leaflets, bipinnate; *leaflets* 12 to 25 pairs, oblong-linear, pointed, with hairy margins, 9 to 12 mm long, 1.5 mm wide, leaflets when touched draw back and fold up together with pinnae and petioles, movements are quickest in young plants; *flower heads* pinkish, ovoid, 9 mm in diameter, on axillary peduncles 12 to 25 mm long, with prickles; *stamens* four; *pods* attached in a cluster, oblong, almost flat, pointed at tip, edges armed with small, outstanding prickles, 1 to 2 cm long, 3 to 5 mm wide, breaking into one-seeded joints; *seeds* one to five, flattened, small, 3 mm in diameter.

The leaves which are sensitive (leaflets are drawn back and folded with the pinnae and petioles) when touched are a distinguishing characteristic of this species. *M. pudica* is common in waste ground up to 1,300 m.

BIOLOGY AND DISTRIBUTION

Plants of the genus *Mimosa* are often grown for their showy or feathery flowers, but the species *M. pudica* and *M. invisa* are also very bad weeds. The genus, in the family Mimosaceae, also known as Leguminosae, is very large and falls into a select group of genera having 300 to 500 species. They may be herbaceous or woody and some are climbers. Some of the species are sensitive to the touch or to any other disturbance. *Mimosa* is from the Greek, "mimic." *Pudica* comes from the Latin word meaning "modest" or "bashful." Many of the common names for this plant around the world mean either ashamed, shy, or asleep. *M. pudica* is cultivated as an annual but it behaves as a perennial in warm

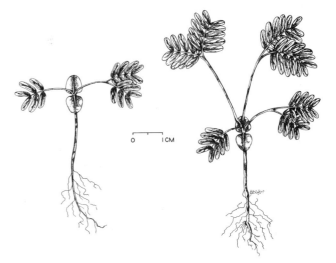

332

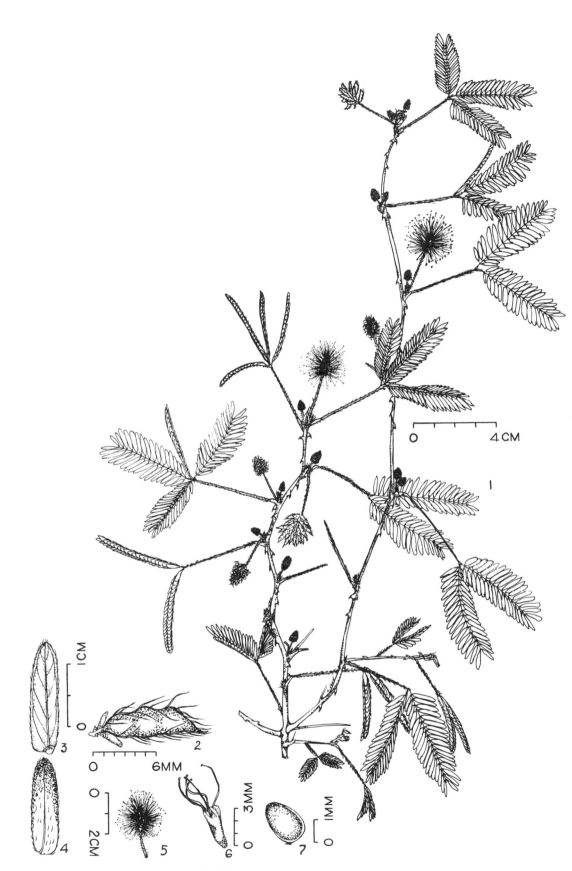

FIGURE 138. *Mimosa pudica* L.

Above: *1*, habit; *2*, pod; *3*, leaflet; *4*, leaflet; *5*, flower head; *6*, individual flower; *7*, seed.
Opposite: Seedlings.

areas. It has been widely introduced and is now pan-tropical.

The plant reproduces by seeds. The bristled seeds travel on the fur of animals and the clothing of man. The weed grows in cultivated areas, lawns, waste places, and frequents settled areas. It grows on a wide variety of soils in many areas. In the Philippines it grows at low to medium elevations. It can stand considerable shading; for that reason it is an important vegetative cover under coconuts (Plucknett 1973).

In the Philippines the plant flowers all year and may produce as many as 675 seeds. Freshly harvested seeds germinate within 2 weeks in moist soil in some areas. In one laboratory test, 80-percent germination was obtained in 4 weeks at alternating temperatures of 20° to 30° C, and many hard seeds, as is the case with many legumes, were produced. Scarification of seeds with sulfuric acid aids germination. Seeds stored in a laboratory for 19 years gave a germination of 2 percent. Efforts have been made to develop spineless types for pasture use, but these lines later became spiny because the seedlings segregated into spiny and spineless forms.

From Figure 139 it may be seen that the species is troublesome in the Caribbean area and South America. On the other side of the world it is reported to be in the fields of India, the mainland of Southeast Asia, the south Pacific Islands, and Australia. It has not been reported as a problem weed in Canada, Africa, Europe, the Near East, or the United States.

AGRICULTURAL IMPORTANCE

From Figure 140 it may be seen that *M. pudica* is a serious weed in corn in Malaysia and a common weed in that crop in Indonesia. It is reported to be a principal weed in sugarcane in Mexico and a common weed in Taiwan. It is a principal weed of rubber in Indonesia, Mexico, and New Guinea and a common weed in Malaysia. It ranks as a principal weed in tea in Indonesia; soybeans in the Philippines; sorghum in Malaysia; and it is a common weed of upland rice in Ceylon, Fiji, Indonesia, the Philippines, and India. Although these facts are not shown on the map, *M. pudica* is a serious weed of tomatoes in the Philippines; a principal weed of coconuts in New Guinea and Trinidad; and a weed of plantation crops such as bananas, papayas, pineapples, coffee, oil palms, citrus, and cotton. It can also become a pest in tropical pastures where its high plant populations and thorny stems make grazing difficult and often deny available forage to the grazing animals. How-

ever, in Fiji it is highly regarded for dairy production on pastures. In such places the dried foliage sometimes is a fire hazard. It is an important pasture plant in the South Pacific where it is a major component of pastures under coconut (Plucknett 1973).

The plant is especially bothersome to man in crops where hand-weeding is practiced, for its thorns penetrate and lacerate the hands, causing painful wounds. Also in dense stands the plants are difficult or painful to walk through because of injury by the thorns to the bare skin.

M. pudica is an alternate host of the parasitic flowering plants *Cassytha filiformis* L. and *Cuscuta sandwichiana* Choisy (Raabe 1965), and of the nematode *Meloidogyne* sp. (Raabe, unpublished; see footnote, *Cyperus rotundus,* "agricultural importance").

COMMON NAMES
OF MIMOSA PUDICA

AUSTRALIA
  sensitive plant
BANGLADESH
  lajjabati
BARBADOS
  sensitive plant
BRAZIL
  dorme dorme
  malicia
  malicia de mulher
CAMBODIA
  paklab
  sampeas
CEYLON
  dedinnaru
  nidi-kumba
    (Sinhalese)
  thodda-chinunki
  thoddal-vadi
  thodda-vadi-kodi
    (Tamil)
COLOMBIA
  sensitive plant
CUBA
  dormidera
DOMINICAN REPUBLIC
  morivivi
FIJI
  sensitive plant
HAWAII
  pua-hilahila
  sensitive plant

INDIA
  lajjavati
  lajkuli (Oriya)
  touch-me-not
INDONESIA
  boedjang kajit (Sundanese)
  daoen kaget-kaget (Indonesian)
  koetjingan (Javanese)
JAMAICA
  dead and awake
  shame-lady
  shamer
MALAYSIA
  mala malu
MAURITIUS
  sensitive plant
MEXICO
  sensitive plant
NICARAGUA
  dormidera
PHILIPPINES
  babain (Ilokano)
  huya-huya,
  kirom-kirom
    (Bisayan)
  makahiya (Tagalog)
  sipug-sipug (Subanun)
  torog-torog (Bikol)
  tuyag-tuyag

PUERTO RICO
  mimosa

TAIWAN
  han-hsiu-tsau

THAILAND
  mai-yarap

TRINIDAD
  shamebush

UNITED STATES
  sensitive plant

VENEZUELA
  dormidera

VIETNAM
  mäc cö

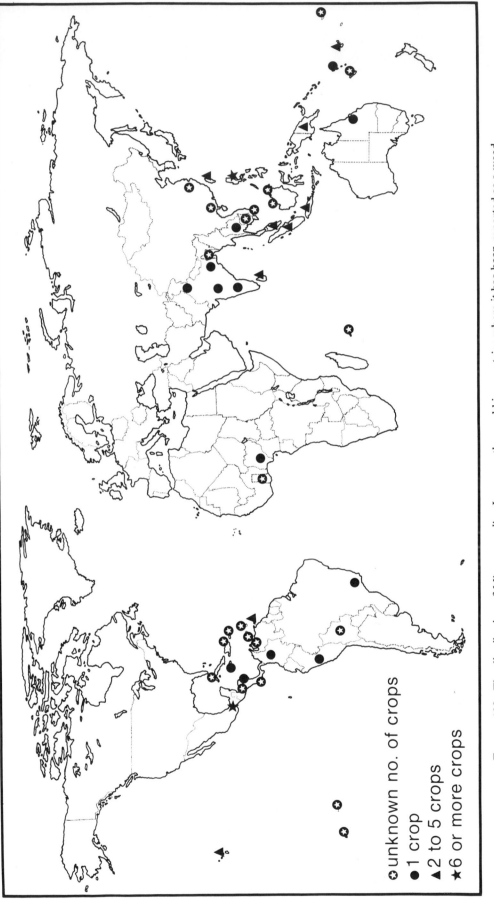

FIGURE 139. The distribution of *Mimosa pudica* L. across the world in countries where it has been reported as a weed.

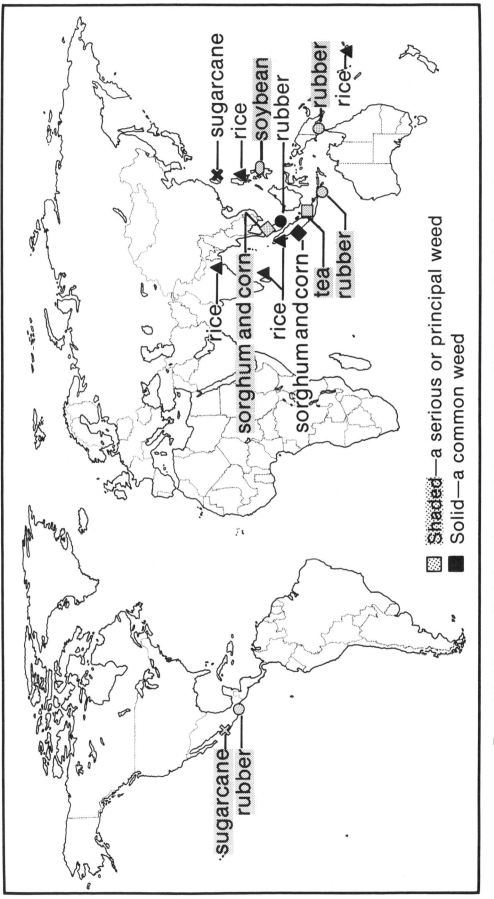

FIGURE 140. *Mimosa pudica* L. is a serious, principal, or common weed in the crops and locations shown above.

Monochoria vaginalis (Burm. f.) Presl and *Monochoria hastata* (L.) Solms (= *M. hastifolia* Presl)

PONTEDERIACEAE, PICKEREL WEED FAMILY

Monochoria refers to the one separate stamen in plants of this genus. *M. vaginalis* is a fleshy annual or perennial herb with a glabrous, shiny appearance. It grows in subaquatic or marshy places. It is native to tropical Asia and Africa. It is a very serious weed in the rice fields of eastern and southern Asia, and is found in 13 countries. In India the young plant is eaten as a pot-herb. A related species, *M. hastata* (L.) Solms, is similar in appearance to *M. vaginalis,* but differs in having a branched, well-developed rhizome which forms a network of plants, leaves with a predominantly sagittate or hastate base, more flowers per spike, and lighter colored flowers. *M. hastata*, though not as common as *M. vaginalis*, can sometimes act as a weed in lowland rice. *M. hastata* is a weed in seven countries.

DESCRIPTIONS

Monochoria vaginalis

M. vaginalis is a smooth tufted, *annual* or *perennial*, aquatic herb with a very short *rhizome*; (Figure 141), 10 to 50 cm tall, stemless; *old plants* often forming large clumps, but these are not connected; *leaves* 2 to 12.5 cm long, 0.5 to 10 cm wide, in very young plants without lamina; leaves of somewhat older plants with a floating linear, or lanceolate blade; leaves of still older plants, ovate-oblong to broadly ovate, sharply acuminate, the base heart-shaped or rounded, shiny, deep green in color, with longitudinal veins; *petioles* soft, hollow, growing from buds at the base, *leaf sheaths* twisted together at the base, slightly reddish when young; *crown* appears bulbous; *inflorescence* spikelike, basally opposite the sheath of the floral leaf, with a large bract arising from a thickened bundle on leaf stalk, about two-thirds of the way up the stalk from the base; *flowers* three to 25, opening simultaneously or in quick succession; on *pedicels* 4 to 25 mm long; *perianth* 11 to 15 mm long; *petals* six, violet or lilac blue, spreading at flowering, afterwards spirally contorted; *stamens* six, one with a lateral obliquely erect tooth; *ovary* with a long style; *capsule* about 1 cm, splitting between the partitions into three valves; *seeds* numerous, longitudinally ribbed.

The spikelike inflorescence which is opposite the floral leaf and the one stamen with a lateral oblique erect tooth are distinguishing characteristics of *M. vaginalis*. The plant may be very variable and for that reason may be misidentified. Backer and Bakhuizen van den Brink, Jr. (1968) described the variability in this manner, "Specimens with few-flowered inflorescences and small, often narrow, proportionally long leaves have wrongly been described as varieties or even separate species. They are either young or weak, or were collected in deep

FIGURE 141. *Monochoria vaginalis* (Burm. f.) Presl. *1*, habit; *2*, flower; *3*, flower vertical section; *4*, ovary, cross section; *5*, stamens; *6 a–c* seedlings; *7*, capsule; *8*, seed.

water." Ohwi (1965) described a variety of *M. vaginalis* in Japan; *M. vaginalis* (Burm. f.) Presl var. *plantaginea* (Roxb.) Solms-Laub (= *Pontederia plantaginea* Roxb. and *Monochoria plantaginea* Roxb. Kunth.). This variety has long petioled basal leaves, shorter petioles on leaves attached to a stem or axis, a three- to seven-flowered inflorescence, and purplish blue flowers. This variety has been reported from Japan, Korea, and China.

Monochoria hastata

M. hastata is a smooth, erect, aquatic, *perennial* herb, 30 to 100 cm tall; with a well-developed, branched *rhizome*; *old plants* form large groups which may be connected by underground stolons; *leaves* of mature plants 5 to 25 cm long, 4 to 20 cm wide, almost always with a sagittate or hastate, rarely with a cordate, base, *basal lobes* diverging, either with acuminate tips or not, borne on long sheathlike petioles; *inflorescence* in spikelike racemes; *flowers* 15 to 60, not expanding simultaneously, lower flower stalks (pedicels) 15 to 30 mm long, higher ones 7 to 20 mm; *perianth* light blue, 15 to 18 mm long; *fruit* enclosed by the perianth, about 1 cm long; *seeds* numerous, light brown, with faint longitudinal lines, blunt at one end, rounding at the other, minute.

These plants are often mistaken for one another, but they can be differentiated. *M. hastata* is larger; also it has a well-developed, branched rhizome, whereas *M. vaginalis* has a short rhizome. *M. hastata* has predominantly sagittate or hastate leaves; *M. vaginalis* has leaves which are usually heart-shaped. *M. hastata* has more flowers in the raceme. Flowers in *M. vaginalis* open nearly simultaneously, those of *M. hastata* do not. The lower pedicels of flowers on *M. hastata* are elongated, whereas all pedicels in *M. vaginalis* are less than 1 cm long.

BIOLOGY

Little is known of the biology of *M. vaginalis*. It is in the family Pontederiaceae along with the dreaded *Eichhornia crassipes*, the water hyacinth. With a group of families comprised of other water-loving plants, it is placed on the taxonomic scale between the Cyperaceae and the Liliaceae. It is thus near the arums (Araceae), the duckweeds (Lemnaceae), the spiderworts (Commelinaceae), and the rushes (Juncaceae).

M. vaginalis may be an annual or pseudoannual in flooded ricefields, but may grow as a perennial in constantly flooded areas. It can be found from 0 to 1,500 m altitude in Indonesia. The plant roots in mud and its upper portions grow above the water.

In Java *M. hastata* grows from 0 to 700 m elevation in freshwater pools, mudflats in rivers, ditches and ricefields, and along canal banks (Backer and Bakhuizen van den Brink, Jr. 1968). In Fiji it grows in shallow water, swampy ground, open drains, rice fields, or in very wet soils (Parham 1958). Many seeds are produced by each plant.

The most helpful work on the biology of *M. vaginalis* was done in Japan by Noda and Eguchi (1965). They studied seedling emergence in both experimental and field plots in early, ordinary (planted at the usual time for rice in the area), and late plantings of rice in southwestern Japan. They also studied seasonal variations in germination and the effects of soil moisture. The greatest number of seedlings emerged in the early planted plots and fields. Seedling emergence at the ordinary and late planting times was only about 15 percent that of the early emergence.

In a second study Noda and Eguchi recorded the seedlings as they emerged from: (1) fields where the water level was held 3 cm above the soil surface, or (2) at the soil surface (saturated), or (3) 20 cm below the soil surface (dried). In (3) there was considerable fluctuation of soil moisture content, and soil moisture probably was below field capacity on the average.

In these tests, 35 percent more seedlings emerged from the saturated soil than they did from submerged or dry soil in the early planting. In the ordinary planting season there was no significant difference among soil moisture conditions although a submerged condition appeared to be most favorable. The differences were not significant in late season but then a dry soil was favored.

A striking change in germination pattern was shown under submerged conditions. Here the majority of the seedlings emerged in a very short time, the peak being between 15 and 25 days; whereas in saturated or dried soil there was a tendency toward gradual emergence through the season. In dry soil the curves of germination were almost flat. It seems likely under these latter conditions that a single herbicide treatment early in the season would be futile. The majority of the weeds might germinate after the herbicide had broken down, permitting a heavy seeding for the next crop.

In Taiwan *M. vaginalis* produced higher fresh weight yields in paddy fields than did any other weed (Lin 1968). This weed produced twice the yield of

Echinochloa crusgalli which ranked second. In competition studies in the Philippines *M. vaginalis* was not as competitive in rice as was *Echinochloa crusgalli* (Lubigan and Vega 1971). In these studies the critical weed population of *M. vaginalis* was 60 plants per sq m; a natural stand of *M. vaginalis* (366 plants per sq m) reduced rice yield by 35 percent. Lubigan and Vega concluded that *M. vaginalis* does not compete seriously for light and that, because it is shallow-rooted, the deeper rooted rice plants may be able to compete more vigorously for nutrients. Further studies of this nature are required for this weed.

Both *M. hastata* and *M. vaginalis* flower throughout most of the year in the Philippines (Merrill 1912). Studies on the morphology of the flowers and seeds of *M. vaginalis* were reported by Juliano (1931). Gupta (1968) suggested that submergence of the fruits after pollination was beneficial for seed development.

AGRICULTURAL IMPORTANCE

As may be seen from Figure 142, the distribution of *M. vaginalis* is simply described. It occurs in Korea and Japan, reaches downward through the Pacific Islands, to the mainland of Southeast Asia, and across to India. With the exception of the species' occurrence in taro in Hawaii, all reports of weediness concern paddy rice. It is a principal weed of rice in Korea, Malaysia (Sarawak), and the Philippines. It is one of the three most serious weeds in rice in Indonesia, Japan, and Taiwan. It is a common weed in rice in Cambodia, Ceylon, China, India, and Thailand. *M. hastata* is a principal weed of rice in Fiji, Malaysia (Sarawak), and the Philippines and a common weed of rice in Indonesia.

The leaves of both species are eaten as food but those of *M. hastata* are preferred to those of *M.*

vaginalis. *M. vaginalis* is used in herbal medicine (Burkill 1935).

COMMON NAMES
Monochoria vaginalis

BANGLADESH
  panee kachu
CAMBODIA
  chrach
HAWAII
  monochoria
  pickerel weed
INDIA
  meerthomari
  nanka
INDONESIA
  bengok (Javanese)
  etjeng lemboet (Sundanese)
  etjeng padi (Indonesian)
JAPAN
  kōnagi

KOREA
  mooldalgebi
PHILIPPINES
  biga-bigaan (Tagalog)
  gabi-gabi (Bisayan)
  gabing-uak
  kalaboa
  lapa-lapa (Ilokano)
  saksaklung (Bontoc)
  upi-upi (Bikol)
TAIWAN
  ya-she-tsau
THAILAND
  ka-kiad
  phak-khiat
VIETNAM
  rau mác lá thon

Monochoria hastata

FIJI
  pickerel weed
HAWAII
  monochoria
INDONESIA
  benggob (Javanese)
  bia-bia (Javanese)
  echeng gede (Sundanese)
  echeng kebo (Sundanese)
  pingo (Javanese)

  wewchan
    (Javanese)
  weweyan
    (Javanese)
MALAYSIA
  chacha layar
  chachang layar
  kangkang ayer
PHILIPPINES
  calaboa (Tagalog)
THAILAND
  pak top

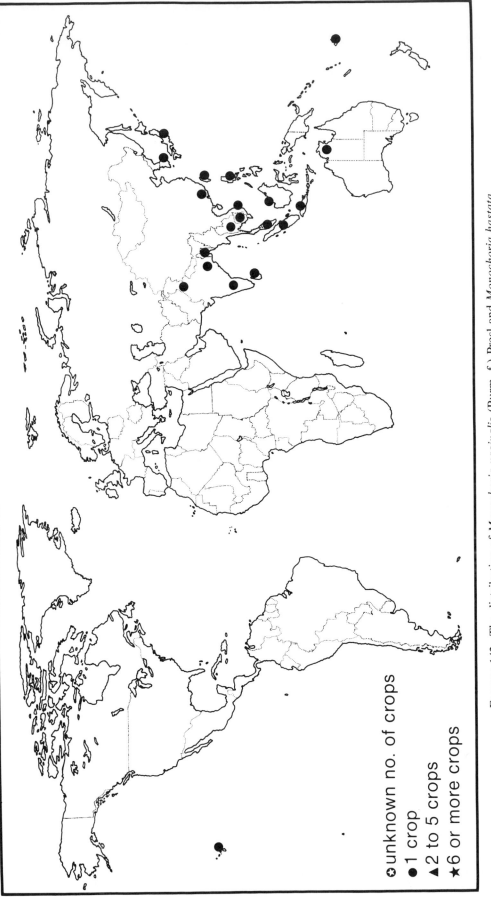

FIGURE 142. The distribution of *Monochoria vaginalis* (Burm. f.) Presl and *Monochoria hastata* (L.) Solms across the world in countries where they have been reported as weeds.

❧ 53 ❧

Oxalis corniculata L.
(= O. repens)

OXALIDACEAE, OXALIS FAMILY

Oxalis corniculata is a slender, cloverlike, semierect, commonly prostrate, spreading perennial. It is native to Europe and America. It is not a large, shade-producing, dominant species but is a constant, space-filling competitor. Its spread is due in part to its forceful ejection of seeds in dry weather. It has been reported to be a weed of 17 crops in 44 countries. It belongs to a select group of species which have been the most successful colonizers of new areas (noncultivated) around the world.

DESCRIPTION

O. corniculata is a creeping, stoloniferous, *perennial* herb (Fig. 143), with aerial stems up to 30 to 50 cm long, often branching at the base, rooting at the nodes, and sometimes covered with long hairs; it may arise from woody rootstocks; *roots* fibrous (no bulbs); *stems* brown, with fine longitudinal lines, young stems densely hairy; *leaves* alternate, three heart-shaped leaflets on slender *petioles* up to 10 cm long, with a winged basal joint (articulation); *leaflets* clover-shaped, up to 12 mm long, 15 mm broad, bright or pale green, almost smooth above, densely shaggy with soft hairs beneath; *flowers* on axillary peduncles which may be 3 to 9 cm long, 1 to 6 on each stalk, central flower developing first, flowers yellow, funnel-shaped, about 1 to 1.5 cm in diameter, bending downward as the fruits develop; *fruits* five-angled, a linear capsule 2 cm long, green, hairy; *seeds* red-brown, flattened, ovoid, 1.5 mm long, with transversely raised ridges, often with a white, membranous, outer integument attached to one side as the seeds emerge from the capsule.

The three heart-shaped leaflets, resembling a cloverleaf; the yellow flowers; and the absence of bulbs are characteristics of this species.

BIOLOGY

O. corniculata is a slender perennial that grows quite vigorously. It is easily recognized by its cloverlike leaves. In the evening the leaves fold down around the leaf stalk. The plant has trailing stems which often root at the node and a long, narrow taproot. The rootstocks of older plants may become woody. Propagation by rootstocks in cultivated land is not important, for few of them survive the plow or the hoe. Thus, the spread is mainly due to high production of fertile seed. The plant flowers in March and April in Iraq and throughout the year in the Philippines. After the plant has flowered, the fruits form into narrow cylindrical capsules which are pointed at the apex. An elastic jacket or aril surrounds the small brown seeds and this serves to eject them from the capsule. Studies in South Africa have shown that fresh seeds will germinate well in light at temperatures just under 21° C.

343

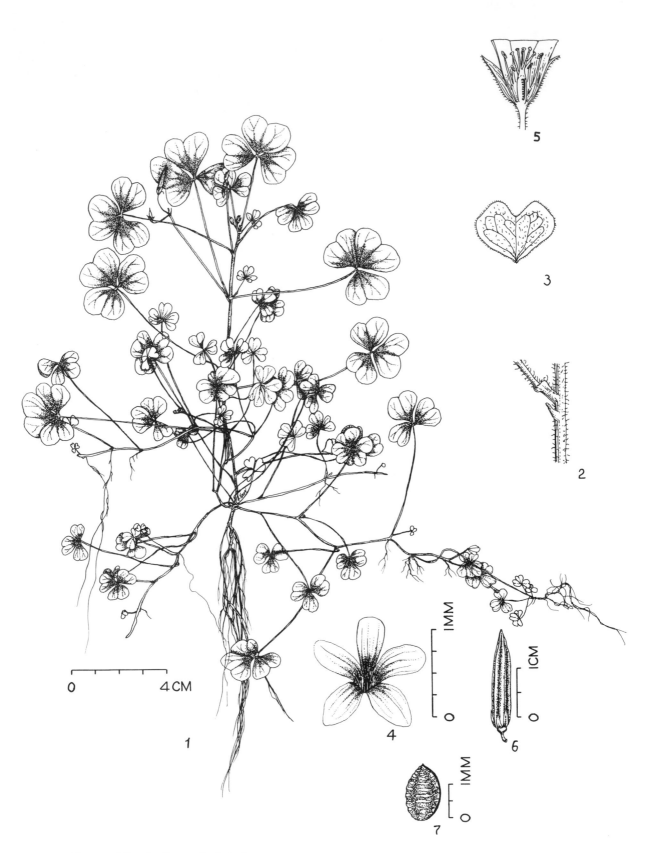

FIGURE 143. *Oxalis corniculata* L.

Above: *1*, habit; *2*, portion of stem with leaf base; *3*, leaflet, enlarged; *4*, flower; *5*, flower, vertical section; *6*, capsule; *7*, seed.

Opposite: Seedlings.

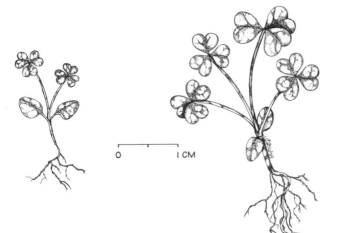

DISTRIBUTION AND HABITAT

O. corniculata is a cosmopolitan weed of the tropical and temperate zones (Figure 144). It is at home in gardens, lawns, arable land, and pastures. It is frequently found also on channel banks and in shady places. Its prostrate habit helps to protect it from being mowed in lawns. In Kenya it is usually found at altitudes over 1,000 m. From Figure 144 it may be seen that it has established itself from the Great Lakes in North America to New Zealand. Its presence on all continents and many islands indicates that its seeds are carried by birds. It is generally distributed to the east and west throughout the world on agricultural lands.

AGRICULTURAL IMPORTANCE

O. corniculata is a principal weed of tea in India and of beans, corn, and potatoes in Mexico. It has been reported to be a common weed of beans, corn, potatoes, rice, tea, and vegetables in Japan; cereals in Ethiopia; rice in India and Indonesia; pastures in Australia; citrus in Spain; coffee in Tanzania; and corn and bananas in Taiwan. It is a weed in bananas in Hawaii and Mexico; cassava in Taiwan; citrus in Iraq and Mexico; coffee in El Salvador, India, Kenya, Mexico, and Venezuela; corn in Ethiopia; pastures in Taiwan; vegetables and pineapples in Hawaii; pyrethrum in Kenya; sugarcane in Mauritius and Peru; vegetables, flower beds, peanuts, corn, soybeans, and root crops in the Philippines; and tea in Ceylon, Indonesia, and Taiwan; it is one of the most serious weed pests in orchid culture in Hawaii.

The weed is often resistant to the hormone type of herbicide and so growers often resort to hand-weeding. In eastern Africa it is believed that this weed in grassland may cause poisoning to cattle. A related species, *O. pes-caprae*, is poisonous to stock animals in New Zealand and Australia.

In some situations the plants may serve as a valuable ground cover in preventing soil erosion in perennial crops such as tea or coconuts.

O. corniculata is an alternate host of the nematodes *Meloidogyne* sp. (Raabe, unpublished; see footnote, *Cyperus rotundus*, "agricultural importance") and *M. incognita* (Kofoid & White) Chitwood (Valdez 1968 as *O. repens*); and of the virus which causes beet curly top (Namba and Mitchell, unpublished; see footnote, *Cynodon dactylon*, "agricultural importance").

COMMON NAMES
OF OXALIS CORNICULATA

ARGENTINA
vinagrillo
vinagrillo rastrero

AUSTRALIA
yellow wood sorrel

BRAZIL
azedinha
três corações
trevo azedo

BURMA
hmô-gyin

CAMBODIA
chantoe phnom kok

CEYLON
hin-embul-embiliya
Indian sorrel
kodippuliyarai
puliyari
rata-embala
(Sinhalese)
sikapu-puliyarai-pillu
(Tamil)

CHILE
hierba de la perdiz

COLOMBIA
acedera
acederilla
chulco

EASTERN AFRICA
yellow sorrel

EGYPT
hamb

EL SALVADOR
trébol

ENGLAND
procumbent yellow
sorrel

FRANCE
pied-de-pigeon
trèfle-jaune

GERMANY
gehörnter Sauerklee

HAWAII
yellow wood sorrel

INDIA
chalmori
Indian sorrel
khatibuti
pulichinta
puliyari
sarutengensi

INDONESIA
daoen asam ketjil (Indonesian)
rempe
semanggen
semanggi goenoeng
(Javanese)
tjalingtjing (Sundanese)

IRAN
toroshak

IRAQ
sheep sorrel

JAMAICA
sorrel
sourgrass

JAPAN
katabami

LEBANON
  hammad
  hhamidah
  hhullwah
MAURITIUS
  oseille filante
  petit trèfle
MEXICO
  agrito
PERU
  vinagrillo

PHILIPPINES
  daraisig (Bikol)
  kanapa (Igorot)
  malabalug-dagis
    (Pampangan)
  marasiksik (Ilokano)
  pikhik (Ivatan)
  salamagi (Bontoc)
  taingang-daga
    (Tagalog)

PUERTO RICO
  vinagrillo
RHODESIA
  yellow oxalis
  yellow sorrel
SOUTH AFRICA
  creeping sorrel
  tuinranksuring
SPAIN
  aigret

TAIWAN
  ja-jyang-tsau
THAILAND
  somkop
UNITED STATES
  creeping wood sorrel
VENEZUELA
  trébol
VIETNAM
  me dât

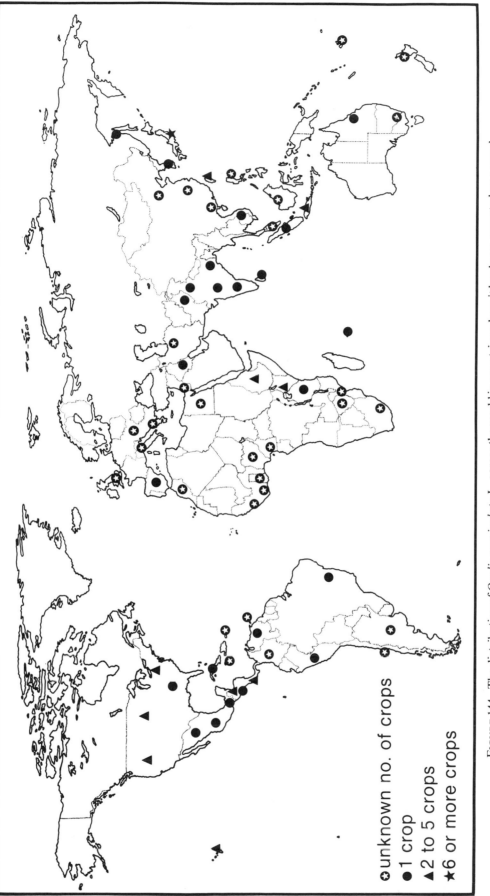

FIGURE 144. The distribution of *Oxalis corniculata* L. across the world in countries where it has been reported as a weed.

❧54❧

Panicum maximum Jacq.

POACEAE (also GRAMINEAE), GRASS FAMILY

Panicum maximum, a native of Africa, is an important weed of sugarcane. It is a robust, tufted, perennial grass that grows to 4 meters in height. Forty-two countries report that it is a weed in 20 of the world's major crops. The young plants are palatable and nutritious, and the species has become an important grass for pastures, hay, and silage in several countries of the world.

DESCRIPTION

P. maximum is a tall, vigorous, tufted, perennial grass (Figure 145) that grows to 4 m in height; it spreads slowly from *fibrous roots* or sometimes by short *rhizomes* or by rooting at the lower nodes, to form large stools; *culms* mostly erect, stout, somewhat flattened, often 1 cm in diameter at the base; *nodes* hairy; *sheaths* smooth or sparsely hairy with tubercle-based hairs, margins above sometimes hairy; *ligule* fringed with hairs, 1 to 3 mm long; *blades* linear, finely pointed, 15 to 100 cm long, 1 to 3.5 cm wide, with numerous erect to spreading branches up to 30 cm long; *spikelets* awnless, alike, symmetrical, short-pedicelled, clustered, oblong, obtuse to slightly acute, plump, 2.5 to 4 mm long, green or purplish, smooth; *glumes* very unequal, the lower broad, very obtuse or nearly or quite straight across (truncate), 0.25 to 0.35 the length of spikelet, the upper as long as the spikelet, five-nerved; *lower floret* male, its *lemma* similar to upper glume, five- to seven-nerved; *upper floret* perfect; upper lemma oblong, up to 3 mm long, finely transversely wrinkled; *seeds* 3 mm long, 1 mm wide, somewhat flattened on one side.

This species is distinguished by its hairy nodes, hair-fringed ligules, loose spreading panicle, and transversely wrinkled upper lemma.

DISTRIBUTION AND BIOLOGY

P. maximum occurs naturally in areas receiving 1,000 to 1,700 mm of rainfall and having a dry season not exceeding 4 months (Figure 146). It grows well on a wide variety of well-drained soils, and, although it is drought-resistant, will not withstand long periods of severe desiccation. Some strains of *P. maximum* prefer wet situations. The species can stand waterlogging or flooding for only short periods. It is very susceptible to even occasional light frosts. The plant will grow well even under trees because it is shade-tolerant.

In some areas the plant flowers all year round. It seeds readily but the heads ripen very unevenly and shatter early. In the Philippines, 9,000 seeds have been counted from one plant, and there the plant blooms mainly during the evening (Javier 1970). The number of fresh seeds which germinate is usually low, but this number may be increased by storing the seeds under dry conditions for 6 months or more. In spite of the low percentage of germination, the plant spreads mainly by seeds and only slowly by vegetative means. In commercial pastures the grass is sometimes propagated by crown divisions. *Melinis minutiflora*, a spreading perennial which forms

FIGURE 145. *Panicum maximum* Jacq. *1*, habit; *2*, portion of inflorescence; *3*, spikelet; *4 a–f*, bracts; *5*, flower; *6*, stamen; *7*, ligule; *8*, node; *9*, seed; *10*, seed, cross section.

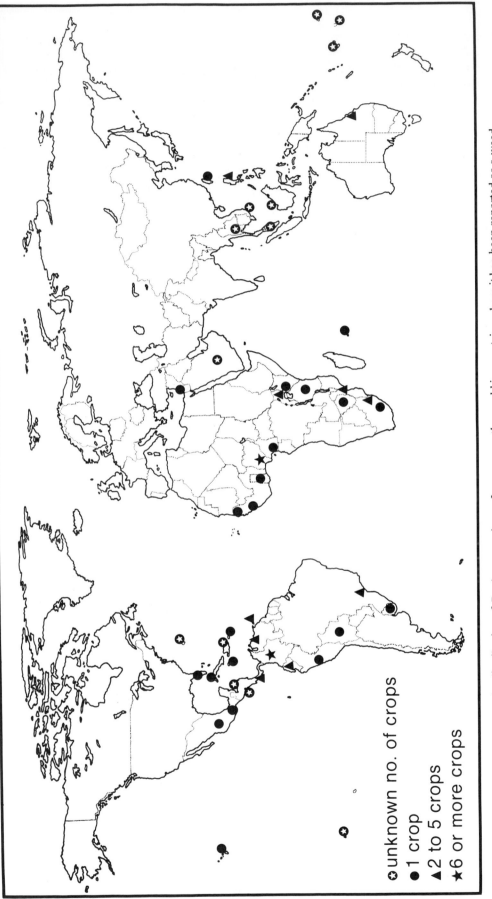

○ unknown no. of crops
● 1 crop
▲ 2 to 5 crops
★ 6 or more crops

FIGURE 146. The distribution of *Panicum maximum* Jacq. across the world in countries where it has been reported as a weed.

large, loose tussocks, is sometimes sown with it to cover the ground more quickly.

The species shows considerable variation in its growth habit and many countries have selected strains and varieties particularly suited for local use. Different botanical types have been described in Australia, Ceylon, Jamaica, Puerto Rico, and in the United States in the Gulf coastal states and Hawaii. The most common type cultivated in Brazil is called *sempreverde* ("always green") and it is distinctive in that the base of the culm is expanded into a bulb. *Panicum maximum* is mainly apomictic; although seeds are formed there is little sexual reproduction and, therefore, small chance of segregation and recombination. Some efforts have been made to breed new lines of *P. maximum*. Javier (1970) described the flowering habits and methods of reproduction. He found two main flowering periods: July to October (the rainy season) and October to December (the end of the wet season).

AGRICULTURAL IMPORTANCE

With its vigorous growth, large size, and ability to adapt to a wide variation in soil moisture, light, and elevation, *P. maximum* is able to flourish in several crops which are grown under widely varying conditions (Figure 147). Of all crops it is most troublesome in sugarcane, where it is ranked number one in Cuba and Hawaii and number three in South Africa. It is a principal weed of that crop in Australia, Costa Rica, Mexico, and Taiwan. The appearance of some strains is so like that of sugarcane that it may be unnoticed until it flowers. It is a serious weed of cotton in Uganda, and a principal weed of oil palms in Costa Rica, corn in Colombia, cotton in Mozambique, cacao in Ecuador, and citrus in Swaziland. Although not shown on the map, it is a principal weed of bananas and coffee in Costa Rica, coconuts in Trinidad, and pineapples in Swaziland. It is also a weed of pineapples and abaca in the Philippines; corn and rice in Mexico; peanuts in Trinidad and Colombia; sisal in Mozambique and Tanzania; tea in Mozambique and Uganda; coconuts in Venezuela;

cotton, oil palms, and sugarcane in Colombia; corn, rice, and cowpeas in Nigeria; citrus in Jamaica and the United States; cacao in Venezuela; and sugarcane in Mauritius and Mozambique.

P. maximum is distributed over much of the warm parts of the world because it has been widely introduced for cattle food. It is cultivated on a large scale in India. Some authorities relate the rapid development of the cattle industry of Latin America to the introduction of grasses of tropical origin, in particular *P. maximum*. When it is young and leafy it can produce a crude protein content of 13 percent, but this percentage falls off rapidly and the grass becomes very coarse with increasing maturity. It dies rapidly under continued close grazing.

P. maximum is an alternate host of hoja blanca virus of rice (Granados and Ortega 1966).

COMMON NAMES
OF PANICUM MAXIMUM

AUSTRALIA
  guinea grass
BRAZIL
  capim colonia guiné
  sempreverde
CEYLON
  gini tana (Sinhalese)
  gino pul (Tamil)
COLOMBIA
  guina
CUBA
  yerba guinea
ECUADOR
  saboya
FIJI
  guinea grass
HAWAII
  guinea grass
INDONESIA
  roempoet banggala (Indonesian)
  soeket londo (Javanese)

MAURITIUS
  fataque
MEXICO
  o privilegio
  zacate guinea
NETHERLANDS
  bengaalsch gras
PERU
  grama castilla
  zaina
PHILIPPINES
  guinea grass
PUERTO RICO
  yerba de guinea
SOUTH AFRICA
  barbegras
THAILAND
  yah guinea
TRINIDAD
  guinea grass
UNITED STATES
  guinea grass
VENEZUELA
  gamelote

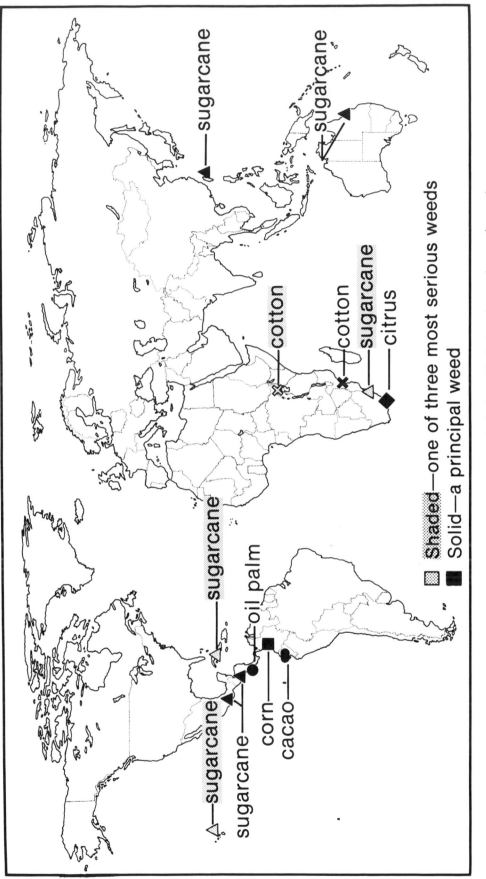

FIGURE 147. *Panicum maximum* Jacq. is a serious or principal weed in the locations and crops shown above.

✺§ 55 ✺

Panicum repens L.

POACEAE (also GRAMINEAE), GRASS FAMILY

Panicum repens is an aggressive, creeping, perennial grass that spreads by coarse, long-lived rhizomes. A native of the Old World, it has spread into the tropics and subtropics, growing from about lat 35° S to 43° N. A weed of 19 crops in 27 countries, it is found mostly in plantations, orchards, or in field crops. Primarily a weed of moist, coastal, sandy soils, it can also grow on heavy upland soils and can tolerate drought. Although it has been considered important for pasturage, it is of relatively low feeding value and palatability compared to more desirable improved grasses.

DESCRIPTION

P. repens is an erect, rather wiry, creeping, *perennial* grass, 30 to 90 cm tall, (Figure 148), growing extensively but not densely; *rhizomes* strongly developed, often swollen or knotty, smooth with brownish or whitish scales, branched, sending out erect culms from the rather distant nodes; *culms* clothed at base with bladeless sheaths; *leaf sheaths* fringed with long hairs along the margin; *ligule* short, fringed with hairs; *blades* two-ranked, 15 to 25 cm long, 1.5 cm wide or less, acuminate, linear, flat or folded, with a rounded base, smooth or somewhat rough, sparsely long-hairy above, smooth below, covered with a whitish bloom, somewhat rigid, with long white hairs behind ligule, midrib prominent and keeled; *panicles* exserted, 7 to 18 cm long, somewhat loose and open, erect or ascending, sometimes nodding, elongated, scattered, one to three at each node, branches stiffly ascending, distant, 2 to 19 cm

long, smooth or slightly rough; *spikelets* two-flowered, pale green or pale yellow, often tinged with purple, oblong-ovate, acute or slightly acuminate, 2.2 to 3 mm long; *first glume* truncate, about one-fifth as long as the spikelet, *second glume* as long as spikelet, seven- to nine-nerved; *first lemma* as long as spikelet, five- to nine-nerved; *first palea* nearly as long as spikelet; *anthers* yellowish orange; *second lemma* shorter than spikelet; *stigmas* purple; *caryopsis* lanceolate, straw-colored.

The sharp-pointed rhizome which is coarse, swollen, and shaped like a ginger rhizome and the blades which are often covered with whitish waxy bloom are the distinguishing characteristics of this species.

BIOLOGY AND DISTRIBUTION

P. repens is a native of the Old World and has spread in the tropics and subtropics to become one of the most serious grass weeds. It grows within a belt of about lat 35° S to 43° N (Figure 149).

P. repens grows in many soils but is found most frequently in sandy soils along seacoasts or in poorly drained heavy soils. It usually cannot stand permanently flooded conditions, but can withstand occasional flooding and can encroach upon and invade ditches, drains, watercourses, fishponds, and floating mats of vegetation from banks and adjacent areas. It is frequent in cultivated lands, grasslands, roadsides, and gardens. It is usually found in low coastal areas and along rivers but it does spread to local inland areas to elevations of 1,500 to 2,000 m in the tropics. It grows best in open sunny areas but can

353

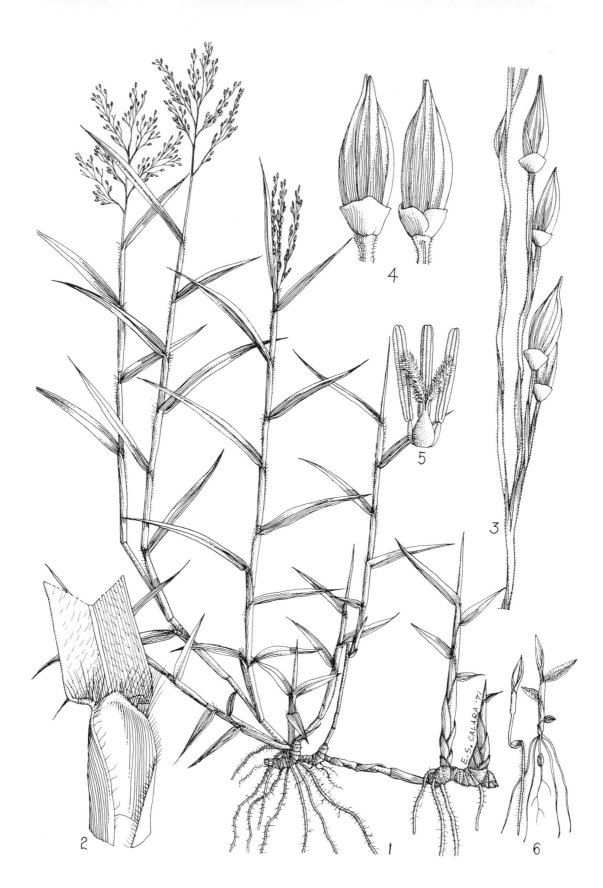

FIGURE 148. *Panicum repens* L. *1*, habit; *2*, ligule; *3*, portion of spike; *4*, spikelet, two views; *5*, flower; *6*, seedling.

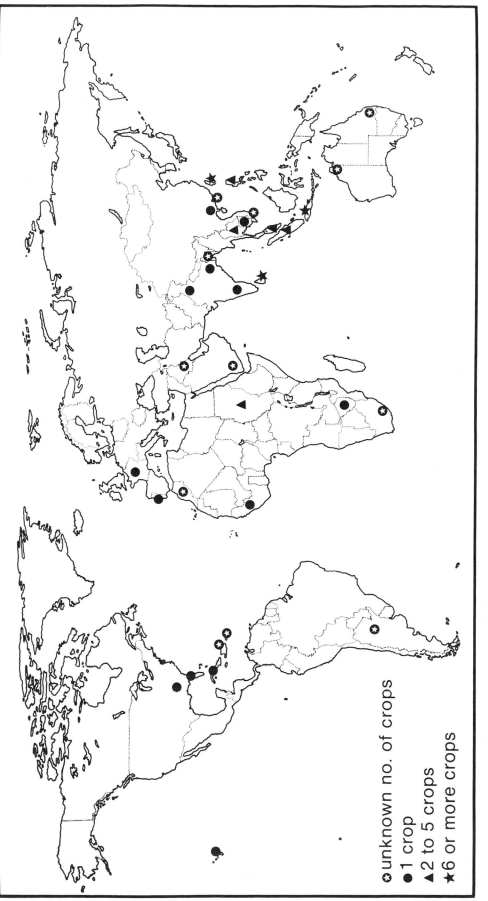

○ unknown no. of crops
● 1 crop
▲ 2 to 5 crops
★ 6 or more crops

FIGURE 149. The distribution of *Panicum repens* L. across the world where it has been reported as a weed.

stand partial shade. It is very drought-resistant, and the rhizomes may live through prolonged dry periods.

The plant spreads mainly by coarse, extensively creeping rhizomes which may extend as far as 7 m from the parent plant. Rhizomes are sharp-pointed and are often swollen and enlarged, resembling the rhizomes of ginger. Bor (1960) reported that deep plowing favored the spread of the weed in cultivated fields. Seeds are produced in a few situations, but we know of no reports of the biology of the seeds.

The plants are variable depending upon growing conditions. In dry soils they may be short and produce few flowers per panicle, whereas under moist conditions they grow tall and produce a full rich panicle.

AGRICULTURAL IMPORTANCE

From Figure 150 it may be seen that *P. repens* is a serious or principal weed of plantations, fruits, and field crops. It is a serious weed of oil palms, cacao, coconuts, and rubber in Malaysia; tea and rice in Ceylon; tea in Indonesia; and orchard crops in Thailand. It is a principal weed of pineapples in Guinea; sugarcane in Hawaii and Taiwan; lowland and upland rice and rubber in Indonesia; and tea in India. It is also a weed of pastures, pineapple, tea, bananas, and irrigated crops in Taiwan; citrus in Taiwan and the United States; rice in Cambodia, France, India, the Philippines, Portugal, Taiwan and Thailand; corn in Ceylon, Indonesia, and Taiwan; peanuts, coconuts, and chilies in Ceylon; cotton and irrigated crops in Sudan; and peanuts and sugarcane in Indonesia. It is frequently reported as a weed of lawns and gardens.

P. repens is an important weed because of its rapid rate of spread and its persistence, these factors being related to its coarse, enlarged rhizomes.

It is often resistant to herbicides. Tillage will not control it; rather, the plant spreads and thrives under such management unless the rhizomes are dragged to the surface and then dried and burned.

The grass may spread into improved pastures, competing and choking out desirable species. It is sometimes reported to be useful for pasturage, although it is lower in protein throughout its growing stages than are many improved grasses and may contain only 3.3 percent crude protein and as much

as 39 percent crude fiber at maturity (Whyte, Moir, and Cooper 1959). Its advantage lies in the facts that it may be quite palatable when young and that it can stand heavy grazing and trampling. Because a number of more desirable and nutritious grasses are available for the tropics and subtropics, however, *P. repens* should no longer be planted for pasture use.

Burkill (1935) reported that in Malaysia *P. repens* stabilizes mine sediments which have been spread by rivers and streams.

It is an alternate host of the rice leafhopper *Thaia oryzivora* (Leeuwangh and Leuamsang 1967), *Ustilago* sp., *Piricularia* sp. (Chandrasrikul 1962), and *P. oryzae* (Paje, Exconde, and Raymundo 1964).

COMMON NAMES
OF PANICUM REPENS

ARGENTINA
  camalote
AUSTRALIA
  torpedo grass
BANGLADESH
  baranda
BURMA
  myet-kha
CAMBODIA
  chhlong
CEYLON
  couch
  etora (Sinhalese)
  inji-pul (Tamil)
FRANCE
  panic rampant
HAWAII
  torpedo grass
  wainaku grass
INDIA
  injipilla
  karigaddi
INDONESIA
  baloengan
  benda laut
  jahean
  jajahean
  lampoejangan
  lampujangan (Sundanese)

rumput jae-jae
suket balungan (Javanese)
suket lempujangan (Indonesian)
MALAYSIA
  creeping panic grass
  kerunong padi
  metubong
  telur ikan
  upat
NETHERLANDS
  victoriagras
PHILIPPINES
  kayana (Manobo)
  luya-luyahan (Tagalog)
PORTUGAL
  escalracho
TAIWAN
  pu-shu-tsau
THAILAND
  yah-chan-ah-kat
  yah chanagard
UNITED STATES
  torpedo grass
  panic rampant
VIETNAM
  cò ống

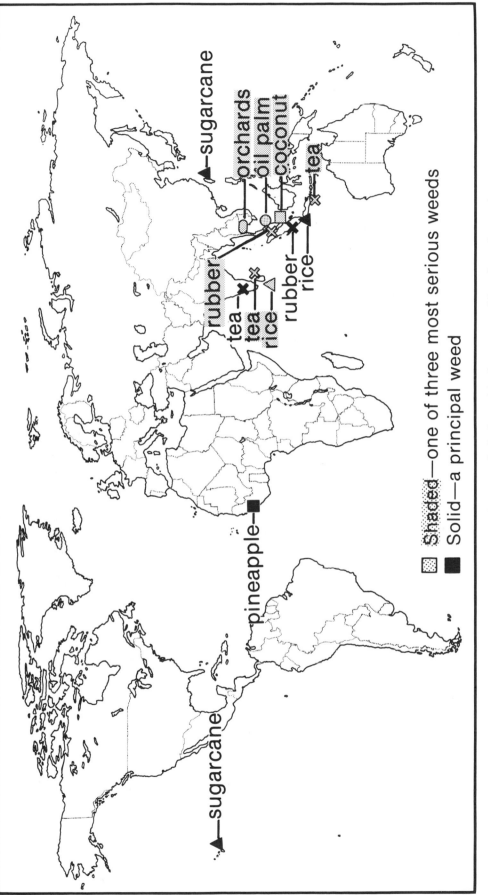

FIGURE 150. *Panicum repens* L. is a serious or principal weed in the crops and locations shown above.

❦56❧

Paspalum dilatatum Poir.

POACEAE (also GRAMINEAE), GRASS FAMILY

Paspalum dilatatum is a perennial grass that spreads rapidly by seeds and by the distribution of vegetative parts. It is a native of South America and is one of the species of grasses which can thrive on soil which is permanently damp or occasionally waterlogged. For some regions it is an important grass for pasture, silage, and erosion control. It is a weed of 14 crops in 28 countries.

DESCRIPTION

P. dilatatum is a *perennial* grass (Figure 151); *culms* strongly tufted, leafy at base, compressed, 40 to 170 cm tall, with a narrow depression at the base, smooth, erect, to ascending, usually bent like a knee (geniculate) at the base, sometimes rooting at the lower nodes; *leaf sheaths* smooth or the lower ones hairy at the back; *inflorescence* erect or nodding, 10 to 25 cm long; *racemes* three to five, rarely nine, widely separated, 4 to 13 cm long; *spikelets* alternate in pairs, the outer pedicel longer than the inner, oval with acute apex, light green or purplish; *upper glume* and *lower lemma* five- to nine-nerved, fringed with long, white, silky hairs; *anthers* and *stigma* dark purple or black; *fruit* about 2.5 mm long, broadly elliptic, minutely papillose-striate.

The distinguishing characteristic of this species is that the paired spikelets have outer pedicels which are longer than the inner ones.

BIOLOGY AND DISTRIBUTION

The species is an example of a perennial grass that produces large quantities of seeds either sexually or apomictically and that spreads mainly by this means.

In plantation crops the plant parts are moved about by mechanical cultivation so that vegetative propagation becomes an additional means of spread. Such weeds, having two systems of propagation, can often adapt to a variety of soil types and climates. If the seeds of this species are fresh, only a small percentage will germinate; if they have had an opportunity to afterripen—for exmple, if they have been exposed on the soil to a hot sun for several days—a much larger percentage will germinate.

Bennett (1944) studied the embryology of *P. dilatatum* and Knight (1955) studied flowering and seed production in response to photoperiod and temperature. Plants were placed in two greenhouses at temperatures of 21° to 26° C during the day. One house was held at 18° to 24° C and the other at 7° to 13° C at night and both had 12-, 14-, and 16-hour photoperiods. A high night temperature and a 14-hour photoperiod provided the best vegetative growth and seed production. No seedheads were produced in an 8-hour photoperiod, and flowering was incomplete at 12 hours. Low night temperatures inhibited seedhead production in long days.

The species prefers wet situations but on low ground may extend from rather dry prairies to marshy meadows. It becomes a serious problem when it encroaches on banks of irrigation ditches. The plant succeeds on light soils or heavy clay soils and is drought-resistant and frost-sensitive. It responds readily to nitrogen fertilization and can compete moderately well with other species under conditions

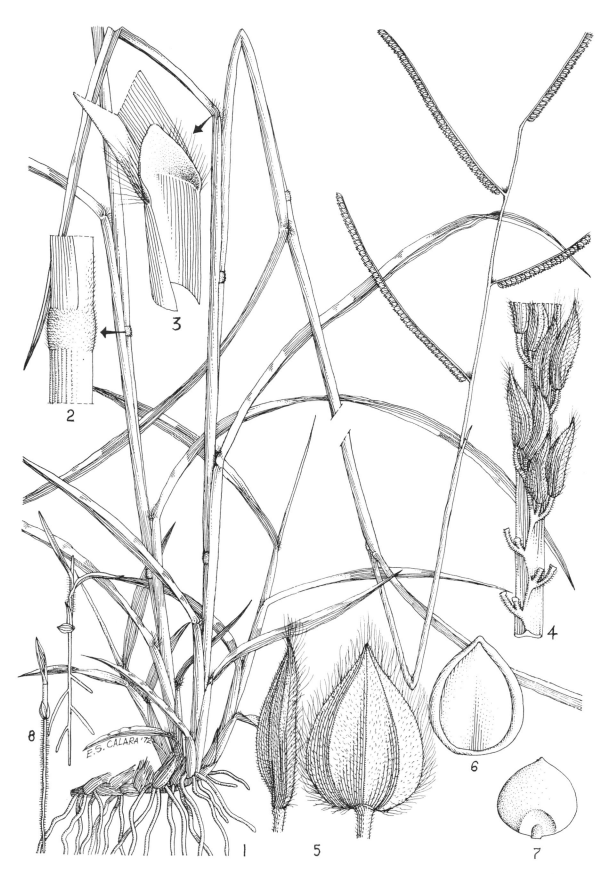

FIGURE 151. *Paspalum dilatatum* Poir. *1*, habit; *2*, node; *3*, ligule; *4*, portion of spike; *5*, caryopsis, two views; *6*, caryopsis, bracts removed; *7*, grain; *8*, seedlings.

of high fertility. Close defoliation either by being grazed or mowed results in lower tiller production and shoot growth.

The plant is found in the northern temperate zone in Great Britain, the Soviet Union, and the United States. In the southern temperate zone it is found in Argentina, Australia, New Zealand, and South Africa. From Figure 152 it may be seen that the weed is generally distributed across the world between these extremes of its range.

AGRICULTURAL IMPORTANCE

P. dilatatum is one of the three most serious weeds in bananas and papayas in the Philippines where it is also troublesome in pineapples and paddy rice. In Australia it is one of the most important weed problems in citrus and often grows along the banks of irrigation channels; it is also a pest of sugarcane, bananas, orchards, and vineyards in Australia. It is a principal weed of bananas in Hawaii and is also a weed of pineapples, papayas, and sugarcane. It is a principal weed of tea and a weed of citrus in the Soviet Union. In Brazil and India it is found in paddy rice, and in New Zealand it is a weed of potatoes and other vegetables and of pastures. In the United States it is a weed of pastures. It is frequently a pest along roads, railroads, and around lights and along runways on airports.

The species is rated good to excellent for pasturage and silage in those warm temperate and subtropical areas (such as parts of South America and Asia) which have extended rainfall. It is an important summer-growing perennial in those pasture and cattle-raising areas of Australia where the lands are irrigated or where they receive a moderately high rainfall (1,200 mm). When planted for pasture the grass requires very careful management, for it spreads slowly and is easily damaged by trampling and grazing. For this reason it does not compete well with most other grasses, especially under low-nitrogen conditions. Sometimes the stands become sod-bound rather early. In some areas it is the custom to seed the grass with legumes such as white clover (*Trifolium repens*) or Rhodes grass (*Chloris gayana*).

P. dilatatum is an alternate host of *Claviceps paspali* Stevens & Hall (Parris 1940) and *Sorosporium paspali* McAlp (Stevens 1925); and of a slug caterpillar (*Paraso bicolor*), a new insect of rice in West Bengal (Chatterjee 1969). It may serve as one of the wild hosts for the parasitic weed *Striga lutea* Lour.

COMMON NAMES
OF PASPALUM DILATATUM

ARGENTINA
   canota
   pasto dulce
   pasto miel
AUSTRALIA
   paspalum
BRAZIL
   capim papua
CEYLON
   miti paspalum tana
      (Sinhalese)
CHILE
   camalote
   chepica gigante
HAWAII
   dallis grass
   paspalum

INDONESIA
   rumput australi (Indonesian)
JAPAN
   shimasuzumenohie
PHILIPPINES
   dallis grass
PUERTO RICO
   yerba dalis
TAIWAN
   da-li-tsau
UNITED STATES
   dallis grass
   paspalum
   watergrass
   water paspalum

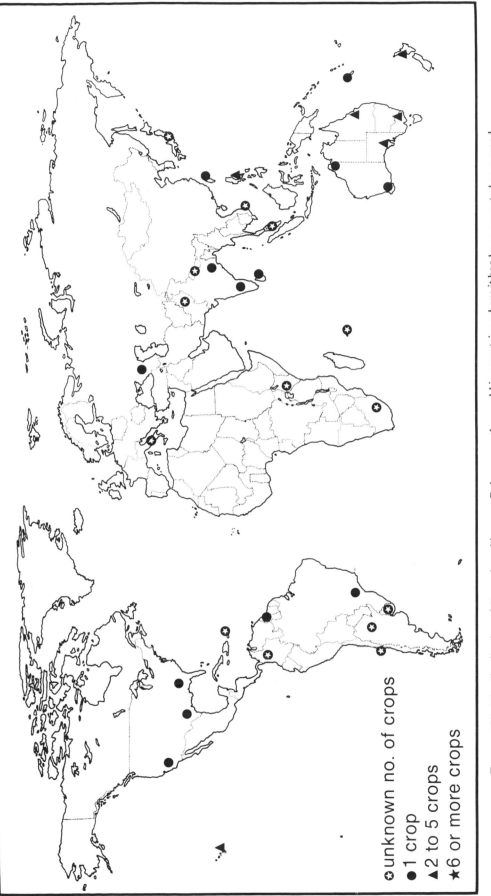

FIGURE 152.   The distribution of *Paspalum dilatatum* Poir. across the world in countries where it has been reported as a weed.

○ unknown no. of crops
● 1 crop
▲ 2 to 5 crops
★ 6 or more crops

❧57❧

Pennisetum clandestinum Hochst.

POACEAE (also GRAMINEAE), GRASS FAMILY

Pennisetum clandestinum is an aggressive, creeping, sod-forming, perennial grass that spreads by stolons and rhizomes. It rarely produces seeds. A native of the highlands of tropical eastern Africa, it has been spread widely throughout the tropics and subtropics as a pasture grass. It has been reported as a weed in 14 crops in 36 countries, mainly in perennial or plantation crops. It is important for both dairy and beef production as pasture or as a soilage crop and is also important as a soil binder. It produces very small, hidden inflorescences on vegetative shoots, hence its specific name *clandestinum*.

DESCRIPTION

P. clandestinum is a low, mat-forming, *perennial* grass (Figure 153); creeping extensively by stout *rhizomes* and long branched *stolons; culms* 30 to 120 cm, prostrate and rooting from the nodes, internodes short, profuse vertical leafy branches arise from the stolons and rhizomes; *leaf sheaths* overlapping, membranous to papery, pale to brown, hairless or hairy; *ligule* a hairy rim; *blades* narrow, spreading, blunt to pointed, 1.25 to 5 cm long, 3 to 4 mm wide, folded at first, later flat, hairless or hairy, the margins rough; the small white or tawny *panicles* are not borne at the top of the culms as in other grasses but are enclosed within short leaf sheaths at the top of short side shoots which resemble regular vegetative shoots, *spikelets* in clusters of two to four and nearly enclosed in the uppermost leaf sheath, terminal spikelet shortly stalked, the others stalkless, each spikelet partly or wholly surrounded by delicate bris-tles up to 15 in number that are unequal in length, inner bristles plumose, spikelets with two florets, slender, narrow, 1 to 2 cm long, whitish below, greenish above; the only parts of the *flower* visible are the stamens which appear as a mass of fine white threads attached to the leaves and consist of filaments, 2.5 cm or more long, with anthers at the tip; *grain* (caryopsis) oblong, brown, 1.5 to 2.5 mm long, the seeds can be found only by dissecting the leaf sheaths.

This species can be distinguished by its extensively creeping rhizomes and stolons which form a dense mat, its culms with overlapping leaf sheaths and by its flowers which, if present at all, appear on leafy, vegetative side shoots with only the stamens visible above the leaf sheaths.

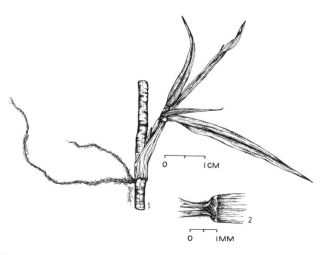

FIGURE 153. *Pennisetum clandestinum* Hochst. ex Chiov.
Above: *1*, habit; *2*, flowering culm; *3*, spikelet; *4*, lower glume; *5*, outer glume; *6*, lemma; *7*, grain.
Opposite: *1*, detail of a young shoot developing from a node of a stolon cutting; *2*, ligule.

BIOLOGY

P. clandestinum is a creeping, sod-forming grass that can spread rapidly to form vigorous pure stands. It is well adapted to the humid tropics or subtropics, especially at higher elevations and in soils of high fertility. In Hawaii the grass grows luxuriantly from sea level to more than 2,000 m (Hosaka 1958), and in Colombia it is found from 1,500 to 3,000 m (Romero, Jeffery, and Revelo 1970; Romero et al. 1970). In Kenya where it is native it is common in pastures, forest clearings, and forest edges above 2,000 m in red loam soils, usually soils of high fertility, and in areas where rainfall is greater than 900 mm (Morrison 1966). The plant grows rapidly under high levels of nitrogen, but it will also respond to phosphorus and sulfur on deficient soils. It has long been associated with volcanic soils or with the red soils of the tropics and subtropics.

In the highlands of eastern Africa, where it is native, mean maximum and mean minimum temperatures range from 2° to 8° C and 16° to 22° C, respectively (Mears 1970). The plant can withstand a light frost but cannot tolerate sustained cold weather. Light frosts will kill only the exposed upper parts of the plant; but the perennial root system will survive and the plant will make rapid growth during the onset of warm weather. The species does not appear to be sensitive to daylength; flowering has been reported from a latitude range of 0° to 38° (Mears 1970). Higher temperatures at low elevations may inhibit flowering; indeed, seed production, always rare in the species, may occur only infrequently at lower elevations. The weed can grow in medium shade where it can form the dominant understory in forest margins and in plantation crops.

Although the grass grows best under moist, humid conditions, well-established plants can withstand a considerable amount of dry weather; however, they do not grow as rapidly during such periods of moisture stress. In studies in Tunisia it was found that maximum water consumption of *P. clandestinum* during the summer was about half that of alfalfa (de la Sayette 1967).

The plant rarely sets seeds except at higher elevations in the tropics or subtropics. It has been described as a facultative apomict (Bor 1960). Several experiments on seed production and flowering have indicated that repeated defoliation is necessary if the plant is to flower (Wilson 1970). This fact has been observed elsewhere; for example, lawns at 1,200 to 1,400 m in Hawaii flower profusely while adjacent unmowed areas do not flower at all. In Australia the plant will produce flowers and seeds throughout the year but most commonly in late autumn and spring (anon. 1954). Male-sterile and hermaphroditic races are known (Wilson 1970). The plant has been known to spread from seeds germinating in dung pats (Mears 1970).

Because it rarely produces seeds, *P. clandestinum* usually propagates vegetatively. It produces strong vigorous stolons and rhizomes which spread rapidly in a circular pattern from the parent plant, colonizing bare ground or encroaching on croplands, grasslands, forests, and waste areas. Stolons root readily at the nodes. The thick, white rhizomes spread through the soil to a depth of 30 to 40 cm or more, producing a tough, dense sod difficult to plow or to penetrate with tillage equipment. Roots of the plant have been found at a depth of almost 5 m in Kenya (Hosegood 1963). Both rhizomes and tillers can survive considerable drying after tillage; however, severe desiccation will kill them.

The plant can withstand severe and repeated defoliation; hence, it is very resistant to overgrazing or mowing. It is often used as a lawn grass and can survive and prosper even if it is mowed weekly. If the plant is infrequently or never defoliated, it grows somewhat upright; if it is frequently defoliated, it forms a dense, tough turf. The grass does well in pastures either under continuous or rotational grazing. In Africa it is considered to be a colonizer of lands which have been cleared for cultivation but on which fire has not been used as a land management tool (Rattray 1960).

DISTRIBUTION AND AGRICULTURAL IMPORTANCE

P. clandestinum is native to tropical eastern Africa, and its most frequently used common name, kikuyu grass, refers to its origin in the regions where the Kikuyu tribe lives. It has spread widely, especially in the past 50 or so years, mainly because it has been used as a pasture grass. Today its usual range extends from about lat 35° N to 37° S, where it grows mostly in the humid tropical highlands or in subtropical areas subject to occasional frosts (Figure 154). In Kenya it has been reported that it is a weed only above 2,000 m and in areas where rainfall exceeds 900 to 1,000 mm.

P. clandestinum is a much underrated but important pasture grass. It is often listed as a weed of pastures and especially of sown pastures where it encroaches and invades the swards, usually finally becoming dominant. However, to label it as a weed

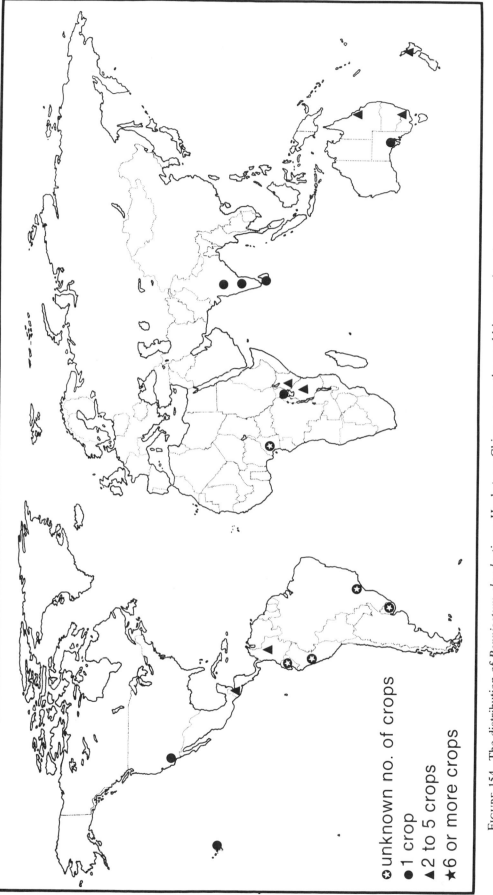

FIGURE 154. The distribution of *Pennisetum clandestinum* Hochst. ex Chiov. across the world in countries where it has been reported as a weed.

⊛ unknown no. of crops
● 1 crop
▲ 2 to 5 crops
★ 6 or more crops

in some pastures may be shortsighted, for the grass can be highly productive; it withstands heavy grazing, tolerates drought stresses, and produces a high quality forage having high levels of crude protein under conditions of good soil fertility. It is used for dairy production in eastern Australia, chiefly in areas with summer rainfall; and it is the most important pasture grass in Hawaii where it can carry as much as 2.5 animal units (1 animal unit = a 454-kg animal) per hectare the year round while producing liveweight gains of 300 to 800 kg per hectare. In Colombia it can carry three or four animals per hectare all year round (Romero, Jeffery, and Revelo 1969). Under good management the grass will choke out weeds and maintain pure stands. Unless frequent, close grazing is practical, legume mixtures with *P. clandestinum* may be difficult to maintain.

The same features which make *P. clandestinum* an important pasture grass also make it an important weed. It can be especially important in plantation or perennial crops because of its tendency to form vigorous, pure stands. It is rarely a serious weed in short-term field crops because it usually is vegetatively propagated. If good tillage is practiced, all that is necessary to control the weed is to quarantine it to field margins.

It has been reported to be a principal weed of pyrethrum in Kenya; sisal in Tanzania; and tea in India; and a weed of pastures in Australia, Ecuador, Kenya, and Taiwan; row crops in Costa Rica; cereals, citrus, sugarcane, and vineyards in Australia; forests and sisal in Kenya; pyrethrum in Tanzania; tea in Ceylon, Costa Rica, Kenya and Uganda; potatoes, wheat, corn, fruit crops and barley in Colombia; irrigated crops in Colombia and the United States (California); and of vegetables, papayas, and bananas in Hawaii. It is considered to be a weed of first importance in Peru and Colombia, where it is rated as the most agressive weed in the highlands.

It is an important grass for use in soil conservation practices such as grassed waterways, spillways, and for other uses as a soil binder. In Brazil it is used as a soilage crop.

Two insect pests, *Sphenophorus venatus vestitus* and *Herpetogramma licarsicalis*, have caused severe injury to *P. clandestinum* in Hawaii (Plucknett 1970).

It is an alternate host of *Piricularia grisea* (Cke) Sacc. (Raabe, unpublished; see footnote, *Cyperus rotundus*, "agricultural importance").

COMMON NAMES
OF PENNISETUM CLANDESTINUM

AUSTRALIA
  kikuyu grass
CEYLON
  kikiyu pul (Tamil)
  kikiyu tana (Sinhalese)
COLOMBIA
  kikuyo
EASTERN AFRICA
  kikuyu grass
HAWAII
  kikuyu grass

INDIA
  kikuyu grass
JAMAICA
  kikuyu grass
PERU
  kikuyo
PHILIPPINES
  kikuyu grass
UNITED STATES
  kikuyu grass
URUGUAY
  kikuyo

✵§58✵

Pennisetum purpureum Schumach., *Pennisetum polystachyon* (L.) Schult., and *Pennisetum pedicellatum* Trin.

POACEAE (also GRAMINEAE), GRASS FAMILY

Pennisetum purpureum is a large, upright, perennial bunchgrass that has been spread widely throughout the world as a fodder and pasture crop. A native of tropical Africa, the grass spreads and colonizes wastelands and grows along marshes and ditchbanks, in forest clearings and margins, and in perennial tropical crops. Although prospering in moist conditions or high rainfall, it can also survive droughts and fires. It has been reported as a weed in nine crops in 25 countries. It is the most productive fodder and soilage crop of the tropics. Two other weedy *Pennisetum* species which resemble *P. purpureum* in growth habit are *P. polystachyon* and *P. pedicellatum*, both of which become dominant in upland tropical hills and croplands after forests have been cleared or when shifting cultivation or subsistence agriculture has been practiced. All three of these weeds can become dominant in fire climax or subclimax savannahs when soil fertility has declined.

DESCRIPTIONS

Pennisetum purpureum

P. purpureum is a tall, robust, densely tufted, *perennial* grass, spreading by short creeping *rhizomes* 15 to 25 cm in length (Figure 155); *culms* 2 to 7 m high, stout, smooth, shortly hairy just below the flowering spike, joints hairy or hairless; *sheaths* smooth and hairless or rough and hairy in their upper parts; *ligule* a very narrow rim, fringed with dense white hairs up to 3 mm long; *blades* finely pointed, 30 to 90 cm long and up to 2.5 cm wide, with a very stout midrib, dull or bluish green or faintly tinged with purple, more or less rough on both sides, hairy or hairless, margins very rough; *inflorescence* a compact, erect, bristly, cylindrical *spike*, 8 to 30 cm long, 1.5 to 3 cm wide, usually yellow or tinged with brown or purple; *spikelets* arranged around a hairy axis, falling from the axis at maturity, each spikelet or group of spikelets surrounded by numerous, rough, dark yellow, brownish, or purplish *bristles* from 5 to 14 mm long, one much longer and up to 4 cm, the inner frequently hairy, spikelets with two florets; *lower floret* male or barren, half to almost as long as the upper floret, *upper floret* perfect, or both florets male or barren, stalkless, narrow, tapering to a fine point, 5 to 7 mm long, hairless, straw-colored or tinged with brown or purple toward the tips; *lower glume* minute or absent; *upper glume* narrow, finely pointed, translucent, one-nerved, up to 2 mm long; *stamens* three; *anthers* 2.5 to 3 mm long, tips very shortly hairy; *styles* joined, with two *stigmas* exserted from tip of flower.

367

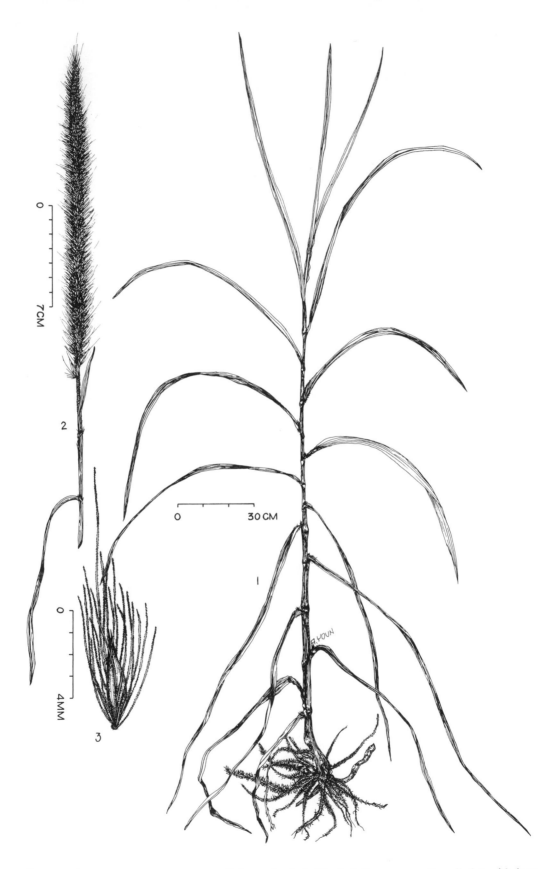

FIGURE 155. *Pennisetum purpureum* Schumach. *1*, habit; *2*, inflorescence; *3*, spikelet with long bristles.

Pennisetum polystachyon

P. polystachyon is a tufted *annual* or *perennial* grass (Figure 156); *culms* slender to fairly stout, 0.5 to 3 m or more, simple, sometimes branched; *blades* narrow, 5 to 45 cm long, 5 to 18 mm wide, smooth to hairy; *inflorescence* a dense *spike*, yellow-brown, 5 to 26 cm long, 1.3 to 2.6 cm wide; *spikelets* alike but surrounded by numerous bristles and falling from the spike with bristles attached; *bristles* densely hairy at the base, unequal, one longer than the others, 1.2 to 2.5 cm long; *spikelets* sessile, about 5 mm long, two-flowered, only the upper floret perfect; *lower glumes* minute or small, rarely, if ever, half as long as the *lower lemma.*

Pennisetum pedicellatum

P. pedicellatum is a tufted *perennial* grass; *culms* erect, woody, with prominent nodes, 60 to 120 cm or more high; *blades* 10 to 25 cm long, 5 to 10 mm wide, accuminate; *inflorescence* a tight *panicle* 1.2 to 2.5 cm long, changing from light purple to light brown as it matures; *spikelets* in fascicles 2.5 to 5 mm long, surrounded by bristles; *spikelets* pedicelled or in groups of two to five with one sessile and the others pedicelled; *lower glume* at least half the length of the *lower lemma.*

P. pedicellatum and *P. polystachyon* can be distinguished from *P. purpureum* by their lower lemmas which are often three-lobed, their upper florets which disjoint readily, and their rachises with decurrent wings on the ribs below the pedicels. The spikelets in *P. pedicellatum,* however, are pedicelled or in groups of two to five, with one sessile and the others pedicelled and with the lower glume at least half the length of the lower lemma; in *P. polystachyon* the spikelets are sessile and the lower glumes are minute or small and rarely, if ever, half as long as the lower lemma.

BIOLOGY

P. purpureum is a large, tufted, perennial grass which is a native of tropical Africa. It has become widespread throughout the tropics and subtropics as a soilage and fodder crop and as a pasture grass. It is well adapted to the humid regions of the tropics, and, although it grows especially well in lowland areas, it also can grow well at elevations as high as 1,500 m or more. It is easily injured by frost.

P. purpureum grows well on a wide range of soils and in many habitats. Although it can persist and compete in dry, sandy soils, it grows best in rich, well-drained soils. It does not grow well in water-logged soils but does flourish in irrigated lands. It is also very drought-resistant and can survive long dry periods such as are experienced in eastern Africa or in those parts of Asia subject to a monsoon climate.

As a weed *P. purpureum* colonizes along roadsides, ditchbanks, watercourses in marshy depressions, and in open places in forests. In moist rich places it forms reed jungles, thereby eliminating the use of such land for cultivation.

In Africa *P. purpureum* is found frequently near perennial swamps and disturbed forest lands (Rattray 1960). In these areas it is often associated with secondary tree growth, particularly in those places where fire or cultivation have occurred and have opened the lands to a savannah-type vegetation. In Uganda *P. purpureum* grasslands are considered to be the fire subclimax of the evergreen forests.

The root system of *P. purpureum* is mainly fibrous; however, the plant may produce short rhizomes. The plant does not spread much from its original site but, by tillering extensively, it widens the stool. When cut, the plant tillers freely from buds on branches or stems as well as from dormant buds near the crown. The fibrous root system is very deep and extensive; when well established it enables the plant to persist for long periods, even during drought.

P. purpureum rarely produces seeds, perhaps because the stigmas emerge from 2 to 4 days before the stamens, thus causing partial sterility and making fertilization possible mainly by cross pollination (Wilsie and Takahashi 1934), a process which is common in the species (Hanson and Carnahan 1956). Fertile hybrids for forage have been produced by crossing *P. purpureum* and *P. typhoides.* Although the plants do not produce seeds readily in most countries, they have been reported to flower freely and produce a good seed crop in August to September from sea level to 1,500 m in El Salvador (Flores and Olive 1952). In Hawaii *P. purpureum* blooms mainly in autumn and winter and seldom during the rest of the year (Wilsie and Takahashi 1934).

For pasture the grass can be propagated from root clump divisions from old plants or from stem cuttings consisting of one to several nodes. When seeds are produced, they often have low viability.

A well-known strain of *P. purpureum* is merkergrass, which has narrower leaves than the more common *P. purpureum* strain known as napiergrass. Although the two strains resemble one another, merker may be differentiated by its blue-green color, its upright early growth, its lighter red pigment in the younger stems, and its production of fewer tillers.

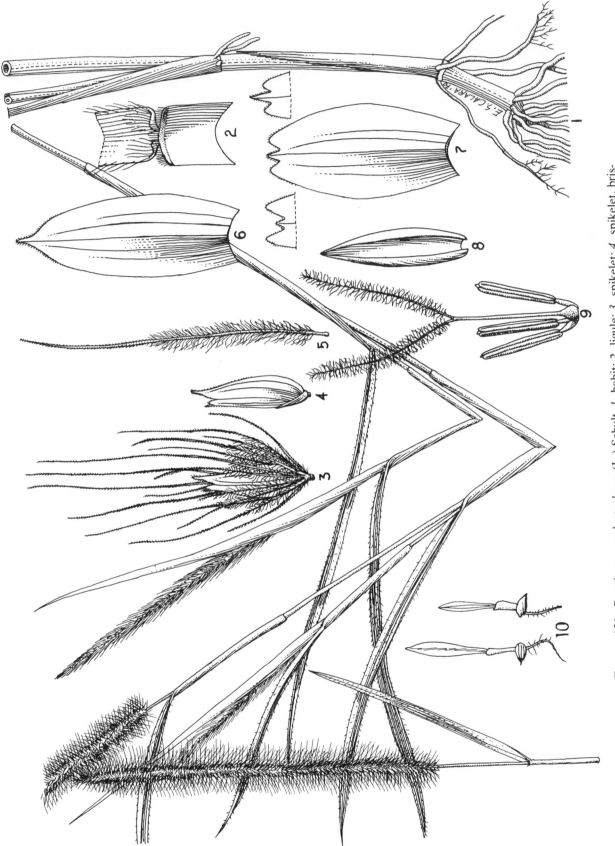

FIGURE 156. *Pennisetum polystachyon* (L.) Schult. *1*, habit; *2*, ligule; *3*, spikelet; *4*, spikelet, bristles removed; *5*, bristle; *6*, bract; *7*, bract; *8*, bract; *9*, flower; *10*, seedlings.

Merker can bloom 60 days after planting and has been observed to reseed (Wilsie and Takahashi 1934).

P. pedicellatum has been introduced into several countries as a pasture grass. In Thailand this species, together with P. polystachyon, has invaded upland areas that have been cleared for agricultural use or for shifting cultivation; in such places weedy grasses are now the dominant vegetation. Severe infestations take land out of crop production.

DISTRIBUTION AND AGRICULTURAL IMPORTANCE

P. purpureum, P. polystachyon, and P. pedicellatum are weeds of the tropics and subtropics and only rarely extend beyond lat 23° N and 23° S (Figure 157). They are important as weeds because they take over both wastelands and cultivated lands, often denying further use by man. All three may be used for pasture or for fodder crops, but only P. purpureum can be considered as important for these uses (Takahashi, Moomaw, and Ripperton 1966). When established in croplands or in forests, P. purpureum can form reed jungles 3 m or more in height.

In Fiji, P. polystachyon has become dominant in hilly, former forest lands that do not support livestock and that are difficult to improve because of the aggressive, competitive growth of this grass (Parham 1955). Under such conditions cultivation does not kill enough weeds to provide control. Because the crops grown in these lands are weed-infested, the farmers obtain poor yields.

P. purpureum is a principal weed of citrus in Trinidad. It is a weed of pastures in Australia and Brazil; coffee in eastern Africa; young cypress plantings in Uganda forests; rubber in Malaysia; tea and sisal in Mozambique; pineapples in the Philippines and Hawaii; oil palms in Colombia; and sugarcane in Florida (United States), where it encroaches into the fields from ditchbanks.

P. purpureum is an alternate host of Cassytha filiformis L. (Raabe 1965); Helminthosporium sacchari (B. de Haan) Butl. (Martin 1940); and Leptosphaeria sacchari B. de Haan, Meloidogyne sp., and Phyllosticta sp. (Parris 1941).

COMMON NAMES

Pennisetum purpureum

ARGENTINA
  pasto elefante

AUSTRALIA
  elephant grass

BRAZIL
  capim elefante

CAMBODIA
  smao kantuy damrey

CEYLON
  nepiar pul (Tamil)
  nepiar tana (Sinhalese)

COLOMBIA
  elefante

EL SALVADOR
  yerba elefante
  zacate elefante

HAWAII
  elephant grass
  merker grass

napier grass

HONDURAS
  elephant grass
  napier grass
  zacate elefante

MALAYSIA
  elephant grass
  napier grass

PERU
  pasto elefante

PHILIPPINES
  elephant grass
  napier grass

PUERTO RICO
  yerba elefante

UNITED STATES
  napier grass

VENEZUELA
  pasto elefante

Pennisetum polystachyon

FIJI
  mission grass

Pennisetum pedicellatum

FIJI
  kyasuwa grass

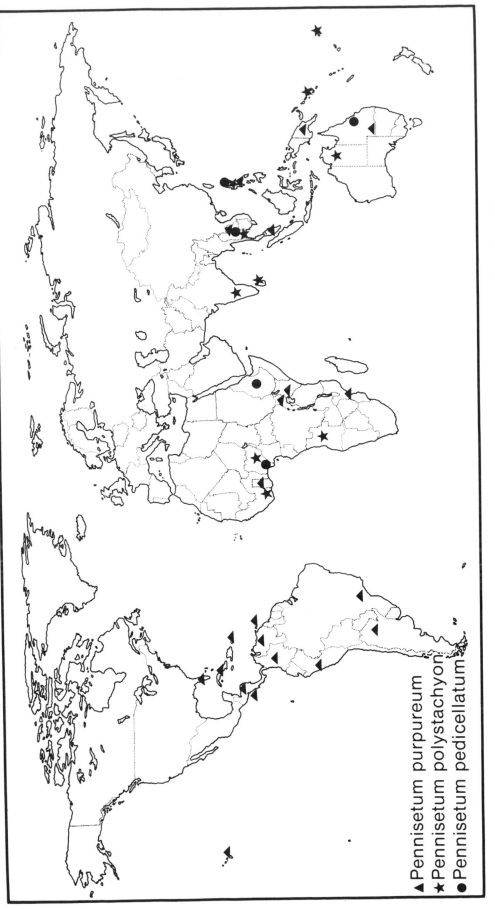

FIGURE 157. The distribution of *Pennisetum purpureum* Schumach., *Pennisetum polystachyon* (L.) Schult., and *Pennisetum pedicellatum* Trin. across the world in countries where they have been reported as weeds.

▲ Pennisetum purpureum
★ Pennisetum polystachyon
● Pennisetum pedicellatum

❧ 59 ❧

Phragmites australis (Cav.) Trin. (= *P. communis* Trin.) and *Phragmites karka* (Retz.) Trin.

POACEAE (also GRAMINEAE), GRASS FAMILY

Phragmites australis is believed by many to be the most widely distributed of all angiosperms. It is a perennial reed with broad, flat, leaf blades and large terminal panicles. It reproduces from vegetative propagules and has a vigorous, branched, rhizome system that runs quickly to new areas in either the water or the substrate. Although a native of the Old World tropics, it is remarkable for being equally at home in the countries of the northern temperate zone and in the torrid swamps of the Nile. At times the reed threatens man's waterways, pastures, and arable fields, but, as with many weeds, it can be a helpful companion. It provides shelter, food for animals, chemicals, fuel, fertilizer, and the raw material for an important papermaking industry in Romania.

DESCRIPTIONS

Phragmites australis

This is a robust, erect, aquatic or subaquatic, *perennial* grass (Figure 158); *culms* 2.5 to 4.5 m (sometimes 6 m) high, strongly tufted, with stout creeping *rhizomes*, often also with *stolons*; *leaf sheaths* loose and overlapping; *ligule* up to 1.5 mm long; *blade* flat, up to 60 cm long, 8 to 60 mm wide, tapering to a spiny point, rigid, glabrous or covered with whitish bloom; *inflorescence* a feathery *panicle*, somewhat nodding, 15 to 50 cm long, tan-

nish, brownish or purplish, rather dense, very many flowered, the branches slender, ascending; *spikelets* several flowered, 10 to 18 mm long, the florets exceeded by the hairs of the rachilla; *first glume* 2.5 to 5 mm long; *second glume* 5.7 mm long; *lemmas* thin, three-nerved, densely and softly hairy, the nerves ending in slender teeth, the middle tooth extending into a straight awn; *grain* slender, dark brown.

This species is distinguished by the long hairs of the rachilla which exceeds the florets and by the densely and softly hairy three-nerved lemma, the middle one of which extends into a straight awn.

P. australis is often confused with the Southeast Asian species, but the two may be differentiated as follows: *P. australis* has a long ligule up to 1.5 mm long, whereas *P. karka* has a ligule which is 0.5 mm long or absent; the lowest lemmas in *P. karka* are much shorter than are those in *P. australis*; and *P. australis* has rhizomes and stolons, *P. karka* has only stolons. For a review on the current taxonomy of this species, see Clayton (1968).

Phragmites karka

P. karka is an erect, strongly tufted, robust, aquatic or subaquatic, *perennial* grass; with creeping *stolons* often up to 20 m long; *culms* 3 to 4 m high, simple or branched; *sheaths* glabrous or sparsely hairy, with fine hairs at the top, leathery; *ligule* want-

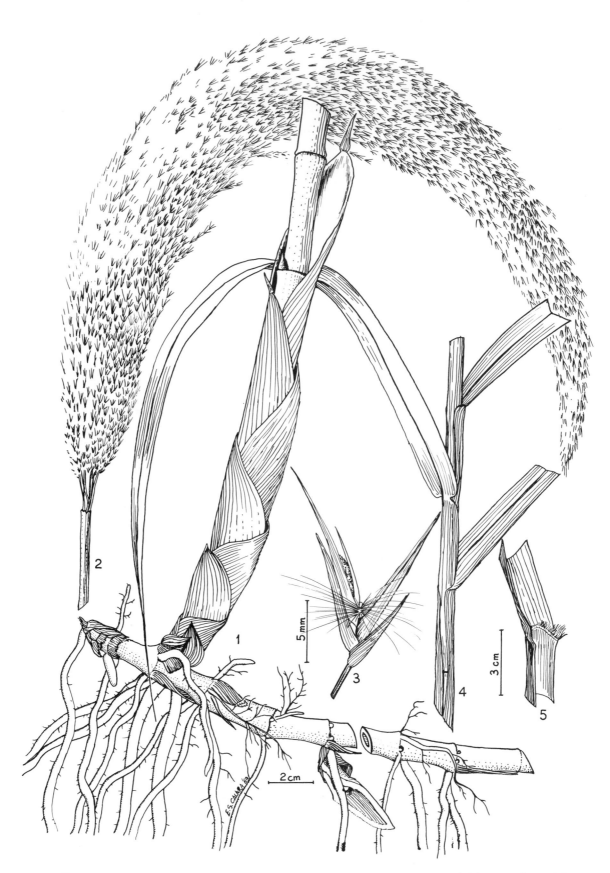

FIGURE 158. *Phragmites australis* (Cav.) Trin. ex Steud. *1*, root system and rhizomes; *2*, panicle; *3*, spikelet; *4*, leaf and leaf sheath; *5*, ligule.

ing or very short, minutely fine-hairy; *blades* coarse, sword-shaped with a broad base and an acute tip, up to 3.5 cm wide, margins smooth or rough, midnerve green; *inflorescence* a long, upright, oblong *panicle* 20 to 70 cm long, with a robust central axis and with wide-spreading, alternate, roughened, threadlike, nodding branches, central axis hairy or smooth near base of branches, branches crowded in groups, repeatedly branched; *pedicels* hairlike, rough or smooth, 2.5 to 10 mm long; *upper glume* 5 mm long; *lower glume* ovate-lanceolate, 2.5 to 4 mm long, three- (rarely five-) nerved; *lemmas* 7 to 9 mm long, tapering; *palea* 2.5 mm long with smooth or slightly roughened nerves; *stamens* two; *stigmas* white or yellowish, exserted laterally from near the base of the spikelet.

DISTRIBUTION AND HABITAT

P. australis is believed by many to be the most widely distributed angiosperm. The members of the genus *Phragmites* are all more or less aquatic. *P. australis*'s extensive range over the world and its vigorous rhizomes which can grow with great speed have made this species a threat to agriculture wherever there are lowlands, ponds, streams, or banks associated with arable fields or pasture. It serves to remind us that many of the most widespread hydrophytes are monocots which may be emergent, submersed, or free-floating plants. *P. australis* can live in the area above lat 70° N as well as in the great swamps along the Nile in Sudan. Close relatives such as *P. mauritiana* and *P. karka* are often found in the warm tropics. The general distribution of *P. australis* across the world is seen in Figure 159.

The plant prefers to grow where the current is less swift and where there are abundant, silted, muddy bottoms. It is a very common species in reed swamp communities, along banks, and in wet cultivated fields. It will grow in fresh or brackish water. It is one of the important emergent aquatic species which form the sudds (floating islands of vegetation) of the hot, humid region along the Nile in the central Sudan. *Cyperus papyrus* is often the first plant to begin to tie together several types of vegetation but it is quickly assisted by *Vossia cuspidata* and *Phragmites australis*. In the temperate zone floating islands are formed in lakes and at the mouths of rivers. These are called plavs and they differ from sudds in that a plav is usually made up of a single species. The most widely known of these plav-forming areas is at the delta of the Danube in Romania.

The waterline of the northwest coast of the Caspian Sea has receded from 3 to 20 km in the last 40 years. One of the first colonizers of the newly exposed soil has been *P. australis* and other associated species, all of which provided the initial phase of a succession which then moved through a meadow stage and finally, as the waters continued to recede, to desert saline communities of annuals. As the land was drained to form polders in Japan, *P. australis* communities also formed in that area. Annuals occupied the exposed areas until the end of the second year when *P. australis* appeared. It became more conspicuous in the third and fourth years, with several types of succession patterns finally being merged into a community dominated by the reed (Iwata and Ishizuka 1967). *P. australis* has also been used to prevent the growth of other weeds on newly exposed polders in the Netherlands.

P. karka is found in low country wetlands and swamps in Afghanistan, Australia, Burma, Ceylon, Egypt, India, Pakistan, the Philippines, Vietnam, and Sudan. It grows well in either fresh or brackish water, inhabits gravel banks on riversides and canals, and is found on humid slopes. It frequently forms dense masses of pure stands over large expanses of shallow water. These masses seriously interfere with fish production and provide shelter for *Anopheles* and other mosquitos which may be carriers of animal (including human) parasites.

The strongest culms are used for roof thatch or awnings and the small ones are sometimes used for pipestems. The plant is a good soil binder on slopes and banks.

PROPAGATION

There is no doubt that vegetative propagules are the most important means by which the species propagates and spreads. Underground parts may float when torn loose by fish, mammals, and turbulence. Spence (1964) has pointed out that for *P. australis* and many other hydrophytes the possible range of suitable habitats for successful seed germination is significantly narrower than the range of habitats which would allow development of vegetative propagules and mature individuals. In this species the seeds would not even germinate in 5 cm of water, yet the mass of emergent plants can grow and spread in 1 m of water. As a matter of fact, there was doubt for a time that seeds of some types could supply new individuals in natural sites. Even in the laboratory there were few reports of germination. Harris and Marshall (1960) in the United States re-

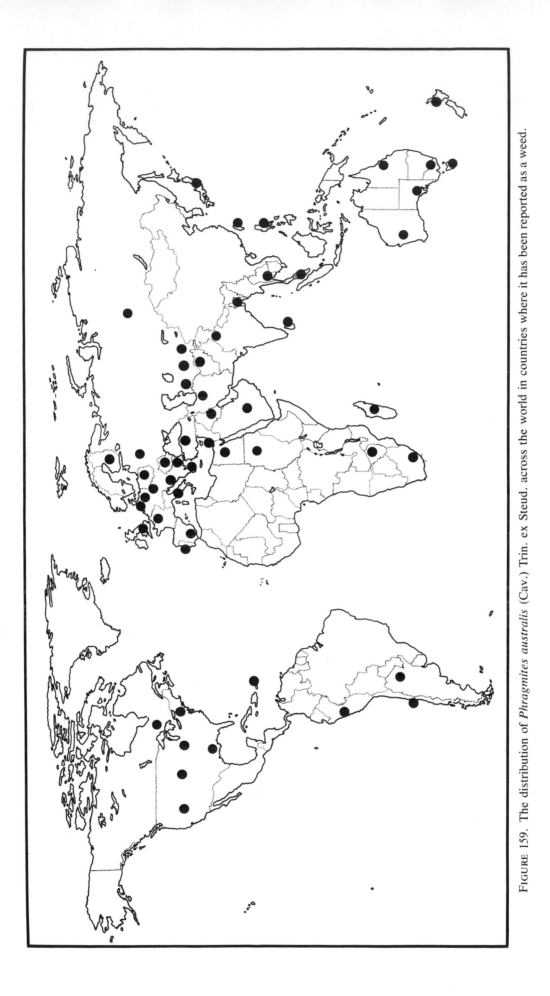

FIGURE 159. The distribution of *Phragmites australis* (Cav.) Trin. ex Steud. across the world in countries where it has been reported as a weed.

ported laboratory germination of seed which had been stored dry in paper bags for 7 months at room temperatures and then had been placed outside in April in closed vials half filled with water. Marked seeds were also placed in natural sites on mud flats where they germinated and established seedlings. Plumes—long, silky hairs in association with the fruit or related structures—assist the wind distribution of the seeds.

BIOLOGY

P. australis is quite remarkable for its strong, tall, aerial culms and leaves and its extensive creeping rhizome system with abundant lateral branching (Sculthorpe 1967). The erect leaves are similar in form and anatomy to the terrestrial grasses of drier sites; but the basal parts are surrounded by much the same environment as a wholly submerged plant, with which they have many morphological features in common. Because its roots must always be in water or in saturated soil, while its shoots are high in the air, the plant has developed air passages to supply the roots and water-covered parts.

P. australis develops some adventitious roots which are long, thick, and unbranched and others which are slender, much branched, and up to 20 to 30 cm long. The former are usually in the substrate and the latter are usually in the water. The creeping rhizomes may reach 20 m and may live from 2 to 3 years. In a dense stand of *P. australis* the underground parts (rhizomes, roots, and the bases of aerial shoots) comprise about 80 percent of the weight of the biomass. Haslam (1969a,b; 1970a,b) made extensive studies of the growth and development of this reed. In speaking of bud development, he reported that the growth rate and final height of aerial shoots are correlated with the width of the emergent bud, which really determines the basal diameter of the stem. At the peak periods of bud emergence there may be severe internal competition for food and small shoots may suffer. The most rapid growth occurs in early bud stage but growth at all times is most rapid in warm, humid habitats and, of course, when arising from buds with the greatest diameter.

To study shoot weight as it influences the size of standing populations of the reed, Gorham and Pearsall (1956) studied several sites in northern Great Britain. Their results suggested that the size of the shoot is most commonly determined by the supply of nutrients, although the summer water level also has

an effect. When water levels were too far below the surface the shoot weight decreased. Buttery and Lambert (1965a,b), also working in Great Britain, studied competition mechanisms between this reed and other plants and concluded that competition in a given site is not for nutrients. They discovered that the main source of nutrient supply for this plant was in the water and not in the substrate.

AGRICULTURAL IMPORTANCE

It should not be surprising that a weed which is so widespread has had a fascinating role in the life of man as he sought to make a home, feed his animals, prepare articles for commerce, and to use it (or war against it) in his irrigation schemes and agricultural fields. Though it may block canals and streams at times, it is probably a greater menace when it blocks drainage ditches because water flow there is not so turbulent and the reed can invade from the sides. Such ditches can be completely blocked in the tropics. In very similar circumstances, however, the plant may serve as an excellent soil binder to prevent erosion and washouts. The species is also a weed of crops in many areas. For example, it is a weed in cotton, corn, and rice in the Soviet Union; of rice in Greece, Portugal, and Taiwan; of sugar beets in Rhodesia and the Netherlands; of sugarcane and waterways in Australia; of wheat in Iran and the Netherlands, and of rape in the Netherlands.

One of the truly successful stories concerning the usefulness of weedy vegetation has been related by Rudescu (1965). Where the mouth of the Danube enters the Black Sea in Romania there is a delta which covers 5,100 square kilometers; 60 percent of this delta is covered with emergent reeds and principal among them is *P. australis*. For many years these reeds were used only for peasant crafts, thatching, windbreaks, and so forth. Then, in 1956, the areas were placed under state protection and now they are "managed" to obtain maximum yields. Most of the reeds are turned into pulp for the production of printing paper. Other products are cemented reed blocks, cardboard, cellophane, synthetic fibers, furfural, alcohol, fuel, insulation materials, and fertilizer. The annual harvest amounts to hundreds of thousands of tons, so that the reeds of the delta have become an important component of the Romanian economy.

Bakker (1960) in the Netherlands devised a way to secure new polders against perennial weed problems by planting the area in *P. australis*. From experience on previous polders it was known that *Cirsium arvense* and *Tussilago farfara* could be serious invaders of newly drained lands and could be very difficult to eradicate when cropping began. Bakker was also aware that *Phragmites australis* could shade the two species so severely that they could not make good growth. He selected the most vigorous strains of *P. australis* growing in his country, harvested and chopped the seedheads, and sowed them from the air as the water receded from the land. Thirty-five thousand hectares of the East Flevoland polder were seeded. The serious weeds were successfully kept under control by competition from the reed until finally, when the land was tiled and drained, *P. australis* was eliminated. In the meantime the reed had promoted the aeration and permeability of the soils.

P. australis growing on the delta of the Volga River in the Soviet Union is used for fodder and cellulose. To understand its management for pasture or for cropping Chuzhova (1968) studied the morphology and physiology of the species' regeneration from vegetative structures in that area. He found that new shoots arise from axillary buds on "vertical" rhizomes either in spring or early autumn. Most of the spring shoots he examined were weak and few succeeded. Those formed in early autumn provided the major part of the new stand in the following season. It has also been observed in the Soviet Union that the use of mole drains in wet meadows results in a decrease in *P. australis*.

The vegetative parts of the reed provide shade, shelter, and food for fish and the seeds provide food for ducks. The vegetative parts are eaten by cattle, goats, and sheep, and are a very important food for muskrats and for pigs in some areas. It is usually satisfactory as fodder only when young.

It was an important source of matting in ancient Egypt and is widely used for that purpose today. It is important in the horticultural trade for mats, shading, and containers in the Netherlands. It is plaited into sandals; the culms are carved into writing pens; it makes excellent thatch; and it is an important raw material in papermaking.

P. australis is an alternate host of *Piricularia oryzae* (Kuribayashi, Ichikawa, and Terazawa, unpublished).*

COMMON NAMES OF PHRAGMITES AUSTRALIS

ARGENTINA
  canizo
  carrizo
AUSTRALIA
  common reed
CAMBODIA
  prabos
CEYLON
  nala-gas (Sinhalese)
CHILE
  carrizo
DENMARK
  tagrør
FINLAND
  jarviruoko
FRANCE
  roseau à balais
  roseau commun
GERMANY
  gemeines Schilfrohr
  Schilfrohr
  Teichrohr
  Teichschilf
IRAN
  common reed
IRAQ
  common reed
  qasad
ITALY
  canna di palude
  canna da spazzole
MALAYSIA
  common reed

  rumput gedabong
NETHERLANDS
  riet
NEW ZEALAND
  phragmites
NORWAY
  takrør
  vannrør
PERU
  caña de India
PHILIPPINES
  bugang
  lupi (Bikol)
  tabunak
  tambo (Tagalog)
  tanobong (Ilokano)
PORTUGAL
  canico
SPAIN
  carrizo
SWEDEN
  bladvass
  vass
THAILAND
  or-lek
TURKEY
  kamis
UNITED STATES
  common reed
  giant reed
VIETNAM
  say

* "Study of physiological specialization in rice blast fungus" is a mimeographed report for 1951-1952 prepared for the Naguno Agricultural Experiment Station 31.

❧ 60 ❧

Pistia stratiotes L.

ARACEAE, ARUM FAMILY

Pistia stratiotes, a hydrophyte occurring in both the Old and New World tropics, is a very pale green, velvety, free-floating, perennial plant which consists of a rosette of leaf blades; a very short stem axis; and when the plant is suspended in water, a long, feathery root system. The place of origin of the species is uncertain. The prominently veined leaves may lie flat on the water or rise into the air to give a fanlike appearance. It multiples rapidly and can block streams, interfere with hydroelectric schemes and fisheries, and cause the breakdown of water control schemes at gates, sluices, and canals. The plants shelter the mosquitoes which carry the parasites responsible for malaria, encephalomyelitis, and rural filariasis. The species was at one time used for human food during periods of famine and it is presently used as feed for stock animals and fowl. It is one of the most widely distributed of all hydrophytes.

DESCRIPTION

P. stratiotes is a free-floating but soon stoloniferous, small, aquatic, *perennial* plant (Figure 160), with a tuft of long, very fibrous roots beneath, primary roots 2 to 7 mm in diameter and 1 m long, very fine, plumosely spreading root hairs; *leaves* obovate-cuneate, erect, few to many, 2.5 to 15 cm long, the basal part somewhat velvety-hairy, becoming thickened by the production of very porous tissue except when stranded on banks; *flowers* bisexual; bracts (spathes) subtending flowers; *spathes* white, densely dotted when dry, finely hairy (pilose)

outside, smooth inside, 7 to 12 mm long, 5 mm wide, short-penduncled in the center of the rosette of leaves; *spadix* bearing the individual flowers is shorter than the spathe, flowering parts minute; *fruit* berrylike (baccate), rupturing irregularly; *seeds* usually numerous, oblong, tapering toward the base, the apex appearing as if cut off at the end, about 2 mm long.

The light yellow-green leaves in the form of a rosette (cabbagelike) which are velvety-hairy and prominently veined below are characteristic of this species.

DISTRIBUTION AND HABITAT

Pistia stratiotes is one of the most widely distributed of all the hydrophytes. From Figure 161 it may be seen that *P. stratiotes* is most troublesome in Africa but is also a serious problem in the waters of southern Asia and the islands and shores of the Caribbean area.

To the good fortune of Australia and the islands of the South Pacific, the species is strangely missing as a weed of these areas, this in spite of the fact that conditions would seem to be ideal. Recent communications with botanists and others in these areas confirm that the plant is only seldom seen as an ornamental.

The free-floating plants are found in reservoirs, ponds, and marshes along the edges of large tropical lakes where they are able to thrive amidst the offshore vegetation and debris; in slow-moving or stagnant waters; and in old wells. The plants are cold

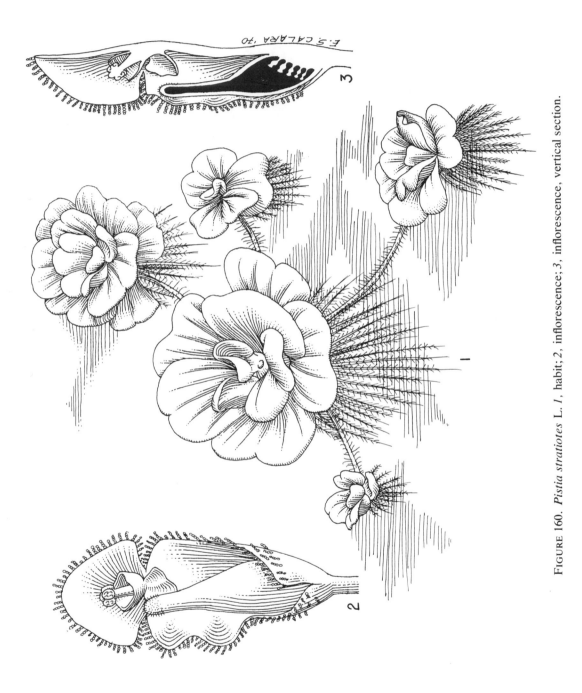

F. S. CALARA '70

FIGURE 160. *Pistia stratiotes* L. *1*, habit; *2*, inflorescence; *3*, inflorescence, vertical section.

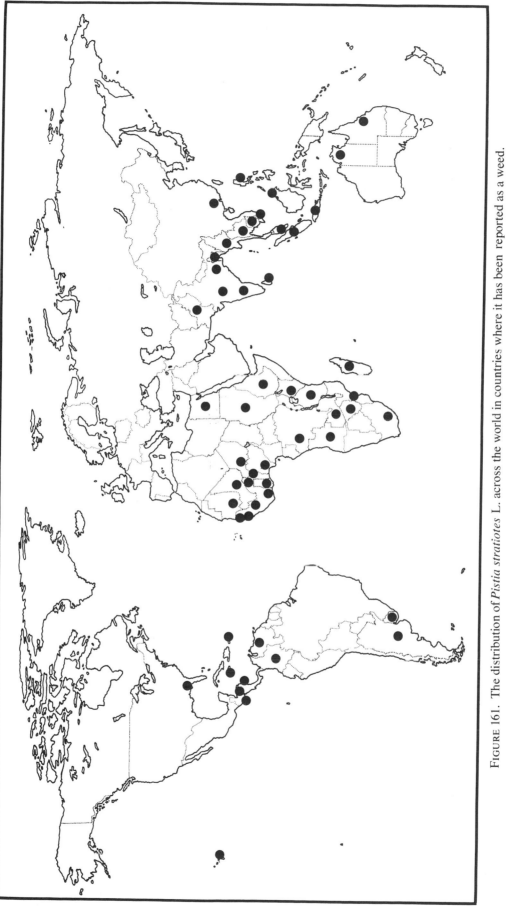

FIGURE 161. The distribution of *Pistia stratiotes* L. across the world in countries where it has been reported as a weed.

sensitive and thus cannot exist far beyond the tropics of Cancer and Capricorn. In the United States, for example, they are not troublesome north of the Gulf states and Arizona. In India and Madagascar they are found above altitudes of 1,000 and 1,200 meters, respectively.

Sculthorpe (1967) has discussed the role of *P. stratiotes* in the formation of dense aggregations of free-floating vegetation known as sudds. Sudd formation is especially common in wide, slow-flowing rivers which cross great expanses of open, flat land. Some of these major rivers are the White Nile, Ganges, Niger, and Brahmaputra. The swampy deltas of such rivers are especially favorable for formation of these floating islands of vegetation. Sudds may also form in extensive swampy areas in mid-country. There are such areas in Bengal in India and in the Okovanggo swamps of South Africa, but the most famous is the Sudd on the White Nile in Sudan, between Juba and Malakal, which has claimed so many lives. Sudd formation also takes place, with *P. stratiotes* among the sudds' important constituents, in large man-made lakes such as the Kariba on the Zambezi River between Rhodesia and Zambia.

Free-floating plants of *Pistia*, with their stoloniferous habit, create compact, thick, floating mats which spread from marginal sites into open water. The rooting medium thus formed is colonized by emergent hydrophytes, mainly grasses and sedges, which often invade from the shore by moving out over the mat. The mat of plants is thus stabilized by the intertwined rhizomes and roots. When the raft is large enough, it may be torn away by wind or flood to begin its journey as a floating island.

Andrews (1945) made extensive observations on the behavior of this species in the Blue Nile and White Nile rivers in Sudan. The plant has the same troublesome qualities as the water hyacinth, *Eichhornia crassipes*. *P. stratiotes* was much more plentiful in the White Nile before the recent explosion of water hyacinth and, because they occupy similar habitats, it is suspected that the *Pistia* populations have been reduced by competition from the larger plant.

Although *P. stratiotes* is common to both the Blue Nile and White Nile rivers, the species seems unable to colonize the canals of the Gezira cotton irrigation scheme (which is on the Blue Nile). The plant gathers in abundance at the Sennar Dam (upstream) but it has not yet threatened the canals. The depth of water does not appear to influence the growth of *Pistia*, but the plants tend to gather among water grasses near margins and in pools heavily infested with other aquatic vegetation. The canals provide few such sites and this may be a part of the explanation for the absence of the species.

Chadwick and Obeid (1966) studied the effects of water pH and nutrition on the growth of *P. stratiotes* and *Eichhornia crassipes*, plants which have a tendency to compete for the same sites. *Eichhornia* gave the greatest dry weight yield at pH 7; whereas *Pistia* performed best at pH 4, would not grow at pH 3, and lost vigor rapidly at any pH over the optimum. Its range of pH tolerance is much more narrow than *Eichhornia*. The pH of most river water is thus more favorable for the growth of *Eichhornia* and this may be an additional advantage in its ability to crowd out *Pistia*. For both plants, on an area basis, an increase in nitrogen level caused a linear increase in both total dry weight yield and number of plants.

Although *P. stratiotes* is usually a free-floating plant, it can survive for extended periods when stranded on very moist muck or sandbars and banks.

REPRODUCTION

In most areas *P. stratiotes* reproduces mainly by vegetative offshoots that are connected to the mother plant by stolons which may be 60 cm in length. The vegetative buds that give rise to these extensions form in a lateral pocket that is derived partly from the leaf sheath and partly from the axial tissue. In Africa, however, it is believed that the plant reproduces principally by seeds.

Flowering and fruiting vary by regions. It is said that the species does not flower in Thailand but that it does flower in Africa, the Philippines, and the United States. In India flowering begins in the hot season and continues up to the rainy season; the fruits appear after the rainy season. A typical plant may produce three to eight flowers in a whorl at the center. The flowers are 1 to 2 cm long, aroid in form, perfect, have perianth lacking, and have a unilocular ovary. When it separates, the spathe first exposes the pistil; then within a few hours the stamens; and soon the flower aborts. The period of time from the appearance of the first flower buds until the flowers open is about 8 days. The flowers fall from the plant within 2 weeks. Reports on seed production vary from none to one per flower in laboratory experiments in the United States, to several seeds per flower in Africa. The seeds are small and float on the water for 2 days, after which they sink and germinate. The seedlings appear at the surface in 5 days.

BIOLOGY

Because *P. stratiotes* very much resembles a small lettuce plant, it is often known as water lettuce. The plants are very light green, very velvety in appearance, with leaves lying flat on the water when light is diffuse or the plants uncrowded. When the rosettes are mature or when the plants are crowded, the leaves rise into the air.

Free-floating hydrophytes are, of course, exposed to falling rain, so that any structural modifications to repel water are beneficial, especially to those plants which may be quickly submerged by heavy monsoon storms. Many hydrophytes have a waxy surface but *Pistia* has short, depressed hairs on both surfaces which trap air, repel water, and thus prevent the epidermis from becoming wet (Sculthorpe 1967).

All of the leaves are succulent and some have on the underside a conspicuous, ovoid swelling filled with spongy parenchyma which gives flotation to the plant. The bladderlike swelling may be several centimeters long and may contain as much as 70 percent air. The tissues of most land plants have only 3 to 9 percent air. When the plants become stranded on mudbanks the floats do not develop.

Elimination of water from the leaves through pores is common in hydrophytes. In *P. stratiotes* this guttation process takes place through apical hydathodes which are located in a protected pocket. Beneath the pore is a cavity lined with thin-walled cells and into this chamber the tracheids of the vein endings open.

Each plant has an abundance of feathery, underwater roots. The primary roots may be 2 to 7 mm in diameter and 90 cm in length. Numerous secondary roots or branches may be 40 mm in length. The number and size of roots often correspond to the size of the plant which is above water. These roots have been relieved of their anchoring function and tend to stabilize the rosettes. Destruction of a portion of the root system disturbs the equilibrium of the plant, whereupon the foliage may become partly or wholly submerged.

ROLE IN HUMAN AFFAIRS

Humans encounter *P. stratiotes* in a variety of activities and it will be necessary to learn much more about its biology and control in the future. Man's health, his mobility, and the production of his food may be threatened or interfered with by this species.

The weed disrupts navigation, plugs grills at hy-droelectric plants, and interferes with fisheries by creating physical barriers and lowering the oxygen content and *p*H of the water. Because plants grow so rapidly in the tropics, the species can seriously interfere with paddy crops. The loss of water needed for agriculture, through transpiration from beds of *P. stratiotes*, has not been accurately measured but is believed to be considerable.

It is of equal concern that the plant serves as a preferred host for several species of mosquitoes which in turn serve as principal vectors of malaria, encephalomyelitis, and rural filariasis. The *Anopheles* mosquito, which carries the parasite responsible for malaria, is frequently associated with *P. stratiotes* because the hydrophyte provides suitable shelter and breeding sites. The *Mansonia* mosquito deposits eggs in large masses on the undersides of the leaves, and larvae subsequently attach themselves to roots of the plant. In many areas of the world it has been shown that removal of this plant species significantly reduces the populations of these mosquitoes in the area.

In Pakistan a blanket of this species on ponds and reservoirs has been shown to cause the decay of submersed hydrophytes such as *Hydrilla*, *Vallisneria*, *Najas*, and *Ceratophyllum* species (Chokder 1965, 1968).

Plants can float into paddy crops such as rice, take root in the soil, and compete with the crop under the shallow water conditions of the field. In such situations the plant behaves much like other weeds of flooded crops.

The plant was used as a famine food in India in 1877 to 1878. It is presently used as feed for pigs and ducks. Various parts of the plant have been used for centuries to cure skin diseases, to treat asthma and dysentery, and to serve as a laxative; its ashes are used for treatment of ringworm.

In several places in the world the plant is periodically devastated by insects, and this information is now being used in biological control studies.

COMMON NAMES
OF PISTIA STRATIOTES

ARGENTINA
repollito del agua

BANGLADESH
tokapana

CAMBODIA
chak thom

COLOMBIA
lechuguilla

EASTERN AFRICA
Nile cabbage
water lettuce

EL SALVADOR
  repollo de sapo
  verdolago de agua

HAWAII
  water lettuce

HONDURAS
  lechuga de agua

INDIA
  akasathamarai

antharathamra
boranjhanji
jalakumbi
kumbi
takapana

INDONESIA
  apoe-apoe
  apon-apon (Javanese)
  ki-apoe (Sundanese)

MALAYSIA
  kiambang

NETHERLANDS
  sla-kroos

NICARAGUA
  lechuga de agua

PHILIPPINES
  water lily

PUERTO RICO
  lamparilla

lechuguilla de vaca

THAILAND
  chok
  jawg

UNITED STATES
  water lettuce

VENEZUELA
  repollo de agua

VIETNAM
  bèo cái

❧ 61 ❧

Plantago major L.
and *Plantago lanceolata* L.

PLANTAGINACEAE, PLANTAGO FAMILY

These species are small, stemless, perennial herbs that are native to Europe. They have a compact crown of round or lanceolate leaves with long petioles and they produce very small, black seeds in capsules. They are temperate-zone plants with extreme ranges to the north and south; yet they have entered the warmest regions to become pests of almost all tropical crops. On a world basis, *P. lanceolata* is said to be one of the 12 most successful noncultivated colonizing species.

They are important world weeds because they are very successful in disturbed agricultural areas where they enter a microsite, become established, spread out in a rosette to occupy valuable space, and then hold it against almost all other species. In this fashion they have invaded many of the world's agricultural fields and have been reported from more than 50 countries. They are thus a constant source of difficulty and man must spend a great deal of time and energy in holding them back. Their lack of size and vigor means that they are seldom reported as a principal weed for a particular crop.

P. major and *P. lanceolata* have been companions of man for a long time. Helbaek (1950, 1958) reported that seeds were found in the stomachs of the Tollund Man and Graubelle Man from peat bogs in northern Europe. The remains dated back to the 3rd and 5th centuries A.D.

DESCRIPTIONS

Plantago major

P. major is a small, *perennial* herb (Figure 162); *stem* short, stout, bearing a rosette of smooth or hairy, spirally arranged leaves; *leaves* entire or obscurely toothed, oblong, broadly ovate or oblong-ovate, 5 to 10 cm long, 2 to 5 cm wide, usually three- to five-veined; *petiole* often as long as the leaf blade or longer; *spikes* 1 to 30 cm long, cylindric, dense to lax or interrupted at the base; *flowers* usually crowded, the bracts small, brownish with a green keel, *corolla* small, yellowish white, triangular, the lobes spreading or reflexed; *anthers* at first lilac, then dirty white; *capsule* three- to 30-seeded, ovoid, about 3 mm long, opening by a line around the fruit above the base; *seed* 1 to 1.5 mm long, variable in shape, dark brown or nearly black, angular, finely marked with wavy threadlike ridges, one surface with a pale scar, sticky when wet; *roots* many, fibrous, adventitious, whitish, up to 1 cm and probably contractile. It is an extremely variable species which frequently has been split into varieties.

The long-petioled leaf blade, its many-seeded capsule that opens around the fruit above the base, and the angular seeds which are marked with wavy threadlike ridges, and a light-colored hilum are distinguishing characteristics of *P. major*.

Plantago lanceolata

P. lanceolata is a small, *perennial* herb (Figure 163); *leaves* lanceolate, acute, narrowed gradually to the leaf base, densely clustered at the top of short, thick rootstocks, 8 to 30 cm long, 2 to 3.5 cm wide, entire or shallowly dentate, with three to five parallel veins, sparsely pubescent or glabrous; *leaf stalk* slender, grooved, the base sometimes surrounded by brownish hairs; *inflorescence* in terminal spikes often much longer than the leaves, sometimes as long as 120 cm, angled, grooved, and ending in cylindrical flower groups, 2 to 10 cm long; *flowers* subtended by dark green-black, ovate-acuminate bracts with membranous points, the lower larger than the upper; *corolla* about 4 mm in diameter, brownish white or yellow with prominent brownish midrib reaching the tip; *capsule* ellipsoid, 3 to 4 mm long, one- to two-seeded, splitting across the middle; *seeds* oblong, 1.5 to 2.5 mm long, with convex side and a scar about midway on the concave side, sticky when wet. It has an erect, thick, short *rhizome*.

The distinguishing characteristics of *P. lanceolata* are the lanceolate leaves, the one- to two-seeded capsule which splits around the middle, and the smooth, boat-shaped seeds with a scar about the middle.

DISTRIBUTION

From Figures 164 and 165 it may be seen that these two species are distributed as weeds through most of the agricultural areas of the world. The genus *Plantago*, which has about 250 species, is now found all over the world; yet it is essentially temperate in its natural distribution. Its wide range is largely made possible by the almost cosmopolitan introduction of *P. major* and *P. lanceolata* into agricultural lands in all climatic zones. Allard (1965) has stated that *P. lanceolata*, on a world basis, is one of the 12 most successful noncultivated colonizing species. *P. major* extends nearly from pole to pole, its northernmost point having been recorded in the Norwegian island of Spitsbergen at lat 77° N (Sagar and Harper 1964). Both species have ventured more freely into the tropics than have other members of the genus, but it may be seen from the distribution maps that they are not often reported as being troublesome weeds along the equator. The competition from taller and more vigorous plants in the heat of the tropics tends to keep the *Plantago*s under control. The species are rarely seen in tropical lowlands. Their distribution across the world is similar, with the exception of *P. lanceolata* seldom being a serious weed in Southeast Asia.

The most complete summary of the biology of these species and their behavior in the temperate zone is that of Sagar and Harper (1964). In their judgment, the most important factor governing the distribution of *P. major* is the disturbance of an area by men or animals. In the temperate zone, both species are found in open grasslands, cultivated fields, around gateways, along lanes and roadsides, and in gardens. *P. major* is more tolerant of compacted soil with the result that it appears frequently in almost pure stands on the edges of pathways. With *Lolium perenne* and *Poa annua*, it is among the species most resistant to treading. In England, for example, it grows best in those pastures which are subject to the most severe treading. *Plantago lanceolata* is less tolerant of the pressure of carts and feet and, indeed, it is said in some areas to be rather sensitive to treading and compaction. Both species have adapted to a wide range of soil types over the world. In England they are not found in acid uplands or peats with very low pH. *P. lanceolata* grows on all basic and neutral grasslands, and both species frequent poor lawn sites. *P. major* is more tolerant of waterlogging, and *P. lanceolata* is more drought resistant. The plants overwinter below the ground if they are in open areas, but they winter as small rosette plants if they are growing in grasslands where they find some cover. In extremely cold seasons or in areas where the leaves become frozen, the plant emerges in spring from underground storage organs. In Europe both species appear above 2,000 meters, whereas in Morocco they have been recorded above 2,100 meters. In summary, the species are so widely distributed over the world that their habitat is probably not restricted by climate.

BIOLOGY

The plants of both species are quite variable everywhere in the world. In Java, the forms of *P. major* vary so much that they sometimes appear to be different species, but they are usually connected by a series of intergrading types in the habitat. At high altitudes the leaves of this species tend to be more lanceolate and rather pubescent and to have shorter, thicker spikes. The leaves of *P. lanceolata* tend to be upright and more linear-lanceolate when the surrounding herbage is tall. Under close grazing its habit is prostrate and its leaves are more ovate.

The plants are certainly perennial, but the longevity of individuals varies with the region and with the

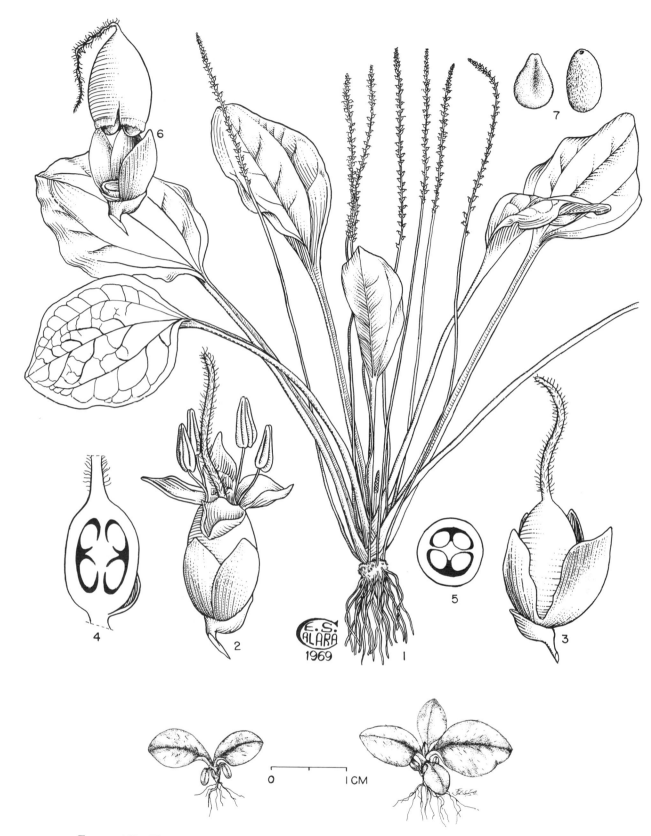

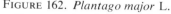

FIGURE 162. *Plantago major* L.

Upper: *1*, plant in flower; *2*, flower; *3*, flower, excised to show ovary; *4*, ovary, vertical section; *5*, ovary, cross section; *6*, capsule dehiscing; *7*, seeds.

Lower: Seedlings.

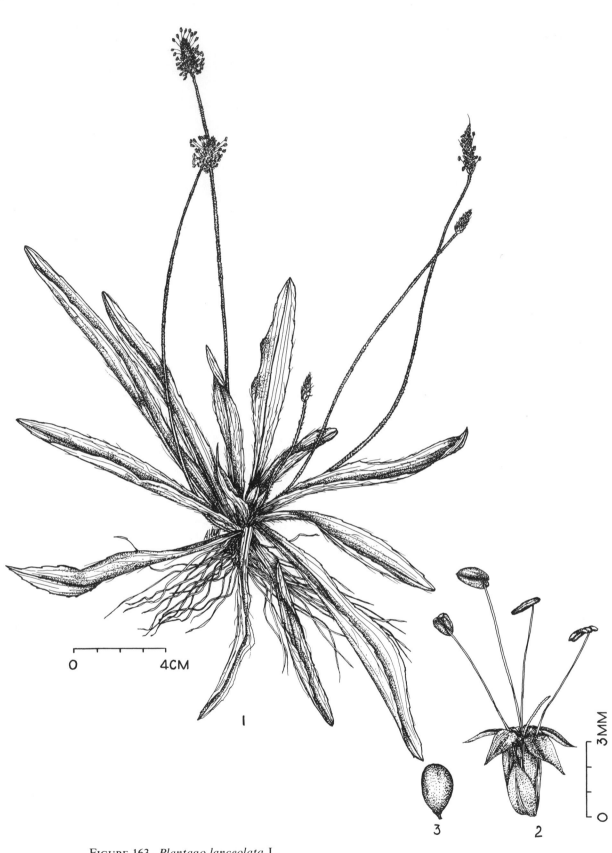

FIGURE 163. *Plantago lanceolata* L.
Above: *1*, plant in flower; *2*, flower; *3*, seed.
Opposite: Seedlings.

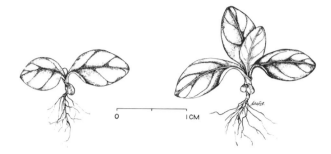

management or disturbance of the site. Some plants of *P. lanceolata* have been known to persist for at least 12 years. Both species have a rosette habit with a short underground stem. The primary roots are replaced early by adventitious roots. In *P. major* some of the later roots are contractile and help to keep the growing point at the soil surface where it is protected. In *P. lanceolata* the main descending roots may reach a depth of 1 meter while a fibrous system develops just under the surface. In experiments, this species was able to regenerate buds from root fragments. Older plants of the species develop a thick rhizome which frequently produces buds and new aerial shoots in the field. The underground organs serve for food storage.

P. major reproduces mainly by seeds and cannot multiply freely by vegetative means. A disturbed or injured crown may, however, give rise to several new growth centers. Normally many seeds are produced. The fruiting body is a capsule with a thin elastic wall which opens by means of a lid to release the seeds. There may be three to 30 seeds per capsule; and a seed production of 14,000 seeds per plant per year has been recorded. The seed shape varies with the number contained in individual capsules.

The situation is not as well defined in *P. lanceolata*. In Great Britain it has been reported that the spread of the species in the field is mainly from the growth of new buds arising on the thick underground stem. This vegetative habit leads to clumping. The plant spreads most rapidly in open areas and forms thick swards where its expansion is inhibited by competition. Dowling (1935) reported that in this area an examination of 400 capsules revealed that 44 percent had two fertile seeds and 56 percent had one. Seed production is highest in open arable land and least in mowed, grazed, and trampled areas. In Mauritius, a climate warmer than Great Britain's produces a different propagation behavior in *P. lanceolata* (Rochecouste 1969). In this area seed production is more important as a means of spread. In older plants the leaf bases become swollen and

buds may be formed in the axils to serve in propagation by vegetative means. This cannot be the principal means of distribution, however, for few of these survive the plow or hoe.

Genetic studies of *P. lanceolata* have revealed that the polymorphic species possesses several sexual forms; and the forms, in turn, have various degrees of sterility (Stout 1919). The plant types represented include hermaphroditic plants with long anthers and a large quantity of pollen having good germination, hermaphroditic plants with short anthers and a small quantity of pollen having poor germination, and female plants with rudimentary stamens and often no anthers. Ross (1960) found that male sterility occurs in *P. lanceolata*. In general, it appears that *P. lanceolata* plants are self-fertile, except when male sterility occurs; also, widespread sterility in the species probably helps to stabilize the heterogeneity of the weed (Brewbaker personal communication*).

In warm regions the plants may flower all year. In the northern temperate zone, *P. lanceolata* may begin to flower in April whereas *P. major* flowers slightly later. The main period of flowering may be June, July, and August but if competition is severe it may end in July. A full life cycle may be accomplished in about 6 weeks. The flowers are wind-pollinated.

Rethke (1946) suggested that the rupture of the fruits of *P. major* is not due to differential rates of drying of the cap and base but rather to a general contraction of the fruit wall as it dries on ripening. In both species a sticky mucilaginous cover forms on the seeds when they are moist, and this cover causes them to stick on soil particles or to adhere to feathers, fur, or skin. Seeds have remained viable in the soil for at least 60 years. In *P. lanceolata*, seed can be ripe and ready to shed 2 or 3 weeks after fertilization. It has been shown experimentally that brown, plump seeds show a higher percentage of germination than do very dark, brown, wrinkled seeds.

There is no active dehiscence in either species and so the seeds must be knocked off by animals, wind, or water; if they are not taken off, they remain attached to the spike until it rots at the base and falls to

* Information obtained from Dr. J. Brewbaker's unpublished files and notes on *Plantago lanceolata* pollen germination and related sterility problems. Dr. Brewbaker kindly allowed us use of his files, thereby adding to the biological information on this important weed. His address is: Department of Horticulture, College of Tropical Agriculture, University of Hawaii, Honolulu, Hawaii 96822, U.S.A.

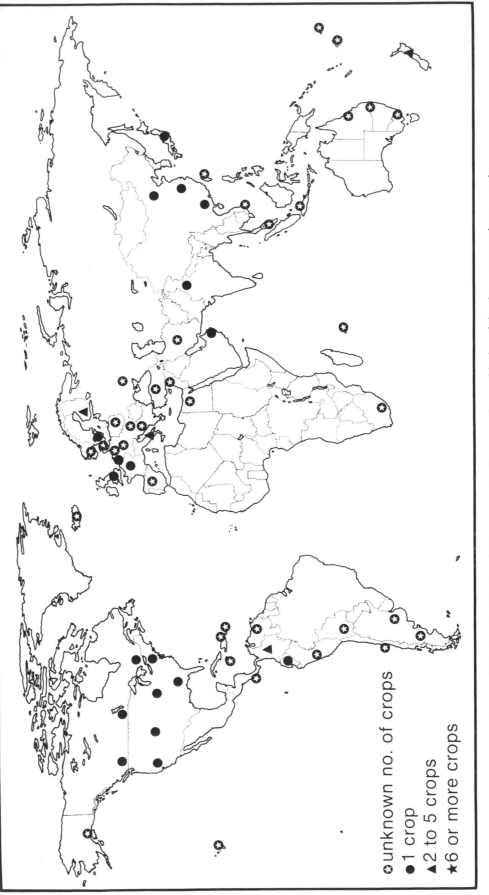

FIGURE 164. The distribution of *Plantago major* L. across the world where it has been reported as a weed.

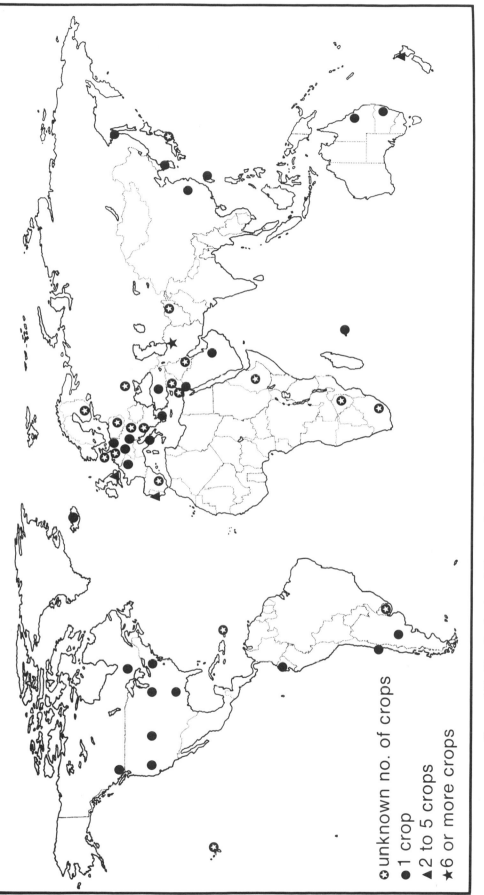

FIGURE 165. The distribution of *Plantago lanceolata* L. across the world where it has been reported as a weed.

the ground the next year. Under competitive conditions *P. lanceolata* may not produce seeds each year. On heavily grazed fields the animals may remove all spikes. Seeds of this species can germinate in autumn of the year of production or in the spring following. In *P. major* the seeds do not germinate in the field until the next growing season.

There is general agreement that many seeds of both species have a dormant period which can be broken by several months of dry storage at room temperature or by a few weeks at 5° C. Alternating temperatures are said to give the best germination but it is not yet clear whether the seeds are responding to the alternation or to the maximum temperature, for high temperatures of 25° to 30° C often give best germination. In laboratory experiments, the seeds of *P. major* germinated best in a long (16-hour) photoperiod. Some believe that this light requirement is dissipated after a short period of soil storage. The necessity for light for the germination of seeds of *P. lanceolata* has not been agreed upon, because the requirement varies so much with the age of the seed under study. Thus, we have a situation in which the seeds of these species can germinate sporadically as they are stimulated or released by the action of heat, cold, water, light, and chemicals. Germination seems best when the seeds are close to the surface and when open, desirable sites are present in the herbage.

AGRICULTURAL IMPORTANCE

Rochecouste (1969) suggested that seed-bearing perennials are the most difficult to control of all weeds in the tropics. The two *Plantago* species discussed here are unique in that more than 50 countries report them as weeds in 26 crops that are as different as alfalfa, coffee, temperate orchards, rice, pastures, cereals, onions, cotton, and small fruits. *Plantago* species are seldom regarded as being principal problems, but the sheer numbers of places, crops, and cultural systems in which they prosper make it necessary to regard them as important world weeds. According to the few reports we have of serious competition, they are always in cereals; they are common in the arable fields of sugar-producing areas; and their seeds frequently contaminate small-seeded legume seedstocks.

Because *P. major* grows best in open arable land, it cannot become established unless the surrounding vegetation is short. It can survive and set seeds in closed grassland communities, however, if it finds

temporary open places there in which to begin life (Sagar and Harper 1964).

The seedlings of *P. lanceolata* in the temperate zone appear in autumn and spring when there are open microhabitats in grasslands. Successful establishment depends on the site remaining open during the crucial transition of seedlings from dependence on seed reserves to independent assimilation (Saxby 1943). We have mentioned that this species is the least tolerant of trampling, but it is also vulnerable to the activities of livestock in other ways. It seems to be more palatable to animals than is *P. major*, and it is not unusual for all flowering spikes in a field to be removed. Sheep use their lower incisors to chisel the crowns of *P. lanceolata* out of the ground. Grazing cattle do not interfere seriously with either species, for the plants flatten out if they are grazed short and the animals then cannot tear the leaves off. In earlier times in England, *P. lanceolata* was sown in sandy soils for sheep-grazing. It is still employed in herbage strips in modern leys to give variety to animals and to supplement mineral supplies, for the species is a good source of calcium, chlorine, phosphorus, potassium, sodium, and cobalt. In Hawaii *P. lanceolata* is the most dominant weed of dry zone pastures at 1,000 to 1,200 m where it persists and increases even during periods of drought when rainfall drops to 250 mm.

For an extensive list of animal feeders or parasites, plant parasites, and diseases of these species, see Sagar and Harper (1964).

COMMON NAMES
Plantago major

ARGENTINA
llantén
torraja cimarrona

AUSTRALIA
large plantain

CANADA
broad-leaved plantain
common plantain
dooryard plantain
whiteman's foot

CHILE
llantén de hojas anchas

COLOMBIA
llantén

DENMARK
koempe-vejbred

DOMINICAN REPUBLIC
llantén

ECUADOR
llantén

EGYPT
lisan el-hamal

ENGLAND
great plantain

FIJI
plantain

FINLAND
piharatamo

FRANCE
grand plantain
plantain majeur
plantain des oiseaux

GERMANY
  Breitwegerich
  großer Wegerich
HAWAII
  broad-leaved plantain
INDONESIA
  daon oerat (Indonesian)
  greges otot (Javanese)
  ki oerat (Sundanese)
ITALY
  centonervi
  cinquenervi
JAMAICA
  English plantain
JAPAN
  to-oba-ko
LEBANON
  broad-leaved plantain
  lisan-ulkalb
  warak-sabun masasah
MALAYSIA
  eker anjing
  plantago
NETHERLANDS
  groote weegbree
  konijnenblad

NEW ZEALAND
  broad-leaved plantain
NORWAY
  grobladkjempe
PHILIPPINES
  lanting
  llantin
PERU
  llantén
PUERTO RICO
  llantén
SOUTH AFRICA
  breeblaar plantago
  ripple-seed plantain
SWEDEN
  groblad
TAIWAN
  che-chyan-tsau
TURKEY
  buyuk sinir otu
UNITED STATES
  broad-leaved plantain
VENEZUELA
  llantén
VIETNAM
  mã dề
YUGOSLAVIA
  velika bokvica

Plantago lanceolata

ARGENTINA
  llantén
  plantén
  siete venas
AUSTRALIA
  ribgrass
  ribwort

CANADA
  English plantain
  narrow-leaved plantain
CHILE
  llantén
  siete venas

DENMARK
  lancetbladet
  vejbred
ENGLAND
  ribwort
FINLAND
  heinaratamo
FRANCE
  bonne femme
  herbe-à-cinq-côtes
  plantain lancéolé
GERMANY
  lanzettlicher
    Wegerich
  Spitz-Wegerich
HAWAII
  buckhorn
  narrow-leaved plantain
IRAN
  bukhosn plantain
  kardi
IRAQ
  athan es-sakhlah
  buckhorn plantain
  narrow-leaved plantain
  ribgrass
  ribwort plantain
  zabad
ITALY
  arnoglossa
  lanciuola
JAPAN
  heraoobako
LEBANON
  adham el kabsh

  narrow-leaved plantain
  buckhorn plantain
  ribgrass
MAURITIUS
  herbe caroline
  herbe couteau
  plantain
NETHERLANDS
  hondetong
  smalle weegbree
NEW ZEALAND
  narrow-leaved plantain
NORWAY
  spisskjempe
  lansettbladkjempe
PORTUGAL
  corrijo
  tanchagem menor
PUERTO RICO
  llantén de pantano
  plantago
SOUTH AFRICA
  buckhorn plantain
  smalblaar plantago
SWEDEN
  spetsgroblad
URUGUAY
  llantén
TURKEY
  bataklik sinir otu
UNITED STATES
  buckhorn plantain
YUGOSLAVIA
  uskolisna bokvica

❧ 62 ❧

Polygonum convolvulus L.

POLYGONACEAE, BUCKWHEAT FAMILY

Polygonum convolvulus, a native of Eurasia, is an annual, herbaceous, creeping and twining plant that is becoming increasingly difficult to control in cereals in several places in the world. Its stems may be 2 m or more in length, and its twining habit causes lodging in grain so that the crop is difficult to harvest by machine. It has a very high seed production, is a serious contaminant of seedstocks several places in the world, and causes grain to heat following harvest. Only limited control is obtained with cultural methods and herbicides because the seeds are long-lived in the soil and can germinate throughout the growing season. The plant is often confused with *Convolvulus arvensis* or other species which have the common name "morning glory." It is a weed of 25 crops in 41 countries. It is ranked as a principal or serious weed in 20 crops of these countries.

DESCRIPTION

Polygonum convolvulus is an *annual* vining herb (Figure 166); *roots* fibrous; *stems* slender, freely branched from the base, trailing on the ground or twining about other plants, smooth to slightly rough, internodes long, 20 to 250 cm long; *leaves* alternate, 2 to 6 cm long, simple, entire, heart-shaped with basal lobes directed backward, pointed at the apex, long petioled, upper leaves lanceolate, with basal lobes directed backward, tapering to a gradual point; *stipule sheaths* with smooth margins; *flowers* in short axillary clusters or in terminal interrupted or spikelike racemes, small, greenish white; *calyx* minutely pubescent, green or purple-tipped, petals ab-

sent, sepals five; *fruiting calyx* nearly oblique-angled to egg-shaped, 4 to 5 mm long, three-angled, with scarcely developed keels; *seeds* (achenes) dull or shiny black, prominently three-angled, ovoid-pyramidal, minutely roughened, about 3 mm long, often covered with a dull brown, minutely roughened hull, sometimes part of the reddish brown calyx may remain attached at the bases of the seeds. The small greenish flowers, black three-angled seeds, and the presence of sheaths encircling the stem at the bases of the leaves are characteristics of this species.

DISTRIBUTION

P. convolvulus is most troublesome in the world in cereal fields, but it is also a problem in many cultivated crops. It is at home on waste grounds, in thickets, and along fences. It is a temperate-zone plant which has acquired the plasticity to invade and colonize areas in most regions of the world. It grows everywhere that crops are cultivated in Europe and North America. It is found from south to north in South America. In Africa it is in Morocco and Tunisia, on the east coast, and in South Africa. It ranges, as a weed, from Japan to Iran, down to India, and into Australia and New Zealand. From Figure 167 it may be seen that it tends to remain outside the humid tropics. In warm areas it does tend to seek high altitudes and cooler valleys, but its capacity for adaptation has permitted it to have a tremendous range around the globe.

FIGURE 166. *Polygonum convolvulus* L. *1*, habit; *2*, portion of stem showing ochrea; *3*, achene; *4*, seed; *5*, seedlings.

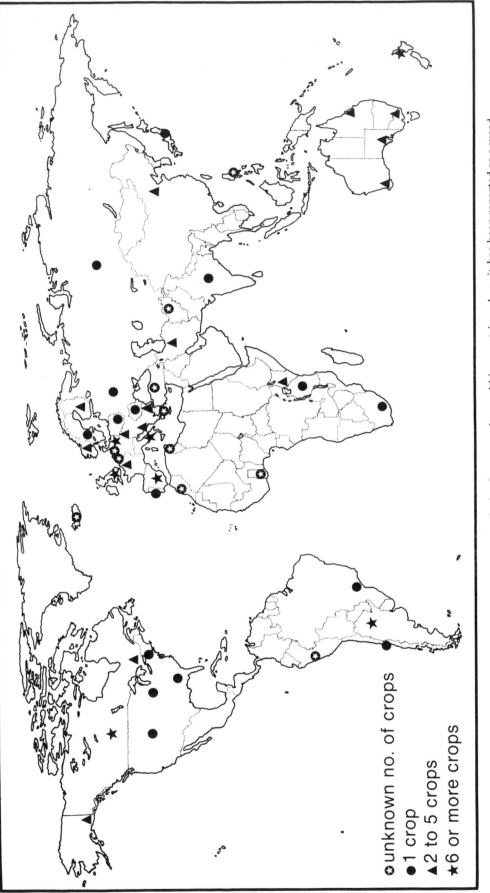

FIGURE 167. The distribution of *Polygonum convolvulus* L. across the world in countries where it has been reported as a weed.

○ unknown no. of crops
● 1 crop
▲ 2 to 5 crops
★ 6 or more crops

PROPAGATION

The fruit of *P. convolvulus* is a nut which provides the only natural means of propagation. Only one nut is produced per flower and it is usually shed with the dried perianth attached. Woodcock (1914) has reported on the anatomy of the seeds of this species.

A single plant that emerges early in the growing season may produce as many as 30,000 seeds. The seeds in the field normally begin to germinate only in the season following production. In wheat fields they have been observed to fall from the mother plant and remain on the soil for 75 days with no sign of germination. There is general agreement that the seeds possess a deep dormancy when mature and that they remain viable in the soil for several years (Justice 1941). New seeds will give only about 3-percent germination upon scarification. Seeds held in dry storage at room temperature will afterripen with the passage of time and eventually give low levels of germination.

Justice (1941) obtained 75-percent germination of seeds stored for 8 weeks at 3° C (37.4° F) in moist peat moss or layers of cotton. The most mature seeds and those kept in storage for some time responded most rapidly to the low temperature treatments. Timson (1966) has discussed the role of the pericarp in the dormancy of the species. As with most seeds possessing a heavy, tight pericarp, the following factors may prevent germination: mechanical pressure or a confinement by the seed covering, some factors which prevent the movement of water and gases, and inhibitors which cannot be released so that the embryo can be freed for metabolic activity. Long-term treatment with water and the application of gibberellic acid and thiourea are ineffective, presumably because they cannot gain entry. Both Justice and Timson found increased germination with the removal of portions of the pericarp. Germination was still low however, and the presence of a chemical inhibitor seems likely.

Experiments in Germany have shown the *P. convolvulus* can withstand storage in liquid manure up to 40 days without injury, whereas the seeds of *Galium aparine* and *Stellaria media* are quickly destroyed. In Denmark, seeds stored in silage for 3 months were still viable, as were those of *Chenopodium album* and *Spergula arvensis*.

BIOLOGY

Under good growing conditions *Polygonum convolvulus* may be decumbent, may produce stems with very long internodes, and may creep long distances. The root system is fibrous. On acid soils small, reddish leaves may be produced. The leaves are similar to *Convolvulus arvensis*, a species with which it is often confused. The latter, however, is a perennial plant and possesses a large funnel-shaped flower. *Polygonum convolvulus* is an annual and has small flowers in the axils of the leaves.

Forsberg and Best (1964) in Canada have given us one of the most helpful studies of the field biology of this species. They prepared large field plots by removing layers of soil to a depth of 19 cm and screening out the *P. convolvulus* seeds. Seeds of known viability were placed at intervals of about 1.3 cm as the soil layers were replaced in order. The germination of seeds from a heavily infested area was also studied during 3 years of summer fallow. Precise knowledge of the initial population of *P. convolvulus* seeds of the area was obtained from extensive sampling at the beginning of the experiment. They also studied rate of growth, flowering, and seed production in experimental plantings.

They found that most seeds emerged from 1.3 to 5 cm, but that some came from a 19-cm depth. Kollar (1968) also studied emergence from different soil depths and found that the practice of trying to "bury" seeds with tillage operations is likely to fail. Most seeds emerge in May and June, and a flush of germination probably coincides with the first very warm period of spring. Seedlings continue to make their appearance all through the growing season, with some emerging in September. Since most herbicide applications are made before mid-June it is obvious that many weed plants will appear subsequently and go on to produce more seeds in the field. One area with an initial average infestation of 3,500 seeds per 30 sq cm showed a germination of only 12 percent of the seeds after three seasons of fallow. Another area with a similar high initial seed population showed only 2.4-percent germination in the 1st year. Witts (1960) obtained similar low germination figures for *P. convolvulus* in the field. In view of these results, Forsberg and Best have warned that an effort should be made to destroy as many plants as possible before mature seeds are formed.

Seeds planted on 15 April and 1 May emerged only 4 days apart. Seeds planted after 15 June emerged in one-half the time required for germination of plantings made on 15 April. For plants which grew from late seedings, shorter intervals of time were required between significant stages of development, and the plants were not as leafy and spreading as those from earlier plantings. The life cycle from sowing to seed production was 15 to 20 days less when seeds were planted late. Some seedlings emerging in mid- to late April eventually produced 30,000 seeds. Individuals planted 2 months later gave 15,000 seeds. It should be remembered that the plant has an indeterminate flowering habit and that immature and mature seeds are present on the same plant.

AGRICULTURAL IMPORTANCE

P. convolvulus is one of the three most important weeds of cereals in Argentina, Canada, and the United States; corn in the Soviet Union; wheat in Canada, Kenya, and South Africa; and sugar beets in Spain. It is a principal weed of barley in Australia, Canada, and Finland; cereals in England and New Zealand; corn in Italy; flax in Australia, Brazil, Canada, and the United States; potatoes in Chile; sugar beets in Czechoslovakia, England, and Germany; beans in England; vegetables in Argentina, Bulgaria, Chile, England, and New Zealand; peas in Bulgaria and New Zealand; wheat in Argentina, Australia, Finland, New Zealand, Tanzania, and the United States; onions in Argentina and England; and sorghum in Italy (Figure 168). In these same crops, *P. convolvulus* is a common weed to be reckoned with in many other countries of the world. It is a weed of cotton in Iran and Spain; sorghum in Bulgaria; sunflowers in Romania and the Soviet Union; orchards in Canada, Finland, and Yugoslavia; potatoes in England, Germany, New Zealand, and the Soviet Union; and of rape, millet, alfalfa, pastures, and several vegetables in many areas of the world.

P. convolvulus is increasing in seriousness because of its deep dormancy and the long life of seeds in the soil; its habit of germinating all through the growing season; and its resistance to herbicides. As other weeds which are more susceptible to herbicides are removed from the field, the competition is effectively reduced, and this species flourishes. Tillage implements and rotations have limited usefulness because of large populations of seeds stored in the soil, the ability of seeds to emerge from great depths, the persistent emergence of new seedlings until late in the season, and the large quantities of seeds produced by the plants which do escape.

During the season the plant climbs upward as it twines on the crop, and in the case of cereals it causes lodging and can interfere with combine-harvesting. When the seed is harvested with cereals it contributes to heating in storage.

With other species of the genus *Polygonum* the weed has been increasing in seriousness in spring and winter cereals in Europe and on the plains of the United States and Canada. Friesen (1965) has given an impressive account of the threat to farmlands in the prairie regions of Canada. Five million six hundred thousand hectares of the prairie are infested, and 70 percent of this area is heavily infested. A study of the Red River Valley in the province of Manitoba revealed that 40 percent of all croplands were heavily infested. Some believe that the importance of this weed to farmers of western Canada is second only to that of *Avena fatua*. Friesen and Shebeski (1960) have shown that 56 and 210 plants per square meter can reduce wheat yields by 15 and 25 percent, respectively.

Dosland and Arnold (1966) have shown that soil moisture supply may be an important factor in the competition of this weed with wheat. In a year when rainfall was low the weed germinated early and made rapid development of leaf area and dry weight. This contributed to the early depletion of moisture reserves in the field; growth of wheat proceeded more slowly; and there was an early loss of leaf area at heading time. All of these factors combined to give a reduced yield of spring wheat. The study showed that in a year when moisture conditions were more favorable the yield of wheat was not significantly affected. The experiments of Gruenhagen and Nalewaja (1966) have also shown serious yield reductions in flax by competition from *Polygonum convolvulus* in North Dakota. Koch (1964) in Germany carried out experiments in cereals over a period of several years and learned that harrowing at a time which preceded crop emergence had the effect of increasing the number of the following species in the fields: *P. convolvulus*, *Avena fatua*, *Galium aparine*, and *Thlaspi arvense*. In general, some weed species were reduced and others remained the same as a result of this operation.

Dospekhov (1967) sprayed monocultures of crops such as rye, potatoes, red clover, and flax with phenoxy herbicides in 8 consecutive years. There was an ecological shift of weed species to resistant types such as *Polygonum convolvulus*, *Spergula arvensis*, and *Polygonum lapathifolium*, but the absolute reduction in total weed population was only 18 percent less than when the experiment was started.

Finally, we have become aware in recent years that contaminated seedstocks have contributed to the spread and seriousness of this weed. Dirty seedstocks serve to resupply the fields with unwanted species and negate our efforts to clean up our crops. A survey of seed lots in Brazil during the decade of

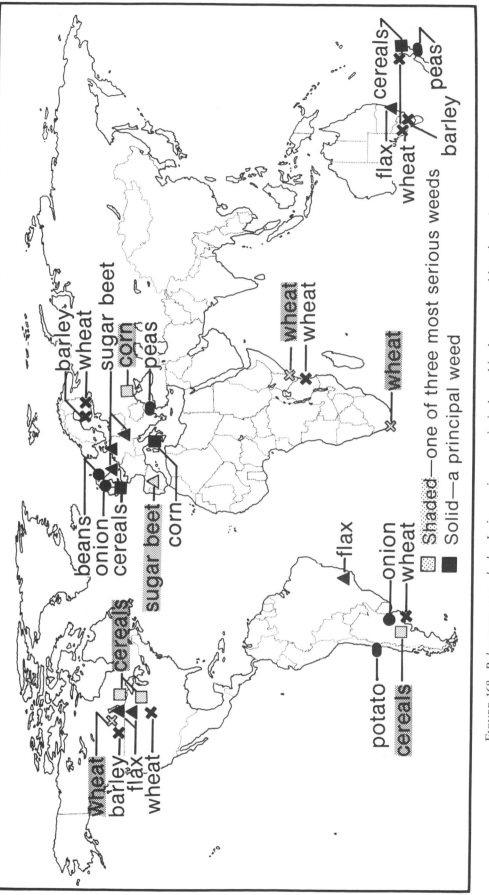

FIGURE 168. *Polygonum convolvulus* L. is a serious or principal weed in the crops and locations shown above.

the 1950s revealed that *P. convolvulus* was one of the three most serious contaminants in wheat seed. A study of weed seeds in the seedstocks of Europe in the period 1950 to 1957 as compared to 1960 to 1961 has been reported by Gooch (1963). *P. convolvulus* increased from 6 to 19 percent in wheat and from 17 to 24 percent in barley. There was a decrease in the seeds of this weed in rye and oats. The species is found in many vegetable seeds; its seeds are among those most frequently found in all seedstocks. Bogdan (1965) found 42 species of weeds in wheat seed lots collected from several places in Kenya. *P. convolvulus* was one of the most frequent contaminants. Harper (1965) has pointed out that seedstocks with quite a high percentage of purity may still include enough weed seeds to create a serious problem in a field. He has calculated that if *P. convolvulus* is present at 1 percent by weight in a sowing of cereals at 124 kilos per hectare, the weed seed will be distributed at the rate of 27 per sq m.

COMMON NAMES
OF POLYGONUM CONVOLVULUS

ARGENTINA
  enredadera

AUSTRALIA
  black bindweed

CANADA
  black bindweed
  climbing bindweed
  wild buckwheat

CHILE
  enredadera

DENMARK
  snerle-pileurt

EASTERN AFRICA
  black bindweed

ENGLAND
  black bindweed

FINLAND
  kiertotatar

FRANCE
  renouée liseron
  vrillée sauvage

GERMANY
  winden-Knöterich
  windenartiger
    Knöterich

IRAN
  pichak band

ITALY
  erba-leprina

JAPAN
  sobakazura

MOROCCO
  faux liseron

NETHERLANDS
  zwaluwtong

NEW ZEALAND
  cornbind

NORWAY
  vindelskjedekne

PORTUGAL
  corriola bastarda

SOUTH AFRICA
  black knotweed
  slingerduisendknoop

SPAIN
  corregüela anual

SWEDEN
  akerbinda

TUNISIA
  renouée liseron

TURKEY
  sarmasik coban
    deǧnei

UNITED STATES
  wild buckwheat

YUGOSLAVIA
  vijušac

Rumex crispus L. and Rumex obtusifolius L.

POLYGONACEAE, BUCKWHEAT FAMILY

Rumex crispus, a native of Europe, is a tall, erect, perennial herb with a thick, fleshy taproot which may extend 150 cm into the soil. It is very widely distributed and is believed by Allard (1965) to be one of the 12 most successful noncultivated colonizing species in the world. Several species of weeds are together commonly known as "docks" and *R. crispus* (curled dock) and *R. obtusifolius* (bitter dock or broad-leaved dock) are the most competitive of these. Although most serious in cereals and pastures, they are found in a wide variety of other crops. One plant of *R. crispus* may produce 60,000 seeds per year; a few seeds were still viable after 80 years of experimental storage in soil; and in one region it was estimated that 1 million seeds were present in the top 15 cm of soil. *R. crispus* is a weed of 16 crops in 37 countries. *R. obtusifolius* is less widely distributed in the world.

DESCRIPTIONS

Rumex crispus

R. crispus is an extremely variable, erect, *perennial* herb (Figure 169); *stems* smooth, sometimes ridged, single or in groups from the root crown, 30 to 160 cm tall, on a somewhat branched, stout, yellow *taproot; leaves* bluish green, lanceolate, prominently curly and wavy along the margins, sometimes flattened, 8 to 25 cm long; *petiole* 2.5 to 5

cm long; above each petiole a delicate sheath surrounds the stem; *inflorescence* a rather dense *raceme*, 30 to 60 cm or less long, intermingled with many linear leaves, rose-colored in the fruiting stage, brown at maturity; *flowers* small, green at flowering, greenish brown at maturity, in dense clusters; *pedicel* 5 to 10 mm long, with a swollen joint; *calyx* of six greenish *sepals*, the three inner, enlarged, heart-shaped (in fruit, called valves), with entire margins, each with a tubercle; *fruit* (achene) enclosed within inner flower parts, about 2 mm long, three-sided with sharp angles, glossy reddish brown. The flowers have no nectar and are wind-pollinated.

This species is distinguished by its crinkly leaf surface and its grain being surrounded by the sepal as a broad wing.

Rumex crispus can be differentiated from any of the other *Rumex* species by the following characteristics: leaves narrowed at the base, leaf margins crinkled and wavy, sepals (valves) with entire margins and with tubercles.

Rumex obtusifolius

Rumex obtusifolius is sometimes confused with *R. crispus*. *R. obtusifolius* is a *perennial* herb with stems to 1 m high; *leaves* broad with heart-shaped base, petioles and veins on undersurface covered with short, blunt, whitish hairs; *sepals* (in fruit, called valves) 0.5 mm long, triangular in outline,

FIGURE 169. *Rumex crispus* (L.) *1*, habit; *2*, fruit surrounded by persistent calyx; *3*, seed; *4*, seedlings.

margins with several long jagged teeth, usually only one sepal with a prominent tubercle; *seeds* three-sided, brown, 2 mm long.

R. obtusifolius differs from *R. crispus* in having a heart-shaped leaf base, sepals (valves) with toothed margins, and only one valve with a tubercle. When these two species hybridize, the teeth of the valves are shorter than in *R. obtusifolius*.

DISTRIBUTION

R. crispus is a troublesome world weed in both arable lands and pastures, but it is also an early colonizer of many disturbed areas both in lowland and upland. It is found on roadsides and neglected lands and in lawns and home gardens; occasional plants are found in many different types of habitats and plant communities. It is present on almost all soil types but less often on peat and rarely on acid soils. Because it is an important and widely distributed weed in the world, a word about its natural distribution may be helpful. It is found throughout Europe and grows above the Arctic Circle in Norway and the Soviet Union. It is found above lat 65° N in North America and also in the more temperate regions of South America. It is in New Zealand and Australia and many places in Asia. It is present in many parts of Africa but is seldom reported to be a problem from that country. Hulten (1950) has prepared maps of the distribution of both species in the Northern Hemisphere. The range of altitude to which *R. crispus* has become adapted is very great. A maritime ecotype grows on the beaches in England; and the species is found at 2,500 m in the Middle East and southwestern United States; at 3,000 m in Iran; and 3,500 m in Argentina.

R. crispus is a weed of agriculture on all continents (see Figure 170). It tends to avoid equatorial areas, but it is a weed in the valleys and high places in those regions. *R. obtusifolius* is also present on all continents but is not as frequently reported as a weed. The latitudinal and altitudinal environments of the two species are similar. *R. obtusifolius* grows better on acid soils than does *R. crispus*.

PROPAGATION

Valuable contributions have been made to our knowledge of the seeds of the two *Rumex* species by Harper (1965), Cavers (1963), and J. Williams (1971). Seed samples of weedy species are frequently quite variable in their germination, and these "polymorphisms" have only recently been ex-amined in detail. Some of them are rather straightforward, as in the case of *Spergula arvensis* where seeds are either smooth or tuberculate. The two forms have different temperature requirements for germination and have a different distribution in Great Britain.

Seed polymorphism exists in seeds borne on different parts of the same plant. In *Rumex crispus* the achenes formed on the proximal ends of branches (near the main stem) are heavier than those formed at the distal ends of the branches. The stage of maturity of the plant may influence the weight of achenes; and achenes of different weights may be produced from year to year. Cavers (1963) and Cavers and Harper (1964) examined the germination requirments of both inland and maritime forms of these two *Rumex* species. Seeds were taken for studies of differences within and between plants and within and between populations. Each sample was tested for germination in three conditions: (1) dark, 20° C, (2) dark with an alternating temperature of 20° C for 16 hours and 10° C for 8 hours, and (3) light and alternating temperature. Among the data obtained in the three different tests, it was possible to detect variations in the response of seeds from different locations. Cavers found that differences between the germination of seeds from different plants in the same habitat were greater than between habitats. Interplant differences were greater than differences in the behavior of seeds borne on different parts of the same plant, plants of different ages, or different degrees of seed ripeness. Thus, in the *Rumex* species, seeds are produced which have a range of different physiological requirements for germination; and we may expect that these requirements will be expressed in a range of potential ecological situations.

A further polymorphism is expressed in the seeds of different sizes growing along the lengths of the branches. The large seeds at the bases of the branches do not differ very much in their germination requirements from the smaller seeds at the distal ends, but they do produce larger, more vigorous seedlings. In tests, the smaller seeds gave seedlings with cotyledons averaging 12.2 mm by 4.2 mm while those seedlings from the larger seeds were 16.3 by 5.1 mm. It is clear from competition studies that vigorous seedlings have an advantage. Thus, we may see another aspect of the way in which a plant may express different ecological potentials through its seeds. J. Williams (1971) has discussed the correlation between hybridity and the germination polymorphism in these two species.

R. crispus and *R. obtusifolius* reproduce mainly

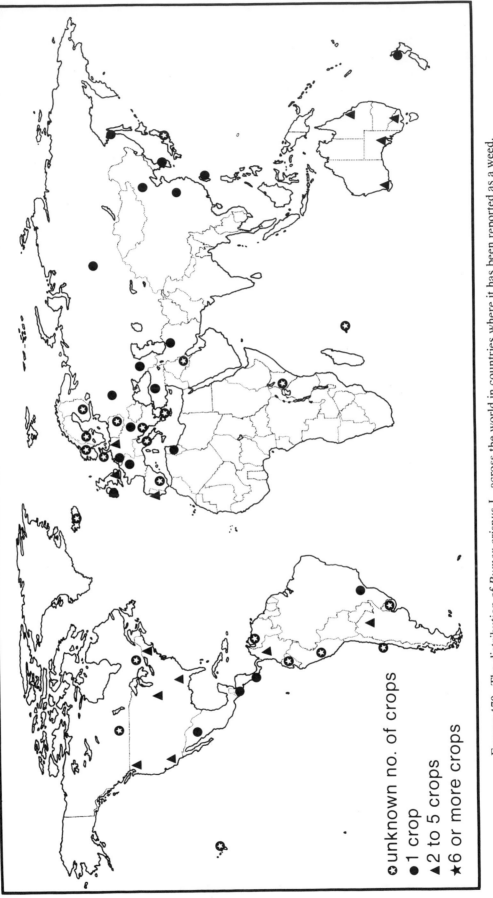

FIGURE 170. The distribution of *Rumex crispus* L. across the world in countries where it has been reported as a weed.

by seeds. Twenty to 50 fruits may be produced per whorl in a panicle. Seeds may number from 100 to more than 60,000 on a single plant. An intermittent germination is to be expected because a population or a plant will produce seeds with different requirements. There are seasonal flushes, nevertheless, and in some regions many seedlings emerge in autumn to form an overwintering rosette. Soil disturbance at any time seems to stimulate some germination. The seedlings grow best in open areas and will not stand crowding by more vigorous species. Seedlings of both species grow slowly and are poor competitors until the taproot systems have been established (Cavers and Harper 1964).

Reports dealing with regeneration from root pieces have been mixed, but the experiments of Healy (1953) in New Zealand and Hudson (1955) in Great Britain have now confirmed that regeneration of shoots can take place only from the uppermost part of the taproot. By studying 1.3-cm segments of entire roots of both species, Healy has shown that only the upper 7.5 cm of the *R. obtusifolius* root and the upper 4 cm of the *R. crispus* root can give rise to shoot buds. Both workers have confirmed that the best regeneration takes place during the early part of the growing season. Both species have a sizeable piece of stem belowground, and if this stem is moved in tillage operations it will quickly give rise to new plants. However, a tillage operation which removes that part of the plant which includes the upper 7 cm of the root, and then destroys it by dessication or freezing, will be an effective control measure because the remainder of the perennial root cannot give rise to new shoot buds.

Monaco and Cumbo (1972) found that plants of *R. crispus* grown from seeds in a growth chamber could produce new plants from rootstocks after 40 days. Plants of *R. obtusifolius* were incapable of regrowth from rootstocks until 51 days, but they produced flower stalks at 35 days. For the latter species it is believed that the food reserves were depleted by the flowering process so that the capability of regeneration from root stocks was delayed.

There is general agreement that the seeds of both species require light, alternating temperatures, or both, for germination. The seeds have been the choice of many workers for studying the phytochrome system (Le Deunff 1971; Le Deunff and Chausatt 1968; Vergnano Gambi 1966; Vincente, Engelhardt, and Silberschmidt 1962; Vincente et al. 1968). Near-red light (6,100 to 7,000 A) stimulates germination and far-red light (> 7,000 A) and darkness inhibit germination of the seeds. The reaction is reversible within limits. For those who wish to germinate seeds experimentally, the results from the seed tests of Steinbauer and Grigsby (1960) may be helpful. For *R. crispus*, 2-month-old seeds placed on moist blotters at 20° C gave 2-percent germination in darkness and 87-percent in light; and at 30° C, 0- and 2-percent germination, respectively. At an alternating temperature of 20° C for 16 hours and 30° C for 8 hours, germination was 24 percent in darkness and 94 percent in light. Scarification with acid or sandpaper greatly increased the percentage of germination in darkness. Seeds of *R. obtusifolius* responded in about the same way.

In a longevity trial started by Beal in 1879, 20 species of seeds were mixed with subsoil, placed in bottles, inverted, and buried to a depth of 50 cm. After 80 years, 2 percent of the *R. crispus* seeds still germinated (Darlington and Steinbauer 1961). In a similar experiment by Duvel the seeds were buried at 1 m. Germination for *R. crispus* at 1 year was 80 percent and at 10 years, 76 percent; germination for *R. obtusifolius* was 94 percent at 3 years and 83 percent at 21 years (Toole and Brown 1946).

The seeds will pass through the digestive tracts of birds and cattle without being harmed but they are destroyed when fed to chickens (Cooper, Maxwell, and Owens 1960). They are short-lived when stored in silage.

The fruits of both species are lightweight and may be moved long distances by wind. They are also spread by moving in water and by adhering with mud to the clothes of man, his machinery, and the fur, feathers, and feet of animals.

BIOLOGY

Maun and Cavers (1969) studied the response of *R. crispus* to daylength by placing 2-year-old plants of this species, which were in a stage of vegetative growth, in 15-, 12-, and 8-hour photoperiods. The day temperature was 26° C and the night temperature was 18° C. The inflorescences were short and decumbent in the 8-hour day, and they were tall, robust, and four times as long in the 15-hour day. Days to anthesis were 70 in the 8-hour day and 35 in the 15-hour day. In some plants fruit weight was 13 times greater and seed production was 16 times greater in the 15-hour day. The weight of individual seeds and percentage of germination were not affected by photoperiod changes. In these experiments the plants used were from two different parents, and there was quite a bit of variation between the two sets of offspring. These experiments tell us that short photoperiods can delay flowering but cannot prevent it, and they explain the short inflorescences seen in late summer and early autumn in fields which have been mowed.

In the temperate region new growth is seen very early on the rosette plants, and in the Northern Hemisphere flowering begins in May and may continue until a hard frost in October or November. Sometimes inflorescences are produced twice a year—in May and again after the first seeds are dropped. This happens more often in *R. crispus* than in *R. obtusifolius*. Plants growing from seed in open areas often flower and set seeds in the 1st year, but if they are crowded they may not flower until the 2nd or 3rd year. The plant, under good growing conditions, may flower 9 weeks after emergence. Some plants die after flowering but some form the winter rosette from which new lateral shoots will arise in the following spring. In sum, it may be seen that the plants, because of their long fleshy taproots, may behave as perennials, but in open areas or arable land may grow as annuals or as perennials. There are no studies of the plant's longevity; but in England some inland plants live at least 3 years and the maritime plants for 5.

Both species have a thick, fleshy, underground stem about 3 to 4 cm long and 5 cm or more in width above the fleshy, vertical taproot. The root of *R. obtusifolius* is much more branched than that of *R. crispus*. The underground stem is the result of root contraction which pulls the crown down into the soil by the time the fifth true leaf is being formed. The roots may extend more than 1.5 m into the soil and some side branches may be 1 m in length. Secondary roots do not thicken until the plant is more mature. New shoots usually form axillary buds on the crown (Cavers and Harper 1964).

In pastures or open, neglected places both species often appear to be patchy and this is because the seeds tend to fall near the plant and germinate there. They can, because of the deep taproot, survive very dry seasons and rather severe freezing. They are sometimes seen in very wet soils but cannot survive there.

Maun and Cavers (1970) grew overwintering plants of *R. crispus* under controlled conditions at soil temperatures of 10°, 15°, 26°, and 35° C to study the effects on top and root growth, seed production, and dormancy. Root growth was best and the largest seeds were produced at 10° C. The highest yield of fruits and seeds was at 26° C. A temperature of 35° C gave a significant reduction in growth rate, plant height, fruit and seed production, and weight per 100 seeds. The total percentage of seeds germinated was

not affected by soil temperature, but seeds which had been ripened at 26° and 35° C germinated faster than those produced at 10° and 15° C. From these experiments it may be seen that *R. crispus* can grow well and produce seeds under a wide range of soil temperatures.

Cavers and Harper (1964, 1967) have contributed a great deal to our knowledge of *R. crispus* and *R. obtusifolius*. To study the weeds' performance in various habitats, Cavers and Harper introduced seeds and transplants of *R. crispus, R. obtusifolius*, and a maritime ecotype of *R. crispus* into seven habitats: (1) arable land, bare soil, (2) a severely trampled area, (3) woods with dense canopy, (4) a lightly grazed meadow, (5) a heavily grazed meadow, (6) a stable plant community at the inland edge of a gravelly beach, and (7) a recently disturbed gravelly beach. Establishment from seeds was possible only on open sites such as the disturbed beach or arable land. Most seeds germinated but survived only 2 to 3 months in a closed grassland community. Transplants survived longer in most places but they had lost weight at the end of the experiment. One of the interesting findings was that the seedlings and transplants had failed on sites where one or more of the three *Rumex* taxa were already abundant. Cavers and Harper suggested that the existing populations living in those habitats in which new seeds or transplants had been unable to survive were relics which had germinated when conditions were more favorable to them; or, alternatively, that the researchers themselves may have chosen habitats which already had reached a saturation point for *Rumex* species when they were selecting areas for the experiment.

Koukol and Dagger (1967) used *Rumex crispus* in studies on the effects of ozone and smog on plants. When exposed to these chemicals the leaves partially turned red because anthocyanin had been stimulated to form. Untreated plants did not form the pigment. The treated plants did not become chlorotic.

AGRICULTURAL IMPORTANCE

R. crispus is a troublesome, widely distributed weed across the world, but it has not often been reported among the most serious. It is a principal weed of wheat and other cereals in Argentina, Great Britain, Mexico, and the Soviet Union; of flax in Brazil; of pastures in Belgium, France, and the United States; and of safflowers in Mexico. It is a weed of beans, orchards, rice, sugar beets, and vineyards in Europe; of citrus, flax, orchards, tea, and vineyards in the Soviet Union; of potatoes and wheat in Colombia; of cereals, pastures, and sugarcane in Australia; of alfalfa, orchards, rice and vegetables in the United States; of pastures in New Zealand; and of wheat in Iran and Turkey. It is the most important of the "dock" weeds in Australia. In Great Britain distribution of both *Rumex* species in seedstocks is monitored by the Seeds Act which requires that the contamination must be declared if it exceeds 0.5 percent by weight. Under the Weeds Act a landowner may be compelled by the Ministry of Agriculture to destroy these species if found on his land.

The very high rate of seed production, a ready germinability coupled with an extremely long seed life in the soil, and a tough, fleshy, storage root all make it possible for the species to spread and persist in agriculture in spite of injuries and disturbances by man and his animals. They enter disturbed areas quickly; and their persistance and success is emphasized in the report by Locket (1946) who found that *R. crispus* is a colonizer of bare chalk. Perhaps their tenacity in agricultural crops is due, however, to a constant resupply of seeds as contaminants in commercial seedstocks. Harper (1965) pointed out that crop seed considered to be of high purity may still bring surprising amounts of weed seeds into the fields each year. A seed supply containing 1 percent by weight of *R. crispus*, if sown at a rate of 18.5 kg per hectare, would bring 12 weed seeds to each square meter of the field. Cavers and Harper (1964) have summarized reports of *R. crispus* as a seed contaminant in the past half century.

The characteristics of these species which make them such successful weeds in agriculture are their ability to establish quickly from seeds, to flower in the first year, and to produce large quantities of seeds, some of which can remain viable for very long periods in the soil. The plants can withstand close grazing and mowing and can quickly enter openings caused by gouging or by dung patches.

Although the plants are not generally considered to be poisonous, they are believed to be responsible for an array of disturbances which have been attributed to the ingestion of foliage or seeds. *R. crispus* is quite toxic to poultry. If large amounts are eaten by cattle, the animals may suffer gastric disturbances and dermatitis. Rumicin from the tops and chrysarobin from the roots are the biologically active substances in the plant.

The Indians of northern California are said to have grown the seeds to make a cooked mush in times of need.

Rumex obtusifolius serves as a weed host for the potato tuber eelworm (*Ditylenchus destructor*) in England (anon. 1968*a*).

COMMON NAMES
Rumex crispus

ARGENTINA
  lengua de vaca
AUSTRALIA
  curled dock
CANADA
  curled dock
  sour dock
  yellow dock
CHILE
  romaza
COLOMBIA
  lengua de vaca
DENMARK
  kruset skraeppe
EASTERN AFRICA
  curled dock
ECUADOR
  lengua de vaca
ENGLAND
  curled dock
FINLAND
  poimuhierakka
FRANCE
  oseille crépu
  patience crépu
  rumex crépu
GERMANY
  kräuser Ampfer
GUATEMALA
  lengua de caballo
  lengua de vaca
IRAN
  curled dock
  sorrel

IRAQ
  curled dock
  hummaithah
  yellow sour dock
ITALY
  romice
  romice crespa
JAPAN
  nagaba-gishi-gishi
MEXICO
  lengua de vaca
NETHERLANDS
  krulzuring
NEW ZEALAND
  curled dock
NORWAY
  krussyre
PERU
  lengua de vaca
PORTUGAL
  labaca crespa
SWEDEN
  drusskräppa
  krussyra
TURKEY
  curly dock
URUGUAY
  lengua de vaca
UNITED STATES
  curly dock
VENEZUELA
  pira de verraco
YUGOSLAVIA
  stavalj

Rumex obtusifolius

AUSTRALIA
  broad-leaved dock
DENMARK
  butbladet skraeppe
ENGLAND
  bitter dock
  broad-leaved dock
FRANCE
  parelle

  patience à feuilles obtuses
  patience sauvage
  rumex à feuilles obtuses

GERMANY
  stumpfblättiger Ampfer

ITALY
  romice comune
  romice a foglie ottuse

NEW ZEALAND
  broad-leaved dock

NORWAY
  storbladet syre
  veisyre

PORTUGAL
  labaças
  labaçol

SPAIN
  acedera de hojas ob-
    tusas

  bijuaca

  lengua de vaca

SWEDEN
  tomtskräppa
  tomtsyra

UNITED STATES
  bitter dock
  broad-leaved dock

❧ 64 ❧

Salvinia auriculata Aublet

SALVINEACEAE, SALVINIA FAMILY

Salvinia auriculata, a fern, is a free-floating hydrophyte that is very widely distributed in the world. It has become so specialized and has undergone such extreme vegetative reduction that it no longer resembles the ferns. It is most notorious for its explosive growth in the rice paddies and waterways of Ceylon, and on Lake Kariba on the Zambezi River in southern Africa, during the last 3 decades. It now threatens the crops and waterways of Java in Indonesia and Kerala state in India. Though the species is a native of South America, it has never seriously interfered with man's activities in that area. The plants cannot survive conditions of drought or cold and they die quickly in salt water. It is reported to be a weed in 22 countries where it fouls irrigation systems and navigable streams and fisheries; interferes with electric power production; and is a direct threat to rice farming. It affects many crops across the world by interfering with the distribution of irrigation water.

DESCRIPTION

S. auriculata is a *free-floating*, rapid-growing, mat-forming, branched, gregarious, *annual* or *perennial* fern (Figure 171); individual plants up to 30 cm long with numerous leaves which usually form a mat to 2.5 cm thick, although they can form very dense mats also; slender horizontal floating *rhizome* producing at each node two short petioled or sessile *fronds* on the upper side and a long filiform feathered *rootlike frond* downward; *leaves* (fronds) produced in groups of three, from the delicate stem, each group with two broadly ovate, entire, undivided,

green, aerial leaves up to 25 mm long situated on the upper side of the stem, with a distinct midrib from the base to the apex, the upper surface covered with close parallel rows of numerous long hairs that terminate in a cagelike, club-shaped tip and that prevent the leaves from getting wet, lower leaf surfaces smooth except for simple hairs (rootlets?) near the midrib; the leaves of young plants on open water float flat on the surface, later with age and crowding they fold against one another with the hairy upper surfaces facing each other; the third, the submerged, rootlike "water" leaf, situated ventrally, is much divided, feathery, brown, up to 25 cm long, resembles and functions as a true root, and bears the *sporocarp* or spore-forming structures; *sporocarps* globose 2 to 3 mm in diameter, on a short stalk 1 mm long, densely hairy, indehiscent, monoecious, seated in clusters on the stem at the base of the rootlike, submerged leaves; *megaspores* and *microspores* about 2 mm long, numerous, globular, covered with minute hairs.

The presence of numerous, cagelike, clubshaped hairs on the upper surfaces of the aerial leaves is very distinctive. This feature imparts an additional buoyancy to the plant and serves to distinguish *S. auriculata* from any other species of the genus.

The African species is sterile and is more robust and vigorous than any other known *Salvinia* species. Mitchell (unpublished*) in Rhodesia believes that it

* Information obtained from Dr. D. S. Mitchell, Salisbury, Rhodesia, in 1972, concerning publications in preparation on ecological surveys of *Eichhornia crassipes* and *Salvinia* species, together with data on the taxonomic position of the latter genus.

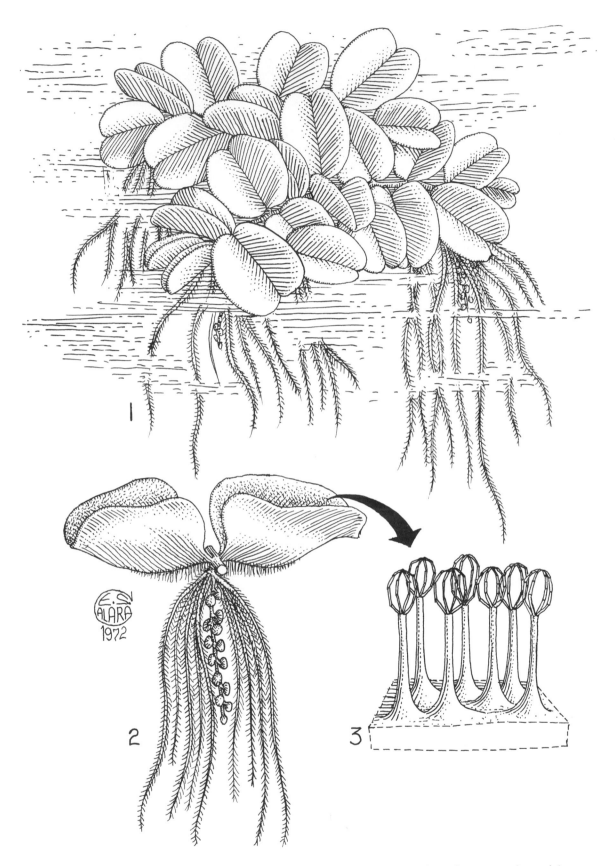

FIGURE 171. *Salvinia auriculata* Aubl. *1*, habit; *2*, portion of plant showing microsporangia, aerial fronds, and rootlike submerged frond; *3*, detail of hairs on upper surface of aerial leaves.

is a hybrid of horticultural origin which has not been described. He has written a new description and is preparing a publication in which he proposes to rename the species *Salvinia molesta*. The genus *Salvinia* is also undergoing a taxonomic revision. Until the above proposed revisions are published there is no choice but to regard the various forms of *S. auriculata* as a single species.

DISTRIBUTION

The world distribution of *S. auriculata* is shown in Figure 172, where it may be seen that cool temperatures have limited the weed's spread into temperate zones. Except for growing in certain locations in Australia and New Zealand, the plant is almost always confined to the thermal tropical zone. The species is found elsewhere in aquariums and garden ponds, and these cultures may very well be the principal mechanisms by which this weed spreads. The genus *Salvinia*, which is one of the most cosmopolitan of the hydrophytes, is made up of 12 species encountered mainly in the tropics, subtropics, and warm temperate regions. Sculthorpe (1967) reported that *S. auriculata* is in southern Europe.

The species is known for its explosive colonization of large expanses of water. Its spread into four regions of the world will be discussed in a later section. It is confined to fresh water and is quickly killed by seawater. Though it normally grows as a free-floating hydrophyte, it also thrives on land in the zone of constant mist near the foot of Victoria Falls in southern Africa (Schelpe 1961). It can survive for short periods on mudbanks but will tolerate very little drying. Mitchell (unpublished)* reported that several nights at temperatures of 10° C caused the death of plants in Rhodesia.

This species grows best in stagnant or slow-flowing water and prefers the small bays of dissected shorelines and the estuaries of small streams. Growing around emergent brush and trees on flooded shorelines it is protected from wave action and multiplies rapidly. If *Salvinia* threatens in areas where man-made lakes are planned, the shorelines should be cleared to discourage the weed and to facilitate control operations if small patches make their appearance. The mechanical tissues of

* "Ecological studies of *Salvinia auriculata* with particular reference to Lake Kariba (Rhodesia and Zambia) and the Chobe River (Botswana)" is the title of a paper presented by Dr. D. Mitchell to the meeting of experts on "Ecology and Control of Aquatic Vegetation" at UNESCO House, Paris, 16-18 December 1968.

this fern are so reduced that it must depend on turgor pressure for its form and organization. The smallest disturbance, therefore, will break off pieces of stems; and even the smallest fragment, if it contains an axillary bud, can give rise to a new plant (Sculthorpe 1967, Mitchell, unpublished [1968]). Plants can withstand the battering of storm waves and high winds, but mature tissues are often seriously damaged under such stress and reproduction follows from axillary buds.

The optimum range of temperature for good growth is 25° to 28° C. The plant can tolerate a wide pH range, being found at pH 5.2 in Malaysia, pH 7.4 in Lake Kariba in Rhodesia, and at pH 6.8 to 9.5 in other waters of Africa. The optimum is thought to be pH 6 to 7.5. The nitrogen supply is very influential in the growth of the species. Twenty-eight percent of the total dry weight of the plant is inorganic ash.

S. auriculata makes such a tight cover that it quickly depletes the oxygen supply beneath the mats. Measurements on Lake Kariba have shown a range of 4.4 to 6.9 mg per ml of dissolved oxygen in open water but only 0.64 and 0.66 mg per ml at the surface and 1 m below the mat, respectively. This is the minimum at which fish will survive. The growth and photosynthesis of phytoplankton and vascular plants are reduced severely as the light is cut off. Sculthorpe (1967) pointed out that the situation worsens as the mats become heavier. The surface of the water is sheltered from the wind and the reduced turbulence retards aeration, prevents mixing, and thus hastens stratification. The small amount of oxygen below the mat may be consumed as the organic matter decays.

On Lake Kariba the species grows in association with *Pistia stratiotes*, *Vossia cuspidata*, *Ceratophyllum demersum*, and species of *Lemna* and *Utricularia*. A second potential danger from *Salvinia auriculata* is that it serves as a mat for the colonization of other species in sudd-formation. On Lake Kariba, *Scirpus cubensis* is one of the most vigorous of the associated plants in the early formation of sudds. *Ludwigia adscendens* sends horizontal stems to a distance of 6 m on the mat. Other associated species are *Ludwigia leptocarpa*, *Polygonum senegalense*, *Panicum repens*, and *Echinochloa pyramidalis*. *Salvinia auriculata* is a base for sudd-formation in Bengal in India. Strangely, one of the colonizers of *S. auriculata* mats in Kerala state, India, and in Lake Kariba is *Ageratum conyzoides*, a terrestrial species discussed as weed number 19 in this volume. The presence of *A. conyzoides* indicates some degree of drainage

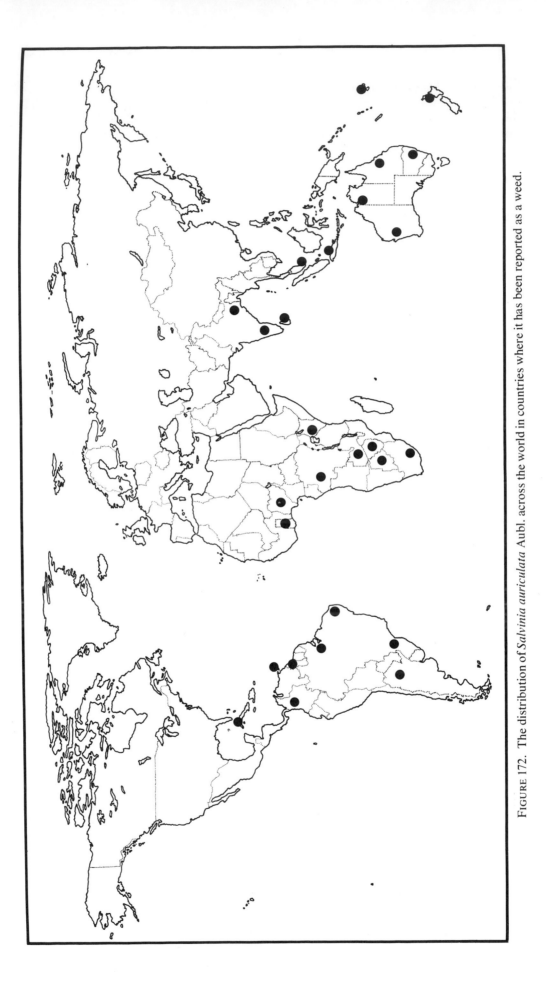

FIGURE 172. The distribution of *Salvinia auriculata* Aubl. across the world in countries where it has been reported as a weed.

in the upper layers of the thick mats (Boughey 1963, Cook and Gut 1971).

Several other species of *Salvinia* are encountered in various parts of the world but none have ever been so troublesome. *S. natans* is perhaps the second most widely distributed, being present in Europe in France and Spain; in Japan and Korea; in Taiwan, New Zealand, Thailand, Burma, Bangladesh, and Indonesia; in Bengal and Assam in India, and in Iraq. *S. cucullata* is found in Bangladesh, Burma, in Bengal and Assam in India, in Thailand, and in Vietnam. *S. hastata* is endemic to Madagascar and eastern Africa from Mozambique north to Kenya. *S. rotundifolia* is reported from Zaïre and Spain, and has become important enough in Florida (United States) that recommendations for its control have been issued. *S. herzogii* is in Argentina, southern Brazil, and New Zealand. *S. biloba* is known only in the vicinity of Rio de Janeiro in Brazil.

BIOLOGY

Detailed accounts of the anatomy, morphology, and embryogeny of *S. auriculata* are given in Bonnet (1955), Loyal and Grewal (1964, 1966, 1967), Herzog (1934), Mitchell (unpublished [1968], 1970), and Sculthorpe (1967).

The plant is without roots and it has been so changed by its reduction to a free-floating habit that it may seem to resemble a liverwort more than it does a fern. It possesses delicate, horizontal, floating stems which bear surface leaves which are light green and submerged leaves which are brownish. The floating leaves are notched at the apex and are broadly elliptic in shape. A long, finely dissected, submerged leaf hangs down from each node. The segments of this leaf serve as roots by absorbing water and nutrients.

There is no lignification of the tissues and as a result the plant must remain turgid for mechanical support of its organs. The plant floats because air is trapped by water-repellent hairs which grow in close, parallel rows on the surface of the floating leaves. The hairs are multicellular and have a basal stalk which divides into three or four spreading arms to form a cage or basketlike structure as the arms recurve and are reunited at the tip. Air is trapped in the cage and the basal area of the rows of hairs. Water droplets are speedily repelled because they cannot penetrate through the hairs and the trapped air bubbles (Sculthorpe 1967).

There are three buds at each node. One produces a submerged, dissected leaf; another, a pair of surface leaves; and the third usually remains dormant. The primary stems rarely branch when plants are tightly packed, but secondary laterals appear when plants are spread out in quiet water.

The plant progresses through three vegetative growth phases which may be confusing to a person trying to identify the plant. These stages are controlled by age, degree of crowding, water turbulence, and other environmental factors. In the first or primary juvenile phase the leaves are flat on the water and small (about 10 mm in diameter). Mitchell (1968) terms this phase the primary invading form, for it characterizes growth in the open and near shores where plants are uncrowded. During the second phase the floating leaves grow to be about 25 mm long and wide and fold upward, giving the structure a keeled shape. In the tertiary phase the leaves are about 38 mm wide and 25 mm long. The terminal bud now forms leaves which are compact, almost vertical, and acutely folded. This phase develops when competition has become severe at the height of the growing season. When growing conditions are optimum the plant can progress through the three phases in as little as 2 or 3 weeks (Sculthorpe 1967, Hattingh 1961).

The significance of spore formation in the reproduction and spread of *S. auriculata* is not yet clear. The submerged leaf has a short petiole which projects vertically downward from the main stem. The distal end is dissected into cylindrical, hairy segments which may be up to 18 mm in length. These serve as "roots" for the plant. The central one to eight segments are fertile and may bear an average of eight sporocarps arranged alternately. Sporocarps are nutlike bodies, equal in size, about 2 to 3 mm long, globular in shape, and covered with minute hairs. Both microsporocarps and megasporocarps are formed, the latter usually being borne singly on a long stalk at the end of a segment. An average of 30 megasporangia and 500 microsporangia are contained in the respective sporocarps. The numbers of the structures may vary greatly, and there is often much underdevelopment and abortion. Sporocarp formation is abundant when the mats are crowded and have been stationary for several weeks (R. Williams 1956).

The development of fertile spores and the subsequent phases of sexual reproduction have not been observed in Ceylon and Africa (Loyal and Grewal 1966). Mitchell (unpublished [1968]) found very few spores formed in the mats at Lake Kariba, and those which he did find were misshapen and almost certainly infertile. In Malaysia it is believed that all reproduction is by vegetative means. Sporocarps are produced but all are microsporocarps. Sutomo (1971) in Indonesia has recently reported spore formation in experiments in East Java, but he did not persist in his observations to determine whether the entire process of sexual reproduction was carried through. Earlier, in India, Biswas and Calder (1936) described the morphological features of the microspores and macrospores and their development in sexual reproduction but did not indicate whether they had observed this in nature.

In vitro germination of spores has not yet been reported in the literature; therefore, we do not understand the species well enough at present to determine whether it reproduces both vegetatively and by sexual means in natural sites.

ROLE IN HUMAN AFFAIRS

The species was originally described from Guyana by Aublet. It has never been a cause of serious concern in waterways of South America, but its uncontrolled growth in two areas of Asia and Africa has had tragic consequences. There is recent evidence that its encroachment on croplands and waterways may be a threat to still other areas of the world. The species cannot survive in salt water, rarely produces by spores, and yet vegetative parts of the plant, fragile though they may be, have been transported all around the world. On a given water body they may be moved by winds, currents, and ships. They may be moved overland for short distances by adhering with mud to the fur and feathers of animals, to the clothing of man, and to the sides and bottoms of boats and the wheels of vehicles. Sometimes the fresh material is used as packing for fish and other products of freshwater lakes and streams. If the plant does produce spores, it is possible that such spores are transported for long distances by birds during their migratory flights. But man, in collecting plants to study or to place in aquariums and fishponds, may be the cause of distribution between continents.

Those who live inland and who seldom visit the seas, lakes, and streams cannot imagine the terror and hopelessness felt by persons whose very food supply and livelihood are tied to the water when tens or hundreds of square kilometers of water become covered with a single weed. In the tropics these weeds interfere with man's food supply, his mobility, his work, and his health. Villages may have to be moved when shores become permanently blocked with weed growth. On Lake Kariba, 15-meter boats

with twin diesel marine engines dare not enter some of the rafts of *Salvinia* (Mitchell, unpublished [1968]). The grids and sluices of electrical generating plants and irrigation systems become plugged. Fishing areas are blocked off or are destroyed by darkness and lack of oxygen; or the weeds may simply prohibit the setting and lifting of nets. Where water flow is retarded, the canals silt up. The weed harbors snails and insects which are vectors of human and animal diseases.

The weed grew out of control in Ceylon during and after World War II, and it multiplied very rapidly in the months after the closing of the new dam at Lake Kariba in Africa in the 1960s. There still was an attitude in some quarters that these were special situations, that *Salvinia* had been widely distributed long before these events took place, and that it had given little trouble. We now know differently. It has recently become a threat in Kerala (India) (Cook and Gut 1971). Steps must be taken to keep it under control in some of the waterways of Malaysia. In some areas around Malang in East Java it is now the number-one weed in rice. To the south of Malang in irrigated sugarcane the species may become the number-two weed in waterways, *Eichhornia crassipes* being number-one (Sutomo 1971). It has broken out in Kenya and South Africa and is contained only by vigilance. It is disturbing that it has now been identified from the Congo River where another free-floating hydrophyte, *Eichhornia crassipes*, has raised havoc for a quarter century. It is reported to be in Ghana but has not yet become a problem on the new Volta Lake. The weed is in Nigeria where the Niger River is filling a tremendous reservoir behind a dam at Kainji.

S. auriculata has infested thousands of hectares of the richest paddy fields along the western shore of Ceylon in the last 2 decades. It is believed that the first plants were brought from India in the 1930s to be studied at the university at Colombo. It escaped confinement and small patches were seen in the suburbs in 1942; by 1943 it was in commercial and experimental rice fields (R. Williams 1956); and by 1952 it had spread widely in western Ceylon and had begun to block canals and streams. Because much of the area is linked by canals, ditches, and streams, the weed was soon in the northwestern and southern provinces; it finally appeared in the central highlands and on the east coast. It is believed that birds, water buffalo, and man hastened the spread of the weed. Twelve years after it had appeared a survey showed that it had infested 800 hectares of waterways and 9,000 hectares of rice paddies. The government declared a "*Salvinia* week" in 1952 to try to clean up several thousand hectares, but those fragments which then were missed were able, in only a few months, to reinfest the areas completely. Some fields were finally abandoned.

In southern Africa, *S. auriculata* was recorded in 1949 at Katombora, 55 km upstream from Victoria Falls on the Zambezi River. A dam was completed below the falls in 1958; and when Lake Kariba was filled in 1963 the lake covered an area 190 km long and 10 to 30 km wide. Sculthorpe (1967) has remarked that "it provided a magnificent sink of still water" for the growth of *Salvinia*. The first fragments were seen on the rising water in 1959. Then the growth "exploded" on an unprecedented scale and in 13 months the fern covered about 10 percent of the lake. The infestation has now receded until 8 to 10 percent of the lake is covered, depending on the season (Mitchell, unpublished [1968]). The inundation of the floor of the valley brought into solution many inorganic and organic nutrients which had had their origin in decaying vegetation, ashes from the burning of debris on cleared land at the shoreline, and from the soil itself (Harding 1966). In addition to the rapid growth of the fern there were luxuriant plankton blooms and large increases in fish production. Today the *Salvinia* mats may reach a thickness of 15 cm as the wind, which may blow from the same direction for days, drives the mats into bays and up onto the shore. Wave action may form ridges to 1 meter in height. Some bays and estuaries are permanently covered, and no fish can exist below them.

The species provides food for some fishes and, if only sparsely distributed, also provides shade and shelter. The *Salvinia* species are eaten by ducks, geese, swans, pigs, and sometimes by cattle, deer, and upland game birds.

The plant appears to suffer little from diseases. Three species of insects seem to be important and are being studied in research now underway on biological control: *Paulinia acuminata* de Geer, *Cyrtobagous singularis* Hulst., and *Samea multiphicalis* Guenee (Bennett 1966).

Recently Hira (1969) reported that *S. auriculata* on Lake Kariba may harbor and transport two snails which are intermediate hosts of *Schistosoma haematobium* and *S. mansoni*, the organisms responsible for sporadic outbreaks of urinary and intestinal schistosomiasis in a resort community on the Zambian shore. The snails attach themselves to the weed mats and are moved about the lake by wind and current. The species is an important plant host of the *Mansonia* mosquitoes which serve as one of the principal vectors of rural filariasis (*Wuchereria malayi*) in Ceylon and elsewhere (Chow, Thevasagayan and Wambeek 1955).

COMMON NAMES
OF SALVINIA AURICULATA

AUSTRALIA
salvinia
EASTERN AFRICA
salvinia
INDIA
African payal

SOUTH AFRICA
kariba weed
water fern
watervaring
THAILAND
chawk hunu

❧ 65 ❧

Setaria verticillata (L.) Beauv.

POACEAE (also GRAMINEAE), GRASS FAMILY

Setaria verticillata is an annual grass that was first described by Linnaeus from a plant collected in Europe. It is easily recognized by the bristly, cylindrical inflorescence that readily detaches and sticks to clothing and to the hairs of animals. In some areas the inflorescence is tinged with purple. It has been reported to be a weed of 18 crops in 38 countries. In some areas the grass is considered to be good for hay and grazing.

DESCRIPTION

S. verticillata is a loosely tufted, *annual* grass (Figure 173), growing up to 1 m tall; *culms* geniculately spreading and ascending, occasionally rooting at the lower nodes, branched, smooth; *sheaths* compressed, hairy or smooth; *ligule* short, membranous, fringed with hairs; *blades* up to 30 cm long and 5 to 18 mm wide, expanded, thin, soft, with well-marked veins, usually loosely hairy, linear to linear-lanceolate with acute apex; *inflorescence* a dense spikelike panicle, bristly, cylindrical, or often broad below and tapering above, up to 15 cm long, 6 to 15 mm broad, often tinged with purple, made up of numerous clusters of spikelets arising in whorls from the main axis; *spikelets* about 2 mm long and 1 mm wide, green or purplish, irregularly clustered together on short stalks and consisting of a single fertile floret, together with a reduced sterile floret, subtended by one to four bristles arising from the base; *bristles* rough, three to nine, the barbs bent or recurved, pale green to purple; *grains* small, 2 to 3 mm long, 1 mm wide, compressed at the back, tightly enclosed by the hardened lemma and palea.

This grass can be recognized by the bristly cylindrical inflorescence which readily becomes detached and sticks to clothing, and by the recurved barbs on the bristles, for all other foxtails have barbs directed upward. It is difficult to distinguish *S. verticillata* from *S. lutescens* and *S. pallidifusca*.

BIOLOGY AND DISTRIBUTION

S. verticillata often grows as a tall annual grass, but it may also grow spreading and prostrate at the base with rooting at the lower nodes. Surely one of its outstanding characteristics is its spread by seed as a result of the recurved bristles in the inflorescence. The seedheads sometimes become twisted and tangled together to make a clump or ball which may be carried about.

The plant flowers from July to November in Iraq. In Hawaii there is a vigorous growth of new seedlings after rains. In one study, about 25-percent germination was obtained at 25°, 30°, and 35° C. Temperatures which were higher and lower than these, as well as several alternating temperatures of different ranges, gave less germination (Andersen 1968).

From Figure 174 it may be seen that the weed is present on all the continents, and it is one of the few that can thrive from Canada in the north to New Zealand in the south. In general the weed grows in cultivated and disturbed places. Because the seedheads stick to clothing, the plant soon finds its way into gardens. It thrives in waste places as well. In many places in the world it seems to prefer damp,

Figure 173. *Setaria verticillata* (L.) Beauv. *1*, plant habit; *2*, spikelets with recurved barbs on the bristles.

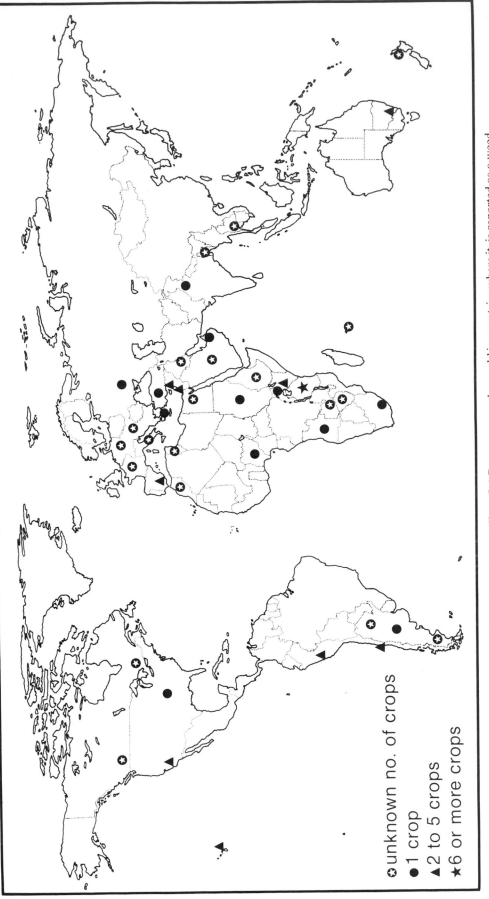

FIGURE 174. The distribution of *Setaria verticillata* (L.) Beauv. across the world in countries where it is reported as a weed.

shady sites. It is notably a troublesome and persistent weed in a crescent which extends from South Africa up the eastern coast of Africa into Europe. In eastern Africa it grows at all altitudes.

AGRICULTURAL IMPORTANCE

S. verticillata is one of the three most serious weeds in corn in Israel and Spain; sorghum in Israel; cotton in Turkey; and sugarcane in Peru. It is a principal weed of cotton in Kenya and Spain; sugar beets, orchards, and vegetables in Spain; sugarcane in South Africa and Tanzania; and corn, pastures, and wheat in Tanzania. It is a common weed in cotton in Peru, Spain, and Sudan; coffee in Kenya; pineapples in Hawaii; and coffee and beans in Tanzania. It is also a weed of cotton in Tanzania; bananas, papayas, and vegetables in Hawaii; sugarcane in Angola and South Africa (Natal); corn and onions in Lebanon; citrus and vineyards in Australia; vegetables in Greece; cotton in Uganda; and alfalfa and pastures in Chile.

In eastern Africa the plant is often found in fallow lands that are reverting to grassland. These areas eventually are covered with perennial grasses, but in the meantime *S. verticillata* has served a purpose of providing good grazing and good hay. Young plants are the best for forage. The species often becomes very conspicuous in areas newly cleared of brush.

When cattle are used to clean crop fields after the harvest, the seedheads of the weed may catch on the animals' coats and be distributed to new areas as the animals move. In Zambia the species is part of a special weed flora that thrives well under irrigation. In Tanzania, repeated treatments with phenoxy herbicides in wheat controlled many broad-leaved weeds, but these have been replaced by annual grasses such as *S. verticillata*.

S. verticillata is an alternate host of *Anaphothrips swezeyi* Moulton (Sakimura 1937); of the nematodes *Meloidogyne* sp. and *Pratylenchus pratensis* (de Man) Filip. (Raabe, unpublished; see footnote, *Cyperus rotundus*, "agricultural importance"); and of the viruses which cause wheat spot mosaic, barley stripe mosaic, and wheat streak mosaic (Namba and Mitchell, unpublished; see footnote, *Cynodon dactylon*, "agricultural importance").

COMMON NAMES
OF SETARIA VERTICILLATA

ARGENTINA
  cola de zorro
  hierba pegajosa
AUSTRALIA
  whorled pigeon grass
CANADA
  bristly foxtail
CHILE
  pega-pega
EASTERN AFRICA
  bristly foxtail
  love grass
EGYPT
  quam el-far
FRANCE
  setaire verticillée
GERMANY
  Borstenhirse
HAWAII
  foxtail
INDONESIA
  kamala (Indonesian)
  oehoe (Sundanese)
IRAQ
  rough bristle grass
KENYA
  love grass
LEBANON
  bristly foxtail
  dukhain

khishin
  rough bristle grass
MOROCCO
  setaire verticillée
NEW ZEALAND
  rough bristle grass
PERU
  rabo de zorro
RHODESIA
  bur grass
SOUTH AFRICA
  bur bristle grass
SPAIN
  almorejo
  carreig
  panissola
  pata de gallina
SUDAN
  lossaig
TANZANIA
  foxtail
  love grass
THAILAND
  yah hang ching-chok
TUNISIA
  setaire verticillée
TURKEY
  kirpi dari
UNITED STATES
  bristly foxtail

❧ 66 ❧

Setaria viridis (L.) Beauv.

POACEAE (also GRAMINEAE), GRASS FAMILY

Setaria viridis is an annual grass that produces abundant seed. It is a native of Europe. Primarily a weed of the temperate zone, it is sometimes found in the cooler subtropics, usually at higher elevations. It is a weed of 29 crops in 35 countries, mostly in cereal and pulse crops. Natural stands are sometimes used for summer pasture.

DESCRIPTION

S. viridis is a tufted *annual* grass (Figure 175); *culms* many, hollow, smooth, erect or ascending, sometimes spreading, simple or branched, 10 to 70 (rarely 100) cm high; fibrous *roots*; *leaf sheath* smooth, with short-hairy margins; *ligule* a fringe of short hairs; *blades* flat, linear-lanceolate with obtuse or rounded base and acute apex, with rough or smooth margins, with or without sparse long hairs above, smooth beneath, 4 to 25 cm long, 5 to 15 mm wide; *inflorescence* a soft, bristly, green or purplish spikelike *panicle*, erect or somewhat nodding, cylindric or narrowed toward the apex, dense, not lobed, 2 to 11 cm long, 1 to 2.3 cm wide; *spikelets* solitary on the pubescent central axis or in lower part in short racemes and often partly poorly developed, falling as a whole (deciduous) at maturity from their small cuplike base leaving the bristles on the axis, 2 to 3 mm long, one-flowered; *bristles* one to three below each spikelet, mostly three to four times the length of the spikelets, obliquely erect, often slightly sinuous, barbs which point upward, at first green, yellowish, or greenish white, sometimes purplish, afterwards dirty yellow; *first glume* one-fourth to one-half as long as the spikelet, oval, obtuse, or shortly acuminate, one- to three-nerved; *second glume* nearly as long as the spikelet, oval elliptic, obtuse, or shortly acuminate, five- to nine-nerved; *sterile lemma* five-nerved, elliptic, slightly rough; *fertile lemma* enclosing a small narrow palea; *anthers* blackish brown; *fruit* (caryopsis) consists of two hard, finely warty scales (lemma and palea) which enclose the grain; *grain* ovate, flat on one side, dull, pale yellow, 2.25 mm long; fruits green, whitish, pale yellow or purplish depending upon maturity, sometimes mottled with small dark spots.

The hollow culm, barbed bristles with barbs pointing upward, and the short-hairy margins of the leaf sheath are distinguishing characteristics of this species. *S. viridis* can be distinguished from *S. glauca* by its rougher leaf surfaces and margins and by the fact that the second glume is as long as the spikelet. It can be distinguished from *S. verticillata* by the upward-pointing barbs of its bristles; the barbs in *S. verticillata* point downward.

BIOLOGY AND DISTRIBUTION

S. viridis is primarily a weed of the temperate zone (Figure 176). It is found within a range of about lat 45° S to 55° N, but is rarely found within the tropical belt from lat 23° S to 23° N except at high elevations. A native of Europe, it has spread to temperate areas of South America, North America, Australia, and Asia.

S. viridis spreads and reproduces by seeds which are produced in great abundance. At one time most harvested crop seeds in the United States contained

FIGURE 175. *Setaria viridis* (L.) Beauv. *1*, habit; *2*, ligule; *3*, barb; *4*, caryopsis, two views; *5*, seed; *6*, seedling.

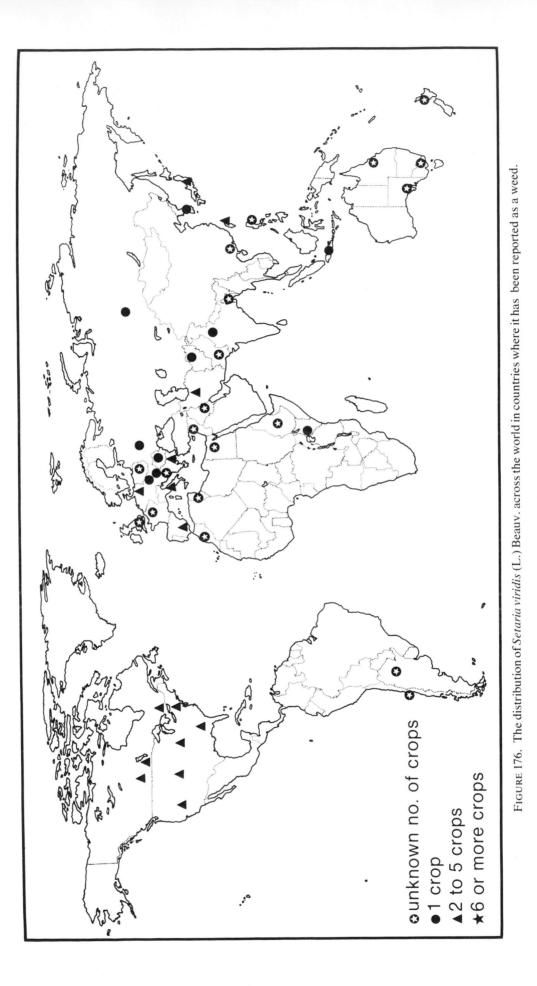

FIGURE 176. The distribution of *Setaria viridis* (L.) Beauv. across the world in countries where it has been reported as a weed.

S. viridis seeds as a contaminant, and there can be no doubt that this problem still exists. Undoubtedly the weed has spread throughout the world both in dirty crop seeds and in the ballast of ships. Ridley (1930) reported that the weed has been reintroduced into England in this way. The barbed bristles on the spikelet may adhere to clothing, wool, fur, or other surfaces. However, the barbed bristles may not be very important in dissemination, for the spikelets usually fall from the panicle when they are mature, leaving the bristles attached to the axis of the head. The species may also be disseminated by floating seeds which can remain buoyant on water for as long as 10 days (Ridley 1930). *S. viridis* may also be dispersed by birds which may eat the seeds and carry them for some distance before excreting or regurgitating them (Ridley 1930).

The weed grows in cultivated fields, gardens, waste places, disturbed areas, and along roads. It is frequently found in fertile soils. Plants may vary widely depending upon growing conditions, and may range in height when mature from less than 10 cm to almost 100 cm. Several varieties with different morphological characteristics have been described (Fairbrothers 1959, Schreiber and Oliver 1971, Hubbard 1915).

The weed grows and produces seeds very quickly, sometimes in 40 days or less. In Canada it is known as a summer annual, germinating readily between mid-May and mid-June under field conditions. With the temperature being maintained at 22.5° C the plant in that country was found to flower the following number of days from seeding: 26 days in an 8-hour photoperiod, 28 days in a 12-hour photoperiod, and more than 60 days at a photoperiod of 16 hours or longer. At 30° C first flowering occurred at 28, 28, 45, and 33 days for 8-, 12-, 16-, and 20-hour photoperiods, respectively (Schreiber and Oliver 1971). During these same studies over a 3-year period, the plants required an average of 37 days from seeding to attain 25-percent flowering, indicating the rapid completion of the life cycle of this weed. The flowering period of the weed in the field ranges from June to September or October.

Although temperature and other conditions for seed germination vary, some generalizations can be made. Freshly harvested seeds will germinate with or without seed coats within a temperature range of 25° to 35° C, with the optimum temperature being within the range of 20° to 30° C (Andersen 1968). Seeds germinate and emerge best from shallow depths of 1.5 to 2.5 cm, with the percentage of seeds which emerge decreasing with increasing depth, and with no emergence from below 12 cm (Dawson and Bruns 1962). Seeds have survived 33 months of dry storage at room temperatures (Andersen 1968).

AGRICULTURAL IMPORTANCE

From Figure 177 it may be seen that *S. viridis* causes major problems in temperate countries. The species grows and competes mainly in cereals, vegetables, or pulse crops. Its importance as a weed relates to its heavy seed production and dense competitive stands which occur largely in spring-sown crops. Because the weed germinates in late spring, it may escape early cultivation and may become prevalent in earlier sown crops. Some form of midseason cultivation or herbicide application may be required to control emerged seedlings in late spring or early summer. Late-season cultivations would probably be ineffective for this weed, because it can complete its life cycle within a very short time.

In addition to competing directly with crop plants, *S. viridis* also causes abnormal or disrupted growth in cabbage and tomato roots (Retig, Holm, and Struckmeyer 1972). In these studies one crop seedling was grown in association with four seedlings of *S. viridis*, resulting in abnormal changes in the anatomy of the crop roots. It was postulated that these effects were caused by diffusible compounds from *S. viridis* seedlings.

S. viridis is a serious weed of corn in Spain and of rice and sunflowers in Iran. It is a principal weed of flax, soybeans, sorghum, and pastures in the United States; barley in Canada; cotton and vegetables in Spain; carrots in the Soviet Union; beans, cereals, rape, sunflowers, wheat, and tomatoes in Canada; sugar beets in Canada, Iran, the Soviet Union, Spain, and the United States; and corn in Canada and the United States. It is a common weed of lowland rice in India and Taiwan; corn in Romania; cereals and cotton in the United States; and soybeans in the Soviet Union. It is also a weed in barley, peas, potatoes, millet, vineyards, orchards, upland rice, sweet potatoes, and irrigated crops.

The grass is sometimes used for pasture.

S. viridis is an alternate host of *Meloidogyne* sp. (Raabe, unpublished; see footnote, *Cyperus rotundus,* "agricultural importance"), *Piricularia oryzae* (Nishikado 1926); and of the virus which produces black-streaked dwarf disease of rice (Shinkai 1957).

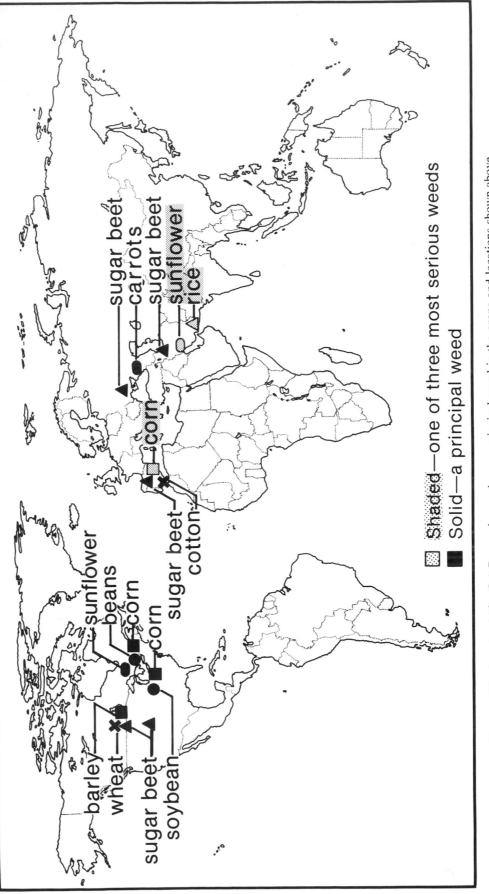

FIGURE 177. *Setaria viridis* (L.) Beauv. is a serious or principal weed in the crops and locations shown above.

Shaded—one of three most serious weeds

Solid—a principal weed

sunflower
beans
corn
corn
sugar beet
soybean

barley
wheat
sugar beet
soybean

sugar beet
carrots
sugar beet
sunflower
rice

corn

sugar beet
cotton

COMMON NAMES
OF SETARIA VIRIDIS

ARGENTINA
  gramilla

AUSTRALIA
  green pigeon grass
  green setaria

BANGLADESH
  shabuz shiallaja

CANADA
  bottle grass
  green bristle grass
  green foxtail
  wild millet

DENMARK
  grøn skaermaks

EGYPT
  deil-el-far

ENGLAND
  green panicum

FRANCE
  sétaire verte
  mierge

GERMANY
  grüne Borstenhirse
  grünes Fennichgras

IRAN
  arzan
  blue foxtail

IRAQ
  bottle grass
  dukhain el-forsheh
  green bristle grass
  green foxtail grass

ITALY
  panicastrella

JAPAN
  enokorogusa
  green panicum

LEBANON
  bottle grass
  green bristle grass

NETHERLANDS
  groene naaldaar

NEW ZEALAND
  green bristle grass

PHILIPPINES
  buntot-pusa

SPAIN
  almorejo

SWEDEN
  grönhirs

TAIWAN
  gou-wei-tsau

UNITED STATES
  bottle grass
  green bottle grass
  green bristle grass
  green foxtail
  pigeon grass
  wild millet

YUGOSLAVIA
  muraika

Sida acuta Burm. f.

MALVACEAE, MALLOW FAMILY

Sida acuta is a small perennial shrub, near-shrub, or herb with woody stems and a deep taproot. It reproduces by seeds. A native of Central America, it has spread throughout the tropics and subtropics. It is a weed of 20 crops in 30 countries, and is troublesome in pastures, plantation crops, cereals, root crops, and vegetables. Its stem fibers are used for coarse cordage.

DESCRIPTION

S. acuta is a small, erect, much branched, *perennial shrub* or *herb* (Figure 178); ranging from 30 to 100 cm in height, with a strong *taproot*; *stem* and *branches* flattened at the extremities, fibrous, almost woody at times; *leaves* alternate, slender, lanceolate, acute, margins toothed, 1.2 to 9 cm or more long, 0.5 to 4 cm wide, lower surface smooth or with sparse, short, branched, starlike (stellate) hairs, with fairly prominent veins; *petiole* 3 to 6 mm long, hairy, with a pair of *stipules*, at least one lanceolate-linear, 1 to 2 mm broad, three- to six-nerved, often curved, finely hairy, the other stipule narrower, one-to four-nerved; *flowers* 1 to 2 cm in diameter, solitary or in densely crowded axillary heads; *pedicels* 3 to 8 mm, slender, jointed near the middle; *calyx* five-lobed; *sepals* pale green, triangular, acute, about 6 mm long; *petals* five, joined at base, 6 to 9 mm or more long, light yellow, yellow, or pale orange, with a shallow notch at the apex; *ovary* eight-celled, one-ovulate; *stamens* of many filaments arising from a tube; *style* divided into six branches; *fruit* a capsule, 3 to 4.5 mm in diameter, rough, consisting of five to eight (rarely more) carpels which break at maturity into equal, one-seeded, segments (mericarp), each of which has two glabrous or nearly glabrous *awns* or beaks 1 to 1.5 mm long; *seeds* small, roughly triangular, 1.5 mm long, with a deep depression on each of the sides, reddish brown or black.

S. acuta is often confused with *S. rhombifolia*, but can be distinguished from that species by its smaller and narrower lanceolate leaves (those of *S. rhombifolia* are rhomboid or diamond shaped), the shorter pedicels of the flower, and by the two awns of the carpels (*S. rhombifolia* has one awn or none).

BIOLOGY AND DISTRIBUTION

S. acuta is mainly a weed of the tropics and subtropics (Figure 179) and is rarely found outside the belt from lat 23° N to 23° S. It is a native of Central America, and has spread to tropical Africa, southern Asia, and Oceania. It has been reported as an equatorial species in Brazil. Although it grows widely throughout the tropics, it can be found at higher elevations. It has been reported up to 1,500 m in Indonesia, at medium and higher elevations in Kenya, and in the foothills of the Andes in Peru.

The weed is frequently found in pastures, wastelands, cultivated lands, roadsides, lawns, and in planted forests. Once the plant becomes established, it is very competitive, holding and denying sites to other plants. It does appear to do best in disturbed habitats.

The plant grows well in many soils, including some

426

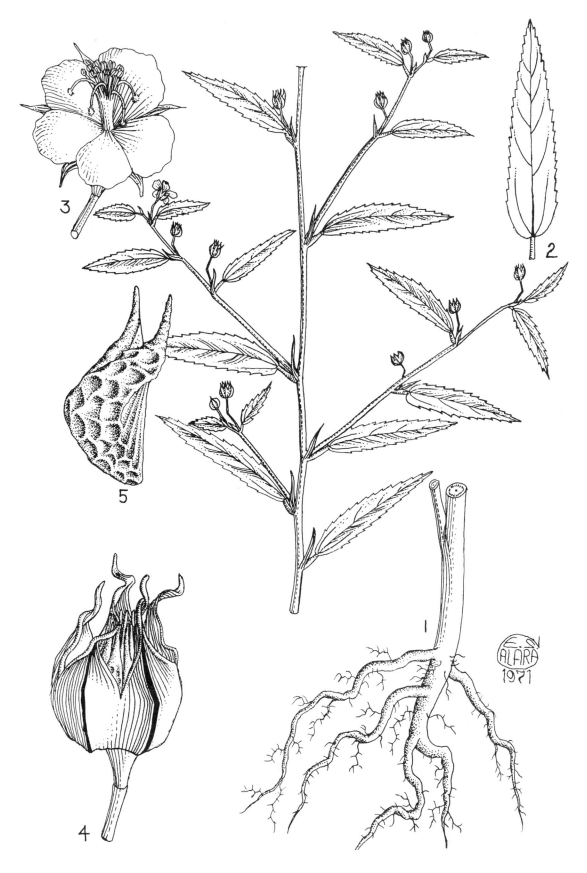

FIGURE 178. *Sida acuta* Burm. f. *1*, habit; *2*, leaf; *3*, flower; *4*, capsule; *5*, seed with two awns.

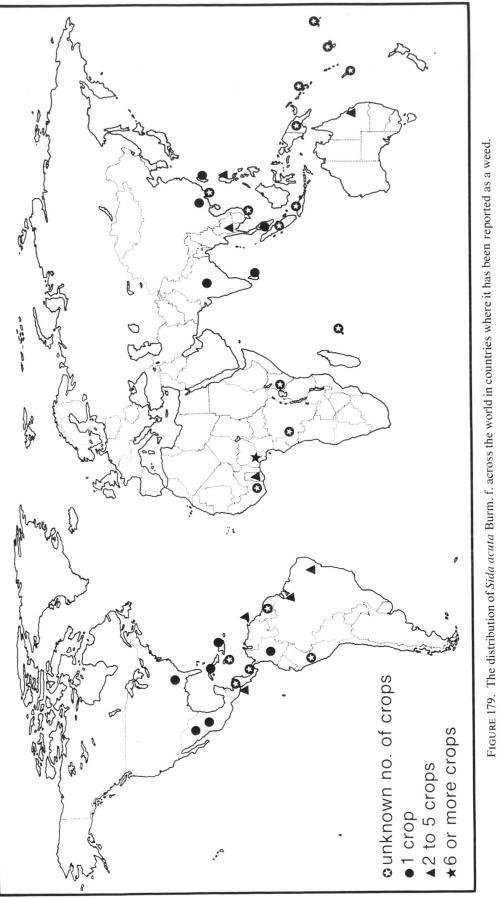

FIGURE 179. The distribution of *Sida acuta* Burm. f. across the world in countries where it has been reported as a weed.

heavy clays, and can tolerate dry as well as high rainfall conditions. In Fiji it is most aggressive in the dry zones. It has a deep taproot which can withstand drought, mowing, or shallow infrequent tillage. Tough woody stems also aid in its persistence.

S. acuta propagates by seeds which are produced in considerable quantity in five- to 12-celled capsules. At maturity the capsule splits fairly easily into triangular one-seeded segments, each of which has two sturdy awns at the tip. These awns cling directly to fur or stick to the mud which is carried on the feet or coats of animals and are, thereby, spread to other areas. In instances when the capsules do not break apart, whole capsules may be disseminated by these means.

Little is known of the biology of *S. acuta*. Juliano (1940) obtained a 54-percent germination of seeds collected a few weeks previously. It appears that the seeds germinate readily under natural conditions; for example, Mune and Parham (1956) reported that in 1943 the weed spread very rapidly over thousands of hectares in Fiji. This spread was undoubtedly due to rapid seed dispersal, germination, and establishment.

The flowers open in early morning and wilt in the afternoon. The species flowers throughout most of the year in Ceylon.

AGRICULTURAL IMPORTANCE

S. acuta is a weed of 15 crops, including plantation and annual field crops and pastures. Its ability to compete with such widely varying crops is probably due to its shrubby perennial nature and to its ability to grow from seeds and to compete as an herb with short-term vegetable, cereal, or field crops.

It is a principal weed of corn in Mexico, sorghum in Thailand, tomatoes in the Philippines, pastures in Australia and Fiji, and onions in Brazil. It is a common weed of tea in Taiwan; peanuts and cassava in Ghana; corn in Ghana and Thailand; coconuts in Trinidad; beans in Brazil; pastures under coconut trees in Ceylon; and pineapples in the Philippines. It is also reported as a weed in sugarcane in Australia, El Salvador, and Trinidad; coffee in Colombia; rubber in Malaysia; upland rice in the Philippines and Nigeria; cotton in El Salvador and Thailand; tea in Ceylon; and corn, cassava, cowpeas, pastures, and sweet potatoes in Nigeria.

Without cultivation or use of herbicides *S. acuta* can be very difficult to control because of its tough fibrous stems and deep taproot. Continued cultivation can kill the plant, especially if practiced long enough to kill most of the seeds remaining in the soil. Ivens (1967) reported that the deep taproot often made cultivation difficult.

In pastures, mowing provides only temporary control because the plants can grow back from the woody stems and perennial taproots. The stems are unpalatable to stock; the animals will eat other plants, thereby increasing the dominance of *S. acuta*. If a pasture is overgrazed, the weed will increase rapidly. Because animals spread the awned seeds and capsules, efforts should be made to prevent or reduce seed production or to restrict grazing during the fruiting periods.

If animals graze in infested areas it would be desirable to hold them in yards to allow ingested seeds to pass before placing them in pastures which are free of the weed. Mune and Parham (1956) suggest a holding period of 96 hours for horses in order that all the seeds may pass before putting the animals in weed-free pastures.

Burkill (1935) reported that the weed is used in native medicine in parts of Asia. The stem fibers can be used as a substitute for jute. Traces of hydrocyanic acid have been found in the leaves, bark, and roots (Burkill 1935).

COMMON NAMES OF SIDA ACUTA

AUSTRALIA
  spinyhead sida
BRAZIL
  cheeseweed
  vassonrinha curraleira
CAMBODIA
  kantreang bay sar
CEYLON
  gas belila
  kesara-belila
    (Sinhalese)
  malai-tanki
  palampasi
  visha-peti (Tamil)
EL SALVADOR
  escoba
  escobilla
  escobilla cabezuda
  escobilla negra
FIJI
  broom weed
  deni vuaka
  Paddy's lucerne
INDONESIA
  galoenggang (Sun-
    danese)
  sadagori (Sundanese)
  sidagoeri (Javanese)
JAMAICA
  broom weed

MALAYSIA
  bunga telur belangkas
  dukong anak
  ketumbar hutan
  lidah ular
  pokok kelulut putch
  sedeguri
  seleguri

PHILIPPINES
  basbasot (Ilokano)
  broom grass
  escuba
  surusighid (Bikol)
  walis-walisan
    (Tagalog)

TAIWAN
  syi-ye-jin-wu-shih-hwa

THAILAND
  mai-kward
  mai-kwat
  yung kwat
  yung pat

TRINIDAD
  ballier savanne
  broom weed

UNITED STATES
  southern sida

VIETNAM
  bái nhon

❦§68?❧

Solanum nigrum L.

SOLANACEAE, NIGHTSHADE FAMILY

Solanum nigrum is a low, spreading or erect annual that has become naturalized in both temperate and tropical regions. It is a native of Europe. It is found in cultivated lands, waste areas, and open woodlands and is considered to be a weed of 37 crops in 61 countries. Mature fruits are sometimes used for making jam. Green fruits may be poisonous to both man and animals.

DESCRIPTION

S. *nigrum* is a low and spreading, sometimes erect, *annual* to *biennial*, somewhat shrubby herb (Figure 180), sometimes woody at the base, 30 to 90 cm tall, having a strong primary root system. *Stems* round or angular, smooth or sparsely hairy; *leaves* alternate, simple, lanceolate-obovate, entire, toothed or irregularly lobed, up to 8 cm long and 5 cm broad, but usually half this size, smooth to hairy on slender petiole; *flowers* perfect, white with yellow centers, in clusters, the peduncle arising directly from the branches; *calyx* of five small lobes, persistent; *corolla* white, 3 to 8 mm in diameter; *stamens* five; fruit a many-seeded globular berry, 5 to 13 mm in diameter, black at maturity; *seeds* about 1.5 mm in diameter, flattened, finely pitted, yellow to dark brown.

It is distinguished by its inflorescence which arises directly from the branches and by its black berry with persistent calyx.

DISTRIBUTION AND BIOLOGY

S. *nigrum* is found from about lat 54° N to 45° S as a common weed in gardens, fields, waste areas, and open forests (Figure 181). It is mainly a weed of moist environments, and thrives in areas of low rainfall only where the land is under irrigation. It is best adapted to soils of high fertility, especially those high in nitrogen and phosphorus.

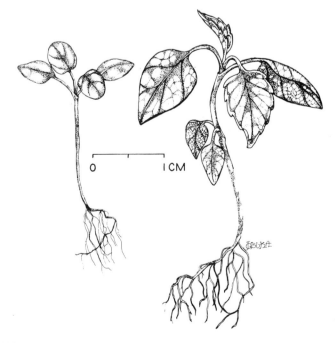

430

FIGURE 180. *Solanum nigrum* L.
Above: *1*, habit; *2*, flower, two views; *3*, stamen, two views; *4*, flower, vertical section; *5*, ovary, cross section; *6*, fruit cluster; *7*, seed, two views.
Opposite: Seedling.

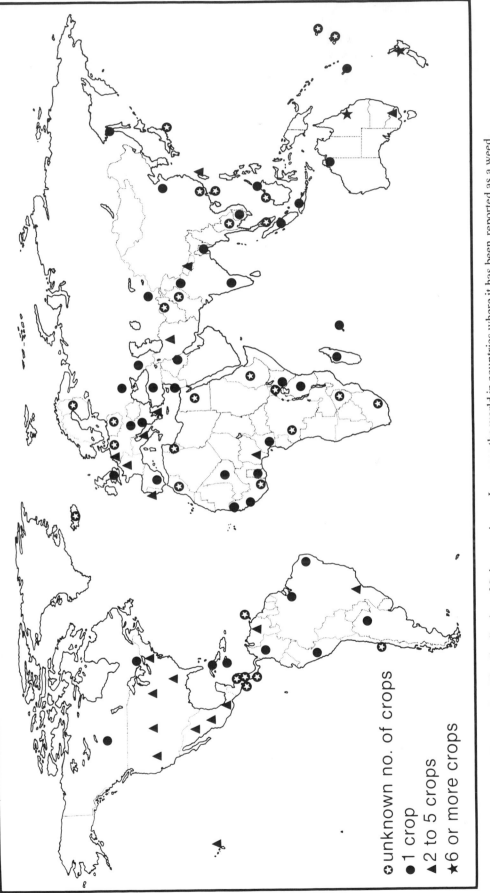

FIGURE 181. The distribution of *Solanum nigrum* L. across the world in countries where it has been reported as a weed.

⊛ unknown no. of crops
● 1 crop
▲ 2 to 5 crops
★ 6 or more crops

The plant reproduces from seeds, of which as many as 178,000 per plant may be produced. The seed seems to germinate best at alternating temperatures of 20° to 30° C. There appears to be little dormancy of the seed for germination was found to be 99 percent in spring after the seeds had been exposed to winter temperatures. Seeds stored at room temperatures for 2 years had 27-percent germination; however, seeds germinated 2 and 0 percent after 8 and 9 years storage, respectively (Andersen 1968).

AGRICULTURAL IMPORTANCE

S. nigrum is reported as a weed in 61 countries and in 37 crops (Figure 182). It is ranked as the most important weed of sugar beets in Iran and as a serious weed of corn, cotton, sorghum, and vegetables in Australia. It is a principal weed of sugarcane in Mauritius and Taiwan; vineyards in the Soviet Union; corn in Italy and Portugal; onions, peas, and potatoes in New Zealand; sunflowers in Iran; tomatoes in Canada and Portugal; sugar beets in Belgium and Italy; vegetables in Belgium and Canada; peanuts and sweet potatoes in Hawaii; sorghum in Italy; and cotton in Greece, Israel, and the Soviet Union.

It is a weed in bananas in Mexico and Taiwan; abaca in Sabah; sugarcane in Argentina, Australia, India, Mexico, Peru, and Tanzania; pineapples in Venezuela; coffee in Mexico and Venezuela; tea in India, Indonesia, and Taiwan; vineyards in Portugal; citrus in Mexico; cereals in New Zealand and Kenya; wheat in Nepal; barley in Nepal and New Zealand; corn in France, Hungary, Mexico, Spain, and Taiwan; soybeans in Canada; tobacco in New Zealand; sugar beets in France and the United States; vegetables in Greece and New Zealand; beans and potatoes in Brazil; onions in Australia; and orchards in Yugoslavia.

In the Great Lakes region of the northern United States, *S. nigrum* and *Abutilon theophrasti* have been the subject of an interesting ecological shift in weed species as a result of herbicide use. Prior to 1962 lima beans were weeded with a combination of dinitrophenol herbicides, cultivation, and hand-hoeing. The two weed species were seldom seen in bean fields. Trifluralin, CDAA (an acetamide, 2-chloro-N,N-diallylacetamide), and chloramben (a benzoic acid) came into use after 1962 and, as a consequence, farmers were no longer required to do much cultivating and weeding with hand tools. *Solanum nigrum* and *Abutilon theophrasti* began to

appear in 1965 to 1966 and are now the principal weeds of lima bean fields.

Solanum nigrum is an alternate host of *Cercospera atro-marginalis* Atk. (Raabe, unpublished; see footnote, *Cyperus rotundus,* "agricultural importance"); and of the following nematodes: *Meloidogyne* sp. (Raabe, unpublished; see footnote, *Cyperus rotundus,* "agricultural importance"), *Rotylenchus reniformis* Linford & Oliveira (Linford and Yap 1940), and sugar beet nematodes (Altman 1968); and of the following important virus diseases: cucumber mosaic, tobacco mosaic, and tomato spotted wilt; other virus diseases include aster yellows, atropa belladonna mosaic, beet curly top, chili (pepper) mosaic, cucumber green mottle mosaic, lucerne mosaic, petunia mosaic, potato A, potato leafroll, potato T, red currant ring spot, tobacco broad ring spot, tobacco itch, tobacco leaf curl, tobacco ringspot, tobacco streak, tobacco yellow dwarf, tomato bunchy top, vaccinium false blossom (Namba and Mitchell, unpublished; see footnote, *Cynodon dactylon,* "agricultural importance"), and potato X (Naperkovskaya 1968).

COMMON NAMES
OF SOLANUM NIGRUM

ARGENTINA
yerba mora
AUSTRALIA
blackberry nightshade
BANGLADESH
gurki
BRAZIL
erva moura
BURMA
haung-laung-nyo
CAMEROONS
black nightshade
CHILE
hierba mora
llague
COLOMBIA
yerba mora
COSTA RICA
yerba mora
DENMARK
sort natskygge
EASTERN AFRICA
black nightshade

EGYPT
enab el-deib
EL SALVADOR
yerba mora
ENGLAND
black nightshade
FIJI
black nightshade
FINLAND
mustakoiso
FRANCE
crève-chien
GERMANY
schwarzer
Nachtschatten
HAWAII
black nightshade
HONDURAS
hierba mora
INDIA
makhoi (Rajasthani)
nunununia (Oriya)
INDONESIA
anti (Indonesian)

leuda
leuntja
leuntja pait (Sundanese)
ranti (Javanese)

IRAN
taj rizi

IRAQ
innaib el-theeb

ITALY
ballerina
erba morella
solano nero

JAMAICA
black nightshade
branched calalu
guma

JAPAN
inuhōzuki

MALAYSIA
terong meranti

MAURITIUS
brède martin

MEXICO
hierba mora
trompillo

MOROCCO
morelle noire

NATAL
black nightshade

NETHERLANDS
zwarte nachtschade

NEW ZEALAND
black nightshade

NORWAY
sort søtvider

PAKISTAN
kanper makoo

PANAMA
pintamora

PERU
yerba mora

PHILIPPINES
kamakamatisan
kunti
malasili

RHODESIA
i xabaxaba (Ndebele)
musaka (Shona)
mutsungutsugu (Ndeo)

SOUTH AFRICA
black nightshade
galbessie
nagskaduuse
nastagal
nastergal

SPAIN
tomatito de moro
tomatitos

SWEDEN
nattskatta

TAIWAN
lung-kwei

THAILAND
toem tok
ya-tomtok

TRINIDAD
agouma

TUNISIA
morelle noire

TURKEY
kopek uzumii

UNITED STATES
black nightshade

VENEZUELA
hierba mora
yocoyoco

YUGOSLAVIA
pomocnica

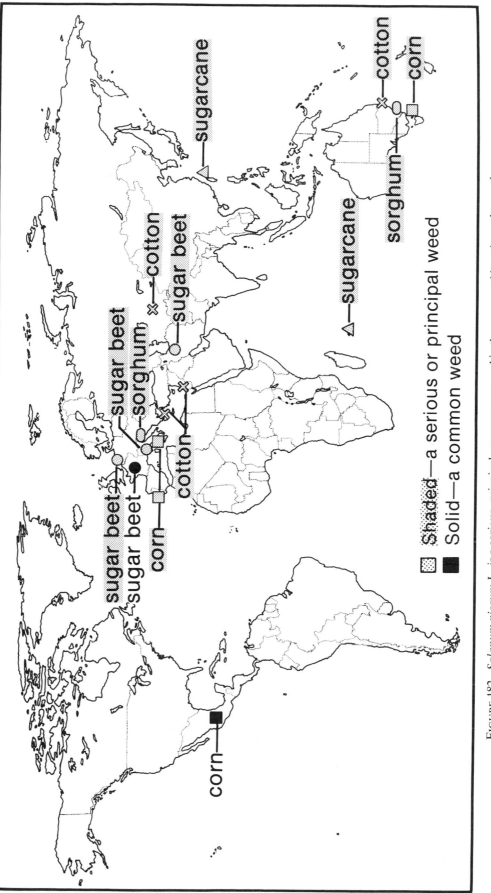

FIGURE 182. *Solanum nigrum* L. is a serious, principal, or common weed in the crops and locations shown above.

❧§69❧

Sonchus oleraceus L.

ASTERACEAE (also COMPOSITAE), ASTER FAMILY

Sonchus oleraceus is a soft, erect, annual or sometimes biennial herb that has become naturalized in 56 countries. A native of Europe and North Africa, it is a problem in field, row, and tree crops in temperate and tropical regions. As the specific name suggests ("akin to potherbs"), the plant is sometimes used as a vegetable; for example, in parts of Africa the leaves are eaten as spinach or are mixed with cereal meals. Livestock sometimes graze on the leaves.

DESCRIPTION

S. oleraceus is an erect, soft, simple or branched, taprooted, *annual* or sometimes *biennial* herb (Figure 183); *stem* smooth, pale bluish green or powdery, sometimes purplish, 30 to 120 cm tall, with milky juice; basal and lower *leaves* stalked, up to 20 cm long, 3 to 6 cm wide, basal leaves originally rosetted, upper leaves sessile and clasping the stem, the leaves divided into broad toothed and spiny segments with the terminal segments being the largest and triangular, and being green on both sides, sometimes tinged with purple; *flowering heads* numerous, pale yellow, 2 to 3 cm broad, with smooth involucre; *seeds* (achene) 2 to 3 mm long, broadest toward the top, tapering to a narrow base, ribbed lengthwise, and transversely wrinkled by the presence of minute warty lumps (tubercles); pappus of numerous silky hairs about 1.5 cm long. The plant may be distinguished by its toothed leaves, hollow stems, and milky sap. It has a fibrous taproot.

PROPAGATION

The plant reproduces by seeds that are produced in large quantities and are carried by wind or water. Seeds germinate readily when nearly mature and may remain viable for 8 years or more. They germinate in both light and dark. Temperatures suitable for germination vary in both range and duration; however, 78-percent germination has resulted with alternating temperatures of 10° to 25° C. Minimum and maximum temperatures for germination have been reported as being below 7° C and above 35° C, respectively (Andersen 1968).

DISTRIBUTION AND AGRICULTURAL IMPORTANCE

S. oleraceus is widely distributed around the

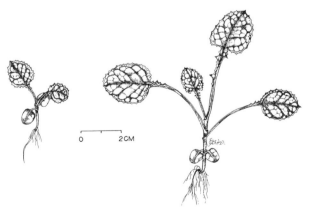

436

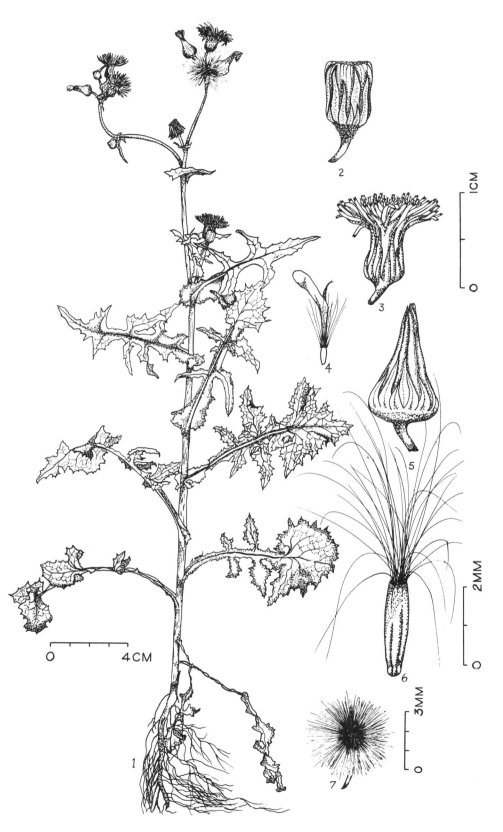

FIGURE 183. *Sonchus oleraceus* L.

Above: *1*, habit; *2*, flower head; *3*, flower head; *4*, disk flower; *5*, flower head; *6*, achene, with feathery pappus; *7*, head with mature achenes and rays of feathery pappuses.

Opposite: Seedling.

world (Figure 184). It has been reported as a principal weed of sugarcane in Brazil, rubber in Mexico, and wheat in Portugal. It has been listed as a weed in the following crops: alfalfa in Argentina; coffee in Angola and Kenya; tea in Brazil, Indonesia (Java), Japan, and Taiwan; macadamia nuts in Hawaii; citrus in Brazil; legumes in South Africa; corn in Lebanon; sugarcane in Australia, India, Mexico, Peru and South Africa; cotton in Egypt, Mozambique, Spain, and Sudan; vegetables in Hawaii and Spain; vineyards in Australia, Greece, Lebanon. Portugal, and Spain; cinchona in Indonesia (Java); potatoes in New Zealand; beans in Brazil and England; onions in Australia, Ireland, and Lebanon; sugar beets in Israel; pastures in Rhodesia and wheat in Portugal and Pakistan.

The weed is adapted to a number of environments and grows well at low and high elevations in the tropics. It is usually found in moist conditions in fields, orchards, roadsides, gardens, or waste areas. It often occurs in irrigated lands.

In Europe the plants are fed to animals and are cooked and used for greens (Neal 1965). Young shoots may be used for salads in Java (Burkill 1935).

S. oleraceus may harbor virus diseases such as lettuce big-vein, aster yellows, beet curly top, datura rugosa leaf curl, tobacco streak, tobacco yellow dwarf, alfalfa mosaic, cucumber mosaic, and lettuce necrotic yellow viruses (Namba and Mitchell, unpublished; see footnote, *Cynodon dactylon,* "agricultural importance"). It has been reported as a host for several nematodes.

COMMON NAMES
OF SONCHUS OLERACEUS

ALASKA
common sow thistle
ARGENTINA
cerraja
nil
AUSTRALIA
common sow thistle
BRAZIL
annual sow thistle

seralha
CANADA
common sow thistle
CHILE
cerrajilla
nilhue
COLOMBIA
cerraja

DENMARK
almindeleg
svinemaelk
EASTERN AFRICA
sow thistle
EGYPT
galawein
godeid
ENGLAND
annual milk thistle
sow thistle
FRANCE
laiteron commun
polais de lièvre
GERMANY
kohlartige Gänsedistel
Kohl-Gänsedistel
Kohl-Saudistel
Gemüse-Gänsedistel
HAWAII
pua-lele
sow thistle
INDIA
corn-sow-thistle
dodhi
INDONESIA
delgijoe ketubar (Javanese)
djombang gerowong (Sundanese)
IRAQ
lubbane
om el-haleeb
ITALY
cicerbita
JAMAICA
sow thistle
JAPAN
common sow thistle
nogeshe
KENYA
sow thistle
LEBANON
libbayn

MAURITIUS
lastron piquant
MEXICO
morraja
MOROCCO
laiteron maraîcher
NATAL
sow thistle
NETHERLANDS
melkdistel
NEW ZEALAND
sow thistle
NORWAY
haredylle
PERU
cerraja
PHILIPPINES
gagatang (Igorot)
PORTUGAL
serralha macia
PUERTO RICO
achicoria silvestre
RHODESIA
sowthistle
SOUTH AFRICA
gewone sydissel
melkdistel
SPAIN
cerraja
llisco
SUDAN
moleita
SWEDEN
kålmolke
TAIWAN
ku-jya-tsai
TUNISIA
laiteron maraîcher
TURKEY
esek
URUGUAY
cerraja
UNITED STATES
annual sow thistle

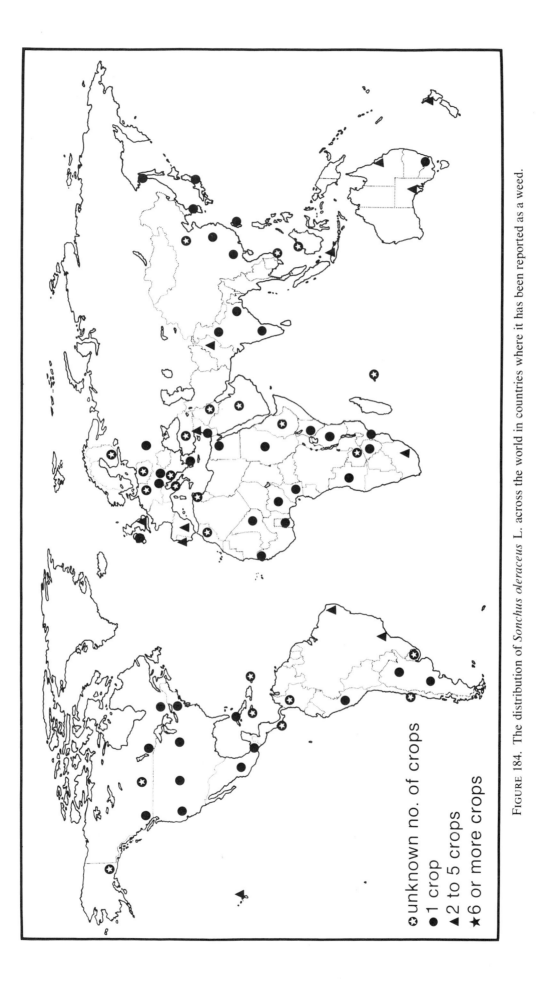

● 1 crop

▲ 2 to 5 crops

★ 6 or more crops

FIGURE 184. The distribution of *Sonchus oleraceus* L. across the world in countries where it has been reported as a weed.

❦70❧

Spergula arvensis L.

CARYOPHYLLACEAE, PINK FAMILY

Spergula arvensis is a rather small, herbaceous, annual plant native to Europe. It is conspicuous by its whorled, threadlike, bright green leaves. In a favorable location it may grow so rapidly that it blankets an entire crop. It is a weed of cereals in almost all areas of the world. *S. arvensis* prefers light, acid soils and sites which have some open spaces. It has been reported to be a weed in 25 crops in 33 countries.

This species was a source of food for man during the 3rd to 5th centuries A.D. in northern Europe. In times of food shortage it was used for making bread more recently in that area. Odum (1965) provided a report of remarkable seed longevity for this species. Excavations of occupation sites dating back to the Iron Age 2,000 years ago yielded seeds which were still able to germinate.

DESCRIPTION

S. arvensis is an erect, ascending or spreading, *annual* herb (Figure 185); *stems* 15 to 60 cm tall, profusely branched below, slender, conspicuously jointed and somewhat sticky, not hairy or only sparsely hairy; *taproot* and secondary roots finely branched; *leaves* appearing as whorls at each joint of the stem, threadlike, bright green, 1.5 to 4 cm long, rounded on the upper surface, and grooved lengthwise on the lower surface, with minute yellowish brown stipules; *flowers* in terminal clusters, small, perfect, more or less flat-topped, often spreading, with the central flower opening first (cyme); *sepals* five, nearly separate, green and white, glan-

dular; *petals* five, white; *stamens* 10 (occasionally five); *pistil* one, with five *styles*, five-valved; *fruit* round, a one-celled *capsule* splitting into five sections containing many seeds; *seeds* thick, lens-shaped, dull black, the surface roughened by minute rounded, protruding bodies, rarely smooth, about 1.5 mm in diameter, with a conspicuous, narrow, light colored wing on the margin.

The threadlike leaves which are arranged in whorls and the black seeds which are usually roughened by minute tubercles and which have a narrow, light colored wing on the margin are distinguishing characteristics of this species.

DISTRIBUTION AND HABITAT

From Figure 186 it may be seen that *S. arvensis* is a cosmopolitan weed that is most widely distributed in the temperate zones but that does enter the tropics and competes with crops at higher elevations. It is on all the continents and is found in most of the cereal-growing areas of the world. As a weed it is found in the most northerly agricultural areas in Finland and Alaska and is found as far south as Tasmania. As a species it is found north of the Arctic Circle and yet may be seen in high places on the equator.

S. arvensis is never a member of a closed community but likes to grow in open places and on arable land. Sometimes it is found in grasslands or clover fields while there are still open areas in the 1st year following a cultivated crop. Hamel and Dansereau (1949) studied its distribution on different types of sites in Canada. They found the weed in increasing

FIGURE 185. *Spergula arvensis* L. *1*, habit; *2*, leaf whorls, enlarged; *3*, flower; *4*, corolla, expanded; *5*, fruit; *6*, seed, two views; *7*, seedlings.

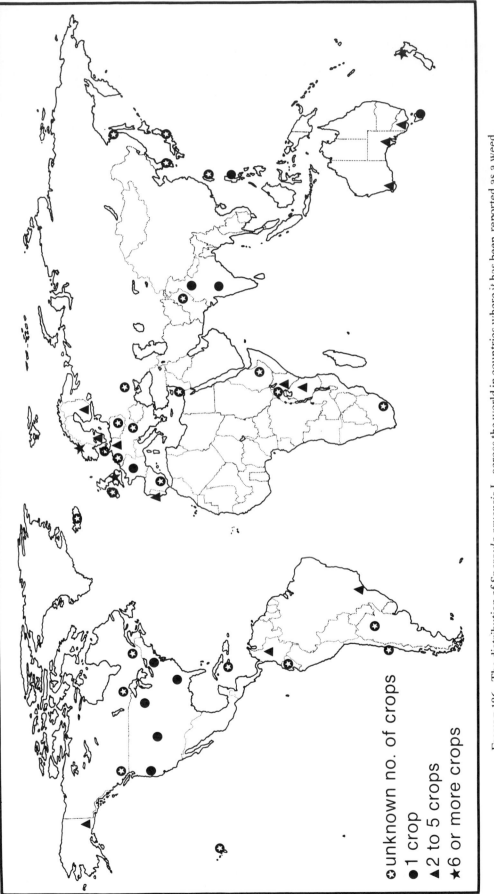

FIGURE 186. The distribution of *Spergula arvensis* L. across the world in countries where it has been reported as a weed.

unknown no. of crops
● 1 crop
▲ 2 to 5 crops
★ 6 or more crops

amounts as they moved from cultivated or hoed areas to cereals to young and old prairies. The greatest density was found in young prairies but the density decreased rapidly in old prairies. The species did not grow in pastures or abandoned lands.

Reports from several places in the world agree that the species prefers an acid soil. A *p*H of 4.6 to 5 seems most favorable; and many workers suggest that an application of lime results in a decline in numbers of this weed. The plant prefers light soils but can grow well on heavy soils. In England it grows very well on peat soils but in Sweden the species does not favor soils having high organic matter. Most roots are in the upper levels of the soil, but a few may penetrate to 30 cm.

The species does not appear to be a companion of any particular crop or set of crops and may be seen growing with equal vigor in wheat, root crops, potatoes, flax, or oats. On some sites the plants tend to clump but usually they are uniformly distributed in a field. Its habit of growth is very responsive to environmental changes or differences. In similar adjacent fields in England the weed was observed to be highly branched in cabbage and peas but unbranched in an oat field (New 1961). The relationship of crop to degree of branching is not a simple one, however, for the weed is branched whenever it grows in root crops but exhibits different degrees of branching when it grows in cereals.

As the weed approaches the tropical zone it must seek higher elevations to be sufficiently vigorous for competition with crops. It is of intermediate importance in cool climate crops at altitudes of 2,400 m. In Kenya the species again prefers acid soils and is most troublesome above 2,500 m. In the Philippines it is found above 2,000 m only and forms very dense mats in vegetable fields if control measures are not adequate. In Jamaica it grows above 1,800 m. The degree of slope of a field is not a factor in distribution or density of the weed. The plants can withstand light frosts and have been found, with flowers, as late as December in England.

BIOLOGY AND PROPAGATION

S. arvensis is a very old companion of man and is another of the seeds used as food by the Tollund Man and the Graubelle Man in the 3rd to 5th centuries A.D. in northwestern Europe (Helbaek 1950, 1958). Syme (1863-1886) reported that the seeds have been used by man more recently to make an inferior bread in times of food shortage in Scandinavia.

One of the principal studies on the biology of *S.*

arvensis is that of New (1961). In the temperate zone it may be expected to flower in about 8 weeks and produce mature seeds in about 10 weeks after germination. Flowering and seeding continue until the plant dies. Late capsules are usually small with few seeds. When growing in openings it may produce two generations in one season. In some areas the weed plants may be cut by machinery at the second or third node level when crops are harvested. New branches may be produced from this stubble in time to flower and to have another release of seeds before frost. The plant can develop to maturity while growing at 12 or 16 hours of daylight or in varying daylengths in the British Isles.

Detailed examinations of the number of stamens, the degree of hairiness of the stems, and the types of seeds have revealed that the species is quite variable. There is no evidence thus far of hybridization with other species or of apomixis. Occasional instances of germination of unshed seed (vivipary) have been found.

The remainder of the known biology of this species deals almost exclusively with the physiology of flowering, with seed formation, and with seed dormancy and germination, and it gives us an excellent opportunity to examine the confusion which exists in the literature about the dormancy of most weed seeds. Harper (1965) pointed out that we have come to expect a germination of 98 to 100 percent in many of our crop seeds, and we are accustomed to having them germinate at about the same time so that we may have "uniform" stands in our fields. In the reports on studies of weed seed dormancy it is often assumed that a species, or the progeny of a single individual, will have in common one, or possibly two, simple requirements or treatments that will cause germination. This is probably not the rule, however, for weedy species usually give very erratic germination. When a specific requirement is quoted for a species it often means that when the seeds were planted in a greenhouse (the normal procedure) a small percentage germinated; but that when seeds were held in cold temperature storage, before sowing for example, a higher percentage of the seeds germinated. If the latter treatment causes about one-half of the seeds to germinate, then this single treatment is erroneously believed to satisfy the characteristic dormancy-breaking requirement of the species and the rest of the seeds are not viable. The fact is that very often additional physical and chemical treatments will bring higher and higher germination percentages. Thus, a substantial body of evidence is accumulating for seed polymorphism in weeds and, as we examine each species carefully, we learn that there are several mechanisms which explain the behavior of dormant seeds of weeds.

Spergula arvensis provides us with just such an interesting example of seed polymorphism (New 1961). In England the seeds of this species are either smooth or papillate (having small projections on the coat) and individual plants bear only one kind of seed. The inheritance of the seed character is controlled by a single gene (two alleles). The two seeds differ in their requirements for germination. The papillate seeds germinate more readily than do the smooth seeds at 21° C and the reverse is true at 13° C. There clearly is a difference in the proportion of these two seed forms in the North and South of Great Britain. Variations in the expression of seed polymorphism may thus be a sensitive indicator of evolutionary change within an invading species.

In England the seeds germinate in the field in mid- or late April and most seedlings emerge from seeds very near the surface. In Canada the plant flowers from June through October and will drop mature seeds from July onward. At least 97 percent selfing occurs in nature. The stamen length is such that pollination takes place when the flowers are closed. There is a small amount of pollination by insects such as syrphids, wasps, and honeybees. Open flowers have an unpleasant smell.

The degree of branching has an effect on the amount of seed produced, but on the average the plants can produce very large amounts of seed in a very short time. A large plant may have 500 capsules and release 7,500 seeds. Capsules produced early in the season may bear 25 seeds but later capsules may contain only five.

Viable seeds have been found in the droppings of horses, sheep, pigs, cattle, and of sparrows, pigeons, and other birds. Short-distance dispersal is by water and by mud on animals, by the feet of humans, and by agricultural machines. The seeds are moved about in commerce with crop seeds. In Denmark seeds were buried 1 meter deep in silage made from sugar beet tops and waste, and, after 2 months, the seeds were no longer viable (anon. 1960*b*).

Many believe that the seeds have no dormancy when mature and may germinate immediately on falling to the ground, while others find that the seeds need special treatments to germinate. Birch (1957) reported from Kenya that fresh seeds germinated faster than seeds which had been buried in soil for 2 years. In laboratory tests germination was found to be best at 20° to 25° C. In all regions it has been reported that seeds germinate in crops throughout the season. Normally there is some viability after 5 years in the soil; but after 10 years only a few seeds can grow. There is a report by Odum (1965) from Denmark, however, which records an exceedingly long life for the seeds of the species. Archaeologically dated samples from several excavation areas revealed that seeds of *Chenopodium album* and *Spergula arvensis* from Iron Age occupation sites could still germinate. These sites were occupied from 100 B.C. to A.D. 400. Champness and Morris (1948) found a density of 4 million seeds per hectare in a grassland site in England. Roberts (1958) estimated that 23 million seeds per hectare were in the top 15 cm of arable land in England. After 2 years of the kind of intensive cultivation required for vegetable culture the number was reduced by one-half.

AGRICULTURAL IMPORTANCE

S. arvensis is one of the three most serious weeds in cereals in Kenya and of wheat and barley in Finland. It is a principal weed of cereals and pastures in Alaska; corn, peas, and wheat in New Zealand; flax in Germany; vegetables in New Zealand, Norway, and the Philippines; barley, oats, potatoes, and wheat in Norway and Sweden; corn and oats in Colombia; oats, potatoes, and sugar beets in Finland; peas in Ireland; rye in Sweden; and wheat in Tanzania. It is also a common weed of several cereal crops in Australia, Colombia, England, India, and Portugal; and of sugar beets in France. It is a weed of cereals, potatoes, rape, onions, tobacco, strawberries, and pastures in New Zealand; corn in Portugal; pastures and vegetables in England; peas in India and Tasmania; flax in Japan; potatoes in Brazil, Colombia, and India; pyrethrum in Kenya; and corn and vegetables in Brazil.

In an extensive survey of weeds in cereals in Finland, *S. arvensis* was found to have a higher average density per field than any other species. It is one of the dominant species in northern and central Sweden where it is said to be favored by the high amounts of pasture in the crop rotations. In New Zealand the use of phenoxy herbicides in cereal fields and pastures for almost 2 decades has taken out many susceptible species of weeds which are now being replaced by weeds tolerant to the herbicides. *S. arvensis* is one of these.

The plant has sometimes been grown for fodder because it can germinate late in the season, mature quickly, and grow into the fall. It is liked by stock in general but particularly by sheep and cattle. It is a favorite of poultry.

Russel (1965) reported that the species is a host for the beet mild yellows virus but apparently not for beet yellows. Since the weed can overwinter successfully in England, it may pass the virus from one crop to the next. Lucerne mosaic will also overwinter in *S. arvensis*.

COMMON NAMES
OF SPERGULA ARVENSIS

AUSTRALIA
corn spurry

BRAZIL
gorga

CANADA
corn spurry

CHILE
linacilla

COLOMBIA
agujillas

anisillo
mardo quea
miona

DENMARK
almendelig spergel

EASTERN AFRICA
spurrey

ECUADOR
alfarillo

ENGLAND
corn spurry

FINLAND
peltohatikka

FRANCE
petite spergoute
spargoute des champs
spourier

GERMANY
Spörgel
Acker-Spark

HAWAII
corn spurry

INDIA
bandhamia
mun-muna
pittpapra

ITALY
renaiola

JAPAN
noharatsumekusa

NETHERLANDS
spurrie

NEW ZEALAND
spurry

NORWAY
linbendel
spergel

PHILIPPINES
devil's gut

PORTUGAL
esparguta

SOUTH AFRICA
sporrie

SWEDEN
åkerspargel

TAIWAN
da-gwa-tsau

UNITED STATES
corn spurry

❧ 71 ❧

Sphenoclea zeylanica Gaertn.

SPHENOCLEACEAE (formerly CAMPANULACEAE),

SPHENOCHLEA FAMILY

Sphenoclea zeylanica is an herbaceous annual that is native to tropical Africa. It is recognized by its cord-like roots, hollow stems, and by its white, terminal, densely spikelike inflorescence. It is distinctive in that it is never reported as a weed in any crop except rice. It is a weed of rice in 17 countries.

DESCRIPTION

S. zeylanica is a fleshy, erect, *annual* marsh herb (Figure 187); *stems* erect, often much-branched, 7 to 150 cm; *leaves* alternate, oblong to lanceolate-oblong, gradually tapering at both ends, apex sometimes acute, smooth, 2.5 to 16 cm long, 0.5 to 1.5 rarely 5 cm wide; *petiole* 0.3 to 30 mm long; *inflorescence* in dense spikes, cylindric, 0.75 to 7.5 cm long, 5 mm in diameter; *peduncle* slender, 1 to 8 cm long; *flowers* sessile, wedge-shaped below, attached longitudinally to the rachis by a linear base; *calyx* five-lobed, triangular (deltoid), semicircular; *corolla* whitish or greenish white, occasionally mauve-tinged, 2.5 to 4 mm long, united slightly more than half way; *stamens* five, alternating with the corolla lobes and free or attached at the corolla base, *filaments* slightly dilated at base; *ovary* two-celled, ovules numerous; *style* short; *fruit* a flattened globose capsule, 4 to 5 mm in diameter opening transversely; *seed* yellowish brown, 0.5 mm long.

This fleshy species is recognized by the bracts that subtend the inflorescence, the flowers of which develop first at the base, then toward the apex; and by the stamens which are inserted halfway between the corolla lobes (Figure 187, part 5).

DISTRIBUTION AND AGRICULTURAL IMPORTANCE

S. zeylanica thrives in almost any kind of damp ground at altitudes below 350 m. It is found along the sides of ponds, ditches, and rivers, and on dry riverbeds. It grows in seasonal swamps or depressions which are periodically inundated and prefers stagnant water sites. It is a gregarious species in the Philippines but is never gregarious in Malaysia. It does not grow on the mud of tidal creeks in Malaysia as it does in Africa.

Little is known of the biology of this species. It reproduces by seeds. Almost every flower on every inflorescence sets fruit; only one or two flowers are open at once on any one head. The plant flowers all year round in the Philippines.

In Java, young plants and tips of older plants are steamed and eaten with rice.

From Figure 188 it may be seen that the species is distributed as a weed across the world in the tropical and subtropical regions. The natural range of the species extends from Iran and Turkestan in the Soviet Union (east of the Caspian Sea) in the north to the Celebes Islands and Timor in the south, but it has not yet been reported as a weed in the agriculture of all of these areas.

S. zeylanica is a serious weed of rice around the

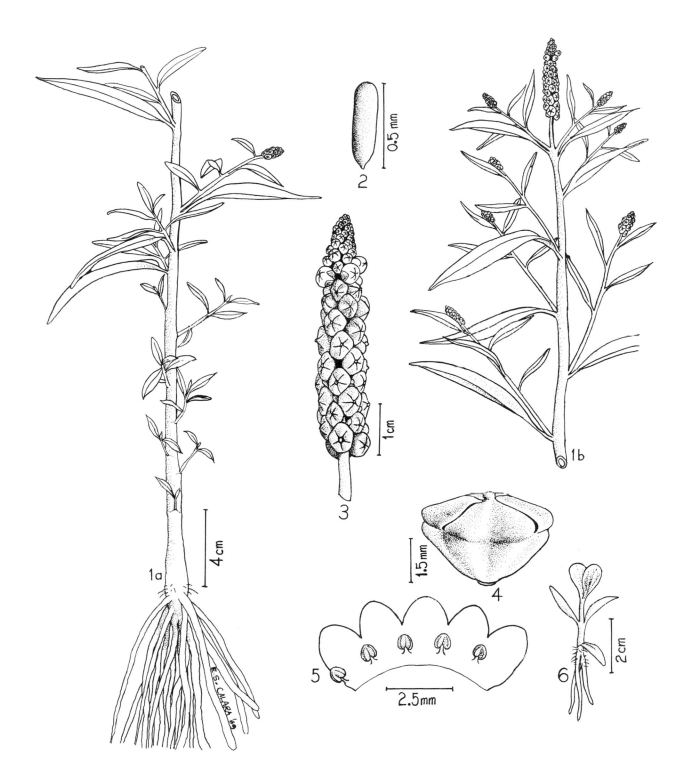

FIGURE 187. *Sphenoclea zeylanica* Gaertn. *1 a–b*, habit; *2*, seed; *3*, inflorescence; *4*, flower; *5*, petal, opened to show stamens; *6*, seedlings.

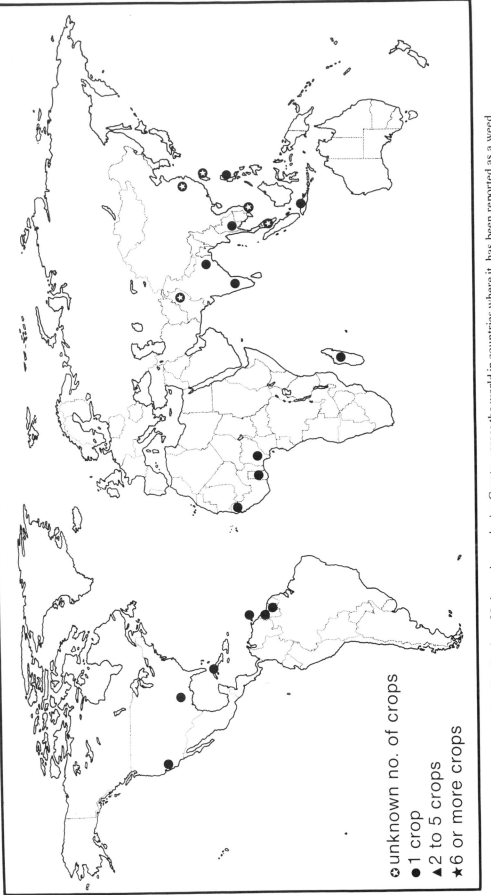

FIGURE 188. The distribution of *Sphenoclea zeylanica* Gaertn. across the world in countries where it has been reported as a weed.

⊛ unknown no. of crops
● 1 crop
▲ 2 to 5 crops
★ 6 or more crops

Caribbean area and in Guyana, India, Pakistan, Southeast Asia, and West Africa. It is a principal weed of rice in the Philippines, Surinam, Trinidad, and the United States. In the past 2 decades there have been about 50 reports in the literature concerning the weediness of this species; and rice is the only crop ever mentioned.

A leaf disease on *S. zeylanica* caused by *Cercosporidium helleri* was observed in paddy fields at Ernakulam, Kerala, India (Ponnappa 1967).

COMMON NAMES
OF SPHENOCLEA ZEYLANICA

INDONESIA
  goenda (Javanese, Indonesian)

NIGERIA
  ekologwe

PAKISTAN
  mirch booti

PHILIPPINES
  mais-mais
  silisilihan (Tagalog)

SURINAM
  pinda grasie

THAILAND
  pak pawd
  phak-pot

UNITED STATES
  gooseweed

VIETNAM
  xà bông

❦72❧

Stellaria media (L.) Cyrill.

CARYOPHYLLACEAE, PINK FAMILY

Stellaria media, an annual, winter annual, or sometimes a perennial herb, is native to Europe. It is one of the most widely distributed weeds in the world. Allard (1965) believes it to be one of the 12 most successful colonizing species among the noncultivated plants. The plant prefers shady, moist places under trees and shrubs where it spreads by rooting at the nodes and forms thick, succulent mats. The green, upright plant can endure extremely cold weather and may flower and seed all winter in the temperate zone. The propagules of the species have no special adaptations for dissemination; yet the weed has traveled to the far corners of the earth. The leaf—shaped like the ear of a mouse—may be deceptive for identification purposes, since the leaves of other closely related species have a similar appearance. *S. media* is distinguished by a small line of hairs down one side of the stem and on the leaf stalks. This species is reported to be a weed in more than 20 crops in 50 countries.

DESCRIPTION

S. media is a weakly tufted, *annual, winter annual*, or sometimes *perennial herb* (Figure 189); *roots* fibrous, shallow, the plants sometimes taking root at the prominent joints; *stems* much branched, erect or ascending from a creeping base, rather weak, minutely pubescent in longitudinal lines bearing a single row of hairs on alternating sides of successive internodes, 20 to 80 cm long; *leaves* opposite, simple, very variable in size in different plants, smooth or fringed with hairs near the base, ovate-elliptic, acute or shortly acuminate, 6 to 30 mm long, 3 to 15 mm wide; *petiole* of lower leaves 5 to 20 mm long and having a line of hairs, petioles of the highest leaves often very short or sessile; *flowers* solitary or in few-flowered, terminal, leafy *cymes* (flat-topped inflorescence with the central flower opening first), white; *pedicels* nearly capillary, ascending, reflexed or recurved, frequently pubescent; *sepals* five, lanceolate-oblong, 3.5 to 6 mm long, blunt to acute, usually with long soft hairs; *petals* five, deeply cleft, white, small, shorter than the sepals, two-parted or absent; *stamens* three to five, rarely more; single *pistil* with three or four styles; *fruit* a many-seeded dry capsule, ovoid, usually a little longer than the sepals, opening by six teeth, breaking into five segments at maturity; *seeds* dark brown, yellowish, or dull reddish brown, nearly circular, slightly elongated toward the notch at the scar, about 1 mm across, the surface covered with conspicuous curved rows of irregular wartlike projections, marginal projections more prominent and toothed in appearance.

The single row of hairs on alternating sides of successive internodes and on the leaf stalks and the irregular wartlike projections on the surface of the seeds are characteristics of *S. media*.

S. media is often confused with two related species, *Cerastium vulgatum* and *C. arvense*, both of which possess chickweed-type leaves. Another, *Stellaria graminea*, has grasslike leaves and is not easily confused with *S. media*.

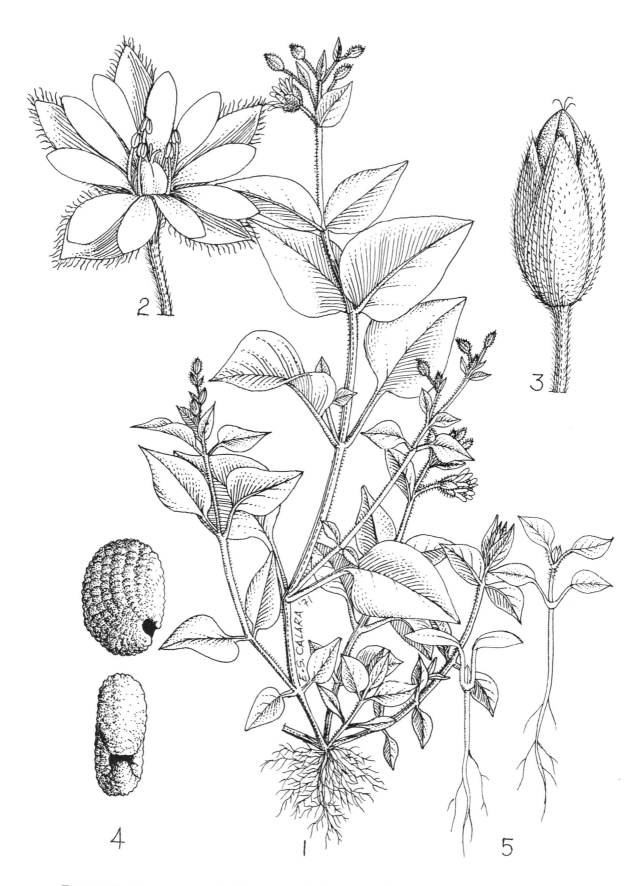

FIGURE 189. *Stellaria media* (L.) Vill. *1*, habit; *2*, flower; *3*, fruit; *4*, seed, two views; *5*, seedlings.

DISTRIBUTION AND HABITAT

S. media is found in cultivated fields, pastures, gardens, shady lawns, roadsides, and neglected lands. Because it prefers cool, moist, shady places, it thrives in orchards, nurseries of all kinds, vineyards, and under shrubs and trees. The seeds germinate well and the seedlings grow quickly in cool, moist fields at the same time as such crops as cabbages are being sown or transplanted. In Great Britain it is believed by some to be the most common broad-leaved weed in both winter and spring cereals (anon. 1968*b*). In Canada, however, the species is more common in British Columbia and in the Eastern Provinces than in the open, dry, windy plains region where it becomes desiccated.

S. media will grow on a very wide range of soils, but it prefers a neutral *p*H and does not like acid conditions. It makes a very luxuriant growth when nitrogen is plentiful. Hamel and Dansereau (1949) in Canada surveyed the distribution of this species on six types of sites in several regions. The sites were hoed or cultivated lands, cereal fields, young and old prairies, pastures, and abandoned lands. The species was found to be confined to the hoed and cultivated lands where it was very plentiful and to the cereal fields where it was found much less frequently. In Colombia it is one of the most aggressive weeds in crops at 2,600 m; and in Java it grows to the same altitude. The species has no special adaptation for the dissemination of seeds, yet this delicate, small-leaved, low-growing, fragile plant has colonized areas all over the world. Its great strength lies in its ability quickly to enter areas disturbed by men and animals and then to become adapted to the habitat. The ecotypes which have resulted from such a wide distribution enable us to find it near the Arctic Circle in Iceland and Alaska, on the mountains and valleys near the equator in Kenya, Tanzania, and Java, and at lat 45° in the Southern Hemisphere. Its distribution across the world may be seen in Figure 190.

BIOLOGY

S. media has been a companion of man for many centuries. Helbaek (1950, 1958) has found the seeds of the species in the stomachs of the Tollund Man and the Graubelle Man in remains preserved in peat bogs in northwestern Europe. The communities existed in the 3rd to 5th centuries A.D.

Plants of this species emerge all through the year. They provide an example of a green plant which can remain upright in winter and tolerate very cold temperatures without cover. They may produce flowers without petals. The species is self-pollinated and cleistogamous (the flowers remain closed). In some areas fertile seeds may be produced and be shed all winter (King 1966). In a warm area such as Java, flowering has also been observed through the year. Goppert (1881) reported that *S. media* can survive temperatures as low as -10° C.

Polunin (1960) has discussed the widespread dissemination of ubiquitous weeds. *S. media*, a plant which is found all over the world, has no special adaptions for long-distance dispersal; whereas other plants, which still are restricted to their original area or to only a few geographical areas, may have special structures to aid in the spread of the propagules, which are more widely carried than are those of *S. media*. Polunin believes that *S. media*'s success is probably due more to the adaptability of the species to the local environment than it is to its means of dissemination. Plants such as *S. media* are quite variable, and can adapt more easily. Nevertheless, climate, soil, or competition from other plants may be very severe and the invaders cannot survive unless they find openings. Small plants such as *S. media* can survive only as long as man keeps on disturbing and opening up the areas. Even after they have colonized a site, they may still depend for survival on disturbances by man and clearings prepared by him to hold other vegetation at bay.

Baker (1951) has pointed out that restrictive factors, such as self-pollination, vegetative reproduction, and so forth, which prevent hybridization, may provide considerable stability among species. This appears to occur among many of the commonest weeds, including *S. media*.

Petersen (1936) has described many ecotypes of this species. Cytological studies have been reported by Sinha and Whitehead (1965) and anatomical and morphological studies, including embryo sac development, have been reported by Pritchard (1964).

PROPAGATION

In regions where *S. media* can grow all through the year several generations may be produced. When the plant has passed into the reproductive phase it usually has flowers and seeds in all stages of development at any one time. Seed production is very high. Champness and Morris (1948) have estimated that 5.5 kg of seeds may be produced per hectare in pastures and 10.8 kg in arable lands. Salisbury (1961) reported a production of 11 to 13 million seeds per hectare.

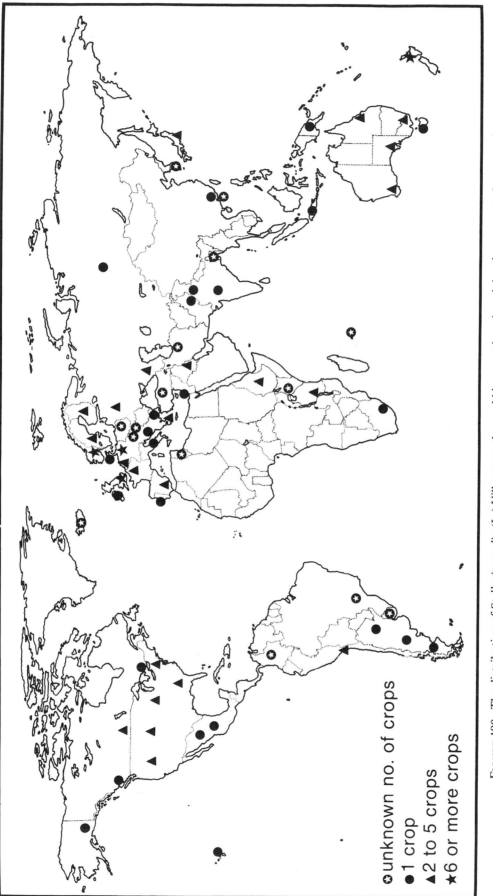

FIGURE 190. The distribution of *Stellaria media* (L.) Vill. across the world in countries where it has been reported as a weed.

⊛ unknown no. of crops
● 1 crop
▲ 2 to 5 crops
★ 6 or more crops

Seeds harvested in August and September in Japan have little or no dormancy, but those harvested in November show increased dormancy. When seeds were buried in the soil a secondary dormancy developed, and this was broken by the alternating soil temperatures during spring days. The procedures of Steinbauer et al. (1955) are suggested for those who require laboratory germination. One-month-old seeds held in alternating temperatures of 20° to 30° C on blotters moistened with 0.2-percent potassium nitrate gave 94-percent germination. In water, germination was 12 percent. Ten-month-old seeds gave 93-percent germination on potassium nitrate and 30-percent on water. Kolk (1947, 1962a) obtained good germination in weak daylight or darkness. The use of alternating temperatures did not improve germination. Seeds germinated well even when soil moisture fell to 30 percent of maximum water capacity in a sandy soil rich in humus. This is one of the few species that is able to emerge at this level of dryness. In this soil, germination was best at the surface or very close to the surface, and seeds buried at 2 cm performed poorly. Shallow cultivation of stubble fields in Sweden in autumn after harvest favored the germination of newly fallen seeds of *S. media*.

In the final tests of the Duvel buried-seed experiment in the United States, *S. media*, which had had 97-percent germination at 1 year, gave 22-percent germination at 10 years (Toole and Brown 1946). Petersen (1936) demonstrated the wide variability and adaptability of this species by gathering and testing seeds of different ecotypes from Arctic alpine, middle latitude, and marine situations in the Northern Hemisphere. In a uniform test of 100 days' duration, some types gave only 5-percent germination; yet others gave 50-percent germination in less than that time.

Rieder (1966) in Germany found that seeds of *S. media* were quickly killed if stored in liquid manure. When the liquid was put on the fields, germination was delayed for a time. Ammonia is believed to be the toxic agent.

AGRICULTURAL IMPORTANCE

S. media is one of the three most serious weeds in cereals in Alaska and England; in forage crops, principally alfalfa, in the United States; in citrus in Spain; and in sugar beets in East Germany. It is a principal weed of cereals in Australia, Belgium, and Japan; wheat in New Zealand and Tanzania; forage crops in New Zealand and Spain; peas in Ireland;

vegetables in Alaska, Belgium, England, the Soviet Union, and Spain; potatoes in Belgium and Chile; pastures in the United States; strawberries in the Soviet Union; sugar beets in Belgium, England, and Spain; and barley, oats, potatoes, sugar beets, rye, vegetables, and wheat in the Scandinavian countries. It is a common weed of citrus and sugarcane in the United States; coffee in Tanzania; peas in New Zealand; sugar beets in France and Israel; hemp in Bulgaria; orchards in Yugoslavia; wheat in the Soviet Union; and vineyards in Australia. *S. media* is also a weed of forest nurseries, bananas, flax, upland rice, and tobacco in several places in the world. It is a weed under macadamia trees in Hawaii and mulberry trees in Japan.

S. media is often a bad weed in perennial crops which overwinter, and in early-planted crops such as peas, spinach, and onions because the weather at planting is often rainy and cool, a condition which favors the germination and rapid growth or regrowth of the species. Frequently it is a serious weed problem before the soil is dry enough to cultivate. In small fruits, vineyards, nurseries, and orchards the weed overwinters and is well established when the warm spring comes. It is one of the most common weeds in spring and winter cereals in northern Europe. Surveys conducted in the Hokkaido region of Japan showed that *S. media* was one of the most common of the 60 species examined in the studies —and this regardless of the soil type or the crop in the field. In East Germany, surveys of weeds in sugar beets for several years prior to 1967 revealed that *S. media*, *Thlaspi arvense*, and *Chenopodium* species made up 50 percent of the entire weed problem. Swietochowski (1967), from 1956 to 1961 in Poland, measured the responses of weed communities to rotations of crops, and to hand, mechanical, and chemical methods of control. *S. media* was one of the most common weeds in his fields. He found that weeding by hand or with chemicals caused *Stellaria media* to increase. Rotations tended to restrict the spread of individual species.

S. media seems such a small weed when compared to some of the great thistles or some of the perennial grasses that many believe it not to be a competitor of crops. Henson (1968) in England studied weed competition in early-planted onions. *S. media*, *Poa annua*, and *Polygonum aviculare*, all "small" weeds, were dominant in his fields. The weeds which were allowed to remain in the field for the first 5.5 weeks after emergence of the onions, and those which appeared more than 7.5 weeks after onion emergence, had no effects on yields. The critical time for weed competition was between the 5.5-to-7.5-week period when the onions were in the two- to three-leaf stage and when they were in the rapid exponential growth phase. At this time it made little difference what the weed density was, for yield reductions were about the same whether there were 160 or 230 weeds per sq m. If competition was allowed to continue from the time the onions emerged through the first 7.5 weeks, the crop plants developed no more leaves, even if the weeds were then removed, and they showed little increase in dry weight by harvest time at 21 weeks.

But the role of weeds in man's crops need not always be that of an adversary. There are farmers in Scandinavia who encourage the growth of *Stellaria media* in orchards because they believe that it brings good fruiting and a better yield. In the recent past (and perhaps even today), one could see along the Rhine in Europe trays and baskets of plants of this species being transported upward on steep inclines to be transplanted under the grapes to give soil cover, to hold the soil, to serve as a buffer to the changes in environment, and to bring a crop of better quality and yield.

This species can be a host for beet yellows virus and beet mild yellowing virus (Russell 1968).

COMMON NAMES
OF STELLARIA MEDIA

ARGENTINA
caapiqui
capiqui
ojo de gringo
pajarera
pamplina
yerba de los caminos
yerba de los canarios

yerba del pajarero

AUSTRALIA
chickweed

CANADA
chickweed

CHILE
bocado de gallina
quilloi quilloi

COLOMBIA
pajarera

DENMARK
fugelgraes

EASTERN AFRICA
chickweed

ENGLAND
chickweed

FRANCE
morgéline
mouron des oiseaux
stellaire intermédiaire

FINLAND
pihatahtimb

GERMANY
Hühnerdarm
Mäusedarm
Vogelmiere

IRAQ
kazazah

ITALY
centocchio

JAPAN
hakobe
kohakobe

NATAL
chickweed

NETHERLANDS
muur

NEW ZEALAND
chickweed

NORWAY
vassarv

PORTUGAL
morugem
morugem branca
morugem vulgar

SOUTH AFRICA
gewone sterremuur

SPAIN
borrisol
yerba gallinera
morrons
pamplina
revola

SWEDEN
våtarv

TUNISIA
mouron des oiseaux

TURKEY
serce dili

URUGUAY
capiqui

UNITED STATES
chickweed
satin-flower
starwort

YUGOSLAVIA
misjakinja

❧73❧

Striga lutea Lour.
(= *S. asiatica* [L.] O. Ktze.)

SCROPHULARIACEAE, FIGWORT FAMILY

Ten families are represented in a group of plants that are called flowering or higher parasites. Six of these are very troublesome and they contain both root and stem parasites. Some of the worst parasites are in *Scrophulariaceae*, and the *Striga*s are chief among the members of this family. About 60 species of *Striga* have been described, but many races of these species which have distinct physiological differences still are undescribed. *Striga lutea* is the most widespread in the world and is the cause of the greatest economic loss in agriculture. Other species of *Striga* may be devastating on a local level but none compare with *S. lutea* as a problem for man and his crops. This species is an annual, growing from 15 to 45 cm high, usually having orange to red flowers, and having rounded stems below the soil and square stems above. It parasitizes root systems and is serious on sorghum, corn, sugarcane, and millet. The tiny seeds resemble dust, have a deep dormancy, and may survive for 20 years in fields. Early research on the weed was conducted in South Africa, and a great deal of the biology was described at that time (Saunders 1933). In the period around 1930 in South Africa it was believed that this weed caused greater losses in sorghum and corn than the total losses from insects and diseases. About 1950 the weed was found on corn and crabgrass, *Digitaria sanguinalis*, in North Carolina (United States). This was the northernmost part of its range to be reported, and the infestation brought about renewed interest in the biology and control of the species.

The most frequently used common name for *S. lutea* is "witchweed." This is also a term used loosely to indicate several *Striga*s across the world.

Striga angustifolia (Don) Saldanha, of Asia and Africa, is mainly a problem in rice, sorghum, and sugarcane. *Striga densiflora* (Benth.) Benth. is mainly a problem in Asia on millet, sorghum, and sugarcane. *Striga hermonthica* Bentham is most serious on corn, sorghum, and sugarcane in Africa. All of the above species are principally associated with members of the grass family. *Striga gesneroides* (Willd.) Vatke. is in both Asia and Africa, has more limited distribution, and is not so much attracted to the grasses. It may be serious on tobacco, *Nicotiana tabacum*, in Africa, but it also parasitizes several legumes as well as certain species of *Ipomoea*.

Striga lutea is a weed of 35 countries but is not recorded as a problem in Europe or South America.

DESCRIPTION

S. lutea is an erect, usually branched, stiff, *annual* herb, 7 to 30 cm (sometimes 50 cm) tall (Figure 191), *stems* slender, rough, four-sided or grooved, densely clothed with rough white hairs, drying green or brown; *leaves* nearly opposite or alternate, narrowly linear or sometimes lanceolate, 6 to 37 mm long, up to 4 mm wide, acute or obtuse, entire, sessile, rough with small prickles or scabrid-hairy; *inflorescence* in terminal spikes, 10 to 15 cm long; *bracts* linear, up to

FIGURE 191. *Striga lutea* Lour. *1*, habit; *2*, leaf, undersurface; *3*, leaf, upper surface; *4*, flower; *5*, portion of stem with the capsule enlarged; *6*, seeds.

8 mm long, obtuse or acute, scabrid-hairy; *bracteoles* similar to bracts but shorter; *flowers* 6 to 9 mm wide, varying from yellow (variety *lutea*), white (variety *albiflora* O. Ktze.) or red (variety *coccinea* O. Ktze.) or pink, sessile, axillary, solitary; *calyx* 5 to 8 mm in length, narrowly tubular, the ribs more than the number of normal lobes, five to as many as 17 ribs (usually 10), one calyx rib terminating at the tip of each lobe, the rest terminating between the lobes, thin (scarious) and dry between the ribs, scabrid-hairy on the ribs and teeth, teeth stiff, subulate or narrowly triangular, up to 2.8 mm long, the uppermost usually smaller; *corolla* two-lipped, corolla tube 6 to 12 mm long, very slender, smooth or slightly pubescent, straight and cylindric, but distinctly curved and inflated at the apex, upper lip of corolla broadly ovate, about 2 mm long and 4 mm broad, lower lip three-lobed, lobes obovate; about 5 mm long by 3 mm broad; *style* about 6 to 8 mm long; *fruit* (capsule) oblong-ovoid or ellipsoid, about 4 mm long by 2 mm wide, usually five-sided, each side terminating in a characteristic spur, *seeds* dustlike, about 0.2 mm long, golden brown, ellipsoid, glabrous, with reticulations and longitudinal lines or ridges.

S. lutea is distinguished by one of its calyx ribs terminating at the tip of each lobe, the rest terminating in the sinuses.

S. hermonthica is larger than *S. lutea*, having larger leaves and flowers, and having five calyx ribs instead of the (usually) 10 found in *S. lutea*. Flower color in *S. hermonthica* is variable but is usually pink to pale pink-white. *S. hermonthica* produces about 60 capsules per plant; each capsule contains about 700 tiny, dark brown, ridged seeds.

DISTRIBUTION AND HABITAT

The world distribution of *S. lutea* is shown in Figure 192. The species is a weed problem on the eastern coast of the United States (in North Carolina) and in Africa, Asia, New Zealand, and Australia. It is the most widely distributed of all the *Striga* species. The weed frequents light soils, but it will grow well and produce seeds on soils ranging from coarse sands to heavy clays; and infestations are found on peat and muck as well. The parasite cannot succeed in wet, high rainfall areas. In South Africa it thrives in low rainfall areas receiving 450 to 500 mm per year and in Nigeria it grows in areas having 650 to 750 mm per year. Robinson (1960) has made temperature measurements in witchweed areas of the United States in order to compare them with temperatures at the Potchefstroom School of Agriculture, South Africa, which is located in an area long infested with the species. In both places there are 150 to 180 days when the temperature of the soil at a 5- to 8-cm depth is above 21° C. A 5- to 8-cm depth is a satisfactory level for germination and parasitization of roots of crops. The plant requires only 90 to 120 days to complete a life cycle. Robinson speculated that the weed could advance still farther north. Lincoln, Nebraska (United States), has 120 days of soil temperatures above 21° C; and, with the variability exhibited by physiological races of the weed, it does seem possible that some could survive there. These matters also raise questions about the warm, low rainfall areas of the Mediterranean, such as the coasts of Spain and Turkey; the Caribbean Islands; and about the countries of Central America and South America. In these areas extreme vigilance is in order.

Until recently it was believed that the plant had never invaded areas with a photoperiod of more than 14.5 hours of daylight. Kust (1964) grew sorghum and *Striga* in a 17-hour day at 2,000 foot candles of light intensity and at a temperature of 28° to 31° C. There was excellent parasitism of sorghum and good production of viable seeds by the weed. Kust feels that length of photoperiod is not a limiting factor to the extension of *S. lutea*'s range.

GROWTH AND DEVELOPMENT

There are a great many reviews and some very detailed reports on most aspects of the biology of *S. lutea* and the witchweed group. Schmucker (1959) described many of the flowering parasites; Stephens (1912) studied the anatomy of *S. lutea* with special emphasis on the haustorium; and Michell (1915) studied the fertilization process and the development of the embryo. Literature reviews which include information on the biology of the species have been published by McGrath et al. (1957), anon. (1957), and Subramanian and Srinivasan (1960).

The growing habit of the plant is quite variable from one crop to another, in different environmental conditions, and between geographical areas. Wild (1954) described the plant as being bushy and 15 to 30 cm tall in corn in Rhodesia. On indigenous veld grasses, however, it is usually spindly, more or less unbranched, and 15 to 22 cm tall. In Burma the plant grows to 50 cm. Normally the mechanical tissues of the stem are not well developed and so it cannot attain great height. As mentioned previously, there are different ecotypes of the species. The weed is

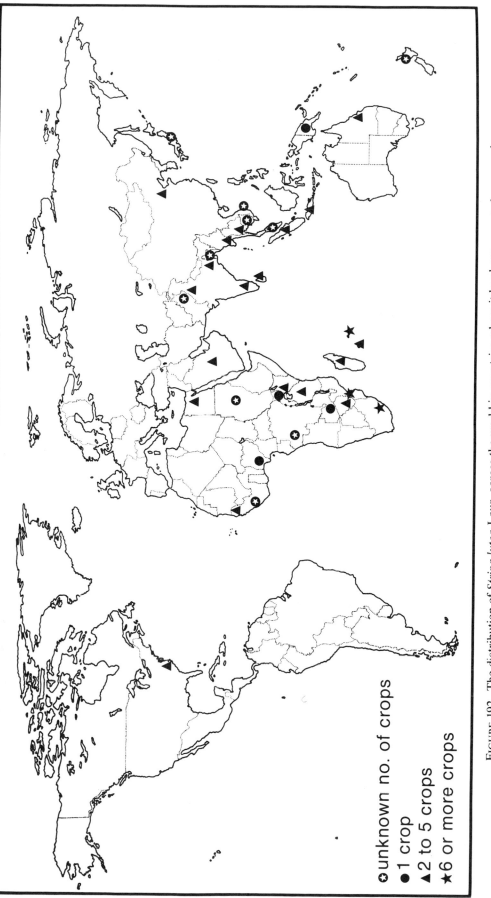

FIGURE 192. The distribution of *Striga lutea* Lour. across the world in countries where it has been reported as a weed.

○ unknown no. of crops
● 1 crop
▲ 2 to 5 crops
★ 6 or more crops

strictly an annual, even when growing on a perennial host.

The seeds of the parasite can germinate only when close to a suitable host. The chemical which is excreted by the host is volatile; its production is influenced by temperature; and it is released as soon as root hairs form on the radicle. This early production occurs before photosynthesis begins in the host. The crop root may be invaded within 2 or 3 weeks after the seeds have been sowed; and, within 3 to 8 weeks after this invasion, a white, almost transparent stem of the parasite emerges through the soil. The time varies with soil type and depth of the seeds. The weed plants continue to invade the host plants and to emerge over a considerable period of the growing season. One month after emergence the flowers open and 1 month after that seeds may ripen. This is a normal course, but some *Striga* plants which attack in late season may hasten the cycle, emerge, and flower at a height of 5 cm. The seed capsules ripen from the bottom of the stem upward.

INVASION AND INJURY
TO THE HOST PLANT

As the seed cracks at the narrow end, the radicle emerges and grows toward the nearest host root. This is a chemotropic response and contact is unnecessary. If the tip of the new root touches the host cells, a bell-like swelling develops and is closely appressed to the host. The root cap cells are absorbed and a profuse growth of root hairs develops near the *Striga* root tip. Ordinarily the *Striga* roots lack hairs, but if a *Striga* root is put near a corn root, for example, it will develop hairs. If the *Striga* root is now withdrawn a short distance, the root hairs will grow to contact the corn and the process can be repeated.

There may be as many *Striga* roots invading as there is space for them on the host root. Saunders (1933) reported one extreme case in which 22 *Striga* seedlings were attached to a 6-mm section of young corn root. The parasite prefers young roots. After they have invaded the host roots, the parasitic roots now move into the host by digesting away the host's root tissue. Small fingerlike papillae spread through the host from the haustorium (the tip of the *Striga* root). Within one day the endodermis is reached, and growth is slowed a bit for this tissue offers some resistance. Some of the papillae go between endodermal cells to enter the vascular tissues of the host (often plugging the conducting cells) and now the conducting systems of parasite and host are joined.

At present there is no evidence that special new tissues develop to transfer food from the host. The material simply seems to pass from the conducting system of the host to that of the parasite. When underground, the *Striga* is totally dependent on the host but, once the parasite has emerged, developed chorophyll, and begun to photosynthesize, it may be dependent on the host only for water (Saunders 1933, Tarr 1962).

If the *Striga* roots invade a host which has immunity, they may be able to effect only partial penetration and may not be able to complete a full parasitic relationship.

There is some evidence that the haustorium has a higher osmotic pressure than the host cells, and by this means it may "suck" water from the conducting tissue. Solomon (1952) showed that if the nitrogen level is low in the soil, the grain yield of sorghum is low and the parasitic attack is more severe. He measured the osmotic pressure of plants under different fertility levels and found that the pressure in parasitic tissues was always much greater than that in the host tissues at a lower concentration of nutrients. At a higher concentration of nutrients, the pressure levels tended to be the same.

Thus the host is robbed of water and nutrients in proportion to the number of parasites it must support. The parasite is known to lower the host's transpiration at the hot part of the day and thus may also reduce the host's level of nutrient uptake.

There is presently no evidence to show that the parasite poisons the host, but some workers feel that often there are too few haustorial attachments to account for all the damage done to the host by loss of water and nutrients alone (Tarr 1962).

THE CHEMICAL STIMULANT

The long search to find and characterize the germination stimulant is a very interesting story; and, although the picture has not yet been completed, the information already at hand may lead us to practical tools for control. Each worker has built on the contributions of his predecessors, so that the properties of the chemicals and their roles in the physiology of the parasitism of particular species have been further defined. Knowledge has been gained which may be used to lure *Striga* seeds into germination with synthetic chemicals. As the pieces have fallen into place, however, we have seen that the phenomenon of the chemical stimulation of seed germination is much broader than we had imagined. Egley (1971) recently reported that chemical compounds contained in many types of natural water bodies and in more than 100 plant species can stimulate *Striga* seeds to germinate.

It has long been known that the chemicals are heat-labile, that prolonged rain or irrigation may dilute them and delay seed germination, and that root washings deteriorate in 3 to 4 days in open flasks in the laboratory. Some species supply the stimulants in sufficient quantity for seed germination of *Striga*, but then for some reason parasitization does not follow. There are many instances of "spontaneous" germination—the opening of the seed and the growth of seedlings in water or without benefit of stimulation from a host species. Such germination may be produced in the laboratory by scarification, fluctuation of temperatures over a wide daily range, and alternate freezing and thawing. In some cases, however, the seeds may imbibe water and break the testa, but then grow no more. The research to characterize the natural stimulants and the attempt to develop synthetic stimulants have been reported or reviewed by Brown (1965), Brown and Edwards (1945), Dowler et al. (1959-1960), Worsham (1961), and Worsham, Moreland, and Klingman (1964). Kust (1966) reported evidence for a germination inhibitor in *Striga* seeds. In a recent report Cook et al. (1966) identified a *Striga* seed germination stimulant from cotton seed extracts and named it "strigol." It is a highly active crystalline substance whose structure has not yet been elucidated. The molecular formula assigned is the same as that for gibberellic acid, a hormone which has a function in seed germination. Strigol does not seem to be identical to any known gibberellin. Gibberellic acid does not stimulate germination of witchweed seeds.

An occasional piece of research seems to synthesize vast quantities of information which may broaden our view of the natural world and stimulate us to think of our problems in very different ways. Such an experiment is that reported by Dale and Egley (1971) on the universality of germination stimulants in the environment. In a survey of 163 species, 118 of them in 57 families could stimulate the germination of *Striga*. They range from the great redwood, *Sequoia sempervirens*, to an emergent aquatic weed, *Typha latifolia*. The stimulants were found in 22 of 29 farm ponds, streams, and lakes. No germination stimulants were found in water from deep wells or water from natural lakes where all vegetation was more than 9 m distant, and none were found in distilled water. This must certainly enlarge our view and explain some contradictory reports about the response of the weed to natural stimulants.

Egley and Dale (1969) and Eplee (1972) have provided encouraging news about the possibility of using synthetic chemicals to lure the *Striga* seeds into germination so that they will destroy themselves. Ethylene gas (C_2H_4) and an ethylene-releasing agent, 2-chloroethane phosphonic acid, have been shown to be strong stimulators of germination. When ethylene is chisel-injected at 3 kg per hectare, it diffuses horizontally for a distance of 1 m; and *Striga* seed germination has been obtained at that distance in light soils. Because of dormancy, lack of preconditioning, or other factors, all *Striga* seeds are not ready to germinate at one time. In experiments, however, 85 to 90 percent of those in the soil did germinate, and this is as good as that expected from the natural stimulants which cannot be applied on a field scale until we know more about them.

GERMINATION OF THE SEEDS

Emergence and flowering occur more quickly and more completely on light soils. Nelson (1958) found that plants emerged in 28 days at 32° C on light soil, in 53 days on a light clay soil, and none penetrated to the surface in heavy clay at this temperature. No flowering occurred on heavy soils below a temperature of 28° C, but plants flowered successfully at 25° C on light soils. The seeds are obviously adapted for dark germination.

Seeds are about 0.31 mm long, 0.16 mm wide, weigh about 0.0045 gm and one plant may bear 50,000 to 500,000 seeds. They are so small as to resemble dust. Most are found in the top 30 cm of the soil, but viable seeds have been found at a depth of 1.5 meters. Plants can emerge from a 30-cm depth even in heavy soils. About 5 percent of the seeds will germinate as they ripen but most have a dormant period of 18 months (Saunders 1933).

The seed cover is a hard brown testa above a single layer of food storage cells which are rich in protein granules and oil. The embryo has little starch but is high in oil. The seeds may be distributed as dust in the wind, on fur, feathers, and clothing, with mud and machines, in fodder, and may lodge between the grain and glume in cereals and be disseminated with the crop seeds. It has been estimated that some infested soils may have 3.5 million seeds per square meter.

Saunders (1933) found in South Africa that the best germination occurs at a high temperature of 30° to 35° C when the stimulants are present. He could get no germination at 20° C and he feels that this explains the lack of parasitic activity on winter cereals. Twenty to 30 percent of oven dry weight seems to be the optimum moisture content for germination in many soils. Too much moisture dilutes or washes the chemicals away, so that irrigation may be used to depress germination in emergency conditions. Kust (1963) stored seeds at -18°, 5°, 24°, and 31° C in relative humidities of 20, 50, and 100 percent from 15 weeks to 26 months. Seeds stored in 100-percent relative humidity at 31° C perished within 20 weeks. At -18°, 5°, and 24° C and at 20-percent humidity the seeds were viable at 26 months. All seeds were very erratic in germination under uniform conditions. Dormancy was broken at 31° C in 6 weeks and in 40 weeks at -18° C. It appears a long viability is possible if humidity is low and the temperature remains below 24° C.

Brown and Edwards (1944, 1946) discovered that the stimulating substances from the host seem to be more potent near the root tips. They suggest that much-branched root systems provide more tips but also expand into a greater volume of the soil where seeds of *Striga* are located. Studying germination potential along a host root system, they found that host seedlings vary greatly in production of the chemicals. In their experiments the seeds germinated slowly and erratically when exposed directly to the host stimulators. If, however, the seeds were preconditioned in the dark, with moisture, at 22° C for about a week, they germinated quickly. This suggests that developmental changes must take place before the seeds can respond well to the stimulants. It also suggests that a period of warm, rainy weather in early season may prepare *Striga* seeds to receive the stimulants and portend a high level of parasitization.

AGRICULTURAL IMPORTANCE

The mere presence of *S. lutea* in a country is a threat and the weed must be considered a serious problem no matter how much acreage is infested. The crops which are damaged most often by *S. lutea* are oats, rice, teosinte, wheat, rye, sorghum, several of the millets (Italian, pearl, and finger), Sudan grass, corn, and sugarcane. Infestation of the last four crops—and occasionally of rice—causes the greatest economic losses. McGrath et al. (1957) presented extensive tables of host plants as well as common names for them and for *Striga* species. A geographical index of the *Striga*s and one of the most extensive reviews of literature ever made for the species are also presented in this publication. Tables and information on host species may also be found in Tarr (1962) and Dale and Egley (1971).

There is little crop loss in the early stages of an infestation and only scattered *Striga* plants appear above the soil surface. In the United States an infestation in corn reduced yields by 20, 60, and 100 percent in the 1st, 2nd, and 3rd years, respectively. As the infestation becomes more intense the crop plants appear as they do in times of severe drought, and their condition is not improved by rainfall or irrigation. The leaves wilt and growth is stunted; the leaves then yellow and the whole plant shrivels and dies. All of this may happen before seeds are formed on the crop. In such an infestation crop plants are often ringed with *Striga* plants appearing above ground. In the center only sparsely distributed plants are seen because competition (underground) is so severe that the mass of germinating seedlings cannot emerge. In a really severe attack the host may be killed so quickly that the witchweed cannot emerge and it too will perish without producing seeds (Saunders 1933).

In Rhodesia the first infestations were noted in 1916. Thirteen years later, despite efforts to control the weed, 30,000 hectares of cultivated land were affected. In Sudan, where there is great dependence on sorghum, there is a long history of *Striga* infestation. In areas of shifting cultivation, conflicting views arose, as the yields became progressively lower, as to whether the land was exhausted or parasitism was preventing crop growth; however, experiments conducted in that area and, later, in other places demonstrated that *Striga* infestations can cause almost complete crop losses while soil fertility is still adequate. Upland rice culture in some areas of southern India has been stopped because of 80 to 90 percent crop losses from *Striga*, and millet-growing in the Rajasthan area farther north is sometimes a complete failure because of the weed. Many of the very severe losses, however, come as a result of primitive agricultural methods which include continuous culture of millet, corn, or sorghum, methods which allow *Striga* to build up until it is no longer economical to pursue such a cropping system. A suitable crop rotation scheme instituted at the first appearance of the weed will seldom allow the infestation to become severe.

Control measures have been studied for almost half a century and these have been reviewed recently by McGrath et al. (1957), Shaw et al. (1962), and Tarr (1962). There are many different methods available; and an integrated control program made up of several of the measures will first contain the weed and then gradually reduce the seed population until only a maintenance program is necessary. The principal objective should be to prevent seed formation by the parasite. This may be done by hand-weeding, burning, cultivating, or using herbicides. Deep plowing is sometimes done to bury the seeds, but they are so long-lived that the process is effective only once. The use of a lister plow can shape the field so that the crop can be planted in a furrow. As the parasitic plants appear they can be buried by throwing soil on them in the course of several cultivations. Phenoxy and benzoic acids and other herbicides applied pre- and post-emergence have given good control in some crops under suitable soil and weather conditions. The advisability of fighting the weed by improving fertility was viewed differently among many workers for a long time; however, experiments in recent years have shown that healthy, well-fertilized crops are indeed more resistant, but that the vigor of the parasite is also enhanced. Mathur and Mathur (1967) designed experiments to test the hypothesis that impoverished soils were the principal cause of such high losses of millet to *S. lutea* in India. They applied nitrogen, phosphorus, and potassium at several levels, alone and in combination. It was found that emergence of the parasite was not discouraged by high fertility, but that lack of nitrogen was the cause of reduced yields no matter how great the *S. lutea* infestation. Egley (1971) in the United States studied sorghum at three levels of nutrition and found that the weed reduced host-shoot yields by 70 percent at low levels of nutrition and 45 percent at high levels. Witchweed produced seeds at all nutrition levels, but the sorghum produced seeds only at the high level. As nutrition increased, the weight of witchweed increased. *S. lutea* could not survive when separated from the host, and Egley believes that this is because the parasite could no longer obtain water. Plants could be kept alive, but did not grow, if they were covered and provided with a high humidity around the aerial shoots. The subterranean parts of the plant are not adapted for procuring nutrients and water from the soil. Some experiments have shown that wider spacing of some crops makes for more vigorous growth and thus helps the plant to withstand the parasitic attack.

Two remaining aids for control which are receiving much attention at present are the use of "trap" or "catch" crops in rotation systems and the use of resistant crop varieties. With a seed viability of up to 20 years in soil, crop rotations will obviously be of little consequence unless they include host species which stimulate germination. For example, a field which had been heavily parasitized by *S. lutea* was planted with corn 14 years later—and the crop was severely damaged (Saunders 1933). Time alone will not clear the field. The crop species used in the rotations to lure the *Striga* seeds into germination have come to be known as "trap" crops, those which

stimulate seed germination but are not themselves parasitized, and "catch" crops, those which stimulate germination and are in turn parasitized by the witchweed. A good example to illustrate such use is given in the experiments of Robinson and Dowler (1966). They used soybeans and field peas as trap crops and corn, sorghum, and millet as catch crops. For 5 years following 1958 each crop was sown in its own field. All five crops were about equally effective. All of them except millet required about 3 years before the weed density was reduced enough to give improved yields; millet gave improvement in 2 years. But after 5 years, eradication was still not complete. In trap-cropping all grasses must be controlled for obvious reasons. In catch-cropping the *S. lutea* must be destroyed in some way before it bears seeds. Much effort has been devoted to a search for resistant strains of corn, sorghum, and sugarcane. Robinson and Stokes (1963) found both tolerant and susceptible varieties in the sugarcane stocks of the United States. They used methyl bromide to fumigate fields planted to different varieties and found that cane growth was much improved for susceptible varieties but was not much improved for tolerant varieties. Sorghum varieties can also be sorted into tolerant and susceptible lines. The resistant varieties are actually invaded by the parasite which then seems to be unable to continue its development. Some witchweed plants are killed when attacking resistant varieties. Tolerance in a strain of sorghum is also indicated when few *Striga* plants appear on the surface. The search for resistant corn varieties has not been successful. One of the difficulties in breeding for resistance in crops is that different strains of the weed may vary in their pathogenicity. This complicates the work because a new resistant variety of a crop planted in a distant area may be successfully attacked by a different strain of the weed.

The question of the effect of the parasite on the quality of juice in sugarcane has not been resolved. In India it is believed the juice is of lower quality and in the United States it is believed that the quality is unaffected.

COMMON NAMES OF STRIGA LUTEA

BURMA
gwin-bin
mogyo-laung-mi
naga-the
pwinbyu
pyang-sa-bin
thagya-laungmi

EASTERN AFRICA
red witchweed

INDIA
bile kasa (Kanarese)
palli poondu (Tamil)
theepalli (Malayalam)

INDONESIA
  radja tawa

ITALY
  erba strega

MALAYSIA
  rumput siku-siku

SOUTH AFRICA
  brandboschjes
  fireweed

  mielie-gift
  redweed
  rooiblom
  rooi-bloometje

  rooiboschje
  vuurbosie
  witchweed

UNITED STATES
  witchweed

❧74❧

Tribulus terrestris L. and *Tribulus cistoides* L.

ZYGOPHYLLACEAE, CALTROP FAMILY

Tribulus terrestris and *T. cistoides* are members of a small genus composed of about 12 species that are native to both warm and temperate regions. The genus name is said to be derived from the Latin *tribo*, "to tear," which relates to the spiny fruits. "Caltrop," the common name of members of the Zygophyllaceae, refers to the resemblance of the fruits to the spiked metal devices used in medieval warfare which were also called by that name. *Tribulus* is one of the very few pantropic genera which thrives in arid conditions (Good 1964). *T. terrestris*, a native of the Mediterranean region, is widespread throughout the world from lat 35° S to 47° N. It is a weed of 21 crops in 37 countries and is an especially serious one in pastures, cotton, corn, and other field crops. *T. cistoides*, a native of tropical America, is less widely distributed, having a range of lat 22° S to 32° N and growing mostly along tropical coasts. Both *T. terrestris* and *T. cistoides* may be poisonous to livestock, causing both photosensitization and nitrate poisoning. The thorny burs of both plants cause injury to the feet and skin of animals, cause reduced grazing in pastures, reduce hay and seed quality, puncture tires of bicycles and motor vehicles, and cause difficulty and pain for man on golf courses, playing fields, or lawns.

DESCRIPTIONS

Tribulus terrestris

T. terrestris is a much branched, prostrate to de-cumbent, mat-forming, *annual* herb (Figure 193); long *taproot*, many fibrous roots; *stems* radiating from a central axis, to 2.4 m long, green to reddish or brownish, striate, pubescent, finely or stiff white-hairy, sometimes sparsely hairy; *leaves* opposite, compound, pinnate, short-petioled, larger leaves up to 6 cm long, smaller up to 3.5 cm long; *leaflets* in five to eight (rarely to 14) pairs, usually unequal with one pair being shorter than the other, *rachis* densely pubescent, leaflets obliquely oblong-lanceolate up to 15 mm long, 5 mm wide, upper surface green, sparsely pubescent, lower surface whitish with dense pubescence; *stipules* linear, up to 10 mm long; *flowers* solitary, axillary, on pubescent pedicels slightly shorter than the leaves below; *sepals* 3 to 5 mm long, acute; *petals* five, pale yellow, 3 to 12 mm long, stamens about as long as the petals; *fruits* split when mature into five indehiscent, approximately triangular, dry, almost woody segments (cocci); *cocci* pubescent or almost glabrous, each with a tuberculate bristly dorsal crest, two tough lateral divergent acute spines above the middle, and two shorter spines near the base, directed downward; *seeds* one to four in each coccus, whitish, flattened, ovate, with a pointed tip.

Tribulus cistoides

T. cistoides is a decumbent or ascending, *perennial* herb; *taproot* long; *stems* 0.3 to 1.4 m long, pubescent; *leaves* opposite, pinnate, unequal,

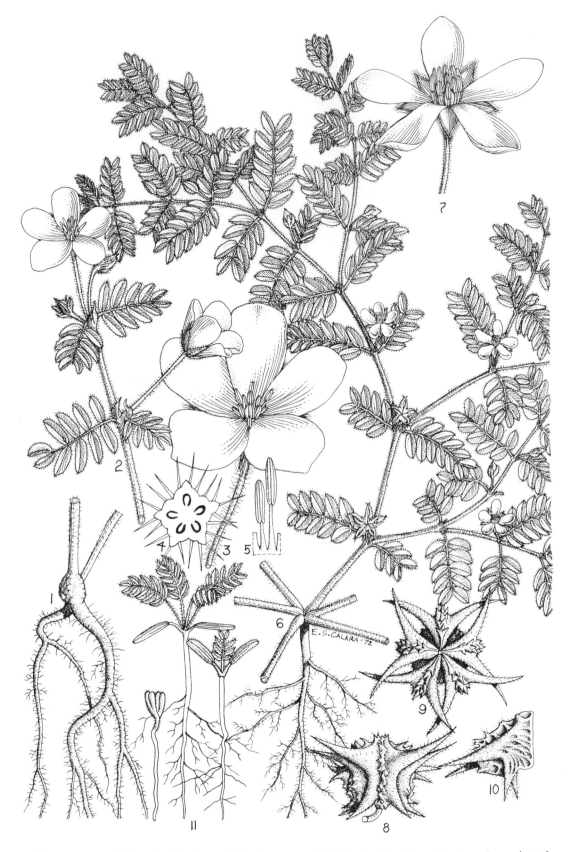

FIGURE 193. *Tribulus cistoides* L. and *Tribulus terrestris* L. *T. cistoides: 1*, root system; *2*, portion of stem; *3*, flower; *4*, ovary, cross section; *5*, stamens. *T. terrestris: 6*, habit; *7*, flower; *8*, fruit, side view; *9*, fruit, top view; *10*, separate carpel, internal view; *11*, seedlings.

2 to 6 cm long, *leaflets* usually in six to eight pairs, sometimes less, subsessile, elliptic, 6 to 22 mm long, 3 to 9 mm wide; *stipules* sickle shaped, pointed, erect, 3 to 6 mm long; *flowers* solitary, in axils of smaller leaves; *pedicels* 2 to 4 cm, longer at fruiting; *sepals* lanceolate, acute, 7 to 11 mm long; *petals* obovate, bright yellow, 10 to 20 mm long, 9 to 16 mm wide; *fruit* splits into four or five segments (cocci) with sharp stout spines of which the two lateral ones are largest, pericarp corky.

T. cistoides closely resembles *T. terrestris* except for its larger flowers (more than 3 cm wide whereas those of *T. terrestris* are less than 2 mm), larger leaflets, and in being perennial. Both may be easily recognized by their radiating, mat-forming, prostrate or nearly prostrate growth and by their spiny fruit segments. *T. cistoides* is a somewhat more showy and taller plant.

DISTRIBUTION AND HABITAT

From Figure 194 it may be seen that *T. terrestris* is distributed over the world from lat 35° S to 47° N. Probably native to the Mediterranean region (some have suggested the fringes of the Sahara Desert), the weed is now widespread in southern Europe, eastern Africa, southern Asia, and Australia. It is found throughout the United States except in the most northerly states, but, with the exception of Trinidad and Argentina, it has not been reported as a weed in Central America and South America. It is seldom a weed in western Africa. It is a weed in all states of Australia.

Seeds of *T. terrestris* may have been disseminated across the world in the wool of European sheep. Certainly the weed is usually reported first near agricultural communities, railroad yards, or coastal towns. It is often found in hay, straw, or manure.

T. terrestris is a weed of the subtropics and warm temperate zone. It is often concentrated at low elevations in coastal areas but does extend far inland under suitable conditions. In eastern Africa it grows as high as 1,800 m in dry areas. In the United States it is found on the coastal plain from Florida to Texas and New York as well as in inland states. It is commonly found in railroad yards and right-of-ways, along roadsides and field margins, in barnyards, near gravel pits and open sandy places, and in ballast heaps, dry waste areas, fallow fields, pastures, cultivated fields, lawns, and playgrounds.

T. terrestris grows best in dry, loose, sandy soils and it prospers near sand dunes or loose blown soil by field margins. However, it also grows in heavier soils, especially if they are fertile and moist. The plant can grow on compacted soils such as those found along the sides of unsurfaced roads or in playgrounds.

T. cistoides is a native of tropical America and is less widespread than *T. terrestris*. It has been reported as a weed in Australia, Jamaica, Puerto Rico, Mexico, Venezuela, Mauritius, Madagascar, Indonesia, and Hawaii. In the United States it is found in hummocks and waste places along the coastal plain from Florida to Texas and Georgia. It is mostly a plant of dry, warm, coastal areas and is found on sand beaches and coastal dunes. It is believed that the seeds were spread to Hawaii by ocean currents and animals. The plant has been reported from lat 22° S to 32° N.

PROPAGATION

T. terrestris, an annual, and *T. cistoides*, a perennial, both reproduce by seeds. The seeds are produced in woody, spiny segments of the fruit. Each fruit usually comprises five segments (cocci) and each coccus contains from one to four seeds. A plant growing by itself without severe competition may produce thousands of seeds. In India one plant was found which produced 1,000 fruits; if these fruits contained an average of two seeds per coccus, the plant would have produced 10,000 seeds.

Fruits are produced as early as 5 or 6 weeks after seed has germinated and growth has begun. The fruits have a bony or woody covering which encases and protects the seeds and which also bears the woody sharp spines. Misra (1962) found that 100 fruits weighed about 8 grams fresh, 4.5 grams dry.

In the field, germination starts in spring or during warm weather soon after the first rains. Germination, studied in India over a 3-year period, was found to begin within about 5 to 7 days after the first late spring or early summer showers (Misra 1962). Laboratory germination was very erratic, ranging from no germination at all with seeds only a few weeks to several years old, to as much as 22 percent with seeds which had been stored in a plastic bag for 16 months at room temperatures and planted 0.6 cm deep in soil in the greenhouse (Andersen 1968). Treatments which have been used with little or perhaps even negative effects include allowing 24 hours of water imbibition on filter paper at 32° C (89.6° F), soaking for 30 minutes in concentrated sulfuric acid, alternating temperatures from 32° C to 5° C, and treating with gibberellic acid. In most cases untreated seeds germinated as well or better than

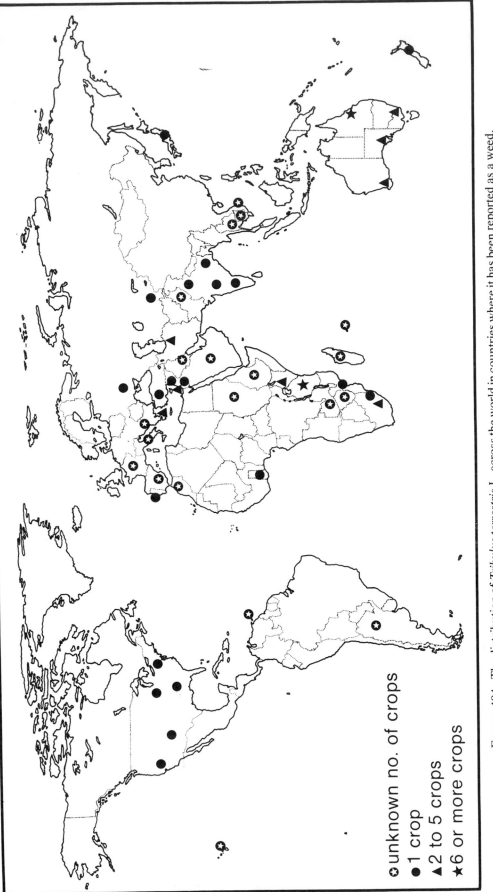

FIGURE 194. The distribution of *Tribulus terrestris* L. across the world in countries where it has been reported as a weed.

treated seeds. Seeds in the field may germinate throughout the growing season. It is clear that much more work must be done to clarify the germination story for *T. terrestris*.

The spiny fruit segments are the weed's primary means of dissemination. The large and small spines are arranged at different angles so that, no matter how the seed falls, at least one of the spines always points upward to meet the unwary foot, hoof, or vehicle tire. The sharp spines easily penetrate leather, rubber tires, skin, or other surfaces, puncturing them and sometimes adhering for some distance before being knocked or picked off.

BIOLOGY

Although *T. terrestris* is a summer annual in most areas, it may become a perennial under suitable tropical conditions. The plant has a deep, somewhat woody taproot that enables it to obtain moisture and to grow under conditions much too rigorous for most plants. A deeper, more extensive root development occurs in loose, open soils and this may explain the weed's association with sandy soils. However, the water requirement of *T. terrestris* is very low compared with that of other plants. In studies in Texas Davis and Wiese (1964) found that *T. terrestris* required only 96 kg of water to produce 1 kg of dry matter as contrasted with sorghum or alfalfa that require about 300 and 840 kg of water, respectively, to produce 1 kg of dry matter. In later experiments Davis, Wiese, and Pafford (1965) found that *T. terrestris* has a very large root volume of 5.3 cubic m and is able to extract 14.1 kg of water per plant in excess of the rainfall received, this amount indicating an exceptional ability on the part of the plant to remove water from soil at very high moisture tension levels. Several authors report that the taproot produces a network of fine rootlets which enables the plant to survive drought. Davis, Johnson, and Wood (1967) have studied root profiles of seven weeds, including *T. terrestris*, in the field in Texas. Three-week-old plants were transplanted into the field in mid-May in single rows, the plants being spaced 15 cm apart in the row. Root measurements were made at 10 weeks after planting (late July) and in October when seeds were fully developed and plants were beginning to dry off. Although taprooted and quite small aboveground, *T. terrestris* was again found to have a very extensive root system, one with a lateral diameter of 6.6 m and a depth of 2.6 m. This large root system developed slowly; from the 10-week sampling date until maturity, the root profile area

had increased by 142 percent. In India the taproots penetrate as deep as 1 m, with a maximum radial spread of 27 cm (Misra 1962).

The seeds of *T. terrestris* do not seem to have specific germination requirements, this situation often being the case with the seeds of highly adaptable, ubiquitous, annual weeds. Some dormancy apparently does exist for seeds have been reported to last for 3 or 4 years or more in the soil. Undoubtedly the woody outer shells of the spiny fruit segments do protect the seeds, perhaps by being impervious to water for some time. Misra (1962) stated that dormancy is lengthened if the fruit segments are buried deeply.

Probably the most complete study of the biology of *T. terrestris* was done by Misra (1962) in the arid zone of India. In India the weed, after it has germinated, grows very rapidly and can cover a large area within a week. Flowering starts about 3 to 5 weeks after germination. By contrast, in Russia flowering may occur only 2 weeks after germination, and late-season germination shortens the period between germination and flowering. In India, maximum shoot growth occurs in 3 to 5 weeks. At the end of the period Misra obtained the following mean plant measurements from 10 plants; root length, 126 cm; shoot length, 91 cm; number of roots, 6.5; number of branches, 11; and number of fruits, 252. One plant was found which covered about 2.5 sq m and which produced 1,000 fruits. Although normally a prostrate plant, *T. terrestris* may grow almost upright when competition and shading occur in dense crop stands.

Flowering occurs from June to September in the United States, from August to October in southern India, and from May to October in Iraq. In California (United States) Goeden and Ricker (1970) found that maximum flower production usually occurs in July or August. Fruit formation occurs during July to September in the United States and from September to December in southern India. Flowers usually remain on the plant about 2 weeks before fruit formation starts. Fruits mature in about 2 weeks, splitting apart into segments soon thereafter.

Little biological information has been reported for *T. cistoides*. Pope (1968) described the growth of *T. cistoides* in Hawaii where it forms pure patches of considerable size near the leeward coasts. The plant begins to flower and set seeds while only a few cm tall, continuing these processes throughout the year. It is known to flower from spring to fall in the subtropics and all year round in the tropics. Flowers open just after sunrise and close at sunset, lasting for about 2 days. The plant's rapid growth allows it to form sizable radial patches very quickly. Water freshets may spread the fruits.

POISONOUS PROPERTIES

Both *T. terrestris* and *T. cistoides* have been reported to be poisonous to livestock, especially sheep. They have both been implicated in hepatogenic photosensitization in sheep. This condition, known as geeldikkop or dikoor in South Africa, has been reported most frequently from that country (Steyn 1934), from Australia where it is known as yellow thick head or bighead, and from the United States where it is known as bighead (Schmutz, Freeman, and Reed 1968).

Beasley (1967) and Watt and Bryer-Brandwijk (1962) researched the disease in some detail. They discovered that the toxic condition is brought about by damage to the sheep's liver by unknown toxins in *T. terrestris*. They believed that the damaged livers cannot filter out chlorophyll or a derivative of chlorophyll, phylloerythrine. Chlorophyll or phylloerythrine is then carried in the blood to skin cells, sensitizing them to sunlight (Gardner and Bennetts 1956). Photosensitization symptoms (dying of the skin, loss of skin) appear on the lips, ears, and around the eyes, followed by swelling of the head. As might be expected with liver damage, the animals also suffer from jaundice. The whites of the eyes become yellow, as do the mucous membranes and affected parts of the skin. Fever is an early symptom. If the eyes become inflamed, blindness may result. If the poisoned animals do live, they may never fully recover. Mortality or sickness may exceed 90 percent, especially among young animals.

In Australia it was shown that photosensitization occurs when feed or pasture is in very short supply and animals are hungry (Blohm 1962). The disease affects animals with unpigmented skin when they eat the leaves and stems of *T. terrestris* and are then exposed to full sunlight. In Yugoslavia black sheep were not affected but white sheep were.

There may be more than one kind of toxicity caused by the ingestion of *T. terrestris*. Kingsbury (1964) stated that the species belongs to a small group of plants which can poison animals because of two or more compounds which are in different chemical groups. The species has been found to have toxic levels of nitrate in its tissues. Ragonese (1955) reported that nitrate-induced asphyxiation occurred in animals in Argentina after they had eaten the weed. Similar nitrate poisoning has been reported from the United States (Schmutz, Freeman, and Reed 1968). Hsu, Odell, and Williams (1968) studied three saponins of *T. terrestris* and found them to be in leaves and roots but not in stems or seeds. The authors doubt that these chemicals play a role in hepatogenic photosensitivity, for saponins usually have a hemolytic effect.

T. cistoides also has been reported to be poisonous to animals in Australia, Colombia, and Venezuela (Blohm, 1962; Gardner and Bennetts, 1956). In Australia the weed was blamed for poisoning sheep that had eaten it when their stomachs were empty (Bailey and Gordon 1887).

Another *Tribulus* species that has been implicated in livestock poisoning is *T. inermis* (Hurst 1942). It was suspected to have caused the death of cattle in Australia. The animals, apparently being blind and deaf, had a staggered gait and walked in a hunched manner; their ears and lower lips twitched. Hurst also mentioned that 300 sheep may have perished from eating leafless stems of *T. inermis*.

BIOLOGICAL CONTROLS

T. terrestris and *T. cistoides* are among those weeds for which biological control appears to be feasible. In recent years a fruit-infesting weevil (*Microlarinus lareynii*) and a stem-and-crown-mining weevil (*M. lypriformis*) were introduced from India into the United States to control *Tribulus terrestris* and *T. cistoides* (Goeden and Ricker 1970). The fruit weevil lays its eggs in the fruits, and the larvae then hatch and destroy the seeds. The stem weevil lays its eggs in the plant crowns. *Microlarinus* spp., which breed only on *Tribulus*, have been known to infest 50 percent of the weed in California. Arizona also has introduced these weevils for biological control (Butler 1965).

Perhaps the most dramatic success in biological control of *T. terrestris* and *T. cistoides* has been obtained in Hawaii where the fruit weevil was established in 1962 and the stem weevil in 1963 (Davis and Krauss 1967a,b; Nakao 1969). Nakao (1966) reported that the fruit weevil had destroyed 75 percent of the seed crop and the stem weevil had destroyed all growth of both weeds on the island of Kauai within a year after having been established on that island. Nakao (1969) suggested that complete control of both *Tribulus* species may be possible.

In Australia, Squires (1965) reported that the larvae of an *Aristotelia* species destroys the seeds of *Tribulus terrestris*.

AGRICULTURAL IMPORTANCE

The problems and losses due to *T. terrestris* in California have been reported by Johnson (1932). These include: (1) flesh and skin punctures, infections, and other mechanical injuries to animals, (2) reduced crop seed quality, (3) increased harvesting costs, and (4) increased costs of land preparation and weed control. Other problems caused by the species in California and elsewhere include livestock poisoning, crop competition, reduced hay and feed quality, poor grazing and poor animal performance (amount of weight gained, health, general status of stock) in infested pastures, and reduced wool quality.

Livestock can suffer injury to their mouths, eyes, digestive tracts, and skin from the sharp woody spines of the fruit segments of both *T. terrestris* and *T. cistoides*. Painful punctures of the feet can also occur, sometimes causing suffering, infection, and lameness, especially to horses. In severely infested pastures, wounds of the mouth may result in a reluctance to graze, followed by loss of condition.

The damage caused by these two species is of serious economic concern in many areas of the world. *T. terrestris*, however, must also be regarded as a troublesome weed of croplands as well (Figure 195). It is a principal weed of cotton, corn, sorghum, vineyards, and other summer crops in Australia, and of cotton in Kenya, Swaziland, and the United States. It is a common weed of cotton and corn in Tanzania, of corn in India, and of cotton in Iran and Turkey. It is also a common weed of corn and sugar beets in Greece, of sugar beets in Israel, of citrus in Mozambique, and of beans, barley, sorghum, and wheat in Tanzania. It is a weed of the following crops in one or more countries: pastures, peanuts, cereals, sugarcane, coffee, onions, orchard crops, fiber crops, fruits, and potatoes.

T. terrestris is a problem in crops because of its ability to extract soil moisture from great depth in the soil. This allows it to offer severe competition under very dry conditions. The weed prospers in alfalfa, clover, hay, and corn fields as well as in overgrazed pastures.

In India *T. terrestris* was the dominant weed in pearl millet (*Pennisetum typhoides*), causing a 15 to 20 percent loss in grain yield. In the arid zone of India, where 4 to 9 percent of the surface of cultivated fields is covered with *T. terrestris*, the presence of the weed indicates overgrazing and declining soil fertility (Misra 1962). It has been estimated that as much as 30 percent of the maize fields of Punjab (India) are infested with the species (Adlakha 1961).

There are indications that the weed affects crops in indirect ways, too. Sen, Chawan, and Sharma (1969) used aqueous extracts of *T. terrestris* to study the effects on *Pennisetum typhoides* and found that germination was not affected but that seedling growth was inhibited. Root extracts were most harmful if they were followed by leaf and stem extracts.

To control these weeds, we must first look to seed destruction or prevention. Because the seeds germinate irregularly throughout the year and because the plants can flower and set seeds in such a short time, single or infrequent cultivations will not be sufficient to reduce seed populations. Because the plant is taprooted, shallow cultivation to sever the taproot just below the soil surface is probably best. Deep-plowing appears to have little advantage because the seeds are long lived in the soil. If fruits have formed, plants should be cut so that seeds cannot ripen.

In India *T. terrestris* is used in folk medicine as a tonic, diuretic, and as an aphrodisiac (Tadulingam and Venkatanarayana 1932).

Tribulus spp. are an alternate primary host for *Cuscuta hyalina* in India (King 1966).

COMMON NAMES

Tribulus terrestris

AUSTRALIA
bendy-eye
bindii
bull's head
caltrops
catshead
double gee
puncture vine
yellow vine

EASTERN AFRICA
caltrops
puncture vine

HAWAII
puncture vine

INDIA
chota gokhru
gokhru
ground bur-nut
kanti
nerinji (Tamil)
nerunjil
palleru (Telugu)

small caltrops

IRAQ
caltrops
gotob
Malta cross
puncture vine

LEBANON
caltrops
kutrah
Malta cross
·puncture vine

SOUTH AFRICA
common dubbeltjie
gewone dubbeltjie

UNITED STATES
burnut
caltrop
ground bur-nut
land caltrop
Mexican sandbur
puncturevine
puncture weed

Tribulus cistoides

AUSTRALIA
  rock rose

HAWAII
  carpet weed
  false puncture vine
  nohu

UNITED STATES
  burnut
  caltrops

VENEZUELA
  abrojo
  flor amarilla

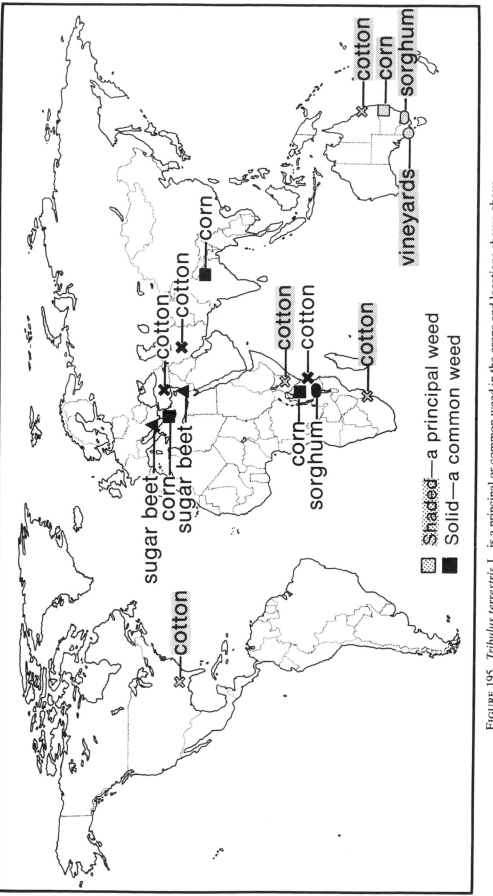

FIGURE 195. *Tribulus terrestris* L. is a principal or common weed in the crops and locations shown above.

❧ 75 ❧

Xanthium spinosum L.

ASTERACEAE (also COMPOSITAE), ASTER FAMILY

Xanthium spinosum, an annual, is a native of South America and has now spread throughout the world from lat 43° S to 50° N. It is a weed in 39 countries. It is the most stable *Xanthium* species and is easily recognized because of the three-pronged spines at the leaf axils. Largely a pasture weed, it grows along roadsides, river or creek flats, floodplains, and disturbed areas. It is frequently found in areas subject to periodic or shallow flooding. The spiny burs of this weed cause discomfort to animals and lower wool quality; the seedlings are poisonous to cattle, goats, horses, poultry, and sheep; and the species is troublesome in cotton, corn, peas, potatoes, sorghum, sugarcane, and vineyards. The toxic principle is hydroquinone which is present in the seeds and translocates throughout the seedling after germination.

DESCRIPTION

X. spinosum is an erect, rigid, much-branched, robust, *annual* herb (Figure 196); *stems* from 30 to 120 cm tall, striate, yellowish or brownish gray, finely pubescent; *leaves* alternate, usually three-lobed but sometimes simple, entire, or two-lobed, more or less lanceolate, central lobe longer than the others, thick in texture, dark gray-green above and slightly rough, with a conspicuous white midrib, the underside clothed with soft, short, white, wooly hairs, from 6 to 10 cm long, 2 cm wide, *petioles* short, about 1 cm long; each *leaf base* armed at the axil by yellow three-pronged spines 2 to 5 cm long, often opposite in pairs; *heads* in axillary clusters or often solitary; *flowers* inconspicuous, greenish, monoecious, male flowers in almost globular heads in axils of uppermost leaves, female flowers in axils of lower leaves, developing into a bur; *bur* two-celled, oblong, nearly egg-shaped, slightly flattened, 10 to 13 mm long, 4 mm wide, pale yellowish, more or less striate, glandular, covered with slender, hooked, glabrous spines from more or less thickened bases, the two apical beaks short and straight; *seeds* two per bur, flattened, thick coated, dark brown or black, the lower germinating first. The cotyledons are linear-lanceolate in shape, and differ in appearance from later leaves.

This species is distinguished by the conspicuous three-pronged spine at the base of the leaf and the egg-shaped bur which is covered with hooked, thorny prickles. Löve and Dansereau (1959) consider *X. spinosum* to be a more stable species than is *X. strumarium*.

DISTRIBUTION AND HABITAT

X. spinosum, a native of South America, grows from lat 43° S to about 50° N (Figure 197). It is widely distributed in the Mediterranean region and Europe, throughout most of Australia, in some coastal African countries, and in southern parts of South America and the United States. It is seldom found in the tropics.

Mainly a pasture or meadow weed, the plant grows along roads and in disturbed areas, and abandoned fields; it is sometimes common around waterholes and along floodplains, canals, ditches,

FIGURE 196. *Xanthium spinosum* L. and *Xanthium strumarium* L.
X. strumarium: 1, bur; *2*, prickle, enlarged; *3*, seedling; *4*, seedling.
X. spinosum: 5, leaf and trifid spines; *6*, bur.

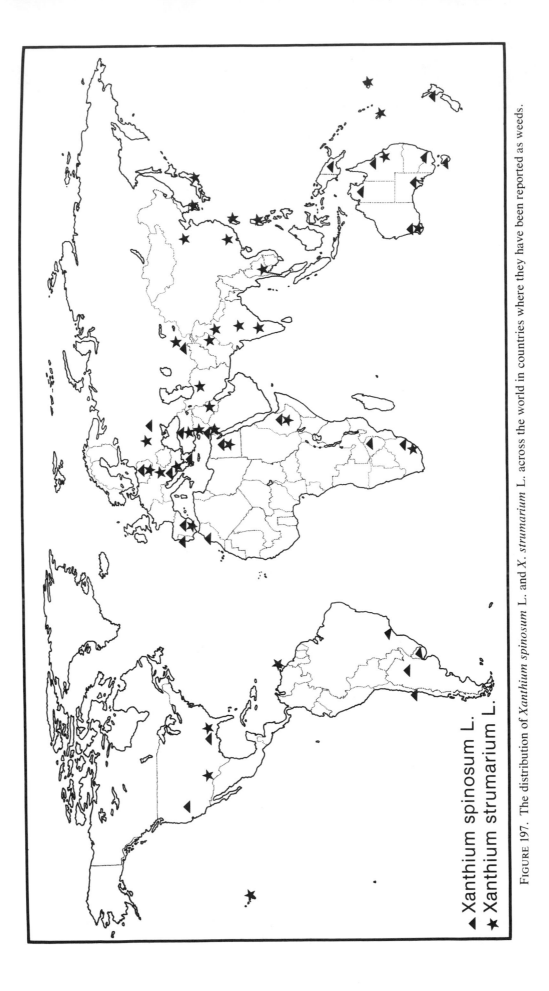

▲ Xanthium spinosum L.
★ Xanthium strumarium L.

FIGURE 197. The distribution of *Xanthium spinosum* L. and *X. strumarium* L. across the world in countries where they have been reported as weeds.

creek flats, river terraces, and other moist places. Often these are the same places where livestock go to obtain water. *X. spinosum* was probably spread around the world because of its prickly burs which cling to the hair or wool of animals, to clothing, or to feedsacks, canvas, and so forth. It is believed to have been introduced to Australia around 1840 on the tails of horses imported from Chile (Whittet 1968). It may have been introduced into South Africa on the wool of sheep. It usually appears first along roads; then animals, vehicles, or water spread the burs from there. The burs or seeds also are dispersed in impure seedstocks and in weedy hay.

X. spinosum is usually a summer annual, thriving in warm, moist soil. It grows best in open, unshaded areas and cannot stand dense crowding or intense competition.

PROPAGATION AND BIOLOGY

X. spinosum reproduces by seeds which are heteromorphic. About 150 seeds are produced per plant. The spiny bur of this plant has two elongated cavities, each containing one seed. The lower seed of the bur cavities has a shorter dormancy and so germinates first. Dormancy of the upper seed has been found to be due to low permeability of the seed coat to oxygen (Crocker 1906, Davis 1930). Higher temperatures of 32° to 38° C or higher oxygen partial pressures can overcome the dormancy. Both upper and lower seeds germinate readily if the seed coat is removed. The lower seed may germinate promptly or within a few months after maturity; the upper seed may not germinate for several months or even years. Such staggered germination means that the seedlings may emerge at any time over a period of several years during favorable growing conditions. The seeds tend to germinate during summer rains or whenever soil moisture is adequate. The plant has a high water requirement. Sometimes seeds germinate out of season, and the resulting plants produce fruits out of season.

Although *Xanthium* species have been used for a long time in photoperiod and flowering research, little has been done with *X. spinosum*. It flowers in January and February in Chile, from February to July in South Australia, and from July to September in the United States. Fruits are produced in September to November in the United States.

BIOLOGICAL CONTROL

Biological control of *X. spinosum* in Australia was reviewed by Wilson (1960). Although no insects

were introduced into Australia to control the weed, several do attack it. A seed fly (*Camaromyia bullans*) is widespread in Queensland and New South Wales. Although it does destroy large quantities of the seeds of the weed, it has had little important general effect on the weed population. Stem borers (*Zygrita diva* Thoms., *Corrhenes paulla* Germ., and *Lixus tasmanicus* Germ.) also attack the plant. Of these *Corrhenes paulla* has caused greatest mortality in some local situations.

A fungus disease caused by *Colletotrichum xanthii* has caused death of *Xanthium spinosum* in summer rainfall areas of New South Wales (Butler 1951, Wilson 1960). Most serious in black soil areas, the disease is favored by wet weather during summer and autumn. Seedling plants may be killed before the four-leaf stage; and in older plants lesions are formed on stems and they may be girdled. Seed production may be only 10 to 20 percent of normal and viability may be reduced. The disease can be established artificially in the field and will then persist naturally.

AGRICULTURAL IMPORTANCE

X. spinosum has been reported as a weed in 13 crops in 39 countries. It is a serious weed of cotton, corn, pastures, and sorghum; a principal weed of cereals; and a common weed of vineyards in Australia. It is a principal weed of potatoes and peas in Argentina and a common weed of sugarcane in South Africa. It is also found in rice, carrots, and wheat in one or more countries.

The importance of *X. spinosum* in animal production may overshadow its effects in crops. The plant invades pastures and grazing lands, causing reduced forage production. It is not palatable to stock so that it will increase at the expense of pasture species. The greatest losses may be to the quality of wool. The spiny burs become entangled in wool and cause discomfort to the animals. *X. spinosum* may also poison livestock.

To prevent movement of the seeds, Orchard (1949) recommended that sheep be shorn before being moved from infested to clean pastures.

X. spinosum has been reported as a host of *Orobanche ramosa* in California (King 1966); and of the fungi zinnia mildew (*Erysiphe cichoracearum*) and of *Verticillium dahliae*. It has been reported to be a medicinal plant in Argentina.

Frequent cultivation of fallow fields can control the weed. The use of 2, 4-dichlorophenoxy acetic acid and related compounds in pastures or the establishment and maintenance of vigorous improved pasture species are recommended control measures.

POISONOUS PROPERTIES
OF XANTHIUM SPECIES

Although the toxic chemical is not definitely known, Kingsbury (1964) has reported that the poisoning symptoms can be reproduced by administering a synthetic form of hydroquinone to animals. The most frequent cases of livestock poisoning in the United States occur with horses and swine, but other animals may suffer as well.

The poisonous compound is present in the seeds. When ingested at 0.3 percent of an animal's body weight, the seeds will cause toxicity; but this situation rarely occurs because the spiny burs are not palatable to animals. When seedlings have two cotyledonary leaves they are palatable and also have the highest toxicity. Plant material ingested at 0.75 to 1.5 percent of the animal's body weight will cause toxicity at this stage. Toxicity decreases rapidly as true leaves are formed. Evidence of poisoning appears in about 12 to 48 hours, the symptoms being nausea, vomiting, lassitude, depression, weakened muscles, and prostration. Severe poisoning may result in convulsions and spasmodic running movements in prostrate animals. Ruminants may not vomit. Death may occur within a few hours or days. Administration of fatty substances such as milk, lard, or linseed oil has been recommended as an antidote (Kingsbury 1964).

Poisoning usually results only when plants with cotyledonary leaves are eaten. This situation occurs most frequently at the edges of ponds, lakes, floodplains, or other bodies of water where shallow flooding followed by recession of the waterline occurs. Under such conditions seeds germinate readily, constantly supplying new generations of potentially poisonous seedlings as dry weather continues. Animals are sometimes attracted to such areas because of their need to obtain drinking water. The problem is accentuated because *Xanthium* seeds do have natural dormancy and germinate over long periods of time.

COMMON NAMES
OF XANTHIUM SPINOSUM

ARGENTINA
abrojillo
abrojo chico
cepa caballo

AUSTRALIA
bathurst bur
spiny burweed

BRAZIL
carrapicho do carneiro

CHILE
cepacaballo
clonqui
concli

EGYPT
shobbeit

MOROCCO
lampourde épineuse

NEW ZEALAND
bathurst bur

SOUTH AFRICA
boetebossie
boeteklis
boxopa
burweed
clotbur
cocklebur
mokwala
pinotiebossie
sehlabahlabane
spiny cocklebur

TURKEY
zincir pitraği

UNITED STATES
dagger cocklebur
daggerweed
spiny burweed
spiny clotbur
thorny burweed

URUGUAY
cepa caballo

YUGOSLAVIA
boca

❧ 76 ❧

Xanthium strumarium L.

ASTERACEAE (also COMPOSITAE), ASTER FAMILY

Xanthium strumarium is an exceedingly variable, annual herb which can be found across much of the world. Many "species" of *Xanthium* which have been reported are probably local variants of *X. strumarium*. Some "species" names applied to some of these include: *orientale, canadense, chinense, occidentale, macrocarpum, longirostre, pungens, echinatum, italicum, Cavanillesii, australe, californicum, saccharatum, pensylvanicum* and *oviforme*. Löve and Dansereau (1959) suggested seven complexes of *X. strumarium*, as follows: strumarium, Cavanillesii, oviforme, echinatum, chinense, hybrid and orientale. Sometimes a weed of crops, it is more frequently regarded as a problem because its leaves may poison livestock and its prickly burs will cling to wool, thereby reducing its quality and requiring special processing steps for their removal. It can become a problem in pastures. Like *X. spinosum* it is found along streams, rivers, in low-lying areas subject to flooding, and in waste places. It has been reported as a weed from lat 33° S to 53° N.

DESCRIPTION

X. strumarium is a coarse, erect, branching, *annual* herb (Figure 196); *stems* 30 to 150 cm tall, tough, with short dark streaks or spots and covered with short hairs which give a coarse texture; *leaves* alternate, triangular-ovate to broadly ovate in shape, 2 to 12 cm long, base often cordate, *petiole* 2 to 8 cm long, margins irregularly toothed or lobed, both surfaces rough-pubescent; *flowers* monoecious, male flowers in inconspicuous, many-flowered heads 5 to 8 mm across, clustered at the tips of branches or axillaries above the female flowers, female flower heads axillary, greenish, two flowers in the head enclosed by the involucre; *fruit* a hard brown, ovoid bur, 1.5 to 2.5 cm long, covered with hooked spines 2 to 4 mm long, and with two terminal beaks, fruits readily stick to clothing and fur, and thus are easily spread; *seeds* (achenes) black, two in each bur, one above the other.

Xanthium is a genus of cosmopolitan weeds that develops numerous races, based mainly on variations in the bur. The genus consists of 2 to 30 species, depending on botanical opinion. Most of these "species" can be placed under *X. strumarium*, according to a recent work by Löve and Dansereau (1959). The many forms of *X. strumarium* vary widely in plant size, fruit shape and size, and in their response to environmental factors.

TAXONOMY

In dealing with *X. strumarium*, a major taxonomic difficulty becomes apparent; that is, what constitutes the species itself and how shall we classify the many other "species" of *Xanthium*. Linnaeus described two species, *X. spinosum* and *X. strumarium*. From that point on, many "species," perhaps as many as 30, were described. Further work led to groupings, recombinations, and new names. Löve and Dansereau (1959), in a tentative reclassification of these names, have suggested that all *Xanthium* types, except for *X. spinosum* which is easily recognized and stable, belong to *X. strumarium* and can be grouped into complexes within the species. These complexes are: strumarium, Cavanillesii, oviforme, echinatum, chinense, hybrid, and orientale. Löve and Dansereau have suggested two subspecies of *X. strumarium*, *strumarium* and *Cavanillesii*. They have grouped some of the former "species" under the two subspecies as varieties, forms, or notomorphs. This classification has been presented as tentative, and further changes or revisions may be presented in the future.

DISTRIBUTION AND HABITAT

Figure 197 shows the world distribution of *X. strumarium* as a weed. It is found from about lat 53° N to 33° S and most frequently in the temperate zone. It is sometimes found in the subtropics and only infrequently in the tropics. It may have originated in the Mediterranean region. World distribution of *X. strumarium* has been discussed by Kaul (1965a).

A weed of wastelands or disturbed lands, *X. strumarium* is often associated with open grounds. It frequents roadsides, railway banks, small streams, and riverbanks, as well as the edges of ponds and freshwater marshes and poorly managed or overgrazed pastures. It cannot stand shading.

The weed can grow on a wide range of soils, from sands to heavy clays, and in a wide range of moisture supply. On rich soils with high moisture and little competition from other plants, it grows tall and luxurious, forming pure stands. In dry, poor soils, if the seedlings have become established, the plant may grow to only a few cm in height, persisting under drought, but still fruiting and setting seeds. This ability to grow under a wide range of conditions results in a constant seed supply if the plant is not controlled.

The spiny fruits provide the main means of dispersal of the weed. By becoming entangled in wool or other hair of animals, and in clothing or in the fabric of cloth feed sacks, tarpaulins, or other materials, the burs travel long distances. The fruits also float, moving downstream with floods or high water to germinate and grow along the edges of rivers and streams or in flat fields or pastures subject to flooding. McMillan (1971) suggested that the Noogoora bur, a form of *X. strumarium* in Australia, may have been carried to that country mixed with cotton seeds imported from Mississippi.

PROPAGATION AND BIOLOGY

The range of response of *Xanthium* to photoperiod is a factor in the wide adaptation of this complex species. Coupled with variable seed germination response, photoperiod plays an important role in survival and maintenance of stands of this annual weed.

As is true of *X. spinosum*, the two seeds in the fruit of *X. strumarium* germinate at different rates. The lower seed germinates more readily than the upper seed, which has a seed coat with low permeability to oxygen. Both seeds show dormancy, but the lower is less dormant than the upper. Kaul (1968) reported that *X. strumarium* seeds have a high water requirement for germination. (For a further discussion of seed germination, see the section on propagation and biology for *X. spinosum*, number 75). Seedling development is very similar to that of *X. spinosum*.

Seeds of *X. strumarium* may germinate the year round in the subtropics, as early as February in Texas, and as late as June in Canada. This wide range in germination with season is largely due to higher temperature effects in warmer climates. Plants which germinate early will have a long period of vegetative growth and become very large before a suitable photoperiod arrives to send them into flowering. Plants which germinate late, when the proper photoperiod is already upon them, may flower when small.

X. strumarium has been used widely as an experimental plant in studies of photoperiod. Salisbury (1969) reviewed these studies in detail.

Xanthium is a short-day plant and usually will not flower if the day length exceeds 14 hours. However, there is evidence of differences in light response between "complex" groups for some plants will flower in a day length as long as 16 hours. At high latitudes, day length is greater than 14 hours during summer; for that reason, *X. strumarium* does not flower until late summer when daylength shortens enough to stimulate flowering. Seeds also ripen late under these conditions—usually in early autumn.

The photoperiod behavior in *Xanthium* has been shown to be a response to the length of the dark period. One dark period of suitable length (about 8 to 10.5 hours, depending upon the plant "complex") is enough to induce flowering when leaves of suitable age are exposed. Cotyledonary leaves do not play a role in flower induction.

Detailed studies of *X. strumarium* growth and behavior in India have been published by Kaul (1959; 1961; 1965*a,b,c;* 1967, 1971).

AGRICULTURAL IMPORTANCE

X. strumarium has been reported as a weed in 11 crops in 28 countries, but it has not been reported as a serious or principal weed in these crops. Crops in which it has been reported as a weed include abaca, corn, cotton, orchard crops, lowland rice, sugar beets, sugarcane, wheat, upland rice, and pastures.

X. strumarium, like *X. spinosum*, is probably most important in livestock production. It becomes a serious weed in pastures which have degenerated because of overgrazing or drought, pastures in which desirable species have died or thinned out. In such open conditions, the seeds, which are moved about by animals or water, germinate with the onset of rains and favorable weather to create pure stands. It is then that the weedy nature of the *Xanthium* is expressed. It is also at this time, when feed is short, that hungry animals may feed on the germinating seedlings of *Xanthium* and become poisoned (see the section on the poisonous properties of *Xanthium* species in the discussion of *X. spinosum,* number 75).

The burs of *X. strumarium* will lodge in wool, reducing the quality and requiring a special acid treatment for their removal (Meadly 1958).

COMMON NAMES
OF XANTHIUM STRUMARIUM

AUSTRALIA
noogoora bur

HAWAII
cocklebur
kikania

INDIA
adhisishi
bada gokhru bhakra
bur weed
chota dhatura

IRAN
cocklebur

IRAQ
bathurst bur
burweed
clotbur
cocklebur
hasach
lizzage
sheepbur

JAPAN
ōnamomi

LEBANON
clotbur
cocklebur
kharika-ul-bahr
sheepbur

MALAYSIA
buah anjang

PAKISTAN
puth kando

SOUTH AFRICA
kankerroos
large cockle bur

TAIWAN
tsai-er
ye-chye

THAILAND
kachab

TURKEY
siraco otu

UNITED STATES
cocklebur
ditchbur
heartleaf cocklebur
sheepbur

PART TWO
The Crops

Cassava, *Manihot esculenta* Crantz

EUPHORBIACEAE, SPURGE FAMILY

Cassava, a tropical root crop, is native to Central and South America. Some of its close relatives are rubber-bearing trees of northeastern Brazil, where about 100 species of *Manihot* are found. The plant was probably domesticated in southern Mexico, Guatemala, and northeastern Brazil.

Cassava is a medium-sized, short-lived perennial shrub which may grow 2 to 4 m in height. The upright stems are knobby, rather woody, smooth, and sometimes branch from the base or laterally, thereby forming an erect branched growth or a spreading habit. The leaves have long petioles and are deeply lobed, with as many as three to seven lanceolate, entire lobes. The fleshy cylindrical roots are harvested for starch and other uses. The tubers, which arise from swellings on adventitious roots near the base of the stem, may reach a weight of 15 to 20 kg and a length of 1 m or more. Shorter and smaller tubers are more useful for commercial purposes and are easier to harvest than are the massive, long tubers produced by some varieties.

Most cassava is produced in tropical lowlands, but the crop sometimes appears as high as 1,500 m near the equator. It cannot stand cold or frost but will tolerate widely varying conditions within a rainfall range of 500 to 5,000 mm. Perhaps it is not commonly known that cassava produces more food under relatively dry conditions than does any other crop. As much as 35 metric tons per hectare have been recovered. For this reason and because cassava is very resistant to plant diseases, it has long been used as a subsistence crop or as an emergency food source in tropical countries.

Because cassava usually is regarded as a subsistence crop of tropical countries, it has received little attention in the literature; however, it is actually one of the major sources of food of the world. It now ranks as the sixth largest source of staple food. For example, the world production in 1968 has been estimated as being over 80 million metric tons grown on 9.4 million hectares of cropland. The annual harvest of cassava is significantly higher than the combined yield of millets and sorghum, and it is greater than the total production of oats and rye in the world. Almost one-half of the land devoted to cassava production is in Africa where average yields are low and total production is about 30 million metric tons. Like some other root crops, the true crop area of cassava is difficult to estimate, for it is grown widely as a backyard or garden crop throughout the tropics. The world's leading producer is Brazil, with 32 percent of the total world production on 1.9 million hectares, followed by Indonesia with 14 percent, Nigeria with 10.4 percent, and the Congo with 8.5 percent. Thailand production is sold in Europe for animal feed. The average yield in Brazil is 14 tons per hectare.

The uses for cassava are many. Perhaps its best known product is tapioca which is prepared by grinding or grating the peeled tubers, washing the starch with running water, and then gently heating the starch to form small pellets. However, cassava also provides fresh or processed animal feed, the feed including such forms as dried chips and silage. In addition, the leaves and stems have been fed to animals in feed mixtures. The leaves have been used to feed silkworms in China (Martin 1970).

Because many varieties contain hydrocyanic acid, cassava is normally cooked or processed in some way before being used as food for the table. The processed products include tapioca, cassava beer, chips, and meals or flours used for bread or as food supplements. Not all varieties contain hydrocyanic acid, those varieties without this substance often being referred to as "sweet" types. Varieties having a low hydrocyanic acid content are often early maturing with crop durations as short as 6 to 12 months; whereas "bitter" types, those having a high hydrocyanic acid content, have crop durations ranging from 1 or 4 years. The bitter types are often planted as emergency food crops.

Cassava has the potential to become a major industrial starch. In Brazil cassava starch is used for foodstuffs, paper, textiles, plywood, adhesives, and ice cream cones. As manioc meal it is eaten with meats, beans, or fish. Manioc meal has even been used as a seal and lubricant in drilling oil wells. Other products which have been investigated include dextrins, pregelatinized starch, glucose, ethyl alcohol, and "vegetable cheese" (Normanha 1970).

CULTURAL REQUIREMENTS

Cassava is very hardy and will give good yields where other crops will fail or be seriously damaged by birds, insects, or diseases. The crop has suffered severe virus attacks in eastern Africa; this has been the most severe disease problem of record.

The crop grows in extremely variable conditions of rainfall, soils, and management regimes. Except at the time they are planted, the plants can tolerate drought and will recover quickly when moisture conditions improve. Crop yields are very high under high soil fertility but are also reasonably good under low fertility. Cassava is reputed to be a soil-exhausting crop, but this may be due to the fact that it produces under low soil fertility. It is often the last crop to be grown in a shifting cultivation system. Although the crop will grow on almost all soils, it does not do well on stony or shallow soils.

Cassava is grown most frequently as an intercrop in perennial crops, or in other short-term food crops. In some areas it may be grown alone, especially as a commercial crop for export. It is propagated by stem cuttings 0.2 to 0.4 m in length. Cuttings are planted in furrows or in holes, usually in rows spaced 1 m apart and with 0.3 to 0.5 m between plants in the row. Cuttings may be planted horizontally, vertically, or at an angle. Though deep cultivation will favor root development, it is seldom practiced. Instead hilling up is sometimes done 3 months or so after planting. Fertilizers are seldom used.

Control of weeds is necessary from the time the crop is planted until it closes in and provides heavy shade. Late cultivation can cause breakage of lateral parts growing both above and below ground.

Harvesting begins at 6 to 12 months after planting in short-term varieties and from 1.5 to 2 years or more in long-term varieties. For table use the crop may be harvested over a period of months (Purseglove 1968).

MAJOR WEEDS OF CASSAVA

Most troublesome weeds of cassava are those which cause problems in annual or short-term, perennial, tropical crops (Figure 198). A number are reported to be problems in various countries and of these, *Ageratum conyzoides*, an annual broad-leaved herb, is reported most frequently. It has been reported to be the number-one weed in Taiwan, a serious weed in Ghana, and as a weed in the crop in Nigeria and Indonesia. Other annual broad-leaved herbs of importance are *Amaranthus spinosus*, a principal weed in Ghana and reported in the crop in Taiwan and Nigeria; and *Tridax procumbens*, which can be annual or perennial, and is a principal weed in Ghana and a common weed in Taiwan and Indonesia. *Euphorbia heterophylla* is a principal weed in Ghana.

The perennial sedge *Cyperus rotundus* is a serious weed in Ghana and a weed in Indonesia and Taiwan. Annual grasses also cause difficulty. Of these perhaps *Eleusine indica* may be most important, for it is a serious weed in Taiwan, a principal weed in Thailand, a common one in Indonesia, and a weed of the crop in Nigeria. *Digitaria adscendens* is a principal weed in Thailand and Taiwan, and *Dactyloctenium aegyptium* is a common weed in Ghana, Taiwan, and Indonesia. *Echinochloa colonum* is a principal weed in Thailand. Other weeds of the crop are *Bidens pilosa*, *Euphorbia hirta*, *Mimosa pudica*, and *Physalis angulata*.

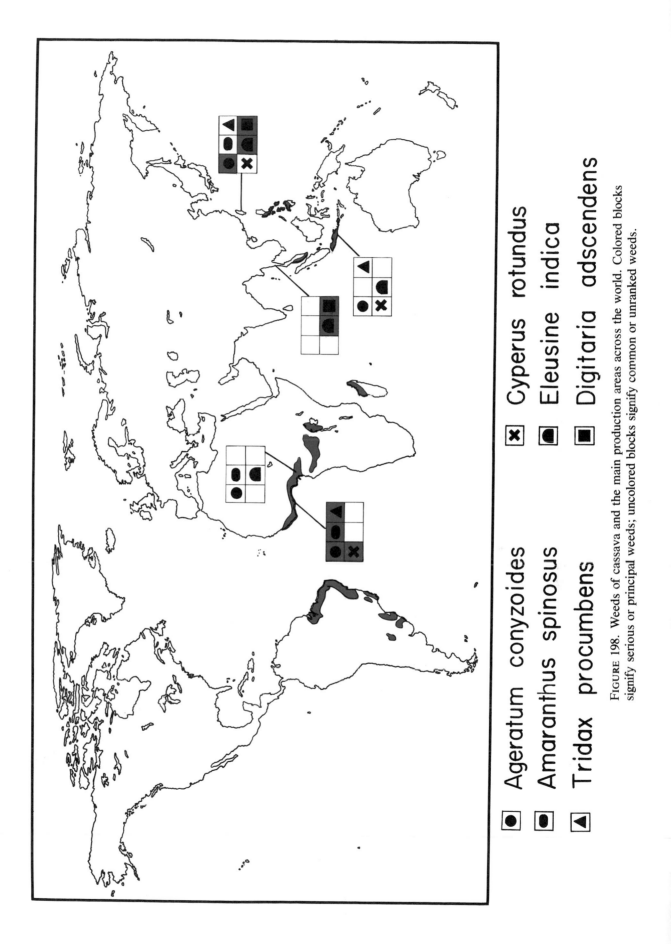

FIGURE 198. Weeds of cassava and the main production areas across the world. Colored blocks signify serious or principal weeds; uncolored blocks signify common or unranked weeds.

☒ Cyperus rotundus
◐ Eleusine indica
◼ Digitaria adscendens

◉ Ageratum conyzoides
◖ Amaranthus spinosus
◄ Tridax procumbens

Coffee, *Coffea arabica* L. and *Coffea canephora* Pierre ex Froehner

RUBIACEAE, COFFEE FAMILY

Brazil produces one-third to one-half of the world's coffee and its neighbor Colombia is the second greatest producer. Mexico, El Salvador, Guatemala, and Costa Rica are important producers. Ecuador, the Dominican Republic, Haiti, Venezuela, Nicaragua, Jamaica, and Trinidad are other exporters. Most of this crop is *Coffea arabica* L., commonly called arabica coffee. It originated in the high forest of Ethiopia and demands more shade and cooler conditions than the hardier *Coffea canephora* Pierre ex Froehner, commonly known as robusta coffee, which had its origin in West Africa. Although Africa is the home of the coffee tree, it produced only 1 percent of the world's crop before 1920; but this figure has now been increased to 20 percent. Most of it is the robusta type which is in great demand for instant coffee. Principal African producers are Angola, Cameroon, Ethiopia, Ivory Coast, Kenya, Malagasy Republic, Tanzania, Uganda, and Zaïre. A small amount of coffee also is produced in India, Indonesia, the Philippines, and the Pacific Islands. Ceylon once had a sizeable coffee industry in the highlands, but that industry was wiped out by leaf rust (*Hemileia vastatrix* Berk. & Br.) at the end of the 19th century. This disease now threatens the industry in Brazil. Statistics are difficult to obtain but some observers believe that most of the world's coffee is produced by smallholders.

Coffee is the principal beverage of continental Europe and the United States. The United States alone imports one-half the world's coffee. In 1962 coffee was the largest single item imported into the United States, accounting for 6 percent by value of the total imports.

CULTURAL REQUIREMENTS

Arabica coffee grows naturally near the equator in Ethiopia as an understory tree in forests growing at 1,500 to 2,000 m elevation. In general, optimum conditions are found near the equator at these elevations. Here temperatures are 15° to 25° C, and rainfall is 1,500 to 2,500 mm annually with good distribution except for a dry period of about 10 weeks during which the first flower buds of the coffee tree appear. The plant cannot tolerate strong winds. Periods of mist and low clouds are favorable to its growth. Coffee may be grown at sea level in the subtropics, but arabica coffee is more subject to leaf rust at these elevations. In Brazil coffee is grown as far south as lat 25° S where it is sometimes killed by frost. The species prefers slightly acid, deep, well-drained, fertile, loam soils. It is an evergreen plant and thus requires subsoil moisture at all times. If the soil is heavy or the water table high, root penetration is poor. The use of shading varies greatly across the world. Much of the coffee in Brazil is grown in full sun because shade trees would compete for soil moisture. Under ideal conditions of growth and careful management shade is not needed; and yields will be higher if the crop is grown in full sunlight. If there is so much light that trees have a tendency to over-bear, or if the rainfall, temperature, and soil of the area are not favorable, then shade must be used. Coffee can be grown under irrigation as in Yemen, for example.

Robusta coffee is not so specific in its requirements and is much more tolerant of adverse conditions and poor management. Many coffee types (varieties) can be grown only in areas where soil and climate are particularly favorable to their particular requirements. One of the most sought after coffees in the world is the "Blue Mountain" variety which was developed in Jamaica but is also grown in Kenya. Only a small amount is produced and it is always in short supply.

The longevity of a plantation obviously depends a great deal on how well it is managed. In Brazil the average life of a coffee tree is said to be 30 to 40 years but some even live to be 100 years old. All misses or blank places in the spacing of coffee trees must be filled early because it is difficult to establish the crop if other trees have become dominant.

New coffee plantings can be intercropped for only about 2 years because coffee is very sensitive to competition. Perennial weeds are particularly harmful if soil moisture is marginal. Because clean weed-ing and tillage may bring on soil erosion problems, a mulch is often used to hold and enrich the soil and to control weeds. Cover crops are not commonly used because they reduce coffee yields (Purseglove 1968, Ochse et al. 1961).

MAJOR WEEDS OF COFFEE

From Figure 199 it may be seen that *Cyperus rotundus*, a perennial sedge, and *Cynodon dactylon*, a perennial grass, are in most of the areas where coffee is grown and they are principal weeds in about half of the locations. The remaining three species shown are all annual broad-leaved weeds and they are all present in Africa, Central America, and South America where most of the world's coffee is grown. They have not been reported as serious weeds in Asia. The perennial grass *Digitaria scalarum* is a very difficult weed to eradicate wherever coffee is grown in eastern Africa. The annual grasses *Eleusine indica* and *Digitaria sanguinalis* are principal weeds of coffee in Brazil and the Philippines. *Paspalum conjugatum* is distributed across the world in Zaïre, Costa Rica, Mexico, New Guinea, and the Philippines, and it is ranked as a principal weed in most of those locations. *Sida acuta* is a weed of coffee in Colombia.

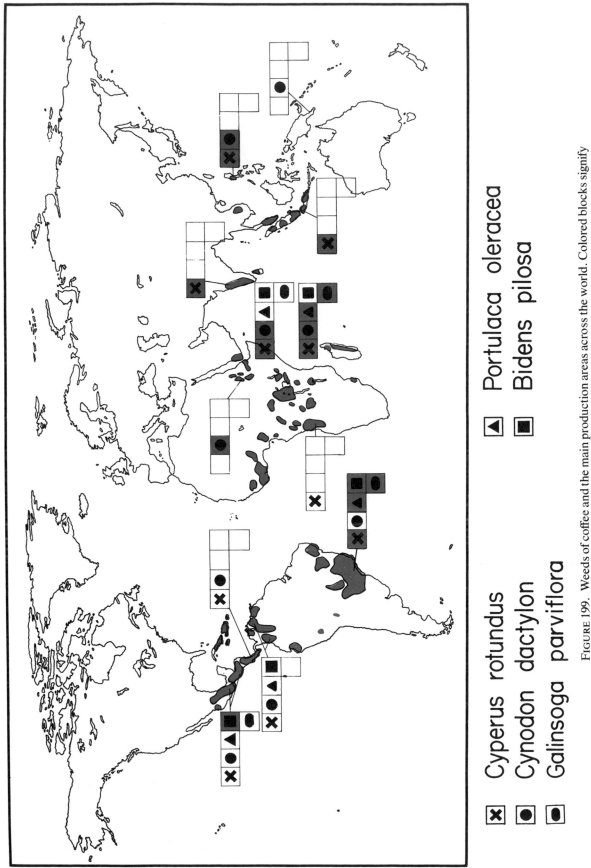

FIGURE 199. Weeds of coffee and the main production areas across the world. Colored blocks signify serious or principal weeds; uncolored blocks signify common or unranked weeds.

⊠ Cyperus rotundus
⊙ Cynodon dactylon
◙ Galinsoga parviflora

◀ Portulaca oleracea
■ Bidens pilosa

Corn, *Zea mays* L.

POACEAE (also GRAMINEAE), GRASS FAMILY

Although corn (maize) is generally considered to be indigenous to the Western Hemisphere, some botanists believe that the species may have arisen in Asia and Africa. It was the main staple food of the people of the Americas for more than 40 centuries. In recent times it has come to rank third in terms of total food calories supplied to the world.

In the United States, the largest producer, the crop was a main food staple for humans in an earlier time, but tremendous increases in yield have now made corn the most important feed grain. One out of every 4 hectares is planted to corn. Through crop improvement, better systems of management, and mechanization, the crop in recent times has expanded into many new areas of the world and yields have risen sharply. In many countries the yields have improved by 25 to 50 percent in the last 2 or 3 decades. By 1960 in the United States, for example, 95 percent of corn acreage had been converted to a crop derived from hybrid seeds.

Corn is an important human subsistence crop of Africa and Asia, but most of the crop enters international trade as feed for animals. Large quantities are also converted to glucose, breakfast foods, starch, oils, dextrins, and alcohol. About 40 percent of the crop is produced in the United States, with Brazil, Mexico, Romania, the Soviet Union, and Yugoslavia being the next leading producers. About 15 percent of the world's corn is harvested in Europe.

CULTURAL REQUIREMENTS

Corn is most intensively grown in the temperate zone, although it is very susceptible to even a few hours of frost; therefore, the principal factor limiting distribution is the length of the frost-free period. On the average across the world the optimal grain yields come from areas with a 120- to 140-day growing season. The crop grows well at a mean summer day temperature of 24° C and a night temperature of 14° C. The seeds germinate and emerge very slowly at cool temperatures. Growth stops at 13° C. In a suitable warm period the plants may emerge only 5 or 6 days after the seeds have been planted. Annual rainfall must be 600 to 1,000 mm with adequate moisture during the rapid growth at midseason. Maize is often grown with irrigation in drier regions. There are now varieties of corn which are much more adaptable to tropical and subtropical conditions. In these areas growth is often limited by the onset of the dry season and by severe disease problems.

Aside from temperature limits the plant is able to accommodate a wide range of conditions and cultural practices. In many areas it can be planted over a period of 30 to 60 days. The crop grows successfully on a very wide range of soils if fertility is high and if nitrogen is available in the later phases of growth. Plant populations may vary from 25,000 to 75,000 plants per hectare and row widths from 25 to 105 cm. Seeds are normally planted at a depth of 2.5 to 10 cm but in arid regions they may be planted at 25 cm. The seeds can absorb moisture from drier soils than most crops. Seed bed preparation may vary from the traditional system of plow-disc operation with four or five later cultivations, to no tillage at all. Fertilizers are applied before or at planting, and may be side-dressed with liquid, dry, slurry, or gaseous fertilizers later.

The root system of corn is interesting. As the plant emerges and begins early growth, its primary root system ceases to function and a secondary root system arises in successively higher whorls beneath the soil surface. Later, brace roots develop at the first, second, and even the third node above the soil. These roots grow downward, branch repeatedly in the surface soil, and become heavy feeders for nutrients.

There have been many improvements in corn growing in recent years, chief among them being the continuing replacement of open-pollinated corn varieties by hybrids. This has made it possible to grow the crop in areas in which there were no available varieties previously. Better control of soil insects, increased use of fertilizers, and improved machinery have increased the efficiency of production and raised yields. Improved drying facilities for picker-shelled corn and more satisfactory storage facilities have made the crop easier to handle.

At maximum dry weight, the grain in most varieties has a moisture content of about 30 to 35 percent. At this point it is physiologically mature, and fields which are used for seeds are harvested as soon as possible. Corn to be used for other purposes should be taken at 26 to 28 percent moisture if it is to be dried artificially, and at a lower moisture content if it is to be placed directly in a crib. Corn is usually taken as soon as it is dry enough to avoid losses from lodging caused by stalk rot and to avoid the wet, stormy weather which builds up in many areas at this season.

In conventional tillage the fields are plowed and a spring-tooth or disk harrow is used for leveling; the crop is cultivated several times, the first when it is about 7.5 cm in height. To accommodate extremes of moisture a lister plow may be used to make a shallow furrow as soil is thrown in both directions. The seeds may be planted in the furrow or up on the ridge.

A chisel is sometimes used in areas where there are critical wind and soil erosion problems. The narrow tines or chisels penetrate to plow depth. The method requires soil dry enough to shatter and a surface clean enough that residues of former crops do not interfere. Chiseling works well following many vegetable crops or soybeans. Reduced or minimum tillage systems are quite variable, but in general both plowing and planting are done in a single operation. A seedbed may be prepared directly in sod or stubble in one pass but in the row area only. Sometimes, however, such systems refer to the practice of leveling with a single harrowing or with sweeps mounted on a cultivator tool bar just ahead of the planter.

Herbicides are frequently used for weed control in all of the above operations. In a zero tillage system, however, they must be used. Here the planting is done in a slit made by a colter or special shoe without turning or stirring the rest of the soil. This practice may be used following row crops or in sod or stubble; the weeds are controlled with herbicides because the area cannot be cultivated. This method, like chiseling, is advantageous in areas of wind and water erosion. In many areas it is used to plant corn immediately after a hay crop is cut or small grains harvested (Inglett 1970).

Corn is especially sensitive to weed competition when it is under stress for moisture or nutrients. Weed control between the rows is not enough to prevent competition with the crop. Moolani, Knake, and Slife (1964) have shown in a 3-year experiment in the United States that the annual broad-leaved weed, *Amaranthus hybridus*, when left in the corn rows, will seriously reduce yields. Weed plants spaced at 100 cm in the row reduced yield by 290 kg per hectare, and weeds at a 2.5-cm spacing in the row reduced the yield to 2,300 kg per hectare.

When phenoxy herbicides were developed, corn became one of the first major crops to benefit from the use of selective chemical weed control. Because of the large acreages involved and variations in growing conditions many new herbicides have since been developed for specific weed problems. Following an early beginning, in some regions the same herbicide was used routinely for many years, and thus it was that in corn culture the first awareness came that an ecological shift of weed species was inevitable with continued use of one type of herbicide on a regular basis. A suitable rotation of crops and herbicides will prevent a shift to weed species that may become increasingly more difficult to control.

MAJOR WEEDS OF CORN

The four species shown in Figure 200 are judged to be the most important in corn across the world because they are so consistently reported to be serious and principal weeds, and also because they are so widespread on all of the continents. *Cyperus rotundus* is the most widely distributed, being in at least 12 or more additional countries not shown on the map. From examining almost 1,000 reports on corn, we learned that *Echinochloa colonum* has been judged a serious or principal weed in every

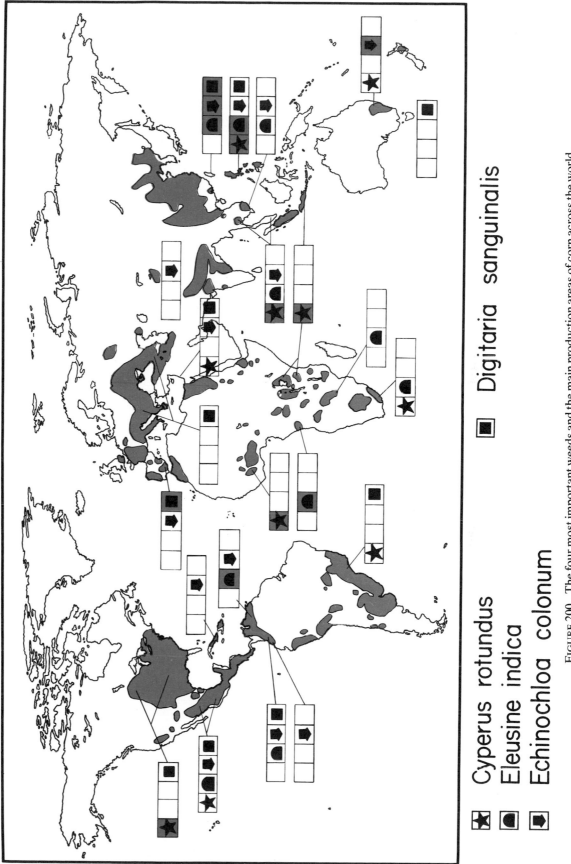

FIGURE 200. The four most important weeds and the main production areas of corn across the world. Colored blocks signify serious weeds; uncolored blocks signify principal weeds.

★ Cyperus rotundus   ■ Digitaria sanguinalis

◖ Eleusine indica

➡ Echinochloa colonum

country that reported it. In addition to the locations shown on the map, *Eleusine indica* and *Digitaria sanguinalis* are common weeds of corn in several countries on four continents.

In Figure 201 are shown two perennial grasses, *Cynodon dactylon* and *Sorghum halepense*, another annual grass, *Echinochloa crusgalli*, and, *Portulaca oleracea*, the only broad-leaved weed to rank high among the weeds of corn on a world basis. Although these weeds are very often reported to be principal weeds of the crop, they are not often ranked among the most serious. They are widespread but some are concentrated in regional areas, as may be seen by the empty boxes in the figure.

Amaranthus retroflexus, *Amaranthus spinosus*, and *Chenopodium album* are next in importance, and all are reported to be principal weeds in at least six countries. Three perennial weeds deserve mention. *Cyperus esculentus* is important in four African countries where it is sometimes a serious or a principal weed; it is a weed in the United States and Mexico as well. Many of the weeds important in corn on a world basis are not a problem in southern and eastern Europe, but *Convolvulus arvensis* is a troublesome weed in France, Greece, Hungary, Portugal, the Soviet Union, Spain, and Yugoslavia. It is also a problem in Latin America, the Near East, North America, and the Philippines. *Cirsium arvense* is also troublesome in several countries of southern and eastern Europe, and in New Zealand and the United States as well.

Rottboellia exaltata deserves special mention because it is a serious weed in Ghana, Rhodesia, the Philippines, and Zambia, and is also found in corn in Colombia, Nigeria, Tanzania, and Venezuela.

Bidens pilosa is widely distributed in South America and Africa, and is found also in Hawaii and Taiwan. *Striga lutea*, a parasitic weed, is reported to be a weed of corn in more than 20 countries. The following are not very widespread but are reported to be principal weeds of corn in three to five countries: *Ageratum conyzoides*, *Agropyron repens*, *Digitaria adscendens*, *Galinsoga parviflora*, *Leptochloa panicea*, and *Setaria verticillata*.

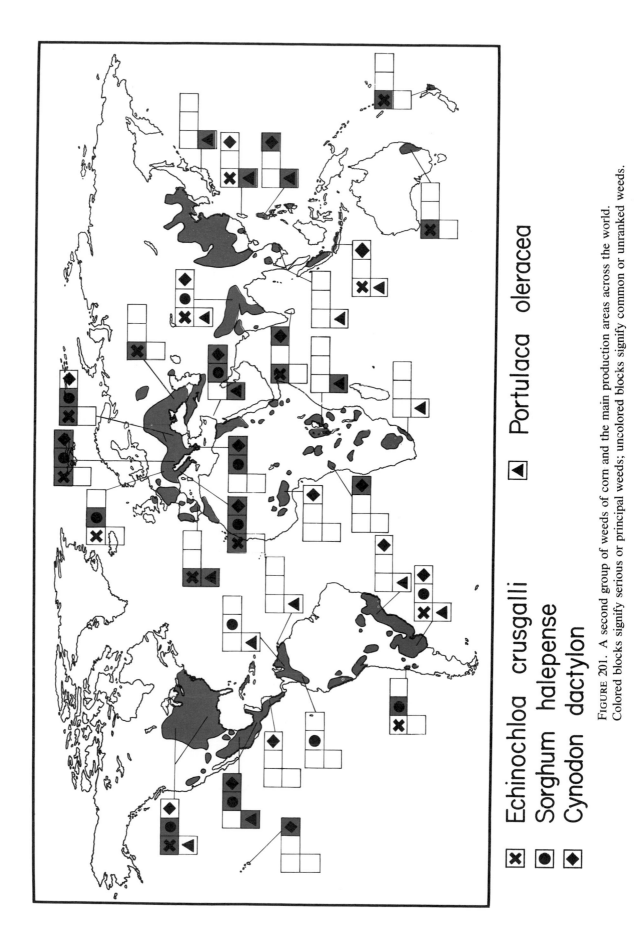

Echinochloa crusgalli   ▣   Portulaca oleracea

☒   Echinochloa crusgalli
◉   Sorghum halepense
◆   Cynodon dactylon

FIGURE 201. A second group of weeds of corn and the main production areas across the world. Colored blocks signify serious or principal weeds; uncolored blocks signify common or unranked weeds.

Cotton, *Gossypium hirsutum* L., *Gossypium arboreum* L., *Gossypium barbadense* L., and *Gossypium herbaceum*

MALVACEAE, MALLOW FAMILY

Three genera of the *Malvaceae* provide fibers of commerce. *Gossypium* is the cotton which is used for clothing, sheeting, cordage, and bags. The bast fibers from the stems of *Hibiscus cannabinus* L., sometimes 3 meters long, are used for sacks, canvas, cordage, and fish nets. This is the kenaf of commerce. The aramina fiber of Brazil and the Congo jute of Zaïre are *Urena lobata*. The stem fibers are used for making carpets and rope. *U. lobata* is a weed in several places in the world and has been declared a noxious weed of Fiji.

The hairs which form on the epidermis of cotton seeds have become the most important vegetable fiber of commerce. The lint is formed from single epidermal cells by extension outward of the external walls. Spiral bands of cellulose are then laid down internally. Cotton began to replace wool as man's most important fiber during the 19th century when machines became available for processing. Competition from synthetic fibers has resulted in a decrease from peak production in the 20th century. Oil from the seed is used to manufacture margarine, salad and cooking oils, and other food products. Inferior grades of oil are used for soap, lubricants, and many other products. The cake or whole seed may be used for cattle feed or supplements and the hulls for animal roughage, fuel, or fertilizer. A fuzz is left on the seed after ginning and this is removed for use in a wide variety of industrial products such as carpets, felt, mattresses, paper, lacquer, photographic film, and explosives.

The United States produces about 25 percent of the world's cotton, followed by the Soviet Union, China, and India. Egypt, Mexico, Sudan, Brazil, and Pakistan are important producers. There is also some production around the Mediterranean Sea, in eastern and western Africa, the Caribbean area, and north of lat 20° S in South America.

The *Gossypium* species are divided into eight sections of which six are wild lintless species. The Old World linted diploid species, *G. herbaceum* and *G. arboreum*, and the New World linted tetraploid species, *G. barbadense* and *G. hirsutum*, are in the remaining two sections. The cultivars of the species are classified according to length of lint, with the medium and medium-long staple types providing more than three-fourths of the world's cotton. They are mainly rain-grown and are produced in the United States, India, the Soviet Union, Brazil, and Mexico. These are cultivars (varieties of cultivated crops) of *G. hirsutum*. The rather coarse, short, staple lints of *G. herbaceum* and *G. arboreum* are rain-grown crops in India, Pakistan, and China. The long staple and extra-long staple varieties of *G. barbadense* are rain-grown or irrigated and they provide about 15 percent of world cotton lint. The largest producers are Egypt, the Soviet Union, Peru, Sudan, the United States, and Uganda.

Cotton is produced from northern Argentina to the central United States in the Western Hemisphere. In the Eastern Hemisphere production extends much farther north—to about lat 45° N near the Black Sea. The crop does not grow well above 1,000 m in India or 1,600 m in tropical Africa. The northern limit in the United States for cotton culture is the latitude at which there is a frost-free period of 200 days. Much of the rain-grown cotton is found in areas receiving 1,000 to 1,500 mm of rainfall annually. A period of dry weather is needed for maturation and harvesting. In dryer areas the crop must be irrigated; in heavy rainfall areas the crop cannot be grown.

CULTURAL REQUIREMENTS

Cotton will not tolerate shading from interplanted crops or from too-heavy stands, and it responds to a lack of sunlight with a delay in fruiting and increased shedding of bolls. The perennial cottons are medium- and short-day plants but the annuals are day neutral. Low temperatures promote vegetative growth and high temperatures favor flowering and fruiting.

Land preparation is held to the minimum necessary for clearing weeds although the seedbed must be firm. In Sudan cotton presently follows a bare fallow, but more economical rotations are being worked out. Short-term crops such as peanuts and beans are sometimes interplanted with cotton. In eastern Africa and elsewhere in smallholder agriculture, cotton is often grown in rotation with annual food crops. In hand-planting, several seeds are placed in a hill and plants are thinned to two when 15 to 25 cm high. The spacing in the row and between the rows is heavily dependent on soil, amount of rainfall, and several other factors in the environment. Mechanical planting dictates a wide row width. After thinning, cultivation is done only when needed for weed control (Cobley 1956, Ochse et al. 1961, and Purseglove 1968).

MAJOR WEEDS OF COTTON

Well over 100 species of plants have been reported to be weeds in cotton. Perhaps more important is the fact that several of the world's worst weeds are troublesome in cotton. Those which are most important are shown in Figure 202, and, because they are so widely distributed, only those places where they are reported to be serious or principal weed problems can be indicated. *Portulaca oleracea* and *C. dactylon* are in 12 or more countries that are not shown, and *Cyperus rotundus* and *Eleusine indica* are each in six or more countries on four continents.

The weeds next in importance are: *Sorghum halepense*, *Echinochloa crusgalli*, *Echinochloa colonum*, *Dactyloctenium aegyptium* (Figure 203). These, too, are often reported to be serious or principal problems but they are not as widespread. Each species is found in the cotton fields of Africa, Asia, Australia, Europe, Latin America, and North America.

Although this fact is not shown on the maps, *Digitaria sanguinalis* is of almost equal importance with some of the species cited above. It is reported to be a serious weed in Spain, Turkey, and Swaziland; a principal weed of cotton in Israel and the United States; and is found in Brazil, Colombia, and Guatemala as well.

Special weed problems are *Trianthema portulacastrum* which is a principal weed in Australia and Nicaragua but is also found in Guatemala and Thailand. *Physalis angulata* is a principal weed of Mexico and Venezuela and *Leptochloa panicea* is a special problem in cotton in Mexico. *Rottboellia exaltata*, a difficult weed to eradicate which is spreading rapidly in several places in the world, is found in cotton in Africa and South America. It is reported to be serious in Rhodesia and Zambia, and is troublesome in Colombia, Ethiopia, Mozambique, Sudan, Uganda, and Venezuela.

Tribulus terrestris is in cotton on five continents. It is a principal weed in Australia, Kenya, Swaziland, and the United States, but is also found in India, Iran, the Soviet Union, Tanzania, and Turkey. Two principal perennial weeds are *Panicum maximum* in Colombia and *Cyperus esculentus* in the United States; both are also important in eastern Africa. *Convolvulus arvensis*, a perennial broad-leaved species, is a principal weed in cotton in several countries at the eastern end of the Mediterranean Sea; *Solanum nigrum*, an annual broad-leaved species, is a principal problem in the same area and in Australia as well.

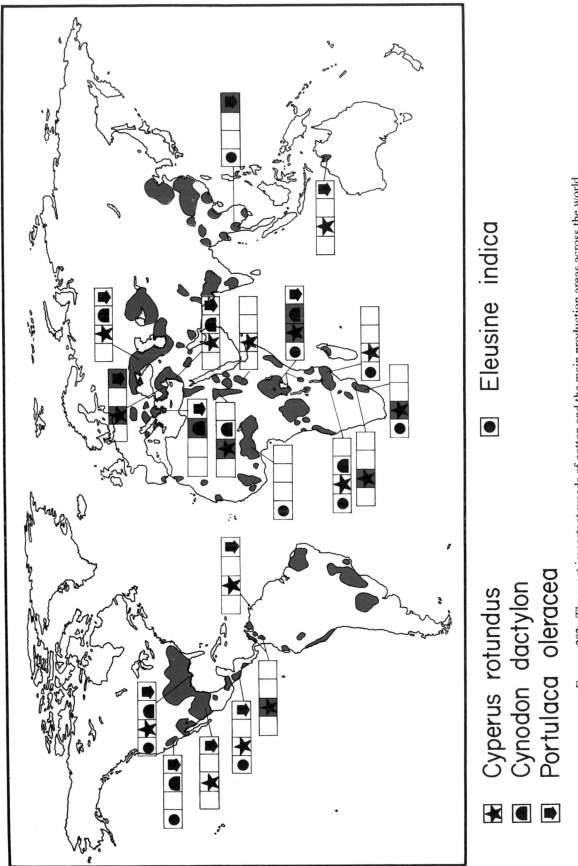

Cyperus rotundus

Cynodon dactylon

Portulaca oleracea

Eleusine indica

FIGURE 202. The most important weeds of cotton and the main production areas across the world. Colored blocks signify serious weeds; uncolored blocks signify principal weeds.

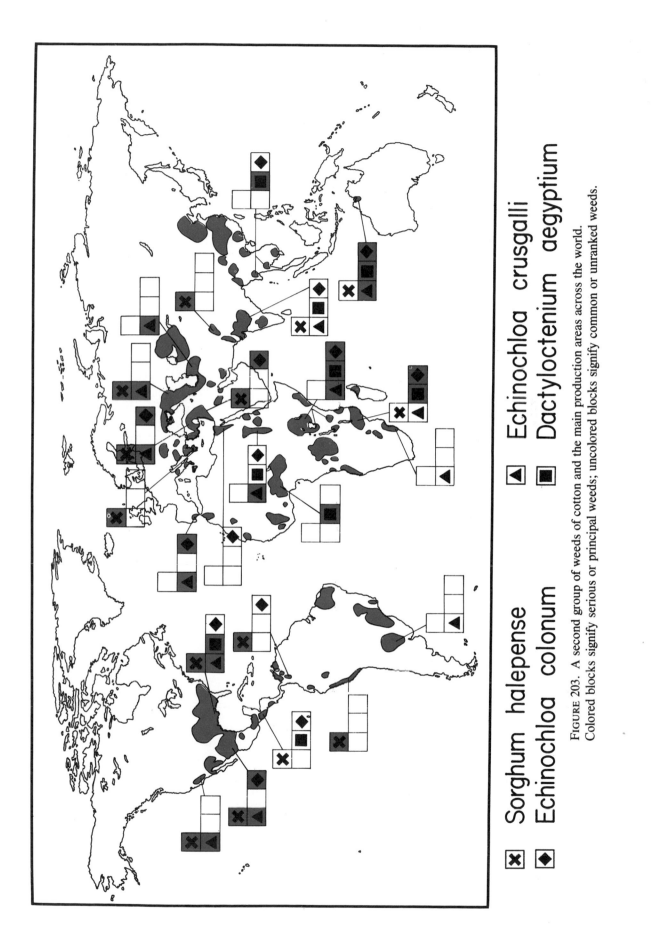

FIGURE 203. A second group of weeds of cotton and the main production areas across the world. Colored blocks signify serious or principal weeds; uncolored blocks signify common or unranked weeds.

⊠ Sorghum halepense	◬ Echinochloa crusgalli
◆ Echinochloa colonum	■ Dactyloctenium aegyptium

Grapes (Vineyards), *Vitis vinifera* L.

VITACEAE, GRAPE FAMILY

The major production from the world's vineyards is in the temperate zone, although grapes are grown on every continent. Many wild species are present in Europe, Asia, and North America where they may be seen as tangled vines at the forest's edge or climbing over bushes and low trees.

About 75 percent of the fruits from the domestic grapes are made into wines and spirits. Wine-making is a science about which we have accumulated much knowledge, but it is an art as well. The ancient craft was practiced with grapes and many other fruits thousands of years before fermentation was described. It is likely that wine now supplies about one-sixth of the calories needed daily by man in countries around the Mediterranean.

More than 50 percent of the crop is produced in Europe, with France and Italy being the largest producers. Each of them has 15 to 20 percent of the world's production. The next largest producers in order of importance are Spain, Turkey, the United States, Argentina, and Portugal, followed by Germany, Greece, Bulgaria, the Soviet Union, Australia, Algeria, Romania, and Yugoslavia. There is little production in Asia. Grapes for local consumption are grown or harvested in the wild in many areas of the world. Twenty percent of the commercial harvest goes to fresh market, with about 5 percent being grown for raisins.

Cultural Requirements

All of the world's great wines and many of the common ones are fermented from *Vitis vinifera*. The variety of grape determines the quality and character of the wine. There may be as many as 5,000 or more varieties across the world and each is usually adapted to a very narrow set of conditions. Within a variety, the subtle differences in taste and bouquet (aroma) depend on the soil, climate, and processing technique. The world's best wines come from areas where soil, moisture, temperature, and hours of sunlight are in the best combination for development of the vines themselves and for the growth and development of the fruit. In France, Germany, and Italy these conditions happen best to be met in hilly regions. On the Rhine River in Germany it is common to see the terraced, steep slopes of the sunny side covered with vineyards, while the opposite bank has none. Such are the delicate preferences of varieties for very particular microclimates. But there is no typical ecology and no standard set of conditions for raising grapes because the varieties differ so greatly in their requirements.

In the Northern Hemisphere the most rapid development takes place in July and August, with food manufacture in the leaf falling sharply toward the end of August. Water and nutrients from the soil are much in demand for about 4 weeks. A mean annual temperature of about 16° C is required for grapes. Grapes cannot be grown in many regions which meet this requirement, however, because of late spring frosts or because of high humidity which encourages development of molds which spoil the fruit. In the United States, for example, the climatic limits are a 170-day growing season and a mean temperature of 16° C from May through September. In other parts of the world some varieties may need a longer season and a higher temperature. Long periods of high temperature and sunny weather do not necessarily make good grapes. At high temperatures sugar production is increased at the expense of acidity and it is the latter which is important in the flavor and bouquet of the wine.

The crop is often grown on very steep hillsides and on many soil types. With suitable rootstocks grapes can be grown on highly calcareous soils. Good drainage is required. The plant requires a good water supply in its vegetative phase and dry sunny weather during ripening and harvesting. This is especially true for raisin grapes. Varieties with deep root systems, once established, can survive and remain green and productive through the very long dry periods around the Mediterranean.

Grapes may be started from cuttings but more often are grafted on suitable rootstocks, placed in nursery beds, and later transplanted to the vineyard. They are supported on trellises, poles, or wires and usually are trimmed twice yearly—in winter and in summer. The trimming is done to limit rather than to increase yield, and it insures a steady growth of uniform quality.

During the late 19th century the aphid *Phylloxera vastatrix* Planchon destroyed grapevines in France and in other parts of Europe, going as far east as the vineyards of Georgia on the Black Sea. The vines were mainly *Vitis vinifera*. The industry was reestablished by grafting the varieties of *Vitis vinifera* on American root stocks of different species, mainly *V. berlandieri*, *V. rupestris*, and *V. riparia*. In some areas of the eastern Mediterranean *V. labrusca* was used.

Many varieties begin to give a good yield in the 5th year from planting, and may produce a crop for another 20 to 35 years thereafter. Peak quality and quantity are reached at about 8 to 15 years. Perennial weed problems tend to build up in a crop of this kind which remains in the field for such a long time. Many reports from across the world tell of agriculturalists repeatedly using the same herbicide to control perennial weeds, particularly grasses, only to discover that removal of the competition has brought in broad-leaved weeds which sometimes pose even a greater problem.

MAJOR WEEDS OF GRAPES

From Figure 204 it may be seen that three principal weeds of vineyards are perennial plants. A grass, *Cynodon dactylon*, is everywhere that grapes are grown and, in the major areas, is always a serious or a principal weed. Another grass, the vigorous *Sorghum halepense*, is in most of the vineyards except those in the northermost areas of Europe and the southern tip of Africa. The broad-leaved weed, *Convolvulus arvensis*, is in all of the major regions of the Northern Hemisphere and is in the vineyards of Argentina, Australia, and Chile as well. Other species which have been reported to be principal weeds of grapes are: *Eragrostis pilosa*, *Portulaca oleracea*, *Solanum nigrum*, and *Sonchus arvensis* in the Soviet Union; *Cyperus rotundus* in Argentina and Portugal; *Chenopodium album* in France and Spain; *Daucus carota* in France; *Emex australis* and *Tribulus terrestris* in Australia; *Sinapis arvensis* in Spain; and *Chrysanthemum coronarium* in Tunisia.

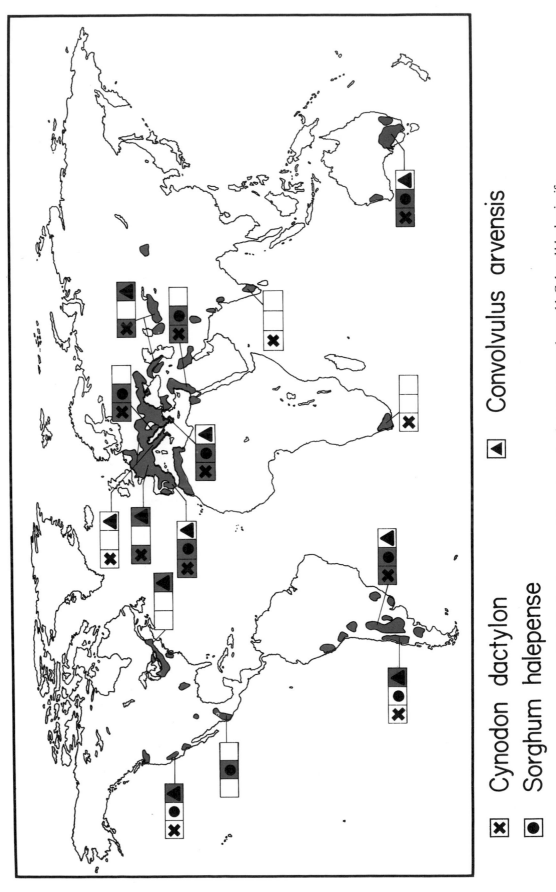

FIGURE 204. Weeds of grapes and the main production areas across the world. Colored blocks signify serious or principal weeds; uncolored blocks signify common or unranked weeds.

⊠ Cynodon dactylon

◉ Sorghum halepense

▲ Convolvulus arvensis

Peanuts, *Arachis hypogaea* L.

LEGUMINOSAE, LEGUME FAMILY

The peanut or groundnut, *Arachis hypogaea* L., a member of a small genus native to central South America, is in all tropical and subtropical countries and now is cultivated in parts of the warm temperate zone where the growing season is at least 4 months long and the temperature is high for a considerable part of the period. This species does not grow in the wild although a closely related wild species, *A. monticola*, is found in northwest Argentina. *A. hypogaea* grows as an annual in cultivation but may be weakly perennial. The wild species of *Arachis* are perennial, are extensively grazed by animals, and all of them ripen their fruits below the ground.

Peanut seeds contain up to 50 percent of an edible, nondrying oil and about 35 percent protein. They are used for human food and both the oil and protein are used in feed industries. A cattle cake is prepared from the meal. The peanut seed is the world's second largest source of vegetable oil (soybeans are first), and it is this product which enters world trade. The oil is used as a substitute for olive oil as a salad and cooking oil. It is also used in the manufacture of margarine and in the manufacture of vegetable ghee by hydrogenation in India; it is used in the preparation of a high quality oil for pharmaceutical purposes and a low quality oil for soap. The major countries, with the exception of the United States, raise peanuts for their oil. Much of the export of nuts and oil goes to Europe with about one-half of it to France.

India is the largest producer, followed by China, and both countries consume most of what they raise. The other countries with large production are Argen-

tina, Brazil, Gambia, Niger, Nigeria, Senegal, Sudan, and the United States. Nigeria is the largest exporter.

CULTURAL REQUIREMENTS

There are two main types of *A. hypogaea*. The Spanish-Valencia type has an erect bunch form, two to six seeded pods, and no seed dormancy; it is a lighter green and is highly susceptible to *Cercospora* leaf spot; it requires only a short season of 90 to 110 days. The Virginia type has true runners and a spreading bunch habit, tends to be indeterminate, and is dark green; its pods are usually two-seeded and its seeds have dormancy; it is moderately resistant to *Cercospora* leaf spot; and it requires a season of 120 to 140 days.

The species is little, if at all, affected by photoperiod; and the time from planting to maximum kernel yield appears to be determined (in the absence of drought) primarily by temperature. The growing period thus is quite different in the various production areas of the world. For example, the upright short-season Spanish-Valencia type matures in 120 days in South Africa, 100 days in central Tanzania, and 90 days in the Sudan zone between Senegal and the Blue Nile in Sudan.

The need to harvest the crop at maturity in a dry season often dictates the time of sowing. If the soil is wet when the plant becomes mature, there is danger of disease and of seeds of the upright type with no dormancy germinating as soon as they mature.

Peanuts are grown from lat 40° S to 40° N of the

equator and much of the crop is in areas of 1,000 mm of annual rainfall. Plants are easily killed by frost. The best soils are loose, sandy loams which are well drained, for the crop does not tolerate waterlogged soils. Soils which crust over prevent peg penetration. In Africa and the Orient, peanuts are grown as a subsistence crop and the buried pods are dug and collected by hand; thus they may be grown on soils not suitable for mechanical harvesting because of heavy texture or liability to compaction. In these situations they are often interplanted with other crops. The crop may be grown on flat or ridged fields. Ridged fields make lifting easier but reduce the number of plants per hectare. Peanuts are often grown late in a rotation with cotton, tobacco, maize, and other cereals. Across the world the crop is grown in many kinds of fields and under different systems and, although herbicides are avilable, much mechanical cultivation and manual weeding still are necessary. All weeding practices must avoid disturbance after pegging has begun (Cobley 1956, Ochse et al. 1961, and Purseglove 1968).

MAJOR WEEDS OF PEANUTS

The distribution of four of the principal weeds of peanuts is shown in Figure 205. *Eleusine indica* has a wider distribution than the other weeds and it is more frequently reported to be important. It is reported to be among the three most important weeds in the crop in Hawaii, Indonesia, Taiwan, and Zambia and it is a serious weed in Malaysia. *Cyperus rotundus*, the world's worst weed, is in all of the great peanut-growing areas and is a principal weed in most of them. *Digitaria sanguinalis* is interesting because it is ranked as a serious weed in Indonesia, Taiwan, and the United States, and is reported to be a princi-

pal weed in every other country except South Africa. *Portulaca oleracea* ranks high because it is so widely distributed and because it has a growing habit like that of the peanut, necessitating the expenditure of much time and energy to keep it at bay.

Although this is not indicated on the map, *Echinochloa colonum* and *Rottboellia exaltata* are almost as important as the above annual grasses. The former is among the three most important weeds of Taiwan; it is a principal weed in Colombia, Israel, and the Philippines; and is a common weed in Australia, Hawaii, India, Indonesia, and Sudan. *R. exaltata* is a serious weed in Rhodesia and Zambia, a principal weed in Sudan, and a problem in Trinidad. Other annual grasses are *Cenchrus echinatus*, an important weed in Australia, Brazil, Hawaii, and the United States; and *Digitaria longiflora*, *Setaria pallidefusca*, and *Dactyloctenium aegyptium*, all of which are principal weeds of peanuts in Gambia. The latter is a weed of peanuts in Indonesia, Israel, and the United States as well.

The perennial grass *Cynodon dactylon* is a principal weed of peanuts in Ceylon, Indonesia, and Israel, and a common weed in Colombia, India, Taiwan, and Trinidad. Another perennial grass, *Sorghum halepense*, is a principal weed of Israel, Pakistan, and the United States and a common weed of India.

At this time grasses or grasslike sedges are the greatest problems in the peanut fields of the world. Among the broad-leaved weeds which have been reported are *Ageratum conyzoides*, a principal weed in Ceylon, Ghana, and Indonesia, and a common weed in Brazil; *Amaranthus spinosus*, a principal weed in Ghana, Hawaii, and the Philippines, and a troublesome one in the United States; *Tridax procumbens* which is present in Ghana, New Guinea, and Taiwan; and *Tribulus terrestris* which is present in Australia, Israel, and South Africa.

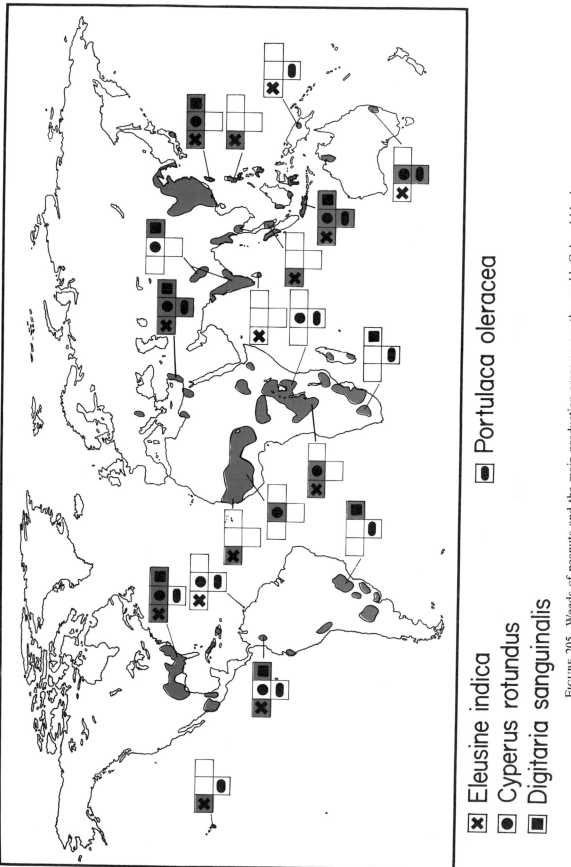

FIGURE 205. Weeds of peanuts and the main production areas across the world. Colored blocks signify serious or principal weeds; uncolored blocks signify common or unranked weeds.

☒ Eleusine indica

◉ Cyperus rotundus

▣ Digitaria sanguinalis

▣ Portulaca oleracea

Pineapples, *Ananas comosus* (L.) Merr.

BROMELIACEAE, PINEAPPLE FAMILY

Although its exact origins are unknown, the pineapple is native to South America where it was grown for food, medicine, and wine long before the discovery of the New World. It is one of the crop plants that was domesticated by prehistoric people. It has become one of the most popular of tropical crops, and has been spread widely in the tropics and subtropics because of the ease with which it can be propagated vegetatively (Collins 1960).

The pineapple is a member of the Bromeliaceae, a large family of more than 1,300 species, which divides into two distinct habitat groups: terrestrial plants which root in soil and epiphytic plants which grow on plants, fences, or other objects for support. They are not parasitic. Pineapples can be grown under a wide range of conditions in the tropics and subtropics, both in regard to soils and rainfall conditions. In South America the genus *Ananas* is found from the warm, wet Amazon basin to the cool semiarid upland regions of southern Brazil. The crop grows well within a temperature range of 15° to 33° C and within a rainfall range of 1,200 mm to over 5,000 mm. Fruit quality is best under high sunlight conditions; therefore, semiarid areas with at least 1,200 mm of well-distributed rainfall are often used successfully for its cultivation. If rainfall is limiting, irrigation is sometimes practical.

Ananas comosus is a highly variable species, and cultivated varieties differ widely in morphology. A short-lived perennial, pineapples are propagated vegetatively by "slips" from the main stem. The "crown" of leaves from the fruit can be used for planting materials, as can "suckers" from the base of the plant; however, these materials are not as desirable as "slips."

The plants produce a multiple fruit which results from the fusion of about 100 flowers. The fruits are borne on long peduncles which arise from terminal buds on the main stem. The fruit ripens 6 to 7 months after flowering. The plants can continue to produce for many years, but fruit size declines with each ratoon crop; for this reason the crop is usually replanted after one or two ratoon crops have been harvested (Plucknett, Evenson, and Sanford 1970). Hormones are used to ensure a more uniform flowering and fruit ripening and also to adjust production to cannery schedules.

Most pineapples are grown as a processed crop and are sold as canned fruit slices, chunks, and juice. Bromelain, a meat tenderizer, is also made from pineapples. Recently the market for fresh fruit has expanded. The most important producing area is Hawaii where the cultivation methods are among the most scientific for any tropical crop.

The principal pineapple-producing areas are Australia, the Caribbean, Hawaii, Mexico, western Africa, and Southeast Asia. The United States (Hawaii) has about 25 percent of the production, followed by Brazil, Malaysia, Mexico, the Philippines, and Taiwan.

CULTURAL REQUIREMENTS

Most pineapples are grown on acid soils. Deep tillage is usually practiced, and preplant fertilizers or soil amendments are incorporated during the tillage operations. After the pineapples have been planted, most fertilizers are applied as foliar sprays.

Many producing areas lay down plastic or paper mulch strips after the soil has been fumigated to control nematodes. The plastic strips, about 1 m in width, are covered with soil at the sides leaving a width of about 0.6 m of mulch which covers the soil. Two rows of pineapples are then planted by hand, using a special trowellike tool to cut through the plastic and to open a hole in the soil into which the slip is inserted. Row spacing from the center of one double row to the next is about 1.6 m. Therefore, over two-thirds of the soil surface is covered with the black plastic mulch which prevents some weed growth, serves as a barrier to the loss of volatile soil fumigants, and contributes to higher soil temperatures.

To prevent damage both to the plants and to the mulch strips and to reduce compaction, pineapple growers install field roads every 40 m or so, thereby ensuring that only men on foot enter the in-field areas. Preemergence herbicides are used to treat the bare soil interrows. As mentioned, most fertilizers which are applied after the pineapples are planted are foliar sprays.

At harvest the fruits are picked by hand and carried out of the field in bags or on conveyor belts mounted on long booms which are suspended over the crop from a machine traveling on the special field roads.

After the ratoon crops have been harvested, men work through the crop, detaching and collecting slips as planting materials for the new fields to be planted.

MAJOR WEEDS OF PINEAPPLE

Weeds which cause problems in pineapples are those which grow through or perforate the plastic mulch; those which are resistant to the preemergence herbicides being used; or those which, if allowed to grow unchecked, can overtop and compete with the crop. Any weed which interferes with the movements of men through the fields is also of concern.

The principal weeds of pineapples are shown in Figure 206. Five of the six are grasses or sedges. Three of these, *Cyperus rotundus*, *Cynodon dactylon*, and *Paspalum conjugatum*, are perennials; whereas two of them, *Eleusine indica* and *Digitaria adscendens*, are annuals. The only broad-leaved species is the annual, *Bidens pilosa*. *Cyperus rotundus* is the principal weed and it is found almost everywhere that pineapples are grown. In some areas, present control measures for this species are so fruitless, yet so expensive, that little effort is made to do anything about the problem. *Cynodon dactylon* and *Eleusine indica* appear to be next in importance. The remaining three weeds are scattered at random in the pineapple regions of the world. Although not shown on the map, *Imperata cylindrica* is also a problem in pineapples in some areas.

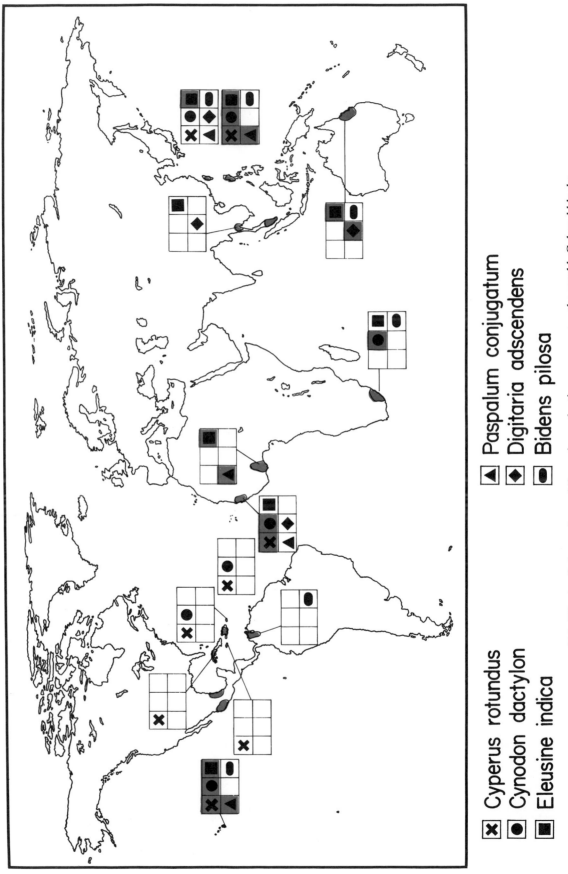

Cyperus rotundus

Cynodon dactylon

Eleusine indica

Paspalum conjugatum

Digitaria adscendens

Bidens pilosa

FIGURE 206. Weeds of pineapples and the main production areas across the world. Colored blocks signify serious or principal weeds; uncolored blocks signify common or unranked weeds.

White or Irish Potatoes, *Solanum tuberosum* L.

SOLANACEAE, NIGHTSHADE FAMILY

This crop was cultivated in the Andean Mountain regions of South America long before the Spanish came to the area. Although it is native to the region of the equator, it evolved in cool, moist, mountain regions and it is now grown almost exclusively in temperate regions or high mountain valleys in warm places. Three-fourths or more of the harvest is in the northern temperate zone.

The white potato, *Solanum tuberosum*, is often known as the Irish potato in contrast to the sweet potato. The world harvest of sweet potatoes and yams is less than one-half that of white potatoes. The total harvest of white potatoes is difficult to know because many gardeners, farmers, and smallholders have their own plots and the crop is consumed by the grower and his family. The world center of production is in Central and Eastern Europe where the tubers are the staple food of millions of people. In some parts of Europe, including parts of the Soviet Union, as much as 40 percent of the crop is fed to livestock. Potatoes are used to make alcoholic beverages and are an important source of industrial alcohol for some areas.

From Figure 207 it may be seen that the principal production area lies in a band from Spain across Europe and Asia to Korea. Almost one-half of the world's production is in the far northern temperate region of Europe and includes the European portion of the Soviet Union. North America and the Far East (mainly China, Korea, and Japan) are the areas of next greatest production. It is strange that the potato's place of origin, Latin America, ranks fourth. The Soviet Union, the country of greatest production, may grow almost one-third of the total world's crop. Poland, West Germany, the United States, East Germany, and France follow in order. There is little production in Africa, Australia, and the Near East. It is believed by some that this crop, under good conditions, can yield more food per hectare than any other. It was an important factor in the rapid increase in population in northwestern and Central Europe in the 19th century. The failure of the crop in Ireland in the mid-19th century, however, was the cause of a famine that took the lives of a million people and caused the emigration of a half million more. Late blight, a disease caused by the fungus *Phytophora infestans*, invaded the cells of the potato leaves and brought about a full destruction of the crops during 1846 and 1847.

CULTURAL REQUIREMENTS

Potatoes require a long, cool growing season and adequate moisture at all times. Growth takes place between 7° and 18° C and if temperatures exceed the upper limit the yields are lowered. Where springs are mild but the summers are hot, or where the season is short and the days are long as in northern Europe, less productive, early-maturing varieties are used. The crop can be grown very successfully on sand if

509

sufficient nutrients and water are added in a uniform manner. The best soils, however, are loams or silts with a good content of organic matter. The crop does not grow well on heavy clays or in wet, undrained soils. A high level of nutrients is required on all soils; and it is customary in heavy production areas to add large amounts of fertilizer because the cost is easily justified by increased yields. The most serious pest of potatoes is late blight, a disease which strikes when temperatures and humidity are high.

MAJOR WEEDS OF POTATOES

The main weeds in potatoes are annuals. When properly grown, the crop quickly closes over the field and smothers much of the weed competition. It is important, however, that early season weed control be practiced because annual grasses or broad-leaved weeds which become established may later appear above the crop and compete intensively for nutrients and moisture when the potato tubers are enlarging. Large weed clumps seriously interfere with harvesting, may cause potato tuber damage, and may increase the wear on machinery. Potatoes should be grown in a 2- or 3-year rotation and, when this practice is followed, the variety in cultural practices for different crops gives opportunity to keep perennial weeds at a low level.

Chenopodium album is reported to be a principal weed of potatoes much more often than is any other species (Figure 207). It is, of course, one of the most widespread weeds of the world in many crops; in potatoes it is in most of the producing areas of the temperate and tropical zones. *Portulaca oleracea*, a low, creeping, vigorous, annual weed, is able to persist in the system of potato culture and is very widespread. It should be noticed that it has not been reported as a principal weed in that great area of production across the Northern Hemisphere, but it has been reported as a troublesome weed in the warmer areas of South America, Africa, and Asia. *Galinsoga parviflora* is a weed of potatoes on four continents. *Stellaria media* is most often a problem in northern Europe but it grows in New Zealand and Chile as well. *Echinochloa crusgalli* is a principal weed in Bulgaria, Poland, and the United States, and is a common one in Iran, New Zealand, and the Soviet Union.

In addition, *Polygonum persicaria* deserves special mention for it is a principal weed in Belgium, Chile, England, and New Zealand; and it is a weed of potatoes in Germany and the United States as well. *Digitaria sanguinalis* is a principal weed in Brazil and the United States; *Bidens pilosa* is a principal weed in Colombia, Mexico, and Mozambique; *Spergula arvensis* is a principal weed in Finland, Norway, and Sweden. *Anagallis arvensis* and *Asphodelus tenuifolius* are principal weeds of potatoes in India. The perennial grass *Agropyron repens* is a principal weed in North America, New Zealand, and several countries of northern Europe; the broad-leaved perennial *Convolvulus arvensis* is a weed of the Soviet Union and northern India, and is common also in England, Iran, and the United States. The perennial sedge *Cyperus esculentus* is a principal weed of potatoes in Canada, South Africa, and the United States.

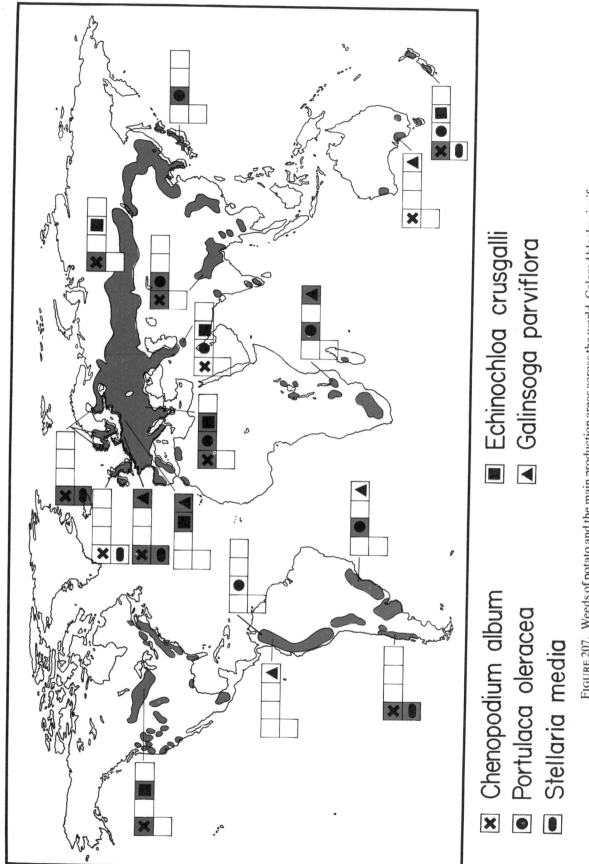

FIGURE 207. Weeds of potato and the main production areas across the world. Colored blocks signify principal weeds; uncolored blocks signify common or unranked weeds.

☒ Chenopodium album   ▣ Echinochloa crusgalli

◉ Portulaca oleracea   ▲ Galinsoga parviflora

▤ Stellaria media

Rice, *Oryza sativa* L.

POACEAE (also GRAMINEAE), GRASS FAMILY

The cereals rice and wheat are the two most important food crops in the world. Rice is especially important, for this crop provides the staple food of more of the world's population than does any other. Ninety percent of the world's rice is grown and eaten in Asia where millions of people subsist almost entirely on this food.

The yield of paddy rice per hectare varies by a factor of four across the world. This is true in part because the crop is planted in many areas and on many soils which are not suitable. On the other hand, the areas with high yields are located mainly in the warm, temperate zones where the high-yielding and fertilizer-responsive *japonica* varieties can be grown. Yields are often lower in tropical areas where the *indica* varieties are used and, although more hardy, these are less responsive to fertilizer and generally have not been improved by plant breeding. The longer days during rice-growing seasons in warm temperate regions, together with often more sophisticated farming techniques, also contribute to higher yields. But this is not all of the story. It has been demonstrated repeatedly in the tropics that good water control, use of fertilizer, suitable pest control, and good farming practices can produce startling results. The more recent introductions of improved rice varieties are causing marked changes and rapidly shifting patterns in rice production.

CULTURAL REQUIREMENTS

The cultivation of rice dates from the earliest ages of man and there is considerable evidence to indicate that it was a staple food in Asia long before recorded history. The plant is native to at least two areas in Asia and Africa and there are now about 25 species distributed in tropical, subtropical, and warm, temperate regions throughout the world. The most commonly cultivated species is *Oryza sativa* of which there are now thousands of varieties. Among these are types which can be grown from upland, rain-fed regions to deep undrained swamps; in water varying in depth from a few cm to 5 to 7 m; and on many different soil types.

The principal systems of cultivation are as follows:

1. DIRECT SOWN UPLAND (upland rice). Except for the system used in areas where rainwater is impounded to grow rice, this system most closely approximates the production methods used for other cereal crops. The land is prepared before the rains come. When the rains begin, the seed is sown. Sowing may be done in rows with drills or other equipment much like the methods used to sow wheat or other cereals; sowing in shifting cultivation is often done by hand with a planting stick. Bunds or levees may be used to retain as much water as possible. In some situations enough water may be impounded in this way to bring about a flooded condition. An assured 3 to 4 months of rainfall is necessary for this kind of culture. Weed control is very difficult in this system.

2. DRY-ESTABLISHED. Fields are mechanically graded and carefully cultivated; the seed is placed

512

in dry soil; irrigation water is supplied; and moisture is carefully controlled for the entire season.

3. WET SOWN. The lowland varieties are sown directly into mud or into standing water. Seeds are often pregerminated. In some places, the fields are cultivated to a depth of 10 to 20 cm about 1 month before planting. Secondary tillage consists of puddling the soil after it had been flooded. The wet soil is worked to a depth of 15 cm, producing a fine mud. This provides the seedbed for rice but also builds a hardpan layer to prevent undue loss of water.

Large-scale fields in Australia and the United States are prepared and flooded; then pregerminated seeds are sown from the air. Control of weeds in this system depends upon use of herbicides and flooding.

4. NURSERY-TRANSPLANTED. This best known of all traditional systems is widely practiced in Asia. Fields for this system are often terraced and very careful water control is maintained. The seeds are sown on a concrete slab, banana leaves, plastic or matting, or in the soil. The seeds may be covered with leaves which are then removed as germination takes place. The minimum period of germination and seedling growth before transplanting is 10 to 15 days. If the young plants have been taken from a soil seedbed, the minimum period of germination and seedling growth before transplanting is 30 to 40 days. Most young rice plants are transplanted by hand and usually more than one seedling is placed in a "hill." Some rice is planted in a square or other definite pattern, but in some areas the rice is planted at random. Until recently most of this rice was hand-weeded or cultivated with small rotary hand machines.

5. FLOATING RICE. In some areas, such as in central Thailand, heavy rains may prevent the control of water; and in these areas floating rice is grown. These varieties are dry-sown and the stems may elongate to as much as 7 m.

6. RATOONING. Cultivated varieties of rice are generally grown as annuals. However, if the field is irrigated after the crop is cut, a second crop may be obtained from the axillary buds which develop at the lower nodes. The *japonica* varieties and *Oryza glaberrima* (often grown in western Africa and called "red rice") are not used for ratoon crops. Experiments in the United States have shown that two-thirds as much nitrogen applied to the ratoon crop will produce about the same grain yield per kilogram weight of fertilizer as was obtained from the first crop. Usually, however,

the ratoon crop cannot be expected to yield more than one-half the yield of the first crop (Plucknett, Evenson, and Sanford 1970).

The total rice production of the world in 1968 to 1969 was 180 million metric tons. It is believed that about 155 million of these were grown in Asia. Mainland China is probably the world's largest producer. India is the world's second-largest producer, growing 57 million metric tons. The next leading producers are Pakistan, Japan, and Indonesia. Each of the last three countries produces more rice than any single continent outside of Asia.

MAJOR WEEDS OF RICE

The following discussion of the major weeds of rice covers all methods of culture and management, including those used for flooded and upland crops.

A great deal is known about the species of weeds in the rice-growing regions of the world. The most important weeds in rice are shown in Figures 208 and 209. *Echinochloa crusgalli* is the most serious weed of rice; it is found almost everywhere that rice is grown, and it is usually ranked as a serious weed. *Echinochloa colonum* is the weed of second importance and it, too, is found almost everywhere that rice is grown. *E. colonum* tends to be clustered along the equator while *E. crusgalli* has a greater range from north to south. *Fimbristylis miliacea* is a very serious weed in southern and eastern Asia and is also serious in the Caribbean area. *Cyperus difformis* will be discussed later. In Figure 209 is shown the distribution of six additional major weeds of rice. These species have much more limited distribution.

In Figure 210 the distribution of *Echinochloa crusgalli* only is shown throughout the rice areas of the world. The weed is present in almost every area, and, on this basis alone and quite aside from its competitiveness, it must be ranked as a very serious weed in rice. *Monochoria vaginalis*, shown in Figure 211, is an example of a weed with a very strict regional distribution. It is an increasing threat to rice production in eastern Asia and is considered to be of major importance in Indonesia, Japan, Korea, the Philippines, Sarawak, and Taiwan. *Cyperus difformis* (Figure 212) is a principal weed of rice in Africa, Asia, Australia, and Europe. It is reported in the Americas only on the western coasts of the United States and Mexico.

Because weeds of upland rice are sometimes different from those of lowland rice, it may be useful to list some of these separately. *Echinochloa colonum* is probably the worst weed of upland rice; next is the annual grass *Eleusine indica*. *Echinochloa crusgalli*, most important in lowland rice, is also important in upland rain-fed rice. Other important weeds are *Cynodon dactylon*, *Portulaca oleracea*, *Digitaria sanguinalis*, *Ageratum conyzoides*, *Amaranthus spinosus* and *Digitaria adscendens*.

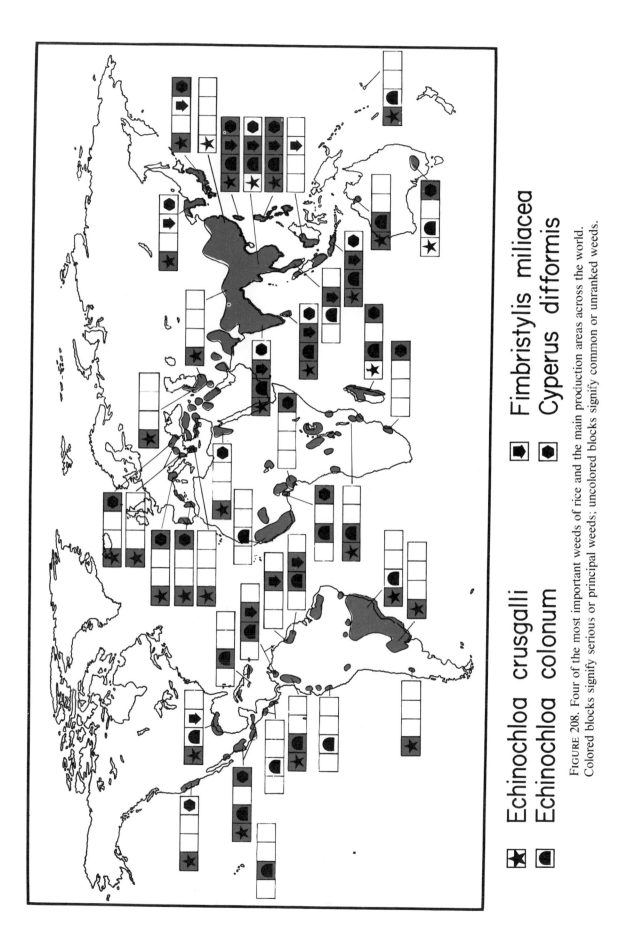

FIGURE 208. Four of the most important weeds of rice and the main production areas across the world. Colored blocks signify serious or principal weeds; uncolored blocks signify common or unranked weeds.

Echinochloa crusgalli

Echinochloa colonum

Fimbristylis miliacea

Cyperus difformis

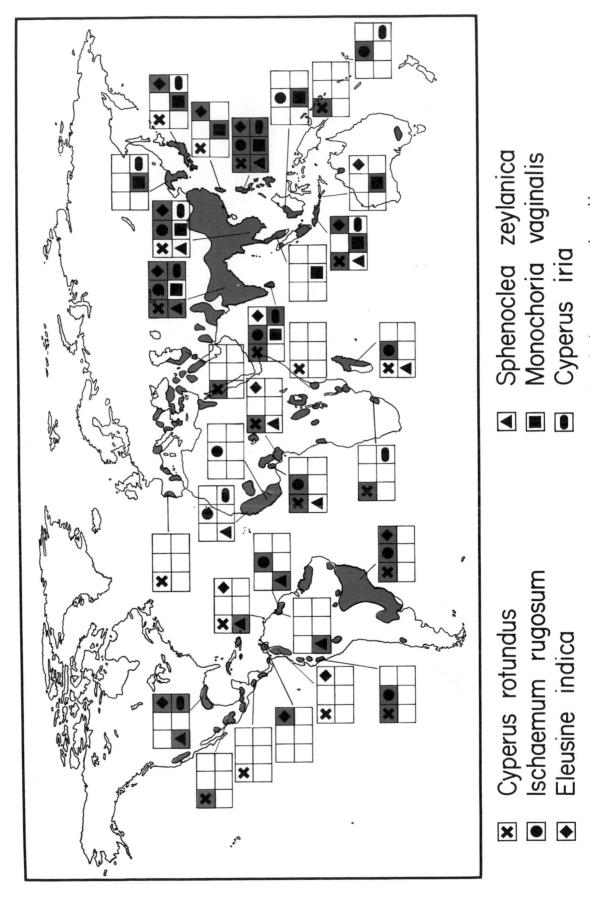

Figure 209. A second group of weeds of rice and the main production areas across the world. Colored blocks signify serious or principal weeds; uncolored blocks signify common or unranked weeds.

⊠ Cyperus rotundus
⊙ Ischaemum rugosum
◆ Eleusine indica

◀ Sphenoclea zeylanica
■ Monochoria vaginalis
⬭ Cyperus iria

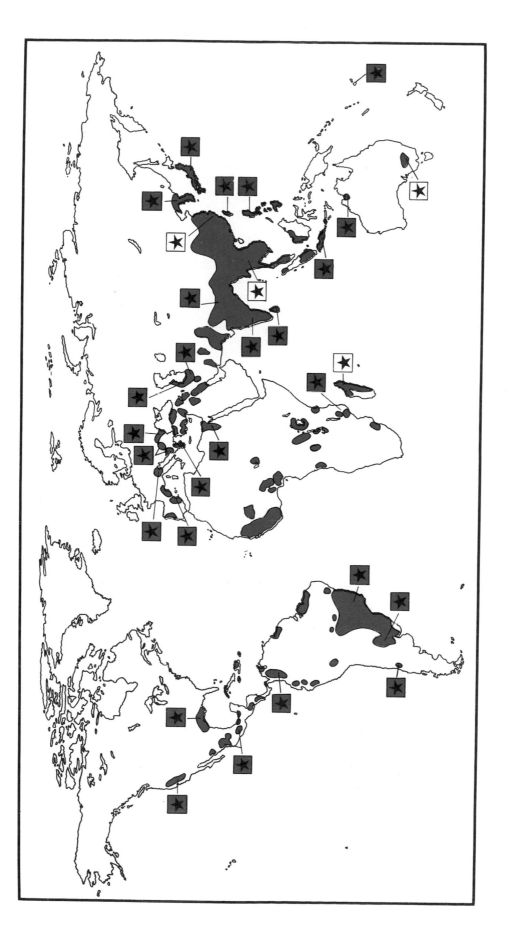

Figure 210. The importance of *Echinochloa crusgalli* in the main rice production areas across the world. Colored blocks signify the species is a serious or principal weed; uncolored blocks signify it is a common or unranked weed.

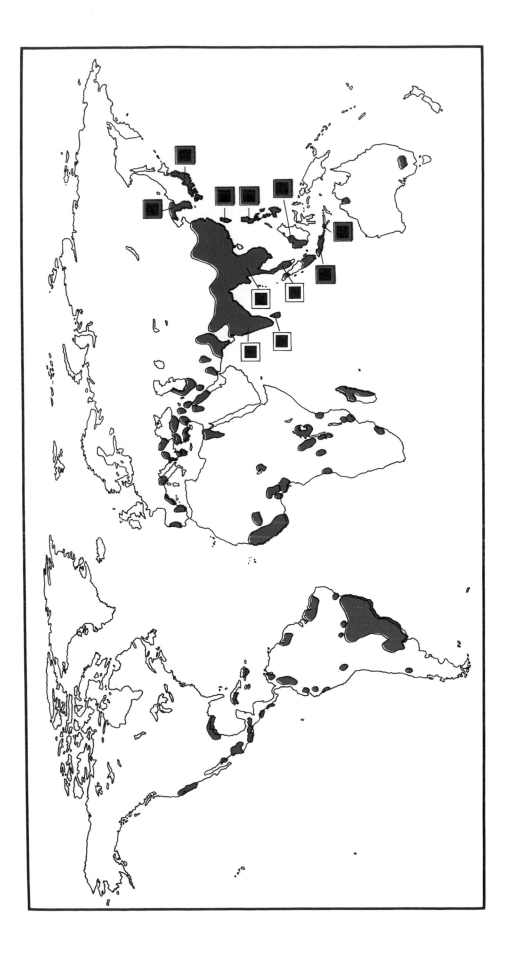

FIGURE 211. The importance of *Monochoria vaginalis* in the main rice production areas across the world. Colored blocks signify the species is a serious or principal weed; uncolored blocks signify it is a common weed.

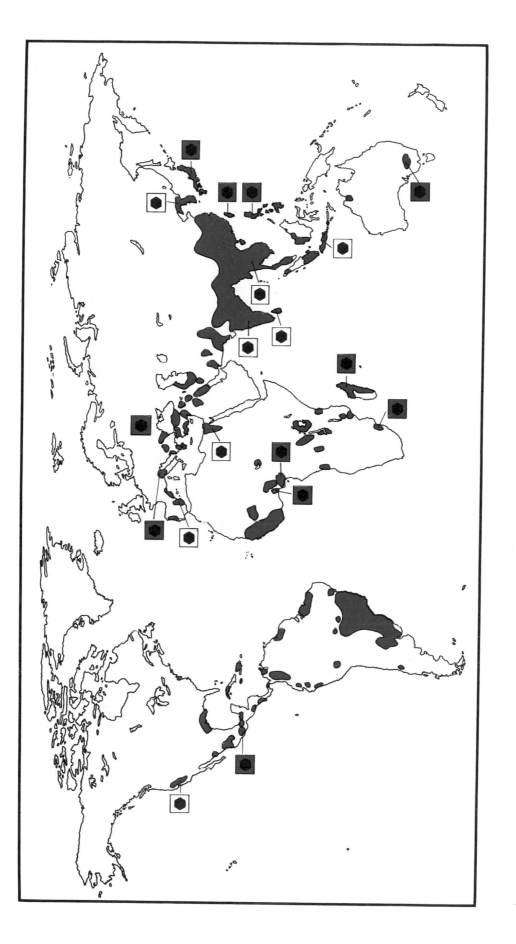

FIGURE 212. The importance of *Cyperus difformis* in the main rice production areas across the world. Colored blocks signify the species is a serious or principal weed; uncolored blocks signify it is a common weed.

Rubber, *Hevea brasiliensis* (Willd. ex Adr. de Juss.) Muell.-Arg.

EUPHORBIACEAE, SPURGE FAMILY

More than 90 percent of the world's natural rubber is produced in southern and eastern Asia. Indonesia and West Malaysia produce about 70 percent, followed by Thailand, Ceylon, South Vietnam, and Sarawak (East Malaysia). In Africa, where about 7 percent is produced, Nigeria is the largest producer with Liberia and Zaïre also being sources. Brazil produces about 1 percent. One-third of all rubber comes from smallholders who raise *Hevea brasiliensis*. *Taraxacum kok-saghyz* (kok-saghyz or Russian dandelion) and *Parthenium argentatum* (Mexican guayule) are also sources of rubber latex but the production from these species is not important on a world scale. In the south of the Soviet Union some effort is being made to bring these plants to significant commercial production.

CULTURAL REQUIREMENTS

H. brasiliensis is a fast-growing tree that reaches to 25 m in plantations and 40 m in the wild. Wild Para rubber trees came from the tropical rain forest of the Amazon basin, where they sometimes grow in periodically flooded areas. The trees grow larger and more vigorous, however, on moist, deep, well-drained loam soils. They will tolerate a pH from 4 to 8 but the optimum is pH 5 to 6. Most plantation rubber is found between lat 15° N and 10° S in hot, humid areas with temperatures of 25° to 35° C and with an annual rainfall of 1,900 to 2,500 mm well distributed over the seasons.

Planting distances vary from about 4.5 by 4.5 m to 9 by 2 m. Wider spacings increase the chance of wind damage. On small holdings annual crops and sometimes bananas are intercropped during the first 3 to 5 years. In some wider plantings coffee or cocoa may be intercropped. The rubber trees are very sensitive to crop and weed competition and intercrops should not be placed within 1.5 meters of the trees. Cultivation must cease at 3.5 years.

South American leaf blight (*Dothidella ulei* P. Henn.) is of major concern and has limited rubber production on plantations in the New World. At present it is found in South America, Central America, and Trinidad. The high-yielding clones of Southeast Asia are very susceptible, but the disease has been under control in that area up to now. Powdery mildew (*Oidium hevea* Steinm.), a disease found on the lower surfaces of leaves, is restricted to Asia and Africa, being particularly serious in Ceylon at higher altitudes (Cobley 1956, Ochse et al. 1961, Purseglove 1968).

In earlier times, when forests were being extensively cleared so that rubber trees could be planted, such techniques as clean-cutting and burning resulted in much erosion and loss of organic matter. Today, when the fields are replanted with newer, high-yielding clones, the old trees are first removed

520

or killed with chemicals or are stumped and treated with creosote. In either case, ground covers now are used and legumes are important in the mixtures. *Calopogonium mucunoides*, *Centrosema pubescens*, and *Pueraria phaseoloides* are important ground-cover species. Extensive replanting of rubber in Malaysia often resulted in hundreds of hectares of leguminous ground covers which were contiguous, and these ground covers allowed insect pests to increase to the point where insecticides were necessary to maintain the cover crop. If the cover crop is unable to stand shading after 4 or 5 years, it begins to die out; then the soil again becomes bare and erosion once more becomes a problem.

MAJOR WEEDS OF RUBBER

To obtain good stands of cover crops, one must often control weeds in the early stages. Short, weedy grasses such as *Axonopus compressus* and *Paspalum conjugatum* are believed to be satisfactory covers in some areas. It must be pointed out, however, that these species are considered to be serious weeds of plantation rubber in many places.

From Figure 213 it may be seen that *Imperata cylindrica*, *Chromolaena odorata*, *Paspalum conjugatum*, *Axonopus compressus,* and *Mikania cordata* are all present in Ceylon, Indonesia, and Malaysia, the major rubber-producing areas. These species are also present in western Africa. Other major weeds not shown on the map are *Melastoma malabathricum* and *Mimosa pudica*. These, too, are present in Ceylon, Indonesia, and Malaysia. Only *melastoma malabathricum* has been reported to be a principal weed of rubber in western Africa.

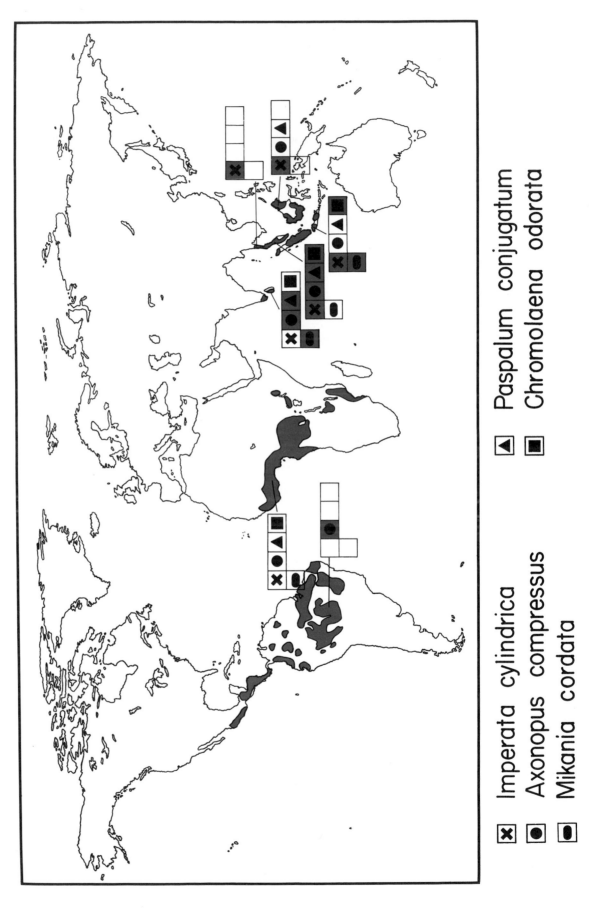

FIGURE 213. Weeds of rubber and the main production areas across the world. Colored blocks signify principal weeds.

Imperata cylindrica Paspalum conjugatum
Axonopus compressus Chromolaena odorata
Mikania cordata

Sorghums, *Sorghum bicolor* (L.) Moench (= *S. vulgare* Pers.) and Millets, *Setaria italica, Panicum miliaceum, Pennisetum glaucum,* and *Echinochloa crusgalli* var. *frumentacea*

POACEAE (also GRAMINEAE), GRASS FAMILY

Sorghums and millets are among the oldest of the important cultivated plants. A sorghum, *Sorghum bicolor* (L.) Moench, was depicted in Egyptian tombs as early as 2200 B.C. The crop was cultivated and used in Syria, Mesopotamia, and China in ancient times. Millets were included in Chinese religious rites as early as 2700 B.C. These cereals are very important sources of food for the world.

Sorghums are canelike grasses which resemble the tall varieties of corn. The stems may be 0.6 to 5 m in length. The leaves are 0.6 to 0.8 m in length and are coated with a waxy, whitish bloom. A heavy head of flowers (which may be loosely arranged or packed closely in their spikelets) is borne at the tip. Seeds are smaller than those of wheat and may be round, flat, or ellipsoid, and are colored red, brown, white, or yellow.

Sorghums are grown for grain (*Sorghum bicolor*), forage (*S. bicolor* var. *Sudanese*, *S. bicolor* var. *saccharatum*, S. almum), syrup (*S. bicolor* var. *saccharatum*), and brooms (*S. bicolor* var. *technicum*). Most of the sorghums are annuals or short-lived perennials; however, *S. halepense*, a weed, and *S. almum* are perennials. Subgroups of the grain sorghums are: milo, kafir, feterita, shallu, kaoliang, durra, hegari, and others. Dwarf types are grown in areas where machine-harvesting is practiced and hybrid sorghums are grown on a field scale. Their food value is similar to that of corn but their protein content is a bit higher and fat content a bit lower. The grain is ground into meal and made into porridge, bread, and cakes. The seeds are sometimes puffed or popped. The crop is of great value for the production of livestock as well as for human consumption.

In Africa and Asia it is common to chew the stems of sweet sorghum. Syrup is made in the following way: the stalks are stripped after the seeds have ripened; the tops of the stalks then are cut off; the stalks themselves are cut close to the ground and are fed through rollers which press out the sap; the sap is then boiled until it reaches the desired thickness. It is possible to make sugar from the already-extracted syrup, but this further process requires expensive alcohol for crystallization. It is not, therefore, a large-scale practice. It is easier and cheaper to use sugarcane.

The sorghums have an enormous variety of uses. The wax from the seeds is similar to carnauba and is used for polishes and carbon paper. Industrial and grain alcohol are made from the seeds. Distillers and manufacturers of malt beverages use large quantities of sorghum. Oil from the seeds is used in salad dressings. Much starch is used in foods, adhesives, and sizings for paper and cloth. Sorghum starch is mixed with other materials and circulated through drilling bits as oil wells are dug; about a ton of starch per well is used in this way to cool and to lubricate the drill and to seal the walls of the well.

The broomcorn sorghums develop a head made up of many long, slender branches in the form of a brush. These are bunched and tied to make small and large brooms.

"Millet" is a term loosely used to describe a wide range of plants from several genera. The word is from the Latin *mille* meaning "thousand," indicating the fertility of the plants. Most of these coarse grasses are small-seeded species and varieties. They are grown as cereals and forages and since ancient times have been confused with the sorghums.

The four principal millets grown in the world are: foxtail, *Setaria italica*; proso or broomcorn millet, *Panicum miliaceum*; pearl millet, *Pennisetum glaucum*; and Japanese sauva or barnyard millet, *Echinochloa crusgalli* var. *frumentacea*. These are four different genera but all are included in the same general tribe *Paniceae*; whereas the sorghums are more closely related to the sugarcane and bluestem grasses in the tribe *Andropogoneae*.

It is believed that foxtail millet is planted on more than 14 million hectares of the arid regions of India. One or more recognized varieties of foxtail millet are grown extensively for forage in Canada, China, Indonesia, Japan, North Africa, southeastern Europe, the Soviet Union, and the United States. In the Old World the crop is grown mainly for human consumption and in the New World it is more often grown for forage. Proso millet is nearly always a grain crop for men and livestock, and southern Russia is an important area of production. The variety is important also in the northern Great Plains of North America. Pearl millet is often used as food by the poor in Africa, Egypt, and India. It is a moisture-loving crop and is grown in India during the rainy season. Japanese millet is produced to a limited extent as a forage plant

and several crops a year may be obtained from it.

Sorghums and millets are crops that can grow successfully in dry, hot, climates and in rather dry soils. This accounts in great part for their wonderful adaptability to severe and exacting soil and climatic conditions. As is true of many other crops, however, sorghums and millets will do much better and give greater yields in moist, fertile soils in areas of high humidity. The important point is that these two crops are much better able to withstand drought than are either corn or wheat, and they can be counted upon to produce in places where corn or wheat are uncertain because of periods of aridity.

The world statistics on sorghums and millets are only approximate because much of the harvest is never recorded. Though China is a large producer, exactly how much that country does raise is uncertain. India and the United States alternate as leading producers, each growing about 25 percent of the world's production. Africa is the next largest producer. South America has little production as a whole, but Argentina and Mexico have the greatest harvests in that area. In recent years sorghum production for animal feed has increased significantly.

CULTURAL REQUIREMENTS

Sorghums and millets are usually grown as row crops, but in some situations may be grown in broadcast- or drill-sown solid stands. Tillage is required to provide a good seedbed and to control weed growth. Seedlings are small; early growth is slow; and plant competition or poor growing conditions during this time can drastically reduce yields.

The crops may be grown under rain-fed or irrigated conditions. Panicle initiation may occur within 45 to 80 days or so depending upon varieties.

The grain is usually harvested by combine threshers; although in subsistence agriculture the panicles may be cut in the field and dried and threshed later. If the crop is to be used for silage or forage, forage harvesters are used. In some areas the crop may be cut and stacked to be fed later to animals in the form of stover.

Like many of the grasses, sorghums will produce tillers after harvest, thus making ratoon-cropping possible in tropical or subtropical regions. One major limitation to ratoon-cropping is the increase in serious weed populations unless careful control measures are practiced (Plucknett, Evenson, and Sanford 1970).

MAJOR WEEDS
OF SORGHUMS AND MILLETS

That information which we have available on the weed problems in the sorghums and millets is difficult to interpret because this group is represented by so many tribes and genera and because the major areas of production are usually in places lacking advanced weed control. Most of the information available is for the sorghums; little has been reported for the millets.

From Figure 214 it may be seen that five of the major weeds are grasses. The annual grasses, *Echinochloa colonum*, *Eleusine indica*, *Echinochloa crusgalli* and *Digitaria sanguinalis* have been reported as serious or principal weeds of sorghum. *Echinochloa colonum* is mainly a weed in Asia and Australia, but *E. crusgalli* will grow in sorghum even at some distance from the tropics. *Eleusine indica* is widely distributed in sorghums and millets across the world.

The perennial grass *Sorghum halepense* is a greater problem in temperate areas. The perennial sedge *Cyperus rotundus* is in the crop in the tropics and subtropics.

Although not shown on the map, the broad-leaved weed *Portulaca oleracea* is widely distributed in sorghums and millets across the world. The perennial grass *Cynodon dactylon* is often reported as a problem weed in the tropics or subtropics. Members of the genus *Striga* can be problem weeds in sorghums.

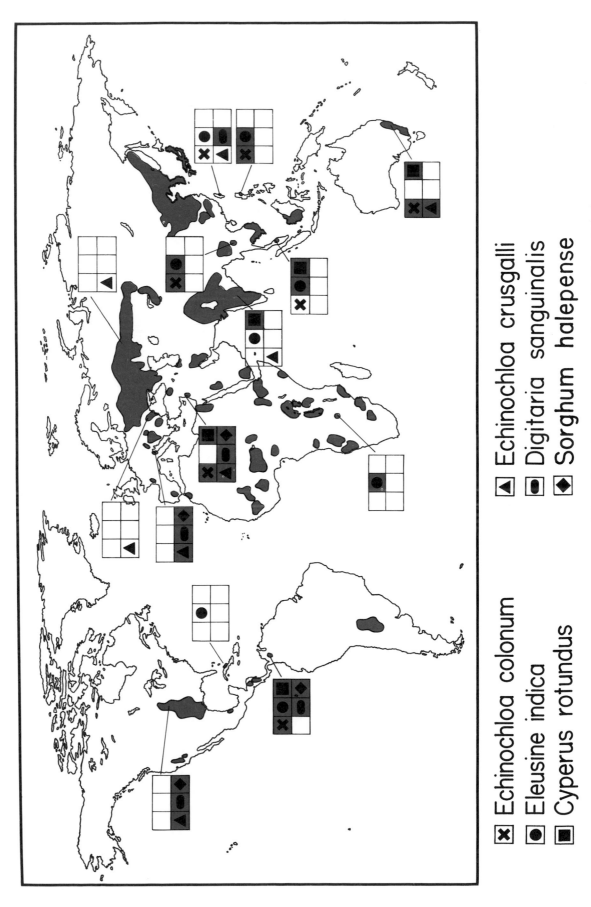

FIGURE 214. Weeds of sorghum and the main production areas across the world. Colored blocks signify serious and principal weeds; uncolored blocks signify common or unranked weeds.

☒ Echinochloa colonum   ◀ Echinochloa crusgalli

◉ Eleusine indica   ◖ Digitaria sanguinalis

◼ Cyperus rotundus   ◆ Sorghum halepense

Soybeans, *Glycine max* (L.) Merr.

LEGUMINOSAE, LEGUME FAMILY

Soybeans are produced for both human and animal food. The beans are especially important as a food source in eastern Asia. They are used both in the fresh and dry state; they may be fermented or may be processed to give flour, oil, sprouts, or a milk which is an important protein supplement for infant feeding and a source of curd and cheese as well. The edible oil is an important food in the Far East and there has been a dramatic increase in its use in Europe and the United States in the last 3 decades. The oil is now used in these areas for margarine, salad oil, and shortening. The oil or its by-products are in demand as raw materials for stabilizers, wetting agents, paints, inks, soaps, plastics, detergents, and disinfectants. The meal remaining after the oil has been extracted is an excellent source of protein for animals and is used by man to manufacture synthetic fibers, adhesives, sizings for textiles, waterproofing, and so on.

The United States has been the largest producer in the past 2 decades and about 75 percent of the crop is used locally. China is the next largest producer; Brazil ranks third and is the leading producer in Latin America. The Soviet Union has recently expanded its acreage and is now a leading producer. Other leading countries are Canada, Indonesia, Japan, and Korea. The plant is the most important food legume of China, Japan, Korea, Malaysia, and Manchuria (Cobley 1956, Ochse et al. 1961, Purseglove 1968).

CULTURAL REQUIREMENTS

The wild form of the soybean plant is unknown. The cultivated forms were grown extensively in China and Manchuria before recorded history. The species is so widely grown over the world now that thousands of varieties have been developed, and it is now possible to select types which will prosper under many different conditions. Some varieties can be grown at lat 52° N which is at about the level of Khabarovsk in the Soviet Union; Brussels, Belgium; or Winnipeg, Canada. The length of day represents one of the most critical factors for flowering. It is a short-day plant and all varieties must have at least 14 to 16 hours of darkness to flower. Most varieties have a very narrow daylength requirement and they are not productive outside of this range. Growing season requirements actually vary from 75 to 100 days.

Because soybeans are susceptible to frost, they cannot be a major crop in northern Europe and can be grown only in the southern part of Canada. Optimum growing conditions are similar to those required by corn. The plant thrives in a warm summer of high humidity although it cannot stand excessive heat. It is not productive where night temperatures fall below 13° C. After the seedlings have become established, the crop is capable of withstanding brief periods of drought; but if it is to produce a good yield, it must have an even supply of moisture. The species is planted on a wide range of soils but it grows best on sandy or clay loams or alluvial soils of good fertility. The nitrogen-fixing bacteria *Rhizobium japonicum* is specific to the root nodules of soybeans, and soybean seeds must be inoculated before being planted in new areas.

Soybeans are often rotated with millet and rice in Asia and they are sometimes used as an intercrop in the early years of some plantation crops. They are also interplanted with corn and other crops. In the United States soybeans are frequently rotated with corn; and the ease with which both crops have been mechanized accounts for the large increases in production. A firm seedbed is required and the spacing depends on the purpose for which the crop is being planted. It is grown in mixtures with corn, Sudan grass, or sorghum to provide hay or forage.

MAJOR WEEDS OF SOYBEANS

The crop is sensitive to weed competition in the cropping systems now used in many of the countries having large production. Herbicides or interrow cultivation must be used until the canopy fills in to shade out weed growth.

Grown most often as a row crop and usually given routine cultivation, soybeans are not plagued with serious perennial weed problems across the world. The practices of cultivation and rotation have resulted in reports of large numbers of weed species, mostly annuals, found in soybeans across the world, but frequently these annuals are serious in only one or two countries. The five species which seem to be most consistently distributed as principal weeds across the world are shown in Figure 215. All of them are grasses or sedges. *Eleusine indica* is almost everywhere that soybeans are grown, being reported among the three most important weeds in Japan and Taiwan, and as a serious weed in Malaysia. *Cyperus rotundus*, the perennial sedge, is one of the most widely distributed in the different soybean areas. The two *Echinochloa* species are in Taiwan and *E. crusgalli* is in the northern part of the United States and the Soviet Union; while *E. colonum* is in the warmer areas of the Philippines and Mexico. *Rottboellia exaltata*, a vigorous annual grass, has been increasing as a threat to several crops in recent years in Asia, Africa, and South America. We are reporting it here as a serious weed in Rhodesia and Zambia as well as in the Philippines on the other side of the world. Three additional annual grasses are reported to be principal weeds of soybeans. They are *Digitaria sanguinalis* in Taiwan and the United States; *Setaria glauca* in the United States; and *Digitaria adscendens*, a species which is reported to be the most serious weed in some fields in Japan. Important broad-leaved weeds in soybeans are *Chenopodium album*, *Ageratum conyzoides*, *Amaranthus spinosus*, *A. hybridus*, *Mimosa pudica*, *Commelina benghalensis*, *Datura stramonium*, and *Physalis angulata*. The perennials *Agropyron repens* and *Cyperus esculentus* are principal weeds in Canada and the United States. *Convolvulus arvensis* and *Sorghum halepense*, also perennials, are also principal weeds in the United States.

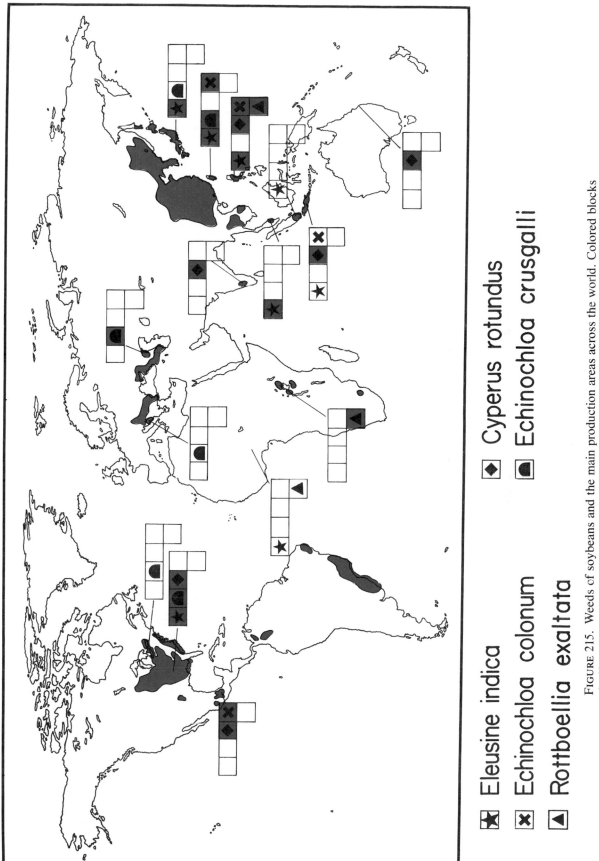

FIGURE 215. Weeds of soybeans and the main production areas across the world. Colored blocks signify serious or principal weeds; uncolored blocks signify common weeds.

★ Eleusine indica   ◆ Cyperus rotundus

✖ Echinochloa colonum   ◖ Echinochloa crusgalli

◀ Rottboellia exaltata

Sugar Beets, *Beta vulgaris* L. var. *altissima* Rossig

CHENOPODIACEAE, GOOSEFOOT FAMILY

It may be seen from Figure 216 that beets grown for sugar are confined to the temperate zones and to the fringe of the subtropical areas where there is a long growing season. Europe produces about 45 percent of the world's sugar beets; the Soviet Union about 40 percent; and the United States about 10 percent. There is no significant production in South America, Africa, or Asia (outside of the Soviet Union).

The species originated in the salty soils along the coasts of Europe, probably near the Mediterranean Sea. Man first selected and cultivated the red-fleshed, edible, sweet-tasting form which we call redbeets or table beets. Then it was discovered that the plant stores some of its reserves in its root in the form of sucrose, so that, in the 18th century in Europe, a massive program of selection and crossbreeding provided varieties with 15 percent sucrose. There are now varieties that yield more than 20 percent sucrose.

CULTURAL REQUIREMENTS

An optimum growing season for beets is 5 or 6 months. Long days and high light intensities are favorable for growth; and in autumn a dry period or cool nights are needed to check the growth of the tops and to accelerate the accumulation of sugar in the roots. Mean summer temperatures of 16° to 21° C (60.8° to 69.8° F) are satisfactory.

A regular water supply during most of the season is necessary if good growth and size are to be obtained. Because the plant sends roots deep into the soil, it can obtain soil moisture more efficiently than many crops. Highest yields are obtained under irrigation or with an annual rainfall of more than 65 cm.

The crop is usually planted in medium to light soils, and the species can tolerate some alkalinity. Soils must be well drained and fertile if good yields are to be obtained. A plentiful supply of potassium is required, for this element tends to increase the proportion of carbohydrate which is stored in the form of sucrose. Sugar beets are grown in the normal fashion of temperate zone row crops. From Figure 216 it may be seen that the principal weeds are all annuals, this being a reflection of the system of mixed agriculture in which the crop is often grown. They enter into rotations with grain, potatoes, fodder beets, or managed pastures and thus the perennial weeds are unable to get a foothold.

MAJOR WEEDS OF SUGAR BEETS

Chenopodium album, an annual species belonging to the same family as the sugar beets, is one of the most frequently reported weeds in this crop (Figure 216). Almost everywhere that this species is found in the crop it is said to be a principal weed. The other three annual broad-leaved weeds, *Sinapis arvensis*, *Echinochloa crusgalli*, and *Amaranthus retroflexus*, seem to be about equally important. These do not

530

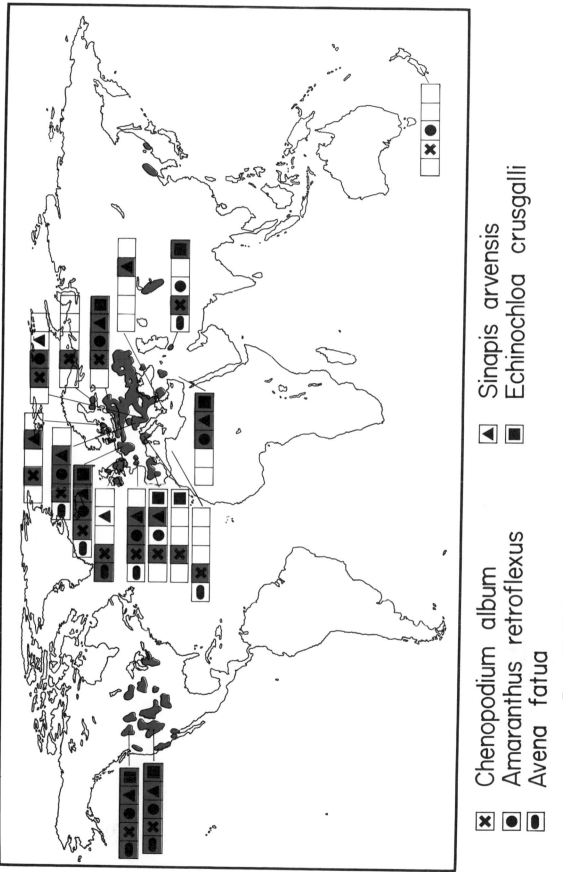

Chenopodium album ▣ Sinapis arvensis
Amaranthus retroflexus ◼ Echinochloa crusgalli
Avena fatua

⊠ ● ⬮

FIGURE 216. Weeds of sugar beets and the main production areas across the world. Colored blocks signify serious or principal weeds; uncolored blocks signify common or unranked weeds.

present regional patterns but are generally scattered over the sugar beet fields of the world. *Avena fatua*, an annual grass, is on four continents.

Matricaria chamomila, *Polygonum aviculare*, *Polygonum convolvulus*, and *Stellaria media* are all principal weeds in several countries of northern Europe. *Polygonum aviculare* is in Spain as well; and *Stellaria media* is in Spain, Israel, and New Zealand. Two additional annual grasses are *Poa annua*, a troublesome weed in Belgium, England, and Germany; and *Setaria viridis*, a principal weed in Canada, Iran, the Soviet Union, Spain, and the United States and one which is in Italy as well. Two perennials, *Agropyron repens* in northern Europe, Spain, and the United States and *Convolvulus arvensis* in Greece, Iran, and the United States, complete the list of major weeds in the world's sugar beet fields.

Sugarcane, *Saccharum officinarum* L.

POACEAE (also GRAMINEAE), GRASS FAMILY

Commercially used sugar is derived from two plants, sugarcane and sugar beet, one a monocot and the other a dicot. About one-half of the world's cane sugar goes into export, whereas beet sugar is largely consumed in the country in which it is grown. The manufacture of raw sugar from both plants is carried out near the site of production. White sugar is produced by a complex refining process which takes place in industrial countries. Because cane sugar usually costs less, beet sugar is often protected by tariffs and subsidies. Sugar beet pulp and sugarcane molasses are valuable by-products in most areas; both pulp and molasses being used for cattle feed and molasses being used in the production of alcohol.

Sugarcane is a tall perennial grass 3 to 4 meters or more in height; the plant grows in clumps or "stools" consisting of many juicy stalks 4 to 5 cm or more in diameter. The leaves are large, from 6 to 8 cm wide and 0.6 to 1.3 m long. Sugarcane is considered to be native to southeast Asia, but it was grown widely throughout the tropics before recorded history. Most of the crop is grown commercially for processed sugar, but some is grown in gardens for chewing. Sugarcane, largely a plantation crop, is produced in some countries by small farmers.

Sugarcane is a plant of the warm tropics or subtropics; and best yields are made possible by frequent heavy rainfall or irrigation interspersed with bright sunshine. This ensures that as the plant is maturing it will accumulate a maximum amount of sugar. Varieties grow for 12 to 24 months before they are harvested. If the season is cut short by frost, the crop is harvested when immature. Minimum rainfall requirements are 1,250 to 1,600 mm per year. In many areas water is carried to the crop by furrows; rarely by overhead sprinklers. Varieties vary a good deal in their temperature requirements, but growth in general is very slow below 15° C (59° F), and activity begins at 21° C (69.8° F). Plants regenerate after being cut but the second, or ratoon, crop, is sometimes not as productive—especially if it is poorly managed. However, as many as 10 or more ratoon crops may be produced (Plucknett, Evenson, and Sanford 1970). The plant will grow in a variety of soils but grows best in deep, well-drained, moisture-holding types. For good yields the soil must be high in fertility.

The countries of Latin America produce about one-half of the world's sugarcane. India produces about 20 percent of the world's total and is the major producer. Brazil is the next leading producer with about 15 percent of the world's total. Cuba, Mexico, Pakistan, and the United States are the next largest producers.

CULTURAL REQUIREMENTS

Sugarcane is planted as a row crop. Deep tillage is desirable to ensure deep root development. Sugarcane is propagated by stem or stalk cuttings (setts) 0.3 to 0.4 m in length, each with three or more nodes. These are planted in furrows, usually by laying them horizontally, sometimes with some overlap. Row width is about 1 to 1.6 m. The setts sprout in about 10 to 14 days. Early growth is quite slow, and for this reason intercropping with short term crops such as vegetables or cereals is sometimes practiced to hold down weed competition as well as to obtain added income. The wide row widths make weed competition a potential problem during the first 4 to 6 months.

As the young plants grow they begin to produce tillers from the base, thus increasing the density and the width of the stool. Under good growth conditions the plants can reach 2 m or more in height by 6 months of age; also, the heavy tillering causes the tops to overlap, thus almost completely shading the interrows. When this occurs (provided good weed control has been practiced) weed growth is no longer a problem. This growth stage is sometimes called "close-in" or the crop is referred to as being "out-of-hand."

Harvesting usually consists of cutting and removing the juicy stalks by hand or by mechanical means. Provided that severe root damage has not occurred during harvest, a ratoon crop may be grown from tillers which arise from the cut stem on the old root system. Management of ratoons follows essentially the same procedures as are used in "plant" crops.

MAJOR WEEDS OF SUGARCANE

From Figure 217 it may be seen that there are six major weeds in sugarcane which have wide distribution across the world. From the number of boxes on the map that are full, or nearly full, it is evident that these weeds are everywhere that cane is grown. With the exception of *Portulaca oleracea*, all are grasses or sedges and three of them are annuals. *Cyperus rotundus*, a perennial sedge, is the major weed and it has the greatest distribution. It is closely followed by *Cynodon dactylon*, a perennial grass.

Figure 218 shows a second group of grass weeds which are serious. *Sorghum halepense*, a perennial grass, is widely distributed in sugarcane in the world. *Panicum maximum*, also a perennial, tends to be troublesome mainly around the Caribbean area and in southern Africa. *Rottboellia exaltata*, an annual, is a dreaded weed of sugarcane. It is serious in southern Asia, northern South America, and eastern and southern Africa. It is, however, on the increase in the world in this and several other crops. It is a vigorous grower and is reducing the acreage of some crops. It mimics sugarcane and thus is often not recognized until it is far along. It develops fiberglasslike needles at the base of the stem and these become embedded in the hands of weeders. In some countries the weeders now refuse to enter the fields when *R. exaltata* is in an advanced stage.

Weeds which are frequently ranked as serious or principal include three annual broad-leaved weeds, *Amaranthus spinosus*, *Ageratum conyzoides*, and *Bidens pilosa*, and the perennial sedge, *Cyperus esculentus*.

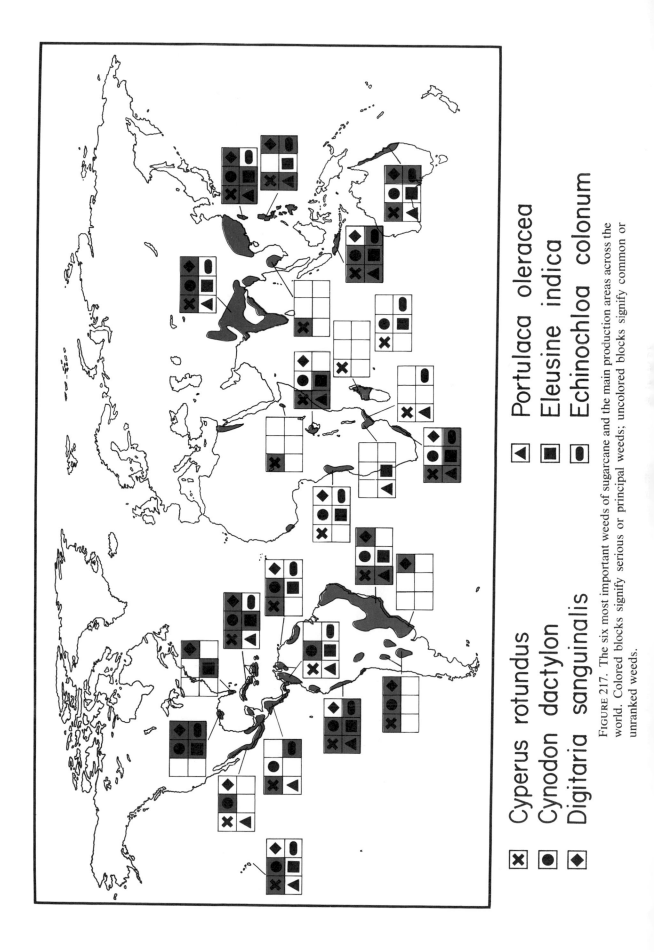

Cyperus rotundus

Cynodon dactylon

Digitaria sanguinalis

Portulaca oleracea

Eleusine indica

Echinochloa colonum

FIGURE 217. The six most important weeds of sugarcane and the main production areas across the world. Colored blocks signify serious or principal weeds; uncolored blocks signify common or unranked weeds.

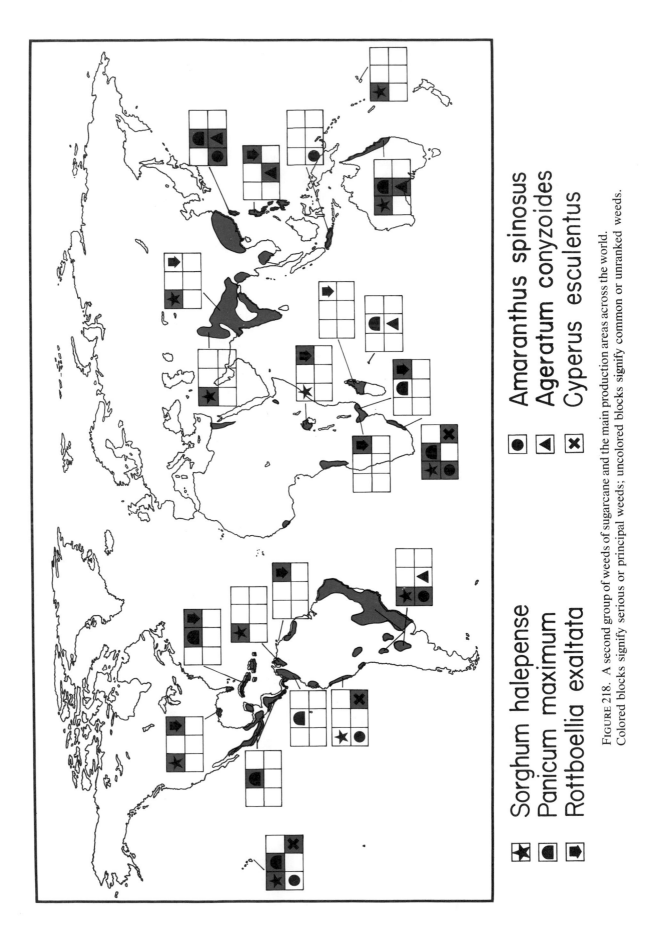

FIGURE 218. A second group of weeds of sugarcane and the main production areas across the world. Colored blocks signify serious or principal weeds; uncolored blocks signify common or unranked weeds.

★ Sorghum halepense ● Amaranthus spinosus
◖ Panicum maximum ◀ Ageratum conyzoides
➡ Rottboellia exaltata ✖ Cyperus esculentus

Tea, *Camellia sinensis* (L.) O. Ktze.

THEACEAE, TEA FAMILY

Tea is mainly grown in the mountainous regions of the tropics (1,300 to 2,000 m) and in the subtropics (Eden 1965). Under natural conditions it is a small evergreen tree which grows to 15 m in height. In commercial cultivation it is pruned as a low-spreading bush 50 to 150 cm tall. It is often grown under shade but in some areas, such as at high elevations in Ceylon, the fields are open. Here it is cool and there is much cloudy weather. In tropical climates growth is continuous. In more temperate regions where there is a definite cold period the plants are pruned annually and the leaves are plucked only during the warm season. Hail can cause serious damage to the bushes.

CULTURAL REQUIREMENTS

The crop requires warm, humid conditions with a minimum of 1,000 to 1,100 mm of rainfall per year. In Ceylon the crop thrives in some areas with 5,000 mm of rain per year. Mean temperatures should not go below 13° C or above 30° C. The crop does not tolerate drought or temperatures more than a few degrees below freezing. Soil requirements are quite specific. They must be acid within certain limits, must not contain more than a trace of available calcium, must be well drained, must be permeable to allow deep root penetration, and should have plenty of available nitrogen.

The flavors which are characteristic of tea are provided by caffeine, polyphenols, and essential oils in the bud and top leaves. Green tea is made from leaves which are steamed and dried without withering and fermenting. To make black tea the leaves are withered, rolled, fermented, and dried. Black tea provides the bulk of the world's supply and, because it requires machine manufacture, it is usually produced on large plantations. Green tea is mainly produced in China and Japan on small holdings and much of it is consumed locally.

The harvest consists of plucking newly grown vegetative shoots composed of the terminal bud and the top two or three leaves surrounding it. Harvesting is often done by women who pluck the leaves with thumb and forefinger and place them in a bag. Pluckers harvest 14 to 35 kg of green shoots per day. These leaves and the bud have the highest content of caffeine and polyphenol and thus yield tea of highest quality. Coarse plucking, which harvests successive leaves down the stem, obtains a reduced quality. Slow growth at high altitudes in the tropics also improves quality.

The length of time from planting to maturity varies from 2 years for stumps to 4 years for seedlings. This period also varies with the elevation. The plucking system and the environmental conditions govern the length of time between growth flushes and the period required for the plucked shoot to produce a new one, a period about 75 to 90 days. Each plant, however, has shoots at varying stages of maturity, and experienced pickers take only those which are at just the right stage. Plucking may be done each 7 to 10 days at low elevations and on a 2-week schedule where it is colder.

About 1.8 kg of green shoots will provide 0.45 kg of made tea. Yields vary greatly with pruning cycles, varieties, locations, fertilizers, and many other factors. Some areas of Assam and Ceylon average about 1,200 kg of made tea per hectare; a plot of clonal tea in Ceylon once yielded 6,000 kg, and yields in Dar Jeeling may be 450 kg of a tea which brings very high prices. Much of the tea produced in India and Ceylon comes from bushes which are 70 to 100 years old, but the economic life of a bush in many areas is 40 to 50 years.

The earliest use of tea was probably medicinal but it has been used as a beverage for 2,000 or 3,000 years. The Mongols carried tea bricks to Central Asia through Siberia. Commercial tea did not reach eastern Europe until after 1650 A.D., a time when coffee drinking was already well established. Great Britain is the largest consumer of tea, taking more than half the world's imports. Per capita consumption is about 4.5 kg per year. The British have always been dominant in the trade and were influential in the development of commercial estates in India and Ceylon. The high tea areas of Ceylon were once planted to coffee. About 1869 the leaf rust, *Hemileia vastatrix*, reached the island and destroyed the coffee. The estates were then planted to tea.

India produces about 33 percent of the world's tea; Ceylon, about 20 percent; and Mainland China, about 15 percent. Next in descending order of production are Japan, Indonesia, Kenya, and Pakistan, but these countries have a very small portion of the world's crop. All of Africa produces only about 5 percent of the world's total. Europe and North America raise no tea; none is grown around the Mediterranean; and there is a small amount in Brazil and on the eastern edge of the Black Sea in Georgia (Soviet Union).

MAJOR WEEDS OF TEA

Virgin forest land is preferred for the planting of tea, but much tea is now started in areas which have been in other crops or in pastures. The weeds of the tropics grow very vigorously and can regenerate quickly after cultivation. They often exceed the growth rates of crop plants and thus compete for moisture, light, and nutrients. In tea it is especially important that weeds be controlled for the first 3 to 4 years, and sometimes through the 6th year, to assure full establishment of the crop at the earliest date. Tea is very susceptible to weed competition. When the canopy has been established, weed problems can be expected to diminish.

Generally it is the perennial grasses of the "couch" type which are most damaging to tea. The stoloniferous and rhizomatous plant materials spread under and between the bushes, with pieces of these organs being spread about during cultivation. In Asia "couch grass" may refer to *Panicum repens*, *Pennisetum clandestinum*, or *Imperata cylindrica*; in eastern Africa, to *Digitaria scalarum*; and in central Africa, to *Cynodon dactylon*.

Nurse crops are sometimes used in new plantings to provide weed control and to protect against erosion. In some places a bulbous herb, *Oxalis latifolia*, is encouraged because it has rapid, renewed growth when the rains come and thus binds the soil. This species is a weed of several crops in other areas of the world.

When plants have become established deep cultivation is no longer permitted in some areas because of damage to surface roots. Weed control must, therefore, be effected either with herbicides or by repeated light scrapings with a hoe or knife. If high rainfall and sloping terrain are present, these conditions necessitate a choice between the competition offered to the crop by weeds against their soil-binding properties which minimize erosion. Chambers (1963) has reviewed management and weed control in tea.

Imperata cylindrica, a perennial grass, is the principal weed of tea across the world (Figure 219). It is ranked as one of the three most important weeds in Ceylon, India, and Indonesia; a principal weed in Japan, Malaysia, Mozambique, Taiwan, and Uganda; and a weed in the crop in Kenya and Tanzania. The severity and distribution of *Cynodon dactylon*, *Paspalum conjugatum*, and *Digitaria scalarum* serve to emphasize that perennial grasses are the major problem for tea. *C. dactylon* is a serious weed in Ceylon and Indonesia; a principal weed in India, the Soviet Union, and Tanzania; and is common elsewhere in eastern Africa, Japan, and Taiwan. *P. conjugatum* is serious in Ceylon, India, Indonesia, and Taiwan; and *P. distichum* and *P. dilatatum* (not shown) are principal weeds in the Soviet Union. *D. scalarum* is confined to eastern Africa and western Asia. The persistence and rapid growth of *Ageratum conyzoides*, the only broad-leaved weed on the map, and the dense stands which it forms make it a problem to be reckoned with almost everywhere that tea is grown.

In addition, *Saccharum spontaneum* is a serious problem in tea and in many other crops in India; and *Digitaria adscendens* is serious in Japan and Taiwan. *Cyperus rotundus*, *Panicum repens*, and *Mikania cordata* are principal weeds in one or more of the important tea-producing countries of Asia.

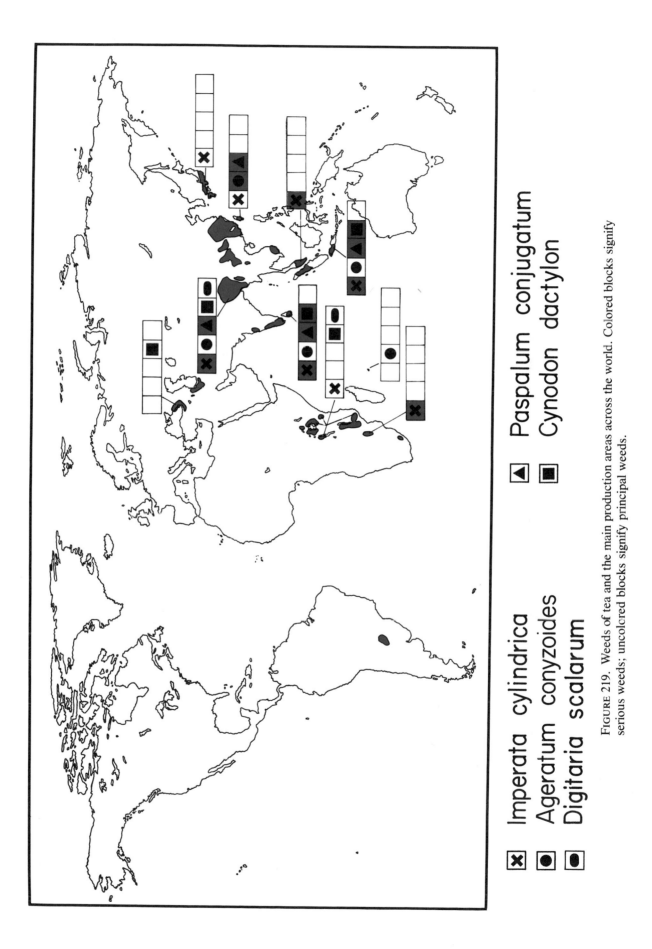

Imperata cylindrica   ▲ Paspalum conjugatum
Ageratum conyzoides   ■ Cynodon dactylon
Digitaria scalarum

FIGURE 219. Weeds of tea and the main production areas across the world. Colored blocks signify serious weeds; uncolored blocks signify principal weeds.

Drymaria cordata is in most of the tea in Asia, as is *Bidens pilosa*; *B. pilosa* has also been reported from Mozambique and Brazil. *Pennisetum clandestinum* is in tea in Ceylon, Costa Rica, India, Kenya, and Uganda.

Wheat, *Triticum aestivum* L. em. Thell. and *Triticum durum* Desf.

POACEAE (also GRAMINEAE), GRASS FAMILY

The place of origin of wheat is a question which has not been resolved, but there is evidence that it was grown 10,000 to 15,000 years B.C. in the Middle East. Wheat has been important in the Americas for the past 400 years, but the civilizations which existed for 4,000 years prior to that were sustained by corn (Quisenberry and Reitz 1967). Presently wheat and rice each provide about 20 percent of the total food calories for the world. Corn and potatoes are next in order of importance. Wheat is the main staple food of about 35 percent of the world's people. Quisenberry and Reitz (1967) have warned that a serious global failure of the wheat crop could lead to catastrophe, for few nations could survive for even 1 year. Much of the world's wheat is grown in the Northern Hemisphere. The Soviet Union has the largest acreage and one which is equal to the rest of Europe and North America combined. The three areas comprise about 90 percent of the world's wheatlands. The Soviet Union is the leading producer followed by the United States and China, and these three produce much more than the following countries which are the next major producers: Canada, France, India, and Italy. The countries of northern Europe have the highest yields per hectare in the world.

Most of the world's wheat (bread wheat) is of the common or hexaploid type, *Triticum aestivum* L. em. Thell. The grain may be hard or soft and is brownish red to white in color. Most of it is used for bread although the softer types are used for crack-ers, cookies, and pastry. Durum wheat, *Triticum durum* Desf., the macaroni wheat, is a tetraploid type grown principally in France, India, Italy, North Africa, the Middle East, the Soviet Union, and in the north central area of the United States and the adjacent portion of Canada. The grain may be white (amber) to reddish brown. It has a flinty grain and is sometimes called hard wheat, although the term is confusing because both species have hard types.

CULTURAL REQUIREMENTS

A growing period of about 90 days is required, but a long history of selection and improvement has resulted in some types which can be grown at high altitudes and in cooler areas where the growing season is short. The crop prefers cool weather. Warm moist weather is optimum for early growth and a minimum temperature of about 15° C (59° F) is required for the grain to ripen. It is a hardy crop and although about three-fourths of the world's acreage is planted in autumn, the plants cannot stand extreme cold. Wheat must be sown in spring in the areas where it is too cold for the plant to be able to survive winter conditions. Wheat cannot prosper if temperatures and humidity are too high; in the United States, for example, the limits of production are at about a mean temperature of 20° C and 1,250 mm of annual rainfall. General rainfall requirements across the world vary a great deal but most of the acreage is in zones where annual rainfall is 250 to 1,000 mm. In some places the crop is irrigated; in semiarid regions it is grown in alternate years with a bare fallow.

The systems for wheat culture vary from subsistence agriculture to large-scale, completely mechanized systems in which field operations follow a 24-hour schedule. The traditional preparation of the seedbed and the use of fertilizer, rotations, and good seedstocks make it possible to have good stands and thus allow the least opportunity for weed growth. High quality, clean seeds provide earlier, more vigorous growth and prevent the reseeding of a field with weeds which are associated with wheat. Rotations often include cultivated row crops or fallow. In recent times the bare fallow has been less favored in all crops because of the tendency of continued cultivation to break down the soil structure. Planting may be done by hand or with the modern precision drill used on large-scale operations. The drill makes it possible to use fewer seeds and, by distributing the seeds more accurately and uniformly, gives maximum control of rate and depth of placement. The result is better stands, higher yields, and more uniform maturity.

MAJOR WEEDS OF WHEAT

After the crop has emerged, it requires few maintenance operations during the growing season. Winter wheat may have to be rolled to firm the soil around the roots if there has been frost-heaving. In wet areas of fields or after a beating rain that has packed the soil, a tillage operation may be carried out with a harrow or rotary hoe to break the crust and to control weeds. A roller may follow to break clumps and firm the seedbed. On small farms in Asia, Africa, South America, and southern Europe, weeding is often done with hand tools. Herbicides are used extensively in large-scale operations. The most difficult weeds are the annual grasses, such as *Avena fatua*, which have life cycles similar to that of wheat, for such grasses may keep reseeding themselves until the land can no longer be cropped. Of equal difficulty in many areas are the deep-rooted perennial weeds, such as *Agropyron repens*, which may persist if the land is continuously cropped in wheat (Delorit and Ahlgren 1967, Peterson 1965).

Among the important weeds of wheat, *Avena fatua* stands out because of its wide distribution across the world and because it has been reported to be a serious or principal weed in so many countries (Figure 220). Two annual weeds, *Chenopodium album* and *Polygonum convolvulus*, are found on four continents, and two more, *Stellaria media* and *Spergula arvensis* (not shown), are mainly a problem in northern Europe but are also troublesome in New Zealand and Tanzania. *Convolvulus arvensis*, a perennial, is a principal weed in the United States but it is also a weed in several countries of Asia where water may be limiting and where the weed can persist because of its very deep root system. Another perennial broad-leaved weed, *Cirsium arvense*, is a principal weed of wheat in Belgium, Canada, Germany, India, New Zealand, and Turkey. *Anagallis arvensis* and *Asphodelus tenuifolius* are reported to be principal weeds in western Asia in the belt from Iran down through India. *Chondrilla juncea* is a serious problem in Australia and also ranks high in Iran. *Sinapis arvensis* is a principal weed in the Northern Hemisphere from Canada through England, Sweden, and into the Soviet Union. *Raphanus raphanistrum* is widespread and reported to be an important weed of wheat in Australia, England, Kenya, and South Africa. The perennial grass *Agropyron repens*, together with *Polygonum aviculare*, *P. persicaria*, *Galium aparine*, several *Galeopsis* species, *Tripleurospermum maritimum*, *Alopecurus myosuroides*, and *Apera spica-venti*, is reported to be a principal weed of wheat in one or more countries of northern Europe or of New Zealand.

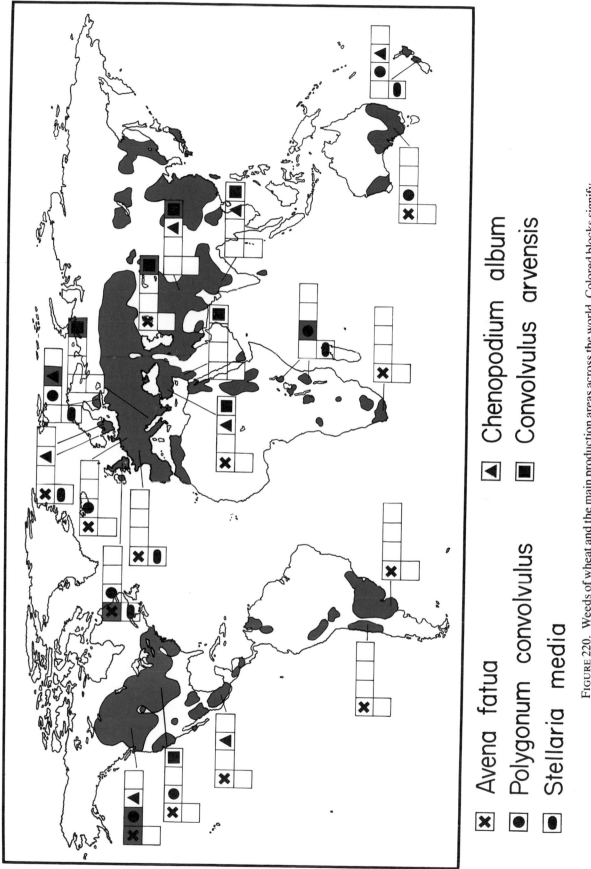

FIGURE 220. Weeds of wheat and the main production areas across the world. Colored blocks signify serious weeds; uncolored blocks signify principal weeds.

Avena fatua   Chenopodium album

Polygonum convolvulus   Convolvulus arvensis

Stellaria media

Appendix A

WORLD LIST OF USEFUL PUBLICATIONS ON WEED DISTRIBUTION, IDENTIFICATION, BIOLOGY, AND CONTROL

These publications are useful for the study of world weeds. They range from small to large, from books to bulletins, and from taxonomy to weed control. Floras are included only where they make a special contribution to the study of weeds or cover areas where we are as yet unable to find satisfactory weed publications.

Please note that special publications on three of the world's most serious weeds are listed in this appendix under a country as follows: *Cyperus rotundus*, India; *Eichhornia crassipes*, United States; and *Imperata cylindrica*, Great Britain and Indonesia.

There is in the world an impressive collection of books about weeds, and the following list is known to be incomplete. The authors are genuinely interested in suggestions for the correction of entries and for information about publications that may be added.

AFGHANISTAN
Kitamura, S. 1960. *Flora of Afghanistan*. Kyoto University, Kyoto. 486 pp. English.
———. 1964. *Plants of West Pakistan and Afghanistan*. Kyoto University, Kyoto. 283 pp. English.

AFRICA
Wild, H. 1961. *Harmful aquatic plants in Africa and Madagascar*.Government Printer, Salisbury. 68 pp. English.

AFRICA, EAST
Karker, K., and D. Napper. 1960. *An illustrated guide to the grasses of Uganda*. Ministry of Natural Resources, Entebbe. 186 pp. English.
Hocombe, S., and R. Yates. 1963. *A guide to chemical weed control in East African crops*. East African Literature Bureau, Dar es Salaam. 84 pp. English.
Ivens, G. 1967. *East African weeds and their control*. Oxford University Press: Nairobi, Addis Ababa, and Lusaka. 244 pp. English.
Lind, E., and A. Tallantire. 1962. *Some common flowering plants of Uganda*. Oxford University Press, London. 257 pp. English.
Napper, D. 1965. *Grasses of Tanganyika*. Bulletin 18. Ministry of Agriculture, Forest, and Wildlife, Dar es Salaam. English.

AFRICA, NORTH
Catalogue of the principal weeds encountered in North Africa. 1963. Publier Société, Usines Chimiques, Rhone-Poulenc. 8 pp. French.

AFRICA, SOUTH

Henderson, M., and J. Anderson. 1966. *Algemene onkruide in Suid-Afrika–Common Weeds of South Africa*. Department of Agriculture Technical Services, Pretoria. 440 pp. Afrikaans and English.

Meredith, D., ed. 1955. *Grasses and pastures of South Africa*. First published by Central News Agency, Johannesburg; later published by Hafner Publishing Co., New York. 771 pp. English.

Phillips, E. 1938. *The weeds of South Africa*. Bulletin 195, Division of Botany series 41. Department of Agriculture, Pretoria. 229 pp. English.

ALASKA (U.S.)

Taylor, R., A. Wilton, and L. Klebesdel. 1958. *Weeds in Alaska*. University of Alaska, Agricultural Extension Service, Palmer. 32 pp. English.

ANGOLA

Teixeira, J. 1965. *Flora infestante das culturas de Angola*. 3 bulletins. Instituto de Investigaçao Agronomica de Angola, Nova Lisboa. 15, 18, and 3 pp. Portuguese.

ARGENTINA

Cabrera, A. 1964. *Las plantas acuaticas*. Editorial Universitaria, Buenos Aires. 94 pp. Spanish.

Godoy, E., ed. 1959. *La maleza en la región Pampeana*. Estación Experimental Agropecuaria de Pergamino, Provincia de Buenos Aires. 88 pp. Spanish. This work is the proceedings of a conference held in 1955.

Marzocca, A. 1957. *Manual de malezas*. Imprenta y Casa Editora Coni, Buenos Aires. 530 pp. Spanish. Out of print.

ASIA

Bor, N. 1960. *Grasses of Burma, Ceylon, India and Pakistan*. Pergamon Publishing Co., Elmsford, New York. 776 pp. English. Out of print.

AUSTRALIA

Bailey, F. 1906. *The weeds and suspected poisonous plants of Queensland*. Queensland Department of Agriculture and Stock, Brisbane. 245 pp. English. Write to the Queensland Herbarium, Indooroopilly, Queensland 4068.

Bill, S., and N. Graham. 1968. *Chemical weed control in irrigation channels and drains*. State Rivers and Water Supply Commission of Victoria, Melbourne. 38 pp. English.

Clarke, G. 1949. *Important weeds of South Australia*. Bulletin 406. Department of Agriculture, South Australia, Adelaide. 119 pp. English.

Commonwealth Scientific and Industrial Research Organisation. 1953. *Standardised plant names: a revised list of standard common names for the more important Australian grasses, weeds, and other pasture plants*. Bulletin 272. Melbourne. 132 pp. English

Everist, S. 1957. *Common weeds of farm and pasture*. Department of Agriculture, South Adelaide. 135 pp. English.

Johnston, A. 1968. The changing pattern of weeds in cereal crops. *Proceedings of the 1st Victorian weed conference* 3: 31-39. English. Write to the Weed Society of Victoria, c/o Keith Turnbull Research Station, Frankston, Victoria 3199.

Maiden, J. 1920. *The weeds of New South Wales*. Part 1. Government Printer, Sydney. 141 pp. English.

Mann, J. 1970. *Cacti naturalised in Australia and their control*. Alan Fletcher Research Station, Department of Lands, Sherwood, Brisbane. 128 pp. English.

Meadly, G. 1965. *Weeds of Western Australia*. Department of Agriculture, Western Australia, South Perth. 173 pp. English.

Orchard, H. 1951. *Weeds of South Australia*. Bulletin 418, series 1. 42 pp. Department of Agriculture, South Australia, Adelaide. English.

Whittet, J. 1968. *Weeds*. New South Wales Department of Agriculture, Government Printer, Sydney. 487 pp. English.

Young, H. 1962. Weed control. *Cane Growers' Quarterly Bulletin* 25(4). Brisbane. 66 pp. English. This is an extensive weed list with some descriptions.

AUSTRIA

Schönbrunner, J. 1954. Übersicht über die wichtigen Ackerunkrauten und deren Bekämpfung [Review of the important field weeds and their control]. Bundesanstalt für Pflanzenschutz, Vienna. 71 pp. German.

BANGLADESH

Ghani, S., and S. Khan. 1967. Economic importance of some of the common water plants of East Pakistan. *Pakistan Journal of Biological and Agricultural Sciences* 10(2): 30-45. Department of Biochemistry, University of Dacca, Dacca. English.

BELGIUM

Slaats, M., and J. Stryckers. 1958. *Les plantes adventices des champs et leur destruction* [Control of arable weeds]. Institut Supérieur Agronomique de l'Etat, Ghent. 90 pp. French; also published in Flemish in 1964.

———. 1965. *Graslandonkruiden en hun bestrijding*

[Control of weeds of grassland]. Rijksland-bouhogeschool, Ghent. 60 pp. Flemish; also published in French in 1964.

Stryckers, J. 1968. *Les végétations aquatiques et palustres et leur destruction*. Ministry of Agriculture, Brussels. 100 pp. French.

BRAZIL

Camargo, P., ed. 1970. *Texto basico de controle químico de plantas daninhas*. University of São Paulo, Escola Superior de Agricultura Luiz de Queiroz, Piracicaba, São Paulo. 257 pp. Portuguese. Five hundred weed species occurring in Brazil are listed.

Sacco, J. 1961. *A flora da sucessão dos campos do Instituto Agronomico do Sul*. Instituto Agronomico do Sul, Pelotas, Rio Grando do Sul. 26 pp. Portuguese.

———. 1960. *Plantas invasora dos arrozais*. Instituto Agronomico do Sul, Pelotas, Rio Grande do Sul. 46 pp. Portuguese.

Warren G., et al. 1973. *Curso intensivo de controle de ervas daninhas* [Intensive course in the control of weeds]. Universidade Federal de Vicosa, Dept. de Fitotecnia, Vicosa, Minas Gerais. 339 pp. Portuguese.

BULGARIA

Andreeva-Fetvadzhieva, N. 1966. *Bor'ba s plevelite* [Weed control]. Zemizdat, Sofia. 201 pp. Bulgarian.

BURMA

see Asia, India, and India and Burma

CAMBODIA

Ho Tong Lip. 1966. *Liste des plantes courantes du Cambodge*. Division des Recherches Agronomiques, Phnom-Penh. 9 pp. Latin, Cambodian, and French.

CANADA

Alex, J. 1965. *Survey of weeds of cultivated land in the prairie provinces*. Experimental Farm, Research Branch, Department of Agriculture, Regina. 68 pp. English. Provides an excellent example of methods that may be used when surveying large areas.

Best, K., and A. Budd. 1964. *Common weeds of the Canadian prairies*. Publication 1136. Department of Agriculture, Ottawa. 70 pp. English. Provides aids to identification by using vegetative characters.

Department of Agriculture. 1930. *Weeds and weed seeds*. Bulletin 137, new series. Ottawa. 72 pp. English.

———. 1969. *Common and botanical names of weeds of Canada*. Publication 1397. Ottawa. 67 pp. English.

Frankton, C., and G. Mulligan. 1970. *Weeds of Canada*. Publication 948. Department of Agriculture, Ottawa. 217 pp. English.

McLeod, J., B. McGugan, and H. Coppel. 1962. *A review of the biological control attempts against insects and weeds of Canada*. Technical Communication 2, Commonwealth Institute of Biological Control, Trinidad. Published by the Commonwealth Agricultural Bureaux, Farnham Royal, Bucks, England. 216 pp. English.

Montgomery, F. 1964. *Weeds of Canada and northern United States*. Ryerson Press, Toronto. 226 pp. English.

CEYLON

see also Asia

Haigh, J., committee of. 1951. *A manual of the weeds of the major crops of Ceylon*. Paradeniya manual 7. Department of Agriculture, Ceylon Government Press, Colombo. 107 pp. English. All extra copies were lost in a flood at Paradeniya.

Senaratna, S. 1956. *The grasses of Ceylon*. Paradeniya manual 8. Department of Agriculture, Ceylon Government Press, Colombo. 279 pp. English.

CHILE

Matthei, O. 1963. *Manual ilustrado de las malezas de la provincia de Nuble*. Escuela de Agronomia, Universidad de Concepción, Chillán. 116 pp. Spanish.

Vallejo, A. de. 1968. *Malezas de Chile*. Boletín Técnico 34, Servicio Agricola y Ganadero, Santiago. 26 pp. Spanish. Gives Latin scientific names and Spanish common names.

Zaviezo, S. 1965. *Manual de control químico de malezas*. Privately published. Write to author at the Instituto de Investigaciones Agropecuarias, Ministerio de Agricultura, Santiago. 160 pp. Spanish.

CHINA

Chi, K. 1948. *Weeds and lawn grasses*. Chung Hwa Book Co., Shanghai. 82 pp. Chinese.

Porterfield, W. 1933. *Wayside plants and weeds of Shanghai*. Kelly and Walsh, Hong Kong and Singapore. 232 pp. English.

Sauer, G. 1947. *A list of plants growing on the Lingnan University campus and vicinity*. Lingnan University, Canton. 112 pp. English.

COLOMBIA

Bristow, J., J. Cardenas, T. Fullerton, and J. Sierra. 1972. *Malezas acuáticas* [Aquatic weeds]. Published jointly by the Instituto Colombiano Agropecuario, Director de Commun. A.A. 7984, Bogota; and by Oregon State University, International Plant Protection Center, Corvallis. 115 pp. Spanish and English.

Cardenas, J. 1969. *Manual de terminología de control de malezas y fisiología vegetal*. Sociedad Colombiana de Control de Malezas y Fisiología Vegetal, Bogota. 74 pp. Spanish.

Cardenas, J., O. Franco, C. Romero, and D. Vargas. 1970. *Malezas de clima frío*. International Plant Protection Center, Oregon State University, Corvallis, U.S.A. 127 pp. Spanish.

Cardenas, J., C. Reyes, J. Doll, and F. Pardo, ed. 1972. *Malezas tropicales* [Tropical weeds]. Volume 1. Instituto Colombiano Agropecuario, Bogota; also available from Oregon State University International Plant Protection Center, Corvallis, U.S.A. 341 pp. English and Spanish.

Reyes, C., and J. Cardenas. 1969. *Catálogo de malezas del Tolima sur*. Publication 001. Instituto Colombiano Agropecuario, Bogota. 66 pp. Spanish.

Sierra, J., A. Vera, T. Fullerton, and J. Cardenas. 1970. *Problemas de malezas en sistemas de riego*. División de Producción Agropecuaria, Instituto Colombiana de la Reforma Agraria, Bogota. 32 pp. Spanish.

Vargas, D., J. Cardenas, and C. Romero. 1971. *Catálogo de semillas de malezas de clima frío*. Division de Investigación, Departamento de Agronomía, Instituto Colombiano Agropecuario, Bogota. 117 pp. Spanish.

CZECHOSLOVAKIA

Deyl, M. 1964. *Plevele poli a zahrad*. Ceskoslovenská Akademie Zemedelskych Ved, Prague. 387 pp. Czech.

Vodak, A., Z. Kropac, and M. Nejedla. 1956. 1. *Semena nebo plody nasich kulturnich rostlin a nejcastejsich plevelu* [Seeds or fruits of crop plants and common weeds]. 2. *Klicni rostliny nasich beznych plevelu* [Seedlings of common weeds]. Ceskoslovenská Akademie Zemedelskych Ved, Prague. The two parts of this book have been published as one volume. 271 pp. Czech.

DENMARK

Ferdinandsen, C. 1918. Undersogelser over danske ukrudtsformationer pa mineral-jorder [Investigations of Danish weed populations on mineral soils]. *Tidsskrift for Planteavl* 25: 629-919. Danish; includes English summary.

Frederiksen, H., P. Grontved, and H. Petersen. 1950. *Ukrudt og ukrudtsbekaempelse* [Weeds and weed control]. Danske Landhusholdningsselskab, Copenhagen. 320 pp. Danish.

Jensen, H. 1969. Content of buried seeds in arable soil in Denmark and its relation to the weed population. *Dansk botanisk Arkiv* 27(2). 56 pp. English.

Jessen, K., and J. Lind. 1922. *Det. danske markukrudts historie* [The history of Danish field weeds]. Danske Videnskabernes Selskab Skrifter, Naturvidenskabelig og Mathematisk Afedling. 8. Copenhagen. 496 pp. Danish.

Mikkelsen, V., and F. Laursen. 1966. Markukrudtet i Danmark omkring 1960. [Field weeds in Denmark at the beginning of the 1960s]. *Botanisk Tidsskrift* 62: 1-26. Danish; includes English summary.

Petersen, H. 1960. *Ukrudtsplanter og ukrudtsbekaempelse* [Weeds and weed control]. Danske Landhusholdningsselskab, Copenhagen. 144 pp. Danish.

Petersen, H., and G. Bakkendrup-Hansen. 1966. *Den gule bog: ukrudtsbekaempelse i gartneri og have brug* [A yellow book: weed control in gardens and lawns]. Almindelig Dansk Gartnerforening, Copenhagen. 99 pp. Danish.

EGYPT

Boulos, L., and N. El-Hadidi. 1966. *Common weeds of Egypt*. Cairo Herbarium, Botanic Gardens, Cairo. 150 pp. English.

Public Work Department. 1932. *A report on the weed flora of the irrigation channels in Egypt*. Government Press, Cairo. 124 pp. English.

Tackholm, V. 1956. *Student's flora of Egypt*. Anglo-Egyptian Bookshop, 165 Rue Mohamed Farid, Cairo. 649 pp. English.

EL SALVADOR

Lagos, J. 1969. *Malas hierbas en plantaciones de algodoneros en El Salvador*. Universidad de El Salvador, San Salvador. 35 pp. Spanish.

Lagos, J., and E. Calles. 1969. *Malas hierbas en cafetales de El Salvador*. Boletino Informativo Suplemento 26, Instituto Salvadoreño de Investigaciones del Café, Nueva San Salvador. 36 pp. Spanish.

ETHIOPIA

Bengtsson, B. 1968. *Cultivation practices and the weed, pest and disease situation in some parts of the Chilalo Awraja*. Swedish International Development Authority (SIDA), Klarabergsgatan

60, S-105 25 Stockholm. 62 pp. English.

Hakansson, S. 1968. *Introductory agrobotanical investigations in grazed areas in the Chilalo Awraja*. Swedish International Development Authority (SIDA), Klarabergsgatan 60, S-105 25 Stockholm. 70 pp. English.

Senni, L. 1951 and 1952. Contribuzione alla conoscenza della flora dell' Africa orientale [Contributions to knowledge of the flora of East Africa]. *Nuovo Giornale Botanica Italiana* 58: 450-461; 59:64-74. Italian and English. This work is an identification of 200 plants collected in Eritrea, Abyssinia, and Italian Somalia, 1937 to 1951.

FIJI

Parham, J. 1955. *Grasses of Fiji*. Bulletin 30. Department of Agriculture, Suva. 166 pp. English.

————. 1958. *The weeds of Fiji*. Bulletin 35. Department of Agriculture, Suva. 202 pp. English.

Parham, J., and T. Mune. 1967. *The declared noxious weeds of Fiji and their control*. Bulletin 48. Department of Agriculture, Suva. 87 pp. English.

FINLAND

Mukula, J. 1964. *Rikkaruohot ja niiden torjunta* [Weeds and their control]. Kirjayhtyma, Helsinki. 140 pp. Finnish.

Mukula, J., M. Raatikainen, R. Lallukka, and T. Raatikainen. 1969. *Composition of weed flora in spring cereals in Finland*. Annales Agriculturae Fenniae, Helsinki. 110 pp. English. This report is included because it provides a good example of one method of making a weed survey of a country.

FRANCE

Detroux, L. 1959. *Les herbicides et leur emploi*. Editions J. Duculot, Gembloux. 220 pp. French.

François, L. 1943. *Semences et premières phases du développement des plantes commensales des végétaux*. Imprimerie Nationale, Paris. 182 pp. French.

Guyot, L., and J. Guillemat. 1962. *Semences et plantules des principales mauvaises herbes*. La Maison Rustique, Paris. 94 pp. French.

Jussiaux, P., and R. Pequignot. 1962. *Mauvaises herbes—techniques modernes de lutte*. La Maison Rustique, Paris. 222 pp. French.

GERMANY

Bursche, E. 1963. *Wasserpflanzen*. Neuman Verlag, Dr. Schminke Allee 19, Radebeul. 124 pp. German.

Burschel, P., and E. Rohrig. 1960. *Unkrautbekampfung in der Forstwirtschaft*. Paul Parey Verlag, Spitalerstrasse 12, Hamburg. 92 pp. German.

Eggebrecht, H. 1957. *Unkräuter im Feldbestand*. Neuman Verlag, Dr. Schminke Allee 19, Radebeul. 264 pp. German.

Hanf, M. 1969. *Die Ackerunkräuter und ihre Keimlinge*. Badische Anilin & Soda-Fabrik A.G., Ludwigshafen am Rhein; also published in 1970 by Bayerischer Landwirtschaftsverlag, Munich. 347 pp. German.

Koch, W. *Unkrautbekampfung*. 1970. E. Ulmer, Stuttgart. 342 pp. German.

Kurth, H. 1960. *Chemische Unkrautbekampfung*. Gustav Fischer Verlag, Jena. 229 pp. German.

Kutschera, L. 1960. *Wurtzelatlas mitteleuropäischer Ackerunkrauter und Kulturpflanzen*. DLG Verlags GMBH, Frankfurt. 574 pp. German.

Luders, W. 1963. *Unkräuter Ungräser*. Landesanstalt für Pflanzenschutz, Stuttgart. 68 pp. German.

Petersen, A. 1953. *Die Gräser als Kulturpflanzen und Unkräuter auf Wiese, Weide und Acker*. Akademie-Verlag, Berlin. 273 pp. German.

Rauh, W. 1953. *Unsere Unkräuter*. C. Winter Universitätsverlag, Heidelberg. 182 pp. German.

Sauer, T., and R. Hocker. 1969. *Unkraut Fibel*. Schering A.G., Berlin. 313 pp. German.

Schindlmayr, A. 1956. *Welches Unkraut ist das?* Franksche Verlagsbuchhandlung, Stuttgart. 237 pp. German.

Snell, K. 1912. Über das Vorkommen von keimfähigen Unkrautsamen im Boden. *Landwirtschaftliche Jahrbuch* 43: 323-347. German.

Tittel, C. 1964. *Unkräuter im Feldbestand*. Neudamm Verlag, Radebeul. 294 pp. German.

Wehsarg, O. 1954. *Ackerunkräuter*. Akademie-Verlag, Berlin. 189 pp. German.

GHANA

Hall, J., P. Pierce, and G. Lawson. 1971. *Common plants of the Volta Lake*. Department of Botany, University of Ghana, Accra. 123 pp. English.

Wills, J., ed. 1962. *Agriculture and land use in Ghana*. Oxford University Press, London. 514 pp. English. Chapter 22, compiled by D. Adams and H. Baker, presents an extensive list of weeds by region.

GREAT BRITAIN

Arber, A. 1963. *Water plants, a study of aquatic angiosperms*. Originally published at the University Press, Cambridge, in 1920; reprinted in 1963 by J. Cramer, Weinheim, Germany. 436 pp. English.

Audus, L. 1964. *The physiology and biochemistry of*

herbicides. Academic Press, London. 555 pp. English.

Bates, G. 1948. *Weed control*. E. & F. N. Spon, London. 235 pp. English.

Brenchley, W. 1920. *Weeds of farm land*. Longmans, Green & Co., London. 239 pp. English.

Chancellor, R. 1962. *The identification of common water weeds*. Technical Bulletin 183. Ministry of Agriculture, Fisheries and Food, Her Majesty's Stationery Office, London. 48 pp. English.

————. 1966. *The identification of weed seedlings of farm and garden*. Blackwell Scientific Publications, Oxford. 88 pp. English.

Evans, S. 1962. *Weed Destruction*. Blackwell Scientific Publications, Oxford. 172 pp. English.

Fryer, J. 1966. *Herbicides in British fruit growing*. Blackwell Scientific Publications, Oxford. 155 pp. English.

Fryer, J., and S. Evans, eds. 1970. *Weed control handbook*. Volume 1, *Principles*. Revised reprint, 5th edition. Blackwell Scientific Publications, Oxford. 494 pp. English.

Fryer, J., and R. Makepeace, eds. 1970. *Weed control handbook*. Volume 2, *Recommendations*. 6th edition. Blackwell Scientific Publications, Oxford. 331 pp. English.

Harper, J., ed. 1960. *The biology of weeds*. Blackwell Scientific Publications, Oxford. 256 pp. English.

Hubbard, C., D. Brown, A. Gray, and R. Whyte. 1944. Imperata cylindrica, *taxonomy, distribution, economic significance, and control*. Imperial Agricultural Bureau Joint Publication No. 7. Imperial Bureau of Pastures and Forage Crops, Aberystwyth, Cardiganshire. 63 pp. English.

Long, H. 1910. *Common weeds of the farm and garden*. Smith, Elder & Co., London. 451 pp. English.

Robson, T. 1968. *The control of aquatic Weeds*. Bulletin 194. Ministry of Agriculture, Fisheries and Food, Her Majesty's Stationery Office, London. 54 pp. English.

Salisbury, E. 1961. *Weeds and aliens*. Collins, London and Glasgow. 384 pp. English.

Tarr, S. 1962. *Diseases of sorghum, Sudan grass, and broom corn*. The Commonwealth Mycological Institute, Kew, Surrey. 380 pp. English. Contains an excellent chapter on parasitic plants.

Woodford, E., ed. 1963. *Crop production in a weed-free environment*. Blackwell Scientific Publications, Oxford. 116 pp. English.

Woodford, E., and G. Sagar, eds. 1960. *Herbicides and the soil*. Blackwell Scientific Publications, Oxford. 96 pp. English.

HAWAII (U.S.)

Haselwood, E., and G. Motter. 1966. *Handbook of Hawaiian weeds*. Hawaiian Sugar Planters' Association, Honolulu. 479 pp. English.

Hilton, H., and D. Jones. 1966. *Weed control manual for the Hawaiian sugar industry*. Hawaiian Sugar Planters' Association, Honolulu. 102 pp. English.

Hosaka, E. 1957. *Common weeds of Hawaii*. University of Hawaii, Cooperative Extension Service, Honolulu. 16 pp. English.

Rotar, P. 1968. *Grasses of Hawaii*. University of Hawaii Press, Honolulu. 364 pp. English.

St. John, H., and E. Hosaka. 1932. *Weeds of the pineapple fields of the Hawaiian Islands*. Research Publication 6. University of Hawaii, Honolulu. 196 pp. English.

Shinbara, B. 1966. *Noxious weed seeds of Hawaii*. Department of Agriculture, Honolulu. 54 pp. English.

HUNGARY

Csapody, V. 1968. *Keimlings Bestimmungsbuch der Dikotyledonen*. Akademiai Kiado, Budapest. 286 pp. German.

Timor, L., and G. Ubrizsy. 1957. Die Ackerunkräuter Ungarns mit besonderer Rucksicht auf die chemische Unkrautbekampfung. *Acta Agrobotanica Hungarica* 7(1-2): 123-155. Academy of Sciences, Budapest. German.

Ubrizsy, G., and A. Gimesi. 1969. *A vegyszeres gyomirtás gyakorlata* [Chemical weed control]. Mezögazdasági Kiado, Budapest. 310 pp. Hungarian.

Ubrizsy, G., and A. Penzes. 1960. Beiträge zur Kenntnis der Flora und der Vegetation Albaniens. *Acta Botanica Hungarica* 6(1-2): 155-170. Academy of Sciences, Budapest. German.

ICELAND

Davidsson, I., and L. Jonsson. 1961. *Illgresi og illgresiseyding* [Weeds and their control]. Útegefandi, Búnadarfélag Islands, Reykjavik. 136 pp. Danish.

INDIA

see also Asia

Joshi, N. 1973. *Manual of weed control*. Researchco Publications, 1865 Trinagar, Delhi 110035. 248 pp. English.

Joshi, N., and S. Singh. 1965. Weeds of agricultural importance of India. *Plant Protection Bulletin* 17

(3-4): 1-32. Department of Agriculture, New Delhi. English.

Ranade, S., and W. Burns. 1925. The eradication of *Cyperus rotundus* L.: a study in pure and applied botany. *Memoirs of the Department of Agriculture of India* 13(5): 99-192. Published for the Imperial Department of Agriculture in India by Thacker, Spink & Co., Calcutta. English.

Sastry, K. 1957. *Common weeds of cultivated and grasslands of Mysore*. Bulletin of the Department of Agriculture, Mysore. Botanical Series 2. Department of Agriculture, Bangalore. 86 pp. English.

Singh, H., and P. Khanna. 1965. *Common weeds of the northwest plains of India*. Indian Council of Agricultural Research, Farmer's Bulletin 5 (new series), New Delhi. 56 pp. English.

Subramanyam, K. 1962. *Aquatic angiosperms*. Botanical Monograph 3. Council of Scientific and Industrial Research, New Delhi. 190 pp. English.

Tadulingam, C., and G. Venkatanarayana. 1955. *Handbook of south Indian weeds*. Government Press, Madras. 488 pp. English.

Thakur, C. 1954. *Weeds in Indian agriculture*. Motilal Banarsidass, Patna. 126 pp. English.

INDIA AND BURMA

Biswas, K., and C. Calder. 1936. *Handbook of common water and marsh plants of India and Burma*. Manager of Publications, Delhi. 216 pp. English. Work is out of print.

INDONESIA

Backer, C. A. 1934. *Handboek ten dienste van de suikerrietcultuur en de rietsuiker-fabricage op Java: Onkruidflora der Javasche suikerriet brondon door*. Prepared by Proefstation voor de Java Suikerindustrie, Pasuruan. 1,333 pp. Dutch. This large edition is seen in many forms. The hardbound text containing weed descriptions is usually separate. The line drawings may be in one or two bound volumes but they were issued as 15 separate bulletins. The Second World War interrupted the completion and publication of the final 220 species. These finally were published in Holland in 1973. *See* Backer, 1973, below.
This set of books, together with Backer and van Slooten, 1924 (listed below), include so many weeds that they provide good coverage for many crops of Java and perhaps of many parts of Southeast Asia.

———. 1973. *Handbook for the cultivation of sugar-cane and manufacturing of cane-sugar in Java*. Volume 7. Atlas of 220 weeds of sugar-cane

fields in Java. Copyright by Indonesian Sugar Experiment Station (BP3G), Pasuruan, Indonesia. 240 pp. English. Edited for Greshoff's Rumphius Fund, Amsterdam, by C. G. G. J. van Steenis. Book may be obtained from the Department of Agricultural Research of The Royal Tropical Institute, Mauritskade 63, Amsterdam.

Backer, C. A., and R. Bakhuizen van den Brink, Jr. 1963, 1965, and 1968. *Flora of Java*. 3 volumes. Wolters-Noodhoff N.V., Groningen. Volume 1, 648 pp.; Volume 2, 641 pp.; Volume 3, 761 pp. English.

Backer, C. A., and D. van Slooten. 1924. *Geillustreed handboek voor Javaansche theeonkruiden en hunne beteeknis voor de cultur*. Prepared by Proefstation voor Thee, Buitenzorg (now Bogor). 597 pp. Dutch. This book discusses weeds of tea.

Soerjani, M. 1970. *Alang-Alang*, Imperata cylindrica, *pattern of growth as related to its problem of control*. Biotrop Bulletin 1. Regional Center for Tropical Biology, P.O. Box 17, Bogor. 88 pp. English.

Voogd, C. de. 1950. *Tanaman apakah gerangan ini?* [Do you know this plant?]. N. V. Uitgeverij W. van Hoeve, Bandoeng, Indonesia, and The Hague, Netherlands. Approximately 100 pp. Indonesian and Dutch editions. A small pocket edition to be used in field identification.

IRAN

Agricultural Research Department. 1971. *Weeds and weed control at the Haft Tappeh Cane Sugar Project 1950-1971*. Ministry of Water and Power, Khuzestan Water and Power Authority, Haft Tappeh Cane Sugar Project. 31 pp. English and Persian.

Behboodi, E. 1962. Weeds and weed control. Ministry of Agriculture. Teheran. 227 pp. Persian.

IRAQ

Hassawy, G., S. Tammimi, and H. Al-Izzi. 1968. *Weeds in Iraq*. Technical Bulletin 167. Botany Division, Ministry of Agriculture, Baghdad. 256 pp. English.

ISRAEL

Bar Daroma, M., M. Horowitz, and S. Osherov. 1968. *Weeds of our fields*. Ministry of Agriculture, Educational Extension Service, Hakirya, Tel-Aviv. 190 pp. Hebrew.

Cohen, G. 1966. *Chemical weed control handbook*. Sifriath Hassadeh Publisher, Tel-Aviv. 236 pp. Hebrew.

Horowitz, M. 1970. *Weeds—Biology and Control*.

Agricultural Publication 86. Ministry of Agriculture, Division of Weed Research, Hakirya, Tel-Aviv. 75 pp. Hebrew.

———. 1972. *Biology of troublesome perennial weeds in Israel:* Cynodon dactylon, Cyperus rotundus, Sorghum halepense. Agricultural Research Organization, Newe Ya'ar. English. 147.

ITALY

Haussman, J., and I. Scurti. 1953. *Le piante infestanti* [Weeds]. Edizione Agricole, Bologna. 305 pp. Italian.

Società Italo-Americana Prodatti Antiparassitari. 1965. *Impariamo a conoscere le malerbe* [How to recognize weeds]. Rome. 65 pp. Italian.

JAPAN

Ihara Chemicals Company. 1969. *How to control paddy field weeds.* Tokyo. 146 pp. Japanese.

Japan Association for the Advancement of Phyto-Regulators. 1968. *Main weeds in Japan.* Tokyo. 13 pp. Japanese and English. An index of Latin scientific names and Japanese and English common names.

Kasahara, Y. 1954. Studies on the weeds of arable land in Japan, with special reference to kinds of harmful weeds, geographical distribution, abundance, life-length, origin and history. *Berichte des Ohara Instituts für Landwirtschaftliche Forschungen* 10: 72-115. English.

———. 1968. *Weeds of Japan illustrated: seeds, seedlings and plants.* Yokendo, Tokyo. 518 pp. Japanese.

Numata, M., and Y. Yashara. 1964. List of weeds: English, German, Japanese and scientific names. *Weed Research* (Japan) 3: 112-152. English, German, and Japanese. This is the report of the Terminology Committee, Weed Society of Japan, 1-26 Nishigahara, Kitaku, Tokyo.

Numata, M., and N. Yoshizawa, eds. 1968. *Weed flora of Japan, illustrated by color.* Japanese Association for the Advancement of Phyto-Regulators, Tokyo. 334 pp. Japanese.

Tsutsui, K. 1970. *Control of weeds, illustrated in color.* Ienohikari Kyokai, 11 Kawaramachi, Ichigaya, Shinjuku-ku, Tokyo. 218 pp. Japanese.

Ueki, K., and S. Matsinaka. 1972. *Outline of weed science.* Yokendo, 5-30-15 Hongo, Bunkyoku, Tokyo. 200 pp. Japanese.

LEBANON

Edgecombe, W. 1970. *Weeds of Lebanon.* American University of Beirut, Beirut. 457 pp. English.

MADAGASCAR

see Africa

MALAWI

Binns, B. 1969. *Weed plants of Malawi.* University of Malawi, Soche Hill College, P.O.B. 5496, Limbe. 24 pp. English.

MALAYSIA

Barnes, D., and M. Chandapillai. 1971. *Sixty weeds of Malaysian plantations.* Privately published. 62 pp. English.

———. 1972. *Common Malaysian weeds and their control.* Ansul (Malaysia) Sendirian Berhad, Kuala Lumpur. 144 pp. English.

Burkill, I. 1935. *A dictionary of the economic products of the Malay Peninsula.* 2 volumes. Crown Agents for Colonies, 4 Millbank, London S.W.1; republished in 1966 on behalf of the governments of Malaysia and Singapore by the Ministry of Agriculture and Cooperatives, Kuala Lumpur, Malaysia. Volume 1, 1,220 pp.; Volume 2, 1,182 pp. English.

Chin Hoong Fong. 1969. *Agricultural and horticultural seeds in Malaysia.* College of Agriculture, Sungai Besi, Selangor. 114 pp. English.

Henderson, M. 1954. *Malayan wild flowers.* Volume 2, *monocotyledons.* Malayan Nature Society; reprinted by Tien Press, Singapore, in 1959. 358 pp. English.

———. 1959. *Malayan wild flowers.* Volume 1, *Dicotyledons.* Malayan Nature Society; reprinted by Tien Press, Singapore, in 1959. 478 pp. English.

Samy, J., A. Wong, and M. Jaafar. 1970. *A handbook on padifield weeds.* Rice Research Centre, Blumbong, Lima. 41 pp. English.

MALTA

Symposium on Parasitic Weeds. 1973. Organized by European Weed Research Council in association with Royal University of Malta. Proceedings compiled at the Agricultural Research Council Weed Research Organization, Begbroke Hill, Oxford; reproduced by Malta University Press. Write to the Secretariat of European Weed Research Council, P.O. Box 14, Wageningen, The Netherlands. 296 pp. English.

MAURITIUS

see also West Indies

Hubbard, C., and R. Vaughan. 1940. *The grasses of Mauritius and Rodriquez.* Waterloo & Sons, London and Dunstable. 127 pp. English.

Rochecouste, E. 1958. *Observations on chemical weed control in Mauritius.* Bulletin 10. Sugar Industry Research Institute, Reduit. 62 pp. English.

———. 1967. *Weed control in sugar cane.* Sugar Industry Research Institute, Reduit. 117 pp. English.

Rochecouste, E., and R. Vaughan. 1959 to present. *Weeds of Mauritius*. Sugar Industry Research Institute, Reduit. English. This is a series of ongoing leaflets.

MOROCCO

Sauvage, C., and J. Veilex. 1970. *Les mauvaises herbes des cultures*. Direction de la Recherche Agronomique, Station Centrale de Phytoécologie, Rabat. 323 pp. French.

NETHERLANDS

Van Dord, D., and P. Zonderwijk. 1964. *Kiemplantentabel van akkeronkruiden* [Seedling plant table of arable weeds]. Plantenziektenkundige Dienste, Wageningen. 80 pp. Dutch; a German edition has also been published.

NEW ZEALAND

Healy, A. 1970. *Identification of weeds and clovers*. Editorial Services, Wellington. 191 pp. English.

Hilgendorf, F., and J. Calder. 1952. *Weeds of New Zealand and how to eradicate them*. 5th ed. Whitcombe & Tombs, Christchurch. 260 pp. English.

Matthews, L. 1960. *Chemical methods of weed control*. New Zealand Department of Agriculture, Wellington. 201 pp. English.

Miller, D. 1970. *Biological control of weeds in New Zealand 1927-48*. Information Series 74. New Zealand Department of Scientific and Industrial Research, Wellington. 103 pp. English.

New Zealand Weed and Pest Control Society. 1969. *Standard common names for weeds of New Zealand*. Published in association with Editorial Services, Box 6443, Wellington. 140 pp. English.

NIGERIA

Egunjobi, J. 1969. *Some common weeds of western Nigeria*. Bulletin of the Research Division, Ministry of Agriculture and Natural Resources, Western State, Ibadan. 40 pp. English.

Rains, A. 1968. *A field key to the commoner genera of Nigerian grasses*. Samaru Miscellaneous Paper 7. Institute of Agricultural Research, Ahmadu Bello University, Samaru, Zaria. 29 pp. English.

Stanfield, D. 1970. *Flora of Nigeria: Grasses*. Ibadan University Press, Ibadan. 118 pp. English.

NORWAY

Hansen, L., T. Rygg, and A. Bylterud. 1968. *Handbok i plantevern* [Handbook of plant protection]. Bondenes Forlag A/S, Oslo. 144 pp. Norwegian.

Korsmo, E. 1935. *Weed seeds*. Gyldendal Norsk Forlag, Oslo. 175 pp. Norwegian, German, and English.

————. 1954. *Anatomy of weeds*. Grøndahl and Søns Forlag, Oslo. 430 pp. English.

————. 1954. *Ugras i natidens jordbruk* [Weeds in present-day agriculture]. Norsk Landbruks Forlag A/S, Oslo. Norwegian.

Vidme, T. 1961. *Ugrasboka* [Weed Book]. Bondenes Forlag A/S, Oslo. 162 pp. Norwegian.

PAKISTAN

see Afghanistan, Asia

PERU

Ferreyra, R. 1970. *Flora invasora de los cultivos de Pucallpa y Tingo Maria*. Botánica Sistemática de la Universidad Nacional Mayor de San Marcos, Lima. 263 pp. Spanish.

Programa Cooperativo de Experimentación Agropecuaria. 1954. *Diccionario de plantas cultivadas, hierbas silvestres y malas hierbas en el Peru*. 56 pp. English and Spanish.

PHILIPPINES

Pancho, J., and M. Guantes. 1963. Seed identification of common weeds in lowland rice fields. *Philippine Agriculturist* 46(7): 481-513. University of the Philippines, College of Agriculture, Los Baños. 33 pp. English.

Pancho, J., M. Vega, and D. Plucknett. 1969. *Some common weeds of the Philippines*. Weed Science Society of the Philippines, College of Agriculturee, Los Baños. 106 pp. English.

Quisumbing, E. 1923. General characters of some Philippine weed seeds. *Philippine Agricultural Review* 16: 298-351. 53 pp. English. This journal became *The Philippine Journal of Agriculture* in 1930. Write to the Department of Agriculture and Commerce, Oriente Building, Plaza Binondo, Manila.

POLAND

Kulpa, W. 1958. *Owoce i nasiona chwastow klucze do oznaczania* [The fruits and seeds of weeds: key to identification]. Panstwowe Wydawnictwo Naukowe, Warsaw. 418 pp. Polish.

PORTUGAL

see also Brazil

Carvalho e Vasconcellos, J. de. 1954. *Plantas vasculares infestantes dos arrozais*. Comissao Reguladora do Comercio de Arroz, Lisbon. 189 pp. Portuguese.

————. 1958. *Ervas infestantes das searas de trigo*. Federaçao Nacional dos Produtores de Trigo, Lisbon. 404 pp. Portuguese.

PUERTO RICO (U.S.)

Velez, I., and J. van Overbeek. 1950. *Plantas indeseables en los cultivos tropicales*. Editorial Universitaria, Rio Piedras. 497 pp. Spanish.

RHODESIA

Wild, H. 1958. *Common Rhodesian weeds*. Government Printer, Salisbury. 220 pp. English.

SOVIET UNION

Araratian, A. 1963. *Sornye rastenija Armenii* [Weeds of Armenia]. Armianskoe Gosudarstvennoe, Yerevan. 257 pp. Armenian.

Baranov, M. 1956. *Sorniaki zernovyx kul'tur i mery bor'by s nimi* [Weeds of the grain cultures and means of combating them]. Kazakhakoe Gosudarstvennoe Izdatel'stvo, Alma-Ata. 100 pp. Russian.

Dorogostaiskaia, E. 1972. *Sornye rastenija krainego severa* [Weeds of the far north of the USSR]. Botanicheski Institut, Akademiya Nauk SSSR. 172 pp. Russian.

Geshtout, Y. 1969. Porodnyj sostav samyx glavnyx sornyx rastenij Kazaxstana i ix cuvstvitel'nost' k xerbicidam [The species composition of the most important weeds in Kazakhstan and their susceptibility to herbicides]. *Vestnik Akademii Nauk Kazaxskoj Sovetskoj Socialisticeskoj Respubliki* 25(9): 45-52. Russian.

Gurskis, V. 1962. *Nezalu apkarosana ar herbicidiem*. Latvijas Valsts Izdevnieciba, Riga. 143 pp. Latvian.

Keller, B., and A. Volkov, general eds. 1934. *Sornye rastenija SSSR* [Weeds of the USSR]. Volume 3. Akademiya Nauk SSSR, Leningrad. 447 pp. Russian.

————. 1935. *Sornye rastenija* SSSR [Weeds of the USSR]. Volumes 1, 2, and 4. Akademiya Nauk SSSR, Leningrad. Volume 1, 323 pp.; Volume 2, 244 pp.; Volume 4, 414 pp. including a 50-page index. Russian.

Kott, S. 1969. *Sornye rastenija i bor'ba s nimi* [Weeds and the fight against them]. Kolos, Moscow. 200 pp. Russian.

Malkov, F. 1936. *Glavnye sornye rastenija Turkmenskoj SSR i mery bor'by s nimi* [Principal weeds of the Turkmenian SSR and means of controlling them]. Supplement 80. Bulletin of Applied Botany, Genetics, and Plant Breeding, The Institute of Plant Industry, Leningrad. 48 pp. Russian.

Rychin, U. 1959. *Sornye rastenija SSSR* [Weeds of the USSR]. Ministry of Education, Moscow. 290 pp. Russian. This is a book of weed plants covering the European part of the Soviet Union.

Vasil'chenko, I., and O. Pidotti. 1970. *Opredelitel' sornyx rastenij rajonov orosaemogo zemledelija* [Manual of weeds found in the regions of irrigation agriculture]. Kolos, Leningrad. 368 pp. Russian.

Volkov, A., and W. Ljubimenko, general eds. 1935. *Rajony raspredelenija samyx glavnyx porod sornyx rastenij v SSSR* [Regions of distribution of the most important weed species in the USSR]. Institute of Botany of the USSR Academy of Science, Moscow. 150 pp. Russian.

SPAIN

Detroux, L., and J. Gostinchar. 1970. *Los herbicidas y su empleo*. Oikos-tau SA. Ediciones, Apartado 5 347, Barcelona. 476 pp. Spanish.

Guell, F. 1970. *Malas hierbas, diccionario clasificatorio ilustrado*. Oikos-tau SA. Ediciones, Apartado 5 347, Barcelona. 216 pp. Spanish.

SUDAN

Andrews, F. 1954. *Weeds of the Gezira*. Bulletin of the Agriculture Publications 13, Ministry of Agriculture, Government of the Sudan, Khartoum. 66 pp. English.

SURINAM

Dirven, J. 1968. Weed flora in field crops on loamy sands in the Old Coastal Plain. *Suriname handboek* 16(3): 123-130. Dutch. Write to Landbouwproef-station, Postbus 160, Paramaribo.

SWEDEN

Granstrom, B. 1962. *Studies on weeds in spring-sown crop*. No. 130. Meddelanden Statens jordbruksforsok, Uppsala. 188 pp. Swedish.

Granstrom, B., and G. Almgard. 1955. *Studies on the weed flora in Sweden*. Publication no. 18. Institute of Plant Husbandry, Royal Agricultural College of Sweden, Uppsala. 22 pp. Swedish.

Ighe, I., and E. Aberg. 1970. *Ogras pa akern* [Field weeds]. LTs Förlag LTK, Stockholm. 100 pp. Swedish.

Kolk, H. 1962. *Viability and dormancy of dry stored weed seeds*. Almqvist & Wiksell Boktryckeri Artiebolag, Uppsala. 192 pp. English.

SWITZERLAND

Geigy Weed Tables. 1968. J. R. Geigy S. A., Basle. Three hundred colored prints with a brief description of biology, geographical distribution, habitat, morphology, and comparison with close relatives. English.

TAIWAN

Hsu, Y. 1968. *Taiwan sugar cane weed control handbook*. Taiwan Sugar Industry, Tainan and Taipei. 259 pp. Chinese.

Lin, C. 1968. *Weeds found on cultivated land in Taiwan*. 2 volumes. College of Agriculture, National Taiwan University, Taipei. Volume 1, 505 pp; Volume 2, 444 pp. Chinese and English.

Sung, T., and H. Chang. 1964. *Weeds found on cultivated land in western Taiwan*. Plant Industry

Series no. 25. Joint Commission for Rural Reconstruction, Taipei. 185 pp. Chinese and English.

TANZANIA
see Africa

THAILAND

Suvatabandhu, K. 1950. *Weeds in paddy field in Thailand*. Technical Bulletin 4. Department of Agriculture, Bangkok. 41 pp. English.

Yongboonkird, U. 1966. *Weeds in cornfields*. Department of Agriculture, Bangkok. 30 pp. Thai.

———. 1971. *Weeds of cotton fields*. Technical Bulletin 12. Crops Division, Department of Agriculture, Bangkok. 54 pp. Thai.

TURKEY

Guneyli, E. 1970. *Weed problems of Turkey*. International Plant Protection Center, Oregon State University, Corvallis, U.S.A. 22 pp. English.

Kurhan, N. 1969. *Turkiyenin baslica yabanciotlari ve zararli oldukhari onemli kultur bitkileri sozlugu* [Dictionary of important Turkish weeds and crops]. Bulletin 45. Türkiye Cumhuriyeti Tarim Bakan Teknik, Weed Control Laboratory, Agricultural Control Institute, Ankara. 75 pp. Turkish.

UGANDA
see Africa

UNITED STATES
see also Alaska, Hawaii, and Puerto Rico

Algren, G., G. Klingman, and D. Wolf. 1951. *Principles of weed control*. John Wiley & Sons, New York. 368 pp. English.

Andersen, R. 1968. *Germination and establishment of weeds for experimental purposes*. Weed Science Society of America, Department of Agronomy, University of Illinois, Urbana. 236 pp. English.

Baker, H., and G. Stebbins, eds. 1965. *The genetics of colonizing species*. Proceedings of the First International Union of Biological Sciences Symposia on General Biology. Academic Press, New York. 588 pp. English.

Barton, L. 1967. *Bibliography of seeds*. Columbia University Press, New York. 858 pp. English.

Beal, W. 1910. *Seeds of Michigan weeds*. Bulletin 260. Michigan Agricultural Experiment Station, Botany Division, East Lansing. 182 pp. English.

Cocannover, J. 1950. *Weeds: guardians of the soil*. Devin-Adair, New York. 179 pp. English.

Crafts, A. 1961. *The chemistry and mode of action of herbicides*. Interscience, New York. 269 pp. English.

Crafts, A., and W. Robbins. 1962. *Weed control*. 3rd ed. McGraw-Hill Book Co., New York. 660 pp. English.

Delorit, R. 1970. *Illustrated taxonomy manual of weed seeds*. Agronomy Publications, River Falls, Wisconsin. 175 pp. English.

Deutsch, A., and A. Poole, eds. 1972. *Manual of pesticide application equipment*. International Plant Protection Center, Oregon State University, Corvallis. 132 pp. English.

Dewey, L. 1896. Migration of Weeds. Pages 263-286 in *Yearbook*. United States Department of Agriculture, Washington, D.C. 23 pp. English.

Dunham, R. 1965. *Herbicide manual for noncropland weeds*. Agriculture Handbook 269. United States Department of Agriculture, Washington, D.C. 90 pp. English.

Eyles, D., J. Robertson, and G. Jex. 1944. *A guide and key to the aquatic plants of the southeastern United States*. Public Health Bulletin 286. United States Department of the Interior, Bureau of Sport Fisheries and Wildlife, Washington, D.C. 150 pp. English.

Fassett, N. 1960. *A manual of aquatic plants*. University of Wisconsin Press, Madison. 405 pp. English.

Fogg, J., Jr. 1945. *Weeds of lawn and garden: a handbook for eastern temperate North America*. University of Pennsylvania Press, Philadelphia. 215 pp. English.

Furtick, R., and R. Romanowsk. 1971. *Weed research methods manual*. International Plant Protection Center, Oregon State University, Corvallis. 80 pp. English.

Georgia, A. 1937. *A manual of weeds*. Macmillan Co., New York. 593 pp. English.

Henkel, A. 1904. *Weeds used in medicine*. Farmers Bulletin 188. United States Department of Agriculture, Washington, D.C. 47 pp. English.

Hitchcock, A. 1950. *Manual of the grasses of the United States*. Miscellaneous Publication 200. United States Department of Agriculture, Washington, D.C.; reprinted in two volumes in 1971 by Dover Publications, New York. 1,051 pp. English.

Holm, L. 1971. Chemical Interactions between Plants on Agricultural Lands. Pages 95-101 in *Biochemical interactions among plants*. Division of Biology and Agriculture, National Academy of Sciences, Washington, D.C. 134 pp. English.

Hotchkiss, N. 1967. *Underwater and floating-leaved plants of the United States and Canada*. Resource Publication 44. United States Department of the Interior, Bureau of Sport Fisheries and

Wildlife, Washington, D.C. 124 pp. English.

Johnson, E. 1932. *The puncture vine in California*. California Agricultural Experiment Station, Berkeley. 42 pp. English.

Kearney, P., and D. Kaufman, eds. 1969. *Degradation of herbicides*. M. Dekker, New York. 394 pp. English.

Klingman, G. 1966. *Weed control as a science*. John Wiley & Sons, New York. 421 pp. English.

Kuijt, J. 1969. *The biology of parasitic flowering plants*. University of California Press, Berkeley. 246 pp. English.

Kummer, A. 1951. *Weed seedlings*. University of Chicago Press, Chicago. 435 pp. English. Out of print but available from University Microfilms, Ann Arbor, Michigan.

Martin, A., and W. Barkley. 1961. *Seed identification manual*. University of California Press, Berkeley. 222 pp. English.

Muenscher, W. 1944. *Aquatic plants of the United States*. Cornell University Press, Ithaca. 374 pp. English.

———. 1960. *Weeds*. Macmillan Co., New York. 560 pp. English.

Musil, A. 1963. *Identification of crop and weed seeds*. Agriculture Handbook 219. United States Department of Agriculture, Washington, D.C. 211 pp. English.

Muzik, T. 1970. *Weed biology and Control*. McGraw-Hill Book Co., New York. 273 pp. English.

National Academy of Sciences. 1968. *Weed control*. Washington, D.C. 471 pp. English. A text on ecology, competition, weed management, herbicide chemistry and physiology, weed control methods, safety, and other aspects of weed control.

Nebraska Department of Agriculture. 1968. *Nebraska weeds*. Weed and Seed Division, Lincoln. 445 pp. English.

Pammel, L. 1911. *Weeds of the farm and garden*. Orange Judd Co., New York. 281 pp. English.

Penfound, W., and T. Earle. 1948. The biology of water hyacinth. *Ecological monographs* 18: 447-472. English.

Robbins, W., M. Bellue, and W. Ball. 1970. *Weeds of California*. Write to the Printing Division, Documents and Publications, P.O. Box 20191, Sacramento, California 95820. 547 pp. English.

Smith, R., and W. Shaw. 1966. *Weeds and their control in rice production*. Agriculture Handbook 292. United States Department of Agriculture, Washington, D.C. 64 pp. English.

Steward, A., L. Dennis, and H. Gilkey. 1963. *Aquatic plants of the Pacific northwest*. Oregon State University Press, Corvallis. 261 pp. English.

United States Department of Agriculture. 1968. *Extent and cost of weed control with herbicides and an evaluation of important weeds, 1968*. Agricultural Research Service Publication 34-102. Washington, D.C. 85 pp. English.

———. 1970. *Selected weeds of the United States*. Agriculture Handbook 366. Washington, D.C. 462 pp. English.

Weed Science Society of America. 1967. *Herbicide handbook of the Weed Society of America*. Department of Agronomy, University of Illinois, Urbana. 293 pp. English.

Weldon, L., R. Blackburn, and D. Harrison. 1969. *Common aquatic weeds*. Agriculture Handbook 352. United States Department of Agriculture, Washington, D.C. 43 pp. English.

Wilkinson, R., ed. 1972. *Research methods in weed science*. Southern Weed Science Society, c/o R. Wilkinson, Experiment, Georgia. 198 pp. English.

VIETNAM

Ho-Minh-Si. 1969. *Weeds in South Viet Nam*. Ministry of Land Reform and Development of Agriculture and Fisheries, Agricultural Research Institute, Forestry Service, 121, Nguyen-binh-Khiem, Saigon. 140 pp. English.

WEST INDIES

Adams, C., L. Kasasian, and J. Seeyave. 1970. *Common weeds of the West Indies*. Faculty of Agriculture, University of the West Indies, St. Augustine, Trinidad. 139 pp. English.

Kasasian, L., and J. Seeyave. 1969. *Weedkillers for Caribbean agriculture*. University of the West Indies, St. Augustine, Trinidad. 44 pp. English.

May and Baker, Limited. 1964. *Weeds of the West Indies and Mauritius*. Dagenham, Essex, England. 76 pp. English.

YUGOSLAVIA

Kojic, M. 1967. *Biologija i suzbijanje korova* [Biology and control of weeds]. Centar Za Unapredenje Poljoprivredne Proizvodnje S.R. Srbije, Belgrade. 245 pp. Serbo-Croatian.

Kovacevic, J. 1970. *Atlas korovskih biljnih vrsta za terenske vjezbe studenata poljoprivrednog fakulteta* [Atlas of weed species for field training of students in the Faculty of Agriculture]. Faculty of Agriculture, University of Zagreb, P. O. Box 95, Zagreb. 65 pp. Serbo-Croatian.

Kovacevic, J., and J. Ritz. 1969. *Korovi i herbicidi*

poznavanje i suzbijanje [Weeds and herbicides: identification and control]. Agronomski Glasnik, Zagreb. 198 pp. Serbo-Croatian.

Sostaric-Pisacic, K., and J. Kovacevic. 1968. *Travnjačka flora i njena poljoprivredna vrijednost* [The grassland flora and its agricultural value]. Znaje, Zagreb. 444 pp. Serbo-Croatian.

Todorovic, D. 1959. *Imenik korovskog bilja Jugoslavije* [List of weeds of Yugoslavia]. Naucna Knjiga, Izdavacko Preduzece, Belgrade. 98 pp. Serbo-Croatian.

Zaplotnik, I. 1955. *Zatiranje plevela* [Weed control]. Zalozila Kmecka Knjiga, Ljubljana. 109 pp. Slovene.

GENERAL

Cardenas, J., and L. Coulston. 1966-1967. *Weed lists*. International Plant Protection Center, Oregon State University, Corvallis. Spanish and English; the Brazil list is in Portuguese. Lists are available for each of the following: Argentina, Brazil, Central America, Chile, Colombia, the Dominican Republic, Mexico, Peru, Puerto Rico, Trinidad, Uruguay, and Venezuela.

Good, R. 1964. *The geography of flowering plants*. Longmans, Green & Co., London. 518 pp. English.

Kasasian, L. 1971. *Weed control in the tropics*. Leonard Hill, 158 Buckingham Palace Road, London, S.W.1. 320 pp. English.

King, L. 1966. *Weeds of the world: biology and control*. John Wiley & Sons, New York. 526 pp. English.

Little, E., ed. 1968. *Handbook of utilization of aquatic plants*. Publication PL:CP/20. Food and Agriculture Organization of the United Nations, Rome. 123 pp. English.

Pijl, L. van der. 1969. *Principles of dispersal in higher plants*. Springer-Verlag, Berlin. 153 pp. English.

Sculthorpe, C. 1967. *The biology of aquatic vascular plants*. Edward Arnold, London. 610 pp. English.

Simmonds, F., ed. 1969. *Proceedings of the First International Symposium on biological control of weeds*. Miscellaneous Publication 1. Commonwealth Institute of Biological Control, Trinidad. 110 pp. English. Obtain from the Commonwealth Agricultural Bureau, Central Sales, Farnham Royal, Slough, SL2-3BN, England.

Weed Science Society of America. 1971. *Technical papers of the FAO International Conference on Weed Control, Davis, California, USA, June 1970*. Department of Agronomy, University of Illinois, Urbana. 668 pp. English.

Whyte, R., T. Moir, and J. Cooper. 1959. *Grasses in agriculture*. Agricultural Studies 42. Food and Agriculture Organization of the United Nations, Rome. 416 pp. English; also printed in Spanish and French.

Appendix B

BOOKS AND SPECIAL PUBLICATIONS ON POISONOUS PLANTS

AFRICA

Quarre, P. 1945. *Contribution a l'étude des plantes toxiques du Katanga*. Comité spécial du Katanga, Elisabethville. 72 pp. French.

Steyn, D. 1934. *The toxicology of plants in South Africa*. Central News Agency, Johannesburg. 631 pp. English.

Verdcourt, B., E. Trump, and M. Church. 1969. *Common poisonous plants of East Africa*. Collins, St. James Place, London. 254 pp. English.

Watt, J., and M. Breyer-Brandwijk. 1962. *The medicinal and poisonous plants of southern and eastern Africa*. E. & S. Livingstone, Edinburgh and London. 1,457 pp. English. An outstanding reference.

Xerhavo, J. 1950. *Plantes médicinales et toxiques de la Côte-d'Ivoire-Haute-Volta*. Frères, Paris. 295 pp. French.

ASIA, SOUTHEAST

Oakes, A., Jr. 1967. *Some harmful plants of Southeast Asia*. Naval Medical School, National Naval Medical Center, Bethesda. 50 pp. English. Write to the Superintendent of Documents, United States Government Printing Office, Washington, D.C.

AUSTRALIA

Bailey, F. 1906. *The weeds and suspected poisonous plants of Queensland*. Queensland Department of Agriculture and Stock, Brisbane. 245 pp. English.

Everist, S. 1964. Review of the poisonous plants of Queensland. *Proceedings of the Royal Society of Queensland* 74(1). 20 pp. English. This article was issued separately.

Ewart, A., and J. Tovey. 1909. *The weeds, poison plants, and naturalised aliens of Victoria*. Ministry of Agriculture, Melbourne. 110 pp. English.

Francis, D., and R. Southcott. 1967. *Plants harmful to man in Australia*. Miscellaneous Bulletin 1. Botanic Garden, Adelaide. 53 pp. English.

Gardner, C., and H. Bennetts. 1956. *The toxic plants of Western Australia*. Periodicals Division, West Australian Newspaper, Perth. 253 pp. English.

Hurst, E. 1942. *The poison plants of New South Wales*. Poison Plants Committee of New South Wales, Department of Agriculture, Sydney. 498 pp. English.

Macadam, J. 1966. *Some poisonous plants in the northwest*. Bulletin P331. Division of Animal Industry, New South Wales Department of Agriculture, State Office Block, Phillip Street, Sydney. 6 pp. English.

Webb, L. 1948. *Guide to the medicinal and poisonous plants of Queensland*. Bulletin 232. Council for Scientific and Industrial Research, 314 Albert Street, Melbourne. 202 pp. English.

BRAZIL

Hoehne, F. 1939. *Plantas e substancias vegetais*

tóxicas e medicinais. Graphicar, S. Paulo-Rio de Janeiro. 355 pp. Portuguese.

Moraes, M. 1941. *Contribuição ao estudo das plantas tóxicas brasileiras*. Serviço de Informação Agrícola, Ministeria da Agricultura, Rio de Janeiro. 106 pp. Portuguese.

CANADA

see also United States

Bruce, E. 1927. Astragalus campestris *and other stock poisoning plants of British Columbia*. Bulletin 88. Department of Agriculture, Ottawa. 44 pp. English.

Department of Agriculture. 1951. *Weeds poisonous to livestock*. Shnitka, Edmonton. 36 pp. English.

Thomson, R., and H. Sifton. 1922. *A guide to the poisonous plants and weed seeds of Canada and the northern United States*. University of Toronto Press, Toronto. 169 pp. English.

COLOMBIA

Investigación sobre el problema conocido en sabanas de Bolívar con el nombre de "caída de los ganados" [Research in the savannahs of Bolivar on the problem known as "cattle drop"]. 1964. Banco Ganadero, Bogota. 30 pp. Spanish.

Perez, E. 1937. *Plantas medicinales y venenosas de Colombia*. Editorial y Litografía Cromos, Bogota. 295 pp. Spanish.

FRANCE

Pontois, M. 1954. *Contribution à l'étude des intoxications végétales chez les animaux domestiques en France*. Imprimerie Nouvelle, Orléans. 77 pp. French.

FRENCH GUIANA

Heckel, E. 1897. *Les plantes médicinales et toxiques de la Guyane française*. Portat Frères, Imprimeurs, Mâcon. 93 pp. French.

GERMANY

Gessner, O. 1953. *Die Gift und Arzneipflanzen von Mitteleuropa* [The poisonous and medicinal plants of Middle Europe]. Carl Winter Universitätsverlag GMBH, Heidelberg. 804 pp. German.

Mildner, T. 1951. *Giftpflanzen in Wald und Flur* [Poisonous plants in forest and meadow]. Goest & Portig, Leipzig. 49 pp. German.

Schonfelder, B. 1962. *Welche Heilpflanze ist das?* [Which medicinal plant is that?]. Frank'sche Verlagshandlung, Stuttgart. 156 pp. German.

GREAT BRITAIN

Forsyth, A. 1968. *British poisonous plants*. 2nd ed. Bulletin 161. Ministry of Agriculture, Fisheries and Food, London. 131 pp. English.

Long, H. 1934. *Poisonous plants on the farm*. Bulletin 75. Ministry of Agriculture and Fisheries, London. 58 pp. English.

GUAM (U.S.)

Keegan, H., and W. Macfarlane, eds. 1963. *Venomous and poisonous animals and noxious plants of the Pacific region*. Pergamon Press, New York; distributed by the Macmillan Co., New York. 456 pp. English. This is a collection of papers from the 10th Pacific Science Congress which was held in Hawaii in 1961. It contains a small section on the poisonous plants of Guam.

INDIA

Chopra, R., R. Badwar, and S. Ghosh. 1965. *Poisonous plants of India*. Volume 1, revised and enlarged. Scientific Monograph 17. Indian Council of Agricultural Research, Manager of Publications, Delhi. 631 pp. English.

————. 1965. *Poisonous plants of India*. Volume 2, revised and enlarged. Indian Council of Agricultural Research, New Delhi. 341 pp. English.

Jain, S. 1968. *Medicinal plants*. National Book Trust, New Delhi. 176 pp. English.

KOREA

To, P., and H. Sim. 1948. *Korean plants: poisonous plants*. Keijo Pharmaceutical College. 170 pp. Korean, Latin, and Japanese. Write to the College of Pharmacy, Seoul National University, Seoul.

MADAGASCAR

Heckel, E. 1903. *Les plantes médicinales et toxiques de Madagascar avec leurs noms et leurs emplois indigènes*. Institut Colonial, Marseille. 148 pp. French.

NEW ZEALAND

Connor, H. 1951. *Poisonous plants in New Zealand*. Bulletin 99. Department of Scientific and Industrial Research, Wellington. 141 pp. English.

Finlay, J. 1964. *Plants in New Zealand poisonous to man*. Pamphlet 147. Department of Health, Wellington. 41 pp. English.

PHILIPPINES

Quisumbing, E. 1947. Philippine plants used for arrow and fish poisons. *Philippine Journal of Science* 77: 127-177. Institute of Science and Technology, Manila. English.

————. 1947. Vegetable poisons of the Philippines. *Philippine Journal of Forestry* 5: 145-171. Department of Agriculture Resources, Manila. English.

RHODESIA

Shone, D., and R. Drummond. 1965. Poisonous plants of Rhodesia. *Rhodesia Agricultural Journal* 62(4): 1-64. English.

SOVIET UNION

Bazhenov, S. 1964. *Veternarnaia toksikologiia* [Veterinary toxicology]. Kolos, Leningrad. 375 pp. Russian.

Gusynin, I. 1962. *Toksikologiia i adovitykh rastenii* [Toxicology of poisonous plants]. Izdvo Selkhos Litry, Shurnalov i Plakatov, Moscow. 623 pp. Russian.

Mikhailovskaia, V. 1962. *I adovitye i vrednye rastenia* [Poisonous and harmful plants]. Izdvo Akademii Nauk BSSR, Minsk. 116 pp. Russian.

TRINIDAD

Stehle, H. 1956. *Poisonous and doubtful plants.* Caribbean Commission, Central Secretariat, Kent House, Port-of-Spain. 482 pp. English. Pages 312-341 cover the plants of Trinidad.

UNITED STATES

see also Canada, Guam

Agricultural Extension Service. 1957. *Principal livestock-poisoning plants of New Mexico ranges.* Circular 274. New Mexico College of Agriculture and Mechanic Arts, Las Cruces, New Mexico. 77 pp. English.

Arnold, H. 1944. *Poisonous plants of Hawaii.* Tongg Publishing Co., Honolulu. 71 pp. English.

Evers, R., and R. Link. 1972. *Poisonous plants of the Midwest and their effect on livestock.* Special Publication 24. College of Agriculture, University of Illinois at Urbana-Champaign. 165 pp. English.

Gilkey, H. 1958. *Livestock-poisoning weeds of Oregon.* Bulletin 564. Agricultural Experiment Station, Oregon State College, Corvallis. 74 pp. English.

Hardin, J., and J. Arena. 1969. *Human poisoning from native and cultivated plants.* Duke University Press, Durham, North Carolina. 167 pp. English.

Harshberger, J. 1920. *Textbook of pastoral and agricultural botany, for the study of the injurious and useful plants of country and farm.* P. Blakiston's Son & Co., Philadelphia. 294 pp. English.

Hosaka, E., and A. Thistle. 1954. *Noxious plants of the Hawaiian ranges.* Extension Bulletin 62. College of Agriculture, University of Hawaii, Honolulu. 39 pp. English.

Kingsbury, J. 1964. *Poisonous plants of the United States and Canada.* Prentice-Hall, Englewood Cliffs, New Jersey. 626 pp. English.

————. 1965. *Deadly harvest: a guide to common poisonous plants.* Holt, Rinehart & Winston, New York. 128 pp. English.

Krochmal, A., R. Walters, and R. Doughty. 1971. *A guide to medicinal plants of Appalachia.* Agricultural Handbook 400. United States Department of Agriculture, Washington, D.C. 291 pp. English.

Morton, J. 1971. *Plants poisonous to people in Florida and other warm areas.* Hurricane House, Miami. 116 pp. English.

Muenscher, W. 1951. *Poisonous plants of the United States.* Macmillan Co., New York. 266 pp. English.

Newsom, I. 1936. *Timber milk vetch* (Astragalus hylophilus) *as a poisonous plant.* Bulletin 425. Colorado Experiment Station, Colorado State College, Fort Collins. 42 pp. English.

Oakes, A. 1962. Poisonous and injurious plants of the U.S. Virgin Islands. Miscellaneous Publication 882. United States Department of Agriculture, Washington, D.C. 97 pp. English.

Pammel, L. 1911. *A manual of poisonous plants.* Torch Press, Cedar Rapids, Iowa. 977 pp. English.

Schmutz, E., B. Freeman, and R. Reed. 1968. *Livestock poisoning plants of Arizona.* University of Arizona Press, Tucson. 176 pp. English.

Sperry, O., J. Dollahite, and G. Hoffman. 1964. *Texas plants poisonous to livestock.* Texas Agricultural Experiment Station, College Station. 57 pp. English.

VENEZUELA

Blohm, H. 1962. *Poisonous plants of Venezuela.* Harvard University Press, Cambridge. 136 pp. English.

YUGOSLAVIA

Lekovito i otrovno bilje. 1951. Narodni Universitet, Belgrade. 90 pp. Serbo-Croatian.

ZAMBIA

Farshawe, A. 1969. *Poisonous plants of Zambia.* Government Printer, Lusaka. English. This is a small pamphlet.

GENERAL

Bernhard-Smith, A. 1923. *Poisonous plants of all countries.* Bailliere, Tindall and Cox, 8, Henrietta Street, Covent Garden, W.C.2, London. 112 pp. English.

Merrill, E. 1943. *Emergency food plants and poisonous plants of the islands of the Pacific.* United States Department of Agriculture, Washington, D.C. 149 pp. English.

Glossary

achene. A small, dry, hard, one-seeded fruit that ripens without bursting its thin outer sheath.

acuminate. Tapering to a slender point; drawn out into a long slender point.

acute. Sharp at the tips; ending in a sharp angle.

adventitious. Originating from other than its usual place; for example, roots arising from a stem.

anatropous. Having the ovule inverted at an early period in its development, so that the micropyle is bent down to the funiculus, to which the body of the ovule is united.

anterior. Toward the front.

antheridium. The male organ of the sexual generation in bryophytes, pteridophytes, and some gymnosperms.

apiculate. Ending in a short pointed tip.

appressed. Lying close to; pressed close to; flat against something.

archegonium. The flask-shaped female sex organ in bryophytes, pteridophytes, and some gymnosperms. It contains the egg, which develops into the sporophyte.

articulate, articulated. Jointed, easily separating at the joints or nodes.

axil. The angle between a branch or a leaf and the stem (axis) from which it arises.

axillary. Situated in, growing from, or pertaining to an axil.

baccate. Pulpy throughout, like a berry; berrylike.

biconvex. Convex on both sides. *See* convex.

bipinnate. Twice pinnate. *See* pinnate.

callus. Soft parenchymatous tissue that forms over any wounded or cut surface of a stem; a hard protuberance; in grasses, the tough often hairy swelling at the base of the lemma or the palea that is often part of the rachis or rachilla.

calyx. The outer whorl of usually green leaf or bractlike structures called sepals.

capsule. A dry dehiscent fruit composed of more than one carpel, having many seeds, and usually opening at maturity by valves or teeth.

caryopsis. The grain or fruit of grasses; a small, one-celled, dry, indehiscent fruit, with a thin pericarp forming the outer surface of the grain.

ciliate. Fringed on the margins with short, usually stiff, hair.

ciliolate. Minutely ciliate.

circumscissile (circumsciss). Dehiscent around the circumference of a capsule in such a way that the top falls away like a lid.

coccus (plural, cocci). One of the parts of a lobed or deeply divided fruit when each part is one-seeded; these parts may be leathery or even dry.

convex. Curved or rounded outward.

cordate. Heart-shaped with the point upward; referring to the base of some leaves.

corolla The petals of a flower; the inner perianth.

crenate. Having the margin cut into rounded scallops; shallowly dentate with rounded teeth.

culm. The jointed stem of a grass or sedge.

cuneate. Wedge-shaped with the narrow part below; triangular with the acute end at the point of attachment.

cyme. A usually broad and more or less flat-topped determinate inflorescence. The central or terminal flowers open first.

dehiscent. Bursting open; splitting apart along definite lines, as a capsule or anther may do; discharges its contents, usually at maturity.

deltoid. Triangular like the Greek letter delta: Δ.

dentate. Toothed, with the teeth more or less perpendicular to the margin of the leaf.

digitate. Spreading like the fingers of a hand; compound, with the members arising together at a common point.

drupe. A fleshy, pulpy, or fibrous fruit with a hard endocarp (the stone) enclosing a single seed; "stone-fruit" of plum, cherry, apricot, peach, etc.

elliptic. Oval and narrowed to both ends.

erose. Irregularly cut along the margin as if eaten or worn away.

exsert. Protruding beyond some enclosing organ or part.

filament. The part of a stamen that supports the anther; any threadlike body.

floret. A small flower, usually one of a cluster, as found in members of the Asteraceae or in grasses.

gamopetalous. Having flower petals that are fused to each other, either at the base only or throughout their length.

geniculate. Bent abruptly like a knee.

glabrous. Having a surface without hairs, although not necessarily smooth.

glaucous. Covered with a bluish or whitish bloom (as on the skin of some grapes or plums) that rubs off.

globose. Spherical, globe-shaped.

glomerule. An inflorescence with flowers crowded in small compact clusters.

glume. One of the two, empty, chaffy bracts at the base of the spikelet in grasses.

hastate. Arrowhead-shaped (triangular), with the basal lobes spreading at nearly right angles.

hilum. The scar or point of attachment of the seed.

inflorescence. A flower cluster.

involucre. A whorl or rosette of bracts, often resembling an ordinary calyx, subtending or supporting a flower cluster or fruit.

keel. A prominent dorsal rib or ridge resembling the keel of a boat; the two lower united petals of a papilionaceous flower.

lanceolate. Shaped like the head of a lance, several times longer than wide, broadest near the base and tapering at both ends.

lemma. In the grasses, the lower of usually two bracts immediately enclosing the flower.

ligule. A thin often scarious projection at the top of the leaf sheath where the sheath and leaf meet; a strap-shaped corolla, as in the ray florets of composites.

mealy. Soft, dry, and friable.

mucro. An abrupt point or tip at the end of an organ.

obovate. Inversely ovate.

obtuse. Blunt or rounded at the tip.

ochrea. A sheathing or tubular stipule.

ovate. Egg-shaped, with the broadest end downward.

ovule. The structure that after fertilization becomes the seed.

palea. The upper or inner of the two bracts that enclose the grass flower.

panicle. A loose, irregularly compound inflorescence with pedicellate flowers.

pappus. The modified calyx limb of composites (Asteraceae) that forms a crown of various characters such as bristles, hairs, scales, teeth, etc., at the summit of the achene.

pectinate. Comb-shaped.

peduncle. The support of an inflorescence or of a solitary flower.

perianth. The floral envelope of a flower.

pericarp. The ripened wall of the ovary when it becomes a fruit.

pilose. Pubescent with long soft hair.

pinna (plural, pinnae). A main division or leaflet of a pinnate or compoundly pinnate frond or leaf.

pinnate. Refers to a compound leaf that has the leaflets arranged on each side of a petiole or rachis.

pistillate. Bearing pistils; usually inferring that stamens are absent.

plumose. Feathery; having fine hairs on each side like the fine barbs on a feather.

posterior. At or toward the back.

propagule. The structure or organ by which plants are propagated; for example, seeds, cuttings, etc.

prothallus. The minute, reduced gametophyte of ferns and their allies. The prothallus bears the sex organs: the archegonia and the antheridia.

pubescent. Generally means hairy, but refers in particular to short soft hair.

raceme. A simple indeterminate inflorescence of pedicelled flowers on a common, more or less elongated axis.

rachilla. A small or secondary rachis; in grasses and sedges, applies especially to the floral axis of a spikelet as opposed to that of a spike.

Rachis. The axis of an inflorescence or of a compound leaf beyond the petiole.

recurved. Curved downward or backward.

reticulate. Being in the form of a network; net-veined.

retuse. Having a shallow notch at the rounded apex, as found in some leaves.

rhizome. Any underground rootlike stem. Leaf shoots grow from the upper surface and roots develop from the lower surface.

rosette. A cluster of closely crowded radiating leaves that appear to arise from the ground.

rugose. Wrinkled.

sagittate. Like an arrowhead, with the basal lobes pointing backward.

scabrid. Rough to the touch.

scape. A peduncle arising at or beneath the surface of the ground, naked or without foliage, and bearing one or more flowers at the summit.

scarious. Thin, dry, scalelike, membranous, not green.

seam. A line of junction; a groove or ridge formed by or between abutting edges.

sepal. One of the leaves of the calyx.

serrate. Toothed with the sharp teeth pointing forward on the margin.

serrulate. Finely serrate.

sessile. Without a stalk of any kind.

sheath. A long, more or less tubular envelope, as the tubular basal portion of a leaf of a grass or sedge which encloses the culm.

spathe. A leaflike, usually more or less concave bract, sometimes colored; a pair of bracts enclosing the inflorescence (spadix), as in aroids, palms, etc.

spikelet. A small or secondary spike; the unit of the inflorescence of sedges and grasses consisting of two glumes and one or more florets.

stamen. One of the anther-bearing organs of the flower.

staminate. Bearing stamens (usually with the implication that the pistil is lacking).

staminode. Sterile, nonfunctional, and often antherless stamens, present in some flowers as accessory structures, some resembling petals.

stigma. The part of the pistil that is adapted to receive pollen.

stipule. An appendage at the base of the petiole or on each side of the petiole.

stolon. A shoot that bends to the ground and takes root; more commonly, a horizontal stem at the surface of the ground, rooting from the nodes and there giving rise to vegetative shoots or culms.

striate. Marked, usually longitudinally, with fine lines or ridges.

strobilus. A conelike structure consisting of a central axis about which are many closely packed, spirally arranged, fertile microsporophylls that bear microsporangia or pollen sacs on their lower surfaces.

subsessile. Almost sessile.

suture. A line of dehiscence in dry fruits or a line of junction or of cleavage of two united organs.

syngenesious. Having stamens joined together by their anthers.

terminal. Growing at the end of a branch or stem.

tomentose. Covered with densely matted hairs.

truncate. Having the end square or even.

tubercle. A small tuber or tuberlike structure, subterranean or otherwise.

utricle. A small inflated achenelike fruit; a small bladder.

villous. Pubescent with long, soft (not matted) hair.

Bibliography

ABDUL WAHAB, A., and E. RICE. 1967. Plant inhibition by johnsongrass and its possible significance in old-field succession. *Bulletin of the Torrey Botannical Club* 94:486-497.

ADAMS, A. 1967. The vectors and alternate hosts of groundnuts rosette virus in Central Province, Malawi. *Rhodesian, Zambian, Malawian Journal of Agricultural Research* 5(2):145-151.

ADLAKHA, P. 1961. Incidence and losses caused by particular weeds in different areas and in different crops and preparation of a weed map. *Proceedings of the Indian Council of Agricultural Research Seminar on Weed Control, Bombay*. Indian Council of Agricultural Research, New Delhi. 12 pp.

AGATI, J., and C. CALICA. 1949. The leaf-gall disease of rice and corn in the Philippines. *Philippine Journal of Agriculture* 14:31-38.

———. 1950. Studies in the host range of the rice and corn leaf-gall virus. *Philippine Journal of Agriculture* 15:249-258.

AGUNDIS, M., and O. VALTIERRA. 1963. El coquillo *Cyperus rotundus* L. una mala hierba del trópico. *Agricultura técnica en Mexico* 2:183-188.

ALCORN, J. 1968. Occurrence and host range of *Ascochyta phaseolorum* in Queensland. *Australian Journal of Biological Sciences* 21(6):1143-1151.

ALEX, J. 1965. *Survey of the weeds of cultivated land in the prairie provinces*. Experimental Farm, Research Branch, Department of Agriculture, Regina. 68 pp.

ALLARD, R. 1965. Genetic systems associated with colonizing ability in predominantly self-pollinated species. Page 49 *in* H. Baker and G. Stebbins, eds. *The genetics of colonizing species*. Academic Press, New York. 588 pp.

ALLEN, S., and W. PEARSALL. 1963. Leaf analysis and shoot production in *Phragmites*. *Oikos* 14:176-189.

ALLSOPP, W. 1960. The manatee: ecology and use for weed control. *Nature* 188:762.

ALLYN, R., and J. NALEWAJA. 1968a. Competitive effects of wild oat in flax. *Weed Science* 16:501-504.

———. 1968b. Competition of wild oat on wheat and barley. *Weed Science* 16:505-508.

———. 1968c. Effect of duration of wild oat competition in flax. *Weed Science* 16:509-512.

ALTMAN, J. 1968. The sugar beet nematode. *Down to Earth* 23(4):27-31.

AMBASHT, R. 1964. Ecology of the underground parts of *Cyperus rotundus* L. *Tropical Ecology* 5:67-74.

ANDERSEN, R. 1968. *Germination and establishment of weeds for experimental purposes*. Weed Science Society of America, Department of Agronomy, University of Illinois, Urbana. 236 pp.

ANDERSON, L., A. APPLEBY, and J. WESELOH. 1960. Characteristics of johnsongrass rhizomes. *Weeds* 8:402-406.

ANDREWS, C. 1967. The initiation of dormancy in developing seeds of *Avena fatua*. Ph.D. Thesis. University of Saskatchewan, Saskatoon. 139 pp.

ANDREWS, F. 1940. A study of nut grass (*Cyperus rotundus* L.) in the cotton soil of the Gezira. 1. The maintenance of life in the tuber. *Annals of Botany*, new series, 4:177-193.

———. 1945. Water plants in the Gezira canals: a study of aquatic plants and their control in the canals of the Gezira cotton area. *Annuals of Applied Biology* 32(1):1-14.

———. 1946. A study of nut grass (*Cyperus rotundus* L.) in the cotton soil of the Gezira. 2. The perpetuation of the plant by means of seed. *Annals of Botany*, new series, 10:15-30.

ANONYMOUS. 1937-1938. The effect of lalang on the growth of young rubber trees. *Journal of the Rubber*

Research Institute of Malaya 8:227-231.

——. 1954. *Kikuyu grass*. Plant Industry Advisory Leaflet 327. Queensland Department of Agriculture and Stock Division, Brisbane. [Extract from *Queensland Agriculture Journal*, August 1954.]

——. 1957. *Witchweed*. Agricultural Research Service Special Report ARS-22-41. United States Department of Agriculture, Washington, D.C. 17 pp.

——. 1959. *Eupatorium odorata*. *Eastern Region Development*, Enugu, Nigeria 3(11):14-15.

——. 1960*a*. *Index of plant diseases in the United States*. Handbook 165. United States Department of Agriculture, Washington, D.C. 531 pp.

——. 1960*b*. Ukrudtsfros spirevne efter opbevaring i ensilage. *Meddelelser fra Statens Forsøkvirksomhed i Plantekultur* 633:2.

——. 1960*c*. *Cuscuta australis*, a parasite of cover plants. *Planters' Bulletin* 49.

——. 1961. Survival of wild oats. *Agriculture* 68:390.

——. 1964. Identification of plants on Malayan rubber estates. Plates 33-40, climbing and scrambling dicotyledons. *Planters' Bulletin* 71:29-40.

——. 1965. Estimated average annual losses to various crop groups caused by weeds and cost of controlling weeds, 1951-1960. Chapter 5 *in Handbook* 291. United States Department of Agriculture, Washington, D.C.

——. 1966. *Stem eelworm on cereals and other farm crops*. Advisory Leaflet 178. Ministry of Agriculture, Fisheries and Food, London. 6 pp.

——. 1967*a*. Covers and fertilizers for immature rubber. *Planters' Bulletin* 89:66-72.

——. 1967*b*. Cover management. *Planters' Bulletin* 89:73-76.

——. 1967*c*. Identification of plants on Malayan rubber estates. Plates 73-80, dicotyledons, herbaceous or half shrubby. *Planters' Bulletin* 90:88-98.

——. 1968*a*. *Potato tuber eelworm*. Advisory Leaflet 372 (revised). Ministry of Agriculture, Fisheries and Food, London. 4 pp.

——. 1968*b*. *Broad-leaved infestations in cereals*. Fisons Agricultural and Technical Information, spring 1968:21-28. Fisons Pest Control, Ltd., Cambridge, England.

ARAI, M. 1961. Ecological studies on weeds in winter cropping on drained paddy fields: a basis for weed control in barley and wheat cultivation. *Journal of the Kanto-Tosan Agricultural Experiment Station* 19:1-182.

ARAI, M., H. CHISAKA, and K. UEKI. 1961. Comparison in ecological characteristics of noxious weeds in winter cropping. *Proceedings of the Crop Science Society of Japan* 30:39-42.

ARAI, M., and M. MIYAHARA. 1960. Physiological and ecological studies on barnyard grass (*Echinochloa crusgalli*). I. On the primary dormancy of seed. 1. Relation of the seed covering to dormancy and effects of temperature and oxygen on breaking dormancy [in Japanese, English summary]. *Proceedings of the Crop Science Society of Japan* 29:130-132.

——. 1962*a*. Physiological and ecological studies on barnyard grass (*Echinochloa crusgalli*). II. On the primary dormancy of seed. 2. On the dormancy breaking of the seed in the soil [in Japanese, English summary]. *Proceedings of the Crop Science Society of Japan* 31:73-77.

——. 1962*b*. Physiological and ecological studies on barnyard grass (*Echinochloa crusgalli*). III. On the secondary dormancy of seed [in Japanese, English summary]. *Proceedings of the Crop Science Society of Japan* 31:186-189.

——. 1962*c*. Physiological and ecological studies on barnyard grass (*Echinochloa crusgalli*). IV. On the death of the seeds in the process of dormancy awakening [in Japanese, English summary]. *Proceedings of the Crop Science Society of Japan* 31:190-194.

——. 1962*d*. Physiological and ecological studies on barnyard grass (*Echinochloa crusgalli*). V. On the germination of the seed [in Japanese, English summary]. *Proceedings of the Crop Science Society of Japan* 31:362-366.

——. 1962*e*. Physiological and ecological studies on barnyard grass (*Echinochloa crusgalli*). VI. On elongation of the plumule through soils after irrigation [in Japanese, English summary]. *Proceedings of the Crop Science Society of Japan* 31:367-370.

ARBER, A. 1963. *Water plants, a study of aquatic angiosperms*. J. Cramer, Weinheim. 436 pp.

ARNOLD, H. 1944. *Poisonous plants of Hawaii*. Tongg Publishing Company, Honolulu. 71 pp.

ARNY, A. 1928. *Quackgrass control*. Extension Circular 25. Minnesota Agricultural Experiment Station, St. Paul. 4 pp.

ARTSCHWAGER, E. 1920. On the anatomy of *Chenopodium album* L. *American Journal of Botany* 7:252-260.

ATWOOD, W. 1914. A physiological study of the germination of *Avena fatua*. *Botanical Gazette* 57:386-410.

AULD, B. 1970. *Eupatorium* weed species in Australia. *Pest Articles and News Summaries* 16(1):82-86.

BACKER, C., and R. BAKHUIZEN VAN DEN BRINK, JR. 1963. *Flora of Java*. Volume 1. Wolters-Noordhoff N.V., Groningen. 648 pp.

——. 1965. *Flora of Java*. Volume 2. Wolters-Noordhoff N.V., Groningen. 641 pp.

——. 1968. *Flora of Java*. Volume 3. Wolters-Noordhoff N. V., Groningen. 761 pp.

BAILEY, F., and P. GORDON. 1887. *Plants reputed poisonous and injurious to stock*. James C. Beal, Government Printer, Brisbane. 112 pp.

BAKER, H. 1951. Hybridization and natural gene flow between higher plants. *Biological Reviews* 26:302-307.

——. 1965. Characteristics and modes of origin of weeds. Pages 147-169 *in* H. Baker and G. Stebbins, eds. *The genetics of colonizing species: proceedings*. Academic Press, New York. 588 pp.

BAKER, H., and G. STEBBINS. 1965. *The genetics of colonizing species*. Academic Press, New York. 588 pp.

BAKKE, A. 1939. Experiments on the control of the European bindweed. Pages 364-440 in Research Bulletin 259. Iowa Agricultural Experiment Station, Ames. 440 pp.

BAKKE, A., and W. GAESSLER. 1945. The effect of reduced light intensity on the aerial and subterranean parts of European bindweed. Plant Physiology 20: 246-257.

BAKKE, A., W. GAESSLER, and W. LOOMIS. 1939. Relation of root reserves to control of European bindweed (Convolvulus arvensis). Pages 112-144 in Research Bulletin 254. Iowa Agricultural Experiment Station, Ames. 440 pp.

BAKKE, A., W. GAESSLER, L. PULTZ, and S. SALMON. 1944. Relation of cultivation to depletion of root reserves in European bindweed at different soil horizons. Journal of Agricultural Research 69:137-147.

BAKKER, D. 1960. A comparative life-history study of Cirsium arvense (L.) Scop. and Tussilago farfara L., the most troublesome weeds in the newly reclaimed polders of the former Zuider Zee. Pages 205-222 in J. Harper, ed. Biology of weeds. Blackwell Scientific Publications, Oxford. 256 pp.

BALLARD, L. 1969. Anagallis arvensis L. Chapter 17 in L. T. Evans, ed. The induction of flowering: some case histories. Macmillan Co. of Australia, South Melbourne. 488 pp.

BAMBER, M. 1909. The cultivation of Passiflora foetida and Mikania scandens to keep down other weeds. Journal of the Royal Botanical Garden 4(16):141.

BANDEEN, J., and K. BUCHHOLTZ. 1967. Competitive effects of quackgrass upon corn as modified by fertilization. Weeds 15:220-227.

BARRALIS, G. 1961. Étude de la distribution des diverses espèces de folles en France [Study of distribution of various species of wild oats in France]. Annales de physiologie végétale 3:39-53.

BASNAYAKE, V. 1966. Eradication of sporadic alang-alang. Planters' Bulletin 87:197-198.

BEASLEY, P. 1967. Photosensitization in sheep. Queensland Agricultural Journal 93:350-352.

BELL, R., W. LACHMAN, R. SWEET, and E. RAHN. 1962. Life history studies as related to weed control in northeastern United States. 1. Nutgrass. Bulletin 364. Rhode Island Agricultural Experiment Station, Kingston. 33 pp.

BENDIXEN, L. 1970. Altering growth form to precondition yellow nutsedge for control. Weed Science 18(5):599-603.

BENEDICT, R. 1941. The gold rush: a fern ally. American Fern Journal 31:27-32.

BENNETT, F. 1966. Investigations on the insects attacking the aquatic ferns, Salvinia spp. in Trinidad and northern South America. Proceedings of the Southern Weed Control Conference 19:497-504.

BENNETT, H. 1944. Embryology of Paspalum dilatatum. Botanical Gazette 106:40-45.

BERGER, G. 1966. Dormancy, growth inhibition, and tuberization of nutsedge (Cyperus rotundus L.) as affected by photoperiod. Ph.D. Thesis. University of California at Riverside, Riverside.

BHARDWAJ, R., and R. VERMA. 1968. Seasonal development of nutgrass (Cyperus rotundus L.) under Delhi conditions. Indian Journal of Agricultural Science 38(6):950-957.

BHARGHAVA, J. 1937. The life history of Chenopodium album L. Proceedings of the Indian Academy of Sciences (section B) 4:179-200.

BHAT, R., and M. KARNIK. 1954. Indigenous cellulosic raw materials for the production of pulp, paper and board. Indian Forest Bulletin 182. Manager of Publications, Delhi. 9 pp.

BHATIA, H. 1970. Grass carps can control aquatic weeds. Indian Farming 20:36-37.

BIRCH, W. 1957. The seeding and germination of some Kenya weeds. Journal of Ecology 45:85-91.

BISWAS, K., and C. CALDER. 1936. Handbook of common water and marsh plants of India. Health Bulletin 24. Malaria Bulletin 11. Manager of Publications, Delhi. 216 pp.

BLACK, C., T. CHEN, and R. BROWN. 1969. Biochemical basis for plant competition. Weed Science 17:338-344.

BLACK, M. 1959. Dormancy studies on the seed of Avena fatua. 1. The possible role of germination inhibitors. Canadian Journal of Botany 37:393-402.

BLOHM, H. 1962. Poisonous plants of Venezuela. Wissenschaftliche Verlagsgesellschaft G.M.B.H., Stuttgart. 136. pp.

BOGDAN, A. 1965. Weeds in Kenya wheat. Weed Research 5(4):351-352.

BONNET, A. 1955. Contribution à l'étude de Hydropteridees: Recherches sur Salvinia auriculata Aublet. Annales des sciences naturelles (botanique) 16:524-600.

BOR, N. 1960. The grasses of Burma, Ceylon, India and Pakistan. 3rd ed. Pergamon Press, London. 767 pp.

BORNKAMM, R. 1961. The competition for light by arable weeds. Flora, oder allgemeine botanische Zeitung, Jena 151(1):126-143.

———. 1963. Manifestation of competition between higher plants and its ideal formulation [in German]. Bericht über das Geobotanische Forschungsinstitut Rübel in Zurich 34:83-107.

BOUGHEY, A. 1963. The explosive development of a floating weed vegetation on Lake Kariba. Adansonia 3:49-61.

BOWDEN, B., and G. FRIESEN. 1967. Competition of wild oats (Avena fatua) in wheat and flax. Weed Research 7:349-359.

BRENCHLEY, W., and K. WARINGTON. 1930. The weed seed population of arable soil. 1. Numerical estimation of viable seeds and observations on their natural dormancy. Journal of Ecology 18:235-272.

BROD, G. 1968. Untersuchungen zur Biologie und Ökologie der Hühner-Hirse, Echinochloa crusgalli.

568 BIBLIOGRAPHY

Weed Research 8:115-127.

BROWN, E., and R. PORTER. 1942. The viability and germination of *C. arvensis* and other perennial weeds. Pages 477-504 *in Research Bulletin* 294. Iowa Agricultural Experiment Station, Ames. 998 pp.

BROWN, R. 1965. The germination of angiospermous parasite seeds. Pages 925-932 *in* W. Ruhland, ed. *Handbuch der Pflanzenphysiologie*. Volume 15. Springer-Verlag, Berlin.

BROWN, R., and M. EDWARDS. 1944. Germination of the seeds of *Striga lutea*. 1. The host effect and the progress of germination. *Annals of Botany*, new series, 8:131-148.

————. 1945. Effects of thiourea and allylthiourea on the germination of the seed of *Striga lutea*. *Nature* 155:455-456.

————. 1946. Germination of the seed of *Striga lutea*. 2. The effect of time of treatment and concentration of the host stimulant. *Annals of Botany*, new series, 10:133-142.

BRUNS, V., and L. RASMUSSEN. 1958. The effects of fresh water storage on the germination of certain weed seeds. 3. Quackgrass, green bristlegrass, yellow bristlegrass, watergrass, pigweed and halogeton. *Weeds* 6:42-48.

BUCHHOLTZ, K. 1968. Use of the split root technique for study of competition between quackgrass and corn. Page 90 *in Abstracts: 1968 meeting*. Weed Science Society of America, Department of Agronomy, University of Illinois, Urbana.

BURKILL, I. 1935. *A dictionary of the economic products of the Malay Peninsula*. Volumes 1 and 2. Governments of the Straits Settlements and Federated Malay States, Crown Agents for the Colonies, London. 2,402 pp.

BURNS, E., and G. BUCHANAN. 1968. Preliminary studies of the competitive effects of cocklebur (*Xanthium pennsylvanicum*) and nutsedge (*Cyperus rotundus*) on early cotton growth. Page 151 *in* R. Upchurch, ed. *Proceedings of the 21st Southern Weed Conference*. Southern Weed Conference, North Carolina State University at Raleigh, Raleigh. 410 pp.

BURT, G., and I. WEDDERSPOON. 1971. Growth of johnsongrass selections under different temperatures and dark periods. *Weed Science* 19:419-423.

BURTON, G. 1966. Breeding better bermuda grass. *Proceedings of the 9th International Grassland Congress, 1965*, 1:93-96.

BUTLER, F. 1951. Anthraconose and seedling blight of Bathurst burr caused by *Colletotrichum xanthii* Halst. *Australian Journal of Agricultural Research* 2(4): 401-410.

BUTLER, G. 1965. Progress report on the biological control of puncture vine with weevils. Pages 17-20 *in* Report 230. Arizona Agricultural Experiment Station, Tucson. 290 pp.

BUTTERY, B., and J. LAMBERT. 1965*a*. Competition between *Glyceris maxima* and *Phragmites communis* in the region of Surlingham Road. 1. The competition mechanism. *Journal of Ecology* 53:163-181.

————. 1965*b*. Competition between *Glyceris maxima* and *Phragmites communis* in the region of Surlingham Road. 2. The fen gradient. *Journal of Ecology* 53:183-195.

BUYCKX, E., and R. TAS. 1958. Essais d'éradication de jacinthe d'eau *E. crassipes*, à l'aide de hélicoptère à turbine. Pages 141-155 *in Proceedings of the 1958 African Weed Control Conference*, Victoria Falls, Southern Rhodesia, July 1958.

BYLTERUD, A. 1965. Mechanical and chemical control of *Agropyron repens* in Norway. *Weed Research* 5:169-180.

CARTER, W. 1939. Populations of *Thrips tabaci*, with reference to virus transmission. *Journal of Animal Ecology* 8(2):261-276.

CAUM, E. 1940. A devastating weed. *Hawaiian Planters' Record* 44:243-249.

CAVERS, P. 1963. The comparative biology of *Rumex obtusifolius* L. and *R. crispus* L. including the variety *trigranulotus*. Ph.D. Thesis. University of Wales, Cardiff.

CAVERS, P., and J. HARPER. 1964. Bilogical flora of the British Isles: *Rumex obtusifolius* and *Rumex crispus*. *Journal of Ecology* 52:737-766.

————. 1967. Studies in the dynamics of plant populations. 1. The fate of seed and transplants introduced into various habitats. *Journal of Ecology* 55:59-71.

CERRIZUELA, E. 1965. Effect of weeds in sugar cane fields (Argentina). *Revista industrial y agricola de Tucumán* 43:1-12.

CHADWICK, M., and M. OBEID. 1966. A comparative study of the growth of *Eichhornia crassipes* and *Pistia stratiotes* in water-culture. *Journal of Ecology* 54:546-575.

CHAKRAVARTY, S. 1963. Weed control in India, a review. *Indian Agriculturist*. 7(1,2): 23-58.

CHAMBERS, G. 1963. The problem of weed control in tea. *World Crops* 15:363-367.

CHAMPNESS, S. 1949. Notes on the buried seed populations beneath different types of ley in their seeding year. *Journal of Ecology* 37:51-56.

CHAMPNESS, S., and K. MORRIS. 1948. The population of buried viable weed seeds in relation to contrasting pasture and soil types. *Journal of Ecology* 36:149-173.

CHANDRASRIKUL, A. 1962. *A preliminary host list of plant diseases in Thailand*. Technical Bulletin 6. Department of Agriculture, Bangkok. 23 pp.

CHAPMAN, L. 1966. Prolific nut grass. *Cane Growers' Quarterly Bulletin* 30:16.

CHATTERJEE, P. 1969. Slug caterpillar—a new pest of rice. *Indian Farming* 19(5):29.

CHOKDER, A. 1965. Control of aquatic vegetation in fisheries. *Agriculture Pakistan* 16:235-247.

————. 1968. Further investigations on control of aquatic vegetation in fisheries. *Agriculture Pakistan* 19:101-118.

CHONA, B., and S. RAFAY. 1950. Studies on the sugar-cane diseases in India. 1. Sugar-cane mosaic virus. 2. The phenomenon of natural transmission and recovery from mosaic of sugar cane. *Indian Journal of Agricultural Science* 20:39-78.

CHOU, T., and F. LIAO. 1964. The introduction and culture of *Cyperus esculentus* and its use in the landscaping of parks. *Acta Horti Sinica* 3(1):83-94.

CHOW, C., E. THEVASAGAYAN, and E. WAMBEEK. 1955. The control of Salvinia—a host plant of *Mansonia* mosquitos. *Bulletin of the World Health Organization* 12:365-369.

CHUZHOVA, A. 1968. *Phragmites communis* regeneration in the delta of the River Volga. *Rastitel'nye Resursy* 4(2):230-236.

CLAYTON, W. 1968. The correct name of the common reed. *Taxon* 17:168-169.

COBLEY, L. 1956. *An introduction to the botany of tropical crops*. Longmans, Green and Co., London. 357 pp.

COFFMAN, F. 1961. *Oats and oat improvement*. American Society of Agronomy, Madison. 650 pp.

COLE, M. 1957. Variation and interspecific relationships of *C. album* L. in Britain. Ph.D. Thesis. University of Southampton, Highfield, Southampton.

———. 1962. Interspecific relationships and intraspecific variation of *C. album* L. in Britain. 2. The chromosome numbers of *C. album* L. and other species. *Watsonia* 5:117-122.

COLLINS, J. 1960. *The pineapple—botany, cultivation, and utilization*. Leonard Hill Books, London. 294 pp.

CONLEY, M. 1939. The seeds of the species of *Chenopodium album* occurring in the United States. M.S. Thesis. University of Notre Dame, South Bend.

COOK, C., and B. GUT. 1971. Salvinia in the state of Kerala, India. *Pest Articles and News Summaries* 17(4):438-447.

COOK, C., L. WHICHARD, B. TURNER, M. WALL, and G. EGLEY. 1966. Germination of witchweed (*Striga lutea* Lour.): isolation and properties of a potent stimulant. *Science* 154(3753):1,189-1,190.

COOPER, J., T. MAXWELL, and A. OWENS. 1960. A study of the passage of weed seeds through the digestive tract of the chicken. *Poultry Science* 39:161-163.

COQUILLAT, M. 1951. Sur les plantes des plus communes à la surface du globe. *Bulletin mensuel de la Société linnéenne de Lyon* 20:165-170.

COSTER, C. 1932. Eenige waarnemingen omtrent groei en bostrijding van alang-alang (*Imperata cylindrica*) [Some observations on the growth and control of *Imperata cylindrica*]. *Tectona* 25:383-402.

———. 1939. Grass in teak Taungya plantations. *Indian Forestry* 65:169-170.

CRAIG, N., and H. EVANS. 1946. Preliminary reports on the progress of weed control investigations in Mauritius. 2. *Mikania scandens* known in Mauritius as "liane marzoge," "liane Pauline," and "liane raisan." *Revue agricole de l'Île Maurice* 25(5):198-203.

CROCKER, W. 1906. Role of seed coats in delayed germination. *Botanical Gazette* 42:265-291.

CRUTTWELL, R. 1968. Preliminary survey of potential biological control agents of *Eupatorium odoratum* in Trinidad. Pages 836-849 *in Proceedings of the 9th British Weed Control Conference*, Brighton, England. Published by British Crop Protection Council. 1,045 pp.

CRUZ, R., C. ROMERO, and J. CARDENAS. 1969. Control of *Cyperus rotundus* in Sinu Valley (Colombia). Pages 60–61 *in Proceedings, Seminar de la Sociedad Colombiana de Control de Malezas y Fisiologia Vegetal* (Bogota).

CRUZ, R., et al. 1971. *El Coquito y su control* [Cyperus rotundus and its control]. Hoja Divulgativa 042. Instituto Colombiano de Agropêcuario, Departamento de Agronomía, Bogotá.

CUMMING, B., and J. HAY. 1958. Light and dormancy in wild oats. *Nature, London* 182:609-610.

CUPAHINA, K. 1963. The palatability of non-leguminous forbs of the Primor'je (Maritime Territory) [in Russian]. *Botanicheskiĭ zhurnal Akademii nauk SSSR* 48(8):1,161-1,167.

CUSSANS, G. 1968. The growth and development of *Agropyron repens* in competition with cereals, field beans and oil seed rape. Pages 131-136 *in Proceedings of the British Weed Control Conference*, Brighton, England. Published by British Crop Protection Council. 1,045 pp.

DALE, J., and G. EGLEY. 1971. Stimulation of witchweed germination by run-off water and plant tissues. *Weed Science* 19:678-681.

DALMACIO, S., and O. EXCONDE. 1967. Host range of *Xanthomonas oryzae* in the Philippines. *Philippine Agriculturist* 51:283-289.

DARLINGTON H., and G. STEINBAUER. 1961. The 80 year period of Dr. Beal's seed viability experiment. *American Journal of Botany* 38:379-381.

DAS, R. 1969. A study of reproduction in *Eichhornia crassipes*. *Tropical Ecology* 10:195-198.

DAVIS, C., and N. KRAUSS. 1967a. Recent introductions for biological control in Hawaii. *Proceedings of the Hawaiian Entomological Society* 19(2):201-207.

———. 1967b. Recent introductions for biological control in Hawaii. *Proceedings of the Hawaiian Entomological Society* 19(3):375-380.

DAVIS, R., W. JOHNSON, and F. WOOD. 1967. Weed root profiles. *Agronomy Journal* 59:555-556.

DAVIS, R., and A. WIESE. 1964. Weed root growth patterns in the field. Pages 367-368 *in* H. Andrews, ed. *Proceedings of the 17th Southern Weed Conference*. Southern Weed Conference, University of Tennessee at Knoxville. 444 pp.

DAVIS, R., A. WIESE, and J. PAFFORD. 1965. Root moisture extraction profiles of various weeds. *Weeds* 13(2):98-100.

DAVIS, W. 1930. The development of dormancy in seeds of cocklebur (Xanthium). *American Journal of Botany* 17(1):77-87.

DAWSON, J., and V. BRUNS. 1962. Emergence of

barnyardgrass, green foxtail, and yellow foxtail seedlings from various soil depths. *Weeds* 10:136-139.

DeLORIT, R., and H. AHLGREN. 1967. *Crop production.* 3rd ed. Prentice-Hall, Englewood Cliffs, New Jersey. 662 pp.

DERSCHEID, L., and R. SHULTZ. 1960. Achene development of Canada thistle and perennial sow thistle. *Weeds* 8:55-62.

DERSCHEID, L., J. STRITZKE, and W. WRIGHT. 1970. Field bindweed control with cultivation, cropping, and chemicals. *Weed Science* 18:590-596.

DERSCHEID, L., and K. WALLACE. 1959. *Control and elimination of thistles.* Circular 147. South Dakota Agricultural Experiment Station, Brookings. 32 pp.

DETMERS, F. 1927. Canada thistle, *Cirsium arvense.* Bulletin 414. Ohio Agricultural Experiment Station, Wooster. 45 pp.

DEXTER, S. 1936. Response of quack grass to defoliation and fertilization. *Plant Physiology* 11:843-851.

———. 1937. Drought resistance of quackgrass under various degrees of fertilization with nitrogen. *Journal of the American Society of Agronomy* 29:568-576.

———. 1942. Seasonal varieties in drought resistance of exposed rhizomes of quackgrass. *Journal of the American Society of Agronomy* 34:1125-1136.

DICKERSON, C. 1964. *Life history studies of barnyardgrass.* M.S. Thesis. University of Delaware, Newark. 56 pp.

DIRVEN, J. 1962. The feeding value of fallow vegetation. Pages 74-75 *in Jaarverslag Landbouwproefstation*, Paramaribo, Surinam, 1961. 98 pp.

DIRVEN, J., I. DULDER, and W. HERMELIJN. 1960. De braakvegetatie rijstvelden in Nickerie [The vegetation of fallow rice fields in the district of Nickerie, Surinam]. *Surinaamse landbouw* 8:1-7.

DIRVEN, J., and H. POERINK. 1955. Weeds in rice and their control in Suriname. *Tropical Agriculture, Trinidad* 32(2):115-123.

DOSLAND, J., and J. ARNOLD. 1966. Leaf area development and dry matter production of wheat and wild buckwheat growing in competition. Page 56 *in Abstracts of the Meeting of the Weed Society of America*, St. Louis. Department of Agronomy, University of Illinois, Urbana. 109 pp.

DOSPEKHOV, B. 1967. The effect of long term application of fertilizers and of crop rotation on weed infestation on arable land [in Russian]. *Izvestiya Timiryazevskoi Sel'sko Khozyaistvennoi Akademii* 3:51-64.

DOWLER, C., G. EGLEY, C. KUST, and P. SAND. 1959-1960. *Witchweed investigations*, Witchweed Laboratory, Whiteville, North Carolina. Annual Reports, Plant Pest Control Division, United States Department of Agriculture, Washington, D.C.

DOWLING, R. 1935. The ovary in the genus *Plantago*. 1. The British species. *Journal of the Linnean Society* (Botany) 50:329-336.

DUNN, S. 1970. Light quality effects on the life cycle of common purslane. *Weed Science* 18:611-612.

DUNN, S., G. GRUENDLING, and A. THOMAS. 1968. Effects of light quality on life cycles of crabgrass and barnyardgrass. *Weed Science* 16:58-60.

DUTTA, S. 1965. Weed control in tea of north east India. *Pest Articles and News Summaries*, section C, 11(1):9-12.

DUVAL-JOUVE, M. 1875. Histotaxie des feuilles de graminées [Histology and taxonomy of leaves of the *Gramineae*]. *Annales des sciences naturelles* (botanique) 1:294-371.

EDEN, T. 1965. *Tea.* 2nd ed. Longmans, London 205 pp.

EDLIN, H. 1969. *Plants and man, the story of our basic food.* Natural History Press, New York. 251 pp.

EGLEY, G. 1971. Mineral nutrition and parasite-host relationship of witchweed. *Weed Science* 19:528-533.

EGLEY, G., and J. DALE. 1969. Stimulation of witchweed (*Striga lutea*) seed germination. Page 339 *in* A. Worsham, ed. *Proceedings of the 22nd Annual Meeting of the Southern Weed Science Society.* Southern Weed Science Society, North Carolina State University at Raleigh. 464 pp.

EHRENDORFER, F. 1965. Dispersal mechanisms, genetic systems, and colonizing abilities in some flowering plant species. Pages 331-351 *in* H. Baker and G. Stebbins, eds. *The genetics of colonizing species.* Academic Press, New York. 588 pp.

EICHINGER, A. 1911. Ueber den Wert einiger tropischer Gräser [The value of some tropical grasses]. *Pflanzer* 7:26-32.

EPLEE, R. 1972. Induction of witchweed seed germination. Page 22 *in Abstracts of the Weed Science Society of America*, meeting, St. Louis, Missouri, February 8-10, 1972. Department of Agronomy, University of Illinois, Urbana.

EUGENIO, C., and M. DEL ROSARIO. 1962. Host range of tobacco mosaic virus in the Philippines. *Philippine Agriculturist* 46:175-197.

EVANS, L. 1969. *Lolium temulentum* L. Chapter 14 *in* L. T. Evans, ed. *The induction of flowering, some case histories.* Macmillan Co. of Australia, South Melbourne. 488 pp.

FAIL, H. 1954. The effect of rotary cultivation on rhizomatous weeds. *Journal of Agricultural Engineering Research* 1:3-15.

———. 1959. Mechanical eradication of couch and twitch. *World Crops* 11:241-244.

FAIRBROTHERS, D. 1959. Morphological variations of *Setaria faberii* and *Setaria viridis*. *Brittonia* 11:44-48.

FILIPJEV, I. 1936. On the classification of the *Tylinchinae*. *Proceedings of the Helminthological Society of Washington* 3:80-82.

FLORES, M., and F. OLIVE. 1952. Forage species of El Salvador. Pages 1434-1439 *in* R. E. Wagner et al., eds. *Proceedings of the Sixth International Grassland Congress.* 2 vols. Pennsylvania State College, State College, Pennsylvania. 1802 pp. total.

FOOD AND AGRICULTURE ORGANIZATION OF

THE UNITED NATIONS. 1957. *Protein requirements*. Nutritional Studies 16. Rome.

FORSBERG, D., and K. BEST. 1964. The emergence and plant development of wild buckwheat (*Polygonum convolvulus*). *Canadian Journal of Plant Science* 44:100-103.

FORSYTH. A. 1968. *British poisonous plants*. 2nd ed. Bulletin 161. Ministry of Agriculture, Fisheries and Food, London. 131 pp.

FRAZIER, J. 1943*a*. Nature and rate of development of the root system of *Convolvulus arvensis*. *Botanical Gazette* 104:417-425.

———. 1943*b*. Amount, distribution, and seasonal trend of certain organic reserves in root systems of field bindweed (*Convolvulus arvensis*). *Plant Physiology* 18:167-184.

FRIEDMAN, T., and M. HOROWITZ. 1970. Phytotoxicity of subterranean residues of three perennial weeds. *Weed Research* 10:382-385.

———. 1971. Biologically active substances in subterranean parts of purple nutsedge. *Weed Science* 19(4):398-401.

FRIESEN, G. 1965. Wild buckwheat control with Tordon. *Down to Earth* 20(4):9-10.

FRIESEN, G., and L. SHEBESKI. 1960. Economic losses caused by weed competition in Manitoba grain fields. 1. Weed species, their relative abundance and their effect on crop yields. *Canadian Journal of Plant Science* 40:457-467.

FURTICK, W., and A. DEUTSCH. 1971. Will the weeds rule the world? *World Crops* 23:150-151.

GAD, A., and F. OSMAN. 1959. Studies on the chemical constitution of Egyptian chufa, *Cyperus esculentus*. *Egyptian Journal of Chemistry* 2:123-124.

GALVEZ, E., H. THURSTON, and P. JENNINGS. 1960. Transmission of hoja blanca of rice by the planthopper *Sugata cubara*. *Plant Disease Reporter* 44:394.

GARDNER, C., and H. BENNETTS. 1956. *The toxic plants of Western Australia*. Periodicals Division, West Australian Newspaper, Perth. 253 pp.

GARESE. P. 1965. *Situacion actual de las problemas de malezas en el país*. [The present situation in regard to weed problems in the country] [in Spanish]. Actas de Reunión de Programación de Malezas. This work is lodged in the library of the Instituto Nacional de Technología Agropêcuaria, Mar del Plata, Argentina.

GARG, D., L. BENDIXEN, and S. ANDERSON. 1967. Rhizome differentiation in yellow nutsedge. *Weeds* 15:124-128.

GARRY, R. 1963. The changing fortunes and future of pepper growing in Cambodia. *Journal of Tropical Geography* 17:133-142.

GAVARRA, M., and A. ELOJA. 1964. Experimental transmission of the abaca mosaic virus by *Taxoptera citricidus* (Kirkaldy). *Philippine Journal of Plant Industry* 29:47-54.

———. 1970. Further studies on the insect vectors of abaca mosaic virus. 2. Experimental transmission of the abaca mosaic virus by *Schizaphis cyperi* (van der Goot) and *S. graminum* Rondani. *Philippine Journal of Plant Industry* 35:89-96.

GAY, P., and L. BERRY. 1959. The water hyacinth: a new problem on the Nile. *Geography Journal* 125:189-191.

GEORGI, C. 1934. Fodder and feeding stuffs in Malaya. Page 35 *in General Series* 17. Department of Agriculture, Straits Settlements and Federated Malay States, Kuala Lumpur.

GEORGIA, A. 1938. *A manual of weeds*. Macmillan Co., New York. 593 pp.

GHOSH, R. 1969. The development of female gametophyte and embryo in *Eupatorium odoratum* L. *Indian Journal of Science and Industry* 3(1):59-60.

GIFFORD, E., and STEWART, K. 1965. Ultrastructure of vegetative and reproductive apices of *C. album*. *Science* 149:75-79.

GIFFORD, E., and H. TEPPER. 1961. Ontogeny of the inflorescence in *C. album*. *American Journal of Botany* 48:657-667.

GOEDEN, R., and D. RICKER. 1970. Parasitization of introduced puncture vine weevils by indigenous *Chalcidoidea* in southern California. *Journal of Economic Entomology* 63(3):827-831.

GOLUB, S., and R. WETMORE. 1948*a*. Studies on the development in the vegetative shoot of *Equisetum arvense*. 1. The shoot apex. *American Journal of Botany* 35:755-767.

———. 1948*b*. Studies on the development in the vegetative shoot of *Equisetum arvense*. 2. The mature shoot. *American Journal of Botany* 35:767-781.

GONGGRIJP. 1952. Het gebruik van oliehoudende onkruidbestrijdingsmiddelen in Indonesia. *Bergcultures* 21(11):215-227.

GOOCH, S. 1963. The occurrence of weed seeds in samples tested by the Official Seed Testing Station 1960-1961. *Journal of the National Institute of Agricultural Botany* 9(3):353-371.

GOOD, R. 1964. *The geography of the flowering plants*. John Wiley & Sons. New York. 518 pp.

GOPPERT, A. 1881. Über Einwirkung neidriger Temperaturen auf die Vegetation. *Gartenflora* 30:1-13, 168-179.

GORHAM, E., and W. PEARSALL. 1956. Production ecology. 3. Shoot production in *Phragmites* in relation to habitat. *Oikos* 7:207-214.

GOULDER, R., and D. BOATMAN. 1971. Evidence that the N supply influences the distribution of a freshwater macrophyte, *Ceratophyllum demersum*. *Journal of Ecology* 59:783-791.

GRANADOS, R., and C. ORTEGA. 1966. Hospederas de los vectores del virus de la hoja blanca. Pruebas de transmisión y reacción de diferentes variadades de arroz al viruz. *Agricultura Tecnica en Mexico* 2(6):276-284.

GRANSTROM, B., and G. ALMGARD. 1955. Studies on the weed flora in Sweden. *Meddelanden. Statens jordbruksförsök* 56:187-209.

GRANT LIPPS, A., and L. BALLARD. 1963. Germination patterns shown by the light-sensitive seed of *Anagallis arvensis*. *Australian Journal of Biological Sciences* 16(3):572-584.

GRAY, B., and C. HEW. 1968. Cover crop experiments in oil palms on the west coast of Malaya. Pages 56-65 *in Oil palm developments in Malaysia*. Incorporated Society of Planters, Kuala Lumpur.

GRUENHAGEN, R., and J. NALEWAJA. 1966. Wild buckwheat competition in flax. Page 56 *in Abstracts of the Meeting of the Weed Society of America*, St. Louis. Department of Agronomy, University of Illinois, Urbana. 109 pp.

GRUMMER, G. 1963. The reaction of couch grass to dry air. *Weed Research* 3:44-51.

GUERIN, P. 1899. The probable causes of the poisonous nature of the darnel (*Lolium temulentum* L.). *Botanical Gazette* 28(2):136-137.

GUHA, H., and G. WATSON. 1958. Effects of cover plants on soil nutrient status and on growth of *Hevea*. 1. Laboratory studies on the mineralization of nitrogen in different soil mixtures. *Journal of the Rubber Research Institute of Malaya* 15:175.

GUPTA, S. 1968. Flowering process and curvature behavior of the inflorescence in *Monochoria vaginalis* Presl. *Tropical Ecology* 9:234-238.

HAKANSSON, S. 1967. Experiments with *Agropyron repens* (L.) Beauv. 1. Development and growth, and the response to burial at different developmental stages. *Lantbrukshögskolans Annaler* 33:827-873.

———. 1968a. Experiments with *Agropyron repens* (L.) Beauv. 2. Production from rhizome pieces of different sizes and from seeds. Various environmental conditions compared. *Lantbrukshögskolans Annaler* 34:3-29.

———. 1968b. Experiments with *Agropyron repens* (L.) Beauv. 3. Production of aerial and underground shoots after planting rhizome pieces of different lengths at varying depths. *Lantbrukshögskolans Annaler* 34: 31-51.

———. 1969a. Experiments with *Agropyron repens* (L.) Beauv. 4. Response to burial and defoliation repeated with different intervals. *Lantbrukshögskolans Annaler* 35:61-78.

———. 1969b. Experiments with *Agropyron repens* (L.) Beauv. 5. Effects of TCA and amitrole applied at different developmental stages. *Lantbrukshögskolans Annaler* 35:79-97.

———. 1969c. Experiments with *Agropyron repens* (L.) Beauv. 6. Rhizome orientation and life length of broken rhizomes in the soil, and reproductive capacity of different underground shoot parts. *Lantbrukshögskolans Annaler* 35:869-894.

———. 1969d. Experiments with *Agropyron repens* (L.) Beauv. 7. Temperature and light effects on development and growth. *Lantbrukshögskolans Annaler* 35:953-987.

———. 1970. Experiments with *Agropyron repens* (L.) Beauv. 9. Seedlings and their response to burial and TCA in soil. *Lantbrukshögskolans Annaler* 36:351-359.

HAKANSSON, S., and E. JONSSON. 1970. Experiments with *Agropyron repens* (L.) Beauv. 8. Responses of the plant to TCA and low moisture contents in the soil. *Lantbrukshögskolans Annaler* 36:135-151.

HAMDOUN, A. 1967. A study of the life history, regenerative capacity, and response to herbicide of *Cirsium arvense*. Master of Philosophy Thesis. University of Reading, Reading. 226 pp.

HAMEL, A., and P. DANSEREAU. 1949. L'aspect écologique du problème des mauvaises herbes. *Bulletin du Service de Biogéographie, Université de Montréal* 5:1-41.

HANF, M. 1941. Keimung und Entwicklung des Klettenlab Krautes (*Galium aparine*) in verscheidener Aussaattiefe. *Angewandte Botanik* 23:152-163.

HANSON, A., and H. CARNAHAN. 1956. *Breeding perennial forage grasses*. Technical Bulletin 1145. United States Department of Agriculture, Washington, D.C.

HARDING, D. 1966. Lake Kariba, the hydrology and development of fisheries. Pages 7-20 *in* R. Lowe-McConnel, ed. *Man-made lakes*. Academic Press, London. 218 pp.

HARKER, K. 1957. A note on *Digitaria scalarum* seed. *East African Agricultural Journal* 23:109.

HARLAN, J. 1970. *Cynodon* species and their value for grazing and hay. *Herbage Abstracts* 40:233-238.

HARPER, J. 1960. Factors involving plant numbers. Chapter 13 *in* J. Harper, ed. *The biology of weeds*. Blackwell Scientific Publications, Oxford. 256 pp.

———. 1965. Establishment, aggression, and cohabitation in weedy species. Pages 243-266 *in* H. Baker and G. Stebbins, eds. *The genetics of colonizing species*. Academic Press, New York. 588 pp.

HARPER, J., and A. CHANCELLOR. 1959. The comparative biology of closely related species living in the same area. 4. *Rumex*: Interference between individuals in populations of one and two species. *Journal of Ecology* 47:679-695.

HARRINGTON, G. 1916. Germination and viability of johnsongrass seed. *Proceedings of the Association of Official Seed Analysts of North America* 9:24-28.

———. 1917. Further studies on germination of johnsongrass seeds. *Proceedings of the Association of Official Seed Analysts of North America* 10:71-76.

HARRIS, S., and W. MARSHALL. 1960. Experimental germination of seed and establishment of seedlings of *Phragmites communis*. *Ecology* 41:395.

HASLAM, S. 1969a. The biology of the reed, *Phragmites communis*. *Annals of Botany* 33:289-301.

———. 1969b. The development of shoots of *P. communis*. *Annals of Botany* 33:695-709.

———. 1970a. Variation in population types of *P. communis*. *Annals of Botany* 34:147-158.

———. 1970b. Development of the annual populations in *P. communis*. *Annals of Botany* 34:571-591.

HATTINGH, E. 1961. The problem of *Salvinia auriculata* Aubl. and associated aquatic weeds on Kariba Lake. *Weed Research* 1:303-306.

HAUSER, E. 1962*a*. Development of purple nutsedge under field conditions. *Weeds* 10:315-321.

———. 1962*b*. Establishment of nutsedge from space-planted tubers. *Weeds* 10:209-212.

HAY, J. 1962*a*. Biology of quackgrass and some thoughts on its control. *Down to Earth* 18:14-16.

———. 1962*b*. Experiments on the mechanism of induced dormancy in wild oats, *Avena fatua*. *Canadian Journal of Botany* 40:191-202.

———. 1968. The changing weed problem on the prairies. *Agricultural Institute Review* 25:17-20.

———. 1970. Weed control in wheat, oats, and barley, Pages 38-47 *in* J. Holstun et al., eds. *Proceedings of the Food and Agriculture Organization of the United Nations*, International Conference on Weed Control, Davis, California, June 1970. Published by the Weed Science Society of America, Department of Agronomy, University of Illinois, Urbana. 668 pp.

HAYDEN, A. 1934. Distribution and reproduction of Canada thistle in Iowa. *American Journal of Botany* 21:355-372.

HEALY, A. 1953. Control of docks. *New Zealand Journal of Science and Technology* (section A) 34(5):473-475.

HEARNE, J. 1966. The Panama Canal's aquatic weed problem. Pages 443-449 *in* R. Upchurch, ed. *Proceedings of the 19th Southern Weed Conference*. Southern Weed Conference, North Carolina State University at Raleigh, Raleigh. 638 pp.

HEINEN, E., and A. AHMED. 1964. *Water hyacinth control on the Nile River, Sudan*. Publication of the Information Production Center, Department of Agriculture, Khartoum.

HEJNY, S. 1957. Eine Studie ueber die Oekologie der Echinochloa-Arten (*Echinochloa crusgalli* (L.) P. Beauv. und *E. coarctata* (Ster.) Koss.). *Biologické práce Slovenskej akademie vied* (5):1-115.

HELBAEK, H. 1950. Tollund-Mandens sidste Måltid [in Danish, English summary]. *Årboger f. nordisk Oldkyndighed og Historie* 1950.

———. 1958. Grauballemandens sidste Måltid [in Danish, with English summary]. *Kuml, Arbog for Tysk arkaelogisk Selskab* (Arhus, Denmark) 1958:83-116.

———. 1960. Comment on *Chenopodium album* as a food plant in prehistory. *Berichte des Geobotanischen Institut* Rubel, Zürich 31:16-19.

HENSON, R. 1968. *Weed competition studies*. Report. National Vegetable Research Station, Wellesbourne, Warwick, England.

HERZOG, R. 1934. Anatomische und experimentelmorphologische Untersuchungen über die Gattung *Salvinia*. *Planta* 22:490-514.

HIGGINS, R. 1967. Loss caused by noxious weeds in potatoes. *Down to Earth* 22(4):2, 24.

HILL, C. 1967. Investigation of the weed problem on Lake Ohakuri (New Zealand). Pages 15-20 *in* V. Chapman and C. Bell, eds. *Rotorua and Waikato water weeds—problems and research for a solution*. Department of Agriculture, Extension Service, University of Auckland, Auckland. 76 pp.

HILL, E., W. LACHMAN, and D. MAYNARD. 1963. Reproductive potential of yellow nutsedge by seed. *Weeds* 11:165-166.

HIRA, P. 1969. Transmission of schistosomiasis in Lake Kariba, Zambia. *Nature* 224:670-672.

HJELMQUIST, H. 1955. The oldest history of cultivated plants in Sweden. *Opera botanica a Societate botanica lundensi* 1(3):168-180.

HODGSON, J. 1964. Variations in ecotypes of Canada thistle. *Weeds* 12:167-171.

———. 1968. *The nature, ecology and control of Canada thistle*. Technical Bulletin 1386. United States Department of Agriculture, Washington, D.C.

———. 1971. *Canada thistle and its control*. Leaflet 523. United States Department of Agriculture, Washington, D.C. 8 pp.

HOLE, R. 1911. On some Indian forest grasses and their ecology. *Imperata arundinacea* Cyrill. *Indian Forest Memoirs* 1(1):91-102.

HOLM, L. 1971. Chemical interactions between plants on agricultural lands. Pages 95-101 *in Biochemical interactions among plants*. National Academy of Sciences, Washington, D.C. 134 pp.

HOLZ, W. 1957. The alkaloid content of *E. palustre* after treatment with growth regulators and the possibilities of its control therewith. *Mitteilungen der biologischen Bundesanstalt für Land- und Forstwirtschaft* 87:51-58.

HO-MINH-SI. 1969. *Weeds in South Viet Nam*. Ministry of Land Reform and Development of Agriculture and Fisheries, Agricultural Research Institute, Forestry Service, 121 Nguyễn-binh-Khiêm, Saigon. 140 pp.

HONESS, B. 1960. A preliminary ecological study of nut grass (*Cyperus rotundus* L.) under Trinidad conditions. Thesis for the Diploma in Tropical Agriculture, Imperial College of Tropical Agriculture, Trinidad.

HOROWITZ, M. 1965. Data on the biology and chemical control of the nutsedge, *Cyperus rotundus*, in Israel. *Pest Articles and News Summaries* 11(4):389-416.

HOROWITZ, M., and T. FRIEDMAN. 1971. Biological activity of subterranean residues of *Cynodon dactylon*, *Sorghum halepense*, and *Cyperus rotundus*. *Weed Research* 11:88-93.

HOSAKA, E. 1958. *Kikuya grass in Hawaii*. Bulletin 389. Cooperative Extension Service, University of Hawaii, Honolulu.

HOSEGOOD, P. 1963. The root distribution of kikuya grass and wattle trees. *East African Agricultural and Forestry Journal* 29(1):60-61.

HSU, C., G. ODELL, and T. WILLIAMS. 1968. Characterization of the saponin fraction of *Tribulus terrestris*. *Proceedings of the Oklahoma Academy of Science* 47:21-24.

HUBBARD, C., D. BROWN, A. GRAY, and R. WHYTE. 1944. Imperata cylindrica, *taxonomy, distribution, economic significance and control*. Imperial Agricultural Bureau Joint Publication 7. Imperial

Bureau of Pastures and Forage Crops, Aberystwyth, Cardiganshire. 63 pp.

HUBBARD, F. 1915. A taxonomic study of *Setaria italica* and its immediate allies. *American Journal of Botany* 2:169-198.

HUDSON, J. 1955. Propagation of plants by root cuttings. 2. Seasonal fluctuation of capacity to regenerate from roots. *Journal of Horticultural Sciences* 30:242-251.

HULTEN, E. 1950. *Atlas över växternas utbredning i norden*. Generalstabens Litografiska Anstalts Förlag, Stockholm. 512 pp.

HUNT, I., and R. HARKESS. 1968. Docks in grassland. *Scottish Journal of Agriculture* 47(3):160-162.

HURST, E. 1942. *The poison plants of New South Wales*. Poison Plants Committee of New South Wales, Department of Agriculture, Sydney. 498 pp.

HUXLEY, P., and A. TURK. 1966. Factors which affect the germination of the seeds of six common East African weeds. *Experimental Agriculture* 2(1):17-25.

INGLETT, G., ed. 1970. *Corn: culture, processing, products*. Avi Publishing Co., Westport, Connecticut. 369 pp.

ITO, K., J. INOUYE, and T. FURUYA. 1968. Comparison of autoecology of purple nutsedge (*Cyperus rotundus*) and yellow nutsedge (*Cyperus esculentus*). *Weed Research* 7:29-34.

IVANOVSKAYA, T. 1943. Influence of different conditions on the germination of seeds of wild oats [in Russian]. *Proceedings of the Lenin Academy of Agricultural Sciences of the U.S.S.R.* 3:28-33.

IVENS, G. 1967. *East African weeds and their control*. Oxford University Press, Nairobi, Lusaka, and Addis Ababa. 244 pp.

IWATA, E., and K. ISHIZUKA. 1967. Plant succession in Hachirogata polder. 1. Ecological studies on the common reed (*P. communis*). *Ecological Review* 17(1):37-46.

JACOMETTI, G. 1912. Le erbe che infestano le risaie italiane. Atti del Congresso Risicolo Internazionale, Vercelli 4:57-91.

JAGOE, J. 1938. The effect of lalang grass (*Imperata arundinacea*) on the growth of coconut palms. *Malayan Agricultural Journal* 26:369-375.

JANGARRD, N., M. SCKERL, and R. SCHIEFERSTEIN. 1971. The role of phenolics and abscisic acid in nutsedge tuber dormancy. *Weed Science* 19:17-20.

JANSEN, L. 1971. Morphology and photoperiodic responses of yellow nutsedge. *Weed Science* 19:210-219.

JAVIER, E. 1970 The flowering habits and mode of reproduction of Guinea grass (*Panicum maximum* Jacq.). Pages 284-289 in M. J. T. Norman, ed. *Proceedings of the Eleventh International Grassland Congress, 1970*. University of Queensland Press, St. Lucia. 956 pp.

JESSON, K., and H. HELBAEK. 1945. Cereals in Great Britain and Ireland in prehistoric and early historic times. *Biologiske Skrifter* 3:1-68.

JOHNSON, A., and S. DEXTER. 1939. The response of quackgrass to variations in height of cutting and rates of application of nitrogen. *Journal of the American Society of Agronomy* 31:67-76.

JOHNSON, B., and K. BUCHHOLTZ. 1958. Factors affecting the bud dormancy of quackgrass rhizomes. Pages 38-39 *in Abstracts of the Weed Society of America*, Memphis, Tennessee. Department of Agronomy, University of Illinois, Urbana.

————. 1961. An *in vitro* method of evaluating the activity of buds on the rhizomes of quackgrass (*Agropyron repens*). *Weeds* 9:600-606.

————. 1962. The natural dormancy of vegetative buds on rhizomes of quackgrass. *Weeds* 10:53-57.

JOHNSON, E. 1932. *The puncture vine in California*. Bulletin 528. University of California Agricultural Experiment Station, Berkeley. 42 pp.

JOHNSON, L. 1935. General preliminary studies on the physiology of delayed germination in *Avena fatua*. *Canadian Journal of Research* (sections C and D) 13:283-300.

JOHNSTON, E. 1962. *Chenopodium album as a food plant in Blackfoot Indian pre-history*. *Ecology* 43:129-130.

JONES, J. 1933. Effects of depth of submergence on control of barnyardgrass, and yield of rice grown in pots. *Journal of the American Society of Agronomy* 25:578-583.

JULIANO, J. 1931. Morphological study of the flower of *Monochoria vaginalis*. *Philippine Agriculturist* 20:177-184.

————. 1940. Viability of some Philippine weed seeds. *Philippine Agriculturist* 29:313-326.

JUSTICE, O. 1941. A study of dormancy in the seeds of *Polygonum*. *Memoirs of the Cornell University Agricultural Experiment Station* 235:41-49.

JUSTICE, O., and M.WHITEHEAD. 1946. Seed production, viability and dormancy in the nutgrasses, *Cyperus rotundus* and *Cyperus esculentus*. *Journal of Agricultural Research* 73:303-318.

KACPERSKA-PALACZ, A., E. PUTALA, and J. VENGRIS. 1963. Developmental anatomy of barnyardgrass seedlings. *Weeds* 11:311-316.

KANODIA, K., and R. GUPTA. 1968. Some useful and interesting supplementary food plants of the arid regions. *Journal d'agriculture tropicale et de botanique appliquée* 15(1-3):71-74.

KASASIAN, L. 1967. Herbicide research and usage in West Indies. *Pest Articles and News Summaries* 13(4):282-290.

————. 1971. *Weed control in the tropics*. Leonard Hill, London. 307 pp.

KAUL, V. 1959. Distribution of *Xanthium strumarium* L. complex in India. Pages 8-9 *in* Part 4, *Proceedings of the 46th Indian Science Congress, Delhi, India*. General Secretary, Indian Science Congress Association, 64 Dilkhusa St., Calcutta 17. 1,217 pp.

————. 1961. Water relations of *Xanthium strumarium* L.

Science and Culture 27:495-497.

————. 1965*a*. Physiological ecology of *Xanthium strumarium* L. 1. Seasonal morphological variants and distribution. *Tropical Ecology* 6:72-87.

————. 1965*b*. Physiological ecology of *Xanthium strumarium* L. 2. Physiology of seeds in relation to its distribution. *Journal of the Indian Botanical Society* 44:365-380.

————. 1965*c*. Physiological ecology of *Xanthium strumarium* L. 3. Effect of edaphic and biotic factors on growth and distribution. *Proceedings of the National Academy of Sciences of India* 35:203-216.

————. 1967. Physiological ecology of *Xanthium strumarium* L. 6. Plasticity and distribution. Journal of the Indian Botanical Society 46:392-396.

————. 1968. Physiological ecology of *Xanthium strumarium* L. 5. Water relations. *Tropical Ecology* 9:88-102.

————. 1971. Physiological ecology of *Xanthium strumarium* L. 6. Effect of climatic factors on growth and distribution. *New Phytologist* 70(4):799-812.

KEELEY, P., R. THULLEN, and J. MILLER. 1970. Biological control studies on yellow nutsedge with *Bactra verutana* Zeller. *Weed Science* 18:393-395.

KENNEDY, P., and A. CRAFTS. 1931. The anatomy of (*Convolvulus arvensis*) wild morning glory or field bindweed. *Hilgardia* 5:591-622.

KEPHART, L. 1923. *Quackgrass.* Farmers' Bulletin 1307. United States Department of Agriculture, Washington, D.C. 29 pp.

KHANNA, P. 1965. A contribution to the embryology of *Cyperus rotundus* L., *Scirpus mucrinatus* L., and *Kyllinga melanospora* Nees. Canadian Journal of Botany 43:1539-1547.

KIEWNICK, L. 1963. Experiments on the influence of seed-borne and soil-borne microflora on the viability of (*Avena fatua*) seeds. 1. The occurence, specific composition, and properties of microorganisms on the seeds [in German]. *Weed Research* 3:322-332.

————. 1964. Experiments on the influence of seed-borne and soil-borne microflora on the viability of (*Avena fatua*) seeds. 2. The influence of microflora on the viability of seeds in the soil [in German]. *Weed Research* 4:31-43.

KING, L. 1966. *Weeds of the world: biology and control.* John Wiley & Sons, Interscience, New York. 526 pp.

KING, R., and H. ROBINSON. 1970. *Chromolaena. Phytologia* 20(3):204.

KINGSBURY, J. 1964. *Poisonous plants of the United States and Canada.* Prentice-Hall, Englewood Cliffs, New Jersey. 626 pp.

KIRCHNER, O. VON, E. LOEW, and C. SCHRÖTER. 1904. Lebensgeschichte der Blütenpflanzen Mitteleuropas [Life history of the flowering plants of Central Europe]. [in German]. Numbers 23-24, volume 4, part 1, sheets 1-11. Verlagsbuchhandlung Eugen Ulmer, Stuttgart.

KISELEV, A., and V. SINYUKOV. 1967. The growth of field horsetail (*Equisetum arvense*) on cultivated and uncultivated fields [in Russian]. *Doklady Tskha* 131:273-277.

KLEBS, G. 1881. Beiträge zur Morphologie und Biologie der Keimung. Pages 536-631 *in* Wilhelm Pfeffer, ed. *Untersuchungen aus dem Botanischen.* Vol. 1. Institut zu Tübingen, Tübingen.

KNIGHT, W. 1955. The influence of photoperiod and temperature on growth, flowering, and seed production of dallisgrass, *Paspalum dilatatum* Poir. *Agronomy Journal* 47:555-559.

KNIGHT, W., and H. BENNETT. 1953. Preliminary report of the effect of photoperiod and temperature on the flowering and growth of several southern grasses. *Agronomy Journal* 45:268-269.

KOCH, W. 1964. Mechanical weed control in cereals. 1. Mode of action of harrows, hoes, chisel-tined cultivators and when to use them. *Zeitschrift für Acker- und Pflanzenbau* 120:369-382.

KOLK, H. 1947. Studies in germination biology in weeds [in Swedish, English summary] *Växtodling* 2:108-167.

————. 1962. Viability and dormancy of dry stored weed seeds. *Växtodling* 18:1-187.

KOLLAR, D. 1968. The germination and viability of weed seeds matured in winter wheat at different soil depths. *Acta Fytotechnica Nitra* 17:103-110.

KOLLMANN, F., and N. FEINBRUN. 1968. A cytotaxonomic study in Palestinian *Anagallis arvensis* L. *Notes from the Royal Botanic Garden, Edinburgh* 28(2):173-186.

KOMMEDAHL, T., J. KOTHEIMER, and J. BERNARDINI. 1959. The effects of quackgrass on germination and seedling development of certain crop plants. *Weeds* 7:1-12.

KOMMEDAHL, T., K. OLD, J. OHMAN, and E. RYAN. 1970. Quackgrass and nitrogen effects on succeeding crops in the field. *Weed Science* 18:29-32.

KORSMO, E. 1925. *Ugress i nutidens jordbruk, biologiske or praktiske undersøkelser.* J. W. Cappelens Forlag A-S, Oslo. 694 pp.

————. 1954. *Ugras i nåtidens jordbruk.* A-S Norsk Landbruks Forlag, Oslo. 635 pp.

KOUKOL, J., and W. DAGGER. 1967. Anthocyanin formation as a response to ozone and smog treatment in *Rumex crispus* L. *Plant Physiology* 42:1023-1024.

KRAUS, E. 1912. *Die gemeine Quecke* (Agriopyrum repens P. B.). No. 220. Deutsche Landwirtschafts-Gesellschaft, Berlin. 152 pp.

KUST, C. 1963. Dormancy and viability of witchweed seeds as affected by temperature and relative humidity during storage. *Weeds* 11:247-250.

————. 1964. Effect of photoperiod on growth of witchweed. *Agronomy Journal* 56:93.

————. 1966. A germination inhibitor in *Striga* seeds. *Weeds* 14:327-329.

LA SAYETTE, P. de. 1967. Evapotranspiration measurements on lucerne. *Annales de l'Institut national de la recherche agronomique Tunisie* 40(4):18.

LAUER, E. 1953. Über die keim Temperatur von Ackerunkräutern und deren Einfluss auf die Zusammensetzung von Unkrautgesselschaften. *Flora, oder allgemeine botanische Zeitung* 140:551-595.

Le DEUNFF, Y. 1971. Mise en évidence du phytochrome chez les semences de *Rumex crispus* L. *Annales de physiologie végétale* 9:201-208.

Le DEUNFF, Y., and R. CHAUSSAT. 1968. Etude de la dormance secondaire des semences chez *Rumex crispus* L. *Annales de Physiologie végétale* 10(4):227-236.

LEE, O. 1964. Weeds may host dwarf mosaic. Pages 44-47 *in Proceedings of the 20th North Central Weed Control Conference.* Purdue University, Lafayette. 97 pp.

LEEUWANGH, J., and P. LEUAMSANG. 1967. Observations on the ecology of *Thaia oryzivora*, a leafhopper found on rice in Thailand. *Food and Agriculture Organization of the United Nations Plant Protection Bulletin* 15(2):30-31.

LeTOURNEAU, D., and H. HEGGENESS. 1957. Germination and growth inhibitors in leafy spurge foliage and quackgrass rhizomes. *Weeds* 5:12-19.

LI SUN-ZEN. 1962. Badania ecologiczne nad chwastnica jednostronna-*Echinochloa crusgalli* (L.) var. *longisetum* Doll. [in Polish]. *Roczniki nauk rolniczych* 86-A-1.

LIN, C. 1968. Weeds found on cultivated land in Taiwan. Volumes 1 and 2. College of Agriculture, National Taiwan University, Taipei. 950 pp.

LINDSAY, D. 1956. Taxonomic and genetic studies on wild oats (*Avena fatua*). *Weeds* 4:1-10.

LINFORD, M., J. OLIVEIRA, and M. ISHII. 1949. *Paratylenchus minutus* n. sp., a nematode parasitic on roots. *Pacific Science* 3(2):111-119.

LINFORD, M., and F. YAP. 1940. Some host plants of the reniform nematode in Hawaii. *Proceedings of the Helminthological Society of Washington* 7:42-44.

LITTLE, F., ed. 1968. *Handbook of utilization of aquatic plants* (a compilation of the world's publications). Document PL: CP/20. Crop Protection Branch, Food and Agriculture Organization of the United Nations, Rome.

LITZENBERGER, S., and HO TONG LIP. 1961. Utilizing *Eupatorium odoratum* L. to improve crop yields in Cambodia. *Agronomy Journal* 53:321-324.

LOCKET, G. 1946. Observations on the colonization of bare chalk. *Journal of Ecology* 33:205-209.

LOUSTALOT, A., T. MUZIK, and H. CRUZADO. 1954. *Studies on nut grass* (Cyperus rotundus L.) *and its control.* Bulletin 52 of the Federal Experiment Station in Puerto Rico. Agriculture Research Service, United States Department of Agriculture, Washington, D.C.

LÖVE, D., and P. DANSEREAU. 1959. Biosystematic studies on *Xanthium*: taxonomic appraisal and ecological status. *Canadian Journal of Botany* 37:173-208.

LOWE, H., and K. BUCHHOLTZ. 1951. Cultural methods for the control of quackgrass. *Weeds* 1:346-351.

LOYAL, D., and R. GREWAL. 1964. Cytological approach to the life history of *Salvinia auriculata*. *Current Science* 33:344-346.

———. 1966. Cytological study on sterility in *Salvinia auriculata* with a bearing on its reproductive mechanism. *Cytologia* 31:330-338.

———. 1967. Some observations on the morphology and anatomy of *Salvinia* with particular reference to *S. auriculata* Aubl. and *S. natans* All. *Research Bulletin of the Punjab University of Science* 18:13-28.

LUBIGAN, R., and M. VEGA. 1971. The effect of different densities and durations of competition of *Echinochloa crusgalli* (L.) Beauv. and *Monochoria vaginalis* (Burm. f.) Presl. on the yield of lowland rice. Pages 19-23 *in Weed Science Report, 1970-1971.* Department of Agricultural Botany, University of the Philippines, College of Agriculture, Los Baños.

MACALPINE, R. 1959. A note on *Mikania scandens*. *Two and a Bud* 6(1):6-8.

McGRATH, H., W. SHAW, L. JANSEN, B. LIPSCOMB, P. MILLER, and W. ENNIS. 1957. *Witchweed* (Striga asiatica)—*a new parasitic plant in the United States.* Special Publication 10. Plant Disease Epidemics and Identification Section, United States Department of Agriculture, Washington, D.C. 142 pp.

McINTOSH, A. 1951. *Annual report of the Department of Agriculture, Malaya, for the year 1949*, Kuala Lumpur, Malaysia. 87 pp.

McINTYRE, G. 1965. Some effects of the nitrogen supply on the growth and development of *Agropyron repens* (L.) Beauv. *Weed Research* 5:1-12.

———. 1967. Environmental control of bud and rhizome development in the seedling of *Agropyron repens*. *Canadian Journal of Botany* 45:1315-1325.

MacLAGAN GORRIE, A. 1950. Dry zone afforestation and agricultural development in Ceylon. *British Agricultural Bulletin* 3:70.

McMILLAN, C. 1971. Photoperiod evidence in the introduction of *Xanthium* (cocklebur) to Australia. *Science* 171:1029-1031.

McWHORTER, C. 1960. *Johnsongrass—some factors affecting its control.* Information Sheet 688. Agriculture Experiment Station, Mississippi State University, State College.

———. 1961a. Morphology and development of johnsongrass from seeds and rhizomes. *Weeds* 9:558-562.

———. 1961b. Carbohydrate metabolism of johnsongrass as influenced by seasonal growth and herbicide treatments. *Weeds* 9(4):563-568.

———. 1971a. Introduction and spread of johnsongrass in the United States. *Weed Science* 19(5)496-500.

———. 1971b. The anatomy of johnsongrass. *Weed Science* 19:385-394.

———. 1971c. Growth and development of johnsongrass ecotypes. *Weed Science* 19:141-147.

———. 1971d. Control of johnsongrass ecotypes. *Weed Science* 19:220-233.

MAGALHAES, A. 1967. Observations on the influence of light on the growth of purple nutsedge, *Cyperus rotundus. Bragantia* 26(9):131-142.

MAIDEN, J. 1920. *The weeds of New South Wales*. Department of Agriculture, Sydney. 141 pp.

MAINSTONE, G., and WONG PHUI WENG. 1966. If *Mikania* invades. *Planter, Kuala Lumpur* 42(1):3-7.

MANSKE, R., and L. MARION. 1942. The alkaloids of *Lycopodium* species. 1. *Lycopodium complanatum* L. *Canadian Journal of Research* 20(5):87-92.

MARTIN, F. 1970. Cassava in the world of tomorrow. Pages 53-58 *in* D. Plucknett, ed. *Tropical root and tuber crops tomorrow*. Proceedings of the 2nd International Symposium on Tropical Root Crops. College of Tropical Agriculture, University of Hawaii, Honolulu. 171 pp.

MARTIN, J. 1940. Pathology. Pages 22-25 *in* Report of the committee in charge of the Experiment Station. *Proceedings of the Sixtieth Annual Meeting of the Hawaiian Sugar Planters' Association*. Honolulu.

MARZOCCA, A. 1957. *Manuel de malezas*. Instituto Nacional de Technología Agropecuaria, Buenos Aires. 530 pp.

MATHUR, R., and B. MATHUR. 1967. Effect of fertilizers on the incidence of *Striga lutea* Lour. and yield of bajra (*Pennisetum typhoides* Stupf. and Hubbard). *Indian Phytopathology* 20:270-272.

MATUMURA, M., N. TAKASE, and I. HIRAYOSHI. 1960. Physiological and ecological studies on germination of *Digitaria* seeds. 1. Difference in response to germinating conditions and dormancy among individual plants [in Japanese, English summary]. *Research Bulletin, Gifu Imperial College of Agriculture* 12:89-96.

MATVEEV, M., and P. TIMOFEEV. 1965. Effect of water-soluble exudates of certain forest and forest-weed species on one-year oak seedlings. *Ukrayins'kyĭ botanichnyĭ zhurnal* 22:28-32.

MAUN, M., and P. CAVERS. 1969. Influence of photoperiod on flowering of *Rumex crispus*. *Agronomy Journal* 61:823.

———. 1970. Influences of soil temperature on reproduction of curly dock. *Weed Science* 18:202-204.

MEADLY, G. 1958, Weeds of Western Australia: docks. *Journal of the Department of Agriculture of Western Australia*, 3rd ser., 7:621-623.

———. 1965. *Weeds of Western Australia*. Department of Agriculture, South Perth. 173 pp.

MEARS, P. 1970. Kikuyu (*Pennisetum clandestinum*) as a pasture grass—a review. *Tropical Grasslands* 4(2):139-152.

MEHRLICK, F., and H. FITZPATRICK. 1935. *Dichotomophthora portulacae*, a pathogen of *Portulaca oleracea*. *Mycologia* 27:543-550.

MERRILL, E. 1912. *A flora of Manila*. Publication 5. Bureau of Science, Manila. 490 pp.

MEYER, B. 1939. The daily cycle of apparent photosynthesis in a submerged aquatic. *American Journal of Botany* 26:755-760.

MEYER, B., and A. HERITAGE. 1941. Effect of turbidity and depth of immersion on apparent photosynthesis on *Ceratophyllum demersum*. *Ecology* 22:17-22.

MEYER, R., and K. BUCHHOLTZ. 1963. Effect of temperature, carbon dioxide, and oxygen levels on quackgrass rhizome buds. *Weeds* 11:1-3.

MICHAEL, P. 1973. Barnyardgrass (*Echinochloa*) in the Asian-Pacific region, with special reference to Australia. Pages 489-493 *in Proceedings of the 4th Asian-Pacific Weed Science Society*, Rotorua. 2 vols. 522 pp.

MICHELL, M. 1915. The embryo sac and embryo of *Striga lutea*. *Botanical Gazette* 59:124-135.

MISRA, D. 1962. *Tribulus terrestris* weed in arid zone farming. *Indian Journal of Agronomy* 7(2):136-141.

MISRA, R. 1969. *Ecological studies of noxious weeds, common to India and America, which are becoming an increasing problem in the upper Gangetic plains*. Part 1. *Cassia tora, Eleusine indica, Portulaca oleracea, Anagallis arvensis, Amaranthus spinosus*. Part 2. *Chenopodium album, Cyperus rotundus, Eleocharis palustris, Eichhornia crassipes*, and *Spirodela polyrhiza*. Final technical report, United States public law 480, grant no. FG-IN-213, project no. A7-CR-106. Department of Botany, Banaras Hindu University, Varanasi-5. 487 pp.

MITCHELL, D. 1970. *Autecological studies of* Salvinia auriculata *Aubl*. Ph.D. Thesis. University of London, London.

MONACO, T., and E. CUMBO. 1972. Growth and development of curly dock and broadleaf dock. *Weed Science* 20:64-67.

MOOLANI, M., E. KNAKE, and F. SLIFE. 1964. Competition of smooth pigweed with corn and soybeans. *Weeds* 12:126-128.

MOORE, R., and C. FRANKTON. 1969. Cytotaxonomy of some *Cirsium* species of the eastern United States, with a key to eastern species. *Canadian Journal of Botany* 47:1257-1275.

MORRISON, J. 1966. The effects of nitrogen, phosphorus and cultivation on the productivity of kikuyu grass at high altitudes in Kenya. *East African Agricultural and Forestry Journal* 31(3):391-397.

MUENSCHER, W. 1939. *Poisonous plants of the United States*. Macmillan Co., New York. 227 pp.

MUIR, F., and G. HENDERSON. 1926. Nematodes in connection with sugar cane root rot in the Hawaiian Islands. *Hawaiian Planters' Record* 30:233-250.

MUKULA, J. 1963. Studies on the biology and control of marsh horsetail (*Equisetum palustre* L.). *Annales Agriculturae Fenniae*, Helsinki (supplementum) 4(2):1-57.

MUKULA, J., M. RAATIKAINEN, R. LALLUKKA, and T. RAATIKAINEN. 1969. Composition of weed flora in spring cereals in Finland. *Annales Agriculturae Fenniae*, Helsinki 8(2):1-110.

MULLVERSTEDT, R. 1963*a* Investigations on the germination of weed seeds as influenced by oxygen partial pressure. *Weed Research* 3:154-163.

———. 1963*b*. Investigations on the causes of increased emergence of weeds following mechanical weed control measures (post emergence). *Weed Research* 3:298-303.

MUNE, T., and J. PARHAM. 1956. The declared noxious weeds of Fiji and their control. Bulletin 31. Department of Agriculture, Suva. 73 pp.

MUZIK, T., and H. CRUZADO. 1953. The effect of 2,4-D on sprout formation in *Cyperus rotundus*. *American Journal of Botany* 40:507-512.

NAKAO, H. 1966. Weed control in Hawaii. Pages 3-7 *in Proceedings of the 18th Annual California Weed Conference*, San Jose. 129 pp.

————. 1969. Biological control of weeds in Hawaii. Pages 93-95 *in Proceedings of the First Asian-Pacific Weed Control Interchange* [June 1967]. East-West Center, Honolulu.

NALEWAJA, J., and W. ARNOLD. 1970. *Weed control methods, losses and costs due to weeds, and benefits of weed control in wheat and other small grains. Proceedings of the Food and Agriculture Organization of the United Nations*. Conference on Weed Control, Davis, California, June 1970. Published by the Weed Science Society of America, Department of Agronomy, University of Illinois, Urbana. 668 pp.

NAPERKOVSKAYA, G. 1968. Weeds-reservoirs of [potato] virus X. *Kartofel' i ovoshchi* 13(8):42.

NEAL, M. 1965. *In gardens of Hawaii*. Special Publication 50. Bernice P. Bishop Museum, Honolulu. 924 pp.

NELSON, R. 1958. The effect of soil type and soil temperature on growth and development of witchweed (*Striga asiatica*) under controlled soil temperatures. *Plant Disease Reporter* 42:152-155.

NEMEC, B., J. BABICKA, and A. OBORSKY. 1937. Occurrence of gold in *Equisetum palustre* and *E. arvense*. *Rozpravy České Akademie Císaře Františka Josefa pro Vědy, Slovesnost a Uměni* 46(1):1-8.

NEW J. 1961. Biological flora of the British Isles: *Spergula arvensis*. *Journal of Ecology* 49:205-215.

NISHIKADO, Y. 1926. Studies on rice blast disease [in Japanese]. Bulletin 15, Plant Pathogenic Fungi and Injurious Insects, Bureau of Agriculture, Mining, Engineering and Forestry, Japan. 211 pp.

NODA, K., and S. EGUCHI. 1965. Studies on ecology of weeds on arable lands: 1. Emergence patterns of annual representative weeds which are commonly found on the paddy rice fields of south-western Japan. *Bulletin of the Kyushu Agricultural Experiment Station* 11:153-170.

NODA, K., K. IBARAKI, W. EGUCHI, and K. OZAWA. 1965. Studies on ecological characteristics of the annual weed, cleaver, and its chemical control on drained paddy fields for wheat plants in temperate Japan. *Bulletin of the Kyushu Agricultural Experiment Station* 11:345-374.

NODA, K., K. OZAWA, and K. IBARAKI. 1968. Studies on the damage to rice plants due to weed competition (Effect of barnyard grass competition on growth, yield and some eco-physiological aspects of rice plants). *Bulletin of the Kyushu Agricultural Experiment Station* 13:345-367.

NORMANHA, E. 1970. General aspects of cassava root production in Brazil. Pages 61-63 *in* D. L. Plucknett, ed. *Tropical root and tuber crops tomorrow*. Proceedings of the 2nd International Symposium on Tropical Root

Crops, College of Tropical Agriculture, University of Hawaii, Honolulu. 171 pp.

OCHSE, J., M. SOULE, M. DIJKMAN, and C. WEHLBURG. 1961. *Tropical and subtropical agriculture*. Vols. 1 and 2. Macmillan Co., New York. 1,446 pp.

ODUM, S. 1965. Germination of ancient seeds. Floristical observations and experiments with archaeologically dated soil samples. *Dansk botanisk Arkiv* 24(2):70.

OHMAN, J., and T. KOMMEDAHL. 1960. Relative toxicity of extracts from vegetative organs of quackgrass to alfalfa. *Weeds* 8:660-670.

————. 1964. Plant extracts, residues, and soil minerals in relation to competition of quackgrass with oats and alfalfa. *Weeds* 12:222-231.

OHWI, J. 1965. *Flora of Japan*. Shibundo, Tokyo. 1,383 pp.

OLESEN, M., and N. LANGKILDE. 1965. The germinativity of undeveloped seeds of *Agropyron repens* and its bearing on the purity analysis. *Proceedings of the International Seed Testing Association* 30:537-545.

OMID, A. 1964. The effect (phytotoxic) of quackgrass (*Agropyron repens*) rhizomes on germination and development of alfalfa. *Dissertation Abstracts* 25(6):3181.

ONO, M. 1966. Physiological and ecological studies on weeds of mulberry fields in Kumamoto district. *Bulletin of the Imperial Sericultural Experiment Station, Japan*. 20(3):259-290.

ORCHARD, H. 1949. Weeds of South Australia. Bathurst burr (*Xanthium spinosum* L.). *Journal of the Department of Agriculture of South Australia* 53:218-220.

OYER, E., G. GRIES, and B. ROGERS. 1959. The seasonal development of johnsongrass plants. *Weeds* 7(1):13-19.

PAJE, E., O. EXCONDE, and S. RAYMUNDO. 1964. Host range of *Piricularia oryzae* in the Philippines. *Philippine Agriculturist* 48:35-48.

PALAVEEVA-KOVACHEVSKA, M. 1966. Certain biological peculiarities of *Cirsium arvense*. *Pastbiščmye Nauki* 2(6)57-68.

PALMER, J. 1958. Studies in the behavior of the rhizomes of *Agropyron repens*. 1. The seasonal development and growth of the parent plant and rhizome. *New Phytologist* 57:145-159.

————. 1962. Effect of soil factors on the orientation of the rhizome. *Physiologia Plantarum* 15:445-451.

PALMER, J., and G. SAGAR. 1963. Biological flora of the British Isles: *Agropyron repens* (L.) Beauv. *Journal of Ecology* 51:783-794.

PALMER, R., and W. PORTER. 1959a. The metabolism of nut grass (*Cyperus rotundus* L.). 1. The influence of various O_2 and CO_2 levels upon germination and respiration. *Weeds* 7:481-489.

————. 1959b. The metabolism of nut grass (*Cyperus rotundus* L.). 2. The respiratory quotient and its relation to storage material and some terminal enzymes. *Weeds* 7:490-503.

————. 1959c. The metabolism of nut grass (*Cyperus*

rotundus L.). 3. Polyphenoloxidase in the dormant tuber. *Weeds* 7:504-510.

PAMMEL, L. 1911. *A manual of poisonous plants*. Torch Press, Cedar Rapids, Iowa. United States. 977 pp.

PANCHO, J. 1964. Seed sizes and production capacities of common weed species in rice fields of the Philippines. *Philippine Agriculturist* 48:307-316.

PANCHO, J., and D. PLUCKNETT. 1971. *Chromolaena odorata* (L.) R. M. King & H. Robinson—a new record of a noxious weed in the Philippines. *Journal of Animal Science* 8(2):143-149.

PANDEY, S. 1968. Adaptive significance of seed dormancy in *Anagallis arvensis* L. *Tropical Ecology* 9(2):171-193.

PARHAM, J. 1955. *The grasses of Fiji*. Bulletin 30. Department of Agriculture, Suva. 166 pp.

———. 1958. *The weeds of Fiji*. Bulletin 35. Department of Agriculture, Suva. 196 pp.

PARIJA, P. 1934. Physiological investigations on water hyacinth (*Eichhornia crassipes*) in Orissa with notes on some other aquatic weeds. *Indian Journal of Agricultural Science* 4:399-429.

PARKER, C. 1972. The *Mikania* problem. *Pest Articles and News Summaries* 18(3):312-315.

PARKER, C., K. HOLLY, and S. HOCOMBE. 1969. Herbicides for nut grass control, conclusions from 10 years of testing at Oxford. *Pest Articles and News Summaries* 15:54-63.

PARRIS, G. 1936. Plant pathology. *Annual Report of the Hawaii Agricultural Experiment Station* 1936:33-40.

———. 1940. A check list of fungi, bacteria, nematodes, and viruses occurring in Hawaii, and their hosts. Supplement 121. *Plant Disease Reporter*. Bureau of Plant Industry, United States Department of Agriculture, Washington, D.C. 91 pp.

———. 1941. Miscellaneous plant diseases. *Annual Report of the Hawaii Agricultural Experiment Station* 1940:73-74.

PARSONS, W. 1963. Water hyacinth, a pest of world waterways. *Journal of the Department of Agriculture of Victoria, Australia* 61:23-27.

PEGG, K., and M. MOFFETT. 1971. Host range of the ginger strain of *Pseudomonas solanacearum* in Queensland. *Australian Journal of Experimental Agriculture and Animal Husbandry* 11:696-698.

PENFOUND, W., and T. EARLE. 1948. The biology of water hyacinth. *Ecological Monographs* 18:447-472.

PENZES, B., and I. TOLG. 1966. Study of growth and feeding of grass carp (*Ctenopharyngodon idella*) in Hungary. *Bulletin français de pisciculture* 39:70-76.

PEPA, M. 1927. A comparative study of the palatability of some common Philippine forages. *Philippine Agriculturist* 15:547-555.

PETERS, R., and S. DUNN. 1971. *Life history studies as related to weed control in the Northeast. 6. Large and small crabgrass*. Bulletin 415, a northwest regional publication of the Northeast Weed Control Conference. Storrs Agriculture Experiment Station, University of Connecticut, Storrs.

PETERSEN, D. 1936. *Stellaria*-Studien. Zur Zytologie, Genetik, Ökologie und Systematik der Gattung *Stellaria* insbesonders der Media Gruppe. *Botaniska Notiser* 281-419.

PETERSON, R. 1965. *Wheat: botany, cultivation, utilization*. Leonard Hill Books, London. 422 pp.

PHILLIPS, W., and F. TIMMONS. 1954. *Bindweed—how to control it*. Bulletin 366. Fort Hays Branch, Kansas Agricultural Experiment Station, Manhattan, Kansas. 40 pp.

PLOWMAN, A. 1906. The comparative anatomy and phylogeny of the *Cyperaceae*. *Annals of Botany* 20:1-37.

PLUCKNETT, D. 1970. Productivity of tropical pastures in Hawaii. Pages A38-A49 *in* M. J. T. Norman, ed. *Proceedings of the Eleventh International Grassland Congress, Brisbane*. University of Queensland Press, St. Lucia. 956 pp.

———. 1973. Management of natural and improved pastures for cattle production under coconut. *Proceedings of the South Pacific Regional Seminar on Raising Cattle under Coconuts*, Apia, Western Samoa [September, 1972].

PLUCKNETT, D., J. EVENSON, and W. SANFORD. 1970. Ratoon cropping. *Advances in Agronomy* 22:285-330.

POITEAU, M. 1808. Sur l'embryon des *Graminées*, des *Cyperacées*, et du *Nelumbo*. *Annales du Musée d'histoire naturelle* 13:473.

POLUNIN, N. 1960. *Introduction to plant geography*. McGraw-Hill Book Co., New York. 640 pp.

PONNAPPA, K. 1967. *Cercosporidium helleri* on *Sphenoclea zeylanica*—a new record from India. *Current Science* 10:273.

POPAY, A., and E. ROBERTS. 1970a. Factors involved in the dormancy and germination of *Capsella bursa-pastoris* (L.) Medic. and *Senecio vulgaris* L. *Journal of Ecology* 58:103-121.

———. 1970b. Ecology of *Capsella bursa-pastoris* (L.) medic. and *Senecio vulgaris* L. in relation to germination behavior. *Journal of Ecology* 58:123-138.

POPE, W. 1968. *Manual of wayside plants of Hawaii*. Charles E. Tuttle Co., Tokyo. 289 pp.

POTHECARY, B., and W. THOMAS. 1968. Control of *Cyperus rotundus* in the Sudan Gezira. *Pest Articles and News Summaries* 14(3):236-240.

PRENTISS, A. 1889. On root propagation of Canada thistle. *Bulletin. Cornell University Agricultural Experiment Station* 15:190-192.

PRITCHARD, H. 1964. A cytochemical study of embryo sac development in *Stellaria media*. *American Journal of Botany* 51:371-378, 472-479.

PROCTOR, V. 1968. Long distance dispersal of seeds by retention in digestive tract of birds. *Science* 160:321-322.

PUCKDEEDINDAN, P. 1966. *A supplementary host list of plant diseases in Thailand*. Technical Bulletin 7. Department of Agriculture, Bangkok. 24 pp.

PURSEGLOVE, J. 1968. *Tropical crops: dicotyledons*. 2

vols. John Wiley & Sons, New York. 719 pp.

QUISENBERRY, K., and L. REITZ. 1967. *Wheat and wheat improvement*. American Society of Agronomy, Madison, Wisconsin. 560 pp.

QUISUMBING, E. 1951. *Medicinal plants of the Philippines*. Technical Bulletin 16. Republic of the Philippines, Department of Agriculture and Natural Resources, Manila. 1234 pp.

RAABE, R. 1965. Checklist of some parasitic phanerogams and some of their hosts on the island of Hawaii in 1963. *Plant Disease Reporter* 49(7):583-585.

RAGONESE, A. 1955. Plantas tóxicas para el ganado en la región central Argentina. *Revista de la Facultad de agronomía*, La Plata 31:133-336.

RAHN, E., R. SWEET, J. VENGRIS, and S. DUNN. 1968. *Life history studies as related to weed control in the Northeast 5. Barnyardgrass*. Bulletin 368. Northeast Regional Publication, Agricultural Experiment Station, University of Delaware, Newark. 45 pp.

RALEIGH, S., T. FLANAGAN, and C. VEATCH. 1962. *Life history studies as related to weed control. 4. Quackgrass*. Bulletin 365. Rhode Island Agriculture Experiment Station, Kingston. 10 pp.

RANADE, S., and W. BURNS. 1925. The eradication of *Cyperus rotundus* L. *Memoirs of the Department of Agriculture in India*, botanical series, 13:99-192.

RAO, J. 1968. Studies on the development of tubers in nutgrass and their starch content at different soil depths. *Madras Agricultural Journal* 55(1):19-23.

RATTRAY, J. 1960. *The grass cover of Africa*. Agriculture Studies 49. Food and Agricultural Organization of the United Nations, Rome. 168 pp.

RAYCHANDHURI, S., W. MISHRA, and A. GHOSH. 1967. *The virus diseases of the rice plant*. International Rice Research Institute. Johns Hopkins Press, Baltimore.

RETHKE, R. 1946. The anatomy of circumscissile dehiscense. *American Journal of Botany* 33:677-683.

RETIG, B., L. HOLM, and B. STRUCKMEYER. 1972. Effects of weeds on the anatomy of cabbage and tomato. *Weed Science* 20(1):33-36.

RICHARDSON, F. 1965. Bromacil, a herbicide suitable for sisal nurseries. *Kenya Sisal Board Bulletin* 54:41-42.

RICHARDSON, W. 1963. Early succession on arable land in Trinidad. *Tropical Agriculture, Trinidad* 40:89-101.

RIDLEY, H. 1930. *The dispersal of plants throughout the world*. L. Reeve & Co., Ashford, Kent. 744 pp.

RIEDER, G. 1966. *The effect of liquid manure on weed distribution and the use of the tetrazolium method for weed seeds*. Ph.D. Thesis. Landwirtschaftliche Hochschule, Universität Hohenheim, Stuttgart-Hohenheim. 119 pp.

RIEPMA, P. 1962. The effect of paraquat on some grasses and other weeds commonly found in rubber plantations. *Journal of the Rubber Research Institute of Malaya* 17(4):141-144.

————. 1965. Weed control with pre-emergence her-

bicides in tropical plantation crops. *World Review of Pest Control* 4(2):64-74.

ROBERTS, H. 1958. Studies on the weeds of vegetable crops 1. Initial effectsof cropping on weed seeds in the soil. *Journal of Ecology* 46:759-768.

ROBERTSON, H., and B. THEIN. 1932. The occurrence of water hyacinth seedlings (*E. crassipes*) under natural conditions in Burma. *Agriculture and Live-Stock in India* 2:383-390.

ROBINSON, B. 1922. The *Mikanias* of northern and western South America. *Contributions from the Gray Herbarium of Harvard University* 64:21-116.

————. 1934. *Mikania scandens* and its near relatives. *Contributions from the Gray Herbarium of Harvard University* 104:55-71.

ROBINSON, E. 1960. Growth of witchweed (*Striga asiatica*) as affected by soil types and temperatures. *Weeds* 8:576-581.

ROBINSON, E., and C. DOWLER. 1966. Investigation of catch and trap crops to eradicate witchweed *Striga asiatica*. *Weeds* 14:275-276.

ROBINSON, E., and I. STOKES. 1963. Influence of *Striga asiatica* (witchweed) on important varieties of sugar cane in the United States. Pages 812-815 *in* J. Williams, ed. *Proceedings of the 11th Congress of the International Society of Sugar Cane Technologists*, Reduit. Elsevier Nederland, Amsterdam. 1,250 pp.

ROCHECOUSTE, E. 1956. Observations on nutgrass (*Cyperus rotundus*) and its control by chemical methods in Mauritius. Pages 1-11 *in Proceedings of the Ninth Congress of the International Society of Sugar Cane Technologists*.

————. 1962a. Studies on the biotypes of *Cynodon dactylon* (L.) Pers. 1. Botanical investigations. *Weed Research* 2(1)1-23.

————. 1962b. Studies on the biotypes of *Cynodon dactylon* (L.) Pers. *Weed Research* 2(2):136-145.

————. 1967. *Weed control in sugar cane*. Mauritius Sugar Industry Research Institute, Reduit. 117 pp.

————. 1969. The problem posed by perennial weeds in tropical plantation crops. Pages 401-406 *in Proceedings of the Second Asian-Pacific Weed Control Interchange, Laguna, Philippines* [June 1969].

ROCHECOUSTE, E., and R. VAUGHAN. 1959. Weeds of Mauritius. *Bidens pilosa* L. Leaflet Series 1. Mauritius Sugar Industry Research Institute, Reduit.

————. 1960. Weeds of Mauritius. *Argemone mexicana*. Leaflet series 6. Mauritius Sugar Industry Research Institute, Reduit.

ROMERO, C., L. JEFFREY, and M. REVELO. 1969. Control y erradicación del kikuyo (*Pennisetum clandestinum* Hochst) en la sabana de Bogotá. *Revista del Instituto Colombiano de Agropêcuario* 4(3):99-116.

ROMERO, C., D. VARGAS, E. LAGOS, J. CARDENAS, G. RIVEROS, and J. DOLL. 1970. *El kikuyo y su control*. Hoja Divulgativa 006. Instituto Colombiano de Agropêcuario, Departamento de Agronomía, Bogotá.

RONOPRAWIRO, S. 1971. The effect of drying on the

survival of nutgrass tubers, *Cyperus rotundus* L. Contribution 1. Pages 101–105 *in Proceedings of the First Indonesian Weed Science Conference, Bogor* [January 1971]. Weed Science Society of Indonesia, Bogor.

ROSS, M. 1960. *Breeding systems in* Plantago. Ph.D. Thesis. Oxford University, Oxford.

ROWNTREE, J. 1940. Grazing vs. burning as an aid to sal regeneration. *Indian Forestry* 66:689-705.

RUDESCU, L. 1965. Neue biologische Probleme bei den *Phragmites* kulturarbeiten im Donaudelta. *Archiv für Hydrobiologie (und Planktonkunde)* 30(2):80-111.

RUSCHEL, A., and D. BRITTO. 1966. Asymbiotic nitrogen fixation in some grasses and in *Cyperus rotundus* by bacteria of the genus *Beijerinckia* Derx. *Pesquisa Agropecuar* 1:65-69.

RUSSELL, G. 1965. The host range of some English isolates of beet yellowing viruses. *Annals of Applied Biology* 55:245-252.

RUSSELL, G. 1968. The distribution of sugar beet yellowing viruses in East Anglia from 1965 to 1968. *British Sugar Beet Review* 37:77-84.

SAGAR, G., and J. HARPER. 1960. Factors affecting the germination and early establishment of plantains. Pages 236-245 *in* J. Harper, ed. *Biology of weeds*. Blackwell Scientific Publications, Oxford. 256 pp.

———. 1964. Biological flora of the British Isles: *Plantago major* L., *Plantago media* L., and *Plantago lanceolata* L. *Journal of Ecology* 52:189-221.

SAGAR, G., and H. RAWSON. 1964. The biology of *Cirsium arvense*. Pages 553-562 *in Proceedings of the 7th British Weed Control Conference*, Brighton, England. Published by British Crop Protection Council.

SAKIMURA, K. 1937. A survey of host ranges of thrips in and around Hawaiian pineapple fields. *Proceedings of the Hawaiian Entomological Society* 9:415-427.

SALGADO, M. 1963. New menace on coconut estates: *Eupatorium odorata* spreads into the coconut estates. *Ceylon Coconut Planters' Review* 3(3):69-70.

SALISBURY, E. 1961. *Weeds and aliens*. Collins, London, 384 pp.

SALISBURY, F. 1969. *Xanthium strumarium* L. Pages 14-61 *in* L. Evans, ed. *The induction of flowering; some case histories*. Macmillan of Australia, South Melbourne. 488 pp.

SANTIAGO, A. 1965. Studies on the autecology of *I. cylindrica* (L.) Beauv. Pages 499-502 *in Ninth International Grassland Congress*, São Paulo.

SAUNDERS, A. 1933. *Studies in phanerogamic parasitism, with particular reference to* Striga lutea Lour. Bulletin 128. South African Department of Agricultural Science, Pretoria. 56 pp.

SAWHNEY, J. 1965. The structure of the rhizomes of *Agropyron repens*, Beauv., *Poa pratensis* L., and *Festuca arundinacea* Schreb. *Dissertation Abstracts* 26(3):1,309.

SAXBY, S. 1943. Weeds—perennial weeds. *New Zealand Journal of Agriculture* 67:407-410.

SCHELPE, E. 1961. The ecology of *Salvinia auriculata*

and vegetation on Kariba Lake. *Journal of South African Botany* 27:181-187.

SCHMUCKER, T. 1959. Hohere parasiten. Pages 480-529 *in* W. Ruhland, ed. *Das Handbuch der Pflanzenphysiologie*. Vol. 11. Springer-Verlag, Berlin.

SCHMUTZ, E., B. FREEMAN, and R. REED. 1968. *The livestock poisoning plants of Arizona*. University of Arizona Press, Tucson. 176 pp.

SCHREIBER, M. 1967. Effect of density and control of Canada thistle on production and utilization of alfalfa pasture. *Weeds* 15:138-142.

SCHREIBER, M., and L. OLIVER. 1971. Two new varieties of *Setaria viridis*. *Weed Science* 19(4):424-427.

SCHWERZEL, P. 1970a. Weed seed production study. *Pest Articles and News Summaries* 16:357.

———. 1970b. Weed phenology and life-span observations. *Pest Articles and News Summaries* 16:511-515.

SCULTHORPE, C. 1967. *The biology of vascular plants*. Edward Arnold, London. 610 pp.

SEN, D., D. CHAWAN, and K. SHARMA. 1969. Preliminary observations on the influence of certain weeds on germination and growth of *Pennisetum typhoideum* Rich. *Proceedings of the Indian Academy of Science* 60 (section B): 111-117.

SENARATNA, S. 1956. *The grasses of Ceylon*. Paradeniya Manual 8. Department of Agriculture, Ceylon Government Press, Colombo. 279 pp.

SETH, A. 1971. Control of *Mikania cordata* (Burm. f.) B. L. Robinson in plantation crops using paraquat. *Weed Research* 11:77-83.

SEXSMITH, J. 1967. Varietal differences in seed dormancy of wild oats. *Weeds* 15:252-255.

———. 1969. Dormancy of wild oat seed under various temperature and moisture conditions. *Weed Science* 17:405-407.

SEXSMITH, J., and U. PITTMAN. 1963. Effect of nitrogen fertilizer on germination and stand of wild oats. *Weeds* 11:99-101.

SHANLEY, B., and O. LEWIS. 1969. The protein nutritional value of wild plants used as dietary supplements in Natal. *Plant Foods for Human Nutrition* 1:253-258.

SHARMAN, B. 1947. The biology and development morphology of the shoot apex in the *Gramineae*. *New Phytologist* 46:20-34.

SHAW, M., and J. WHITE. 1965. Damage to barley by larvae of the wheat bulb fly, *Leptohylemyia coarctata* (Fall.), associated with couch grass. *Plant Pathology* 18(4):192.

SHAW, W., D. SHEPHERD, E. ROBINSON, and P. SAND. 1962. Advances in witchweed control. *Weeds* 10:182-192.

SHER, S. 1954. Observations on plant-parasitic nematodes in Hawaii. *Plant Disease Reporter* 38:687-689.

SHINKAI, A. 1955. Host range of rice stripe disease. *Annals of the Phytopathological Society of Japan* 20:100.

———. 1956. Host range of rice stripe disease (further

report). [in Japanese]. *Annals of the Phytopathological Society of Japan* 21:47.

————. 1957. Host range and problems concerning transmission of rice black-streaked dwarf. *Annals of the Phytopathological Society of Japan* 22:34.

SIDERIS, C. 1931a. Pathological and histological studies on pythiaceous rots of various agricultural plants. *Phytopathologische Zeitschrift* 3:137-161.

————. 1931b. Taxonomic studies in the family *Pythiaceae* 1. *Nematosporangium*. *Mycologia* 23:252-295.

————. 1932. Taxonomic studies in the family *Pythiaceae*. 2. *Pythium*. *Mycologia* 24:14-61.

SIERRA, J. 1973. Some aspects of the biology of *Cyperus rotundus*. Pages 1-26 in *Proceedings of the Second Indonesian Weed Science Conference*, Jogjakarta.

SINGH, J. 1968. Growth of goosegrass in relation to certain environmental factors. *Tropical Ecology* 9:78-87.

SINGH, J., and R. MISRA. 1969. Influence of the direction of slope and reduced light intensities on the growth of *Eleusine indica*. *Tropical Ecology* 10:27-33.

SINGH, K. 1968a. Thermoresponse of *Portulaca oleracea* seeds. *Current Science* 37(17):506-507.

————. 1968b. Effect of exposure to sub-freezing temperature (-10 C) on survival and subsequent growth of *Anagallis arvensis*. *Tropical Ecology* 9(1):72-77.

————. 1969. Seed dormancy and its control by germination inhibitor in *Anagallis arvensis* L. var. *caerulea* Gren. et Godr. *Proceedings of the National Institute of Science of India*, series B, 35:161-171.

SINHA, R., and F. WHITEHEAD. 1965. Meiotic studies of British populations of *Stellaria media* (L.) Vill., *S. neglecta* Weihe, and *S. pallida* (Dumort) Pire. *New Phytologist* 64:343-345.

SINHA, T., and E. THAKUR. 1967. Control of nutgrass weed by cultivation. *Indian Journal of Agronomy* 12(2):121-125.

SJOSTEDT, S. 1959. Germination biology of cleavers, *Gallium aparine* L. [in Swedish, English summary]. *Växtodling* 10:87-105.

SLYKHUIS, J. 1967. *Agropyron repens* and other perennial grasses as hosts of bromegrass mosaic virus from the U.S.S.R. and the United States. *Food and Agriculture* 15:65-66.

SMALL, J. 1918. The origin and development of the *Compositae*. 9. Fruit dispersal in the *Compositae*. *New Phytologist* 17:200-229.

SMARTT, J. 1961. Weed competition in leguminous grain crops in Northern Rhodesia. *Rhodesia Journal of Agriculture* 58:267-273.

SMITH E., and G. FICK. 1937. Nut grass eradication studies. 1. Relation of the life history of nut grass, *Cyperus rotundus* L., to possible methods of control. *Journal of the American Society of Agronomy* 29:1007-1013.

SMITH, E., and E. MAYTON. 1938. Nut grass eradication studies. 2. Eradication by certain tillage treatments. *Journal of the American Society of Agronomy* 30:18-21.

————. 1942. Nut grass eradication studies. 3. Control on several soil types by tillage. *Journal of the American Society of Agronomy* 34:151-159.

SOERJANI, M. 1970. *Alang-alang*, Imperata cylindrica *(L.) Beauv., pattern of growth as related to its problem of control*. BIOTROP Bulletin 1. Regional Center for Tropical Biology, Bogor.

SOERJANI, M., and O. SOEMARWOTO. 1969. Some factors affecting the germination of alang-alang (*I. cylindrica*) rhizome buds. *Pest Articles and News Summaries* 15:376-380.

SOLOMON, S. 1952. Studies in the physiology of phanerogamic parasitism with special reference to *Striga lutea* and *S. densiflora* on sorghum. *Proceedings of the Indian Academy of Sciences* 35:122-131.

SONNEVELD, F. 1953. Survey of the ecology and control of horsetail [in Dutch]. Pages 29-35 in *Verslagen van het Centraal Instituut voor Landbouwkundig Onderzoek, 1952*, Wageningen, The Netherlands. 136 pp.

SPARROW, D. 1958. Nut grass (*Cyperus rotundus*) and its control. *Pesticide Abstracts and News Summary* 4(3):135-145.

SPENCE, D. 1964. The macrophytic vegetation of freshwater lochs, swamps, and fens. Pages 306-425 in J. Burnett, ed. *The vegetation of Scotland*. Oliver & Boyd, Edinburgh.

SQUIRES, V. 1965. A note on *Aristotelia* sp. (Lep, Gelechiidae) attacking *Tribulus terrestris* L. *Journal of the Entomological Society of New South Wales* 2:1-2.

STAHLER, M. 1948. Shade and soil moisture as factors in competition between selected crops and field bindweed, *Convolvulus arvensis*. *Journal of the American Society of Agronomy* 40:490-502.

STAMPER, E. 1957. The problem of johnsongrass. *Proceedings of the Southern Weed Conference* 10:149-152.

STANDIFER, L., W. NORMAND, and T. RIZK. 1966. An effect of light on basal bulb formation in purple nutsedge. *Proceedings of the Southern Weed Control Conference* 19:550-553.

STANTON, T. 1955. *Oat identification and classification*. Technical Bulletin 1100. United States Department of Agriculture, Washington, D.C. 206 pp.

STEINBAUER, G., and B. GRIGSBY. 1960. Dormancy and germination of docks. *Proceedings of the Association of Official Seed Analysts of North America* 50:112-117.

STEINBAUER, G., B. GRIGSBY, L. CORREA, and P. FRANK. 1955. A study of methods for obtaining laboratory germination of certain weed seeds. *Proceedings of the Association of Official Seed Analysts of North America* 45:48-52.

STEPHENS, E. 1912. The structure and development of the haustorium of *Striga lutea*. *Annals of Botany* 26:1067-1076.

STEVENS, F. 1925. *Hawaiian fungi*. Bulletin 19, Bernice P. Bishop Museum, Honolulu. 189 pp.

STEVENS, N., and C. SHEAR. 1929. *Botryosphaeria* and *Physalosporea* in the Hawaiian Islands. *Mycologia* 21:313-320.

STEYN, D. 1934. *The toxicology of plants in South Africa*. Central News Agency, South Africa.

STOLLER, E., D. NEMA, and V. BHAN. 1972. Yellow nutsedge tuber germination and seedling development. *Weed Science* 20:93-97.

STOUT, A. 1919. Intersexes in *Plantago lanceolata*. *Botanical Gazette* 68:109-133.

STRYCKERS, J., and M. PATTOU. 1963. The biology and distribution of wild oats (*Avena*) species in Belgium [in Flemish]. Pages 1063-1086 *in 15th International Symposium Fytofarm*. Fytiatrie, Gent.

SUBRAMANIAN, C., and A. SRINIVASAN. 1960. *A review of the literature on the phanerogamous parasites*. Monograph 10.24. Indian Council of Agriculture Research, New Delhi. 96 pp.

SUTOMO, M. 1971. Preliminary report on studies of *Salvinia* species. The growth rate of *S. auriculata* and its influence on the growth of rice. Pages 95-99 *in Proceedings of the First Indonesian Weed Science Conference*, Bogor [January 1971].

SVEŘEPOVÁ, G. 1968. Zur Zytotaxonomie der Art *Anagallis arvensis* L. [On the cytotaxonomy of the species *Anagallis arvensis* L.] [in German] *Preslia* 40(2):143-146.

SWAIN, D. 1967. Controlling barnyard grass in rice. *Agricultural Gazette of New South Wales* 78(8):473-475.

SWIETOCHOWSKI, B. 1967. The effect of cultivation and chemical treatments on changes in field plant communities. [in Polish]. *Pamiętnik Państwowego Instytutu naukowego gospodarstwa wiejskiego w Pulawach* 28:3-17.

SYME, J. 1863-1886. *English botany; or coloured figures of British plants*. 3rd ed. 12 vols. in 5. This work is lodged in the library at Kew Gardens, London.

TABITA, A., and J. WOODS. 1962. History of hyacinth (*E. crassipes*) control in Florida. *Hyacinth Control Journal* 1:19-22.

TADULINGAM, C., and G. VENKATANARA-YANA. 1932. *A handbook of some south Indian weeds*. Government Press, Madras.

TAKAHASHI, M., J. MOOMAW, and J. RIPPER-TON. 1966. *Studies of napiergrass. 3. Grazing management*. Bulletin 128. Hawaii Agricultural Experiment Station, University of Hawaii, Honolulu.

TANGL, H. 1959. Erdmandel (*Cyperus esculentus*) als futtermittel [in Hungarian, German summary]. *Állattenyésztés* 8(3):267-272.

TARR, S. 1962. Parasitic flowering plants. Pages 280–310 *in Diseases of sorghum, sudan grass and broom corn*. The Commonwealth Mycological Institute, Kew, Surrey. 380 pp.

TAYLOR, P. 1955. The genus *Anagallis* in tropical and South Africa. *Kew Bulletin* 3:321-350.

TAYLORSON, R. 1967. Seasonal variation in sprouting and available carbohydrate in yellow nutsedge tubers. *Weeds* 9:646-653.

TAYLORSON, R., and C. McWHORTER. 1969. Seed dormancy and germination in ecotypes of johnsongrass. *Weed Science* 17:359-361.

TEMPANY, H. 1951. *Imperata* grass, a major menace in the wet tropics. *World Crops* 3:143-146.

TENG, S. 1932. Fungi of Nanking, II. *Contributions from the Biological Laboratory of the Science Society of China*, Botanical series, 8(1):5-48.

THOMAS, P. 1967*a*. A preliminary study on the dormancy of *Cyperus esculentus* tubers. *Pest Articles and News Summaries* 13:329-333.

———. 1967*b*. A rhizome germinating technique for glasshouse propagation. *Pest Articles and News Summaries* 13(3):221-222.

———. 1970. A survey of the weeds of arable lands in Rhodesia. *Rhodesia Agricultural Journal* 67(2):34-37.

THOMAS, P., and I. HENSON. 1968. Influence of climate and soil moisture on tuber dormancy of *Cyperus esculentus*. *Pest Articles and News Summaries* 14(3):271-276.

THOMAS, P., and P. SCHWERZEL. 1968. A cotton/weed control experiment. Pages 737-743 *in Proceedings of the 9th British Weed Control Conference*, Brighton, England. Published by British Crop Protection Council.

THOMAS, W. 1970. Some effects of weeds on cotton growth and development. Pages 207-217 *in* L. Hughes and M. Siddig, eds. *Cotton growth in the Gezira environment*. Sudan Government Press, Khartoum. 320 pp.

THURSTON, J. 1959. A comparative study of the growth of wild oats (*Avena fatua* and *Avena ludoviciana*) and of cultivated cereals with varied nitrogen supply. *Annals of Applied Biology* 47:716-739.

———. 1960. Dormancy in weed seeds. Pages 69-82 *in* J. Harper, ed. *The biology of weeds*. Blackwell Scientific Publications, Oxford. 256 pp.

———. 1961. The effect of depth of burying and frequency of cultivation on survival and germination of seeds of wild oats (*Avena fatua* L., and *A. ludoviciana* Dur.). *Weed Research* 1:19-31.

———. 1962*a*. An international experiment on the effect of age and storage conditions on viability and dormancy of *Avena fatua* seeds. *Weed Research* 2:122-129.

———. 1962*b*. The effect of competition from cereal crops on the germination and growth of *Avena fatua* in naturally infested fields. *Weed Research* 2:192-207.

———. 1963. Biology and control of wild oats. Pages 235-253 *in Rothamsted Experimental Station Report for 1962*, Harpenden, England. 316 pp.

———. 1966. Survival of seeds of wild oats (*Avena fatua* L. and *Avena ludoviciana* Dur.) and charlock (*Sinapis arvensis* L.) in soil under leys. *Weed Research* 6:67-80.

TILDESLEY, W. 1931. *A study of the changes in the growth and food reserves in the underground parts of* Sonchus arvensis *and* Agropyron repens *during the growing season 1931-1932*. Ph.D. Thesis. University of

Manitoba, Winnipeg.

TIMMONS, F. 1949. Duration of viability of bindweed seed under field conditions and experimental results in the control of bindweed seedlings. *Agronomy Journal* 41:130-133.

TIMMONS, F., and V. BRUNS. 1951. Frequency and depth of shootcutting in eradication of certain creeping perennial weeds. *Agronomy Journal* 43:371-375.

TIMSON, J. 1966. The germination of *Polygonum convolvulus* L. *New Phytologist* 65:423-428.

TOGASHI, K. 1942. *A survey of the plant diseases of cultivated plants in Iwate Prefecture in 1941*. Bulletin 8. Iwate Agricultural Experiment Station, Morioka. 25 pp.

TOOLE, E., and E. BROWN. 1946. Final results of the Duvel buried seed experiment. *Journal of Agricultural Research* 72:201-210.

TRIPATHI, R. 1967. Ecology of *Cyperus rotundus* L. 2. Tuber sprouting in relation to temperature. *Proceedings of the National Academy of Sciences of India* (B)37:409-412.

———. 1968a. Comparison of competitive ability of certain common weed species. *Tropical Ecology* 9(1):37-41.

———. 1968b. Ecology of *Cyperus rotundus* L. 1. Effects of exposure to sub-freezing temperature on survival and subsequent growth. *Tropical Ecology* 9:239-242.

———. 1969a. Ecology of *Cyperus rotundus* L. 3. Population of tubers at different depths of soil and their sprouting response to air drying. *Proceedings of the National Academy of Sciences of India*, B, 39:140-142.

———. 1969b. Ecology of *Cyperus rotundus* L. 4. Effects of two different temperatures on growth and reproduction. *Proceedings of the National Academy of Sciences of India*, B, 39:161-163.

———. 1969c. Ecology of *Cyperus rotundus* L. 5. Reproduction by seeds. *Proceedings of the National Academy of Sciences of India*, B, 39:164-167.

TRIPPI, V., and R. TIZIO. 1961. Posible diseminación de "tamascal," *Cyperus rotundus*, mediante tubérculos de "papa" semilla e influencia de la brotación de esta sobre la brotación de la maleza. Phyton 16(2):141-145.

TULLIS, E. 1941. Diseases of rice. Page 163 *in 53rd Annual Report of the Texas Agricultural Experiment Station*, College Station.

TUMBLESON, M., and R. KOMMEDAHL. 1961. Reproductive potential of *Cyperus esculentus* by tubers. *Weeds* 9:646-653.

———. 1962. Factors affecting dormancy in tubers of *Cyprus Esculentus*. *Botanical Gazette* 123:186-190.

TURNER, D. 1966. A study of the effects of rhizome length, soil nitrogen and shoot removal on the growth of *Agropyron repens*. *Proceedings of the British Weed Control Conference* 8:538-545.

———. 1969. The effects of shoot removal on the rhizome carbohydrate reserves of couch grass, *Agropyron repens* (L.) Beauv. *Weed Research* 9:27-36.

UBRIZSY, G. 1968. Long-term experiments on the flora-changing effect of chemical weed killers in plant communities [in Hungarian]. *Acta agronomica Academiae scientiarum hungaricae* 17:171-193.

UEKI, K. 1965. *Physiological and ecological studies on cleaver* (G. aparine) *control*. Ph.D. Thesis. Kyoto University, Kyoto.

———. 1969. *Studies on the control of nutsedge* (Cyperus rotundus): *on the germination of the tuber*. Pages 355-370 *in Proceedings of the Second Asian-Pacific weed control interchange*. University of the Philippines, Los Baños. 424 pp.

UEKI, K.,and N. SHIMIZU. 1967a. Studies on the termination of dormancy in the seeds of cleavers (*Galium aparine*). 1. Histological and physiological changes during dormancy. *Weed Research (Japan)* 6:26-30.

———. 1967b. Studies on the termination of dormancy in the seeds of cleavers (*Galium aparine*). 2. The effects of some chemicals on the breaking of dormancy. *Weed Research (Japan)* 6:30-33.

———. 1970. Studies on the breaking of dormancy in barnyard grass seeds. 1. The effects of some chemicals on the breaking of dormancy. *Proceedings of the Crop Science Society of Japan* 38(2):261-272.

UNITED STATES DEPARTMENT OF AGRICULTURE. 1962. A survey of extent and cost of weed control and specific weed problems. Joint Report ARS 34-23. United States Department of Agriculture, Agricultural Research Service, Washington, D.C. 65 pp.

VAILLANT, A. 1967. Chemical control of annual weeds in rice. *World Crops* 19(6):38-44.

VALDEZ, R. 1968. Survey, identification and host-parasite relationships of root-knot nematodes occurring in some parts of the Philippines. *Philippine Agriculturist* 51:802-824.

VAN RIJN, P. 1968. Ecological aspects of weed control in cotton in the Ord River valley, Western Australia. 1. Conditions affecting germination of weeds. *Australian Journal of Experimental Agriculture and Animal Husbandry* 8:620-624.

VAYSSIERE, P. 1957. Les mauvaises herbes en Indo-Malasie. *Journal d' agriculture tropicale et de botanique appliquée* 4(9-10):392-401.

VEGA, M., and J. SIERRA. 1970. Population of weed seeds in a lowland rice field. *Philippine Agriculturist* 54:1-7.

VENGRIS, J. 1955. Plant nutrient competition between weeds and corn. *Agronomy Journal* 47:213-215.

———. 1962. The effect of rhizome length and depth of planting on the mechanical and chemical control of quackgrass. *Weeds* 10:71-74.

VENGRIS, J., A. KACPERSKA-PALACZ, and R. LIVINGSTON. 1966. Growth and development of barnyardgrass in Massachusetts. *Weeds* 14:299-301.

VERGNANO GAMBI, O. 1966. Influence of light and temperature on the germination of some *Rumex* seeds. *Webbia* 21(1):461-474.

VICKERY, J. 1935. The leaf anatomy and vegetative characters of the indigenous grasses of New South Wales. 1. Andropogoneae, Zoysieae, Tristegineae.

Proceedings of the Linnean Society of New South Wales 60:340-373.

VIEL, P. 1963. Stubble burning trails for destruction of wild oats in Côtes du Nord. *Phytoma* 15:32-33.

VINCENTE, M., M. ENGELHARDT, and K. SILBERSCHMIDT. 1962. The influence of temperature on the germination response to light of seeds of *Rumex obtusifolius*. *Phyton* 19(2):163-167.

VINCENTE, M., A. NORONHA, K. SILBERSCHMIDT, and M. MENEGHINI. 1968. Successive reversion of the effect of temperature on germination of *Rumex obtusifolius* L. by far-red light. *Phyton* 25(1):11-13.

VINDUSKA, L. 1967. Weeds and the sugar beet eelworm (*Heterodera schachtii* Schmidt). *Ochrana Rostlin* 3(40):219-224.

VIOLA, C. 1947. El razon de propagación de las malezas principales de la llana de Catania [The propagation ratio of the principal weeds of the Catania plain] [in Spanish, English summary]. *Annali della sperimentazione agraria* 1:371-394.

VON HOFSTEN, C. 1947. Investigations of germination biology in some weed species [in Swedish, English summary]. *Växtodling* 2:91-107.

VON KRIES, A. 1963. *Horsetail* (Equisetum palustre L.): *a monograph of practical use in agriculture*. Ph.D. Thesis. Institut für Kulturtechnik und Grünlandwirtshaft, Technische Universität, Berlin. 133 pp.

VOTILA, I., and J. MUKULA. 1961. Marsh horsetail, a poison to livestock [in Swedish]. *Tidskrift för lantmän Andelsförening* 17-18:423-429.

WACQUANT, J., and L. PASSAMA. 1971. On the relationship between preferential rootlet absorption and the selective absorption of Ca, Mg, K, and Na in various populations of *A. arvensis* L. and other species. *Séances Académie des Sciences*, series D, 272:711-714.

WANG, C. 1969. Weed competition studies. Pages 164-189 *in Proceedings of the Second Asian-Pacific Weed Control Interchange*. University of the Philippines, Los Baños. 424 pp.

WATANAKUL, L. 1964. *A study on the host range of tungro and orange leaf virus of rice*. M.S. Thesis. College of Agriculture, University of the Philippines, Los Baños.

WATSON, G., WONG PHUI WENG, and R. NARAYANAN. 1964a. Effects of cover plants on soil nutrient status and on growth of *Hevea*. 3. A comparison of leguminous creepers with grasses and *Mikania cordata*. *Journal of the Rubber Research Institute of Malaya* 18:80.

———. 1964b. Effects of cover plants on soil nutrient status and on growth of *Hevea*. 4. Leguminous creepers compared with grasses, *Mikania cordata* and mixed indigenous covers on four soil types. *Journal of the Rubber Research Institute of Malaya* 18:23.

WATT, J., and M. BREYER-BRANDWIJK. 1932. *The medicinal and poisonous plants of southern Africa*. E. &

S. Livingstone, Edinburgh and London. 314 pp.

WAX, L., E. STOLLER, F. SLIFE, and R. ANDERSON. 1972. Yellow nutsedge control in soybeans. *Weed Science* 20:194-201.

WELBANK, P. 1959. Survey of competitive effects of common weed species. Page 84 *in Rothamsted Experimental Station Report for 1959*, Harpenden, England. 288 pp.

———. 1960. Toxin production from *Agropyron repens*. Pages 158-164 *in* J. Harper, ed. *Biology of weeds*. Blackwell Scientific publications, Oxford. 256 pp.

———. 1963a. Toxin production during decay of *Agropyron repens* and other species. *Weed Research* 3:205-214.

———. 1963b. A comparison of competitive effects of some common weed species. *Annals of Applied Biology* 51:107-125.

WELDON, L., and R. BLACKBURN. 1967. Water lettuce—nature, problem, and control. *Weeds* 15:5-9.

WHITTET, J. 1968. *Weeds*. 2nd ed. New South Wales Department of Agriculture, Sydney. 486 pp.

WHYTE, R., T. MOIR, and J. COOPER. 1959. *Grasses in agriculture*. Food and Agriculture Organization of the United Nations, Rome. 416 pp.

WILD, H. 1954. Rhodesia witchweeds (*Striga* and *Electra* species). *Rhodesia Agricultural Journal* 51:330-342.

———. 1958. *Common Rhodesian weeds*. Government Printer, Salisbury. 220 pp.

———. 1961. *Harmful aquatic plants in Africa and Madagascar*. Publication 73. Conseil Scientifique pour l'Afrique au Sud du Sahara. Joint Commission for Technical Cooperation in Africa South of the Sahara/Scientific Council for Africa South of the Sahara project 14. Govenment Printer, Salisbury. 68 pp.

WILKINSON, R. 1963. Effects of light intensity and temperature on growth of water star grass, coontail and duckweed. *Weeds* 11(4):287-289.

———. 1964. Effects of red-light intensity on growth of water star grass, coontail, and duckweed. *Weeds* 12(4):312-313.

WILLIAMS, E. 1968. Preliminary studies of germination and seedling behavior in *Agropyron repens* and *Agrostis gigantea*. *Proceedings of the British Weed Control Conference* 9:119-124.

———. 1970a. Effects of decreasing the light intensity on growth of *Agropyron repens* (L.) Beauv. in the field. *Weed Research* 10:360-366.

———. 1970b. Studies on the growth of seedlings of *Agropyron repens* (L.) Beauv. and *Agrostis gigantea* Roth. *Weed Research* 10:321-330.

———. 1971a. Effects of light intensity, photoperiod, and nitrogen on the growth of seedlings of *Agropyron repens* (L.) Beauv. and *Agrostis gigantea* Roth. *Weed Research* 11:159-170.

———. 1971b. Germination of seeds and emergence of seedlings of *Agropyron repens* (L.) Beauv. *Weed Research* 11:171-181.

WILLIAMS, E., and P. ATTWOOD. 1971. Seed production of *Agropyron repens* (L.) Beauv. in arable crops in England and Wales in 1969. *Weed Research* 11:22-30.

WILLIAMS, J. 1963. Biological flora of the British Isles: *Chenopodium album* L. *Journal of Ecology* 51(3):711-725.

————. 1966. Variation in the germination of several *Cirsium* species. *Tropical Ecology* 7:1-7.

————. 1971. Seed polymorphism and germination. 2. The role of hybridization in germination polymorphism of *Rumex crispus* and *R. obtusifolius*. *Weed Research* 11:12-21.

WILLIAMS, O. 1957. The effect of irrigated pasture in the rice rotation on seed populations of *Echinochloa crus-pavonis*. *Journal of the Australian Institute of Agricultural Science* 23:331-333.

WILLIAMS, R. 1956. *Salvinia auriculata*: the chemical eradication of a serious aquatic weed in Ceylon. *Tropical Agriculture* 33:145-149.

WILLS, G., and G. BRISCOE. 1970. Anatomy of purple nutsedge. *Weed Science* 18:631-635.

WILSIE, C., and M. TAKAHASHI. 1934. *Napier grass* (Pennisetum purpureum) *a pasture and green fodder crop for Hawaii*. Bulletin 72. Hawaii Agricultural Experiment Station, University of Hawaii, Honolulu. 17 pp.

WILSON, F. 1960. *A review of the biological control of insects and weeds in Australia and Australian New Guinea*. Technical Communication 1. Commonwealth Institute of Biological Control, Ottawa. Published by Commonwealth Agricultural Bureau, Bucks, England. 68 pp.

WILSON, G. 1970. Method and practicability of kikuyu grass seed production. Pages 312-315 *in Proceedings of the Eleventh International Grassland Congress*, Brisbane.

WITTS, K. 1960. The germination of *Polygonum* species in the field and glass-house. *Journal of Ecology* 48:213-216.

WONG, PHUI WENG. 1966. Weed control by sequential applications of Weedazol TL and a contact herbicide. *Planters' Bulletin, Rubber Research Institute of Malaya* 87:203-207.

WONG, R. 1964. Evidence for the presence of growth inhibitory substances in *Mikania cordata* (Burm. f.) B. L. Robinson. *Journal of the Rubber Research Institute of Malaya* 18(5):231-242.

WOOD, H. 1954. Wild oat control. Pages 122-123 *in Proceedings of the Eleventh North Central Weed Control Conference*, Fargo, North Dakota.

WOODCOCK, E. 1914. Observations on the development and germination of the seed in certain *Polygonaceae*. *American Journal of Botany* 1:454-464.

WORSHAM, A. 1961. *Germination of* Striga asiatica *(witchweed) seed and studies on the chemical nature of the germination stimulant*. Ph.D. Thesis. North Carolina State College, Raleigh. 112 pp.

WORSHAM, A., D. MORELAND, and G. KLINGMAN. 1964. Characterization of the *Striga asiatica* (witchweed) germination stimulant from *Zea mays* L. *Journal of Experimental Botany* 15:556-567.

WUNDERLICH, W. 1964. Water hyacinth control in Louisiana. *Hyacinth Control Journal* 3:4-7.

WYCHERLEY, P., and M. CHANDAPILLAI. 1969. Effects of cover plants. *Journal of the Rubber Research Institute of Malaya* 21(2):140-157.

YABUNO, T. 1966. Biosystematic study of the genus *Echinochloa*. *Japanese Journal of Botany* 19(2):277-323.

YAMADA, W., T. SHIOMI, and H. YAMAMOTO. 1956. Studies on the stripe disease of rice plant. 3. Host plants, incubation period in the rice plant and retention of over-wintering of the virus in the insect, *Delphacodes striatella* Fallen [in Japanese, English summary]. *Special Bulletin. Okayama Prefectural Agriculture Experiment Station* 55:35-36.

YOUNG, H. 1962. Weed control. *Cane Growers' Quarterly Bulletin* 25(4):1-66.

ZELENCHUK, T., and S. GELEMEI. 1967. Effect of water extracts of plants on seed germination and early growth of meadow grasses [in Russian] *Byulleten' Moskovskogo obshchestva ispytatelei prirody* 72(2):93-105.

Index of Common Weed Names

General Index

Abaca: common or other weeds in, 45, 68, 94, 137, 151, 271, 277, 294, 331, 351, 433, 481; principal weeds in, 52, 59, 137, 214; serious weeds in, 68
Abaca mosaic, 23, 45, 97
Abutilon theophrasti, increase of, with herbicide use, 433
Aecidium tithymali, 271
Afghanistan, *Avena sterilis* present in, as a weed, 107
Agave sisalana. See Sisal
Ageratum conyzoides, 146–152, 187; as alternate host, 151; as a colonizing species, 148; competition with crops, 148, 151; confusion of, with *A. houstonianum*, 146, 148; countries in which found as a serious weed, 151; crops in which found as a common or other weed, 151, 486, 504, 538; crops in which found as a principal weed, 151, 494, 504, 514, 528, 534; crops in which found as a serious weed, 486, 534; dispersal of, 148; distribution map of, 150; economic uses of, 151; illustrated, 147; map of, in crops, 152; origin of, 146; range of, 148; role of, in sudd formation, 411
Ageratum houstonianum, 146–151; as alternate host, 151; competition with crops, 151; confusion of, with *Ageratum conyzoides*, 146, 148; illustrated, 149
Ageratum mexicanum, 146
Agropyron repens, 131, 153–168; as alternate host, 167; chemical inhibition of crop growth by, 165, 166; competi-

tion with crops, 165, 166; contamination of crop seeds by, 163; control of, 159, 166, 167; control of, by freezing, 156; countries in which found as a principal weed, 163, 165; countries in which found as a serious weed, 163; crops in which found as a principal weed, 163, 165, 494, 510, 528, 532, 542; crops in which found as a serious weed, 163, 542; distribution map of, 155; economic uses of, 167; ecotypes of, 156; effect of tillage on, 156, 166, 167; Hakansson's work on, 156–163, 167; illustrated, 154; map of, in crops, 164; origin of, 153; poor growth of, in tropics, 156; quantity of organic matter produced by, 165, 166; range of, 153; similarity of, to *Triticum repens*, 153
Agrostemma githago, 174
Alaska: principal weeds of, 88, 163, 198, 444, 454; serious weeds of, 88, 454
Alfalfa: common or other weeds in, 102, 187, 198, 223, 392, 398, 407, 419, 438, 471; principal weeds in, 165, 203; serious weeds in, 59, 454
Alfalfa mosaic, 438
Algeria: *Avena sterilis* present in, as a weed, 107; principal weeds of, 88
Allium cepa. See Onion
Allium sativum. See Garlic
Alopecurus aequalis, 289
Alopecurus myosuroides, 174; crops in which found as a principal weed, 542

Alsike clover, common or other weeds in, 223
Altica carduorum, 224
Amaranthus hybridus, 114–118, 492; as alternate host, 118; competition with crops, 114, 118; countries in which found as a principal weed, 114; crops in which found as a common or other weed, 114; crops in which found as a principal weed, 114, 528; distribution map of, 116; economic uses of, 114, 118; illustrated, 115; map of, in crops, 117; origin of, 114
Amaranthus retroflexus, crops in which found as a principal weed, 494, 530
Amaranthus spinosus, 119–124; as alternate host, 121; countries in which found as a common or other weed, 121; countries in which found as a principal weed, 121; countries in which found as a serious weed, 121; crops in which found as a common or other weed, 121, 486; crops in which found as a principal weed, 121, 486, 494, 504, 514, 528, 534; crops in which found as a serious weed, 121, 534; distribution map of, 123; economic uses of, 119; ecotypes of, 121; illustrated, 120; origin of, 119; map of, in crops, 124; range of, 119; toxicity of, 121
Anagallis arvensis, 169–175; chromosome number of, 172; competition with crops, 174; countries in which found as a common or other weed, 174;

countries in which found as a principal weed, 174; crops in which found as a common or other weed, 174; crops in which found as a principal weed, 174, 510, 542; dispersal of, 169; distribution map of, 171; economic uses of, 174; germination inhibitor from, 173; killing of, by freezing, 174; illustrated, 170; range of, 169; subspecies *arvensis*, 169; subspecies *foemina*, 169; subspecies *phoenicia*, 169; toxicity of, 169, 174; varieties of, 172, 173, 174
Ananas comosus. See Pineapples
Anaphothrips swezeyi, 419
Aneilema malabaricum, 225
Aneilema nudiflorum, 225
Anemone brown ring, 83
Anemone mosaic, 90, 151, 198
Angola: common or other weeds of, 121, 131, 194, 227, 234, 261, 271, 419, 438; principal weeds of, 131, 151, 227, 280; serious weeds of, 28, 50, 131
Apera spica-venti, crops in which found as a principal weed, 542
Aphelenchoides fragariae, 151
Aquatic weeds, recent dramatic increase of, 210
Arabia, *Avena sterilis* present in, as a weed, 107
Arabian peninsula, principal weeds of, 102
Arachis hypogaea. See Peanuts
Aramina fiber, 496
Argemone mexicana, 176–179; biological control of, 179; contaminant of alfalfa seed,